Numerical Algebra, Matrix Theory, Differential-Algebraic Equations and Control Theory

Peter Benner • Matthias Bollhöfer •
Daniel Kressner • Christian Mehl •
Tatjana Stykel
Editors

Numerical Algebra, Matrix Theory, Differential-Algebraic Equations and Control Theory

Festschrift in Honor of Volker Mehrmann

Springer

Editors
Peter Benner
Max Planck Institute for Dynamics of Complex Technical Systems
Magdeburg, Germany

Matthias Bollhöfer
TU Braunschweig
Carl-Friedrich-Gauß-Fakultät
Institute Computational Mathematics
Braunschweig, Germany

Daniel Kressner
École Polytechnique Fédérale de Lausanne (EPFL)
MATHICSE
Lausanne, Switzerland

Christian Mehl
TU Berlin
Institute of Mathematics
Berlin, Germany

Tatjana Stykel
University of Augsburg
Institute of Mathematics
Augsburg, Germany

ISBN 978-3-319-38431-3 ISBN 978-3-319-15260-8 (eBook)
DOI 10.1007/978-3-319-15260-8

Mathematics Subject Classification (2010): 15A18, 15A21, 15B48, 49M05, 65F08, 65F15, 65H10, 65L80, 93B36, 93B40, 93D09, 93D15

Springer Cham Heidelberg New York Dordrecht London

Softcover reprint of the hardcover 1st edition 2015

Cover illustration: by courtesy of Tatjana Stykel. Handwritten notes by Volker Mehrmann, characterizing his research work

Printed on acid-free paper

Springer International Publishing AG Switzerland is part of Springer Science+Business Media (www.springer.com)

Preface

This book is dedicated to Volker Mehrmann, our mentor, co-author, colleague, and friend. Each chapter in this book highlights one of the many topics Volker has worked on. Inevitably, there are omissions. In this preface, we therefore make an attempt to not only connect the book chapters to Volker's bibliography but to also provide additional details on topics that receive less attention in the subsequent chapters.

M-Matrices and H-Matrices and Their Friends

A complex or real square matrix A is called a Z*-matrix* if its off-diagonal entries are nonpositive. Equivalently, $A = \alpha I_n - B$ holds for some scalar $\alpha \in \mathbb{R}$ and a nonnegative matrix B. If, additionally, α is not smaller than $\rho(B)$, the spectral radius of B, then A is called an M*-matrix*. Arising frequently, for example, from the discretization of partial differential equations, matrices with such structures have many desirable properties that facilitate, among other things, the design and analysis of iterative methods for solving linear systems.

Ostrowski coined the term M-matrix and proposed the following generalization: A matrix A is called an H*-matrix* if its comparison matrix C (defined via $c_{ii} = |a_{ii}|$ and $c_{ij} = -|a_{ij}|$ for $i = 1, \dots, n$, $j \neq i$) is an M-matrix. Incomplete factorizations for such H-matrices are the topic of the first scientific paper by Volker Mehrmann, a 1980 publication [A1] jointly with Richard S. Varga and Edward B. Saff.

Apart from M- and H-matrices, there is a myriad of other matrix classes with similarly desirable properties, but different targets in mind. In his 1982 dissertation [T2] and subsequent publications [A2, A3, A5], Volker contributed to this scene by proposing and analyzing the concepts of R- and V-matrices. One major motivation for these new concepts was to find a unified treatment for M-matrices and Hermitian positive semidefinite matrices.

Let the number $l(B)$ denote the smallest real eigenvalue of a matrix B, with the convention $l(B) = \infty$ if B has no real eigenvalue. A matrix A is called an

Fig. 1 Young Volker doing Math

ω-matrix if this number is finite and satisfies a monotonicity property for all principal submatrices of A. If, moreover, $l(A) \geq 0$, then A is called a τ-matrix. Since their introduction by Engel and Schneider, there has been significant interest in studying ω- and τ-matrices because they represent generalizations of several important matrix classes, such as Hermitian matrices, totally nonnegative matrices, and M-matrices. In [A4], Volker showed for $n = 4$ that every $n \times n$ τ-matrix is stable by establishing an eigenvalue inequality conjectured by Varga. More than a decade later, in 1998, Olga Holtz disproved the conjecture that this property holds for general n by constructing a whole family of unstable τ-matrices.

In subsequent collaborations with Daniel Hershkowitz and Hans Schneider on the matrix classes discussed above, Volker investigated linear preserver problems [A9], eigenvalue interlacing properties [A11], and a generalization of sign symmetric matrices [A12]. Joint work [A23] with Ludwig Elsner establishes the convergence of block iterative methods for block generalizations of Z-matrices and M-matrices that arise from the discretization of the Euler equations in fluid flow computations. This 1991 paper also seems to mark a transition of Volker's work into other areas.

Hamiltonian and Symplectic Matrices

Let $J = \begin{bmatrix} 0 & I_n \\ -I_n & 0 \end{bmatrix}$. A matrix $H \in \mathbb{R}^{2n\times 2n}$ is called *Hamiltonian* if $(JH)^T = JH$ and a matrix $S \in \mathbb{R}^{2n\times 2n}$ is called *symplectic* if $S^TJS = J$. Hamiltonian and symplectic matrices seem to have a certain fascination for Volker that is shared by several of his co-authors. On the one hand, there are important applications, most notably in optimal and robust control. On the other hand, there is rich algebraic structure: the set of Hamiltonian matrices forms a Lie algebra, and the set of symplectic matrices forms the corresponding Lie group. Additionally to being

Der SR-Algorithmus zur Bestimmung der Eigenwerte einer Matrix

Diplomarbeit
vorgelegt von
Volker Mehrmann

Fakultät für Mathematik
der Universität Bielefeld

im Mai 1979

-2-

Inhaltsverzeichnis

Fig. 2 Cover and table of contents of Volker's *Diplomarbeit* (equivalent to a M.Sc. thesis), written in German and pre-TEX

symmetric with respect to the real line, the spectra of H and S are symmetric with respect to the imaginary axis and the unit circle, respectively.

The analysis and numerical solution of eigenvalue problems for matrices and matrix pencils with structure is a major thread in Volker Mehrmann's bibliography, from the first works, his 1979 "Diplomarbeit" [T1] at Bielefeld (see Fig. 2 for a copy of the cover and table of contents), until today. One major challenge in this field had been the development of algorithms that would compute the eigenvalues (and invariant subspaces) of Hamiltonian/symplectic matrices in a numerically stable, efficient, and structure-preserving manner. Chapter 1 by Bunse-Gerstner and Faßbender as well as Chap. 4 by Benner explain the substantial contributions Volker made in addressing this challenge for small- to medium-sized dense matrices. There have been numerous extensions of the numerical algorithms resulting from this work, most notably to large-scale eigenvalue problems, see Chap. 2 by Watkins, and matrix pencils with similar structures, see Chap. 4 as well as Chap. 5 by Poloni. Their extension to structured matrix polynomials will be discussed in more detail below.

Fig. 3 Volker and Angelika Bunse-Gerstner in San Francisco during what seems to be an enjoyable (of course vegetarian!) dinner at Greens Restaurant in the San Francisco harbor area, 1990

There is a fascinating, more theoretical side to the story on structured eigenvalue problems. Still motivated by algorithmic developments, the 1991 paper with Greg Ammar [A21] discussed the existence (or rather the lack thereof) of structured Hessenberg forms for Hamiltonian and symplectic matrices. This paper is rather unique in not only thanking Angelika Bunse-Gerstner (see Fig. 3 for a picture of her and Volker around this time) for helpful discussions but also "the German police for a speeding ticket during one discussion". As summarized in Chap. 6 by Mehl and Xu, the work of Volker and his co-authors then evolved towards a more general picture of structured canonical forms of structured matrices. Perturbation theory plays a major role in understanding the potential benefits from structure in the presence of uncertainty in the matrix entries, for example, due to roundoff error. Of particular interest is the perturbation behavior of eigenvalues that are critical in a certain sense, such as purely imaginary eigenvalues of Hamiltonian matrices; see Chap. 8 by Bora and Karow for a summary of results in this direction. Chapter 13 by Ran and Rodman gives a more general survey of the stability of matrix analysis problems with respect to perturbations.

Matrix Equations

Continuous-time linear-quadratic optimal control problems give rise to *algebraic Riccati equations* of the form

$$F + A^T X + XA - XGX = 0, \tag{0.1}$$

where $A \in \mathbb{R}^{n \times n}$, and $F, G \in \mathbb{R}^{n \times n}$ are symmetric and, often, positive semidefinite. Under certain conditions, solutions X of (0.1) can be obtained from n-dimensional

Fig. 4 Volker organized numerous workshops: one of the first was the *5-day Short Course on Large Scale Scientific Computing* in Bielefeld, 1992, jointly organized with Angelika Bunse-Gerstner. First encounters of many in this picture

invariant subspaces of the $2n \times 2n$ Hamiltonian matrix

$$\begin{bmatrix} A & G \\ F & -A^T \end{bmatrix}.$$

As explained in Chaps. 4 and 5, such quadratic matrix equations and their links to structured matrices/matrix pencils have been a major theme in Volker's work, ever since his 1986 joint work [A6] with Angelika Bunse-Gerstner and his 1987 habilitation thesis [T3, B1]. Among others, this work has led to robust numerical algorithms for solving optimal and robust control problems.

In a collaboration with Mihail M. Konstantinov, Petko H. Petkov, and others, theoretical properties of matrix equations have been investigated, establishing a general framework for deriving local and nonlocal perturbation bounds. This work, which has also been extended to the perturbation analysis of eigenvalue problems, is described in Chap. 7.

Differential-Algebraic Equations

A *differential-algebraic equation* (DAE) arises when one imposes algebraic constraints on the states of a physical system that is modelled with differential equations. DAEs in all flavors are probably the most central theme of Volker

Fig. 5 Volker also (co-)organized many major international conferences, including the *6th Conference of the International Linear Algebra Society (ILAS)*, held in 1996 Chemnitz (Germany)

Mehrmann's work, and several chapters of this book are therefore dedicated to this topic.

Chapter 16 summarizes the results of the long lasting and very fruitful collaboration with Peter Kunkel. This collaboration gave birth to the *strangeness index* and the monograph [B5], Volker's most-cited research publication at the time of writing.

The interaction between algebraic and differential constraints in a DAE of higher index may lead to further hidden constraints, which often result in difficulties during the numerical solution. As discussed in Chap. 17 by Scholz and Steinbrecher, these difficulties can be avoided by regularization techniques. Volker and his co-authors made several important contributions to this topic, in particular for DAEs modelling mechanical systems, electrical circuits, and flow problems as well as for hybrid DAEs. Nonlinear DAEs for electrical circuits are discussed in more detail in Chap. 18 by Reis and Stykel.

In the context of control theory, *descriptor systems* arise when the dynamics are described by DAEs. Chapter 15 by Nichols and Chu summarizes the work of Volker and his co-authors on regularization and disturbance decoupling. More recent developments include model reduction [C38], the computation of state reachable points [S10], and a behavorial approach [A103, A104] for descriptor systems.

Smooth decompositions of parameter-dependent matrices are frequently needed in the treatment of DAEs and descriptor systems with time-varying coefficients. An early and rather influential work in this area, the 1991 paper [A22] investigates the analytic singular value decomposition and proposes an algorithm based on differential equations. This and subsequent work is discussed in Chap. 11 by Van

Fig. 6 Volker enjoying life (*left*) and with Ralph Byers (*right*) during an excursion at the *14th International Symposium on Mathematical Theory of Networks and Systems (MTNS)*, 2000, in Perpignan (France)

Vleck, which also highlights an application to the computation of Lyapunov and Sacker-Sell spectral intervals.

The presence of time delays complicates the analysis and solution of DAEs significantly. In [A155, A173], the solvability and regularization of such DAEs are investigated. Chapter 19 by Linh and Thuan summarizes the work of Volker and his co-authors on robust stability concepts for DAEs, with and without time delays.

Applications in economy, biology, or chemistry frequently lead to differential equations that feature a nonnegative solution. Characterizing this nonnegativity property and preserving it during the time discretization turns out to be quite a challenge for DAEs, which has been addressed in a number of recent publications [A162, S4, S5]. This connects to earlier work on Perron-Frobenius theory for matrix pencils [A70, A127] and sign controllability [A32].

Other Topics in Control Theory

Apart from optimal/robust control and the treatment of descriptor systems, Volker takes a general interest in the role of numerical methods in control. A 1995 report [M10] with Chunyang He and Alan J. Laub carries the provocative title *Placing plenty of poles is pretty preposterous* and points out a fundamental numerical limitation of pole placement, which was analyzed in more detail in joint work with Hongguo Xu [A44, A52, A60]. A more general picture of the sensitivity

Fig. 7 A very typical situation in Volker's academic life: proof-reading a paper or thesis, often done at home

of computational control problems is given in [A96]. As explained in Chap. 19 as well as Chap. 20 by Kressner and Voigt, distance concepts play an important role in the numerical verification of properties for control systems. This includes the notorious distance to singularity, for which the question *Where is the nearest non-regular pencil?* [A54] still has no satisfactory answer.

Chapter 21 by Baumann, Heiland, and Schmidt summarizes the work of Volker and his co-authors on model reduction techniques for control systems governed by partial differential equations, based on directly discretizing the input/output map.

On the software side, Volker Mehrmann has played a key role in the design and creation of SLICOT [C21], a Fortran library of numerical algorithms in systems and control theory. His contributions to SLICOT are described in several SLICOT working notes [M19, M20, M28, M29, M21, M30, M31, M35, M38] and include – among others – benchmark collections for Lyapunov equations and linear control systems, following up on such collections for algebraic Riccati equations [M7, M8, A45].

Polynomial Eigenvalue Problems

A square matrix polynomial of degree d takes the form

$$P(\lambda) = A_0 + \lambda A_1 + \cdots + \lambda^d A_d, \tag{0.2}$$

for complex or real $n \times n$ coefficient matrices $A_0, \ldots, A_d$ with $A_d \neq 0$. Motivated by applications, such as the vibration analysis of mechanical systems, the treatment of the polynomial eigenvalue problem $P(\lambda)x = 0$ is a classical topic in (numerical) linear algebra. By far the most popular approach to solving such eigenvalue problems is *linearization*, which turns (0.2) into an equivalent matrix pencil $\mathscr{A} - \lambda\mathscr{E}$ of size dn.

Fig. 8 Volker in Szeged (Hungary) while on an excursion during the *Conference on Applied Linear Algebra (in Honor of Richard Varga)* in Palić (Serbia and Montengro, then), October 2005 (Courtesy of Daniel Szyld)

It seems that Volker's interest in polynomial eigenvalue problems was triggered around 2000 by matrix polynomials of the form $K + \lambda G + \lambda^2 M$ with $K = K^T$, $G = -G^T$, and $M = M^T$ positive definite. Such problems arise, for example, from gyroscopic mechanical systems and the computation of singularities in elasticity. In a joint work [A83] with David Watkins, a structure-preserving Arnoldi method for this problem was developed, based on a linearization for which $\mathscr{A} - \lambda\mathscr{E}$ is a skew-Hamiltonian/Hamiltonian pencil. This and some follow-up work is described in Chap. 2.

In a 2004 paper [C33] with Andreas Hilliges and Christian Mehl, a matrix polynomial with a rather different structure came up from the vibration analysis of rail tracks under periodic excitation: a matrix polynomial of the form

$$A_1 + \lambda A_0 + \lambda^2 A_1^T \text{with } A_0 = A_0^T.$$

This structure was called palindromic, and a structure-preserving linearization having the form $\mathscr{A} - \lambda\mathscr{A}^T$ was proposed for the solution of the corresponding eigenvalue problem. This early work turned out to be quite influential and has led to a lot of follow-up work concerning the analysis and numerical solution of polynomial eigenvalue problems with palindromic and alternating structures; see Chap. 3 by Lin and Schröder as well as Chap. 12 by Mackey, Mackey, and Tisseur.

The two examples above naturally led to the consideration of a more general strategy for deriving structured linearizations of structured polynomials. However, at that time, linearizations were usually invented on an ad hoc basis and no clear pattern was visible. Clarity to this question was brought in two 2006 papers [A108, A109] with D. Steven Mackey, Niloufer Mackey, and Christian Mehl. This work provides a whole vector space of linearizations, offering ample opportunities to look for (well-conditioned) structured linearizations; see Chap. 12.

From a numerical perspective, it is highly desirable to deflate singular parts as well as zero and infinite eigenvalues before solving the (linearized) polynomial eigenvalue problem. To achieve this, joint work [A120] with Ralph Byers and Hongguo Xu proposes the concept of trimmed linearizations based on (structured)

Fig. 9 Volker, advisor of many Ph.D., Master, and *Diplom* students. A list of his doctoral descendants can be found after the list of Volker's publications on page xxxvii. The picture shows Volker together with Dario Bauso at the Ph.D. defense of Puduru Viswanadha Reddy at Tilburg University (The Netherlands), November 2011 – they both served as members of the thesis committee

staircase forms. The related problem of linearizations for higher-order DAEs is investigated in [A111, A123].

Polynomial eigenvalue problems are a special case of the more general class of nonlinear eigenvalue problems $T(\lambda)x = 0$, where T is a holomorphic matrix-valued function. Volker wrote a widely appreciated survey [A101] with Heinrich Voss on such eigenvalue problems and participated in the creation of the NLEVP benchmark collection [A163].

The Helmut Wielandt Project

Helmut Wielandt (1910–2001) was a German mathematician, who is most renowned for his work on finite groups, but who also made fundamental contributions to operator and matrix theory. His mathematical diaries were made available by his family after his death. Together with Wolfgang Knapp and Peter Schmid, Volker Mehrmann raised funding from the DFG to transcribe these diaries and make them publicly available.[1] Among others, this project involved Hans Schneider, with whom Volker published the thought-provoking German essay [A89] on the dilemma Wielandt faced in Nazi Germany.

Two joint publications [A97, A167] with Olga Holtz and Hans Schneider follow up on observations made by Wielandt on commutativity properties. Consider two complex square matrices A and B satisfying $AB = \omega BA$, where ω is a primitive qth root of unity. Then Potter's theorem states that $A^q + B^q = (A + B)^q$. In [A97], Wielandt's matrix-theoretic proof of this theorem from his diaries is reproduced and translated. Moreover, a counterexample is given, showing that the converse of

[1]See http://www3.math.tu-berlin.de/numerik/Wielandt/.

Fig. 10 Volker at an event of the DFG Research Center MATHEON "Mathematics for key technologies" in Berlin, 2007. Volker was the MATHEON vice-chair 2002–2008 and serves as MATHEON chair since 2008 (Courtesy of Kay Herschelmann Photographie)

Potter's theorem does not hold in general; a matter that was discussed in more detail later on in joint work [A132] with Raphael Loewy. The paper [A167] follows up on a letter from Issai Schur to Helmut Wielandt written in 1934 concerning the question when a matrix-valued function commutes with its derivative.

Miscellaneous

The above discussion does not exhaust the research topics Volker Mehrmann has worked on. A (certainly) incomplete list of further topics is as follows:

- Linear algebra aspects in adaptive finite element methods for PDE eigenvalue problems, see Chap. 9 by Międlar
- Low-rank matrix and tensor approximation, see Chap. 14 by Friedland and Tammali
- Large-scale linear systems, see Chap. 10 by Bollhöfer, and large-scale eigenvalue problems [A68, A116]
- Matrix completion problems [C13, A36, A39]
- Inverse eigenvalue problems [A72, A94]
- Simultaneous diagonalization [A31]
- Sparse (approximate) solutions [A130, A138, S12]
- Industrial applications, such as train traffic simulation and optimization [A118, C31].

Volker Mehrmann's engagement in promoting mathematics in applications and industry is witnessed in the publications [C27, B6, B8, B9, B13].

Apart from the research monographs [B1, B2, B5], Volker also co-authored two German textbooks: on numerical analysis [B3] with Matthias Bollhöfer and on linear algebra [B10] with Jörg Liesen.

Fig. 11 Volker at the *Mathematische Forschungsinstitut Oberwolfach*, 2009. The print on his T-shirt tells it all and might pass for Volker's philosophy of life: "Math makes you happy!" (Courtesy of *Bildarchiv des Mathematischen Forschungsinstituts Oberwolfach*)

Summary

Without any doubt, Volker Mehrmann is among the most prolific researchers in numerical linear algebra and numerical analysis, reaching out to a diversity of application areas, most notably in systems and control theory. On the one hand, this is witnessed by the sheer length of his publication list (see the following pages) and the diversity of research topics covered. On the other hand, every collaborator of Volker has personally experienced his unmatched energy and enthusiasm about doing mathematics. We, the editors of this book, have had the fortune to enjoy Volker's generous and unconditional support during our Ph.D. studies and all subsequent collaborations. Volker would be the perfect role model for us if only we understood how he is able to achieve all this research output in addition to his major committments and responsibilities in research policy and administration. It is tempting to assume that there exist at least two copies of him – a conjecture that was supported by a machine translation, which translated Volker's last name "Mehrmann" into "multi man". In any case, we wish Volker (and possibly his

Fig. 12 The "multi-man" 2008 (*left*) and 2012 (*right*) in front of his favorite tool (*Left picture* courtesy of TU Berlin/Pressestelle/Ulrich Dahl, *right picture* © 2012 Fernando Domingo Aldama (Ediciones EL PAÍS, SL) All rights reserved)

copies) many more productive decades and we are looking forward to many more fruitful discussions and collaborations with him.

Magdeburg, Germany — Peter Benner
Braunschweig, Germany — Matthias Bollhöfer
Lausanne, Switzerland — Daniel Kressner
Berlin, Germany — Christian Mehl
Augsburg, Germany — Tatjana Stykel
December 2014

Publications of Volker Mehrmann (as of December 18, 2014)

Books

B1. Mehrmann, V.: The Autonomous Linear Quadratic Control Problem. Theory and Numerical Solution, *Lecture Notes in Control and Information Sciences*, vol. 163. Berlin etc.: Springer-Verlag (1991)

B2. Konstantinov, M. M., Gu, D. W., Mehrmann, V., Petkov, P. H.: Perturbation Theory for Matrix Equations. Amsterdam: North Holland (2003)

B3. Bollhöfer, M., Mehrmann, V.: Numerische Mathematik. Eine projektorientierte Einführung für Ingenieure, Mathematiker und Naturwissenschaftler. Wiesbaden: Vieweg (2004)

B4. Benner, P., Mehrmann, V., Sorensen, D. C. (eds.): Dimension Reduction of Large-Scale Systems, *Lecture Notes in Computational Science and Engineering*, vol. 45. Berlin: Springer (2005)

B5. Kunkel, P., Mehrmann, V.: Differential-Algebraic Equations. Analysis and Numerical Solution. Zürich: European Mathematical Society Publishing House (2006)

B6. Grötschel, M., Lucas, K., Mehrmann, V. (eds.): Produktionsfaktor Mathematik. Wie Mathematik Technik und Wirtschaft bewegt. Berlin: Springer; München: acatech–Deutsche Akademie der Technikwissenschaften (2009)

B7. Bini, D. A., Mehrmann, V., Olshevsky, V., Tyrtyshnikov, E. E., van Barel, M. (eds.): Numerical Methods for Structured Matrices and Applications. The Georg Heinig Memorial Volume. Basel: Birkhäuser (2010)

B8. Grötschel, M., Lucas, K., Mehrmann, V. (eds.): Production Factor Mathematics. Berlin: Springer (2010)

B9. Lery, T., Primicerio, M., Esteban, M. J., Fontes, M., Maday, Y., Mehrmann, V., Quadros, G., Schilders, W., Schuppert, A., Tewkesbury, H. (eds.): European Success Stories in Industrial Mathematics. Springer (2011)

B10. Liesen, J., Mehrmann, V.: Lineare Algebra. Ein Lehrbuch über die Theorie mit Blick auf die Praxis. Wiesbaden: Vieweg+Teubner (2011)

B11. Arendt, W., Ball, J. A., Behrndt, J., Förster, K.-H., Mehrmann, V., Trunk, C. (eds.): Spectral Theory, Mathematical System Theory, Evolution Equations, Differential and Difference Equations. Basel: Birkhäuser (2012)

B12. Biegler, L. T., Campbell, S. L., Mehrmann, V. (eds.): Control and Optimization with Differential-Algebraic Constraints. Philadelphia, PA: Society for Industrial and Applied Mathematics (SIAM) (2012)

B13. Deuflhard, P., Grötschel, M., Hömberg, D., Horst, U., Kramer, J., Mehrmann, V., Polthier, K., Schmidt, F., Schütte, C., Skutella, M., Sprekels, J. (eds.): MATHEON, DFG Research Center Mathematics for Key Technologies in Berlin. Zürich: European Mathematical Society (EMS) (2014)

Journal Articles

A1. Varga, R. S., Saff, E. B., Mehrmann, V.: Incomplete factorizations of matrices and connections with H-matrices. SIAM J. Numer. Anal. **17**, 787–793 (1980)
A2. Mehrmann, V.: On a generalized Fan inequality. Linear Algebra Appl. **58**, 235–245 (1984)
A3. Mehrmann, V.: On classes of matrices containing M-matrices and Hermitian positive semidefinite matrices. Linear Algebra Appl. **58**, 217–234 (1984)
A4. Mehrmann, V.: On some conjectures on the spectra of τ-matrices. Linear Multilinear Algebra **16**, 101–112 (1984)
A5. Mehrmann, V.: On the LU decomposition of V-matrices. Linear Algebra Appl. **61**, 175–186 (1984)
A6. Bunse-Gerstner, A., Mehrmann, V.: A symplectic QR like algorithm for the solution of the real algebraic Riccati equation. IEEE Trans. Autom. Control **31**, 1104–1113 (1986)
A7. Byers, R., Mehrmann, V.: Symmetric updating of the solution of the algebraic Riccati equation. Methods Oper. Res. **54**, 117–125 (1986)
A8. Elsner, L., Mehrmann, V.: Positive reciprocal matrices. Linear Algebra Appl. **80**, 200–203 (1986)
A9. Hershkowitz, D., Mehrmann, V.: Linear transformations which map the classes of ω-matrices and τ-matrices into or onto themselves. Linear Algebra Appl. **78**, 79–106 (1986)
A10. Hershkowitz, D., Mehrmann, V., Schneider, H.: Matrices with sign symmetric diagonal shifts or scalar shifts. Linear Algebra Appl. **80**, 216–217 (1986)
A11. Hershkowitz, D., Mehrmann, V., Schneider, H.: Eigenvalue interlacing for certain classes of matrices with real principal minors. Linear Algebra Appl. **88–89**, 373–405 (1987)
A12. Hershkowitz, D., Mehrmann, V., Schneider, H.: Matrices with sign symmetric diagonal shifts or scalar shifts. SIAM J. Algebraic Discrete Methods **8**, 108–122 (1987)
A13. Mehrmann, V.: A symplectic orthogonal method for single input or single output discrete time optimal quadratic control problems. SIAM J. Matrix Anal. Appl. **9**(2), 221–247 (1988)
A14. Mehrmann, V., Tan, E.: Defect correction methods for the solution of algebraic Riccati equations. IEEE Trans. Autom. Control **33**(7), 695–698 (1988)
A15. Bunse-Gerstner, A., Byers, R., Mehrmann, V.: A quaternion QR-algorithm. Numer. Math. **55**(1), 83–95 (1989)
A16. Bunse-Gerstner, A., Mehrmann, V.: The HHDR algorithm and its application to optimal control problems. RAIRO, Autom. Prod. Inf. Ind. **23**(4), 305–329 (1989)
A17. Bunse-Gerstner, A., Mehrmann, V., Watkins, D.: An SR algorithm for Hamiltonian matrices based on Gaussian elimination. Methods Oper. Res. **58**, 339–357 (1989)
A18. Mehrmann, V.: Existence, uniqueness, and stability of solutions to singular linear quadratic optimal control problems. Linear Algebra Appl. **121**, 291–331 (1989)
A19. Mehrmann, V., Krause, G. M.: Linear transformations which leave controllable multiinput descriptor systems controllable. Linear Algebra Appl. **120**, 47–64 (1989)
A20. Kunkel, P., Mehrmann, V.: Numerical solution of differential algebraic Riccati equations. Linear Algebra Appl. **137–138**, 39–66 (1990)
A21. Ammar, G., Mehrmann, V.: On Hamiltonian and symplectic Hessenberg forms. Linear Algebra Appl. **149**, 55–72 (1991)
A22. Bunse-Gerstner, A., Byers, R., Mehrmann, V., Nichols, N. K.: Numerical computation of an analytic singular value decomposition of a matrix valued function. Numer. Math. **60**(1), 1–39 (1991)
A23. Elsner, L., Mehrmann, V.: Convergence of block iterative methods for linear systems arising in the numerical solution of Euler equations. Numer. Math. **59**(6), 541–559 (1991)
A24. Flaschka, U., Mehrmann, V., Zywietz, D.: An analysis of structure preserving numerical methods for symplectic eigenvalue problems. RAIRO, Autom. Prod. Inf. Ind. **25**(2), 165–189 (1991)
A25. Kunkel, P., Mehrmann, V.: Smooth factorizations of matrix valued functions and their derivatives. Numer. Math. **60**(1), 115–131 (1991)

A26. Bunse-Gerstner, A., Byers, R., Mehrmann, V.: A chart of numerical methods for structured eigenvalue problems. SIAM J. Matrix Anal. Appl. **13**(2), 419–453 (1992)
A27. Bunse-Gerstner, A., Mehrmann, V., Nichols, N. K.: Regularization of descriptor systems by derivative and proportional state feedback. SIAM J. Matrix Anal. Appl. **13**(1), 46–67 (1992)
A28. Kunkel, P., Mehrmann, V.: Errata: Numerical solution of differential algebraic Riccati equations. Linear Algebra Appl. **165**, 273–274 (1992)
A29. Ammar, G., Benner, P., Mehrmann, V.: A multishift algorithm for the numerical solution of algebraic Riccati equations. Electron. Trans. Numer. Anal. **1**, 33–48 (1993)
A30. Ammar, G., Mehrmann, V., Nichols, N. K., Van Dooren, P.: Numerical linear algebra methods in control signals and systems - Preface. Linear Algebra Appl. **188**, 1 (1993)
A31. Bunse-Gerstner, A., Byers, R., Mehrmann, V.: Numerical methods for simultaneous diagonalization. SIAM J. Matrix Anal. Appl. **14**(4), 927–949 (1993)
A32. Johnson, C. R., Mehrmann, V., Olesky, D. D.: Sign controllability of a nonnegative matrix and a positive vector. SIAM J. Matrix Anal. Appl. **14**(2), 398–407 (1993)
A33. Mehrmann, V.: Divide and conquer methods for block tridiagonal systems. Parallel Comput. **19**(3), 257–279 (1993)
A34. Mehrmann, V., Rath, W.: Numerical methods for the computation of analytic singular value decompositions. Electron. Trans. Numer. Anal. **1**, 72–88 (1993)
A35. Bunse-Gerstner, A., Mehrmann, V., Nichols, N. K.: Regularization of descriptor systems by output feedback. IEEE Trans. Autom. Control **39**(8), 1742–1748 (1994)
A36. Elsner, L., He, C., Mehrmann, V.: Minimizing the condition number of a positive definite matrix by completion. Numer. Math. **69**(1), 17–23 (1994)
A37. Kunkel, P., Mehrmann, V.: Canonical forms for linear differential-algebraic equations with variable coefficients. J. Comput. Appl. Math. **56**(3), 225–251 (1994)
A38. Kunkel, P., Mehrmann, V.: A new look at pencils of matrix valued functions. Linear Algebra Appl. **212–213**, 215–248 (1994)
A39. Elsner, L., He, C., Mehrmann, V.: Minimization of the norm, the norm of the inverse and the condition number of a matrix by completion. Numer. Linear Algebra Appl. **2**(2), 155–171 (1995)
A40. Kunkel, P., Mehrmann, V.: Generalized inverses of differential-algebraic operators. SIAM J. Matrix Anal. Appl. **17**(2), 426–442 (1996)
A41. Kunkel, P., Mehrmann, V.: Local and global invariants of linear differential-algebraic equations and their relation. Electron. Trans. Numer. Anal. **4**, 138–157 (1996)
A42. Kunkel, P., Mehrmann, V.: A new class of discretization methods for the solution of linear differential-algebraic equations with variable coefficients. SIAM J. Numer. Anal. **33**(5), 1941–1961 (1996)
A43. Mehrmann, V.: A step toward a unified treatment of continuous and discrete time control problems. Linear Algebra Appl. **241–243**, 749–779 (1996)
A44. Mehrmann, V., Xu, H.: An analysis of the pole placement problem I: The single-input case. Electron. Trans. Numer. Anal. **4**, 89–105 (1996)
A45. Benner, P., Laub, A. J., Mehrmann, V.: Benchmarks for the numerical solution of algebraic Riccati equations. IEEE Contr. Syst. Mag. **17**(5), 18–28 (1997)
A46. Benner, P., Mehrmann, V., Xu, H.: A new method for computing the stable invariant subspace of a real Hamiltonian matrix. J. Comput. Appl. Math. **86**(1), 17–43 (1997)
A47. Byers, R., Geerts, T., Mehrmann, V.: Descriptor systems without controllability at infinity. SIAM J. Control Optim. **35**(2), 462–479 (1997)
A48. Byers, R., He, C., Mehrmann, V.: The matrix sign function method and the computation of invariant subspaces. SIAM J. Matrix Anal. Appl. **18**(3), 615–632 (1997)
A49. Byers, R., Kunkel, P., Mehrmann, V.: Regularization of linear descriptor systems with variable coefficients. SIAM J. Control Optim. **35**(1), 117–133 (1997)
A50. Kunkel, P., Mehrmann, V.: The linear quadratic optimal control problem for linear descriptor systems with variable coefficients. Math. Control Signals Syst. **10**(3), 247–264 (1997)

A51. Kunkel, P., Mehrmann, V., Rath, W., Weickert, J.: A new software package for linear differential-algebraic equations. SIAM J. Sci. Comput. **18**(1), 115–138 (1997)
A52. Mehrmann, V., Xu, H.: An analysis of the pole placement problem. II: The multi-input case. Electron. Trans. Numer. Anal. **5**, 77–97 (1997)
A53. Benner, P., Mehrmann, V., Xu, H.: A numerically stable, structure preserving method for computing the eigenvalues of real Hamiltonian or symplectic pencils. Numer. Math. **78**(3), 329–358 (1998)
A54. Byers, R., He, C., Mehrmann, V.: Where is the nearest non-regular pencil? Linear Algebra Appl. **285**(1–3), 81–105 (1998)
A55. He, C., Hench, J. J., Mehrmann, V.: On damped algebraic Riccati equations. IEEE Trans. Autom. Control **43**(11), 1634–1637 (1998)
A56. Hench, J. J., He, C., Kučera, V., Mehrmann, V.: Dampening controllers via a Riccati equation approach. IEEE Trans. Autom. Control **43**(9), 1280–1284 (1998)
A57. Kirkland, S., Mehrmann, V., Michler, G., Shader, B.: 6th conference of the international linear algebra society (ILAS), Chemnitz, Germany, August 14–17, 1996. Linear Algebra Appl. **275–276**, 632 (1998)
A58. Kunkel, P., Mehrmann, V.: Regular solutions of nonlinear differential-algebraic equations and their numerical determination. Numer. Math. **79**(4), 581–600 (1998)
A59. Mehrmann, V.: Chemnitz 1996 conference report. Linear Algebra Appl. **275–276**, 627–629 (1998)
A60. Mehrmann, V., Xu, H.: Choosing poles so that the single-input pole placement problem is well conditioned. SIAM J. Matrix Anal. Appl. **19**(3), 664–681 (1998)
A61. Ammar, G., Mehl, C., Mehrmann, V.: Schur-like forms for matrix Lie groups, Lie algebras and Jordan algebras. Linear Algebra Appl. **287**(1–3), 11–39 (1999)
A62. Benner, P., Mehrmann, V., Xu, H.: A note on the numerical solution of complex Hamiltonian and skew-Hamiltonian eigenvalue problems. Electron. Trans. Numer. Anal. **8**, 115–126 (1999)
A63. Bhatia, R., Bunse-Gerstner, A., Mehrmann, V., Olesky, D. D.: Special issue celebrating the 60th birthday of L. Elsner. Linear Algebra Appl. **287**(1–3), 382 (1999)
A64. Bunse-Gerstner, A., Byers, R., Mehrmann, V., Nichols, N. K.: Feedback design for regularizing descriptor systems. Linear Algebra Appl. **299**(1–3), 119–151 (1999)
A65. Bunse-Gerstner, A., Mehrmann, V.: L. Elsner and his contributions to core, applied and numerical linear algebra. Linear Algebra Appl. **287**(1–3), 3–10 (1999)
A66. Chu, D., Mehrmann, V.: Disturbance decoupled observer design for descriptor systems. Syst. Control Lett. **38**(1), 37–48 (1999)
A67. Chu, D., Mehrmann, V., Nichols, N. K.: Minimum norm regularization of descriptor systems by mixed output feedback. Linear Algebra Appl. **296**(1–3), 39–77 (1999)
A68. Elsner, U., Mehrmann, V., Milde, F., Römer, R. A., Schreiber, M.: The Anderson model of localization: A challenge for modern eigenvalue methods. SIAM J. Sci. Comput. **20**(6), 2089–2102 (1999)
A69. Lin, W.-W., Mehrmann, V., Xu, H.: Canonical forms for Hamiltonian and symplectic matrices and pencils. Linear Algebra Appl. **302–303**, 469–533 (1999)
A70. Mehrmann, V., Olesky, D. D., Phan, T. X. T., van den Driessche, P.: Relations between Perron-Frobenius results for matrix pencils. Linear Algebra Appl. **287**(1–3), 257–269 (1999)
A71. Mehrmann, V., Xu, H.: Structured Jordan canonical forms for structured matrices that are Hermitian, skew Hermitian or unitary with respect to indefinite inner products. Electron. J. Linear Algebra **5**, 67–103 (1999)
A72. Arav, M., Hershkowitz, D., Schneider, H., Mehrmann, V.: The recursive inverse eigenvalue problem. SIAM J. Matrix Anal. Appl. **22**(2), 392–412 (2000)
A73. Benner, P., Byers, R., Fassbender, H., Mehrmann, V., Watkins, D.: Cholesky-like factorizations of skew-symmetric matrices. Electron. Trans. Numer. Anal. **11**, 85–93 (2000)
A74. Chu, D., Mehrmann, V.: Disturbance decoupling for descriptor systems by measurement feedback. Electron. J. Linear Algebra **7**, 152–173 (2000)

A75. Chu, D., Mehrmann, V.: Disturbance decoupling for descriptor systems by state feedback. SIAM J. Control Optim. **38**(6), 1830–1858 (2000)
A76. Konstantinov, M. M., Mehrmann, V., Petkov, P. H.: On properties of Sylvester and Lyapunov operators. Linear Algebra Appl. **312**(1–3), 35–71 (2000)
A77. Mehl, C., Mehrmann, V., Xu, H.: Canonical forms for doubly structured matrices and pencils. Electron. J. Linear Algebra **7**, 112–151 (2000)
A78. Mehrmann, V., Xu, H.: Numerical methods in control. J. Comput. Appl. Math. **123**(1–2), 371–394 (2000)
A79. Chu, D., Mehrmann, V.: Disturbance decoupling for linear time-invariant systems: A matrix pencil approach. IEEE Trans. Autom. Control **46**(5), 802–808 (2001)
A80. Konstantinov, M. M., Mehrmann, V., Petkov, P. H.: Perturbation analysis of Hamiltonian Schur and block-Schur forms. SIAM J. Matrix Anal. Appl. **23**(2), 387–424 (2001)
A81. Kunkel, P., Mehrmann, V.: Analysis of over- and underdetermined nonlinear differential-algebraic systems with application to nonlinear control problems. Math. Control Signals Syst. **14**(3), 233–256 (2001)
A82. Kunkel, P., Mehrmann, V., Rath, W.: Analysis and numerical solution of control problems in descriptor form. Math. Control Signals Syst. **14**(1), 29–61 (2001)
A83. Mehrmann, V., Watkins, D.: Structure-preserving methods for computing eigenpairs of large sparse skew-Hamiltonian/Hamiltonian pencils. SIAM J. Sci. Comput. **22**(6), 1905–1925 (2001)
A84. Apel, T., Mehrmann, V., Watkins, D.: Structured eigenvalue methods for the computation of corner singularities in 3D anisotropic elastic structures. Comput. Methods Appl. Mech. Eng. **191**(39–40), 4459–4473 (2002)
A85. Benner, P., Byers, R., Mehrmann, V., Xu, H.: Numerical computation of deflating subspaces of skew-Hamiltonian/Hamiltonian pencils. SIAM J. Matrix Anal. Appl. **24**(1), 165–190 (2002)
A86. Benner, P., Mehrmann, V., Xu, H.: Perturbation analysis for the eigenvalue problem of a formal product of matrices. BIT **42**(1), 1–43 (2002)
A87. Bollhöfer, M., Mehrmann, V.: Algebraic multilevel methods and sparse approximate inverses. SIAM J. Matrix Anal. Appl. **24**(1), 191–218 (2002)
A88. Freiling, G., Mehrmann, V., Xu, H.: Existence, uniqueness, and parametrization of Lagrangian invariant subspaces. SIAM J. Matrix Anal. Appl. **23**(4), 1045–1069 (2002)
A89. Mehrmann, V., Schneider, H.: Anpassen oder nicht? Die Geschichte eines Mathematikers im Deutschland der Jahre 1933–1950. Mitteilungen der Deutschen Mathematiker Vereinigung **10**(2), 20–26 (2002)
A90. Mehrmann, V., Watkins, D.: Polynomial eigenvalue problems with Hamiltonian structure. Electron. Trans. Numer. Anal. **13**, 106–118 (2002)
A91. Benner, P., Kressner, D., Mehrmann, V.: Structure preservation: A challenge in computational control. Future Gener. Comp. Sy. **19**(7), 1243–1252 (2003)
A92. Hwang, T.-M., Lin, W.-W., Mehrmann, V.: Numerical solution of quadratic eigenvalue problems with structure-preserving methods. SIAM J. Sci. Comput. **24**(4), 1283–1302 (2003)
A93. Konstantinov, M. M., Mehrmann, V., Petkov, P. H.: Perturbed spectra of defective matrices. J. Appl. Math. **2003**(3), 115–140 (2003)
A94. Loewy, R., Mehrmann, V.: A note on the symmetric recursive inverse eigenvalue problem. SIAM J. Matrix Anal. Appl. **25**(1), 180–187 (2003)
A95. Frommer, A., Mehrmann, V., Nabben, R.: Special section dedicated to the GAMM Workshop - Applied and Numerical Linear Algebra with Special Emphasis on Numerical Methods for Structured and Random Matrices - Preface. Linear Algebra Appl. **380**, 1–2 (2004)
A96. Higham, N. J., Konstantinov, M., Mehrmann, V., Petkov, P. H.: The sensitivity of computational control problems. IEEE Control Syst. Mag. **24**(1), 28–43 (2004)

A97. Holtz, O., Mehrmann, V., Schneider, H.: Potter, Wielandt, and Drazin on the matrix equation $AB = \omega BA$: new answers to old questions. Am. Math. Mon. **111**(8), 655–667 (2004)

A98. Kunkel, P., Mehrmann, V.: Index reduction for differential-algebraic equations by minimal extension. Z. Angew. Math. Mech. **84**(9), 579–597 (2004)

A99. Kunkel, P., Mehrmann, V., Stöver, R.: Symmetric collocation for unstructered nonlinear differential-algebraic equations of arbitrary index. Numer. Math. **98**(2), 277–304 (2004)

A100. Mehl, C., Mehrmann, V., Xu, H.: On doubly structured matrices and pencils that arise in linear response theory. Linear Algebra Appl. **380**, 3–51 (2004)

A101. Mehrmann, V., Voss, H.: Nonlinear eigenvalue problems: A challenge for modern eigenvalue methods. Mitt. Ges. Angew. Math. Mech. **27**(2), 121–152 (2004)

A102. Gäbler, A., Kraume, M., Paschedag, A., Schlauch, S., Mehrmann, V.: Experiments, modeling and simulation of drop size distributions in stirred liquid/liquid systems. Chem.-Ing.-Tech. **77**(8), 1091–1092 (2005)

A103. Ilchmann, A., Mehrmann, V.: A behavioral approach to time-varying linear systems. I: General theory. SIAM J. Control Optim. **44**(5), 1725–1747 (2005)

A104. Ilchmann, A., Mehrmann, V.: A behavioral approach to time-varying linear systems. II: Descriptor systems. SIAM J. Control Optim. **44**(5), 1748–1765 (2005)

A105. Kunkel, P., Mehrmann, V.: Characterization of classes of singular linear differential-algebraic equations. Electron. J. Linear Algebra **13**, 359–386 (2005)

A106. Kunkel, P., Mehrmann, V., Stöver, R.: Multiple shooting for unstructured nonlinear differential-algebraic equations of arbitrary index. SIAM J. Numer. Anal. **42**(6), 2277–2297 (2005)

A107. Bora, S., Mehrmann, V.: Linear perturbation theory for structured matrix pencils arising in control theory. SIAM J. Matrix Anal. Appl. **28**(1), 148–169 (2006)

A108. Mackey, D. S., Mackey, N., Mehl, C., Mehrmann, V.: Structured polynomial eigenvalue problems: Good vibrations from good linearizations. SIAM J. Matrix Anal. Appl. **28**(4), 1029–1051 (2006)

A109. Mackey, D. S., Mackey, N., Mehl, C., Mehrmann, V.: Vector spaces of linearizations for matrix polynomials. SIAM J. Matrix Anal. Appl. **28**(4), 971–1004 (2006)

A110. Mehrmann, V.: Numerical solution of structured eigenvalue problems. Bol. Soc. Esp. Mat. Apl., SeMA **34**, 57–68 (2006)

A111. Mehrmann, V., Shi, C.: Transformation of high order linear differential-algebraic systems to first order. Numer. Algorithms **42**(3–4), 281–307 (2006)

A112. Mehrmann, V., Stykel, T.: Descriptor systems: A general mathematical framework for modelling, simulation and control. at - Automatisierungstechnik **54**(8), 405–415 (2006)

A113. Benner, P., Byers, R., Mehrmann, V., Xu, H.: A robust numerical method for the γ-iteration in H_∞ control. Linear Algebra Appl. **425**(2–3), 548–570 (2007)

A114. Byers, R., Mehrmann, V., Xu, H.: A structured staircase algorithm for skew-symmetric/symmetric pencils. Electron. Trans. Numer. Anal. **26**, 1–33 (2007)

A115. Chu, D., Liu, X., Mehrmann, V.: A numerical method for computing the Hamiltonian Schur form. Numer. Math. **105**(3), 375–412 (2007)

A116. Fritzsche, D., Mehrmann, V., Szyld, D. B., Virnik, E.: An SVD approach to identifying metastable states of Markov chains. Electron. Trans. Numer. Anal. **29**, 46–69 (2007)

A117. Kunkel, P., Mehrmann, V.: Stability properties of differential-algebraic equations and spin-stabilized discretizations. Electron. Trans. Numer. Anal. **26**, 385–420 (2007)

A118. Bavafa-Toosi, Y., Blendinger, C., Mehrmann, V., Steinbrecher, A., Unger, R.: A new methodology for modeling, analysis, synthesis, and simulation of time-optimal train traffic in large networks. IEEE Trans. Autom. Sci. Eng. **5**(1), 43–52 (2008)

A119. Brualdi, R. A., Mehrmann, V., Schneider, H.: LAA is 40 years old - Preface. Linear Algebra Appl. **428**(1), 1–3 (2008)

A120. Byers, R., Mehrmann, V., Xu, H.: Trimmed linearizations for structured matrix polynomials. Linear Algebra Appl. **429**(10), 2373–2400 (2008)

A121. Hamann, P., Mehrmann, V.: Numerical solution of hybrid systems of differential-algebraic equations. Comput. Methods Appl. Mech. Eng. **197**(6–8), 693–705 (2008)
A122. Kunkel, P., Mehrmann, V.: Optimal control for unstructured nonlinear differential-algebraic equations of arbitrary index. Math. Control Signals Syst. **20**(3), 227–269 (2008)
A123. Losse, P., Mehrmann, V.: Controllability and observability of second order descriptor systems. SIAM J. Control Optim. **47**(3), 1351–1379 (2008)
A124. Losse, P., Mehrmann, V., Poppe, L. K., Reis, T.: The modified optimal H_∞ control problem for descriptor systems. SIAM J. Control Optim. **47**(6), 2795–2811 (2008)
A125. Mehrmann, V.: Ralph Byers 1955–2007. Linear Algebra Appl. **428**(11–12), 2410–2414 (2008)
A126. Mehrmann, V., Liesen, J.: Gene Golub 1932–2007. GAMM Rundbrief (2008)
A127. Mehrmann, V., Nabben, R., Virnik, E.: Generalisation of the Perron-Frobenius theory to matrix pencils. Linear Algebra Appl. **428**(1), 20–38 (2008)
A128. Mehrmann, V., Xu, H.: Explicit solutions for a Riccati equation from transport theory. SIAM J. Matrix Anal. Appl. **30**(4), 1339–1357 (2008)
A129. Mehrmann, V., Xu, H.: Perturbation of purely imaginary eigenvalues of Hamiltonian matrices under structured perturbations. Electron. J. Linear Algebra **17**, 234–257 (2008)
A130. Jokar, S., Mehrmann, V.: Sparse solutions to underdetermined Kronecker product systems. Linear Algebra Appl. **431**(12), 2437–2447 (2009)
A131. Linh, V. H., Mehrmann, V.: Lyapunov, Bohl and Sacker-Sell spectral intervals for differential-algebraic equations. J. Dyn. Differ. Equations **21**(1), 153–194 (2009)
A132. Loewy, R., Mehrmann, V.: A note on Potter's theorem for quasi-commutative matrices. Linear Algebra Appl. **430**(7), 1812–1825 (2009)
A133. Mackey, D. S., Mackey, N., Mehl, C., Mehrmann, V.: Numerical methods for palindromic eigenvalue problems: Computing the anti-triangular Schur form. Numer. Linear Algebra Appl. **16**(1), 63–86 (2009)
A134. Mehl, C., Mehrmann, V., Ran, A. C. M., Rodman, L.: Perturbation analysis of Lagrangian invariant subspaces of symplectic matrices. Linear Multilinear Algebra **57**(2), 141–184 (2009)
A135. Mehl, C., Mehrmann, V., Xu, H.: Structured decompositions for matrix triples: SVD-like concepts for structured matrices. Oper. Matrices **3**(3), 303–356 (2009)
A136. Mehrmann, V., Schröder, C., Watkins, D.: A new block method for computing the Hamiltonian Schur form. Linear Algebra Appl. **431**(3–4), 350–368 (2009)
A137. Mehrmann, V., Wunderlich, L.: Hybrid systems of differential algebraic equations - Analysis and numerical solution. J. Process Contr. **19**(8), 1218–1228 (2009)
A138. Jokar, S., Mehrmann, V., Pfetsch, M. E., Yserentant, H.: Sparse approximate solution of partial differential equations. Appl. Numer. Math. **60**(4), 452–472 (2010)
A139. Mackey, D. S., Mackey, N., Mehl, C., Mehrmann, V.: Jordan structures of alternating matrix polynomials. Linear Algebra Appl. **432**(4), 867–891 (2010)
A140. Mehl, C., Mehrmann, V., Xu, H.: Singular-value-like decomposition for complex matrix triples. J. Comput. Appl. Math. **233**(5), 1245–1276 (2010)
A141. Ahmad, S. S., Mehrmann, V.: Perturbation analysis for complex symmetric, skew symmetric, even and odd matrix polynomials. Electron. Trans. Numer. Anal. **38**, 275–302 (2011)
A142. Alam, R., Bora, S., Karow, M., Mehrmann, V., Moro, J.: Perturbation theory for Hamiltonian matrices and the distance to bounded-realness. SIAM J. Matrix Anal. Appl. **32**(2), 484–514 (2011)
A143. Benner, P., Byers, R., Losse, P., Mehrmann, V., Xu, H.: Robust formulas for H_∞ optimal controllers. Automatica **47**(12), 2639–2646 (2011)
A144. Carstensen, C., Gedicke, J., Mehrmann, V., Miedlar, A.: An adaptive homotopy approach for non-selfadjoint eigenvalue problems. Numer. Math. **119**(3), 557–583 (2011)
A145. Friedland, S., Mehrmann, V., Miedlar, A., Nkengla, M.: Fast low rank approximations of matrices and tensors. Electron. J. Linear Algebra **22**, 1031–1048 (2011)
A146. Kunkel, P., Mehrmann, V.: Formal adjoints of linear DAE operators and their role in optimal control. Electron. J. Linear Algebra **22**, 672–693 (2011)

A147. Linh, V. H., Mehrmann, V.: Approximation of spectral intervals and leading directions for differential-algebraic equation via smooth singular value decompositions. SIAM J. Numer. Anal. **49**(5), 1810–1835 (2011)
A148. Linh, V. H., Mehrmann, V.: Spectral analysis for linear differential-algebraic equations. Discrete Contin. Dyn. Syst. **SI**, 991–1000 (2011)
A149. Linh, V. H., Mehrmann, V., Van Vleck, E. S.: *QR* methods and error analysis for computing Lyapunov and Sacker–Sell spectral intervals for linear differential-algebraic equations. Adv. Comput. Math. **35**(2–4), 281–322 (2011)
A150. Mackey, D. S., Mackey, N., Mehl, C., Mehrmann, V.: Smith forms of palindromic matrix polynomials. Electron. J. Linear Algebra **22**, 53–91 (2011)
A151. Mehl, C., Mehrmann, V., Ran, A. C. M., Rodman, L.: Eigenvalue perturbation theory of classes of structured matrices under generic structured rank one perturbations. Linear Algebra Appl. **435**(3), 687–716 (2011)
A152. Mehrmann, V., Miedlar, A.: Adaptive computation of smallest eigenvalues of self-adjoint elliptic partial differential equations. Numer. Linear Algebra Appl. **18**(3), 387–409 (2011)
A153. Mehrmann, V., Schröder, C.: Nonlinear eigenvalue and frequency response problems in industrial practice. J. Math. Ind. **1**, 18 (2011)
A154. Betcke, T., Mehl, C., Mehrmann, V., Reichel, L., Rump, S. M.: Special issue dedicated to Heinrich Voss's 65th birthday – Preface. Linear Algebra Appl. **436**(10, SI), 3793–3800 (2012)
A155. Ha, P., Mehrmann, V.: Analysis and reformulation of linear delay differential-algebraic equations. Electron. J. Linear Algebra **23**, 703–730 (2012)
A156. Heiland, J., Mehrmann, V.: Distributed control of linearized Navier-Stokes equations via discretized input/output maps. Z. Angew. Math. Mech. **92**(4), 257–274 (2012)
A157. Martens, W., von Wagner, U., Mehrmann, V.: Calculation of high-dimensional probability density functions of stochastically excited nonlinear mechanical systems. Nonlinear Dyn. **67**(3), 2089–2099 (2012)
A158. Mehl, C., Mehrmann, V., Ran, A. C. M., Rodman, L.: Perturbation theory of selfadjoint matrices and sign characteristics under generic structured rank one perturbations. Linear Algebra Appl. **436**(10), 4027–4042 (2012)
A159. Mehrmann, V., Poloni, F.: Doubling algorithms with permuted Lagrangian graph bases. SIAM J. Matrix Anal. Appl. **33**(3), 780–805 (2012)
A160. Mehrmann, V., Schröder, C., Simoncini, V.: An implicitly-restarted Krylov subspace method for real symmetric/skew-symmetric eigenproblems. Linear Algebra Appl. **436**(10), 4070–4087 (2012)
A161. Ahmad, S. S., Mehrmann, V.: Backward errors for eigenvalues and eigenvectors of Hermitian, skew-Hermitian, H-even and H-odd matrix polynomials. Linear Multilinear Algebra **61**(9), 1244–1266 (2013)
A162. Baum, A.-K., Mehrmann, V.: Numerical integration of positive linear differential-algebraic systems. Numer. Math. **124**(2), 279–307 (2013)
A163. Betcke, T., Higham, N. J., Mehrmann, V., Schröder, C., Tisseur, F.: NLEVP, a collection of nonlinear eigenvalue problems. ACM Trans. Math. Softw. **39**(2), 28 (2013)
A164. Du, N. H., Linh, V. H., Mehrmann, V., Thuan, D. D.: Stability and robust stability of linear time-invariant delay differential-algebraic equations. SIAM J. Matrix Anal. Appl. **34**(4), 1631–1654 (2013)
A165. Emmrich, E., Mehrmann, V.: Operator differential-algebraic equations arising in fluid dynamics. Comput. Methods Appl. Math. **13**(4), 443–470 (2013)
A166. Friedland, S., Mehrmann, V., Pajarola, R., Suter, S. K.: On best rank one approximation of tensors. Numer. Linear Algebr. **20**(6), 942–955 (2013)
A167. Holtz, O., Mehrmann, V., Schneider, H.: Matrices that commute with their derivative. On a letter from Schur to Wielandt. Linear Algebra Appl. **438**(5), 2574–2590 (2013)
A168. Mackey, D. S., Mackey, N., Mehl, C., Mehrmann, V.: Skew-symmetric matrix polynomials and their Smith forms. Linear Algebra Appl. **438**(12), 4625–4653 (2013)

A169. Mehl, C., Mehrmann, V., Ran, A. C. M., Rodman, L.: Jordan forms of real and complex matrices under rank one perturbations. Oper. Matrices **7**(2), 381–398 (2013)
A170. Mehrmann, V., Poloni, F.: A generalized structured doubling algorithm for the numerical solution of linear quadratic optimal control problems. Numer. Linear Algebra Appl. **20**(1), 112–137 (2013)
A171. Mehrmann, V., Poloni, F.: Using permuted graph bases in H_∞ control. Automatica **49**(6), 1790–1797 (2013)
A172. Carstensen, C., Gedicke, J., Mehrmann, V., Miedlar, A.: An adaptive finite element method with asymptotic saturation for eigenvalue problems. Numer. Math. **128**(4), 615–634 (2014)
A173. Ha, P., Mehrmann, V., Steinbrecher, A.: Analysis of linear variable coefficient delay differential-algebraic equations. J. Dynam. Differential Equations pp. 1–26 (2014)
A174. Kunkel, P., Mehrmann, V., Scholz, L.: Self-adjoint differential-algebraic equations. Math. Control Signals Syst. **26**(1), 47–76 (2014)
A175. Linh, V. H., Mehrmann, V.: Efficient integration of strangeness-free non-stiff differential-algebraic equations by half-explicit methods. J. Comput. Appl. Math. **262**, 346–360 (2014)
A176. Mehl, C., Mehrmann, V., Ran, A. C. M., Rodman, L.: Eigenvalue perturbation theory of symplectic, orthogonal, and unitary matrices under generic structured rank one perturbations. BIT **54**(1), 219–255 (2014)
A177. Mehrmann, V.: Preface: Memorial issue for Michael Neumann and Uriel Rothblum. Linear Algebra Appl. **447**, 1 (2014)
A178. Mehrmann, V., Scholz, L.: Self-conjugate differential and difference operators arising in the optimal control of descriptor systems. Oper. and Matrices **8**(3), 659–682 (2014)

Contributions to Books and Proceedings

C1. Mehrmann, V.: A symplectic orthogonal method for single input or single output discrete time optimal control problems. In: Linear Algebra in Signals, Systems, and Control, Proc. SIAM Conf., Boston/Mass. 1986, pp. 401–436 (1988)
C2. Bunse-Gerstner, A., Byers, R., Mehrmann, V.: Numerical solution of algebraic Riccati equations. In: S. Bittanti (ed.) The Riccati Equation in Control, Systems and Signals, Como 1989, pp. 107–116 (1989)
C3. Elsner, L., Mehrmann, V.: Priority vectors for matrices of pairwise comparisons. In: Methods Oper. Res., vol. 59, pp. 15–26. Athenäum Verlag GmbH, Frankfurt am Main (1989)
C4. Bunse-Gerstner, A., Mehrmann, V., Nichols, N. K.: On derivative and proportional feedback design for descriptor systems. In: Signal Processing, Scattering and Operator Theory, and Numerical Methods, Proc. Int. Symp. Math. Theory Networks Syst., Vol. III, Amsterdam/Neth. 1989, Prog. Syst. Control Theory 5, pp. 437–446 (1990)
C5. Kunkel, P., Mehrmann, V.: Numerical solution of differential algebraic Riccati equations. In: Signal Processing, Scattering and Operator Theory, and Numerical Methods, Proc. Int. Symp. Math. Theory Networks Syst., Vol. III, Amsterdam/Neth. 1989, Prog. Syst. Control Theory 5, pp. 479–486. Birkhäuser (1990)
C6. Ammar, G., Mehrmann, V.: A geometric perspective on condensed forms for Hamiltonian matrices. In: Computation and Control II, Proc. 2nd Conf., Bozeman/MT (USA) 1990, Prog. Syst. Control Theory 11, pp. 1–11 (1991)
C7. Bunse-Gerstner, A., Mehrmann, V., Nichols, N. K.: Regularization of descriptor systems. In: 30th IEEE Conference on Decision and Control (CDC), Vols. 1–3, pp. 1968–1969 (1991)
C8. Kunkel, P., Mehrmann, V.: Smooth decompositions and derivatives. In: 13th IMACS World Congress on Computation and Applied Mathematics, Dublin Irland, pp. 1141–1142 (1991)
C9. Mehrmann, V.: Divide and conquer algorithms for tridiagonal linear systems. In: W. Hackbusch (ed.) Parallel Algorithms for PDEs, Proceedings of the 6th GAMM-Seminar, Kiel, January 19–21, 1990, pp. 188–199. Vieweg–Verlag (1991)

C10. Elsner, L., Mehrmann, V.: Convergence of block iterative methods for matrices arising in fluid flow computations. In: Iterative Methods in Linear Algebra. Proceedings of the IMACS International Symposium, Brussels, Belgium, 2–4 April, 1991, pp. 391–394. Amsterdam: North-Holland (1992)
C11. Mehrmann, V.: Preconditioning of blockstructured linear systems with block ILU on parallel computers. In: Incomplete Decomposition (ILU) - Algorithms, Theory, and Applications. Proceedings of the 8th GAMM-Seminar, Kiel, January 24–26, 1992, pp. 88–95. Wiesbaden: Vieweg (1993)
C12. Bunse-Gerstner, A., Mehrmann, V., Nichols, N. K.: Output feedback in descriptor systems. In: P. Van Dooren, B. Wyman (eds.) Linear Algebra for Control Theory, *The IMA Volumes in Mathematics and its Applications*, vol. 62, pp. 43–53. Springer New York (1994)
C13. Elsner, L., He, C., Mehrmann, V.: Completion of a matrix so that the inverse has minimum norm. Application to the regularization of descriptor control problems. In: P. Van Dooren, B. Wyman (eds.) Linear Algebra for Control Theory, *The IMA Volumes in Mathematics and its Applications*, vol. 62, pp. 75–86. Springer New York (1994)
C14. He, C., Mehrmann, V.: Stabilization of large linear systems. In: L. Kulhavá, M. Kárný, K. Warwick (eds.) European IEEE Workshop CMP'94, Prague, September 1994, pp. 91–100 (1994)
C15. Kunkel, P., Mehrmann, V.: Analysis und Numèrik linearer differentiell-algebraischer Gleichungen. In: J. Herzberger (ed.) Wissenschaftliches Rechnen. Eine Einführung in das Scientific Computing, pp. 233–278. Akademie Verlag Berlin (1995)
C16. Byers, R., He, C., Mehrmann, V.: On the matrix sign function method for the computation of invariant subspaces. In: 1996 IEEE International Symposium on Computer-Aided Control System Design, pp. 71–76 (1996)
C17. Kunkel, P., Mehrmann, V., Rath, W., Weickert, J.: GELDA: Ein Softwarepaket zur Lösung linearer differentiell-algebraischer Gleichungen mit beliebigem Index. In: W. Mackens, S. M. Rump (eds.) Software Engineering im Scientific Computing, pp. 242–248. Vieweg+Teubner Verlag (1996)
C18. Benner, P., Mehrmann, V., Xu, H.: A new method for the Hamiltonian eigenvalue problem. In: Proc. European Control Conf. ECC 97, Paper 785. BELWARE Information Technology, Waterloo, Belgium, 1997 (1997). CD-ROM
C19. Benner, P., Byers, R., Mehrmann, V., Xu, H.: Numerical methods for linear quadratic and H_∞ control problems. In: Dynamical Systems, Control, Coding, Computing Vision. New Trends, Interfaces, and Interplay, pp. 203–222. Basel: Birkhäuser (1999)
C20. Benner, P., Byers, R., Mehrmann, V., Xu, H.: Numerical solution of linear-quadratic control problems for descriptor systems. In: O. Gonzalez (ed.) 1999 IEEE International Symposium on Computer-Aided Control System Design, pp. 64–69 (1999)
C21. Benner, P., Mehrmann, V., Sima, V., Van Huffel, S., Varga, A.: SLICOT – a subroutine library in systems and control theory. In: Applied and Computational Control, Signals, and Circuits. Vol. 1, pp. 499–539. Basel: Birkhäuser (1999)
C22. Konstantinov, M. M., Petkov, P. H., Christov, N. D., Gu, D. W., Mehrmann, V.: Sensitivity of Lyapunov equations. In: Advances in Intelligent Systems and Computer Science, pp. 289–292. World Scientific and Engineering Academy and Society (1999)
C23. Konstantinov, M. M., Petkov, P. H., Christov, N. D., Mehrmann, V., Barraud, A., Lesecq, S.: Conditioning of the generalized Lyapunov and Riccati equations. In: American Control Conference, vol. 4, pp. 2253–2254 (1999)
C24. Kunkel, P., Mehrmann, V., Rath, W.: Numerical solution of variable coefficient DAEs and descriptor systems. In: Scientific Computing in Chemical Engineering II: Simulation, Image Processing, Optimization, and Control, pp. 244–252. Springer (1999)
C25. Mehrmann, V., Xu, H.: On invariant subspaces of Hamiltonian matrices. In: IEEE International Symposium on Computer-Aided Control System Design, pp. 40–45 (1999)
C26. Konstantinov, M. M., Mehrmann, V., Petkov, P. H., Gu, D. W.: Additive matrix operators. In: 13th Spring Conference of the Union of Bulgarian Mathematicians, Borovets, Bulgaria 2001, pp. 169–175 (2001)

C27. Deuflhard, P., Mehrmann, V.: Numerical analysis and scientific computing in key technologies. In: DFG Research Center Mathematics for Key Technologies, pp. 9–19. Berlin: Berliner Mathematische Gesellschaft (BMG) (2002)
C28. Bollhöfer, M., Mehrmann, V.: Some convergence estimates for algebraic multilevel preconditioners. In: Fast Algorithms for Structured Matrices: Theory and Applications, pp. 293–312. Providence, RI: American Mathematical Society (AMS); Philadelphia, PA: Society for Industrial and Applied Mathematics (SIAM) (2003)
C29. Mehrmann, V.: Numerical methods for eigenvalue and control problems. In: Frontiers in Numerical Analysis, pp. 303–349. Berlin: Springer (2003)
C30. Apel, T., Mehrmann, V., Watkins, D.: Numerical solution of large scale structured polynomial or rational eigenvalue problems. In: Foundations of Computational Mathematics: Minneapolis 2002 (FoCM 2002), pp. 137–156. Cambridge: Cambridge University Press (2004)
C31. Bavafa-Toosi, Y., Mehrmann, C. B. V., Ohmori, V., Steinbrecher, A., Unger, R.: Time-optimal train traffic in large networks based on a new model. In: 10th IFAC/IFORS/IMACS/IFIP Symposium on Large Scale Systems: Theory and Applications, pp. 729–734 (2004)
C32. Benner, P., Byers, R., Mehrmann, V., Xu, H.: A robust numerical method for optimal H_∞ control. In: 43rd IEEE Conference on Decision and Control (CDC), Vols. 1–5, pp. 424–425. Omnipress, Madison, WI (2004)
C33. Hilliges, A., Mehl, C., Mehrmann, V.: On the solution of palindromic eigenvalue problems. In: ECCOMAS 2004 - European Congress on Computational Methods in Applied Sciences and Engineering, p. 11 (2004)
C34. Konstantinov, M. M., Mehrmann, V., Petkov, P. H., Gu, D. W.: A general framework for the perturbation theory of general matrix equations. In: 30th International Conference Applications of Mathematics in Engineering and Economics AMEE 04, Sozopol, Bulgaria 2004, pp. 36–72 (2004)
C35. Benner, P., Kressner, D., Mehrmann, V.: Skew-Hamiltonian and Hamiltonian eigenvalue problems: Theory, algorithms and applications. In: Conference on Applied Mathematics and Scientific Computing, pp. 3–39. Dordrecht: Springer (2005)
C36. Gäbler, A., Schlauch, S., Paschedag, A. R., Kraume, M., Mehrmann, V.: Transient drop size distributions in stirred liquid liquid dispersions. In: M. Sommerfeld (ed.) 11th Workshop on Two-Phase Flow Predictions (2005)
C37. Konstantinov, M. M., Mehrmann, V., Petkov, P. H., Gu, D. W.: Perturbation theory for general matrix equations. In: Applications of Mathematics in Engineering and Economics, pp. 36–72. Sofia: Softtrade (2005)
C38. Mehrmann, V., Stykel, T.: Balanced truncation model reduction for large-scale system in descriptor form. In: P. Benner, V. Mehrmann, D. C. Sorensen (eds.) Dimension Reduction of Large-Scale Systems, pp. 83–115. Berlin: Springer (2005)
C39. Mehrmann, V., Stykel, T.: Differential equations and stability. In: L. Hogben (ed.) Handbook of Linear Algebra, Discrete Mathematics and Its Applications, pp. 55/1–55/16. CRC Press LLC, Boca Raton, FL (2006)
C40. Henning, L., Kuzmin, D., Mehrmann, V., Schmidt, M., Sokolov, A., Turek, S.: Flow control on the basis of a FEATFLOW-MATLAB coupling. In: R. King (ed.) Active Flow Control, *Notes Numer. Fluid Mech. Multidiscip. Des.*, vol. 95, pp. 325–338 (2007). DOI 10.1007/978-3-540-71439-2_20
C41. Kunkel, P., Mehrmann, V.: Optimal control for linear descriptor systems with variable coefficients. In: IEEE Conference NLASSC 2007 9.-11.1.07 Kharagpur, India (2007)
C42. Schlauch, S., Kraume, M., Mehrmann, V.: Numerical simulation of drop-size distributions in stirred liquid-liquid systems using a compartment model approach. In: 4th International Berlin Workshop on Transport Phenomena with Moving Boundaries 27.–28.9.2007, Berlin, Germany, pp. 215–226 (2007)
C43. Benner, P., Losse, P., Mehrmann, V., Poppe, L., Reis, T.: γ-iteration for descriptor systems using structured matrix pencils. In: 18th Int. Symp. Math. Theory Networks Syst. (2008)

C44. Denkena, B., Günther, G., Mehrmann, V., Möhring, H.-C., Steinbrecher, A.: Kalibrierverfahren für hybride Parallelkinematiken. In: U. Heisel, H. Weule (eds.) Fertigungsmaschinen mit Parallelkinematiken – Forschung in Deutschland, pp. 183–224. Shaker-Verlag (2008)
C45. Byers, R., Mackey, D. S., Mehrmann, V., Xu, X.: Symplectic, BVD, and palindromic eigenvalue problems and their relation to discrete-time control problems. In: Collection of Papers Dedicated to the 60-th Anniversary of Mihail Konstantinov, pp. 81–102. Publ. House RODINA, Sofia (2009)
C46. Knyazev, A., Mehrmann, V., Osborn, J., Xu, J.: Linear and nonlinear eigenproblems for PDEs. Oberwolfach Rep. **6**(3), 2025–2114 (2009)
C47. Mehrmann, V., Miedlar, A.: Adaptive solution of elliptic PDE-eigenvalue problems. Proc. Appl. Math. Mech. **9**(1), 583–584 (2009)
C48. Baum, A.-K., Mehrmann, V.: Positivity inheritance for linear problems. Proc. Appl. Math. Mech. **10**(1), 597–598 (2010)
C49. Heiland, J., Baumann, M., Walle, A., Mehrmann, V., Schäfer, M.: Simulation and control of drop size distributions in stirred liquid/liquid systems. In: 4th International Conference on Population Balance Modelling, pp. 627–646. Berlin, Germany (2010)
C50. Heiland, J., Mehrmann, V., Schmidt, M.: A new discretization framework for input/output maps and its application to flow control. In: R. King (ed.) Active Flow Control II, pp. 357–372. Springer, Berlin (2010)
C51. King, R., Mehrmann, V., Nitsche, W.: Active flow control – A mathematical challenge. In: Production Factor Mathematics, pp. 73–80. Berlin: Springer (2010)
C52. Mehrmann, V., Miedlar, A.: Error bounds for non-selfadjoint PDE eigenvalue problems. Proc. Appl. Math. Mech. **10**(1), 551–552 (2010)
C53. Heiland, J., Mehrmann, V., Schmidt, M.: Systematic discretization of input/output maps and control of partial differential equations. In: Mathematics in Science and Technology. Mathematical Methods, Models and Algorithms in Science and Technology, pp. 45–70. Hackensack, NJ: World Scientific (2011)
C54. Kunkel, P., Mehrmann, V.: Optimal control for linear descriptor systems with variable coefficients. In: Numerical Linear Algebra in Signals, Systems and Control, pp. 313–339. New York, NY: Springer (2011)
C55. Linh, V. H., Mehrmann, V.: Spectral analysis for linear differential-algebraic equations. In: Proceedings for the 8th AIMS Conference on Dynamical Systems, Differential Equations and Applications, Dresden, Germany, May 25 - 28, 2010, DCDS Supplement, pp. 991–1000 (2011)
C56. Campbell, S. L., Kunkel, P., Mehrmann, V.: Regularization of linear and nonlinear descriptor systems. In: L. T. Biegler, S. L. Campbell, V. Mehrmann (eds.) Control and Optimization with Differential-Algebraic Constraints, no. 23 in Advances in Design and Control, pp. 17–36 (2012)
C57. Linh, V. H., Mehrmann, V.: Spectra and leading directions for linear DAEs. In: L. T. Biegler, S. L. Campbell, V. Mehrmann (eds.) Control and Optimization with Differential-Algebraic Constraints, no. 23 in Advances in Design and Control, pp. 59–78 (2012)
C58. Du, N. H., Linh, V. H., Mehrmann, V.: Robust stability of differential-algebraic equations. In: Surveys in Differential-Algebraic Equations I, pp. 63–95. Berlin: Springer (2013)
C59. Knyazev, A., Mehrmann, V., Xu, J. (eds.): Numerical solution of PDE eigenvalue problems, vol. 10, pp. 3221–3304. European Mathematical Society Publishing House, Zürich; Mathematisches Forschungsinstitut Oberwolfach, Oberwolfach (2013)
C60. Lemke, M., Miedlar, A., Reiss, J., Mehrmann, V., Sesterhenn, J.: Model reduction of reactive processes. In: R. King (ed.) Active Flow and Combustion Control 2014, pp. 397–413. Springer (2015)
C61. Mehrmann, V., Van Dooren, P.: Basic numerical methods for the analysis and design of linear time invariant dynamical systems. In: Encyclopedia of Systems and Control. Springer Verlag, Berlin (2015)

Miscellaneous

M1. Bunse-Gerstner, A., Mehrmann, V., Nichols, N. K.: Derivative feedback for descriptor systems. Materialien LVIII, Universität Bielefeld, FSP Mathematisierung, Bielefeld (1989)

M2. Bunse-Gerstner, A., Mehrmann, V. (eds.): 5-day Short Course on Nonlinear Control. Bielefeld (1990)

M3. Geerts, T., Mehrmann, V.: Linear differential equations with constant coefficients: A distributional approach. Preprint Report 90-073, Sonderforschungsbereich 343, Diskrete Strukturen in der Mathematik, Universität Bielefeld (1990)

M4. Bunse-Gerstner, A., Mehrmann, V. (eds.): 5-day Short Course on Large Scale Scientific Computing. FSP Mathematisierung, Universität Bielefeld (1992)

M5. Mehrmann, V.: Die transportable Unterprogrammbibliothek LAPACK. Rundbrief Numerische Software der Fachgruppe 2.2.2. der Gesellschaft für Informatik **6**, 10–11 (1992)

M6. Mehrmann, V.: Workshop on matrix theory. Ergänzungsreihe 1992 E92-004, Sonderforschungsbereich 343 Diskrete Strukturen in der Mathematik, Universität Bielefeld, D-4800 Bielefeld (1992)

M7. Benner, P., Laub, A. J., Mehrmann, V.: A collection of benchmark examples for the numerical solution of algebraic Riccati equations I: Continuous-time case. Preprint SPC 95-22, Forschergruppe Scientific Parallel Computing, Fakultät für Mathematik, TU Chemnitz–Zwickau, D-09107 Chemnitz (1995)

M8. Benner, P., Laub, A. J., Mehrmann, V.: A collection of benchmark examples for the numerical solution of algebraic Riccati equations II: Discrete-time case. Preprint SPC 95-23, Forschergruppe Scientific Parallel Computing, Fakultät für Mathematik, TU Chemnitz–Zwickau, D-09107 Chemnitz (1995)

M9. Bollhöfer, M., He, C., Mehrmann, V.: Modified block Jacobi preconditioners for the conjugate gradient method. Part I: The positive definite case. Preprint SPC 95-7, Forschergruppe Scientific Parallel Computing, Fakultät für Mathematik, TU Chemnitz–Zwickau, D-90107 Chemnitz (1995)

M10. He, C., Laub, A. J., Mehrmann, V.: Placing plenty of poles is pretty preposterous. Preprint SPC 95-17, Forschergruppe Scientific Parallel Computing, Fakultät für Mathematik, TU Chemnitz–Zwickau, D-09107 Chemnitz (1995)

M11. Mehrmann, V.: Book review: Numerical Methods for Large Eigenvalue Problems by Y. Saad. Jahresbericht der DMV 97 **3**, 62–63 (1995)

M12. Mehrmann, V., Nichols, N. K.: Mixed output feedbacks for descriptor systems. Preprint SPC 95-37, Forschergruppe Scientific Parallel Computing, Fakultät für Mathematik, TU Chemnitz–Zwickau, D-09107 Chemnitz (1995)

M13. Benner, P., Mehrmann, V.: Collections of benchmark examples in control. WGS Newsletter 10, NICONET e.V. (1996)

M14. Hench, J. J., He, C., Mehrmann, V.: A note on damped algebraic Riccati equations. Preprint 1872, Institute of Information Theory and Automation (1996)

M15. Mehrmann, V.: Book review: Algebraic Riccati Equations by P. Lancaster and L. Rodman. SIAM Review **38**(4), 694–695 (1996)

M16. Mehrmann, V., Penzl, T., Varga, A.: A new SLICOT implementation and documentation standard. WGS Newsletter 10, NICONET e.V. (1996)

M17. Benner, P., Mehrmann, V., Xu, H.: A new method for computing the stable invariant subspace of a real Hamiltonian matrix or breaking van Loan's curse? Preprint SFB 393/97-01, Sonderforschungsbereich 393, Numerische Simulation auf massiv parallelen Rechnern, Fakultät für Mathematik, TU Chemnitz–Zwickau, D-09107 Chemnitz (1997)

M18. Konstantinov, M. M., Mehrmann, V., Petkov, P. H., Gu, D. W.: Sensitivity of general Lyapunov equations. Technical report 98-15, Department of Engineering, Leicester University, Leicester, GB (1998)

M19. Kressner, D., Mehrmann, V., Penzl, T.: CTDSX– a collection of benchmark examples for state-space realizations of continuous-time dynamical systems. SLICOT Working Note SLWN1998-9, available from www.slicot.org (1998)
M20. Kressner, D., Mehrmann, V., Penzl, T.: DTDSX– a collection of benchmark examples for state-space realizations of discrete-time dynamical systems. SLICOT Working Note SLWN1998-10, NICONET e.V. (1998). Available from www.slicot.org
M21. Mehrmann, V., Penzl, T.: Benchmark collections in SLICOT. SLICOT Working Note SLWN1998-5, NICONET e.V. (1998). Available from www.slicot.org
M22. Mehrmann, V., Xu, H.: Lagrangian invariant subspaces of Hamiltonian matrices. Preprint SFB 393/98-25, Sonderforschungsbereich 393, Numerische Simulation auf massiv parallelen Rechnern, Fakultät für Mathematik, TU Chemnitz, D-09107 Chemnitz (1998)
M23. Petkov, P. H., Konstantinov, M. M., Mehrmann, V.: DGRSVX and DMSRIC: Fortran 77 subroutines for solving continuous-time matrix algebraic Riccati equations with condition and accuracy estimates. Preprint SFB 393/98-16, Sonderforschungsbereich 393, Numerische Simulation auf massiv parallelen Rechnern, Fakultät für Mathematik, TU Chemnitz, D-09107 Chemnitz (1998)
M24. Benner, P., Byers, R., Mehrmann, V., Xu, H.: Numerical computation of deflating subspaces of embedded Hamiltonian pencils. Preprint SFB 393/99-15, Sonderforschungsbereich 393, Numerische Simulation auf massiv parallelen Rechnern, Fakultät für Mathematik, TU Chemnitz, D-09107 Chemnitz (1999)
M25. Bollhöfer, M., Mehrmann, V.: Nested divide and conquer methods for the solution of large sparse linear systems. Preprint SFB 393/98-5, Sonderforschungsbereich 393, Numerische Simulation auf massiv parallelen Rechnern, Fakultät für Mathematik, TU Chemnitz (1999)
M26. Bollhöfer, M., Mehrmann, V.: A new approach to algebraic multilevel methods based on sparse approximate inverses. Preprint SFB 393/98-5, Sonderforschungsbereich 393, Numerische Simulation auf massiv parallelen Rechnern, Fakultät für Mathematik, TU Chemnitz (1999)
M27. Konstantinov, M. M., Mehrmann, V., Petkov, P. H., Gu, D. W.: Structural properties and parametrizations of Lyapunov and Lyapunov-like operators. Technical report 99-6, Department of Engineering, Leicester University, Leicester, GB (1999)
M28. Kressner, D., Mehrmann, V., Penzl, T.: CTLEX– a collection of benchmark examples for continuous-time Lyapunov equations. SLICOT Working Note SLWN1999-6, NICONET e.V. (1999). Available from www.slicot.org
M29. Kressner, D., Mehrmann, V., Penzl, T.: DTLEX– a collection of benchmark examples for discrete-time Lyapunov equations. SLICOT Working Note SLWN1999-7, NICONET e.V. (1999). Available from www.slicot.org
M30. Mehrmann, V., Sima, V., Varga, A., Xu, H.: A MATLAB MEX-file environment of SLICOT. SLICOT Working Note SLWN 1999-11, NICONET e.V. (1999). Available from www.slicot.org
M31. Petkov, P. H., Konstantinov, M. M., Gu, D. W., Mehrmann, V.: Numerical solution of matrix Riccati equations: A comparison of six solvers. SLICOT Working Note SLWN1999–10, NICONET e.V. (1999). Available from www.slicot.org
M32. Benner, P., Byers, R., Mehrmann, V., Xu, H.: A unified deflating subspace approach for classes of polynomial and rational matrix equations. Preprint SFB 393/00-05, Sonderforschungsbereich 393, Numerische Simulation auf massiv parallelen Rechnern, Fakultät für Mathematik, TU Chemnitz, D-09107 Chemnitz (2000)
M33. Ipsen, I., Mehrmann, V.: Conference report on SIAM Applied Linear Algebra meeting, Raleigh, N.C., October 2000. SIAM News **34**, 19–20 (2000)
M34. Konstantinov, M. M., Mehrmann, V., Petkov, P. H.: On fractional exponents in perturbed matrix spectra of defective matrices. Preprint SFB 393/98-7, Sonderforschungsbereich 393, Numerische Simulation auf massiv parallelen Rechnern, Fakultät für Mathematik, TU Chemnitz, D-09107 Chemnitz (2000)

M35. Petkov, P. H., Konstantinov, M. M., Gu, D. W., Mehrmann, V.: Condition and error estimates in the solution of Lyapunov and Riccati equations. SLICOT Working Note SLWN2000-1, NICONET e.V. (2000). Available from www.slicot.org
M36. Blendinger, C., Mehrmann, V., Steinbrecher, A., Unger, R.: Numerical simulation of train traffic in large networks via time-optimal control. Preprint 722, Institut für Mathematik, TU Berlin (2001)
M37. Hwang, T.-M., Lin, W.-W., Mehrmann, V.: Numerical solution of quadratic eigenvalue problems for damped gyroscopic systems. Preprint, National Center for Theoretical Science National Tsing Hua University, Hsinchu 300, Taiwan, ROC (2001)
M38. Gu, D. W., Konstantinov, M. M., Mehrmann, V., Petkov, P. H., Xu, H.: DRCEXC – a collection of benchmark examples for robust control design of continuous-time dynamical systems. version 1.0. SLICOT Working Note SLWN2002-08, NICONET e.V. (2002). Available from www.slicot.org
M39. Kunkel, P., Mehrmann, V., Seufer, I.: GENDA: A software package for the numerical solution of general nonlinear differential-algebraic equations. Preprint 730, Institut für Mathematik, TU Berlin, Str. des 17. Juni 136, D-10623 Berlin (2002)
M40. Mehrmann, V.: Five Berlin institutions join forces to create major new mathematics research center. SIAM News **36** (2003)
M41. Arnold, M., Mehrmann, V., Steinbrecher, A.: Index reduction in industrial multibody system simulation. Preprint 146, MATHEON, DFG Research Center Mathematics for Key Technologies in Berlin (2004)
M42. Benner, P., Byers, R., Mehrmann, V., Xu, H.: Robust numerical methods for robust control. Preprint 06-2004, Institut für Mathematik, TU Berlin, D-10623 Berlin (2004)
M43. Mehrmann, V., Shi, C.: Analysis of higher order linear differential-algebraic systems. Preprint 17-2004, Institut für Mathematik, TU Berlin (2004)
M44. Kunkel, P., Mehrmann, V., Seidel, S.: A MATLAB toolbox for the numerical solution of differential-algebraic equations. Preprint 16-2005, Institut für Mathematik, TU Berlin (2005)
M45. Mehrmann, V.: Book review: Matrix Riccati Equations in Control and Systems Theory by H. Abou-Kandil, G. Freiling, V. Ionescu, G. Jank. SIAM Review **46**(4), 753–754 (2005)
M46. Mehrmann, V., Petkov, P. H.: Evaluation of numerical methods for discrete-time H_∞ optimization. Preprint 08-2005, Institut für Mathematik, TU Berlin (2005)
M47. Kunkel, P., Mehrmann, V.: Necessary and sufficient conditions in the optimal control for general nonlinear differential-algebraic equations. Preprint 355, MATHEON, DFG Research Center Mathematics for Key Technologies in Berlin (2006)
M48. Kunkel, P., Mehrmann, V., Schmidt, M., Seufer, I., Steinbrecher, A.: Weak formulations of linear differential-algebraic systems. Preprint 16-2006, Institut für Mathematik, TU Berlin (2006)
M49. Kuzmin, D., Mehrmann, V., Schlauch, S., Sokolov, A., Turek, S.: Population balances coupled with the CFD-code FEATFLOW. Preprint 12-2006, Institut für Mathematik, TU Berlin (2006)
M50. Mehrmann, V.: Book review: Matrix Analysis for Scientists and Engineers by A. J. Laub. IEEE Control Syst. Mag. **26**(2), 94–95 (2006). DOI 10.1109/MCS.2006.1615276
M51. Brüll, T., Mehrmann, V.: STCSSP: A FORTRAN 77 routine to compute a structured staircase form for a (skew-)symmetric/(skew-)symmetric pencil. Preprint 31-2007, Institut für Mathematik, TU Berlin (2007)
M52. Linh, V. H., Mehrmann, V.: Spectral intervals for differential-algebraic equations and their numerical approximation. Preprint 402, MATHEON, DFG Research Center Mathematics for Key Technologies in Berlin (2007)
M53. Losse, P., Mehrmann, V., Poppe, L. K., Reis, T.: Robust control of descriptor systems. Preprint 47-2007, Institut für Mathematik, TU Berlin (2007)
M54. Friedland, S., Mehrmann, V.: Best subspace tensor approximations. Technical report 0805.4220v1, arXiv (2008)

Theses

T1. Mehrmann, V.: Der SR-Algorithmus zur Berechnung der Eigenwerte einer Matrix. Diplomarbeit, Universität Bielefeld (1979)
T2. Mehrmann, V.: On classes of matrices containing M-matrices, totally nonnegative and Hermitian positive semidefinite matrices. Ph.D. thesis, Fakultät für Mathematik der Universität Bielefeld (1982)
T3. Mehrmann, V.: The linear quadratic control problem: Theory and numerical algorithms. Habilitation, Universität Bielefeld (1987)

Submitted

S1. Mehrmann, V.: Index concepts for differential-algebraic equations. Preprint 3-2012, Institut für Mathematik, TU Berlin (2012)
S2. Mehrmann, V., Poloni, F.: Robust control via the computation of permuted graph bases. Preprint 19-2012, Institut für Mathematik, TU Berlin (2012)
S3. Mehrmann, V., Saha, M.: The Frobenius-Jordan form of nonnegative matrices. Preprint 21-2012, Institut für Mathematik, TU Berlin (2012)
S4. Baum, A.-K., Mehrmann, V.: Positivity characterization of nonlinear DAEs. Part I: Decomposition of differential and algebraic equations using projections. Preprint 18-2013, Institut für Mathematik, TU Berlin (2013)
S5. Baum, A.-K., Mehrmann, V.: Positivity characterization of nonlinear DAEs. Part II: A flow formula for differential-algebraic equations using projections. Preprint 20-2013, Institut für Mathematik, TU Berlin (2013)
S6. Benner, P., Losse, P., Mehrmann, V., Voigt, M.: Numerical solution of eigenvalue problems for alternating matrix polynomials and their application in control problems for descriptor systems. Preprint MPIMD/13-24, MPI Magdeburg (2013)
S7. Emmrich, E., Mehrmann, V.: Analysis of operator differential-algebraic equations arising in fluid dynamics. Part II. The infinite dimensional case. Preprint 28-2013, Institut für Mathematik, TU Berlin (2013)
S8. Ahmad, S. S., Mehrmann, V.: Backward errors and pseudospectra for structured nonlinear eigenvalue problems. Preprint 1-2014, Institut für Mathematik, TU Berlin (2014)
S9. Das, P., Mehrmann, V.: Numerical solution of singularly perturbed convection-diffusion-reaction problems with two small parameters. Preprint 13-2014, Institut für Mathematik, TU Berlin (2014)
S10. Datta, S., Mehrmann, V.: Computation of state reachable points of descriptor systems. Preprint 2-2014, Institut für Mathematik, TU Berlin (2014)
S11. Datta, S., Mehrmann, V.: Computation of state reachable points of linear time invariant descriptor systems. Preprint 17-2014, Institut für Mathematik, TU Berlin (2014)
S12. Kutyniok, G., Mehrmann, V., Petersen, P.: Regularization and numerical solution of the inverse scattering problem using frames. Preprint 26-2014, Institut für Mathematik, TU Berlin (2014)
S13. Mackey, D. S., Mackey, N., Mehl, C., Mehrmann, V.: Möbius transformations of matrix polynomials. Preprint 1051, MATHEON, DFG Research Center Mathematics for Key Technologies in Berlin (2014)
S14. Mehl, C., Mehrmann, V., Ran, A. C. M., Rodman, L.: Eigenvalue perturbation theory of structured real matrices and their sign characteristics under generic structured rank-one perturbations. Submitted for publication, Linear Algebra Appl. (2014)

S15. Mehl, C., Mehrmann, V., Wojtylak, M.: On the distance to singularity via low rank perturbations. Preprint 1058, MATHEON, DFG Research Center Mathematics for Key Technologies in Berlin (2014)

S16. Mehrmann, V., Poloni, F.: An inverse-free ADI algorithm for computing Lagrangian invariant subspaces. Preprint 14-2014, Institut für Mathematik, TU Berlin (2014)

S17. Mehrmann, V., Thuan, D.-D.: Stability analysis of implicit difference equations under restricted perturbations. Preprint 16-2014, Institut für Mathematik, TU Berlin (2014)

Volker Mehrmann's Doctoral Descendants (1992–2015)

- Reinhard Nabben, *Konvergente Iterationsverfahren für unsymmetrische Blockmatrizen*, Universität Bielefeld, 1992
  - Christian Mense, *Konvergenzanalyse von algebraischen Mehr-Gitter-Verfahren für M-Matrizen*, TU Berlin, 2008
  - Florian Goßler, *Algebraische Mehrgitterverfahren mit F-Glättung*, TU Berlin, 2013
  - André Gaul, *Recycling Krylov Subspace Methods for Sequences of Linear Systems: Analysis and Applications* (with J. Liesen), TU Berlin, 2014
- Ulrike Flaschka, *Eine Variante des Lanczos-Algorithmus für große, dünn besetzte symmetrische Matrizen mit Blockstruktur*, Universität Bielefeld, 1992
- Dieter Pütz, *Strukturerhaltende Interpolation glatter Singulärwertzerlegungen*, RWTH Aachen, 1994
- Werner Rath, *Feedback Design and Regularization for Linear Descriptor Systems with Variable Coefficients*, TU Chemnitz, 1996
- Peter Benner, *Contributions to the Numerical Solution of Algebraic Riccati Equations and Related Eigenvalue Problems*, TU Chemnitz, 1997
  - Ulric Kintzel, *Polar Decompositions and Procrustes Problems in Finite Dimensional Indefinite Scalar Product Spaces*, TU Berlin, 2005
  - Viatcheslav Sokolov, *Contributions to the Minimal Realization Problem for Descriptor Systems*, TU Chemnitz, 2006
  - Hermann Mena, *Numerical Solution of Differential Riccati Equations Arising in Optimal Control Problems for Parabolic Partial Differential Equations*, EPN Quito, Ecuador, 2007.
  - Ulrike Baur, *Control-Oriented Model Reduction for Parabolic Systems*, TU Berlin, 2008
  - Jens Saak, *Efficient Numerical Solution of Large Scale Algebraic Matrix Equations in PDE Control and Model Order Reduction*, TU Chemnitz, 2009
  - Sabine Hein, *MPC/LQG-Based Optimal Control of Nonlinear Parabolic PDEs*, TU Chemnitz, 2010

- ► Zoran Tomljanović, *Dimension Reduction for Damping Optimization of Vibrating Systems* (with N. Truhar and Z. Drmač), University of Zagreb, 2011
- ► Mohammad Sahadet Hossain, *Model Reduction for Time-Varying Descriptor Systems*, TU Chemnitz, 2011
- ► Philip Losse, *The $\mathcal{H}_\infty$ Optimal Control Problem for Descriptor Systems*, TU Chemnitz, 2011
- ► Thomas Mach, *Eigenvalue Algorithms for Symmetric Hierarchical Matrices*, TU Chemnitz, 2012
- ► Tobias Breiten, *Interpolatory Methods for Model Reduction of Large-Scale Dynamical Systems*, Otto-von-Guericke-Universität Magdeburg, 2013.
- ► Matthias Voigt, *On Linear-Quadratic Optimal Control and Robustness of Differential-Algebraic Systems,* Otto-von-Guericke-Universität Magdeburg, 2015

- Jörg Weickert, *Applications of the Theory of Differential-Algebraic Equations to Partial Differential Equations of Fluid Dynamics*, TU Chemnitz, 1997
- Matthias Bollhöfer, *Algebraic Domain Decomposition*, TU Chemnitz, 1998

- ► Peter Stange, *Beiträge zu effizienten Algorithmen basierend auf Rang-1 Aufdatierungen*, TU Braunschweig, 2011
- ► André Bodendiek, *Moment Matching Based Model Order Reduction in Computational Electromagnetism*, TU Braunschweig, 2013

- Thilo Penzl, *Numerische Lösung großer Lyapunov-Gleichungen*, TU Chemnitz, 1998
- Brahim Benhammouda, *Numerical Solution of Large Differential-Algebraic Equations of Systems on Massively Parallel Computers*, TU Chemnitz, 1998
- Christian Mehl, *Compatible Lie and Jordan Algebras and Applications to Structured Matrices and Pencils*, TU Chemnitz, 1998
- Uwe Schrader, *Invers-isotone Diskretisierungsmethoden für invers-isotone lineare und quasilineare Zwei-Punkt-Randwertaufgaben*, TU Chemnitz, 2001
- Ulrich Elsner, *Static and Dynamic Graph Partitioning*, TU Chemnitz, 2002
- Tatjana Stykel, *Analysis and Numerical Solution of Generalized Lyapunov Equations*, TU Berlin, 2002

- ► Falk Ebert, *On Partitioned Simulation of Electrical Circuits Using Dynamic Iteration Methods*, TU Berlin, 2008.

- Daniel Kreßner, *Numerical Methods and Software for General and Structured Eigenvalue Problems*, TU Berlin, 2004

- ► Christine Tobler, *Low-Rank Tensor Methods for Linear Systems and Eigenvalue Problems*, ETH Zürich, 2012
- ► Cedric Effenberger, *Robust Solution Methods for Nonlinear Eigenvalue Problems*, École Polytechnique Fédérale de Lausanne, 2013
- ► Meiyue Shao, *Dense and Structured Matrix Computations – the Parallel QR Algorithm and Matrix Exponentials*, École Polytechnique Fédérale de Lausanne, 2014

- Chunchao Shi, *Linear Differential-Algebraic Equations of Higher-Order and the Regularity or Singularity of Matrix Polynomials*, TU Berlin, 2004
- Ingo Seufer, *Generalized Inverses of Differential-Algebraic Equations and their Discretization*, TU Berlin, 2005
- Andreas Steinbrecher, *Numerical Solution of Quasi-Linear Differential-Algebraic Equations and Industrial Simulation of Multibody Systems*, TU Berlin, 2006
- Sonja Schlauch, *Modeling and Simulation of Drop Size Distributions in Stirred Liquid-Liquid Systems*, TU Berlin, 2007
- Simone Bächle, *Numerical Solution of Differential-Algebraic Systems Arising in Circuit Simulation*, TU Berlin, 2007
- Michael Schmidt, *Systematic Discretization of Input/Output Maps and other Contributions to the Control of Distributed Parameter Systems*, TU Berlin, 2007
- Christian Schröder, *Palindromic and Even Eigenvalue Problems – Analysis and Numerical Methods*, TU Berlin, 2008
- Elena Virnik, *Analysis of Positive Descriptor Systems*, TU Berlin, 2008
- Lena Wunderlich, *Analysis and Numerical Solution of Structured and Switched Differential-Algebraic Systems*, TU Berlin, 2008
- Peter Hamann, *Analyse regelungstechnischer Methoden zur modellbasierten Regelung von Toroidgetrieben*, TU Berlin, 2009
- Tobias Brüll, *Dissipativity of Linear Quadratic Systems*, TU Berlin, 2011
- Agnieszka Miedlar, *Inexact Adaptive Finite Element Methods for Elliptic PDE Eigenvalue Problems*, TU Berlin, 2011
- Jan Heiland, *Decoupling and Optimization of Differential-Algebraic Equations with Application in Flow Control*, TU Berlin, 2014
- Ann-Kristin Baum, *A Flow-on-Manifold Formulation of Differential-Algebraic Equations: Applications to Positive Systems*, TU Berlin, 2014
- Phi Ha, *Analysis and Numerical Solutions of Delay Differential-Algebraic Equations*, TU Berlin, 2015

Acknowledgments

First of all, we are grateful to a number of people who helped in reviewing the individual chapters of the Festschrift. These include several chapter authors, but also Ulrike Baur, Leo Batzke, Beatrice Meini, Reinhard Nabben, Yuji Nakatsukasa, and André Uschmajew.

We would like to thank Dario Bauso, Angelika Bunse-Gerstner, Heike Faßbender, Volker Mehrmann (though not knowing of it), Reinhard Nabben, Puduru Viswanadha Reddy, and Daniel Szyld for providing some of the images and additional input used in the preface.

We highly appreciate the support provided by Janine Holzmann, Andreas Steinbrecher, and Heiko Weichelt in creating a hopefully complete publication list of Volker Mehrmann.

Our thanks also go to Martin Peters for the easygoing negotiations with Springer on the "Mehrmann Festschrift" project. Last but not least, we would like to express our gratitude to Ruth Allewelt from Springer for her support and endless patience when accompanying the process of realizing this book project.

Contents

Part I
Numerical Algebra

Chapter 1
Breaking Van Loan's Curse: A Quest for Structure-Preserving Algorithms for Dense Structured Eigenvalue Problems

Angelika Bunse-Gerstner and Heike Faßbender

Abstract In 1981 Paige and Van Loan (Linear Algebra Appl 41:11–32, 1981) posed the open question to derive an $\mathcal{O}(n^3)$ numerically strongly backwards stable method to compute the real Hamiltonian Schur form of a Hamiltonian matrix. This problem is known as Van Loan's curse. This chapter summarizes Volker Mehrmann's work on dense structured eigenvalue problems, in particular, on Hamiltonian and symplectic eigenproblems. In the course of about 35 years working on and off on these problems the curse has been lifted by him and his co-workers. In particular, his work on SR methods and on URV-based methods for dense Hamiltonian and symplectic matrices and matrix pencils is reviewed. Moreover, his work on structure-preserving methods for other structured eigenproblems is discussed.

1.1 Introduction

Matrix eigenvalue problems with special structure of the matrix as well as of its eigenvalues and eigenvectors occur in numerous applications where the special structure reflects specific physical properties of the underlying problem. Numerical eigenvalue methods which are able to exploit the special structure may considerably reduce the amount of storage and computing time compared to general purpose methods. Moreover they may also preserve the problem's special characteristics for the computed solution, whereas in general-purpose methods they might be lost due to rounding errors. The development and investigation of structure preserving

A. Bunse-Gerstner
Fachbereich Mathematik/Informatik, Zentrum für Technomathematik, Universität Bremen, 28334 Bremen, Germany
e-mail: bunse-gerstner@math.uni-bremen.de

H. Faßbender (✉)
Institut Computational Mathematics, AG Numerik, Technische Universität Braunschweig, Pockelsstr. 14, 38106 Braunschweig, Germany
e-mail: h.fassbender@tu-braunschweig.de

P. Benner et al. (eds.), *Numerical Algebra, Matrix Theory, Differential-Algebraic Equations and Control Theory*, DOI 10.1007/978-3-319-15260-8_1

eigenvalue methods is one of the research topics which Volker Mehrmann has been pursuing since the beginning of his career.

In this chapter we will review (some of) his work on dense structured eigenvalue problems. In particular, the work on Hamiltonian and symplectic eigenproblems will be considered. Such eigenproblems appear quite often in various areas of application, see (most of) the references at the end of this chapter for a wide range of specific examples.

A real matrix $H \in \mathbb{R}^{2n\times 2n}$ is Hamiltonian matrix if $HJ = (HJ)^T$ with

$$J := \begin{bmatrix} 0 & I \\ -I & 0 \end{bmatrix} \in \mathbb{R}^{2n\times 2n} \tag{1.1}$$

where the $n \times n$ identity matrix is denoted by I. It is easy to see that any Hamiltonian matrix can be written in block form as

$$H = \begin{bmatrix} A & G \\ Q & -A^T \end{bmatrix}, \qquad G = G^T, \quad Q = Q^T, \tag{1.2}$$

where $A, G, Q \in \mathbb{R}^{n\times n}$. Eigenvalues of Hamiltonian matrices always occur in pairs $\{\lambda, -\lambda\}$, if $\lambda \in \mathbb{R} \cup \imath\mathbb{R}$, or in quadruples $\{\lambda, -\lambda, \overline{\lambda}, -\overline{\lambda}\}$, if $\lambda \in \mathbb{C}\backslash(\mathbb{R} \cup \imath\mathbb{R})$. In linear quadratic control problems, an important application for the Hamiltonian eigenvalue problem, the stable invariant subspace $\mathcal{X}$ of H has to be computed. This is the invariant subspace belonging to all eigenvalues with negative real parts, see e.g. [47]. General-purpose solvers like the QR algorithm will compute eigenvalues which in general do not display this $\{\lambda, -\lambda\}$ – eigenvalue pairing. It imposes unavoidable unstructured errors onto the computed eigenvalues, so that each eigenvalue of an eigenvalue pair or quadruple might be altered in a slightly different way. The correct information on the stable invariant subspace might then be lost. This is similar to the well-known effect that only a symmetry preserving (and exploiting) eigenvalue method will compute just real eigenvalues for a symmetric eigenproblem, otherwise unavoidable rounding errors can result in the computation of complex eigenvalues.

A real matrix $S \in \mathbb{R}^{2n\times 2n}$ is a symplectic matrix if $S^T JS = J$. Symplectic matrices are nonsingular since $S^{-1} = J^T S^T J$ and their eigenvalues occur in reciprocal pairs, i.e. if λ is an eigenvalue of S with eigenvector x, then λ^{-1} is also an eigenvalue of S with left eigenvector $(Jx)^T$. It is well-known that the set of all symplectic matrices $\mathfrak{S}$ forms a multiplicative group (even more, $\mathfrak{S}$ is a Lie group), while the set $\mathfrak{H}$ of all Hamiltonian matrices forms a Lie algebra. Moreover, symplectic similarity transformations preserve the Hamiltonian structure:

$$(S^{-1}HS)J = S^{-1}HJS^{-T} = S^{-1}J^T H^T S^{-T} = [(S^{-1}HS)J]^T. \tag{1.3}$$

A Cayley transform turns a symplectic matrix into a Hamiltonian one and vice versa. This explains the close resemblence of the spectra of Hamiltonian and symplectic matrices. Unfortunately, despite the close relationship the symplectic eigenproblem is much more difficult than the Hamiltonian one. In particular, while for Hamiltonian matrices the structure is explicit (see (1.2)), for symplectic matrices it is given only implicitly. Moreover, perturbation results are quite different for Hamiltonian and symplectic matrices. It turns out that there is no or little difference between the unstructured condition number $\kappa(\lambda)$ and the structured one $\kappa^{\mathsf{Hamil}}(\lambda)$ in the case of Hamiltonian matrices (see, e.g., [27, 39]). Here, the unstructured condition number is defined as

$$\kappa(\lambda) = \lim_{\varepsilon\to 0} \frac{1}{\varepsilon} \sup\{|\hat{\lambda} - \lambda| : E \in \mathbb{C}^{n\times n}, \|E\|_2 \le \varepsilon\}, \tag{1.4}$$

where $\hat{\lambda}$ is the eigenvalue of the perturbed matrix $A + E$ closest to the eigenvalue λ of the $n \times n$ matrix A, and the structured condition number is given by

$$\kappa^{\mathsf{struct}}(\lambda) := \lim_{\varepsilon\to 0} \frac{1}{\varepsilon} \sup\{|\hat{\lambda} - \lambda| : E \in \mathsf{struct}, \|E\|_2 \le \varepsilon\}. \tag{1.5}$$

Also, $\kappa(\mathcal{X})$ and $\kappa^{\mathsf{Hamil}}(\mathcal{X})$ are equal for the important case that $\mathcal{X}$ is the stable invariant subspace of H. If λ is a simple eigenvalue of a symplectic matrix S, then so is $1/\lambda$. There is no difference between $\kappa(\lambda)$ and $\kappa(1/\lambda)$, the unstructured eigenvalue condition numbers, but the structured ones differ (see Chap. 8 for more on this). Hence, the two matrix structures differ significantly in this aspect. While it is absolutely necessary to use a structure-preserving algorithm for computing invariant subspaces of symplectic matrices, the merits of structure preservation for Hamiltonian matrices are of a more subtle nature and not always relevant in applications.

As for the Hamiltonian eigenproblem, if a standard eigensolver is used for a symplectic eigenproblem the computed eigenvalues do not necessarily appear in pairs due to rounding errors.

For the Hamiltonian and the symplectic eigenproblem it is therefore of interest to develop structure-preserving methods as this will enforce computed eigenvalues that come in pairs or quadruples.

The following definition from Wikipedia[1] states the main idea of structure preservation, even though it refers to linguistics. Thc 'Dccp Structure' of the problem has to be preserved:

> The Structure Preservation Principle is a stipulation proposed by Noam Chomsky as part of the Generative-Transformational Grammar. Under the Structure Preservation Principle, Deep Structures should be preserved by a movement transformation, which simply rephrases the sentence.
>
> The following is an example of this Principle:
>
> Fabio strangled Prince Jamal.
>
> can be transformed into:
>
> Prince Jamal was strangled by Fabio.
>
> and this Principle is fulfilled.

[1] http://en.wikipedia.org/wiki/Structure_preservation_principle

> Both sentences hold the same meaning, because their Deep Structure remains equal. Only their Surface Structure changes, this is, just the arrangement of the words changes. Of course, auxiliary words like by are needed for the rearrangement to work.

That is, the deep, underlying structure of a matrix problem (= sentence) which often displays relevant physical properties induced by the original problem to be solved, should be preserved by any transformation applied to the matrix.

Preserving structure can help for any structured eigenproblems to preserve the physically relevant symmetries in the eigenvalues and eigenvectors of the matrix and may improve the accuracy and efficiency of an eigenvalue computation. An ideal method tailored to the matrix structure would

- be strongly backward stable in the sense of Bunch described in [17], i.e., the computed solution is the exact solution corresponding to a nearby matrix with the same structure;
- be reliable, i.e., capable to solve all eigenvalue problems in the considered matrix class; and
- require $\mathcal{O}(n^3)$ floating point operations (flops), preferably less than a competitive general-purpose method.

While for skew-Hamiltonian matrices (that is, $(NJ)^T = -NJ$ for $N \in \mathbb{R}^{2n\times 2n}$) it was possible to derive such a method [53], it has been a long-standing open problem to develop an ideal method for the Hamiltonian eigenvalue problem. As symplectic similarity transformations preserve the Hamiltonian and symplectic structure, it is straightforward to aim for methods making only use of symplectic similarity transformations. This has the potential to construct strongly backward stable methods. It would be perfect to have a numerical method which works in analogy to the QR algorithm. It should use similarity transformations with matrices which are symplectic and – for numerical stability – at the same time unitary, to drive the Hamiltonian matrix to a Hamiltonian analogue of the Schur form. A Hamiltonian matrix H is said to be in Hamiltonian Schur form if

$$H = \begin{bmatrix} T & R \\ 0 & -T^* \end{bmatrix} =$$ 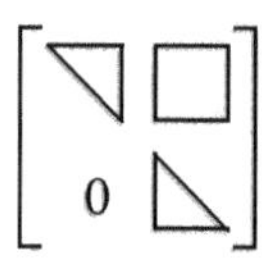 $$, \qquad T, R \in \mathbb{C}^{n\times n} \tag{1.6}$$

where T is upper triangular and $R^H = R$. It is in real Hamiltonian Schur form if in addition $T \in \mathbb{R}^{n\times n}$ is quasi-upper triangular and $R^T = R \in \mathbb{R}^{n\times n}$. It was in fact proved [41, 42, 50] that a real orthogonal and symplectic matrix S exists such that $S^T HS$ is in real Hamiltonian Schur form, if and only if every purely imaginary eigenvalue λ of H has even algebraic multiplicity, say $2k$, and if for any basis $X_k \in \mathbb{C}^{2n\times 2k}$ of the maximal invariant subspace for H corresponding to λ the matrix $X_k^H JX_k$ is congruent to $J \in \mathbb{R}^{2k\times 2k}$. Moreover, S can be chosen such that T has only eigenvalues in the open left half plane. One was hoping for an symplectic orthogonal QR-like algorithm to compute this Hamiltonian Schur form. Unfortunately, the numerical computation of the Hamiltonian Schur form via strongly backward stable $\mathcal{O}(n^3)$ methods turned out to be an extremely difficult problem. It was quite a success, when such a symplectic QR algorithm was presented in the Householder-Prize-winning Ph.D. thesis of Byers [26] for the very special case of a Hamiltonian matrix $H = \begin{bmatrix} A & G \\ Q & -A^T \end{bmatrix}$ with $\text{rank}(G) = 1$ or $\text{rank}(Q) = 1$. For the general case it has been an open problem since its introduction. It

> proved difficult to solve, however, so much so that it came to be known as *Van Loan's curse*

(see [48]). Unfortunately, it finally turned out [1] that the aim for symplectic orthogonal QR-like methods to solve Hamiltonian or symplectic eigenproblems is in general hopeless, due to the lack of an orthogonal symplectic reduction to a Hessenberg-like form. So far there is no method known that meets all three requirements for an ideal method satisfactorily.

Section 1.2 reviews Volker Mehrmann's work on the SR based algorithm which make use of (orthogonal and) symplectic transformations including the negative result [1]. Nevertheless, Volker Mehrmann and his co-authors did not give up in trying to beat Van Loan's curse. Section 1.4 reviews URV-based methods [14, 16, 29] which are considered a break-through even so some details still need to be clarified. Some of Volker Mehrmann's work on other structured eigenproblems is summarized in Sect. 1.3.

1.2 SR Methods for Dense (Generalized) Hamiltonian and Sympletic Eigenproblems

1.2.1 The SR Decomposition

It appears that one of the cornerstones for Volker Mehrmann's work on structure preserving methods was already laid in his Diplomarbeit [45]. This thesis investigated the SR algorithm for solving the standard eigenvalue problem which was one of the algorithms considered in a very basic form by Della-Dora [30]. The SR factorization of a matrix $A \in \mathbb{R}^{2n\times 2n}$ is given by $A = SR$ where $S \in \mathbb{R}^{2n\times 2n}$ is symplectic and $R \in \mathbb{R}^{2n\times 2n}$ is J-triangular. A matrix $R = \begin{bmatrix} R_{11} & R_{12} \\ R_{21} & R_{22} \end{bmatrix} \in \mathbb{R}^{2n\times 2n}$ where $R_{ij} \in \mathbb{R}^{n\times n}$ for $i, j = 1, 2$ is an (upper) J-triangular matrix if R_{11}, R_{12}, R_{22} are upper triangular matrices and R_{21} is a strict upper triangular matrix, that is,

$R =$ 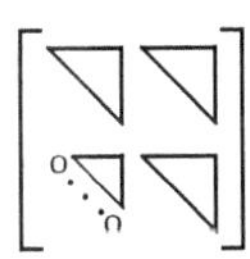 .

Note that this is a permuted upper triangular matrix, i.e. PRP^T is an upper triangular matrix, where P is the permutation matrix

$$P = [e_1, e_3, \ldots, e_{2n-1}, e_2, e_4, \ldots, e_{2n}] \in \mathbb{R}^{2n \times 2n}. \tag{1.7}$$

Here the ith unit vector is denoted by e_i.

In [30] and later in [57, 58] very general matrix decompositions are considered as basis for the computation of matrix eigenvalues. For a matrix group $\mathcal{G}$ a decompositon $A = GR$ is considered, where R is essentially an upper triangular matrix and $G \in \mathcal{G}$. In analogy to the QR algorithm based on the QR decomposition [30] proposed a very basic GR iteration based on such a general GR decomposition. The SR algorithm follows the standard GR iteration: It begins with a matrix $B \in \mathbb{K}^{n \times n}$ ($\mathbb{K}$ denotes either $\mathbb{R}$ or $\mathbb{C}$) whose eigenvalues and invariant subspaces are sought. It produces a sequence of similar matrices B_i that (hopefully) converge to a form exposing the eigenvalues. The transforming matrices for the similarity transformations $B_i = G_i^{-1} B_{i-1} G_i$ are obtained from a GR decomposition $p_i(B_{i-1}) = G_i R_i$ in which p_i is a polynomial and R_i is upper triangular. If p_i has degree 1 one speaks of a single step, if the degree is 2, it is a double step.

Almost every matrix A can be decomposed into a product of a symplectic matrix S and a J-triangular matrix R. The decomposition is unique up to a trivial (that is, symplectic and J-triangular) factor.

Theorem 1 *(for part (a) see [31, Theorem 11], [18, Theorem 3.8], for part (b) see [22, Proposition 3.3]) Let $A \in \mathbb{R}^{2n \times 2n}$ be nonsingular.*

(a) There exists a symplectic matrix S and a J-triangular matrix R such that $A = SR$ if and only if all leading principal minors of even dimension of $PA^T JAP^T$ are nonzero with P as in (1.7). The set of $2n \times 2n$ SR decomposable matrices is dense in $\mathbb{R}^{2n \times 2n}$.

(b) Let $A = SR$ and $A = \tilde{S}\tilde{R}$ be SR factorizations of A. Then there exists a trivial matrix D such that $\tilde{S} = SD^{-1}$ and $\tilde{R} = DR$. Here D is a trivial matrix, if D is symplectic and J-triangular. That is, D is trivial if and only if it has the form

$$D = \begin{bmatrix} C & F \\ 0 & C^{-1} \end{bmatrix},$$

where $C = \operatorname{diag}(c_1, \ldots, c_n)$, $F = \operatorname{diag}(f_1, \ldots, f_n)$.

There are several ways to define a unique SR decomposition. Volker Mehrmann considered two cases in [45]:

1. Restrict the symplectic group $\mathfrak{S}$ to the subgroup $\mathfrak{S}^\star$ with the additional constraint that the column sum of each column of a matrix $S^\star \in \mathfrak{S}^\star$ is one, use the entire group $\mathfrak{R}$ of J-triangular matrices R.

2. Use the entire symplectic group $\mathfrak{S}$ for S, but restrict the group $\mathfrak{R}$ of upper J-triangular matrices

$$R = \begin{bmatrix} R_{11} & R_{12} \\ R_{21} & R_{22} \end{bmatrix} =$$

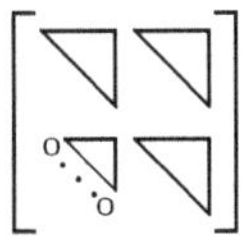

to the subgroup $\mathfrak{R}^\star$ of J-triangular matrices $R^\star$ with the additional constraint $r_{jj} > 0$ and $r_{n+j,n+j} = \pm r_{jj}$ for $j = 1, \ldots, n$ (this implies that the diagonal elements of the tridiagonal matrix $R_{11}^\star$ are all positive and that the diagonal elements of $R_{22}^\star$ are of the same absolute value as their counterparts in $R_{11}^\star$), and in addition $r_{j,n+j} = 0$ for $j = 1, \ldots, n$ (this implies that not only $R_{21}^\star$, but also $R_{12}^\star$ is a strict upper triangular matrix), i.e.

$$R^\star = \begin{bmatrix} R_{11}^\star & R_{12}^\star \\ R_{21}^\star & R_{22}^\star \end{bmatrix} =$$

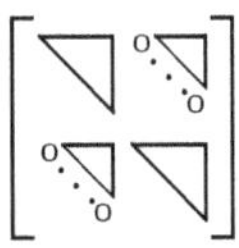

$$, (R_{11}^\star)_{jj} > 0, (R_{22}^\star)_{jj} = \pm(R_{11}^\star)_{jj}.$$

As shown in [45] (under the assumption that all leading principal minors of even dimension of $PA^T JAP^T$ are nonzero as in Theorem 1(a)), an SR decomposition $A = SR^\star$ with $S \in \mathfrak{S}, R^\star \in \mathfrak{R}^\star$ does always exist and is unique. An SR decomposition $A = S^\star R$ with $S^\star \in \mathfrak{S}^\star, R \in \mathfrak{R}$ exists and is unique if and only if there exists an SR decomposition $A = S'R'$ with $S' \in \mathfrak{S}, R' \in \mathfrak{R}$ and $\sum_{i=1}^{2n} s'_{ij} \neq 0$ for $j = 1, \ldots, n$. Algorithms for computing these SR decompositions are derived, based on symplectic transvections $I \pm uu^T J$. It is observed that an initial reduction to upper Hessenberg form reduces the computational costs significantly such that the cost for an SR decomposition of an upper Hessenberg matrix in case 1 is comparable to that of a QR decomposition, see, e.g., [37], but in case 2 the costs are about twice as much. The SR algorithm as well as single and double shift iteration steps are discussed. The additional constraint for existence and uniqueness in case 1 does not pose a real problem in the SR algorithm as there will always be a shift which allows the computation of the SR decomposition of $p(A)$. Convergence theorems are presented (for a more general thorough analysis of the convergence of GR algorithms see [58]). An implementation of the algorithms in PL/1 and some numerical examples complete Volker Mehrmann's first encounter with the SR decomposition.

1.2.2 *The SR Algorithm for Hamiltonian Eigenproblems*

During his PhD studies his interests were drawn to other subjects. Inspired by work of Bunse-Gerstner on the SR decomposition [18, 19] and Byers on Hamiltonian and symplectic algorithms for the algebraic Riccati equation [25], Volker Mehrmann started to make use of the SR algorithm as a means of a structure-preserving algorithm for the (generalized) Hamiltonian and symplectic eigenvalue problem.

As symplectic similarity preserves the Hamiltonian structure (see (1.3)) it seems natural to devise an SR algorithm for the Hamiltonian eigenproblem as this would enforce that each iterate is of Hamiltonian structure again (and therefore enforce the eigenvalue structure): If H_i is the current iterate, then a spectral transformation function q is chosen and the SR decomposition of $q(H_i)$ is formed, if possible:

$$q(H_i) = SR.$$

Then the symplectic factor S is used to perform a similarity transformation on H to yield the next iterate,

$$H_{i+1} = S^{-1} H_i S.$$

As only symplectic similarity transformations are performed, the SR algorithm preserves the Hamiltonian structure. The same holds true for a symplectic structure of a starting matrix.

As observed in [22], all Hamiltonian matrices up to a set of measure zero can be reduced to Hamiltonian J-Hessenberg form

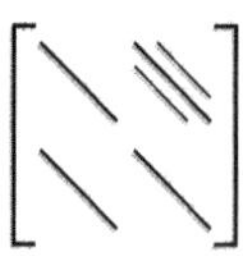
, (1.8)

where each block is an $n \times n$ matrix. Due to the Hamiltonian structure, the $(1,1)$ and the $(2,2)$ block are identical up to sign, while the $(1,2)$ block is symmetric. Hence almost any Hamiltonian matrix can be represented by $4n - 1$ parameters.

The Hamiltonian J-Hessenberg form is preserved by the SR algorithm. An implementation of the (implicit bulge-chasing) SR algorithm based on symplectic Givens, Householder and Gauss transformations is presented in [22]. The symplectic Givens and Householder transformations are orthogonal (and symplectic), while the symplectic Gauss transformations are nonorthogonal. The Gauss transformations are computed such that among all possible transformations of that form, the one with the minimal condition number is chosen. A standard implementation of the SR algorithm will require $\mathcal{O}(n^3)$ flops in each iteration step. Noting that a Hamiltonian J-Hessenberg matrix is determined by $4n - 1$ parameters, one step of the SR algorithm for H can be carried out in $\mathcal{O}(n)$ flops. The (optional) accumulation of the transformations takes another $\mathcal{O}(n^2)$ flops. Moreover, the Hamiltonian structure which will be destroyed in the numerical process due to roundoff errors when working with a Hamiltonian (J-Hessenberg) matrix, can easily be forced. The general convergence theory for GR methods [58] implies that the SR algorithm for Hamiltonian matrices is typically cubically convergent. In [24] a variant of the SR algorithm presented in [22] based solely on symplectic Gauss transformation is developed.

The SR algorithm described above was modified by Volker Mehrmann to be applicable to symplectic eigenproblems. An initial reduction to symplectic J-Hessenberg form is necessary. A J-Hessenberg matrix $A = \left[\begin{smallmatrix} A_{11} & A_{12} \\ A_{21} & A_{22} \end{smallmatrix}\right]$ is such that A_{11}, A_{21}, A_{22} are upper triangular matrices and A_{12} is an upper Hessenberg matrix, that is

$A =$  .

Note that PAP^T is an upper Hessenberg matrix.

1.2.3 *Defect Correction*

It is not recommended to use the SR algorithm just by itself for solving a Hamiltonian eigenproblem, as it is potentially unstable (due to the unavoidable use of symplectic Gauss transformations). In [22] and [24], the goal is to solve the linear quadratic control problem via the solution of a real algebraic Riccati equation $-XGX + XA + A^T X + F = 0$, where $A, G, F \in \mathbb{R}^{n \times n}$, $F = F^T$ is positive semidefinite and $G = G^T$ is positive definite. If $Y, Z, \Lambda \in \mathbb{R}^{n \times n}$ are matrices such that

$$\begin{bmatrix} A & G \\ F & -A^T \end{bmatrix} \begin{bmatrix} Y \\ Z \end{bmatrix} = \begin{bmatrix} Y \\ Z \end{bmatrix} \Lambda$$

holds and Y is invertible, then $X = -ZY^{-1}$ solves the above Riccati equation, if Λ contains all n eigenvalues of $H = \left[\begin{smallmatrix} A & G \\ F & -A^T \end{smallmatrix}\right]$ with negative real part, i.e. if $\left[\begin{smallmatrix} Y \\ Z \end{smallmatrix}\right]$ spans the stable invariant subspace of H. The purpose of the SR algorithm considered in [22] and [24] is therefore to compute the stable invariant subspace of H. The invariant subspace computed via the SR algorithm should be refined using a defect correction method or by using it as a good starting estimate for Newton's method. See Chap. 4 in this book for more.

1.2.4 *A Symplectic Orthogonal Method*

In analogy to Byers' Hamiltonian QR algorithm [26] an orthogonal SR decomposition which avoids the potentially unstable symplectic Gaussian transformations needed for the SR algorithm in [22] was developed in [46] in the context of single input or single output discrete optimal control problems where very special symplectic eigenproblems have to be solved. In particular, complex-valued conjugate symplectic eigenvalues problems $Mx = \lambda x$ with

$$M = \begin{bmatrix} M_{11} & M_{12} \\ M_{21} & M_{22} \end{bmatrix} \in \mathbb{C}^{2n \times 2n}, \quad \text{rank}(M_{21}) = 1 \tag{1.9}$$

where $M^H JM = J$ or generalized eigenproblems of the form

$$\mathcal{A} - \lambda\mathcal{B} = \begin{bmatrix} A & 0 \\ F & I \end{bmatrix} - \lambda \begin{bmatrix} I & -G \\ 0 & A^H \end{bmatrix} \in \mathbb{C}^{2n \times 2n}, \quad \text{rank}(F) = 1, \tag{1.10}$$

where $F = F^H$ and $G = G^H$ are positive semidefinite and $\mathcal{A}^H J \mathcal{A} = \mathcal{B}^H J \mathcal{B}$ are considered. The orthogonal SR decomposition used in this context decomposes a complex conjugate symplectic matrix into the product *SR* where S is orthogonal and conjugate symplectic ($S^T S = I$ and $S^H JS = S$) and R is a symplectic J-triangular matrix with positive diagonal elements, that is,

$R =$ 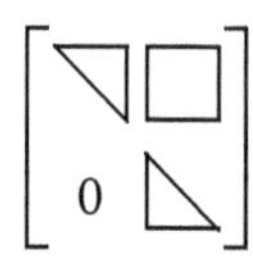 , $r_{ii} > 0.$

For the very special conjugate symplectic matrices (1.9), an initial orthogonal symplectic reduction to a special symplectic Hessenberg form can be achieved which stays invariant under an SR iteration using the just described orthogonal SR decomposition. A QZ-like variant for the generalized problem (1.10) is developed. This paper answered a research question posed by Byers in [25, 26] where the analogous continuous time optimal control problem was considered which involved computing the stable invariant subspace for a single-input Hamiltonian matrix. Implicit single and double shift steps are discussed in [46] as well as deflation and the choice of shifts. The proposed algorithm is (just as the analogue one proposed in [25, 26]) strongly stable in the sense of [17].

1.2.5 Hamiltonian and Symplectic Hessenberg Forms

The question whether an efficient SR algorithm based on the decomposition into an orthogonal and symplectic matrix S (in the real case) and a suitable upper triangular matrix R does also exist for general Hamiltonian or symplectic matrices was considered in [1]. The initial reduction to a suitable Hessenberg form is essential for any GR-type algorithm in order to decrease the computational costs of each iteration step by (at least) one order of magnitude. In particular, it is desirable that such initial reduction (as well as the GR-type decomposition used in the iteration steps) can be performed using orthogonal or unitary matrices. In a paper by Ammar and Mehrmann [1] characterizations are given for the Hamiltonian matrices that can be reduced to Hamiltonian Hessenberg form and the symplectic matrices that can be reduced to symplectic Hessenberg form by orthogonal symplectic similarity transformations. The reduction of a Hamiltonian or symplectic matrix B to its Hessenberg-like form $Q^T BQ$, where Q is orthogonal and symplectic is considered. It turns out that the components of the first column of Q must satisfy a system of n quadratic equations in $2n$ unknowns. Consequently, such a reduction is not always possible. Hence, a general efficient SR algorithm based on the decomposition into an orthogonal and symplectic matrix S and a suitable upper triangular matrix R for general Hamiltonian or symplectic matrices does not exist. An approach to the computation of Lagrangian invariant subspaces (see Chap. 4 for a definition and more on this subject) of a Hamiltonian or symplectic matrix in case the corresponding eigenvalues are known is presented. These problems were a very hot topic in the research community at that time and Volker Mehrmann got so deeply involved in trying to answer these long-standing questions that he discussed them almost everywhere with his co-workers. Therefore parts of the discussions resulting in [1] took part during a car trip with Greg Ammar to a conference and they were so intense, that this lead to a speeding ticket by the German police (probably the only speeding ticket ever acknowledged in a mathematical paper).

1.2.6 The SR Algorithm for the Symplectic Eigenproblem

In [36] it is observed that in analogy to the Hamiltonian case discussed above a symplectic J-Hessenberg matrix can be represented by $4n - 1$ parameters. An SR algorithm based on the decomposition into a symplectic matrix S and a J-triangular matrix R working on the $4n - 1$ parameters only is derived. A version for the generalized symplectic eigenproblem is presented and the numerical properties of these methods are analyzed. It is shown that the resulting methods have significantly worse numerical properties than their corresponding analogues in the Hamiltonian case.

The idea of a parameterized SR algorithm for the symplectic eigenproblem was picked up by other researchers. Banse and Bunse-Gerstner [2–5] presented a new condensed form for symplectic matrices, called symplectic butterfly form. This $2n \times 2n$ condensed matrix is symplectic, contains $8n - 4$ nonzero entries, and, similar to the symplectic J-Hessenberg form of [36], it is determined by $4n - 1$ parameters. It can be depicted by

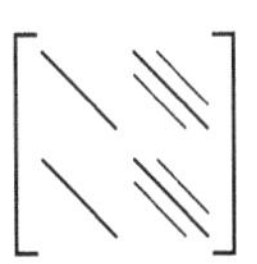

(1.11)

For every symplectic matrix M, there exist numerous symplectic matrices S such that $B = S^{-1}MS$ is in symplectic butterfly form. The SR algorithm preserves the butterfly form in its iterations. As in the Hamiltonian case, the symplectic structure, which will be destroyed in the numerical process due to roundoff errors, can easily be restored in each iteration for this condensed form. There is reason to believe that an SR algorithm based on the symplectic butterfly form has better numerical properties than the one based on the symplectic J-Hessenberg form. For a detailed discussion of the symplectic butterfly form see [7, 8, 10, 11, 32–34]. Structure-preserving SR and SZ algorithms based on the symplectic butterfly form are developed for solving small to medium size dense symplectic eigenproblems. A symplectic Lanczos algorithm based on the symplectic butterfly form is presented that is useful for solving large and sparse symplectic eigenproblems.

The SR-based methods have inspired the derivation of Lanczos-like algorithms for large sparse Hamiltonian and symplectic eigenproblems, [6–9, 35, 49, 54, 56] and Chap. 2.

1.3 Structure-Preserving Methods for Other Structured Eigenproblems

Volker Mehrmann (and his co-authors) also considered other structured eigenproblems.

1.3.1 *The HHDR Algorithm*

In [23] a structure-preserving algorithm for the generalized eigenproblem $F - \lambda G$, where $F, G \in \mathbb{C}^{n\times n}$ are Hermitian and $F - \lambda G$ is a regular (but possibly indefinite) pencil (i.e., $\det(F - \lambda G) \neq 0$) is considered. Like most eigenproblems tackled by Volker Mehrmann in his early career this eigenproblem arises from a linear quadratic optimal control problem. The algorithm proposed is based on the HDR decomposition, that is a factorization $F = HR$ where $H \in \mathbb{C}^{n\times n}$ is a nonsingular matrix such that $H^{-1}DH^{-H} = \tilde{D}$ where D and $\tilde{D}$ are diagonal and R is upper triangular. It exploits the symmetry structure and thus requires less storage and work than the general QZ algorithm. Unfortunately the algorithm is not in general numerically stable. Precautions to avoid numerical instabilities can be taken, but may slow down convergence.

1.3.2 *A Quaternion QR Algorithm*

The eigenvalue problem for matrices with quaternion entries is considered in [20]; it arises naturally from quantum mechanical problems that have time reversal symmetry. The division ring $\mathbf{H}$ of quaternions is the real algebra generated by the unit elements $1, i, j$ and k, with identity 1 and the relations

$$\begin{aligned} i^2 &= j^2 = k^2 = -1, \\ ij &= -ji = k, \\ jk &= -kj = i, \\ ki &= -ik = j. \end{aligned}$$

It is observed that for $\Gamma \in \mathbf{H}^{n\times n}$ there is a decomposition as $\Gamma = \Upsilon + j\Psi$ where $\Upsilon, \Psi \in \mathbb{C}^{n\times n}$. The quaternion matrix $\Gamma = \Upsilon + j\Psi$ can also be represented as the $2n \times 2n$ complex matrix

$$\hat{\Gamma} = \begin{bmatrix} \Upsilon & -\overline{\Psi} \\ \Psi & \overline{\Upsilon} \end{bmatrix}. \tag{1.12}$$

The map $\Gamma \to \hat{\Gamma}$ is an algebra isomorphism of $\mathbf{H}^n$ onto the algebra of $2n \times 2n$ complex matrices of the form of (1.12). A number λ in the closed upper half complex plane $\mathbb{C}^+$ is an eigenvalue of a quaternion matrix $\Gamma \in \mathbf{H}^{n\times n}$ if $\Gamma x = \lambda x$ for a non-zero eigenvector $x \in \mathbf{H}^n$. Eigenvalues and eigenvectors of matrices of the form $\hat{\Gamma}$ appear in complex conjugate pairs. If λ is an eigenvalue of $\hat{\Gamma}$ with eigenvector $y \in \mathbb{C}^n$, then $\overline{\lambda}$ is also an eigenvalue with eigenvector Jy. The eigenvalues of Γ are just the ones of $\hat{\Gamma}$ with nonnegative imaginary part. The standard QR algorithm is extended to quaternion and antiquaternion matrices. It calculates a quaternion version of the Schur decomposition using quaternion unitary similarity transformations. Following a finite step reduction to a Hessenberg-like condensed form, a sequence of implicit QR steps reduces the matrix to triangular form. By preserving quaternion structure, the algorithm calculates the eigenvalues of a nearby quaternion matrix despite rounding errors using only about half of the work and storage of the unstructured QR algorithm.

1.3.3 *The Chart Paper*

The work on structure-preserving eigenvalue methods lead to an overview of structure preserving methods [21] for real and complex matrices that have at least two of the following algebraic structures:

- Orthogonal ($A^T A = I, A \in \mathbb{C}^{n\times n}$),
- Unitary ($A^H A = I, A \in \mathbb{C}^{n\times n}$),
- Symmetric ($A = A^T, A \in \mathbb{C}^{n\times n}$),
- Hermitian ($A = A^H, A \in \mathbb{C}^{n\times n}$),
- Skew symmetric ($A = -A^T, A \in \mathbb{C}^{n\times n}$),

- Skew Hermitian ($A = -A^H, A \in \mathbb{C}^{n\times n}$),
- Symplectic (or J-orthogonal) ($S^T JS = J, S \in \mathbb{C}^{2n\times 2n}$),
- Conjugate symplectic (or J-unitary) ($S^H JS = J, S \in \mathbb{C}^{2n\times 2n}$),
- Hamiltonian (or J-symmetric) ($JH = (JH)^T, S \in \mathbb{C}^{2n\times 2n}$),
- J-Hermitian ($JS = (JS)^H, S \in \mathbb{C}^{2n\times 2n}$),
- J-skew symmetric ($JS = -(JS)^T, S \in \mathbb{C}^{2n\times 2n}$),
- J-skew Hermitian ($JS = -(JS)^H, S \in \mathbb{C}^{2n\times 2n}$).

If $A \in \mathbb{R}^{m\times m}$ some of these structures coalesce, e.g. Hermitian and symmetric, but others have distinct features. A real skew Hermitian matrix has eigenvalues that occur in $\pm$ pairs while a complex skew Hermitian matrix may not. As stated in the abstract of [21]: In the complex case numerically stable algorithms were found that preserve and exploit both structures of 40 out of the 66 pairs studied. Of the remaining 26, algorithms were found that preserve part of the structure of 12 pairs. In the real case algorithms were found for all pairs studied. The algorithms are constructed from a small set of numerical tools, including orthogonal reduction to Hessenberg form, simultaneous diagonalization of commuting normal matrices, Francis' QR algorithm, the quaternion QR-algorithm, and structure revealing, symplectic, unitary similarity transformations.

The paper [21] is the first evidence of Volker Mehrmann's numerous critical endeavours to unify theory and presentation. The paper triggered a lot of work in generating algorithms for the considered classes of problems, and inspired further papers classifying structures and methods related to eigenvalue calculations (see, e.g., [38, 43, 44]).

1.4 Van Loan's Curse: URV-Based Methods for Dense (Generalized) Hamiltonian and Sympletic Eigenproblems

As shown in [1] (and briefly discussed above) a modification of standard QR-like methods to solve (generalized) Hamiltonian or symplectic eigenproblems is in general hopeless, due to the missing reduction to a Hessenberg-like form. Nevertheless, Volker Mehrmann and his co-authors did not give up in trying to derive a method for the Hamiltonian and symplectic eigenproblem that is numerically (strongly) backward stable, has a complexity of $\mathcal{O}(n^3)$ or less and at the same time preserves the Hamiltonian or symplectic structure.

When performing eigenvalue computations one is usually restricted to similarity transformations for matrices and equivalence transformations for pencils, since only these preserve all the spectral properties. The first key observation made in [15] was that a non-equivalence transformation for the original problem can be used which leads to an equivalence transformation for a related problem. From the eigenvalues of this related problem the desired eigenvalues can easily be computed. In particular, for Hamiltonian matrices H it is well-known that if $\lambda \neq 0$ is a simple eigenvalue of H, then λ^2 is a nondefective eigenvalue of the skew Hamiltonian matrix H^2 of multiplicity 2. This fact has been used in the square reduced method of Van Loan [53]. The eigenvalues λ derived via the computed eigenvalues of H^2 may, however, suffer a loss of half of the possible accuracy. The eventual trick to break Van Loan's curse is to use a non-equivalence transformation applied to the Hamiltonian matrix. There exist orthogonal and symplectic matrices Q_1, Q_2 such that

$$\begin{aligned} Q_1^T H^2 Q_1 &= \begin{bmatrix} -H_{11}H_{22}^T & H_{11}H_{12}^T - H_{12}H_{11}^T \\ 0 & -H_{22}H_{11}^T \end{bmatrix} \\ Q_2^T H^2 Q_2^T &= \begin{bmatrix} -H_{22}H_{11}^T & H_{12}^T H_{22} - H_{22}^T H_{12} \\ 0 & -H_{11}^T H_{22} \end{bmatrix} \end{aligned} \tag{1.13}$$

where $H_{ij} \in \mathbb{R}^{n\times n}$, H_{11} is upper triangular and H_{22}^T is quasi-upper triangular. This transformation can be viewed as a symplectic version of the URV decomposition. URV decompositions of a matrix into a product of two unitary matrices U, V and an upper triangular matrix R were first introduced by Stewart in order to achieve a compromise between accuracy and computational cost between the QR decomposition and the singular value decomposition for rank and null space computations, see [51, 52].

In order to compute the eigenvalues of H it suffices to compute those of $-H_{11}H_{22}^T$ [15]. This can be done without forming the products. The periodic QR algorithm applied to $-H_{11}H_{22}^T$ yields real orthogonal transformation matrices $U, V \in \mathbb{R}^{n\times n}$ such that

$$\hat{H} = U^T H_{11} V V^T H_{22}^T U, \qquad \hat{H}_{22}^T := (U^T H_{22} V)^T \tag{1.14}$$

are quasi-upper triangular, while

$$\hat{H}_{11} = U^T H_{11} V \tag{1.15}$$

is upper triangular. Next, the 1×1 or 2×2 eigenvalue problems arising from explicitly multiplying out the diagonal blocks of $\hat{H}_{11}$ and $\hat{H}_{22}$ are solved. This determines n eigenvalues $\mu_i, i = 1, \ldots, n$. Finally the eigenvalues of H are computed by $\lambda_i = \sqrt{\mu_i}$ and $\lambda_{n+i} = -\sqrt{\mu_i}$ for $i = 1, \ldots, n$. See also [12] for a short summary of the algorithm for the Hamiltonian eigenproblem and some numerical examples.

In [15] it is also shown how to use the idea outlined above for the generalized eigenvalues of real Hamiltonian pencils. Via the Cayley transformation a symplectic matrix (or matrix pencil) is transformed to a Hamiltonian matrix pencil. Eigenvalues and the stable invariant subspace of this Hamiltonian matrix pencil are then computed with the method proposed. As the Cayley transformation preserves invariant subspaces, the stable invariant subspace of the original symplectic problem can be read off from the stable invariant subspace of the Hamiltonian matrix pencil. The eigenvalues of the original symplectic problem are obtained via the inverse

Cayley transformation. Hence, in [15] algorithms for computing the eigenvalues of Hamiltonian and symplectic pencils and matrices are derived and analyzed.

A detailed error analysis reveals that the method is numerically backward stable and preserves the structure (i.e., Hamiltonian or symplectic). In the case of a Hamiltonian matrix (as outlined above) the method is closely related to the square reduced method of Van Loan, but in contrast to that method which may suffer from a loss of accuracy, the new method computes the eigenvalues to full possible accuracy.

Based on this algorithm a new backward stable, structure preserving method of complexity $\mathcal{O}(n^3)$ for computing the stable invariant subspace of a real Hamiltonian matrix (and the stabilizing solution of the continuous-time algebraic Riccati equation) is presented in [14]. The method is based on the relationship between the invariant subspaces of the Hamiltonian matrix H and the extended matrix

$$B = \begin{bmatrix} 0 & H \\ H & 0 \end{bmatrix}.$$

If H has no eigenvalues on the imaginary axis, then an orthogonal matrix $Q \in \mathbb{C}^{4n \times 4n}$ exists such that

$$Q^T BQ = \begin{bmatrix} R & D \\ 0 & -R^T \end{bmatrix}$$

is in Hamiltonian Schur form (that is, R is a quasi upper-triangular matrix and $D = D^T$) and all eigenvalues of R have positive real part. Q can be computed using U and V from (1.14) and (1.15) and the transformation of

$$\begin{bmatrix} 0 & -H_{22}H_{11}^T \\ -H_{11}H_{22}^T & 0 \end{bmatrix}$$

(with H_{11} and H_{22} from (1.13)) to real Schur form. The resulting algorithm needs about $159n^3$ flops for the computation of an invariant subspace, while $203n^3$ flops are needed for the computation of the same invariant subspace via the Schur vector method [40] based on the standard QR-algorithm. The storage requirement for the algorithm proposed in [14] is about $9n^2$, a little more than the $8n^2$ required for the Schur vector method.

In [13] an alternative derivation for the method given in [15] via an embedding in skew-Hamiltonian matrices and an example of a structure-preserving iterative refinement algorithm for stable invariant subspaces of Hamiltonian matrices is given. Moreover, Van Loan's algorithm [53] for computing the eigenvalues of a Hamiltonian matrix H via the skew-Hamiltonian matrix $N = H^2$ is reformulated to compute the skew-Hamiltonian Schur form of N. In addition, explicit formulas/bounds for the structured eigenvalue condition numbers and a relation between the structured and unstructured condition numbers for stable invariant subspaces of Hamiltonian matrices are given (see Chap. 8 for more).

An extension of these ideas leads to algorithms for complex Hamiltonian ($(HJ)^H = HJ$ for $H \in \mathbb{C}^{2n\times 2n}$) and skew-Hamiltonian ($(NJ)^H = -NJ$ for $N \in \mathbb{C}^{2n\times 2n}$) matrices [16]. For any complex Hamiltonian matrix H it holds that $\imath H = N$ is a skew-Hamiltonian matrix. Decomposing N into its real and imaginary parts, embedding it into a matrix of double size

$$\begin{bmatrix} N & 0 \\ 0 & \overline{N} \end{bmatrix}$$

and using a simple similarity transformation yields a real skew-Hamiltonian matrix $\mathcal{N} \in \mathbb{R}^{4n\times 4n}$. The eigenvalues of $\mathcal{N}$ can be computed by Van Loan's algorithm [53] which yields the real skew-Hamiltonian Schur form. This gives without any further computations the eigenvalues of H. A stable invariant subspace of H can be computed from the skew-Hamiltonian Schur form. Since Van Loan's method is strongly backward stable, the computed eigenvalues of N are the exact eigenvalues of a real skew-Hamiltonian matrix near to $\mathcal{N}$ and $\mathcal{N}$ is similar to $\mathrm{diag}(N, \overline{N})$. The symmetry of the spectrum of N is preserved.

The ideas of [14] have also been extended to compute deflating subspaces of matrix pencils $N - \lambda H, N, H \in \mathbb{C}^{2n\times n}$ where H is Hamiltonian ($HJ = (HJ)^H$) and N is skew-Hamiltonian ($NJ = -(NJ)^H$). The algorithms proposed circumvent problems with skew-Hamiltonian/Hamiltonian matrix pencils that lack structured Schur forms by embedding them into matrix pencils that always admit a structured Schur form. For the embedded matrix pencils, the algorithms use structure-preserving unitary matrix computations and are strongly backwards stable, i.e., they compute the exact structured Schur form of a nearby matrix pencil with the same structure.

If the skew-Hamiltonian/Hamiltonian pencil is not regular, the singular part and the part associated with higher index singular blocks must be deflated first before the method discussed above can be applied. Hence, an important remaining issue is a structure preserving method to compute the structural invariants under congruence associated with the infinite eigenvalues and the singular part of the pencil. This can be done by the structured staircase algorithm proposed in [28].

In [29] a different URV-based method for computing the Hamiltonian Schur form of a Hamiltonian matrix K that has no purely imaginary eigenvalues is proposed. First it is observed that if U is an orthogonal symplectic matrix such that $U^T K^2 U$ is in real skew-Hamiltonian Schur form

$$U^T K^2 U = \begin{bmatrix} R & X \\ 0 & X^T \end{bmatrix}$$

with

$$R = \begin{bmatrix} R_{11} & \cdots & R_{1,\ell} \\ & \ddots & \vdots \\ & & R_{\ell,\ell} \end{bmatrix}$$

where $R_{jj} \in \mathbb{R}^{n_j \times n_j}$, $1 \leq n_j \leq 2$ for $j = 1, \ldots, \ell$ is in real Schur form, i.e. either 2×2 with a pair of non-real complex conjugate eigenvalues, or 1×1 real, and the magnitude of the real parts of the square roots of the eigenvalues of $R_{i,i}$ $(i = 1, \ldots, \ell)$ is decreasing. Let $H = U^T K U$. Then the columns of $H \left[\begin{smallmatrix} I_{n_1} \\ 0 \end{smallmatrix}\right]$ form an invariant subspace of H^2 associated with the eigenvalues of R_{11}. By an appropriate orthogonalization process this invariant subspace of H^2 can be transformed to a $2n \times n_1$ or $2n \times 2n_1$ quasi-upper-triangular matrix, i.e., an orthogonal-symplectic matrix Q is determined such that the Hamiltonian matrix $Q^T H Q$ can be partitioned as

$$Q^T H Q = \left[\begin{array}{cc|cc} \hat{F}_1 & * & * & * \\ 0 & \hat{F} & * & \hat{G} \\ \hline 0 & 0 & -\hat{F}_1^T & 0 \\ 0 & \hat{H} & * & -\hat{F}^T \end{array}\right]$$

where, depending on the multiplicity of eigenvalues, $\hat{F}_1$ is either an $n_1 \times n_1$ or $2n_1 \times 2n_1$ matrix in real Schur form, $\left[\begin{smallmatrix} \hat{F} & \hat{G} \\ \hat{H} & -\hat{F}^T \end{smallmatrix}\right]$ is Hamiltonian, and $\left[\begin{smallmatrix} \hat{F} & \hat{G} \\ \hat{H} & -\hat{F}^T \end{smallmatrix}\right]^2$ is again in real skew-Hamiltonian Schur form. Once this form has been computed, the same procedure is applied recursively to the matrix $\left[\begin{smallmatrix} \hat{F} & \hat{G} \\ \hat{H} & -\hat{F}^T \end{smallmatrix}\right]$. Note that the process has to perform two tasks simultaneously, i.e., to compute an orthogonal basis of the invariant subspace, and at the same time to perform the transformation in such a way, that the real skew-Hamiltonian Schur form of H^2 is not destroyed during this process. This can be achieved by orthogonal symplectic transformations. A nice interpretation of this method which might be easier to comprehend than the original derivation for readers familiar with the QR algorithm can be found in [55]. Numerical results given in [29] indicate that if no eigenvalues of H are close to the imaginary axis then the method computes the exact Hamiltonian Schur form of a nearby Hamiltonian matrix and thus is numerically strongly backward stable. The method is of complexity $\mathcal{O}(n^3)$.

A modification of the method of [29] for the computation of the Hamiltonian real Schur form is presented in [48] which avoids some of the difficulties that may arise when a Hamiltonian matrix has tightly clustered groups of eigenvalues. A detailed analysis of the method is presented and several numerical examples demonstrate the superior behavior of the method.

1.5 Concluding Remarks

The work of Volker Mehrmann reviewed here led to a new school of structure-preservation which influenced a great number of researchers. He is a driving force in a lot of work on related structured eigenproblems, in particular on structured polynomial eigenproblems which is reviewed in Chap. 12. The palindromic eigenvalue problem which generalizes the symplectic eigenproblem is dealt with in Chap. 3. Canonical forms of structured matrices and pencils are reviewed in Chap. 6, while more on structured eigenvalue perturbation theory can be found in Chap. 8. The work also influenced the numerical solution of Riccati equations, as well as other related matrix equations, see Chap. 4. His never ending effort to transfer the latest knowledge on structure-preservation (and other topics) to engineers working on real-life problems by providing adequate software and benchmark examples and countless talks has been crowned by success. By now, the ideas have been picked up by engineering communities and find their way into (commercial) standard software.

References

1. Ammar, G., Mehrmann, V.: On Hamiltonian and symplectic Hessenberg forms. Linear Algebra Appl. **149**, 55–72 (1991)
2. Banse, G.: Eigenwertverfahren für symplektische Matrizen zur Lösung zeitdiskreter optimaler Steuerungsprobleme. Z. Angew. Math. Mech. **75**(Suppl. 2), 615–616 (1995)
3. Banse, G.: Symplektische Eigenwertverfahren zur Lösung zeitdiskreter optimaler Steuerungsprobleme. PhD thesis, Fachbereich 3 – Mathematik und Informatik, Universität Bremen, Bremen (1995)
4. Banse, G.: Condensed forms for symplectic matrices and symplectic pencils in optimal control. Z. Angew. Math. Mech. **76**(Suppl. 3), 375–376 (1996)
5. Banse, G., Bunse-Gerstner, A.: A condensed form for the solution of the symplectic eigenvalue problem. In: Helmke, U., Menniken, R., Sauer, J. (eds.) Systems and Networks: Mathematical Theory and Applications, pp. 613–616. Akademie Verlag, Berlin (1994)
6. Benner, P., Faßbender, H.: An implicitly restarted symplectic Lanczos method for the Hamiltonian eigenvalue problem. Linear Algebra Appl. **263**, 75–111 (1997)
7. Benner, P., Faßbender, H.: The symplectic eigenvalue problem, the butterfly form, the SR algorithm, and the Lanczos method. Linear Algebra Appl. **275/276**, 19–47 (1998)
8. Benner, P., Faßbender, H.: An implicitly restarted symplectic Lanczos method for the symplectic eigenvalue problem. SIAM J. Matrix Anal. Appl. **22**(3), 682–713 (2000). (electronic)
9. Benner, P., Faßbender, H., Stoll, M.: A Hamiltonian Krylov-Schur-type method based on the symplectic Lanczos process. Linear Algebra Appl. **435**(3), 578–600 (2011)
10. Benner, P., Faßbender, H., Watkins, D.S.: Two connections between the SR and HR eigenvalue algorithms. Linear Algebra Appl. **272**, 17–32 (1998)
11. Benner, P., Faßbender, H., Watkins, D.S.: SR and SZ algorithms for the symplectic (butterfly) eigenproblem. Linear Algebra Appl. **287**(1–3), 41–76 (1999)
12. Benner, P., Kressner, D., Mehrmann, V.: Structure preservation: a challenge in computational control. Future Gener. Comput. Syst. **19**, 1243–1252 (2003)
13. Benner, P., Kressner, D., Mehrmann, V.: Skew-Hamiltonian and Hamiltonian eigenvalue problems: theory, algorithms and applications. In: Proceedings of the Conference on Applied Mathematics and Scientific Computing, pp. 3–39. Springer, Dordrecht (2005)

14. Benner, P., Mehrmann, V., Xu, H.: A new method for computing the stable invariant subspace of a real Hamiltonian matrix. J. Comput. Appl. Math. **86**(1), 17–43 (1997)
15. Benner, P., Mehrmann, V., Xu, H.: A numerically stable, structure preserving method for computing the eigenvalues of real Hamiltonian or symplectic pencils. Numer. Math. **78**(3), 329–358 (1998)
16. Benner, P., Mehrmann, V., Xu, H.: A note on the numerical solution of complex Hamiltonian and skew-Hamiltonian eigenvalue problems. ETNA, Electron. Trans. Numer. Anal. **8**, 115–126 (1999)
17. Bunch, J.R.: The weak and strong stability of algorithms in numerical linear algebra. Linear Algebra Appl. **88/89**, 49–66 (1987)
18. Bunse-Gerstner, A.: Matrix factorizations for symplectic QR-like methods. Linear Algebra Appl. **83**, 49–77 (1986)
19. Bunse-Gerstner, A.: Symplectic QR-like methods. Habilitationsschrift, Universität Bielefeld, Bielefeld (1986)
20. Bunse-Gerstner, A., Byers, R., Mehrmann, V.: A quaternion QR algorithm. Numer. Math. **55**(1), 83–95 (1989)
21. Bunse-Gerstner, A., Byers, R., Mehrmann, V.: A chart of numerical methods for structured eigenvalue problems. SIAM J. Matrix Anal. Appl. **13**(2), 419–453 (1992)
22. Bunse-Gerstner, A., Mehrmann, V.: A symplectic QR like algorithm for the solution of the real algebraic Riccati equation. IEEE Trans. Automat. Control **31**(12), 1104–1113 (1986)
23. Bunse-Gerstner, A., Mehrmann, V.: The HHDR algorithm and its application to optimal control problems. RAIRO Automat.-Prod. Inform. Ind. **23**(4), 305–329 (1989)
24. Bunse-Gerstner, A., Mehrmann, V., Watkins, D.S.: An SR algorithm for Hamiltonian matrices based on Gaussian elimination. In: XII Symposium on Operations Research (Passau, 1987). Volume 58 of Methods of Operations Research, pp. 339–357. Athenäum/Hain/Hanstein, Königstein (1989)
25. Byers, R.: Hamiltonian and symplectic algorithms for the algebraic Riccati equation. PhD thesis, Department Computer Science, Cornell University, Ithaca (1983)
26. Byers, R.: A Hamiltonian QR-algorithm. SIAM J. Sci. Stat. Comput. **7**, 212–229 (1986)
27. Byers, R., Kressner, D.: Structured condition numbers for invariant subspaces. SIAM J. Matrix Anal. Appl. **28**(2), 326–347 (2006). (electronic)
28. Byers, R., Mehrmann, V., Xu, H.: A structured staircase algorithm for skew-symmetric/symmetric pencils. ETNA, Electron. Trans. Numer. Anal. **26**, 1–33 (2007)
29. Chu, D., Liu, X., Mehrmann, V.: A numerical method for computing the Hamiltonian Schur form. Numer. Math. **105**(3), 375–412 (2007)
30. Della-Dora, J.: Sur quelques Algorithmes de recherche de valeurs propres. Thése, L'Université Scientifique et Medicale de Grenoble (1973)
31. Elsner, L.: On some algebraic problems in connection with general eigenvalue algorithms. Linear Algebra Appl. **26**, 123–138 (1979)
32. Faßbender, H.: Error analysis of the symplectic Lanczos method for the symplectic eigenvalue problem. BIT **40**(3), 471–496 (2000)
33. Faßbender, H.: Symplectic Methods for the Symplectic Eigenproblem. Kluwer Academic/Plenum, New York (2000)
34. Faßbender, H.: The parameterized SR algorithm for symplectic (butterfly) matrices. Math. Comput. **70**(236), 1515–1541 (2001)
35. Ferng, W.R., Lin, W.-W., Wang, C.-S.: The shift-inverted J-Lanczos algorithm for the numerical solutions of large sparse algebraic Riccati equations. Comput. Math. Appl. **33**(10), 23–40 (1997)
36. Flaschka, U., Mehrmann, V., Zywietz, D.: An analysis of structure preserving numerical methods for symplectic eigenvalue problems. RAIRO Automat.-Prod. Inform. Ind. **25**(2), 165–189 (1991)
37. Golub, G.H., Van Loan, C.F.: Matrix Computations, 4th edn. Johns Hopkins University Press, Baltimore (2012)

38. Higham, N.J., Mackey, D.S., Mackey, N., Tisseur, F.: Symmetric linearizations for matrix polynomials. SIAM J. Matrix Anal. Appl. **29**(1), 143–159 (2006/2007). (electronic)
39. Karow, M., Kressner, D., Tisseur, F.: Structured eigenvalue condition numbers. SIAM J. Matrix Anal. Appl. **28**(4), 1052–1068 (2006). (electronic)
40. Laub, A.J.: A Schur method for solving algebraic Riccati equations. IEEE Trans. Automat. Control AC-**24**, 913–921 (1979)
41. Lin, W.-W., Ho, T.-C.: On Schur type decompositions for Hamiltonian and symplectic pencils. Technical report, Institute of Applied Mathematics, National Tsing Hua University, Taiwan (1990)
42. Lin, W.-W., Mehrmann, V., Xu, H.: Canonical forms for Hamiltonian and symplectic matrices and pencils. Linear Algebra Appl. **301–303**, 469–533 (1999)
43. Mackey, D.S., Mackey, N., Mehl, C., Mehrmann, V.: Vector spaces of linearizations for matrix polynomials. SIAM J. Matrix Anal. Appl. **28**(4), 971–1004 (2006). (electronic)
44. Mackey, D.S., Mackey, N., Tisseur, F.: Structured mapping problems for matrices associated with scalar products. I. Lie and Jordan algebras. SIAM J. Matrix Anal. Appl. **29**(4), 1389–1410 (2007)
45. Mehrmann, V.: Der SR-Algorithmus zur Berechnung der Eigenwerte einer Matrix. Diplomarbeit, Universität Bielefeld, Bielefeld (1979)
46. Mehrmann, V.: A symplectic orthogonal method for single input or single output discrete time optimal quadratic control problems. SIAM J. Matrix Anal. Appl. **9**(2), 221–247 (1988)
47. Mehrmann, V.: The Autonomous Linear Quadratic Control Problem, Theory and Numerical Solution. Volume 163 of Lecture Notes in Control and Information Sciences. Springer, Heidelberg (1991)
48. Mehrmann, V., Schröder, C., Watkins, D.S.: A new block method for computing the Hamiltonian Schur form. Linear Algebra Appl. **431**(3–4), 350–368 (2009)
49. Mehrmann, V., Watkins, D.: Structure-preserving methods for computing eigenpairs of large sparse skew-Hamiltonian/Hamiltonian pencils. SIAM J. Matrix Anal. Appl. **22**, 1905–1925 (2000)
50. Paige, C.C., Van Loan, C.F.: A Schur decomposition for Hamiltonian matrices. Linear Algebra Appl. **41**, 11–32 (1981)
51. Stewart, G.W.: An updating algorithm for subspace tracking. IEEE Trans. Signal Proc. **40**, 1535–1541 (1992)
52. Stewart, G.W.: Updating a rank-revealing ULV decomposition. SIAM J. Matrix Anal. Appl. **14**(2), 494–499 (1993)
53. Van Loan, C.F.: A symplectic method for approximating all the eigenvalues of a Hamiltonian matrix. Linear Algebra Appl. **61**, 233–251 (1984)
54. Watkins, D.S.: On Hamiltonian and symplectic Lanczos processes. Linear Algebra Appl. **385**, 23–45 (2004)
55. Watkins, D.S.: On the reduction of a Hamiltonian matrix to Hamiltonian Schur form. Electron. Trans. Numer. Anal. **23**, 141–157 (2006). (electronic)
56. Watkins, D.S.: The Matrix Eigenvalue Problem. Society for Industrial and Applied Mathematics (SIAM), Philadelphia (2007)
57. Watkins, D.S., Elsner, L.: Chasing algorithms for the eigenvalue problem. SIAM J. Matrix Anal. Appl. **12**, 374–384 (1991)
58. Watkins, D.S., Elsner, L.: Convergence of algorithms of decomposition type for the eigenvalue problem. Linear Algebra Appl. **143**, 19–47 (1991)

Chapter 2
Large-Scale Structured Eigenvalue Problems

David S. Watkins

Abstract Eigenvalue problems involving large, sparse matrices with Hamiltonian or related structure arise in numerous applications. Hamiltonian problems can be transformed to symplectic or skew-Hamiltonian problems and then solved. This chapter focuses on the transformation to skew-Hamiltonian form and solution by the SHIRA method. Related to, but more general than, Hamiltonian matrices are alternating and palindromic pencils. A SHIRA-like method that operates on alternating (even) pencils $M - \lambda N$ and can be used even when N is singular, is presented.

2.1 Introduction

This chapter discusses eigenvalue problems for which the matrices are large and sparse and have additional Hamiltonian or Hamiltonian-like structure. The plan of the chapter is as follows. In Sect. 2.2 we define the structures that we are going to consider and establish their elementary properties and the relationships among them. In Sect. 2.3 we present an example that illustrates some of the structures. Starting from a quadratic eigenvalue problem $\lambda^2 Ax + \lambda Gx + Kx = 0$ with alternating (even) structure, we convert it to a first-degree eigenvalue problem $(M - \lambda N)z = 0$ that also has alternating structure, then we convert the latter to a Hamiltonian eigenvalue problem. In Sect. 2.4 we consider the problem of exploiting Hamiltonian structure. One can work directly with the Hamiltonian problem, or one can convert it to a skew-Hamiltonian or symplectic eigenvalue problem. We choose to focus on the skew-Hamiltonian case. We show that Krylov subspace methods applied to skew-Hamiltonian matrices automatically produce isotropic subspaces and therefore preserve the structure automatically. We then describe a specific method of this type, the skew-Hamiltonian implicitly-restarted Arnoldi (SHIRA) algorithm of Mehrmann and Watkins [18].

D.S. Watkins (✉)
Department of Mathematics, Washington State University, Pullman, WA 99164-3113, USA
e-mail: watkins@math.wsu.edu

P. Benner et al. (eds.), *Numerical Algebra, Matrix Theory, Differential-Algebraic Equations and Control Theory*, DOI 10.1007/978-3-319-15260-8_2

In Sect. 2.5 we present a second example, a descriptor system that yields an alternating eigenvalue problem of the form $(M - \lambda N)z = 0$. The big difference between this example and the previous one is that here the matrix N is singular, and as a consequence the SHIRA method cannot be applied to this problem. In Sect. 2.6 we develop a variant of SHIRA called even-IRA, due to Mehrmann, Schröder, and Simoncini [17], that can deal with singular N. Some implementation details and difficulties with the computation of eigenvectors and invariant subspaces are also discussed. In Sect. 2.7 we sum things up.

2.2 The Structures

The structures we will discuss include Hamiltonian, skew-Hamiltonian, symplectic, even, odd, alternating, and palindromic. We begin by defining the terms. Some of these were already discussed in Chap. 1; we repeat the definitions here for the reader's convenience.

This author will make his life easy by considering only real matrices, while admitting that there are important applications for which the matrices are complex. The matrix $J \in \mathbb{R}^{2n\times 2n}$ defined by

$$J = \begin{bmatrix} 0 & I \\ -I & 0 \end{bmatrix} \tag{2.1}$$

plays an important role here. The matrix $H \in \mathbb{R}^{2n\times 2n}$ is *Hamiltonian* if $(JH)^T = JH$, i.e. JH is symmetric. The matrix $K \in \mathbb{R}^{2n\times 2n}$ is *skew-Hamiltonian* if $(JK)^T = -JK$, i.e. JK is skew-symmetric. The matrix $S \in \mathbb{R}^{2n\times 2n}$ is *symplectic* if $S^T JS = J$. The set of symplectic matrices in $\mathbb{R}^{2n\times 2n}$ is a Lie group. Its corresponding Lie algebra (tangent space, infinitesimal group) is the set of Hamiltonian matrices, and its Jordan algebra is the set of skew-Hamiltonian matrices [9].

Matrix polynomials are discussed in Chap. 12; we will mention them only briefly here. Consider a matrix polynomial (also called a *matrix pencil*)

$$P(\lambda) = \lambda^k A_k + \lambda^{k-1} A_{k-1} + \cdots + \lambda A_1 + A_0, \tag{2.2}$$

where $A_0, \ldots, A_k \in \mathbb{R}^{m\times m}$. P is called *even* if $A_j^T = (-1)^j A_j$ for $j = 0, \ldots, k$ and *odd* if $A_j^T = (-1)^{j+1} A_j$ for $j = 0, \ldots, k$. Whether P is even or odd, its coefficients alternate between symmetric and skew-symmetric. P is called *alternating* if it is either even or odd. In Sect. 2.3 we will introduce a matrix polynomial $\lambda^2 A + \lambda G + K$, where $A = A^T$, $G = -G^T$, and $K = K^T$. This is a quadratic alternating pencil. We will also introduce polynomials of the form $M - \lambda N$, where $M = M^T$ and $N = -N^T$. These are alternating pencils of degree one. Whether we speak of a matrix pencil $M - \lambda N$ or a *matrix pair* (M, N), we are speaking of the same object.

A matrix pencil $H - \lambda K$ is called *skew-Hamiltonian/Hamiltonian (SHH)* if K is skew-Hamiltonian and H is Hamiltonian. This term appeared in the title of [18] and several other papers published around the year 2000. If one multiplies such a pencil by J, one obtains $JH - \lambda JK$, which is an (even) alternating pencil. Nowadays we prefer to speak of alternating pencils.

The polynomial P in (2.2) is called *palindromic* if $A_{k-j} = A_j^T$, $j = 0, \ldots, k$. For example, a quadratic palindromic polynomial has the form $\lambda^2 A_0^T + \lambda A_1 + A_0$, where $A_1 = A_1^T$. A linear palindromic polynomial has the form $C + \lambda C^T$.

2.2.1 Properties of Structured Matrices and Pencils

Let P be a matrix pencil, as in (2.2). Then a complex number λ is an *eigenvalue* of P if $\det(P(\lambda)) = 0$. P is called *regular* if not every $\lambda \in \mathbb{C}$ is an eigenvalue. We will consider only regular pencils. If λ is an eigenvalue, then there are nonzero vectors $v, w \in \mathbb{R}^m$, such that $P(\lambda)v = 0$ and $w^T P(\lambda) = 0$. The vector v is called a *right eigenvector*, and w^T is called a *left eigenvector*, of P associated with the eigenvalue λ. The set of all eigenvalues of P is called the *spectrum* of P.

2.2.2 Alternating Case

Proposition 1 *Let P be an alternating pencil. If v is a right eigenvector of P associated with eigenvalue λ, then v^T is a left eigenvector associated with eigenvalue $-\lambda$. Thus the eigenvalues of P appear in $\{\lambda, -\lambda\}$ pairs.*

Proof Transposing the equation $P(\lambda)v = 0$ we obtain $v^T P(-\lambda) = 0$. □

Since P is assumed real, the eigenvalues also appear in $\{\lambda, \overline{\lambda}\}$ pairs. If λ is purely imaginary, then $\{\lambda, -\lambda\} = \{\lambda, \overline{\lambda}\}$. If λ is neither real nor purely imaginary, then $\{\lambda, \overline{\lambda}, -\lambda, \ -\overline{\lambda}\}$ is a quadruple of distinct eigenvalues.

Let $H \in \mathbb{R}^{2n\times 2n}$ be a Hamiltonian matrix. Then the eigenvalues of H are the same as the eigenvalues of the pencil $H - \lambda I$. Multiplying this by J we get an equivalent pencil $JH - \lambda J$, which also has the same eigenvalues. From the definition of a Hamiltonian matrix we have that JH is symmetric. Since J is skew symmetric, we see that this is an alternating pencil. This shows that any Hamiltonian matrix can be transformed to an alternating pencil. Therefore the spectrum of a Hamiltonian matrix exhibits the same symmetries as the that of an alternating pencil.

Proposition 2 *If λ is an eigenvalue of the Hamiltonian matrix $H \in \mathbb{R}^{2n\times 2n}$, then so are $-\lambda$, $\overline{\lambda}$, and $-\overline{\lambda}$.*

We have shown that every Hamiltonian matrix can be converted to an alternating pencil $M - \lambda N = JH - \lambda J$. The converse is false. Consider an alternating pencil

$M - \lambda N$ with $M, N \in \mathbb{R}^{m \times m}$. Obviously we will not be able to convert this to a Hamiltonian eigenvalue problem if m is odd. Notice that in this case the skew symmetric N must be singular, so there is at least one infinite eigenvalue. Clearly the alternating pencil concept is more general than the Hamiltonian matrix concept.

If m is even and N is nonsingular, we do get a converse. This is because any such N has a factorization $N = R^T JR$, where R can even be taken upper triangular. R can be computed by a direct method in $O(n^3)$ flops. We were pleased to prove this result [4], but we found out later that Bunch [8] had beaten us to the punch. Using the decomposition $N = R^T JR$, we can transform $M - \lambda N$ to the equivalent pencil $R^{-T}MR^{-1} - \lambda J$, which is also alternating. This can be transformed further to $H - \lambda I$, where $H = J^T R^{-T} MR^{-1}$. H is Hamiltonian, since $JH = R^{-T}MR^{-1}$ is symmetric.

2.2.3 Palindromic Case

Proposition 3 *Let P be a palindromic pencil. If v is a right eigenvector of P associated with eigenvalue λ, then v^T is a left eigenvector associated with eigenvalue λ^{-1}. Thus the eigenvalues of P occur in $\{\lambda, \lambda^{-1}\}$ pairs.*

Proof If we transpose the equation $P(\lambda)v = 0$, the coefficients get reversed. If we then multiply by λ^{-k}, we obtain $v^T P(\lambda^{-1}) = 0$. □

If λ is on the unit circle, then $\{\lambda, \overline{\lambda}\} = \{\lambda, \lambda^{-1}\}$. If λ is neither real nor on the unit circle, $\{\lambda, \overline{\lambda}, \lambda^{-1}, \overline{\lambda}^{-1}\}$ is a quadruple of distinct eigenvalues.

The same eigenvalue symmetry holds for symplectic matrices. Let $S \in \mathbb{R}^{2n \times 2n}$ be symplectic. The defining equation $S^T JS = J$ implies that $S^T = JS^{-1}J^{-1}$. Thus S^T is similar to S^{-1}. The λ, λ^{-1} eigenvalue symmetry follows immediately.

Proposition 4 *If λ is an eigenvalue of the symplectic matrix $S \in \mathbb{R}^{2n \times 2n}$, then so are λ^{-1}, $\overline{\lambda}$, and $\overline{\lambda}^{-1}$.*

In light of the common spectral symmetry it is reasonable to ask whether a symplectic matrix can be transformed to a palindromic pencil, and conversely. The answer is not as straightforward as one might hope, but we will give a partial solution. Let S be a symplectic matrix for which 1 is not an eigenvalue, and thus $(S - I)^{-1}$ exists. If we define $C = J(S - I)^{-1}S$, then $C^T = J(S - I)^{-1}$, as we find by straightforward algebra, making use of the defining equation $S^T JS = J$. The pencil $S - \lambda I$ is equivalent to the pencil $C^T S - \lambda C^T = C - \lambda C^T$. This last is not palindromic according to the definition given above, but the trivial substitution $\mu = -\lambda$ transforms it to a palindromic pencil $C + \mu C^T$, Thus we will refer to $C - \lambda C^T$ as *palindromic* as well. (Sometimes the term *antipalindromic* is used.) For more general results see Schröder [24, 25].

2.2.4 Connecting the Alternating and Palindromic Cases

The *Cayley transform* $\lambda \to (\lambda + 1)/(\lambda - 1)$ is an involution that maps the unit circle to the imaginary axis and back. We can use this and related maps to make a connection between the alternating and palindromic eigenvalue problems.

Let $M - \lambda N$ be an alternating pencil. Specifically, suppose it is even: $M = M^T$ and $N = -N^T$. Let v be an eigenvector with associated eigenvalue $\lambda \neq 1$: $Mv = \lambda N v$. Then $(M + N)v = (\lambda + 1)Nv$ and $(M - N)v = (\lambda - 1)Nv$. Combining these equations we obtain

$$(M + N)v = \frac{\lambda + 1}{\lambda - 1}(M - N)v. \tag{2.3}$$

Letting $C = M + N$, we see that (2.3) has the (anti)palindromic form $Cv = \mu C^T v$. We can also go in the other direction: $(C + C^T)v = (\mu + 1)C^T v$, $(C - C^T)v = (\mu - 1)C^T v$, and

$$(C + C^T)v = \frac{\mu + 1}{\mu - 1}(C - C^T)v.$$

This is alternating. In fact it's our original pencil except for an irrelevant factor 2.

Consider the map $\mu = (\lambda + \tau)/(\lambda - \tau)$, where τ is a positive constant that we insert for added flexibility. The inverse map is $\lambda = \tau(\mu + 1)/(\mu - 1)$. Let H be Hamiltonian, and choose τ so that it is not an eigenvalue of H. Straightforward algebra shows that the matrix

$$S = (H + \tau I)(H - \tau I)^{-1}$$

is symplectic. Going the other way, if S is a symplectic matrix that does not have 1 as an eigenvalue, then

$$H = \tau(S + I)(S - I)^{-1}$$

is Hamiltonian.

To make this last Cayley transform work, we had to assume that 1 is not in the spectrum of S. In the pencil scenario we do not have the same restriction. An eigenvalue 1 in the original pencil corresponds to an eigenvalue ∞ in the transformed pencil. Again the pencil viewpoint is more flexible.

2.3 First Example

In the study of anisotropic elastic materials, especially the study of singularities at corners and crack tips [11, 12, 23], a quadratic eigenvalue problem emerges. Lamé equations are written in spherical coordinates with the origin at the point of interest, and the radial variable is separated from the angular variables, resulting in an infinite-dimensional eigenvalue problem. This is then typically approximated by the finite element method to yield a quadratic eigenvalue problem

$$\lambda^2 Ax + \lambda Gx + Kx = 0, \tag{2.4}$$

where the matrices are real, large, and sparse. Moreover $A = A^T$, $G = -G^T$, and $K = K^T$. A is a mass matrix and $-K$ is a stiffness matrix, so both are nonsingular and even positive definite. This is an alternating pencil. In this application we want to find a few of the eigenvalues that are smallest in magnitude, and these are typically real.

In [1, 2, 18, 19] we studied methods for solving this and related problems. The most common approach to solving quadratic eigenvalue problems is to *linearize* the problem, that is, to write it as a first-degree eigenvalue problem of double size. In the case of (2.4), one way to do this is to introduce auxiliary variables $y = \lambda Ax$ and write (2.4) as a system of two equations

$$\lambda y + \lambda Gx + Kx, \qquad \lambda x = A^{-1}y,$$

which are of first degree in λ. These can be combined into a single matrix equation

$$\lambda \begin{bmatrix} G & I \\ -I & 0 \end{bmatrix} \begin{bmatrix} x \\ y \end{bmatrix} - \begin{bmatrix} -K & 0 \\ 0 & A^{-1} \end{bmatrix} \begin{bmatrix} x \\ y \end{bmatrix} = 0, \tag{2.5}$$

which is a generalized eigenvalue problem of first degree. A nice feature of this linearization is that it preserves the structure: Defining

$$N = \begin{bmatrix} G & I \\ -I & 0 \end{bmatrix} \quad \text{and} \quad M = \begin{bmatrix} -K & 0 \\ 0 & A^{-1} \end{bmatrix},$$

we see that $N = -N^T$ and $M = M^T$, so the pencil $\lambda N - M$ is alternating. In [19] we showed how to linearize alternating pencils of higher degree in a way that preserves the alternating structure. Since then a great deal has been written about linearizations, for example [14, 15]. We will not get into this subject, as it is covered in Chap. 12.

The matrix N is clearly nonsingular, so we can transform this alternating eigenvalue problem into a Hamiltonian problem, as we observed in Sect. 2.2. For this we need a factorization $N = Z^T JZ$, and in this case one is available for free:

$$\begin{bmatrix} G & I \\ -I & 0 \end{bmatrix} = \begin{bmatrix} I & -\frac{1}{2}G \\ 0 & I \end{bmatrix} \begin{bmatrix} 0 & I \\ -I & 0 \end{bmatrix} \begin{bmatrix} I & 0 \\ \frac{1}{2}G & I \end{bmatrix}. \tag{2.6}$$

The alternating pencil $M - \lambda N = M - \lambda Z^T JZ$ is equivalent to $Z^{-T} MZ^{-1} - \lambda J$, which is equivalent to the standard eigenvalue problem for the Hamiltonian matrix

$$H = -JZ^{-T} MZ^{-1} = \begin{bmatrix} 0 & -I \\ I & 0 \end{bmatrix} \begin{bmatrix} I & \frac{1}{2}G \\ 0 & I \end{bmatrix} \begin{bmatrix} -K & 0 \\ 0 & A^{-1} \end{bmatrix} \begin{bmatrix} I & 0 \\ -\frac{1}{2}G & I \end{bmatrix}. \tag{2.7}$$

Its inverse is also Hamiltonian, and it is no less accessible:

$$H^{-1} = ZM^{-1}Z^T J = \begin{bmatrix} I & 0 \\ \frac{1}{2}G & I \end{bmatrix} \begin{bmatrix} (-K)^{-1} & 0 \\ 0 & A \end{bmatrix} \begin{bmatrix} I & -\frac{1}{2}G \\ 0 & I \end{bmatrix} \begin{bmatrix} 0 & I \\ -I & 0 \end{bmatrix}. \tag{2.8}$$

2.4 Exploiting Hamiltonian Structure

We consider the problem of finding the eigenvalues of smallest magnitude, and associated eigenvectors, of a Hamiltonian matrix H by Krylov subspace methods. It is well known [3, 30] that such methods are best at finding eigenvalues on the periphery of the spectrum, so it makes sense to turn the spectrum inside out by working with H^{-1} instead of H. We remark that the inverse of a Hamiltonian matrix is always Hamiltonian. This and other basic properties are summarized in the following proposition. The proofs are elementary.

Proposition 5 (Hamiltonian Matrices)

(a) H is Hamiltonian if and only if HJ is symmetric.
(b) The set of Hamiltonian matrices is a vector space over the real numbers.
(c) If H is Hamiltonian, then so is H^T.
(d) If H is Hamiltonian and nonsingular, then H^{-1} is Hamiltonian.

In the case of our example from the previous section, working with H^{-1} presents no special difficulties. To execute a Krylov subspace method, one needs to be able to perform matrix-vector multiplications: $x \to H^{-1}x$. Looking at (2.8), we see that we can do this if we have a Cholesky factorization of $-K$. We compute this factorization once and use it repeatedly. Thus the limitation of this approach is that the Cholesky factor of $-K$ must be sparse enough that we have room to store it.

Suppose we have some prior information about the location of the desired eigenvalues; say we know they are near some target value $\tau \in \mathbb{R}$. Then it makes sense to work with the shifted matrix $H - \tau I$ and its inverse. Unfortunately the shift destroys the Hamiltonian structure. Recall that if H has eigenvalues near τ, then it must have matching eigenvalues near $-\tau$. Thus it might be better to work with the matrix $K = (H - \tau I)(H + \tau I)$ and its inverse. This matrix is not Hamiltonian, but it does have a related structure: it is skew-Hamiltonian. For reference we list a few basic facts, all of which are easily proved. Recall that K is called *skew-Hamiltonian* if JK is skew-symmetric.

Proposition 6 (skew-Hamiltonian Matrices)

(a) K is skew-Hamiltonian if and only if KJ is skew-symmetric.
(b) K is skew-Hamiltonian if and only if $K^T J = JK$.
(c) The set of skew-Hamiltonian matrices is a vector space over the real numbers.
(d) If K is skew-Hamiltonian, then so is K^T.
(e) If K is skew-Hamiltonian and nonsingular, then K^{-1} is skew-Hamiltonian.
(f) The identity matrix is skew-Hamiltonian.
(g) If H is Hamiltonian, then H^2 is skew-Hamiltonian.
(h) If K is skew-Hamiltonian, then K^m is skew-Hamiltonian for all $m \geq 0$.

From these properties we see right away that the matrix $K = (H - \tau I)(H + \tau I) = H^2 - \tau^2 I$ is skew-Hamiltonian, as claimed, and so is its inverse.

Another possiblity is to perform a Cayley transform $S = (H + \tau I)(H - \tau I)^{-1}$ and work with the symplectic operator S. Still other possibilities are to work with the operator $H(H - \tau I)^{-1}(H + \tau I)^{-1}$ or $H^{-1}(H - \tau I)^{-1}(H + \tau I)^{-1}$, both of which are easily shown to be Hamiltonian.

So far we have assumed that our target τ is real, but complex targets can also be used. One easily shows that

$$K = (H - \tau I)(H - \overline{\tau} I)(H + \tau I)(H + \overline{\tau} I)$$

is skew-Hamiltonian, as is its inverse, and

$$S = (H + \tau I)(H + \overline{\tau} I)(H - \tau I)^{-1}(H - \overline{\tau} I)^{-1}$$

is symplectic.

From these considerations we see that we have the option of attacking a Hamiltonian problem directly or transforming it to a different Hamiltonian problem or to one that is either skew-Hamiltonian or symplectic. Thus we are in a position to take advantage of Krylov subspace methods that preserve any of these three structures. To get an idea how to build such methods, we consider the following easily verified facts.

Proposition 7 *Let $V \in \mathbb{R}^{2n \times 2n}$ be symplectic.*

(a) If H is Hamiltonian, then so is $V^{-1}HV$.
(b) If K is skew-Hamiltonian, then so is $V^{-1}KV$.
(c) If S is symplectic, then so is $V^{-1}SV$.

Thus all of the structures of interest are preserved under symplectic similarity transformations. Recall [30] that Krylov subspace methods, such as the Arnoldi and Lanczos processes, if carried to completion, perform similarity transformations of the matrix to upper Hessenberg form. For example, if we apply the Arnoldi process to a matrix A, we get

$$AV = VB,$$

where B is upper Hessenberg. The vectors produced by the Arnoldi process form the columns of V. Of course, we don't normally carry the process to completion; we just compute a few columns of V. This is a small piece of a similarity transformation:

$$AV_m = V_{m+1}B_{m+1,m},$$

where V_m consists of the first m columns of V.

In light of Proposition 7, if we can produce a Krylov subspace method that builds columns of a symplectic matrix, we will have a method that preserves the structures of interest. And what is the property of the first columns of a symplectic matrix? A subspace $\mathscr{S}$ of $\mathbb{R}^{2n}$ is called *isotropic* if $x^T J y = 0$ for all $x, y \in \mathscr{S}$. The defining equation $S^T JS = J$ shows that the first n columns of a symplectic matrix span an isotropic subspace (and so do the last n). Thus what is needed is a Krylov subspace method that produces isotropic subspaces.

Structured Krylov subspace methods for the Hamiltonian and symplectic cases were developed by Benner and Fassbender [6, 7]. See also [29, 30]. We will not discuss those methods here; instead we focus on the skew-Hamiltonian case.

The interesting thing about skew-Hamiltonian matrices is that they automatically produce isotropic subspaces.

Proposition 8 *Let $K \in \mathbb{R}^{2n\times 2n}$ be skew-Hamiltonian. Then for every $v \in \mathbb{R}^{2n}$ and every non-negative integer m, the Krylov subspace*

$$\mathscr{K}_m(K, v) = span\{v, Kv, K^2v, \ldots, K^{m-1}v\}$$

is isotropic.

Proof For every i and j, $(K^i v)^T J(K^j v) = v^T (K^T)^i JK^j v = v^T JK^{i+j} v = 0$ because JK^{i+j} is skew-symmetric. □

It follows that any Krylov subspace method, when applied to a skew-Hamiltonian matrix, will automatically preserve the structure. But there is a catch: In floating-point arithmetic the isotropy will gradually be lost due to roundoff errors. Therefore one must enforce it. If we consider the Arnoldi process, for example, the jth step has the form

$$\hat{v}_{j+1} = Kv_j - \sum_{i=1}^{j} v_i b_{ij},$$

where the b_{ij} are chosen so that $\hat{v}_{j+1}$ is orthogonal to $v_1, \ldots, v_j$. (In practice one might like to do this orthogonalization step twice, but we omit that detail here.) The modification we made in [18] was simply this:

$$\hat{v}_{j+1} = Kv_j - \sum_{i=1}^{j} v_i b_{ij} - \sum_{i=1}^{j} Jv_i c_{ij},$$

where the c_{ij} are chosen so that $\hat{v}_{j+1}$ is orthogonal to $Jv_1, \ldots, Jv_j$. In exact arithmetic all c_{ij} will be zero by Proposition 8, but in floating point arithmetic they are tiny numbers that provide just the needed correction to keep the spaces isotropic. This modification, together with implicit restarts, led to the Skew-Hamiltonian implicitly restarted Arnoldi (SHIRA) method, which we used in [1, 2, 18, 19] to solve (2.4) and other polynomial eigenvalue problems.

What happens if we fail to enforce isotropy? To answer this we must look into the structure of skew-Hamiltonian matrices. Recalling that a Hamiltonian matrix H has eigenvalues in $\{\lambda, -\lambda\}$ pairs. It follows that if H has a simple eigenvalue λ, then the skew-Hamiltonian matrix H^2 will have a two-dimensional eigenspace corresponding to λ^2, and similarly for $(H - \tau I)(H + \tau I)$ and $(H - \tau I)^{-1}(H + \tau I)^{-1}$. In fact it is true of skew-Hamiltonian matrices in general that they have even-dimensional eigenspaces [30]. A Krylov subspace should contain only one copy of a multiple eigenvalue in principle, but roundoff errors can allow a second copy to creep in. This is what happens if we do not enforce isotropy. If we look for, say, ten eigenvalues, we don't get ten. We get only five in duplicate. In contrast, if we do enforce isotropy, we actually get ten distinct eigenvalues.

A related complication is the computation of eigenvectors. An eigenvector v of H^2 is normally not an eigenvector of H. We have

$$v = c_+ v_+ + c_- v_-, \tag{2.9}$$

where v_+ and v_- are eigenvectors of H associated with $+\lambda$ and $-\lambda$, respectively. Typically both c_+ and c_- will be nonzero. It follows that if one wants eigenvectors, one must do some extra work. In [18] we advocated the use of inverse iteration, but other actions are possible. If v is an eigenvector of H^2 associated with eigenvalue λ^2, then normally $w_+ = (H + \lambda I)v$ and $w_- = (H - \lambda I)v$ will be eigenvectors of H associated with $+\lambda$ and $-\lambda$, respectively. A generalization of this procedure that produces stable invariant subspaces was presented in [10]. However this fails when either w_+ or w_- is zero, i.e. c_+ or c_- is zero in (2.9). In general we can expect this method to do a poor job of reproducing an eigenvector whenever that vector is poorly represented in the linear combination (2.9). We will revisit this question in Sect. 2.6.

2.4.1 Rational SHIRA

A variant that deserves mention is the rational SHIRA algorithm of Benner and Effenberger [5], which adapts the Rational Krylov method of Ruhe [20–22] to the skew-Hamiltonian case. The SHIRA method as we have presented it assumes that a target shift τ is chosen and then kept fixed throughout the computation. The rational SHIRA method allows for efficient changes of shift and is therefore useful in situations where one begins with a crude shift and wishes to refine it in the course of the computation.

2.5 Second Example

The linear-quadratic optimal control problem for a descriptor system minimizes a cost functional

$$\int_0^{\infty} \left(x^T Qx + 2u^T Sx + u^T Ru\right) dt$$

subject to the descriptor system

$$\begin{aligned} E\dot{x} &= Ax + Bu, \\ y &= Cx. \end{aligned}$$

Here $x(t)$, $u(t)$, and $y(t)$ are the state, control (input), and measured output, respectively. In the cost functional, $Q = Q^T$ and $R = R^T$. If the system arises from discretization of a system of partial differential equations, the matrices are large and sparse.

The solution to this problems yields a feedback controller so that the closed-loop system is stable. We can solve this problem by analyzing the even pencil

$$M - \lambda N = \begin{bmatrix} 0 & A & B \\ A^T & C^T QC & C^T S \\ B^T & S^T C & R \end{bmatrix} - \lambda \begin{bmatrix} 0 & E & 0 \\ -E^T & 0 & 0 \\ 0 & 0 & 0 \end{bmatrix}.$$

Omitting all technicalities, we simply state that the existence of the stabilizing controller can be checked by finding the eigenvalues near the imaginary axis. To compute the optimal controller one must find the deflating subspace associated with all of the eigenvalues in the open left half plane [16]. However, in many applications a good approximation can be obtained from the subspace associated with the eigenvalues near the imaginary axis [26].

A big difference between this and our previous example is that here the matrix N is singular, and therefore the methods of Sect. 2.4 are not applicable. Several other examples with this property are listed in [17].

2.6 Dealing with Singular N

Let $M, N \in \mathbb{R}^{m\times m}$ with $M = M^T$, $N = -N^T$, as before, and consider the even pencil $M - \lambda N$. Mehrmann, Schröder, and Simoncini [17] devised a procedure that has some of the characteristics of SHIRA [18] but does not rely on N being nonsingular. Suppose τ is a target that is not an eigenvalue, and we want to find the eigenvalues near τ. The equation $Mv = \lambda Nv$ can be transformed to $(M - \tau N)v = (\lambda - \tau)Nv$ and finally $(M - \tau N)^{-1}Nv = (\lambda - \tau)^{-1}v$. Thus the eigenvalues of the pencil that are close to τ get mapped to peripheral eigenvalues of $(M - \tau N)^{-1}N$. Of course, this transformation destroys the structure. If we want to find eigenvalues near τ, and we want to respect the structure, we must also look for the corresponding eigenvalues near $-\tau$. This suggests we use $(M+\tau N)^{-1}N(M-\tau N)^{-1}N$, for which

$$(M + \tau N)^{-1}N(M - \tau N)^{-1}Nv = \frac{1}{\lambda^2 - \tau^2}v.$$

This is one of the operators we will use; call it K:

$$K = (M + \tau N)^{-1}N(M - \tau N)^{-1}N. \tag{2.10}$$

If τ is either real or purely imaginary, K is a real matrix. If one wants to work with a more general complex target while staying in real arithmetic, one can use the more complicated operator

$$K = (M + \tau N)^{-1}N(M - \tau N)^{-1}N(M + \overline{\tau}N)^{-1}N(M - \overline{\tau}N)^{-1}N \tag{2.11}$$

instead. For simplicity we will focus mainly on the simpler K defined by (2.10), but much of the discussion carries over easily to the more complicated K given by (2.11).

Proposition 9 *Let α, $\beta \in \mathbb{C}$ be any two numbers that are not eigenvalues of the pencil $M - \lambda N$. Then $(M - \alpha N)^{-1}N$ and $(M - \beta N)^{-1}N$ commute.*

Proof if $\alpha = \beta$ there is nothing to prove. If $\alpha \neq \beta$ we can use the evident identity

$$N = \frac{1}{\alpha - \beta}\left[(M - \beta N) - (M - \alpha N)\right]. \tag{2.12}$$

If we replace the sandwiched N in $(M - \alpha N)^{-1}N(M - \beta N)^{-1}N$ by the equivalent expression in (2.12), we obtain

$$(M - \alpha N)^{-1}N(M - \beta N)^{-1}N = \frac{1}{\alpha - \beta}\left[(M - \alpha N)^{-1} - (M - \beta N)^{-1}\right]N.$$

If we then repeat this procedure with the reversed operator $(M - \beta N)^{-1}N(M - \alpha N)^{-1}N$, we get the exact same result. □

Proposition 9 shows that the order of the two factors in the definition of K in (2.10) is irrelevant, and similarly for the four factors in the K defined in (2.11).

Proposition 10 *Let K be defined by either (2.10) or (2.11). Then*

(a) $K^T N = NK$.
(b) For all positive integers m, NK^m is skew-symmetric.

Proof These results are fairly obvious. For example, for part (b), consider NK, where K is given by (2.11):

$$NK = N(M + \tau N)^{-1}N(M - \tau N)^{-1}N(M + \overline{\tau}N)^{-1}N(M - \overline{\tau}N)^{-1}N.$$

The number of skew-symmetric N factors is odd, each $M - \alpha N$ factor has a matching $M + \alpha N = (M - \alpha N)^T$ factor, and the order of these factors is irrelevant. It is therefore clear that $(NK)^T = -NK$. The same argument works for NK^m for any m. A more formal approach would prove (a) first and then deduce (b) from (a). □

A subspace $\mathscr{S}$ of $\mathbb{R}^m$ is called *N-neutral* if $x^T Ny = 0$ for all $x, y \in \mathscr{S}$. It is not hard to build N-neutral spaces; for example, every one-dimensional subspace of $\mathbb{R}^m$ is N-neutral.

Proposition 11 *Let K be defined by either (2.10) or (2.11). Then for every $v \in \mathbb{R}^m$ and every non-negative integer m, the Krylov subspace*

$$\mathscr{K}_m(K, v) = span\{v, Kv, K^2v, \ldots, K^{m-1}v\}$$

is N-neutral.

Proof For every i and j, $(K^i v)^T N(K^j v) = v^T(K^T)^i NK^j v = v^T NK^{i+j} v = 0$ because NK^{i+j} is skew-symmetric. □

Notice that this proof is identical to that of Proposition 8. In the special case $N = J$, these results reduce to results from Sect. 2.4: N-neutrality becomes isotropy, and the operators K considered here become skew-Hamiltonian.

Proposition 11 shows that any Krylov subspace method applied to K will produce an N-neutral space in principle. However, it was found [17] that if one applies, say, the Arnoldi process to K, the N-neutrality is gradually lost due to roundoff errors. Thus one must enforce N-neutrality explicitly. This leads to the *even Arnoldi process* (called *even* because the pencil $M - \lambda N$ is even), for which the jth step has the form

$$\hat{v}_{j+1} = Kv_j - \sum_{i=1}^{j} v_i b_{ij} - \sum_{i=1}^{j} Nv_i c_{ij}. \tag{2.13}$$

The b_{ij} are chosen so that $\hat{v}_{j+1}$ is orthogonal to $v_1, \ldots, v_j$ (as usual), and the c_{ij} are chosen so that $\hat{v}_{j+1}$ is orthogonal to $Nv_1, \ldots, Nv_j$. The jth step is completed by taking $b_{j+1,j} = \| \hat{v}_{j+1} \|_2$ and $v_{j+1} = \hat{v}_{j+1}/b_{j+1,j}$. The c_{ij} corrections serve to guarantee that the space span$\{v_1, \ldots, v_{j+1}\} = \mathscr{K}_{j+1}(K, v_1)$ is N-neutral. By Proposition 11, all of the c_{ij} will be zero in exact arithmetic, but in practice they are tiny nonzero numbers that serve to enforce the structure.

It is a routine matter to incorporate implicit restarts of either the standard [13] or Krylov-Schur [27, 28] type, as these will clearly preserve N-neutrality. (In [17] they use Krylov-Schur.) Doing so, we obtain a method that computes an N-neutral space that is invariant under K and is associated with the largest eigenvalues of K. This is called *even-IRA*.

The eigenspaces of K all have even dimension. Consider for simplicity the version of K defined by (2.10). Each eigenvalue $\mu = 1/(\lambda^2 - \tau^2)$ corresponds to two eigenvalues $\pm\lambda$ of the pair (M, N). The eigenspace of K associated with μ will be the sum of the eigenspaces of (M, N) associated with λ and $-\lambda$, and this will have even dimension.

Consider the generic case, in which all eigenvalues are simple eigenvalues of (M, N). If v_+ and v_- are eigenvectors associated with $+\lambda$ and $-\lambda$, respectively, the eigenspace of K associated with $\mu = 1/(\lambda^2 - \tau^2)$ will be the set of all linear combinations $c_+v_+ + c_-v_-$.

An N-neutral invariant subspace $\mathscr{V}$ produced by the Arnoldi process on K cannot possibly contain this entire eigenspace, as the following proposition shows.

Proposition 12 *Let $\lambda \neq 0$ be a simple eigenvalue of the pair (M, N), and let v_+ and v_- be right eigenvectors associated with $+\lambda$ and $-\lambda$, respectively. Then $v_-^T N v_+ \neq 0$. Thus v_+ and v_- cannot belong to the same N-neutral subspace.*

Proof Transposing the equation $(M + \lambda N)v_- = 0$, we find that v_-^T is a left eigenvector associated with the eigenvalue λ. Since v_+ and v_-^T are right and left eigenvectors associated with the same simple eigenvalue, they must satisfy $v_-^T N v_+ \neq 0$ (true for any pencil). □

Thus the N-neutral invariant subspace $\mathscr{V}$ can contain at most a one-dimensional cross section of each two-dimensional eigenspace of K. This implies that if we compute, say, a ten-dimensional invariant subspace, we will get ten distinct eigenvalues, corresponding to 20 eigenvalues of (M, N). For this it is important that we enforce explicitly the N-neutrality. If we do not, we get five eigenvalues in duplicate instead of ten distinct eigenvalues.

2.6.1 *Implementation*

To begin with, we should mention that the enforcement of N-neutrality is less straightforward in this algorithm than it is in SHIRA because $Nv_1, \ldots, Nv_j$ are generally not orthonormal. We direct the reader to [17] for two methods of N-neutrality enforcement. Apart from that, the application of Arnoldi and restarts in this context is routine.

2.6.1.1 Applying the Operator

In order to apply the Arnoldi process to K, we need to be able to effect matrix-vector multiplications $x \to Kx$, and for this we need LU or similar factorizations of $M - \tau N$ and $M + \tau N$. This is the main limitation on the utility of this approach.

Notice, however, that if $M - \tau N = LU$, then $M + \tau N = (M - \tau N)^T = U^T L^T$. For complex τ we have $M - \overline{\tau} N = \overline{L}\,\overline{U}$ and $M + \overline{\tau} N = U^* L^*$. Thus a single LU decomposition is all that is ever needed. Notice also that if τ is purely imaginary, then $M - \tau N$ is Hermitian, so a symmetric decomposition $M - \tau N = LDL^*$ can be used.

For a more specific illustration, let us reconsider the quadratic example from Sect. 2.3. Starting from the quadratic eigenvalue problem $(\lambda^2 A + \lambda G + K)x = 0$, we began by linearizing the problem. Here we find it convenient to use a different linearization. Letting $y = \lambda x$, we get two first-order equations

$$\lambda A y + \lambda G x + K x = 0, \qquad -Ay + \lambda A x = 0$$

which we write as the single matrix equation

$$\begin{bmatrix} -K & \\ & -A \end{bmatrix} \begin{bmatrix} x \\ y \end{bmatrix} - \lambda \begin{bmatrix} -G & A \\ -A & 0 \end{bmatrix} \begin{bmatrix} x \\ y \end{bmatrix} = 0. \tag{2.14}$$

Letting

$$M = \begin{bmatrix} -K & \\ & -A \end{bmatrix} \quad \text{and} \quad N = \begin{bmatrix} -G & A \\ -A & 0 \end{bmatrix},$$

we have an alternating pencil $M - \lambda N$. For any τ we have

$$M - \tau N = \begin{bmatrix} -K - \tau G & -\tau A \\ \tau A & -A \end{bmatrix} = - \begin{bmatrix} I & \tau I \\ 0 & I \end{bmatrix} \begin{bmatrix} Q(\tau) & \\ & A \end{bmatrix} \begin{bmatrix} I & 0 \\ -\tau I & I \end{bmatrix},$$

where

$$Q(\tau) = \tau^2 A + \tau G + K.$$

Thus

$$\begin{aligned} (M - \tau N)^{-1} &= - \begin{bmatrix} I & 0 \\ \tau I & I \end{bmatrix} \begin{bmatrix} Q(\tau)^{-1} & \\ & A^{-1} \end{bmatrix} \begin{bmatrix} I & -\tau I \\ 0 & I \end{bmatrix} \\ &= - \begin{bmatrix} I & 0 \\ \tau I & I \end{bmatrix} \begin{bmatrix} Q(\tau)^{-1} & \\ & I \end{bmatrix} \begin{bmatrix} I & -\tau A \\ 0 & I \end{bmatrix} \begin{bmatrix} I & \\ & A^{-1} \end{bmatrix} \end{aligned}$$

and

$$(M - \tau N)^{-1} N = - \begin{bmatrix} I & 0 \\ \tau I & I \end{bmatrix} \begin{bmatrix} Q(\tau)^{-1} & \\ & I \end{bmatrix} \begin{bmatrix} I & -\tau A \\ 0 & I \end{bmatrix} \begin{bmatrix} G & A \\ -I & 0 \end{bmatrix}.$$

This is a bit simpler than the operator we derived in [18] for this application. We can apply this operator provided we have an LU or similar decomposition of the matrix $Q(\tau) = \tau^2 A + \tau G + K$. If $Q(\tau) = LU$, then $Q(-\tau) = Q(\tau)^T = U^T L^T$, $Q(\overline{\tau}) = \overline{L}\,\overline{U}$, and $Q(-\overline{\tau}) = U^* L^*$. If τ is purely imaginary, then $Q(\tau)$ is Hermitian, and we can use a symmetric decomposition $Q(\tau) = LDL^*$.

2.6.1.2 Eigenvector Computation

Suppose we are working with an operator K of the form (2.10), and we have computed the invariant subspace $\mathscr{V}$ of dimension j corresponding to the j eigenvalues of K of largest modulus. We have

$$KV = VB,$$

where V has j columns, which form an orthonormal basis for $\mathscr{V}$. We say that V *represents* $\mathscr{V}$.

Each eigenvalue μ of B corresponds to a pair $\pm\sqrt{\tau^2 + 1/\mu}$ of eigenvalues of (M, N). The eigenvector computation requires additional effort. Let

$$W_\pm = (M \pm \tau N)^{-1} N V. \tag{2.15}$$

Then

$$VB = KV = (M \mp \tau N)^{-1} N W_\pm. \tag{2.16}$$

Clearing the inverses from (2.15) and (2.16) and combining the resulting equations into a single equation, we obtain

$$M \begin{bmatrix} V & W_\pm \end{bmatrix} \begin{bmatrix} & B \\ I & \end{bmatrix} = N \begin{bmatrix} V & W_\pm \end{bmatrix} \begin{bmatrix} I & \pm\tau B \\ \mp\tau I & I \end{bmatrix}. \tag{2.17}$$

The eigenvalues of B are the largest eigenvalues of K, and these are normally nonzero. Thus B is normally nonsingular. Assuming this, (2.17) implies that $\begin{bmatrix} V & W_+ \end{bmatrix}$ and $\begin{bmatrix} V & W_- \end{bmatrix}$ represent deflating subspaces for (M, N). In fact they represent the same space, which will normally have dimension $2j$ and contain eigenvectors corresponding to j plus/minus pairs of eigenvalues near $\pm\tau$. In principle we can (normally) extract all of this information from either $\begin{bmatrix} V & W_+ \end{bmatrix}$ or $\begin{bmatrix} V & W_- \end{bmatrix}$. Each eigenvector $\begin{bmatrix} x \\ y \end{bmatrix}$ of the reduced pencil

$$\begin{bmatrix} I & \pm\tau B \\ \mp\tau I & I \end{bmatrix} - \lambda \begin{bmatrix} & B \\ I & \end{bmatrix} \tag{2.18}$$

yields an eigenvector $z = VBy + W_{\pm}x$ of the original pencil. In [17] it is argued that for reasons of numerical accuracy the eigenvectors associated with eigenvalues near τ should be extracted from $\begin{bmatrix} V & W_+ \end{bmatrix}$, and those near $-\tau$ should be extracted from $\begin{bmatrix} V & W_- \end{bmatrix}$. While this is undoubtedly a good practice, one must realize that it does not guarantee success. The success of this method depends upon $W_{\pm}$ containing sufficient information beyond what is contained in V. Technically $\begin{bmatrix} V & W_{\pm} \end{bmatrix}$ has to have full rank, and in practice the rank should be "robustly" full. If not, accurate results are not guaranteed, regardless of which of $\begin{bmatrix} V & W_{\pm} \end{bmatrix}$ is used. Some of the vectors $z = VBy + W_{\pm}x$ can be zero or very small in magnitude. A zero z is useless, and a tiny z will be inaccurate due to cancellation.

Such outcomes are not merely a possibility; they are inevitable, as one sees by looking at the simplest case, $j = 1$. The matrix V has a single column, v, which is an eigenvector of K associated with some eigenvalue $\mu = 1/(\lambda^2 - \tau^2)$. Thus v has the form $v = c_+v_+ + c_-v_-$, where $v_{\pm}$ are eigenvectors of (M, N) associated with eigenvalues $\pm\lambda$. Normally both c_+ and c_- will be nonzero, and neither one will be particularly small. This is good. But consider what happens when, say, $c_- = 0$. Then v is proportional to v_+, so it is an eigenvector of (M, N) associated with the eigenvalue λ. $W_{\pm}$ has a single column $w_{\pm} = (M \pm \tau N)^{-1}Nv = 1/(\lambda \pm \tau)v$. The vectors v, w_+, and w_- are all proportional to v_+. The space represented by $\begin{bmatrix} v & w_+ \end{bmatrix}$ and $\begin{bmatrix} v & w_- \end{bmatrix}$ has dimension 1, and there is no possibility of extracting any information about v_- from it. If c_- is not zero but merely small, the matrices $\begin{bmatrix} v & w_{\pm} \end{bmatrix}$ will have full rank, but just barely. v_- will be computed inaccurately, as cancellation will take place in the attempt to uncover the tiny vector c_-v_-.

It is important to realize that the relationship of c_+ to c_- is set by the starting vector for the Krylov process and remains invariant throughout, including restarts. If v_- is poorly represented in the spectral decomposition of the initial vector, there is no way it can be recovered accurately. One obvious remedy is to apply inverse iteration to compute those eigenvectors that have not been resolved adequately.

The dangers described here can be expected to occur very rarely. Normally a starting vector (chosen at random, for example) will have a significant component in the direction of every eigenvector, and the vectors of interest will be extracted without difficulty. Nevertheless, one should be aware that failures are possible.

A second shortcoming of the eigenvector extraction method described here is that the structure is not fully exploited. Once we get to the level of (2.18), the structure is gone, and there is no way (as far as this author can see) to fix it. But perhaps a clever solution is just around the corner.

2.7 Conclusions

We considered the problem of finding a few eigenvalues of a large, sparse matrix having Hamiltonian or related structure. A Hamiltonian problem can be treated directly or transformed to a skew-Hamiltonian or symplectic problem. Here we focused on the skew-Hamiltonian case. The SHIRA method [18, 19] yields eigenvalues efficiently. Extra work is required to obtain eigenvectors.

Related structures are even (alternating) and palindromic pencils, which are related to Hamiltonian and symplectic matrices, respectively, but are more general. A palindromic pencil can be transformed to an even pencil (and conversely) by a Cayley transform. An even pencil $M - \lambda N$ can be transformed to a Hamiltonian matrix if N is nonsingular, and SHIRA can be applied. Regardless of whether or not N is singular, the SHIRA-like method even-IRA [17] can be applied directly to the even pencil to obtain the eigenvalues. Eigenvectors can be computed inexpensively by a method that usually works well but can occasionally fail. Further work may yield a more satisfactory method of computing eigenvectors.

References

1. Apel, T., Mehrmann, V., Watkins, D.: Structured eigenvalue methods for the computation of corner singularities in 3D anisotropic elastic structures. Comput. Methods Appl. Mech. Eng. **191**, 4459–4473 (2002)
2. Apel, T., Mehrmann, V., Watkins, D.: Numerical solution of large-scale structured polynomial or rational eigenvalue problems. In: Cucker, F., DeVore, R., Olver, P., Suli, E. (eds.) Foundations of Computational Mathematics. London Mathematical Society Lecture Note Series, vol. 312, pp. 137–157. Cambridge University Press, Cambridge (2004)
3. Bai, Z., Demmel, J., Dongarra, J., Ruhe, A., van der Vorst, H. (eds.): Templates for the Solution of Algebraic Eigenvalue Problems. SIAM, Philadelphia (2000)
4. Benner, P., Byers, R., Fassbender, H., Mehrmann, V., Watkins, D.: Cholesky-like factorizations of skew-symmetric matrices. Electron. Trans. Numer. Anal. **11**, 85–93 (2000)
5. Benner, P., Effenberger, C.: A rational SHIRA method for the Hamiltonian eigenvalue problems. Taiwan. J. Math. **14**, 805–823 (2010)
6. Benner, P., Fassbender, H.: An implicitly restarted symplectic Lanczos method for the Hamiltonian eigenvalue problem. Linear Algebra Appl. **263**, 75–111 (1997)
7. Benner, P., Fassbender, H.: An implicitly restarted symplectic Lanczos method for the symplectic eigenvalue problem. SIAM J. Matrix Anal. Appl. **22**, 682–713 (2000)
8. Bunch, J.R.: A note on the stable decomposition of skew-symmetric matrices. Math. Comput. **38**, 475–479 (1982)
9. Hall, B.C.: Lie Groups, Lie Algebras, and Representations, an Elementary Introduction. Graduate Texts in Mathematics, vol. 222. Springer, New York (2003)
10. Hwang, T.-M., Lin, W.-W., Mehrmann, V.: Numerical solution of quadratic eigenvalue problems with structure-preserving methods. SIAM J. Sci. Comput. **24**, 1283–1302 (2003)
11. Kozlov, V.A., Maz'ya, V.G., Rossmann, J.: Spectral properties of operator pencils generated by elliptic boundary value problems for the Lamé system. Rostock. Math. Kolloq. **51**, 5–24 (1997)
12. Leguillon, D.: Computation of 3D-singularities in elasticity. In: Costabel, M., et al. (eds.) Boundary Value Problems and Integral Equations in Nonsmooth Domains. Lecture Notes in Pure and Applied Mathematics, vol. 167, pp. 161–170. Marcel Dekker, New York (1995). Proceedings of the Conference, Held at the CIRM, Luminy, 3–7 May 1993
13. Lehoucq, R.B., Sorensen, D.C., Yang, C.: ARPACK Users' Guide: Solution of Large-Scale Eigenvalue Problems with Implicitly Restarted Arnoldi Methods. SIAM, Philadelphia (1998)
14. Mackey, D.S., Mackey, N., Mehl, C., Mehrmann, V.: Structured polynomial eigenvalue problems: good vibrations from good linearizations. SIAM J. Matrix Anal. Appl. **28**(4), 1029–1051 (2006)

15. Mackey, D.S., Mackey, N., Mehl, C., Mehrmann, V.: Vector spaces of linearizations for matrix polynomials. SIAM J. Matrix Anal. Appl. **28**(4), 971–1004 (2006)
16. Mehrmann, V.: The Autonomous Linear Quadratic Control Problem, Theory and Numerical Solution. Lecture Notes in Control and Information Sciences, vol. 163. Springer, Heidelberg (1991)
17. Mehrmann, V., Schröder, C., Simoncini, V.: An implicitly-restarted Krylov subspace method for real symmetic/skew-symmetric eigenproblems. Linear Algebra Appl. **436**, 4070–4087 (2009)
18. Mehrmann, V., Watkins, D.: Structure-preserving methods for computing eigenpairs of large sparse skew-Hamiltoninan/Hamiltonian pencils. SIAM J. Sci. Comput. **22**, 1905–1925 (2001)
19. Mehrmann, V., Watkins, D.: Polynomial eigenvalue problems with Hamiltonian structure. Electron. Trans. Numer. Anal. **13**, 106–118 (2002)
20. Ruhe, A.: Rational Krylov algorithms for nonsymmetric eigenvalue problems. In: Golub, G., Greenbaum, A., Luskin, M. (eds.) Recent Advances in Iterative Methods. IMA Volumes in Mathematics and Its Applications, vol. 60, pp. 149–164. Springer, New York (1994)
21. Ruhe, A.: Rational Krylov algorithms for nonsymmetric eigenvalue problems, II: matrix pairs. Linear Algebra Appl. **197–198**, 283–295 (1994)
22. Ruhe, A.: Rational Krylov algorithms for nonsymmetric eigenvalue problems, III: complex shifts for real matrices. BIT **34**, 165–176 (1994)
23. Schmitz, H., Volk, K., Wendland, W.L.: On three-dimensional singularities of elastic fields near vertices. Numer. Methods Partial Differ. Equ. **9**, 323–337 (1993)
24. Schröder, C.: A canonical form for palindromic pencils and palindromic factorizations. Preprint 316, DFG Research Center MATHEON, Mathematics for key technologies in Berlin, TU Berlin (2006)
25. Schröder, C.: Palindromic and even eigenvalue problems – analysis and numerical methods. PhD thesis, Technical University Berlin (2008)
26. Sorensen, D.C.: Passivity preserving model reduction via interpolation of spectral zeros. Syst. Control Lett. **54**, 347–360 (2005)
27. Stewart, G.W.: A Krylov-Schur algorithm for large eigenproblems. SIAM J. Matrix Anal. Appl. **23**, 601–614 (2001)
28. Stewart, G.W.: Addendum to “a Krylov-Schur algorithm for large eigenproblems”. SIAM J. Matrix Anal. Appl. **24**, 599–601 (2002)
29. Watkins, D.S.: On Hamiltonian and symplectic Lanczos processes. Linear Algebra Appl. **385**, 23–45 (2004)
30. Watkins, D.S.: The matrix eigenvalue problem. SIAM, Philadelphia (2007)

Chapter 3
Palindromic Eigenvalue Problems in Applications

Wen-Wei Lin and Christian Schröder

Abstract We list a number of practical applications of linear and quadratic palindromic eigenvalue problems. This chapter focuses on two applications which are discussed in detail. These are the vibration analysis of rail tracks and the regularization of the solvent equation. Special purpose algorithms are introduced and numerical examples are presented.

3.1 Introduction

In this chapter we discuss applications of palindromic eigenvalue problems (PEPs), a special structure of eigenvalue problems that is also introduced in Chaps. 2 and 12. Let us recall that a polynomial eigenvalue problem $P(\lambda)x = \sum_{i=0}^{k} \lambda^i A_i x = 0$ with real or complex $n \times n$ matrices A_i and with the property $A_i = A_{k-i+1}^\top$ for all $i = 0 : k$ is called palindromic (or, more precisely, $\top$-palindromic, but we will omit the "$\top$" for simplicity). Most prevalent in applications are the linear and the quadratic case, which are of the form

$$Ax = (-\lambda)A^\top x, \quad \text{and} \quad (\lambda^2 A^\top + \lambda B + A)x = 0, \quad \text{with} \quad B = B^\top. \tag{3.1}$$

It is easy to see (e.g., by transposing (3.1) and dividing by λ^k) that the spectrum of a palindromic eigenvalue problem has a reciprocal pairing, that is the eigenvalues come in pairs $(\lambda, 1/\lambda)$. Such a pair reduces to a singleton whenever $\lambda = 1/\lambda$,

W.-W. Lin
Department of Applied Mathematics, National Chiao Tung University, No.1001 University Road, Hsinchu 30013, Taiwan, Republic of China
e-mail: wwlin@math.nctu.edu.tw

C. Schröder (✉)
Institut für Mathematik, Technische Universität Berlin, Sekretariat MA 4-5, Straße des 17. Juni 136, D-10623 Berlin, Germany
e-mail: schroed@math.tu-berlin.de

P. Benner et al. (eds.), *Numerical Algebra, Matrix Theory, Differential-Algebraic Equations and Control Theory*, DOI 10.1007/978-3-319-15260-8_3

that is for $\lambda = \pm 1$. Note that in case of real matrices A_i the reciprocal pairing is in addition to the complex conjugate pairing. So, in the real case the eigenvalues come in quadruples $(\lambda, \overline{\lambda}, 1/\lambda, 1/\overline{\lambda})$, which reduces to a reciprocal pair for real nonunimodular eigenvalues (that is $\lambda \in \mathbb{R}$, $\lambda \neq \pm 1$), to a complex conjugate pair for unimodular nonreal eigenvalues (that is $|\lambda| = 1, \lambda \neq \pm 1$) and to a singleton for $\lambda = \pm 1$. In many applications the absence or presence of unimodular eigenvalues is an important property.

With the basics out of the way let us now turn to applications. A rich source of palindromic eigenvalue problems is the area of numerical systems and control theory, an area that belongs to the core interests of Volker Mehrmann. A list of PEPs in this area can be found in [29] and the references therein. The linear-quadratic optimal control problem was already mentioned in Chap. 2: this control problem gives rise to a structured linear eigenvalue problem which is equivalent to a palindromic one via the Cayley transformation. Another application is the optimal H_∞ control problem that, when solved with the so-called γ-iteration method, gives rise to two linear even eigenvalue problems in every iteration of that method. In both of these cases the invariant subspace corresponding to the stable eigenvalues inside the unit circle has to be computed. A third problem from systems theory is the test for passivity of a linear dynamical system that may be implemented by finding out whether a certain palindromic pencil has unimodular eigenvalues or not [7].

Other applications of palindromic eigenvalue problems that we only mention in passing include the simulation of surface acoustic wave (SAW) filters [36, 37], and the computation of the Crawford number of a Hermitian pencil [16] (where the latter is actually a $*$-palindromic eigenvalue problem, obtained by replacing $A_i^\top$ by A_i^* in the definition).

In the remainder of this chapter we will focus on two applications in more detail. First is the simulation of rail track vibrations in Sect. 3.2. This is the application that started the whole field of palindromic eigenvalue problems. We show the derivation of the eigenvalue problem, briefly review algorithms for general polynomial palindromic eigenvalue problems and then discuss a special purpose algorithm exploiting the sparse block structure of the matrices arising in the rail problem.

Second we discuss the regularization of the solvent equation which itself has applications in parameter estimation, see Sect. 3.3. This problem is not a palindromic eigenvalue problem in itself, but the algorithm we describe for its solution requires the repeated solution of many PEPs.

We will use the following notation. $\mathfrak{R}$ and $\mathfrak{I}$ denote real and imaginary part, respectively. We use I_n (or just I) for the identity matrix of order n. We denote by $\overline{A}$, $A^\top$, and A^* the conjugate, the transpose, and the conjugate transpose of a matrix A, respectively. The symbol $\rho(A)$ denotes the spectral radius of a matrix. For a vector x we denote by $\|x\|$ its standard Euclidean norm. For a matrix A, $\|A\|_2 := (\rho(A^*A))^{1/2}$ denotes the spectral norm, whereas $\|A\|_F := (\sum_{i,j} |a_{ij}|^2)^{1/2}$ denotes the Frobenius norm. We define for each $m, n \in \mathbb{N}$ the operator $\mathrm{vec}(\cdot) : \mathbb{C}^{m,n} \to \mathbb{C}^{mn}$ that stacks the columns of the matrix in its argument, i.e., for $A = [a_1, \dots, a_n]$

$$\mathrm{vec}(M) := [a_1^\top, a_2^\top, \dots, a_n^\top]^\top.$$

It is well-known that $\mathrm{vec}(AXB) = (B^\top \otimes A)\mathrm{vec}(X)$ for each triple of matrices A, X, B of compatible size, where $\otimes$ denotes the Kronecker product, e.g., [18].

3.2 Rail Track Vibration

With new Inter-City Express trains crossing Europe at speeds up to 300 kph, the study of the resonance phenomena of the track under high frequent excitation forces becomes an important issue. Research in this area does not only contribute to the safety of the operations of high-speed trains, but also to the design of new train bridges. As shown by Wu and Yang [35], and by Markine, de Man, Jovanovic and Esveld [27], an accurate numerical estimation to the resonance frequencies of the rail plays an important role in the dynamic response of the vehicle-rail-bridge interaction system under different train speeds as well as the design of an optimal embedded rail structures. However, in 2004 the classical finite element packages failed to deliver even a single correct digit for the resonance frequencies.

As reported by Ipsen [20], this problem was posed by the Berlin-based company SFE GmbH to researchers at TU Berlin. So, Hilliges, Mehl and Mehrmann [17] first studied the resonances of railroad tracks excited by high-speed trains in a joint project with this company. Apart from the provided theoretical background for the research of the vibration of rail tracks the outcome was two-fold: (a) the traditionally used method to resolve algebraic constraints was found to be ill-conditioned and was replaced by a well-conditioned alternative, and (b) the arising quadratic eigenvalue problem was observed to have the reciprocal eigenvalue pairing. A search for a structure of the matrix coefficients that corresponds to the eigenvalue pairing finally resulted in the palindromic form (3.1). Then, searching for a structure preserving numerical algorithm for palindromic quadratic eigenvalue problems (PQEPs), D.S. Mackey, N. Mackey, Mehl and Mehrmann [26] proposed a structure preserving linearization with good condition numbers. It linearizes the PQEP (1) to a linear PEP of the form

$$\left(\lambda \begin{bmatrix} A^\top & A^\top \\ B - A & A^\top \end{bmatrix} + \begin{bmatrix} A & B - A^\top \\ A & A \end{bmatrix}\right) \begin{bmatrix} x \\ \lambda x \end{bmatrix} = 0. \qquad (3.2)$$

In the same paper [26] a first structure-preserving numerical algorithm for the linear PEP is presented: in a Jacobi-like manner the matrix A is iteratively reduced to anti-triangular form by unitary congruence transformations. The eigenvalues are then given as ratios of the anti-diagonal elements. Later, more algorithms for linear PEPs (QR-like, URV-like, or based on the "ignore structure at first, then regain it" paradigm) were developed by a student of Mehrmann [31] and Kressner, Watkins, Schröder [21]. These algorithms typically perform well for small and dense linear palindromic EVPs. An algorithm for large sparse linear palindromic EVPs is discussed in [29]. From the fast train model, D.S. Mackey, N. Mackey, Mehl and Mehrmann [24, 25] first derived the palindromic polynomial eigenvalue problems (PPEP) and systematically studied the relationship between PQEP/PPEP and a special class of "good linearizations for good vibrations" (loosely from the casual subtitle of [24]). Based on these theoretical developments of the PQEP/PPEP [17, 24–26], Chu, Hwang, Lin and Wu [8], as well as, Guo and Lin [15] further proposed structure-preserving doubling algorithms (SDAs) from two different approaches for solving the PQEP which are described in the following.

In conclusion, a great deal of progress has been achieved since the first works in 2004. Ironically, the mentioned well-conditioned resolution of algebraic constraint in that first paper [17] alone (i.e., without preserving the palindromic structure) was enough to solve the eigenvalue problem to an accuracy sufficient in industry. Still, the story of the palindromic eigenvalue problem is a good example of an academic industrial cooperation where (opposite to the usual view of knowledge transfer from academia into industry) a question from industry sparked a whole new, still growing and flourishing research topic in academia. Moreover, the good experience led to further joint projects with the same company [28].

3.2.1 Modeling

To model the rail track vibration problem, we consider the rail as a 3D isotropic elastic solid with the following assumptions: (i) the rail sections between consecutive sleeper bays are identical; (ii) the distance between consecutive wheels is the same; and (iii) the wheel loads are equal. Based on the virtual work principle, we model the rail by a 3D finite element discretization with linear isoparametric tetrahedron elements (see Fig. 3.1) which produces an infinite-dimensional ODE system for the fast train:

$$\tilde{M}\ddot{u} + \tilde{D}\dot{u} + \tilde{K}u = \tilde{F}, \tag{3.3}$$

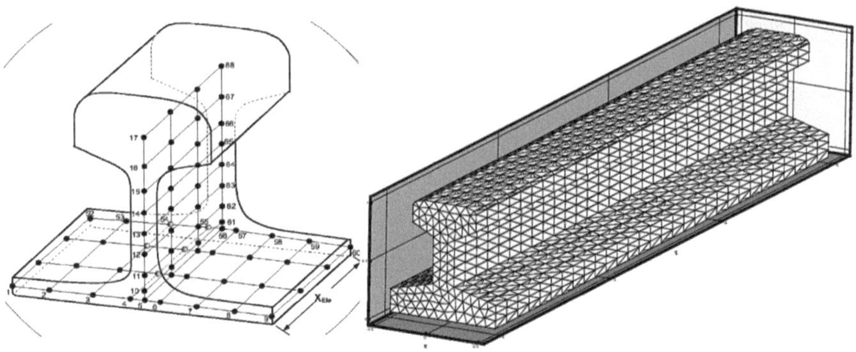

Fig. 3.1 Finite element rail models. *Left*: consisting of three coupled shells, used in industry. *Right*: tetrahedral, used in [8]

where $\tilde{M}$, $\tilde{K}$ and $\tilde{D}$ are block tridiagonal matrices, representing mass, stiffness and damping matrices of (3.3), respectively. The external excitation force $\tilde{F}$ is assumed to be periodic with frequency $\omega > 0$. In practice, we consider $\tilde{D}$ is a linear combination of $\tilde{M}$ and $\tilde{K}$ of the form $\tilde{D} = c_1\tilde{M} + c_2\tilde{K}$ with $c_1, c_2 > 0$. Furthermore, we assume that the displacements of two boundary cross sections of the modeled rail have a ratio λ. Under these assumptions, the vibration analysis of rail tracks induces two real symmetric matrices M and K given by

$$M = \begin{bmatrix} M_0 & M_1^\top & 0 & \cdots & M_1 \\ M_1 & \ddots & \ddots & & \vdots \\ 0 & \ddots & \ddots & \ddots & 0 \\ \vdots & & \ddots & \ddots & M_1^\top \\ M_1^\top & \cdots & 0 & M_1 & M_0 \end{bmatrix}_{m\times m}, \quad K = \begin{bmatrix} K_0 & K_1^\top & 0 & \cdots & K_1 \\ K_1 & \ddots & \ddots & & \vdots \\ 0 & \ddots & \ddots & \ddots & 0 \\ \vdots & & \ddots & \ddots & K_1^\top \\ K_1^\top & \cdots & 0 & K_1 & K_0 \end{bmatrix}_{m\times m}, \tag{3.4}$$

where each block in M and K is of the size $q \times q$. Let M_t be the block tridiagonal part of M, and M_c be the $m \times m$ matrix with M_1 on the upper-right corner and zero blocks else where. Then we can write $M = M_t + M_c + M_c^\top$. Correspondingly, we have $K = K_t + K_c + K_c^\top$ and $D = D_t + D_c + D_c^\top$, where K_t, K_c, D_t, D_c are defined analogously.

Letting $u = xe^{\iota\omega t}$ in the spectral model (3.3), where ι denotes the imaginary unit, leads to a PQEP of the form [8, 15]:

$$P(\lambda)x = (\lambda^2 A^\top + \lambda B + A)x = 0, \tag{3.5}$$

Fig. 3.2 Typical distribution of eigenvalues

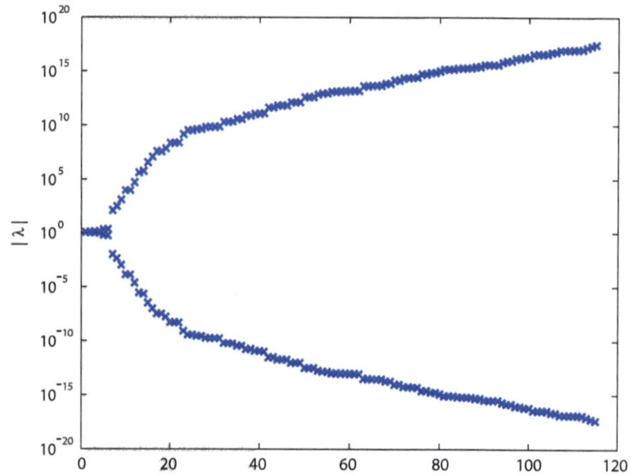

where

$$B = K_t + \iota\omega D_t - \omega^2 M_t = \begin{bmatrix} \blacksquare & \blacksquare & & O \\ \blacksquare & \blacksquare & \ddots & \\ & \ddots & \ddots & \blacksquare \\ O & & \blacksquare & \blacksquare \end{bmatrix} \in \mathbb{C}^{n\times n}.$$

$$A = K_c + \iota\omega D_c - \omega^2 M_c = \begin{bmatrix} O & \cdots & \blacksquare \\ \vdots & & \vdots \\ O & \cdots & O \end{bmatrix} \in \mathbb{C}^{n\times n}$$

with $n = mq$.

The PQEP problem in (3.5) is typically badly scaled and some numerical difficulties need to be addressed: (i) the problem size n can be 30,000–100,000 (typically, m is 50–100, and q is 700–1,000); (ii) it is needed to compute all finite, nonzero eigenvalues and associated eigenvectors for all frequencies ω between 100 and 5,000 Hz; (iii) many of eigenvalues are zero and infinity; (iv) the range of eigenvalues $|\lambda|$ is typically in $[10^{-20}, 10^{20}]$ (see Fig. 3.2).

To solve the PQEP in (3.5), one may use an initial deflating procedure for zero and infinite eigenvalues to obtain a deflated $q \times q$ dense PQEP [17]

$$P_d(\lambda)x_d \equiv (\lambda^2 A_d^\top + \lambda B_d + A_d)x_d = 0. \tag{3.6}$$

On the other hand, one can solve the original block-banded PQEP (3.5) directly.

3.2.2 SDA for General PQEPs

To solve (3.6) we rewrite $P_d(\lambda)$ in (3.6) as

$$P_d(\lambda) = \lambda^2 A_d^\top + \lambda B_d + A_d = (\lambda A_d^\top + X_d) X_d^{-1} (\lambda X_d + A_d) \tag{3.7}$$

assuming that X_d is non-singular. It follows that $P_d(\lambda)$ can be factorized as (3.7) for some non-singular X_d if and only if X_d satisfies the nonlinear matrix equation (NME)

$$X_d + A_d^\top X_d^{-1} A_d = B_d. \tag{3.8}$$

As shown in [25], there are many solutions to the NME (3.8). Each of them enable to facilitate the factorization of $P_d(\lambda)$. Assume that there are no eigenvalues on the unit circle. Then, by (3.7) we can partition the spectrum into $\Lambda_s \oplus \Lambda_s^{-1}$ with Λ_s containing the stable eigenvalues (inside the unit circle). We call a solution $X_{d,s}$ of (3.8) a stabilizing solution if the spectrum of $X_{d,s}^{-1} A_d$ is the same as that of Λ_s. The structure-preserving algorithm (SDA) in [15] can then be applied to solve the NME (3.8) and subsequently the PQEP (3.6).

Algorithm 1 (SDA_CHLW)
Let $A_0 = A_d,\ B_0 = B_d,\ P_0 \equiv 0.$
For $k = 0, 1, \ldots,$ *compute*

$$\begin{aligned} A_{k+1} &= A_k (B_k - P_k)^{-1} A_k, \\ B_{k+1} &= B_k - A_k^\top (B_k - P_k)^{-1} A_k, \\ P_{k+1} &= P_k + A_k (B_k - P_k)^{-1} A_k^\top, \end{aligned}$$

if no break down occurs.

For the convergence of Algorithm 1, we have the following theorem.

Theorem 1 ([8]) *Let* $X_{d,s}$ *and* $\hat{X}_{d,s}$ *be the stabilizing solutions of NME* (3.8) *and the dual NME* $\hat{X}_d + A_d \hat{X}_d^{-1} A_d^\top = B_d$*. Then the sequences* $\{A_k\}, \{B_k\}, \{P_k\}$ *generated by Algorithm 1 satisfy*

(i) $\limsup_{k\to\infty} \sqrt[2^k]{\|B_k - X_{d,s}\|} \le \rho(X_{d,s}^{-1} A_d)^2,$

(ii) $\limsup_{k\to\infty} \sqrt[2^k]{\|A_k\|} \le \rho(X_{d,s}^{-1} A_d),$

(iii) $\limsup_{k\to\infty} \sqrt[2^k]{\left\| B_k - P_k - \hat{X}_{d,s} \right\|} \le \rho(X_{d,s}^{-1} A_d)^2.$

provided that all the required inverses of $B_k - P_k$ *exist.*

3.2.3 SDA for Block-Banded PQEPs

We now apply the solvent approach directly to the original block-banded PQEP (3.5). To this end, as in (3.7), we first factorize the PQEP (3.5) into

$$P(\lambda) = \lambda^2 A^\top + \lambda B + A = (\lambda A^\top + X)X^{-1}(\lambda X + A) \tag{3.9}$$

and then solve the nonlinear matrix equation (NME)

$$X + A^\top X^{-1} A = B. \tag{3.10}$$

There are two advantages of the solvent approach in (3.9) over the deflation approach in (3.6) [2, 26]. First, the deflation procedure is used for the sake of efficiency, which involves the inverses of two potentially ill-conditioned matrices. Second, in the deflation approach, the eigenvalues of the smaller PQEP range in modulus from ε to ε^{-1}, where ε is close to 0, while in the solvent approach the eigenvalues of $\lambda X + A$ range in modulus from ε to 1.

The success of the solvent approach depends on the existence of a stabilizing solution of (3.10) and an efficient method for its computation.

From the classic Poincaré-Bendixson Theorem we obtain the following result.

Theorem 2 ([15]) *Let K_t, K_c, M_t, M_c be given as in* (3.4)*, and set $D_t = c_1 M_t + c_2 K_t$, $D_c = c_1 M_c + c_2 K_c$ with $c_1, c_2 > 0$. Then the PQEP* (3.5) *has no eigenvalues on the unit circle.*

Based on a deep result on linear operators [10] one can prove the following existence theorem.

Theorem 3 ([15]) *Under the assumptions in Theorem 2, the NME* (3.10) *has a unique stabilizing solution, and the solution is complex symmetric. Moreover, the dual equation of* (3.10)

$$\hat{X} + A\hat{X}^{-1}A^\top = B \tag{3.11}$$

also has a unique stabilizing solution and the solution is complex symmetric.

The SDA as in Algorithm 1 with $A_0 = A$, $B_0 = B$, $P_0 = 0$ can then be applied to solve the NME (3.10) and the dual NME (3.11). In order to distinguish it from Algorithm 1, we call this procedure as Algorithm 2.

Algorithm 2 (SDA)
Let $A_0 = A$, $B_0 = B$, $P_0 \equiv 0$.
For $k = 0, 1, \ldots$, compute A_{k+1}, B_{k+1} and P_{k+1} as in Algorithm 1.

In contrast to Theorem 1, the following theorem shows that the Algorithm 2 is well-defined and no break down occurs (i.e., $B_k - P_k$ is always invertible). Moreover, B_k and $B_k - P_k$ converge quadratically to the unique stabilizing solutions of NME as well as the dual NME, respectively.

Theorem 4 ([15]) *Let X_s and $\hat{X}_s$ be the stabilizing solutions of NME* (3.10) *and the dual NME* (3.11), *respectively. Then*

(i) *The sequences $\{A_k\}$, $\{B_k\}$, $\{P_k\}$ generated by Algorithm 2 are well-defined.*
(ii) $\limsup_{k\to\infty} \sqrt[2^k]{\|B_k - X_s\|} \le \rho(X_s^{-1}A)^2$,
(iii) $\limsup_{k\to\infty} \sqrt[2^k]{\|A_k\|} \le \rho(X_s^{-1}A)$,
(iv) $\limsup_{k\to\infty} \sqrt[2^k]{\left\|B_k - P_k - \hat{X}_s\right\|} \le \rho(X_s^{-1}A)^2$.

where $\|\cdot\|$ is any matrix norm.

At the first sight, Algorithm 2 (the solvent approach applied to the original PQEP (3.5)) would be very expensive. However, the complexity of Algorithm 2 can be reduced drastically by using the special structure of the matrix A as in (3.5). Let $B_k = B - R_k$. Then by induction it is easily seen that the matrices in the sequences $\{A_k\}$, $\{R_k\}$, $\{P_k\}$ have the special forms

$$A_k = \begin{bmatrix} & & E_k \\ & 0 & \\ 0 & & \end{bmatrix}, \quad R_k = \begin{bmatrix} 0 & & & \\ & \ddots & & \\ & & 0 & \\ & & & F_k \end{bmatrix}, \quad P_k = \begin{bmatrix} G_k & & & \\ & 0 & & \\ & & \ddots & \\ & & & 0 \end{bmatrix} \tag{3.12}$$

where the $q \times q$ matrices E_k, F_k and G_k can be determined by the following simplified algorithm in which

$$B = \begin{bmatrix} H_0 & H_1^\top & & \\ H_1 & \ddots & \ddots & \\ & \ddots & \ddots & H_1^\top \\ & & H_1 & H_0 \end{bmatrix}_{m\times m} \tag{3.13}$$

is given in (3.5) with $H_0 = K_0 + \iota\omega M_0, \quad H_1 = K_1 + \iota\omega D_1 - \omega^2 M_1$.

Algorithm 3 (SDA_GL; a sparse version of Algorithm 2)
Let $E_0 = H_1,\ F_0 = 0,\ G_0 = 0$. For $k = 0, 1, \ldots$, compute

$$\begin{bmatrix} S_{k,1} & T_{k,1} \\ S_{k,2} & T_{k,2} \\ \vdots & \vdots \\ S_{k,m} & T_{k,m} \end{bmatrix} = \left(B - \begin{bmatrix} G_k & & & \\ & 0 & & \\ & & \ddots & \\ & & 0 & \\ & & & F_k \end{bmatrix} \right)^{-1} \begin{bmatrix} E_k & 0 \\ 0 & \vdots \\ \vdots & 0 \\ 0 & E_k^\top \end{bmatrix}, \tag{3.14}$$

where all matrix blocks on the left side of (3.14) *are $q \times q$.*

Then compute

$$E_{k+1} = E_k S_{k,m}, \quad F_{k+1} = F_k + E_k^\top S_{k,1}, \quad G_{k+1} = G_k + E_k T_{k,m}. \tag{3.15}$$

Note that the linear systems in (3.14) can be solved by the Sherman–Morrison–Woodbury formula. The details can be found in [15].

After the solvent X_s is computed, we can compute all eigenpairs. Let $B = U^H R$ be the QR-factorization in a sparse way. Multiplying U to A and X_s from the left, respectively, we have

$$UA = \begin{bmatrix} 0_{n-q} & \tilde{H}_1^\top \\ 0 & \Phi_1 \end{bmatrix}, \; UX_s = \begin{bmatrix} X_1 & X_2 \\ 0 & \Phi_2 \end{bmatrix}, \tag{3.16}$$

where $X_1 = R(1:n-q, 1:n-q)$ and $X_2(1:n-3q, 1:q) = 0$. In view of the factorization of $P(\lambda) = (\lambda A^\top + X_s)X_s^{-1}(\lambda X_s + A)$, the nonzero stable eigenpairs (λ_s, z_s) of $P(\lambda)$ are those of $\lambda X_s + A$ and can be computed by the generalized eigenvalue problem

$$\Phi_1 z_{s,2} = -\lambda_s \Phi_2 z_{s,2}, \tag{3.17}$$

and set

$$z_{s,1} = -X_1^{-1}(X_2 z_{s,2} + \lambda_s^{-1} \tilde{H}_1^\top z_{s,2}), \; z_s = \begin{bmatrix} z_{s,1} \\ z_{s,2} \end{bmatrix}, \tag{3.18}$$

for $s = 1, \cdots, q$.

We now compute all left eigenvectors of $\lambda\Phi_2 + \Phi_1$ by

$$y_s^\top \Phi_1 = -\lambda_s y_s^\top \Phi_2, \tag{3.19}$$

for $s = 1, \cdots, q$. The finite unstable eigenpairs (λ_u, z_u) of $P(\lambda)$ satisfy

$$P(\lambda_u) z_u \equiv P(1/\lambda_s) z_u = \frac{1}{\lambda_s^2} \left(A^\top + \lambda_s X_s\right) X_s^{-1} \left(X_s + \lambda_s A\right) z_u = 0. \tag{3.20}$$

From (3.16) and (3.19), it follows that

$$(A^\top + \lambda_s X_s) U^\top \begin{bmatrix} 0 \\ y_s \end{bmatrix} = \left(\begin{bmatrix} 0 & 0 \\ \tilde{H}_1 & \Phi_1^\top \end{bmatrix} + \begin{bmatrix} \lambda_s X_1^\top & 0 \\ \lambda_s X_2^\top & \lambda_s \Phi_2^\top \end{bmatrix} \right) \begin{bmatrix} 0 \\ y_s \end{bmatrix} = 0. \tag{3.21}$$

From (3.20) the eigenvector z_u corresponding to $\lambda_u = \lambda_s^{-1}$ can be found by solving the linear system

$$(X_s + \lambda_s A) z_u = X_s \left(U^\top \begin{bmatrix} 0 \\ y_s \end{bmatrix} \right) = \begin{bmatrix} 0 \\ \Phi_2^\top y_s \end{bmatrix}. \tag{3.22}$$

Premultiplying (3.22) by U, the finite unstable eigenpairs (λ_u, z_u) of $P(\lambda)$ can be computed by

$$\begin{bmatrix} \zeta_{u,1} \\ \zeta_{u,2} \end{bmatrix} = U \begin{bmatrix} 0 \\ \Phi_2^\top y_s \end{bmatrix}, \quad z_{u,2} = (\Phi_2 + \lambda_s \Phi_1)^{-1} \zeta_{u,2}, \tag{3.23}$$

$$z_{u,1} = X_1^{-1} \left[\zeta_{u,1} - \left(X_2 + \lambda_s \tilde{H}_1^\top\right) z_{u,2}\right], \quad z_u = \begin{bmatrix} z_{u,1} \\ z_{u,2} \end{bmatrix}, \tag{3.24}$$

for $u = 1, \cdots, q$. The total computational cost for eigenpairs of $P(\lambda)$ is $\frac{154}{3} mq^2$ flops which is the same as the initial deflation procedure.

We quote some numerical results from [15] with $(q, m) = (705, 51)$. The matrices M and K are given by (3.4) and we take $D = 0.8M + 0.2K$. To measure the accuracy of an approximate eigenpair (λ, z) for $P(\lambda)$ we use the relative residual

$$\text{RRes} = \frac{\|\lambda^2 A^\top z + \lambda B z + A z\|_2}{(|\lambda|^2 \|A\|_F + |\lambda| \|B\|_F + \|A\|_F) \|z\|_2}. \tag{3.25}$$

In Table 3.1 we give $\|F_{k+1} - F_k\|_2 / \|F_k\|_2$ for $(q, m) = (705, 51)$, and for $\omega =$ 100, 1,000, 3,000, 5,000, respectively, computed by Algorithm 3. The convergence behavior of F_k is roughly the same as indicated by Theorem 4.

To demonstrate the accuracy of Algorithm 3, in Fig. 3.3, we plot the relative residuals (3.25) of approximate eigenpairs computed by Algorithm 3 (SDA_GL) and those of the other existing methods SA_HLQ [8] as well as Algorithm 1 (SDA_CHLW) [19] for $\omega = 1{,}000$ and $(q, m) = (705, 51)$.

In Fig. 3.3, we see that Algorithm 3 (SDA_GL) has significantly better accuracy for stable eigenpairs. This is because that SA_HLQ [8] and Algorithm 1 (SDA_CHLW) [19] are structure-preserving methods only applied for the deflated PQEP (3.7). The deflation procedure possibly involves the inverses of two poten-

Table 3.1 $\|F_{k+1} - F_k\|_2 / \|F_k\|_2$ for different ω values with $(q, m) = (705, 51)$

k	$\omega = 100$, $\rho = 0.9593$	$\omega = 1{,}000$, $\rho = 0.8745$	$\omega = 3{,}000$, $\rho = 0.7925$	$\omega = 5{,}000$, $\rho = 0.7406$
1	$1.1e-01$	$1.0e-01$	$7.0e-02$	$5.7e-02$
2	$2.8e-02$	$1.2e-02$	$1.0e-02$	$8.8e-03$
3	$4.7e-03$	$3.6e-03$	$1.5e-03$	$7.8e-04$
4	$2.1e-03$	$4.2e-04$	$3.8e-05$	$6.4e-06$
5	$5.7e-04$	$5.8e-06$	$2.2e-08$	$4.3e-10$
6	$4.0e-05$	$1.1e-09$	$7.7e-15$	$2.9e-19$
7	$1.9e-07$	$3.5e-17$	0	
8	$4.6e-12$	0		
9	0			

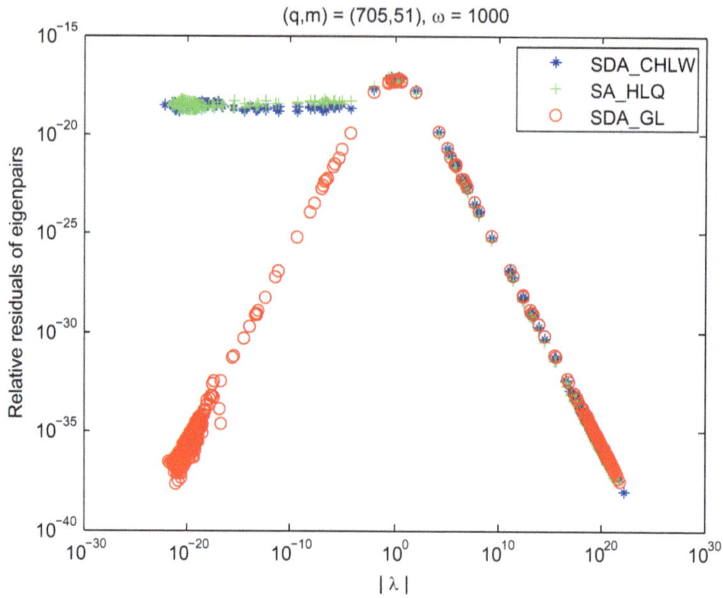

Fig. 3.3 Relative residuals of eigenpairs with $(q, m) = (705, 51)$

tially ill-conditioned matrices so that SA_HLQ [8] and SDA_CHLW may lose the accuracy of eigenpairs when we transform the approximate deflated eigenpairs to the ones of the original PQEP (3.5).

We efficiently and accurately solve a PQEP arising from the finite element model for fast trains by using the SDA_GL (Algorithm 3) in the solvent approach. Theoretical issues involved in the solvent approach are settled satisfactorily. The SDA_GL has quadratic convergence and exploits the sparsity of the PQEP.

3.3 Regularization of the Solvent Equation

Here we consider the nonlinear matrix equation

$$X + A^\top X^{-1} A = B, \tag{3.26}$$

where $A, B \in \mathbb{R}^{n,n}$ with $B > 0$ (i.e., B is Hermitian and positive definite). Note that this is the solvent equation we already saw in (3.8). Here, we are interested in making sure that there is a solution $X \in \mathbb{R}^{n,n}$, $X > 0$. It is known (e.g., [9]) that such a solution exists if and only if the matrix Laurent polynomial

$$Q(\lambda) = \lambda A^\top + B + \lambda^{-1} A$$

is regular (i.e., the matrix $Q(\lambda)$ is non-singular for at least one value of $\lambda \in \mathbb{C}$) and $Q(\lambda) \geq 0$ (i.e., $Q(\lambda)$ is Hermitian and positive semi-definite) for each complex value λ on the unit circle. Moreover, a stabilizing solution X (i.e., one with $\rho(X^{-1}A) < 1$; as it is needed in applications) exists if and only if $Q(\lambda) > 0$ for each unimodular λ. Assuming positive definiteness of $Q(\lambda)$ for at least one such λ, the last condition is equivalent to stating that Q has no generalized eigenvalues on the unit circle.

In practice, often the coefficients A and B are affected by errors, e.g., because they come out of data measurements, or their determination involves some form of linearization, truncation, or other such simplifications. Then it may well be the case that the original intended matrix equation admits a solution, whereas the perturbed one – which is available in practice – does not.

In this section we present a method to compute perturbations $\tilde{A} = A + E$, $\tilde{B} = B + F$, with $\|E\|$ and $\|F\|$ small, such that Eq. (3.26) (with A, B replaced by $\tilde{A}, \tilde{B}$) is solvable. This is achieved by removing all generalized eigenvalues of $\tilde{Q}(\lambda) = \lambda\tilde{A}^\top + \tilde{B} + \lambda^{-1}\tilde{A}$ from the unit circle. The presented method is described in [6] (with an application in parameter estimation, see below) and is based upon similar methods in [3, 7, 12, 32–34] (used there to enforce passivity, dissipativity, or negative imaginariness of an LTI control system). Other related methods that aim to move certain eigenvalues to or from certain regions by perturbing the matrix coefficients in an eigenvalue problem include [13, 14] (where pseudo-spectral methods are used) and [1, 5] by Mehrmann et. al.

We note that λQ is a palindromic matrix polynomial, and that $Q(\lambda^{-1}) = Q(\lambda)^\top$. Thus the eigenvalues of Q come in reciprocal pairs. As a consequence unimodular eigenvalues cannot just leave the unit circle under small perturbations of A and B. For this to happen two of them have to move together, merge, and then split off into the complex plane. Suppose that Q has ℓ unimodular eigenvalues $\lambda_j = e^{\imath\omega_j}$ with normalized eigenvectors v_j, $\|v_j\| = 1$, $j = 1, 2, \ldots, \ell$.

The method to compute E and F is iterative. In a single iteration the unimodular eigenvalues shall be moved to $e^{\imath\tilde{\omega}_j}$, $j = 1, 2, \ldots, \ell$ on the unit circle. We assume that "the $\tilde{\omega}_j$ are closer together than the ω_j" and that $|\tilde{\omega}_j - \omega_j|$ is small for all j. More on how to chose $\tilde{\omega}_j$ will be discussed later.

In order to relate the change of the unimodular eigenvalues to small perturbations of A and B, we use the following first-order perturbation result.

Theorem 5 ([6]) *Let $A, B \in \mathbb{R}^{n,n}$ with $B = B^\top$ and let $Q(\lambda) = \lambda A^\top + B + \lambda^{-1}A$ have a simple unimodular generalized eigenvalue $\lambda_j = e^{\imath\omega_j}$, with eigenvector v_j. Let $\sigma_j := -2\Im(\lambda_j v_j^* A^\top v_j)$. Furthermore, let $\tilde{Q}(\lambda) := \lambda(A+E)^\top + B + F + \lambda^{-1}(A+E)$ be a sufficiently small perturbation of $Q(\lambda)$, with $F = F^\top$. Then $\sigma_j \neq 0$ and $\tilde{Q}$ has a generalized eigenvalue $\tilde{\lambda}_j = e^{\imath\tilde{\omega}_j}$ such that*

$$\sigma_j(\tilde{\omega}_j - \omega_j) = -\Re(2e^{\imath\omega_j} v_j^* E^\top v_j + v_j^* F v_j) + \hat{\Phi}(E, F). \tag{3.27}$$

for some function $\hat{\Phi}(E, F)$ with $\hat{\Phi}(E, F) = o(\|E, F\|)$.

Usually such perturbation results are used to find out where eigenvalues move when the matrices are perturbed. We will use it the other way round: we know where we want the eigenvalues to move to and use the result to find linear constraints to the perturbation matrices.

Moreover, we wish to allow only perturbations in the special form

$$(E, F) = \sum_{i=1}^{m} (E_i, F_i)\delta_i \tag{3.28}$$

for some $\delta_i \in \mathbb{R}$, where $(E_i, F_i) \in \mathbb{R}^{n,n} \times \mathbb{R}^{n,n}$, with $F_i = F_i^\top$ for each $i = 1, 2, \ldots, m$, is a given basis of allowed modifications to the pair (A, B).

For instance, if $n = 2$, a natural choice for this perturbation basis is

$$\left(\left[\begin{smallmatrix}1&0\\0&0\end{smallmatrix}\right],0\right), \left(\left[\begin{smallmatrix}0&0\\0&1\end{smallmatrix}\right],0\right), \left(\left[\begin{smallmatrix}0&1\\0&0\end{smallmatrix}\right],0\right), \left(\left[\begin{smallmatrix}0&0\\1&0\end{smallmatrix}\right],0\right), \left(0,\left[\begin{smallmatrix}1&0\\0&0\end{smallmatrix}\right]\right), \left(0,\left[\begin{smallmatrix}0&0\\0&1\end{smallmatrix}\right]\right), \left(0,\left[\begin{smallmatrix}0&1\\1&0\end{smallmatrix}\right]\right). \tag{3.29}$$

This choice gives all possible perturbations on the entries of each matrix that preserve the symmetry of B. However, if necessary we can enforce some properties of E and F like being symmetric, Toeplitz, circular, or having a certain sparsity structure by choosing $(E_i, F_i)_{i=1}^m$ suitably.

Using the vec-operator, we can rewrite (3.27) as

$$\sigma_j(\tilde{\omega}_j - \omega_j) \approx -\Re([2e^{\imath\omega_i}, 1] \otimes v_j^\top \otimes v_j^*) \begin{bmatrix} \mathrm{vec}(E^\top) \\ \mathrm{vec}(F) \end{bmatrix},$$

and (3.28) as

$$\begin{bmatrix} \mathrm{vec}(E^\top) \\ \mathrm{vec}(F) \end{bmatrix} = \begin{bmatrix} \mathrm{vec}(E_1^\top) & \cdots & \mathrm{vec}(E_m^\top) \\ \mathrm{vec}(F_1) & \cdots & \mathrm{vec}(F_m) \end{bmatrix} \begin{bmatrix} \delta_1 \\ \vdots \\ \delta_m \end{bmatrix}.$$

Together we obtain a system of ℓ linear equations in m unknowns

$$\mathscr{A}\delta = \mathscr{B}, \tag{3.30}$$

where $\delta = [\delta_1, \ldots, \delta_m]^\top$ and

$$\mathscr{A} = \begin{bmatrix} -\Re([2e^{\imath\omega_1}, 1] \otimes v_1^\top \otimes v_1^*) \\ \vdots \\ -\Re([2e^{\imath\omega_\ell}, 1] \otimes v_\ell^\top \otimes v_\ell^*) \end{bmatrix} \begin{bmatrix} \mathrm{vec}(E_1^\top) & \cdots & \mathrm{vec}(E_m^\top) \\ \mathrm{vec}(F_1) & \cdots & \mathrm{vec}(F_m) \end{bmatrix}, \quad \mathscr{B} = \begin{bmatrix} \sigma_1(\tilde{\omega}_1 - \omega_1) \\ \vdots \\ \sigma_\ell(\tilde{\omega}_\ell - \omega_\ell) \end{bmatrix}.$$

So, any sufficiently small perturbation (3.28) satisfying (3.30) moves the unimodular eigenvalues approximately to the wanted positions. We are interested in the smallest such perturbation. To this end we assume that the system (3.30) is under-determined, $m > \ell$, but of full rank. Hence, we can use a simple QR factorization to compute its minimum-norm solution, given by $\delta = QR^{-T}\mathscr{B}$, where $\mathscr{A}^\top = QR$ denotes a thin QR factorization. Note that for the system to be solved efficiently it is sufficient that ℓ is small and the matrix product $\mathscr{A}$ is efficiently formed (e.g., using the sparsity of E_i, F_i); m, on the other hand, may be large.

Using several steps of this procedure, the unimodular eigenvalues are made to coalesce into pairs in sufficiently small steps and then leave the circle. To sum up, our regularization algorithm is as follows.

Algorithm 4
Input: *$A, B = B^\top \in \mathbb{R}^{n,n}$ such that $Q(\lambda) = \lambda A^\top + B + \lambda^{-1}A$ is regular, $\{E_i, F_i\}_{i=1}^m$*
Output: *$\tilde{A}, \tilde{B} = \tilde{B}^\top \in \mathbb{R}^{n,n}$ such that $\tilde{Q}(\lambda) = \lambda\tilde{A}^\top + \tilde{B} + \lambda^{-1}\tilde{A}$ has no unimodular eigenvalues.*

1. *Set $\tilde{A} = A, \tilde{B} = B$*
2. *Compute the unimodular generalized eigenvalues $\lambda_j = e^{\imath\omega_j}$ of $\tilde{Q}(\lambda)$, $j = 1, 2, \ldots, \ell$ and the associated eigenvectors v_j. If there is none, terminate the algorithm. Also compute $\sigma_j = -2\Im(\lambda_j v_j^* \tilde{A}^\top v_j)$.*
3. *Determine suitable locations for the perturbed generalized eigenvalues $\tilde{\omega}_j$.*
4. *Assemble the system* (3.30) *and compute its minimum-norm solution δ.*
5. *Set $\tilde{A} = \tilde{A} + \sum_{i=1}^m \delta_i E_i$, $\tilde{B} = \tilde{B} + \sum_{i=1}^m \delta_i F_i$ and repeat from step 2.*

A few remarks are in order. Although the perturbation in every single iteration is minimized, this does not imply that the accumulated perturbation is also minimal. In numerical experiments the norm of the accumulated perturbation decreased and finally seemed to converged when more and more steps of Algorithm 4 were used that each move the unimodular eigenvalues by a smaller and smaller distance, see Fig. 3.4 (top right plot) below. Second, there is nothing to prevent non-unimodular eigenvalues from entering the unit circle in the course of the iterations. This is not a problem since they are moved off again in the following few iterations. Finally, step 2 consists of solving a quadratic palindromic eigenvalue problem where the eigenvalues on the unit circle are wanted. For general eigenvalues methods it is difficult to decide whether a computed eigenvalue is really on or just close by the unit circle. Here, structured methods that compute the eigenvalues in pairs can show their strengths.

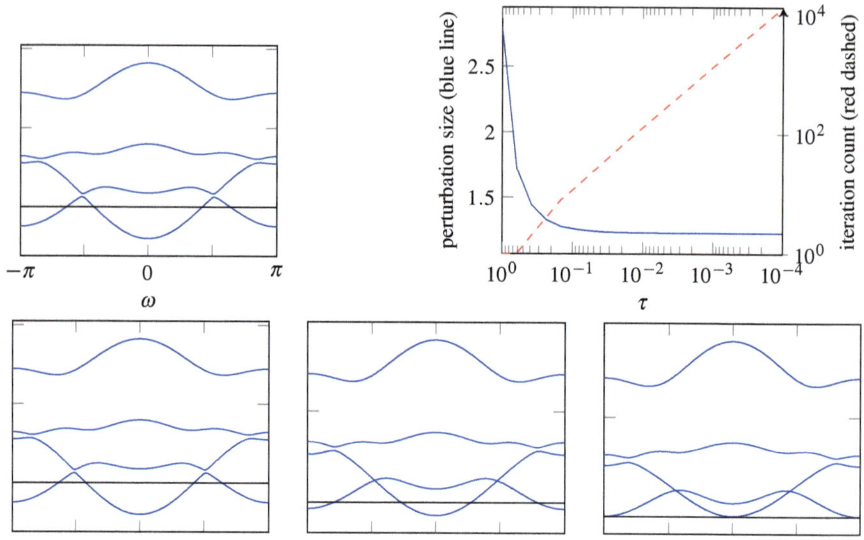

Fig. 3.4 *Top left*: Spectral plot (i.e., eigenvalues of $Q(e^{\imath\omega})$ on the y-axis plotted over $\omega \in [-\pi, \pi]$) for (3.31); *top right*: dependence of size of cumulative perturbation and of number of needed iterations on τ; *bottom row*: Spectral plot after 1, 3, and 6 iterations using $\tau = 0.2$

3.3.1 An Example and Spectral Plots

In order to get a better insight of what is happening and to explain how to choose the $\tilde{\omega}_j$ it is best to look at an example. We start from the matrices [6]

$$A = \begin{bmatrix} 1 & 0 & 0 & 0 \\ 0 & 1 & 1 & 0 \\ 0 & 1 & -1 & 0 \\ 0 & 0 & 0 & -1 \end{bmatrix}, \quad B = \begin{bmatrix} 3 & 2 & 1 & 0 \\ 2 & 3 & 2 & 1 \\ 1 & 2 & 3 & 2 \\ 0 & 1 & 2 & 3 \end{bmatrix}. \tag{3.31}$$

The top left plot of Fig. 3.4 shows the eigenvalues of the Hermitian matrix $Q(e^{\imath\omega})$ for these A, B on the y-axis as they vary with $\omega \in [-\pi, \pi]$. Obviously the plot is 2π-periodic and symmetric (because $Q(e^{\imath\omega}) = Q(e^{\imath(\omega+2\pi)}) = (Q(e^{-\imath\omega}))^{\top}$). For instance, one sees from the graph that all the lines lie above the x-axis for $\omega = \pi/2$, so $Q(e^{\imath\pi/2})$ is positive definite. Instead, for $\omega = 0$ (and in fact for most values of ω) there is a matrix eigenvalue below the imaginary axis, thus $Q(e^{\imath 0})$ is not positive definite. There are four points in which the lines cross the x-axis and these correspond to the values of ω for which $Q(e^{\imath\omega})$ is singular, i.e., for which $e^{\imath\omega}$ is a generalized eigenvalue of Q on the unit circle. We label them $\omega_1, \omega_2, \omega_3, \omega_4$ starting from left.

Notice that in our example the lines corresponding to different matrix eigenvalues come very close to each other, but never cross. This is not an error in the graph, but an instance of a peculiar phenomenon known as *eigenvalue avoidance*, see, e.g., [22, p. 140] or [4].

Recall that the overall goal is to find a perturbation that renders $Q(\lambda)$ positive definite on the whole unit circle. This perturbation will have to move up the two bumps that extend below the x-axis. For this to happen the two central intersections ω_2 and ω_3 have to move towards each other, until they coalesce and then disappear (i.e., the curve does not cross the x-axis anymore). The other two intersections ω_1 and ω_4 moved towards the borders of the graph, coalesce at $\omega = \pi$ and then disappear as well. In particular, we see that the intersections ω_i in which the slope of the line crossing the x-axis is positive ($i = 1, 3$) need to be moved to the *left*, and the ω_i for which it is negative ($i = 2, 4$) need to be moved to the *right*.

Moreover, the sign of the slope with which the line crosses the x-axis is known in the literature as *sign characteristic* of the unimodular generalized eigenvalue [11], and it is well known that only two close-by generalized eigenvalues with opposite sign characteristics can move off the unit circle through small perturbations.

These slopes are easily computable, in fact the are given by σ_j in Theorem 5. In order to obtain the $\tilde{\omega}_j$ that are moved "to the correct direction" and that are not too far away from the ω_j, we use $\tilde{\omega}_j = \omega_j - \tau \operatorname{sign}(\sigma_j)$, where τ is a step size parameter. Other choices are discussed in [7].

For our example, Algorithm 4 with a step size of $\tau = 0.2$ needs 6 iterations. The spectral plots for the intermediate polynomials $\tilde{Q}(\lambda)$ after 1, 3, and 6 iterations are shown in the lower row of Fig. 3.4. One can see that the topmost eigenvalue curves are almost unchanged, while the bottom ones are modified slightly at each iteration, and these modifications have the overall effect of slowly pushing them upwards. After six iterations, the resulting palindromic Laurent matrix polynomial $Q(\lambda)$ is positive definite on the whole unit circle (as shown in the bottom-right graph), and thus it has no more unimodular generalized eigenvalues. The resulting matrices returned by the algorithm are

$$\tilde{A} \approx \begin{bmatrix} 0.816 & 0.183 & 0.0379 & -0.0565 \\ 0.183 & 0.915 & 0.775 & 0.152 \\ 0.0379 & 0.775 & -0.647 & -0.173 \\ -0.0565 & 0.152 & -0.173 & -0.922 \end{bmatrix}, \quad \tilde{B} \approx \begin{bmatrix} 3.16 & 1.67 & 0.956 & 0.0913 \\ 1.67 & 3.28 & 1.62 & 1.13 \\ 0.956 & 1.62 & 3.41 & 1.55 \\ 0.0913 & 1.13 & 1.55 & 3.13 \end{bmatrix}.$$

The relative magnitude of the obtained perturbation is

$$\frac{\|[\tilde{A} - A, \tilde{B} - B]\|_F}{\|[A, B]\|_F} = 0.157.$$

Such a large value is not unexpected, since the plot in Fig. 3.4 extends significantly below the real axis, thus quite a large perturbation is needed.

The step size τ plays a role here. With a smaller value of τ, one expects the resulting perturbation to be smaller, since the approximated first-order eigenvalue locations are interpolated more finely; on the other hand, the number of needed steps should increase as well. This expectation is supported by the top right plot in Fig. 3.4 where we report the resulting values for different choices of τ in this example.

Note that a naive way to move all eigenvalue curves above the x-axis is to add a multiple of the identity matrix to B. This results in an upwards-shift of all eigenvalue curves. In this example the minimum eigenvalue occurs for $\omega = 0$ and is -0.675 (correct to the digits shown) which amounts to a relative perturbation in $\|[A, B]\|_F$ of 0.161. So, the perturbation found by Algorithm 4 is better in the sense that (i) it is smaller and (ii) it perturbs the upper eigenvalue curves less.

3.3.2 Parameter Estimation

The application for enforcing solvability of (3.26) treated in [6] is the parameter estimation for econometric time series models. In particular, the *vector autoregressive model with moving average*; in short *VARMA(1,1)* model [23] is considered. Given the parameters $\Phi, \Theta \in \mathbb{R}^{d,d}$, $c \in \mathbb{R}^d$ and randomly drawn noise vectors $\hat{u}_t \in \mathbb{R}^d$ (independent and identically distributed, for instance, Gaussian), this stochastic process produces a vector sequence $(\hat{x}_t)_{t=1,2,\ldots}$ in $\mathbb{R}^d$ by

$$\hat{x}_t = \Phi\hat{x}_{t-1} + c + \hat{u}_t - \Theta\hat{u}_{t-1}, \quad t = 2, 3, \ldots.$$

The task is to recover (i.e., estimate) the parameters Φ, Θ, and c from an observed finite subsequence $(\hat{x}_t)_{t=1,\ldots,N}$. In [6] such an estimator is described, that only uses (approximations of) the mean $\mu := \lim_{t\to\infty} \mathbb{E}(x_t)$ and the autocorrelation matrices $M_k := lim_{t\to\infty}\mathbb{E}((x_{t+k} - \mu)(x_t - \mu)^\top)$. Here, $\mathbb{E}$ denotes the expected value and x_t is the random variable that $\hat{x}_t$ is an instance of. Using μ and M_k the parameters can be obtained as [30]

$$\Phi = M_{k+1}M_k^{-1} \text{ for any } k \geq 1; \quad c = (I - \Phi)\mu; \quad \Theta = -AX^{-1}, \tag{3.32}$$

where X solves the solvent equation (3.26) for $A := M_1^\top - M_0\Phi^\top$ and $B := M_0 - \Phi M_1^\top - M_1\Phi^\top + \Phi M_0\Phi^\top$. Note that X is guaranteed to exist since it can be interpreted as the covariance matrix of u (the random variable that $\hat{u}_t$ are instances of).

Since μ and M_k are unknown, the estimator approximates them by the finite-sample moments

$$\hat{\mu} := \frac{1}{N}\sum_{t=1}^{N}\hat{x}_t, \quad \hat{M}_k := \frac{1}{N-k}\sum_{t=1}^{N-k}(\hat{x}_{t+k} - \hat{\mu})(\hat{x}_t - \hat{\mu})^\top, \tag{3.33}$$

(which converge to the true values for $N \to \infty$) and then replace μ and M_k by $\hat{\mu}$ and $\hat{M}_k$, respectively, in (3.32) giving rise to approximations $\hat{A}$, $\hat{B}$, $\hat{X}$, $\hat{\Theta}$, $\hat{\Phi}$, $\hat{c}$. Unfortunately, the finite-sample moments (3.33) converge rather slowly to the true asymptotic ones, i.e., substantial deviations are not unlikely. Therefore, one may encounter the situation described above, where the solvent equation (3.26) for $\hat{A}$, $\hat{B}$ admits no solution that satisfies all the required assumptions. The regularization technique presented above can then be used to obtain solutions in cases in which the estimator would fail otherwise; the robustness of the resulting method is greatly increased.

3.4 Conclusion

We saw that the palindromic eigenvalue problem has numerous applications ranging from control theory via vibration analysis to parameter estimation. Often the stable deflating subspace is wanted, but other times just the question whether unimodular eigenvalues exist is of interest.

The rail track vibration problem was discussed and an efficient algorithm for its solution presented. Another presented algorithm aims to remove the unimodular eigenvalues via small perturbations, a task that is useful for passivation and parameter estimation.

References

1. Alam, R., Bora, S., Karow, M., Mehrmann, V., Moro, J.: Perturbation theory for Hamiltonian matrices and the distance to bounded-realness. SIAM J. Matrix Anal. Appl. **32**(2), 484–514 (2011). doi:10.1137/10079464X. http://dx.doi.org/10.1137/10079464X
2. Antoulas, T.: Approximation of Large-Scale Dynamical Systems. SIAM, Philadelphia (2005)
3. Benner, P., Voigt, M.: Spectral characterization and enforcement of negative imaginariness for descriptor systems. Linear Algebra Appl. **439**(4), 1104–1129 (2013). doi:10.1016/j.laa.2012.12.044
4. Betcke, T., Trefethen, L.N.: Computations of eigenvalue avoidance in planar domains. PAMM Proc. Appl. Math. Mech. **4**(1), 634–635 (2004)
5. Bora, S., Mehrmann, V.: Perturbation theory for structured matrix pencils arising in control theory. SIAM J. Matrix Anal. Appl. **28**, 148–169 (2006)
6. Bruell, T., Poloni, F., Sbrana, G., Christian, S.: Enforcing solvability of a nonlinear matrix equation and estimation of multivariate arma time series. Preprint 1027, The MATHEON research center, Berlin (2013). http://opus4.kobv.de/opus4-matheon/frontdoor/index/index/docId/1240. Submitted
7. Brüll, T., Schröder, C.: Dissipativity enforcement via perturbation of para-hermitian pencils. IEEE Trans. Circuits Syst. I **60**(1), 164–177 (2013). doi:10.1109/TCSI.2012.2215731
8. Chu, E.K.W., Hwang, T.M., Lin, W.W., Wu, C.T.: Vibration of fast trains, palindromic eigenvalue problems and structure-preserving doubling algorithms. J. Comput. Appl. Math. **219**(1), 237–252 (2008). doi:10.1016/j.cam.2007.07.016. http://dx.doi.org/10.1016/j.cam.2007.07.016

9. Engwerda, J.C., Ran, A.C.M., Rijkeboer, A.L.: Necessary and sufficient conditions for the existence of a positive definite solution of the matrix equation $X + A^* X^{-1} A = Q$. Linear Algebra Appl. **186**, 255–275 (1993). doi:10.1016/0024-3795(93)90295-Y. http://dx.doi.org/10.1016/0024-3795(93)90295-Y
10. Gohberg, I., Goldberg, S., Kaashoek, M.: Basic Classes of Linear Operators. Birkhäuser, Basel (2003)
11. Gohberg, I., Lancaster, P., Rodman, L.: Indefinite Linear Algebra and Applications. Birkhäuser, Basel (2005)
12. Grivet-Talocia, S.: Passivity enforcement via perturbation of Hamiltonian matrices. IEEE Trans. Circuits Syst. **51**, 1755–1769 (2004)
13. Guglielmi, N., Kressner, D., Lubich, C.: Low-rank differential equations for hamiltonian matrix nearness problems. Technical report, Universität Tübingen (2013)
14. Guglielmi, N., Overton, M.L.: Fast algorithms for the approximation of the pseudospectral abscissa and pseudospectral radius of a matrix. SIAM J. Matrix Anal. Appl. **32**(4), 1166–1122 (2011). doi:10.1137/100817048. http://dx.doi.org/10.1137/100817048. ISSN: 0895-4798
15. Guo, C.H., Lin, W.W.: Solving a structured quadratic eigenvalue problem by a structure-preserving doubling algorithm. SIAM J. Matrix Anal. Appl. **31**(5), 2784–2801 (2010). doi:10.1137/090763196. http://dx.doi.org/10.1137/090763196
16. Higham, N.J., Tisseur, F., Van Dooren, P.M.: Detecting a definite Hermitian pair and a hyperbolic or elliptic quadratic eigenvalue problem, and associated nearness problems. Linear Algebra Appl. **351/352**, 455–474 (2002). doi:10.1016/S0024-3795(02)00281-1. http://dx.doi.org/10.1016/S0024-3795(02)00281-1. Fourth special issue on linear systems and control
17. Hilliges, A., Mehl, C., Mehrmann, V.: On the solution of palindromic eigenvalue problems. In: Proceedings of the 4th European Congress on Computational Methods in Applied Sciences and Engineering (ECCOMAS), Jyväskylä (2004)
18. Horn, R.A., Johnson, C.R.: Matrix Analysis, 1st edn. Cambridge University Press, Cambridge (1985)
19. Huang, T., Lin, W., Qian, J.: Structure-preserving algorithms for palindromic quadratic eigenvalue problems arising from vibration of fast trains. SIAM J. Matrix Anal. Appl. **30**(4), 1566–1592 (2009). doi:10.1137/080713550. http://dx.doi.org/10.1137/080713550
20. Ipsen, I.C.: Accurate eigenvalues for fast trains. SIAM News **37**(9), 1–2 (2004)
21. Kressner, D., Schröder, C., Watkins, D.S.: Implicit QR algorithms for palindromic and even eigenvalue problems. Numer. Algorithms **51**(2), 209–238 (2009)
22. Lax, P.D.: Linear Algebra and Its Applications. Pure and Applied Mathematics (Hoboken), 2nd edn. Wiley-Interscience [Wiley], Hoboken (2007)
23. Lütkepohl, H.: New Introduction to Multiple Time Series Analysis. Springer, Berlin (2005)
24. Mackey, D.S., Mackey, N., Mehl, C., Mehrmann, V.: Structured polynomial eigenvalue problems: good vibrations from good linearizations. SIAM J. Matrix Anal. Appl. **28**(4), 1029–1051 (electronic) (2006). doi:10.1137/050628362. http://dx.doi.org/10.1137/050628362
25. Mackey, D.S., Mackey, N., Mehl, C., Mehrmann, V.: Vector spaces of linearizations for matrix polynomials. SIAM J. Matrix Anal. Appl. **28**(4), 971–1004 (electronic) (2006). doi:10.1137/050628350. http://dx.doi.org/10.1137/050628350
26. Mackey, D.S., Mackey, N., Mehl, C., Mehrmann, V.: Numerical methods for palindromic eigenvalue problems: computing the anti-triangular Schur form. Numer. Linear Algebra Appl. **16**(1), 63–86 (2009). doi:10.1002/nla.612. http://dx.doi.org/10.1002/nla.612
27. Markine, V., Man, A.D., Jovanovic, S., Esveld, C.: Optimal design of embedded rail structure for high-speed railway lines. In: Railway Engineering 2000, 3rd International Conference, London (2000)
28. Mehrmann, V., Schröder, C.: Nonlinear eigenvalue and frequency response problems in industrial practice. J. Math. Ind. **1**(1), 7 (2011). doi:10.1186/2190-5983-1-7. http://www.mathematicsinindustry.com/content/1/1/7
29. Mehrmann, V., Schröder, C., Simoncini, V.: An implicitly-restarted Krylov subspace method for real symmetric/skew-symmetric eigenproblems. Linear Algebra Appl. **436**(10), 4070–4087 (2012). doi:10.1016/j.laa.2009.11.009. http://dx.doi.org/10.1016/j.laa.2009.11.009

30. Sbrana, G., Poloni, F.: A closed-form estimator for the multivariate GARCH(1,1) model. J. Multivar. Anal. **120**, 152–162 (2013). doi:10.1016/j.jmva.2013.05.005
31. Schröder, C.: Palindromic and even eigenvalue problems – analysis and numerical methods. Ph.D. thesis, Technical University Berlin (2008)
32. Schröder, C., Stykel, T.: Passivation of LTI systems. Preprint 368, DFG Research Center MATHEON, TU Berlin (2007). www.matheon.de/research/list_preprints.asp
33. Voigt, M., Benner, P.: Passivity enforcement of descriptor systems via structured perturbation of Hamiltonian matrix pencils. Talk at Meeting of the GAMM Activity Group? Dynamics and Control Theory? Linz (2011)
34. Wang, Y., Zhang, Z., Koh, C.K., Pang, G., Wong, N.: PEDS: passivity enforcement for descriptor systems via hamiltonian-symplectic matrix pencil perturbation. In: 2010 IEEE/ACM International Conference on Computer-Aided Design (ICCAD), San Jose, pp. 800–807 (2010). doi:10.1109/ICCAD.2010.5653885
35. Wu, Y., Yang, Y., Yau, E.: Three-dimensional analysis of train-rail-bridge interaction problems. Veh. Syst. Dyn. **36**, 1–35 (2001)
36. Zaglmayr, S.: Eigenvalue problems in SAW-filter simulations. Master's thesis, Institute of Computer Mathematics, Johannes Kepler University, Linz (2002)
37. Zaglmayr, S., Schöberl, J., Langer, U.: Eigenvalue problems in surface acoustic wave filter simulations. In: Progress in Industrial Mathematics at ECMI 2004. Mathematics in Industry, vol. 8, pp. 74–98. Springer, Berlin (2006)

Chapter 4
Theory and Numerical Solution of Differential and Algebraic Riccati Equations

Peter Benner

Abstract Since Kalman's seminal work on linear-quadratic control and estimation problems in the early 1960s, Riccati equations have been playing a central role in many computational methods for solving problems in systems and control theory, like controller design, Kalman filtering, model reduction, and many more. We will review some basic theoretical facts as well as computational methods to solve them, with a special emphasis on the many contributions Volker Mehrmann had regarding these subjects.

4.1 Introduction

The algebraic and differential Riccati equations (AREs/DREs) play a fundamental role in the solution of problems in systems and control theory. They have found widespread applications in applied mathematics and engineering, many of which can be found in the monographs [1, 36, 68, 86]. In this chapter, we focus on Riccati equations associated to control problems, as these have always inspired Volker Mehrmann's work, and he has mainly focused on the resulting *symmetric* Riccati equations – symmetric in the sense that the associated Riccati operators map symmetric (Hermitian) matrices onto symmetric (Hermitian) matrices. Hence, also the solutions to the Riccati equations we will consider are expected to be symmetric (Hermitian). A class of nonsymmetric AREs that arises, e.g., in queuing theory, certain fluid flow problems, and transport theory (see, e.g., [36]) is of importance as well, but for conciseness, we will omit these AREs even though Volker has also contributed to this area [81].

In most of the literature on AREs and DREs, the motivation is taken from the classical linear-quadratic regulator (LQR) problem. This was the topic of Volker's habilitation thesis [74], where, building upon earlier work by Bender and Laub [6, 7], he extended the LQR theory to so-called descriptor systems, i.e., systems with

P. Benner (✉)
Max Planck Institute for Dynamics of Complex Technical Systems, Sandtorstraße 1, D-39106 Magdeburg, Germany
e-mail: benner@mpi-magdeburg.mpg.de

P. Benner et al. (eds.), *Numerical Algebra, Matrix Theory, Differential-Algebraic Equations and Control Theory*, DOI 10.1007/978-3-319-15260-8_4

dynamics described by differential-algebraic equations (DAEs). Much of Volker's early work on AREs culminated in this thesis which later became the appraised book [76]. We therefore also start by formulating the LQR problems in continuous- and discrete time and how they relate to AREs. These optimal control problems can be formulated in various levels of generality. The setting we consider is:

minimize

$$\mathcal{J}(x^0, u) = \frac{1}{2}\int_0^\infty \left(y(t)^T Q y(t) + 2y(t)^T S u(t) + u(t)^T R u(t)\right) dt \tag{4.1}$$

subject to the linear time-invariant (LTI) system

$$E\dot{x}(t) = Ax(t) + Bu(t), \qquad x(0) = x^0, \tag{4.2a}$$

$$y(t) = Cx(t), \tag{4.2b}$$

in the continuous-time case and

minimize

$$\mathcal{J}(x^0, u) = \frac{1}{2}\sum_{k=0}^{\infty} \left(y_k^T Q y_k + 2y_k^T S u_k + u_k^T R u_k\right) \tag{4.3}$$

subject to the discrete-time LTI system

$$Ex_{k+1} = Ax_k + Bu_k, \qquad x_0 = x^0, \tag{4.4a}$$

$$y_k = Cx_k, \tag{4.4b}$$

in the discrete-time case. In both settings, $A, E \in \mathbb{R}^{n\times n}$, $B \in \mathbb{R}^{n\times m}$, $C \in \mathbb{R}^{p\times n}$, $Q \in \mathbb{R}^{p\times p}$, $R \in \mathbb{R}^{m\times m}$, and $S \in \mathbb{R}^{p\times m}$. Furthermore, we assume Q and R to be symmetric. In both cases, the initial state $x^0 \in \mathbb{R}^n$ can be chosen freely if E is nonsingular and is constrained to a manifold in the descriptor case, see the Chap. 16 for more details on this. In the continuous-time case, u is considered as an element of an appropriate function space like k times continuously differentiable functions $C^k(0,\infty)$ or square-integrable functions $L_2(0,\infty)$. No further constraints are imposed in this setting. In discrete-time, u represents a sequence $(u_k)_{k=0}^{\infty}$ that should be (square-)summable in an appropriate sense. A formulation in complex arithmetic is possible, and most of the results considered here remain valid (cf. [68] for a detailed treatment of real and complex AREs), but as most practical applications are formulated using real data, we stick to this case here.

Under fairly general conditions, the LQR problems have solutions of the form $u(t) = \mathcal{F}_c(x(t))$ and $u_k = \mathcal{F}_d(x_k)$, respectively. As they appear in feedback form, that is, the current control input depends on the current state information, this observation lays the basis for *modern (feedback) control* [54]. Possible sufficient conditions to obtain such feedback solutions are: E is nonsingular, Q and R are positive definite, S is "small" enough such that $\begin{bmatrix} Q & S \\ S^T & R \end{bmatrix}$ is positive semidefinite, and the matrix triples (E, A, B) and (E, A, C) are stabilizable and detectable, respectively. Here, *stabilizable* means that $\operatorname{rank}[A - \lambda E, B] = n$ for all $\lambda \in \{z \in \mathbb{C} \mid \Re(z) \geq 0\}$ in the continuous-time case and for all $\lambda \in \{z \in \mathbb{C} \mid |z| \geq 1\}$ in the discrete-time case, while detectability of (E, A, C) is equivalent to stabilizability of (E^T, A^T, C^T). Under these conditions, the LQR problems have unique solutions. These are given by the feedback laws

$$u(t) = -R^{-1}(B^T X_c E + S^T C)x(t) =: -F_c x(t), \qquad t \geq 0, \tag{4.5}$$

in the continuous-time case and

$$u_k = -(R+B^T X_d B)^{-1}(B^T X_d A+S^T C)x_k =: -F_d x_k, \quad k = 0, 1, \ldots, \tag{4.6}$$

in the discrete-time case. Now, the relation of the LQR problem to AREs becomes evident as X_c is the unique *stabilizing* solution to the *(generalized) continuous-time algebraic Riccati equation* (CARE)

$$\begin{aligned} 0 = \mathscr{R}_c(X) := C^T QC + A^T XE + E^T XA- \\ (E^T XB + C^T S)R^{-1}(B^T XE + S^T C), \end{aligned} \tag{4.7}$$

while X_d is the unique *stabilizing* solution of the *(generalized) discrete-time algebraic Riccati equation* (DARE)

$$\begin{aligned} 0 = \mathscr{R}_d(X) := C^T QC + A^T XA - E^T XE- \\ (A^T XB + C^T S)(R + B^T XB)^{-1}(B^T XA + S^T C). \end{aligned} \tag{4.8}$$

The solutions X_c and X_d are *stabilizing* in the sense that the feedback solutions generated by inserting (4.5) and (4.6) into (4.2) and (4.4), respectively, are *asymptotically stable*, i.e., $x(t), x_k \to 0$ for $t, k \to \infty$ and all initial values $x^0 \in \mathbb{R}^n$. Under the given assumptions, CAREs and DAREs have exactly one stabilizing solution, despite the fact that there may exist many other solutions [68]. These stabilizing solutions are symmetric and positive semidefinite, the latter property again uniquely identifies the solutions X_c, X_d in the respective solution sets.

In his work, Volker has often considered the case that E is singular. This is in particular the case in his habilitation thesis and the resulting book [76]. In general, in this case the relation between the AREs (4.7) and (4.8) and the feedback solutions to the LQR problems is lost. Several modifications of the AREs to re-establish this connection have been suggested in the literature. They usually require special conditions, and the resulting generalized AREs are often not easily solvable numerically. Only recently, efficient methods for a class of generalized

CAREs with singular E have been suggested in [35]. Similarly to [6, 7], Volker shows in [76, § 5] how the LQR problem for descriptor systems (i.e., E singular) can be reduced to a problem with nonsingular E for which the Riccati solutions exist. Nevertheless, this procedure requires strong conditions, in particular quite complicated consistency conditions for the initial values, that are often not satisfied in practice. This observation, among others, led Volker to work on alternative approaches to solve the LQR problem for descriptor systems, avoiding AREs completely. This is the topic of [63, 64, 66, 67], where also extensions to systems with time-varying coefficients are considered and the theory of *strangeness-free* DAEs [65] is applied to the LQR problem. We will not further discuss these fundamental contributions to optimal control for DAEs, as they do not directly relate to Riccati equations anymore. Recently, a solution concept for LQR problems based on purely algebraic considerations was derived in the Ph.D. thesis of Matthias Voigt [88]. This theory relates the feedback solution of the LQR problem for descriptor systems to dissipation inequalities, even matrix pencils (see the Chaps. 6 and 12), and the stabilizing solutions of Lur'e equations. This work may complete three decades of the quest for a general algebraic solution theory for the LQR problem that does not require a special index concept or restrictions on the index of the underlying DAE. Notably, Matthias is Volker's academic grandchild!

In [76], Volker also considers LQR problems with finite time-horizon, i.e., the cost functionals are replaced by finite integrals $\int_0^T$, where $0 < T < \infty$, or finite sums $\sum_{k=0}^{K-1}$, $K \in \mathbb{N}$. In this situation, usually also the final states $x(T)$, x_K are penalized in the cost functional. Often, the LQR problem is used for stabilization problems so that it is desired that the final state is close to zero, which suggests a positive definite quadratic weighting function. This is particularly the case when the modeling is done in such a way that x represents the deviation of the state of a dynamical system from a desired state. In the continuous-time case, the LQR problem can then be formulated as follows:

minimize

$$\mathscr{J}(x^0, u) = \frac{1}{2}\left(x(T)^T M x(T) + \int_0^T \left(y(t)^T Q y(t) + 2y(t)^T S u(t) + u(t)^T R u(t) \right) dt \right)$$

subject to (4.2)

with $M = M^T \in \mathbb{R}^{n \times n}$ positive semidefinite. Note that for consistency, one could write the penalty term at T in terms of $y(T)$ instead of $x(T)$. As we can replace $y(T)$ by $Cx(T)$ using (4.2), this is merely a notational issue.

Under similar conditions as in the infinite-time horizon case, the unique solution of the finite-time LQR problem is then given by the feedback control

$$u(t) = -R^{-1}(B^T X(t) E + S^T C) x(t) =: -F_c(t) x(t), \qquad t \geq 0, \tag{4.9}$$

where $X(t)$ is the unique solution of the DRE

$$-E^T \dot{X}(t)E = \mathscr{R}_c(X(t)) \tag{4.10}$$

with terminal condition $E^T X(T)E = M$ and $\mathscr{R}_c$ as in (4.7). The existence and uniqueness of the DRE solution is considered in [1, 86]. The theory can also be extended to time-varying systems, i.e., systems with $A = A(t)$, etc. In general, if E is singular, again the relation of the solution to the finite-time LQR problem and the DRE is lost, but is re-established using a reduction procedure described by Volker in [76, § 3]. We will re-visit this technique, as well as a numerical procedure to solve (4.10) as suggested by Mehrmann and Kunkel in [62], in the next section.

The finite horizon discrete-time LQR problem is also treated in [76]. This leads to the solution of *difference Riccati equations* of the form

$$-E^T X_k E = \mathscr{R}_d(X_{k+1}) \tag{4.11}$$

with terminal condition $E^T X_K E = M$. As in the other cases considered so far, a reduction to the case E nonsingular is necessary to establish the link between the LQR problem and (4.11). Such a reduction procedure is suggested again in [76, § 3]. A numerical procedure to solve (4.11) is sketched in [76, § 19]. As (4.11) is solved backwards starting from X_K, advancing from X_{k+1} to X_k can be achieved by simply evaluating $\mathscr{R}_d(X_{k+1})$ using, e.g., a Cholesky decomposition of $R + B^T X_{k+1} B$, and then solving a linear system of equations using, e.g., the LU decomposition of E. Due to its conceptual simplicity, we will not discuss this approach here any further.

The remainder of this chapter is organized as follows. As already mentioned above, we will focus on Volker's contributions to the solution of DREs in the next section. The relation between CAREs and DAREs is explored in Sect. 4.3. The bulk of this chapter summarizes Volker's contributions to the numerical solution of AREs and comprises Sect. 4.4, while we briefly touch upon Volker's passion to solve control problems avoiding AREs in Sect. 4.5. Final remarks are given in Sect. 4.6.

4.2 Numerical Methods for Differential Riccati Equations

In [62], the DRE (4.10) is considered with $S = 0$, and in the simplified representation

$$-E^T \dot{X}(t)E = F + A^T X(t)E + E^T X(t)A - E^T X(t)GX(t)E \tag{4.12}$$

with terminal condition $E^T X(T)E = M$, where even time-dependence of A, E, F, G is allowed under the assumption of sufficient smoothness.

The contributions of [62], and [76, § 3,19] are twofold. First of all, conditions are derived under which the solution of the finite-time horizon LQR problem is given via the feedback control law (4.9), defined by the solution of the DRE. In case of singular E, this requires a reduction procedure resulting in a DRE with nonsingular E matrix. We will sketch this procedure briefly. Second, a numerical solution method for (4.12) is derived that resolves some singularity issues when (4.12) is solved in vectorized form as a DAE. In the notation derived later on by Kunkel and Mehrmann [65], it is assumed that the DAE $E\dot{x}(t) = Ax(t)$ is strangeness-free (see the Chap. 16 for details), (E, A) is a regular matrix pair, and $\operatorname{rank}(E)$ is constant on $[0, T]$. For the ease of presentation, in the following we will only consider the time-invariant case, but the derivations in the time-varying case in [62] are formally completely analogous.

4.2.1 Reduction to a Standard LQR Problem

Basically, the singular value decomposition of E,

$$E = U \begin{bmatrix} \Sigma & 0 \\ 0 & 0 \end{bmatrix} V^T \tag{4.13}$$

is the key step in the reduction procedure. Here, $U, V \in \mathbb{R}^{n\times n}$ are orthogonal, $\Sigma \in \mathbb{R}^{r\times r}$ is nonsingular, and $r = \operatorname{rank}(E)$. Not that we do not necessarily need a reduction of Σ to diagonal form, a URV-type decomposition with a nonsingular Σ suffices. Therefore, we do not assume symmetry of Σ in the following. If we now insert the SVD of E in (4.2), multiply the first equation with U^T from the left, and using

$$U^T AV = \begin{bmatrix} A_{11} & A_{12} \\ A_{21} & A_{22} \end{bmatrix}, \quad U^T B = \begin{bmatrix} B_1 \\ B_2 \end{bmatrix}, \quad CV = \begin{bmatrix} C_1, & C_2 \end{bmatrix},$$

as well as the change of basis $\left[\begin{smallmatrix} x_1 \\ x_2 \end{smallmatrix}\right] = V^T x$ with the partitioning implied by the SVD (4.13), the descriptor system (4.2) becomes

$$\Sigma \dot{x}_1(t) = A_{11}x_1(t) + A_{12}x_2(t) + B_1u(t), \tag{4.14a}$$

$$0 = A_{21}x_1(t) + A_{22}x_2(t) + B_2u(t), \tag{4.14b}$$

$$y(t) = C_1x_1(t) + C_2x_2(t). \tag{4.14c}$$

The strangeness-freeness assumption implies that A_{22} is nonsingular, so that we can solve the second of these equations for x_2 and insert the result in the first and third equations. This leads to the standard LTI system

$$\Sigma \dot{x}_1(t) = \underbrace{(A_{11} - A_{12}A_{22}^{-1}A_{21})}_{=:\,\hat{A}}\, x_1(t) + \underbrace{(B_1 - A_{12}A_{22}^{-1}B_2)}_{=:\,\hat{B}}\, u(t), \tag{4.15}$$

$$y(t) = \underbrace{(C_1 - C_2A_{22}^{-1}A_{21})}_{=:\,\hat{C}}\, x_1(t) + \underbrace{(-C_2A_{22}^{-1}B_2)}_{=:\,\hat{D}}\, u(t). \tag{4.16}$$

Now also performing the change of basis in the cost functional results in a standard finite-time horizon LQR problem which can be solved via the DRE

$$-\Sigma^T\dot{\hat{X}}(t)\Sigma = \hat{Q} + \Sigma^T\hat{X}\hat{A} + \hat{A}^T\hat{X}\Sigma - \left(\hat{B}^T\hat{X}\Sigma + \hat{S}^T\right)^T\hat{R}^{-1}\left(\hat{B}^T\hat{X}\Sigma + \hat{S}^T\right) \tag{4.17}$$

with terminal condition $\Sigma^T\hat{X}\Sigma = \hat{M}$, where $\hat{M}$ results from the terminal condition of the original LQR problem after the change of coordinates. Note that when $B_2 \neq 0$, $u(T)$ appears in the terminal penalty term which changes the nature of the problem. As M can usually be chosen freely, it is therefore convenient to assume in the partitioning of $V^TMV = \begin{bmatrix} M_{11} & M_{12} \\ M_{12}^T & M_{22} \end{bmatrix}$ that $M_{12} = 0$, $M_{22} = 0$. The coefficient matrices $\hat{Q}, \hat{S}, \hat{R}$ in (4.17) can be read off from the cost functional after applying the change of coordinates and inserting the modified output equation (4.16) in the first integrand:

$$\begin{aligned} y(t)^TQy(t) &= (\hat{C}x_1(t) + \hat{D}u(t))^TQ(\hat{C}x_1(t) + \hat{D}u(t)) \\ &= x_1(t)^T\underbrace{\hat{C}^TQ\hat{C}}_{=:\hat{Q}}x_1(t) + 2x_1(t)^T\underbrace{\hat{C}^TQ\hat{D}}_{=:\hat{S}}u(t) + u(t)^T\hat{D}^TQ\hat{D}u(t). \end{aligned}$$

With $\hat{R} = R + \hat{D}^TQ\hat{D}$, and the assumption that $S = 0$, the LQR theory for LTI systems yields the feedback solution via the DRE (4.17). In order for this reduction procedure to work, it is of course necessary that the modified matrices defining the LTI system and cost functional inherit the properties like positive (semi)definiteness of $\hat{Q}, \hat{R}$ as well as the stabilizabilty and detectability properties. In a numerical procedure, this needs to be checked before the solution of the original LQR problem can be derived from that of the reduced LQR problem. While often, these properties indeed carry over, a more severe restriction is caused by the consistency condition implied by (4.14b) for $t = 0$: if $\begin{bmatrix} x_1^0 \\ x_2^0 \end{bmatrix} = V^Tx^0$, then (4.14b) implies

$$x_2^0 = -A_{22}^{-1}\left(A_{21}x_1^0 + B_2u(0)\right).$$

If $B_2 \neq 0$, this is a restriction on the possible controls, while for $B_2 = 0$, it yields a consistency condition for the initial values. Whether or not this restricts the applicability of the reduction approach to the LQR problem for descriptor systems certainly depends on the application. It should also be noted that under certain conditions on the system matrices, higher-index DAE problems can be reduced to regular index-1 problems using output feedback. This topic is discussed in more detail in Chap. 15.

Remark 1 The reduction procedure using the SVD (4.13) of E can also be applied directly to the DRE (4.12) without considering the LQR background. This basically leads again to the DRE (4.17) with additional consistency and solvability conditions, see [62, Section 3].

4.2.2 Numerical Solution of DREs

In [62] and [76, § 19], it is suggested to solve the DRE (4.12) by *vectorization*. This uses the "vec" operator

$$\mathrm{vec} : \mathbb{R}^{\ell\times k} \to \mathbb{R}^{\ell k} \; : X \mapsto \mathrm{vec}(X),$$

which stacks the columns of a matrix on top of each other. Vectorization of expressions like AXB is simplified by using the Kronecker product $\otimes$, see, e.g., [61, 69]. In particular, the formulas

$$\mathrm{vec}(AXB) = \left(B^T \otimes A\right) \mathrm{vec}(X), \quad \left(A^T \otimes B^T\right) = (A \otimes B)^T \tag{4.18}$$

are useful for our purposes. Applying vectorization to (4.12) and using (4.18) yields the system of ordinary differential equation (ODEs) or DAEs

$$\begin{aligned} -(E \otimes E)^T \mathrm{vec}(\dot{X}(t)) = \mathrm{vec}(F) + \left((E \otimes A)^T + (A \otimes E)^T\right) \mathrm{vec}(X(t)) \\ -(E \otimes E)^T (X(t) \otimes X(t)) \mathrm{vec}(G). \end{aligned} \tag{4.19}$$

The terminal condition yields

$$(E \otimes E)^T \mathrm{vec}(X(T)) = \mathrm{vec}(M). \tag{4.20}$$

Together, (4.19) and (4.20) form an initial value problem (in reverse time) for a system of ODEs/DAEs if E is nonsingular/singular with quadratic nonlinearity. (In case $n = 1$, a classical Riccati differential equation is obtained, thus the name "differential Riccati equation".)

In case of E being nonsingular, this initial value problem can be solved by any integration method for ODEs. As (4.19) is then a system of n^2 ODEs, already one time step of an explicit integrator would require $\mathcal{O}(n^4)$ floating point operations (flops) in general for the matrix vector products. In an implicit integration technique, linear systems of equations need to be solved at a computational cost of $\mathcal{O}(n^6)$ flops. Therefore, this approach is limited to very small dimensions n, even if one exploits the symmetry of the DRE as suggested in [62] and only works with the $n(n+1)/2$ necessary equations, ignoring the redundant other ones (which is possible, but the data handling is cumbersome). After this initial work on numerical methods for DREs by Kunkel/Mehrmann, it was suggested in [48] to re-write ODE integrators in matrix form. This reduces the cost to $\mathcal{O}(n^3)$ flops in dense arithmetic. Due to the usual inherent stiffness of DREs, in the literature, mostly implicit integrators are discussed. In [48], in particular the backward differentiation formulas (BDF) were considered. This was followed up in [53], where an efficient implementation of the matrix-valued BDF for DREs was described. Later, this was extended to large-scale problems employing sparsity in the coefficient matrices in [29]. Also other popular implicit integration techniques were investigated, e.g., the Rosenbrock methods in matrix form in [30, 70] by members of Volker's academic descendants family.

The case that E is singular is much more involved, though. Let $\tilde{E} := (E \otimes E)^T$ and $\tilde{A} := (E \otimes A)^T + (A \otimes E)^T$. Then it is shown in [62] that even in the simplest form of a strangeness-free DAE $E\dot{x} = Ax$, the matrix pencil $\lambda\tilde{E} - \tilde{A}$ is singular. If the nilpotency index of (E, A) is 1, then so is the nilpotency index of $\lambda\tilde{E} - \tilde{A}$, and the DRE (4.19) can be solved by simply omitting the singular "zero blocks". A procedure to achieve this is suggested in [62, Section 5]. Apart from the SVD (4.13) and the associated reduction, it includes consistency checks and requires additional algebraic equations to be solved. The reduced DRE is then solved by standard DAE integrators for index-1 systems. Also, a variant for time-dependent coefficients $A(t), E(t), \ldots$ is provided in [62]. If the nilpotency index of the regular matrix pencil $\lambda E - A$ is larger than 1, then in addition to singularity of $\lambda\tilde{E} - \tilde{A}$, this matrix pencil may have even a larger nilpotency index, and also different kinds of singular blocks. Therefore, a numerical procedure for this situation cannot be derived easily, and this is an open problem up to now to the best of this author's knowledge. An interesting question for further research would be whether the singularity problem can be avoided in matrix-valued DRE solvers. Recently, a first step in this direction was proposed by another of Volker's many Ph.D. students, Jan Heiland. In his thesis [58], he suggests a matrix-valued implicit Euler scheme for a special index-2 DRE, which he calls a differential-algebraic Riccati equation and which arises in the finite-time horizon LQR problem for flow control problems. Further research in the area of matrix-valued solvers for DREs or differential-algebraic Riccati equations is certainly needed.

4.3 A Unified Treatment of CAREs and DAREs?

A unified treatment of discrete- and continuous-time algebraic Riccati equations is discussed in [77]. We will recall the main result from this paper related to AREs: a then new result on the solution of DAREs using the idea of a unified treatment of continuous- and discrete-time control problems. As this result was derived in complex arithmetic, we will (in contrast to the other sections) in this chapter also formulate the results in complex arithmetic. For this, we denote by M^H the complex conjugate transpose of a matrix or vector M, while $\overline{\mu}$ denotes as usual the complex conjugate of the scalar $\mu \in \mathbb{C}$.

In addition to the Hamiltonian and symplectic matrices introduced already in Chap. 1, we will need the following structures in this section.

Definition 1 Let $J = \left[\begin{smallmatrix} 0_n & I_n \\ -I_n & 0_n \end{smallmatrix}\right] \in \mathbb{R}^{2n\times 2n}$ with $I_n, 0_n$ the identity and zero matrices of order n.

(a) A matrix pencil $\lambda K - H$ is *Hamiltonian* iff $KJH^H + HJK^H = 0$.
(b) A matrix pencil $\lambda T - S$ is *symplectic* iff $TJT^H = SJS^H$.

In this and the next section, a special class of n-dimensional subspaces in $\mathbb{C}^{2n}$ plays a prominent role.

Definition 2 Let $\mathscr{L} \subset \mathbb{C}^{2n}$ and $\dim(\mathscr{L}) = n$. Then $\mathscr{L}$ is *Lagrangian* if $x^H Jy = 0$ for all $x, y \in \mathscr{L}$.

A typical example of a Lagrangian subspace is the graph subspace $\left[\begin{smallmatrix} I_n \\ M \end{smallmatrix}\right]$ for a Hermitian matrix M, see the Chap. 5 for more on the use of graph subspaces.

If we now consider the continuous- and discrete-time LQR problems from Sect. 4.1 with $S = 0$ for simplicity, and the associated AREs, then we have the following well-known observation: let $E = I_n$, and define $F := C^H QC$, $G = BR^{-1}B^H$, and

$$H = \begin{bmatrix} A & G \\ F & -A^H \end{bmatrix}, \qquad \lambda T - S := \lambda \begin{bmatrix} I_n & -G \\ 0_n & A^H \end{bmatrix} - \begin{bmatrix} A & 0_n \\ F & I_n \end{bmatrix}. \tag{4.21}$$

Then H is Hamiltonian, $\lambda T - S$ is symplectic. We then have the following result, which is a collection of results that can be found, e.g., in [68]:

Proposition 1 *Let $E = I_n$, $S = 0$, and H, S, T as in* (4.21).

(a) Assume that $X_c = X_c^H$ is a solution to the CARE (4.7)*, then the columns of $\left[\begin{smallmatrix} I_n \\ -X_c \end{smallmatrix}\right]$ span a Lagrangian H-invariant subspace. On the other hand, if the columns of $\left[\begin{smallmatrix} U \\ V \end{smallmatrix}\right]$ span a Lagrangian H-invariant subspace with $U \in \mathbb{C}^{n\times n}$ invertible, then $X_c = -VU^{-1}$ is a Hermitian solution to the CARE* (4.7).
(b) Let A be nonsingular. Assume that $X_d = X_d^H$ is a solution to the DARE (4.8)*, then the columns of $\left[\begin{smallmatrix} I_n \\ -X_d \end{smallmatrix}\right]$ span a Lagrangian deflating subspace of $\lambda T - S$. On the other hand, if the columns of $\left[\begin{smallmatrix} U \\ V \end{smallmatrix}\right]$ span a Lagrangian deflating subspace of $\lambda T - S$ with $U \in \mathbb{C}^{n\times n}$ invertible, then $X_d = -VU^{-1}$ is a Hermitian solution to the DARE* (4.8).

Note that the CARE (DARE) solutions requested in the LQR problems are associated to the stable invariant (deflating) subspaces of H ($\lambda T - S$), that is, those associated to the n eigenvalues in the open left half of the complex plane (inside the open unit disk).[1]

This result can be used in order to link the CARE and DARE in the following way, employing the generalized Cayley transformation: given $\mu \in \mathbb{C}$ with $|\mu| = 1$,

[1]This spectral dichotomy of the Hamiltonian matrix and symplectic pencil exists under the assumptions used in Sect. 4.1, i.e., there are no eigenvalues on the boundaries of these regions, that is the imaginary axis in the continuous-time and the unit circle in the discrete-time case.

$\mu^2 \neq -1$. Then the *generalized Cayley transformation* of a matrix pencil $\lambda L - M$ is given by the matrix pair

$$\mathscr{C}_\mu(L, M) := ((L - \mu M), (\mu L + M)). \tag{4.22}$$

With this, Volker proved the following result:

Lemma 1 ([77, Lemma 2])

(a) If $\lambda T - S$ is a symplectic pencil, then $\mathscr{C}_\mu(T, S) = (K, H)$ defines a Hamiltonian pencil $\lambda K - H$.
(b) If $\lambda K - H$ is a Hamiltonian pencil, then $\mathscr{C}_{-\mu}(K, H) = (T, S)$ defines a symplectic pencil $\lambda T - S$.

As the Hamiltonian matrix H from (4.21) also defines a Hamiltonian pencil $\lambda I - H$, this lemma relates the Hamiltonian matrix and associated CARE to a symplectic pencil via Lemma 1(b), to which, using some cumbersome algebraic manipulations and some further assumptions on the Cayley shift μ, a DARE can be associated. Vice versa, the symplectic pencil associated to the DARE (4.8) can be related to a Hamiltonian pencil via Lemma 1(a). As it is easy to see that the Cayley transformation leaves deflating subspaces invariant, we thus can use numerical methods for computing deflating subspaces of Hamiltonian pencils to solve DAREs via Cayley transformation and Proposition 1. An algorithm for the Hamiltonian pencil eigenproblem with the desirable properties of being numerically backward stable, requiring only $\mathscr{O}(n^3)$ flops, and being structure-preserving in the sense that the symmetry of the spectrum is preserved exactly, was then derived later in [28]. This algorithm extends the method for Hamiltonian matrices based on the symplectic URV decomposition presented in the same paper and discussed in detail in Sect. 1.4. Unfortunately, this algorithm only computes eigenvalues, but no deflating subspaces. Already the extension of the method for Hamiltonian matrices to compute also invariant subspaces turned out to be rather complicated [27], so an extension for the Hamiltonian pencil algorithm was never derived. But it turned out a bit later that the better structure to consider was that of skew-Hamiltonian/Hamiltonian pencils (see the Chap. 2), or even more general, that of even/odd pencils, or palindromic pencils (see the Chap. 3). Numerically stable and efficient algorithms respecting these structures have been derived by Volker and co-workers in the last decades, see various chapters in this book and the recent overview [24].

The main contribution of [77] was to extend the use of the generalized Cayley transformation to more general situations than those discussed in Proposition 1 and Lemma 1. In particular, $E \neq I_n$ is allowed, and the quite restrictive assumption of A being nonsingular in the discrete-time case is dropped. In the situations where $E \neq I_n$, the matrix pencils associated to the CARE and DARE are no longer Hamiltonian and symplectic, respectively. Nevertheless, the same spectral symmetries are observed (which is evident when one considers, as done later, the associated even and palindromic pencil structures). A number of interesting relations of the Cayley transformed pencils are derived in [77], and it is shown that controllability properties of the linear-time invariant systems associated to the Hamiltonian and symplectic matric pencils before and after Cayley transformation remain invariant. This lead Volker to suggest an *implication scheme* which states that if one can prove "A ⇒ B" for a continuous-time system, one can prove the analogous result for the discrete-time system obtained via Cayley transformation (and vice versa). This eventually lead to the proof of the following result for the existence of DARE solutions which generalizes Proposition 1(b) and is proved using this implication scheme applied to a variant of Proposition 1(a).

Theorem 1 ([77, Theorem 14(b)]) *Consider the symplectic matrix pencil*

$$\lambda T - S := \lambda \begin{bmatrix} A & 0_n \\ F & I_n \end{bmatrix} - \begin{bmatrix} I_n & -BR^{-1}B^H \\ 0_n & A^H \end{bmatrix}.$$

Let the columns of $\left[\begin{smallmatrix} U \\ V \end{smallmatrix}\right]$ *span an n-dimensional deflating subspace of* $\lambda T - S$ *with* $U, V \in \mathbb{C}^{n\times n}$, $V^H U = U^H V$, *and not containing eigenvectors corresponding to infinite eigenvalues. Suppose there exists* $\mu \in \mathbb{C}$, $|\mu| = 1$, *such that* $I_n - \mu A$ *and* $\mu I + A - BR^{-1}B^H(\mu I_n - A^H)^{-1}F$ *are invertible, as well as*

$$\Psi(\mu) := R + B^H(A - \mu I_n)^{-H}F(A - \mu I_n)^{-1}B$$

is definite. Assume further that (A, B) *is controllable (i.e.,* $\operatorname{rank}([A - \lambda I_n,\ B]) = n$ *for all* $\lambda \in \mathbb{C}$*).*

Then U *is invertible and* $X_d := -VU^{-1}$ *solves the DARE*

$$0 = F + A^H XA - X + A^H XB(R + B^H XB)^{-1}B^H XA.$$

This result does not require A to be invertible. The nonsingularity assumptions in the theorem are needed to apply the generalized Cayley transformation with shift μ to $\lambda T - S$ and to convert the resulting Hamiltonian pencil to a Hamiltonian matrix, for which a variant of Proposition 1(a) can then be used to prove the assertions in the continuous-time case. The proof then follows from the implication scheme.

Highlighting this result gives a glimpse on the often observed interdisciplinary work of Volker Mehrmann, here linking systems and control theory with matrix analysis. It also demonstrates his keen interest in deriving fundamental and general principals that allow to solve classes of problems rather than just specific instances of a given problem.

4.4 Numerical Methods for AREs

We note that this section is based in large parts on [10] which concentrated on the CARE and continued the effort of providing surveys of ARE solvers given as in [4] and by Volker with co-workers [40]. We use an abridged and updated version of the survey given there, with pointers to corresponding methods for DAREs which appear throughout Volker's work. In this section, we restrict ourselves again to AREs in real arithmetic. Also, we will now always assume that E is nonsingular such that there is a clear relation of the LQR problem and the related ARE as outlined in the Introduction. For the ease of presentation, we will use the following simplified ARE versions. We consider CAREs of the form

$$0 = \mathscr{R}_c(X) \;:=\; F + A^T X + XA - XGX, \tag{4.23}$$

and DAREs

$$0 = \mathscr{R}_d(X) = F + A^T XA - X - (A^T XB)(R + B^T XB)^{-1}(A^T XB)^T, \tag{4.24}$$

where in the LQR setting, $F := C^T QC$ and $G := BR^{-1}B^T$. The solution of (4.23) yielding the optimal feedback control then is the unique *stabilizing* solution X_c, i.e., $A - GX_c$ is stable in the sense that all its eigenvalues are in the open left half plane $\mathbb{C}^-$. Similarly, in the discrete-time case, the stabilizing solution X_d yields a stable *closed-loop matrix* $A - B(R + B^T X_d B)^{-1} B^T X_d A$ in the sense that all its eigenvalues are inside the open unit disk.

It should be noted that in the non-descriptor, non-singular case, i.e., when E and R are invertible, it is always possible to rewrite CAREs and DAREs in this simplified way. In practice, though, this should only be done in case all transformations are well-conditioned. See, e.g., [4, 9, 31, 76] for algorithms working in the more general formulations and avoiding inversions and matrix products in forming the coefficients as far as possible and necessary.

4.4.1 Methods Based on the Hamiltonian and Symplectic Eigenproblems

As discussed in the previous section, solving AREs can be achieved using methods to compute certain invariant or deflating subspaces of Hamiltonian matrices or symplectic pencils. If such a subspace $\left[\begin{smallmatrix} U \\ V \end{smallmatrix}\right]$ is Lagrangian with $U \in \mathbb{R}^{n \times n}$ invertible, the formula

$$X = -VU^{-1} \tag{4.25}$$

yields a symmetric solution to the CARE or DARE, respectively. Of course, for numerical stability the condition number of U should be small, which can be achieved by certain scaling procedures (see [9] for a comparison of several scaling strategies) or, more recently, using the principle of permuted graph matrices introduced by Mehrmann and Poloni, see the Chap. 5 for details of this approach.

The required solutions for solving the LQR problems are obtained if $\begin{bmatrix} U \\ V \end{bmatrix}$ corresponds to the stable eigenvalues of the corresponding Hamiltonian matrix or symplectic pencil as explained in the previous section. Due to spectral symmetries of Hamiltonian matrices H explained in Chap. 1, i.e., with λ also $-\overline{\lambda}$ is an eigenvalue of H, and the analogous property of symplectic pencils (with λ also $1/\overline{\lambda}$ is an eigenvalue), a sufficient condition for the existence of n-dimensional stable invariant subspaces is spectral *dichotomy*, i.e., no eigenvalues lie on the boundary of the respective stability region. This dichotomy property is usually satisfied under assumptions that lead to the unique solvability of LQR problems via AREs and is thus assumed throughout this section.

First, we will consider methods for CAREs based on solving the Hamiltonian eigenproblem for H as in (4.21). Using Proposition 1(a), in order to solve the CARE it is sufficient to find a nonsingular matrix $T \in \mathbb{R}^{2n \times 2n}$ such that

$$T^{-1}HT = \begin{bmatrix} H_{11} & H_{12} \\ 0 & H_{22} \end{bmatrix}, \tag{4.26}$$

where the eigenvalues of $H_{11} \in \mathbb{R}^{n \times n}$ are all in $\mathbb{C}^-$. Partitioning T analogously, the columns of $\begin{bmatrix} T_{11} \\ T_{12} \end{bmatrix}$ span the stable Lagrangian H-invariant subspace, and the desired stabilizing solution X_c of the CARE (4.23) is obtained via (4.25) setting $U = T_{11}$ and $V = T_{12}$.

Most algorithms for solving matrix eigenproblems, i.e., for computing eigenvalues and -vectors or invariant subspaces of some matrix $M \in \mathbb{R}^{\ell \times \ell}$, are based on the following approach:

***Generic procedure** to compute invariant subspaces:*

1. *Compute an initial transformation matrix $S_0 \in \mathbb{R}^{\ell \times \ell}$ in order to reduce M to some condensed form, i.e., compute*

$$M_0 := S_0^{-1} M S_0. \tag{4.27}$$

2. *Then construct a sequence of similarity transformations such that in each step*

$$M_{j+1} := S_{j+1}^{-1} M_j S_{j+1}, \qquad j = 0, 1, 2, \ldots, \tag{4.28}$$

the reduced form is preserved and moreover, if we define $T_j := \prod_{k=0}^{j} S_k$, then $\lim_{j\to\infty} T_j = T$ and $\lim_{j\to\infty} M_j = M_*$ exist and eigenvalues and eigenvectors and/or M-invariant subspaces can be read off from M_* and T.

The purpose of the initial reduction to a condensed form and the preservation of this form throughout the iteration is twofold: first, such a reduction is usually necessary in order to satisfy the complexity requirements – an iteration step (4.28) on a reduced form can usually be implemented much cheaper than for a full matrix; second, using such a reduced form it is usually easier to track the progress of the iteration and detect if the problem can be decoupled (*deflation*) into smaller subproblems that can then be treated separately. For details see [56, Chapters 7–8].

For numerical stability, one usually requires all S_j and thus T_j, T to be orthogonal. Under this requirement, the real Schur decomposition of H is a reasonable choice in (4.26), and this was suggested in the seminal paper by Laub [71]. A disadvantage is that the structure of the Hamiltonian matrix is already destroyed in the initial Hessenberg reduction, leading to the loss of spectral symmetry in finite precision arithmetic. In the worst case, this may lead to violation of the spectral dichotomy as perturbations may move the eigenvalue to or across the boundary of the stability region. This can be avoided by using symplectic similarity transformations as then all iterates in the generic procedure above remain Hamiltonian and thus, the spectral symmetry is preserved. As already seen in Chap. 1, implementing the above procedure satisfying both demands, symplectic *and* orthogonal similarity transformations is only possible under certain assumptions due to the lack of an efficient procedure for computing the Hamiltonian Schur form, also called *Van Loan's curse*.

We will now briefly discuss the numerical methods for the Hamiltonian and symplectic eigenproblems that can be used to solve CAREs and DAREs. As most of them have already been discussed in Chap. 1, we keep this discussion short and highlight only Volker's contributions in more or less chronological order.

4.4.1.1 The SR Algorithm

In [41], Bunse-Gerstner and Mehrmann suggest to use the SR algorithm to implement the generic procedure to compute the stable H-invariant subspace and then to solve the CARE via (4.25). The SR algorithm is described in some detail in Sect. 1.2. Its obvious advantage is the preservation of the Hamiltonian structure due to the exclusive use of symplectic similarity transformations. Another advantage is that it is fast: it requires only $\mathcal{O}(n)$ flops per iteration and generically converges with a cubic rate. As non-orthogonal transformations are required, numerical stability is lost and the computed CARE solution may not be as accurate as desired. A possible remedy for this problem is *defect correction* (see Sect. 4.4.2), which can be used to improve the accuracy of an approximate CARE solution.

Given that the method is not numerically backward stable, some variations have been suggested that compute the approximation to the CARE solution even faster by giving up orthogonal transformations altogether, see [42]. Another idea is to iterate directly on the approximation of the CARE rather than computing the Lagrangian invariant subspaces explicitly. For this, an updating procedure is suggested in [47].

Moreover, an SR algorithm for symplectic eigenvalue problems was also derived by Volker and co-workers, see [55] for a first variant and Chap. 1 for further developments of this method.

4.4.1.2 The Hamiltonian QR Algorithm

The ideal algorithm for the Hamiltonian and symplectic eigenvalue problems would be a method that computes the Hamiltonian Schur form using the generic method described above, where only symplectic and simultaneously orthogonal similarity transformations are used. The existence of the Hamiltonian Schur form under rather generic assumptions, in particular under the spectral dichotomy condition we assume here, was proved in [82]. The resulting quest for a Hamiltonian QR algorithm became known under the name *Van Loan's curse* and is described in some detail in Chap. 1. In summary, the lack of the existence of a Hamiltonian or symplectic QR algorithm of cubic complexity under the same general assumptions that allow for structured Schur forms is due to the non-existence of structured Hessenberg-like forms that stay invariant under structured QR iterations. Byers [43, 44] was able to derive such a Hamiltonian Hessenberg form for the special case that $\operatorname{rank}(F) = 1$ or $\operatorname{rank}(G) = 1$ in (4.21). This allows to compute the Hamiltonian Schur form using orthogonal and symplectic similarity transformations in $\mathcal{O}(n^3)$ flops and to solve the CARE (4.23) and the associated continuous-time LQR problem via (4.25).

Under the same rank assumptions on G or F for the symplectic pencil in (4.21), Volker was able to derive an analogous QZ-type algorithm [75]. This algorithm then was probably the most involved eigenvalue algorithm of QR-type, with a technically demanding sequence of elementary eliminations in the bulge-chasing process!

In the following, we will discuss the Hamiltonian QR algorithm and Volker's contribution to stop the search for a Hamiltonian Hessenberg form in the general situation. We omit the symplectic case as it is technically much more involved and refer to the original paper [75] for details.

As already noted, the Hamiltonian QR algorithm should be based on orthogonal and symplectic similarity transformations. This implies a special structure.

Lemma 2 ([82]) *If* $U \in \mathbb{R}^{2n \times 2n}$ *is orthogonal and symplectic, then*

$$U = \begin{bmatrix} U_1 & U_2 \\ -U_2 & U_1 \end{bmatrix}, \qquad U_1, U_2 \in \mathbb{R}^{n \times n}. \tag{4.29}$$

Moreover, as the intersection of two matrix groups, orthogonal symplectic matrices form a matrix group $\mathcal{US}_{2n}$ with respect to matrix multiplication.

As elements of $\mathcal{US}_{2n}$ are determined by the $2n^2$ parameters given by the entries of U_1, U_2, only these parameters need to be stored and updated throughout a sequence of similarity transformations.

Now, the following theorem raises the hope that it is possible to find an algorithm based on symplectic *and* orthogonal similarity transformations for solving CAREs.

Theorem 2 ([82]) *If H is Hamiltonian and $\Lambda(H) \cap \imath\mathbb{R} = \emptyset$, then there exists $U \in \mathcal{US}_{2n}$ such that*

$$U^T HU = \begin{bmatrix} \hat{H}_{11} & \hat{H}_{12} \\ 0 & -\hat{H}_{11}^T \end{bmatrix}, \qquad \hat{H}_{11}, \hat{H}_{12} \in \mathbb{R}^{n\times n}, \tag{4.30}$$

where $\hat{H}_{11}$ is in real Schur form and $\Lambda(\hat{H}_{11}) = \Delta$ (the stable part of the spectrum of H).

Partitioning U from (4.30) as in (4.29), we have from (4.25) that the stabilizing solution of the CARE (4.23) is given by $X_c = U_2 U_1^{-1}$.

Remark 2 The decomposition given in (4.30) is called the *Hamiltonian Schur form*. It can be shown that such a form may also exist if eigenvalues on the imaginary axis are present. They have to satisfy certain properties, the most obvious one is that their algebraic multiplicity needs to be even; see [72] and Chap. 6.

As the QR algorithm is considered to be the best method for solving the dense non-symmetric eigenproblem, it is straightforward to strive for a symplectic variant of the QR algorithm converging to the Hamiltonian Schur form given in (4.30). A framework for such an algorithm can easily be derived. Denote the iterates of such an algorithm by H_j. If we choose the QR decomposition performed in each step, i.e., $p_j(H_j) = S_{j+1}R_{j+1}$, such that all S_{j+1} are symplectic and orthogonal, then it follows that all iterates $H_{j+1} = S_{j+1}^T H_j S_{j+1}$ are Hamiltonian. Unfortunately, such a *symplectic QR decomposition* does not always exist. Sets of matrices in $\mathbb{R}^{2n\times 2n}$ for which it exists are described in [39]. In particular, it is also shown there (see [44] for a constructive proof) that if M is symplectic, then there exists $S \in \mathcal{US}_{2n}$ such that

$$M = SR = S \begin{bmatrix} R_{11} & R_{12} \\ 0_n & R_{11}^{-T} \end{bmatrix} =$$

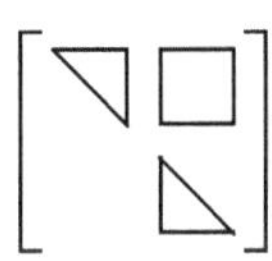

(4.31)

where $R_{11}, R_{12} \in \mathbb{R}^{n\times n}$. Uniqueness of this decomposition can be achieved by requiring all diagonal entries of R_{11} to be positive.

As the matrix R in (4.31) is permutationally similar to an upper triangular matrix and the Hamiltonian Schur form is similar to the real Schur form using the same permutations, it can be shown under mild assumptions that such a Hamiltonian QR algorithm converges to Hamiltonian Schur form if it exists. Moreover, as only similarity transformations in $\mathcal{US}_{2n}$ are used, the algorithm can be shown to be strong backward stable in the sense of Bunch [38].

Byers shows in [44] that if the rational function p_j is chosen to be the *Cayley shift* $c_k(t) := (t - \mu_k)(t + \mu_k)^{-1}$, where μ_k is an approximate real eigenvalue of H, or $d_k(t) := (t - \mu_k)(t - \overline{\mu_k})(t + \mu_k)^{-1}(t + \overline{\mu_k})^{-1}$, where μ_k is an approximate complex eigenvalue of H, then $p_j(H_j)$ is symplectic, and hence, the symplectic QR decomposition of $p_j(H_j)$ exists. In case $\pm\mu_k$ are exact eigenvalues of H and hence of H_j, then deflation is possible, and we can proceed with the deflated problem of smaller dimension without ever being forced to invert a singular matrix. In this way, a double or quadruple Hamiltonian QR step can be implemented.

Unfortunately, the so derived algorithm is of complexity $\mathcal{O}(n^4)$ as each symplectic QR decomposition requires $\mathcal{O}(n^3)$ flops and usually, $\mathcal{O}(n)$ iterations are required (based on the experience that for each eigenvalue, 1–2 iterations are needed). The missing part that would bring the computational cost down to $\mathcal{O}(n^3)$ is an initial reduction analogous to the Hessenberg reduction in the QR algorithm that

- is invariant under the similarity transformation performed in each step of the Hamiltonian QR algorithm (the *Hamiltonian QR step*);
- admits an implementation of the Hamiltonian QR step using only $\mathcal{O}(n^2)$ flops.

In [44] Byers shows that such a form exists.

Definition 3 A Hamiltonian matrix $H \in \mathbb{R}^{2n\times 2n}$ is in *Hamiltonian Hessenberg form* if

$$H = \begin{bmatrix} H_{11} & H_{12} \\ H_{21} & -H_{11}^T \end{bmatrix} = \tag{4.32}$$

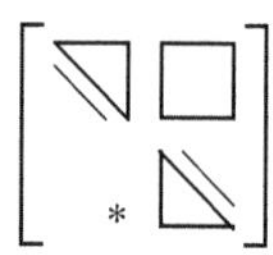

where $H_{ij} \in \mathbb{R}^{n\times n}$, $i, j = 1, 2$, H_{11} is upper Hessenberg, and $H_{21} = \varphi e_n e_n^T$ with $\varphi \in \mathbb{R}$ and e_n being the nth unit vector. The Hamiltonian Hessenberg matrix H is unreduced if $h_{i+1,i} \neq 0$, $i = 1, \ldots, n-1$, and $\varphi \neq 0$.

Byers [44] shows that if H_j is in Hamiltonian Hessenberg form and the rational function p_j is chosen as a Cayley shift, then H_{j+1} is in Hamiltonian Hessenberg form again and the Hamiltonian QR step can be implemented in $\mathcal{O}(n^2)$ flops.

The crux of this algorithm is the initial reduction of a Hamiltonian matrix to Hamiltonian Hessenberg form. Byers shows how this can be achieved if one of the off-diagonal blocks of the Hamiltonian matrix H in (4.21) has rank 1. (This is related to control systems of the form (4.2) having only one input ($m = 1$), i.e., *single-input systems* and/or only one output ($p = 1$), i.e., *single-output systems*.) But unfortunately no algorithm is known for reducing a general Hamiltonian matrix to Hamiltonian Hessenberg form. But the situation is even worse. Analogous to the standard QR algorithm where the QR step is performed on unreduced Hessenberg matrices (possibly zeros on the subdiagonal are used for deflation, i.e., splitting the problem in two or more subproblems consisting of unreduced Hessenberg matrices), the Hamiltonian QR algorithm works for unreduced Hamiltonian Hessenberg matrices. The following theorem due to Ammar and Mehrmann [3] shows that the situation is in general hopeless with respect to the existence of the unreduced Hamiltonian Hessenberg form.

Theorem 3 *If $H \in \mathbb{R}^{2n\times 2n}$ is Hamiltonian, then there exists an orthogonal and symplectic matrix transforming H to unreduced Hamiltonian Hessenberg form if and only if the nonlinear set of equations*

$$x^T x = 1 \qquad \text{and} \qquad x^T J H^{2k-1} x = 0 \quad \text{for} \quad k = 1, \dots, n-1,$$

has a solution that is not *contained in an H-invariant subspace of dimension n or less.*

Obviously, if JH is positive definite, such a vector cannot exist, showing that there really exist situations in which the unreduced Hamiltonian Hessenberg form does not exist. A similar result holds in the symplectic case. Therefore, other approaches have been investigated during the last decade.

4.4.1.3 The Multishift Algorithm

From Theorem 3 we know that the reduction to Hamiltonian Hessenberg form which is necessary to efficiently implement the Hamiltonian QR algorithm is in general not possible. Nevertheless, the same paper [3] suggested a possible alternative method that became the topic of my diploma thesis [8] and eventually lead to the paper [2].

The basis of this idea is that by orthogonal symplectic similarity transformations, the following reduction due to Paige and Van Loan [82] can be achieved.

Theorem 4 *Let $H \in \mathbb{R}^{2n\times 2n}$. Then there exists $U \in \mathcal{US}_{2n}$ such that*

$$U^T H U = \begin{bmatrix} H_{11} & H_{12} \\ H_{21} & H_{22} \end{bmatrix} = \quad (4.33)$$

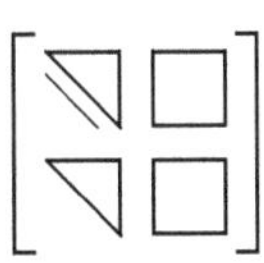

where $H_{11} \in \mathbb{R}^{n\times n}$ is upper Hessenberg and $H_{21} \in \mathbb{R}^{n\times n}$ is upper triangular. The transformation matrix U can be chosen such that

$$U = \left[\begin{array}{cc|cc} 1 & 0 & 0 & 0 \\ 0 & \tilde{U}_1 & 0 & \tilde{U}_2 \\ \hline 0 & 0 & 1 & 0 \\ 0 & -\tilde{U}_2 & 0 & \tilde{U}_1 \end{array}\right]. \tag{4.34}$$

If in addition, H is Hamiltonian, then

$$U^T H U = \begin{bmatrix} H_{11} & H_{12} \\ H_{21} & -H_{11}^T \end{bmatrix} = \quad (4.35)$$

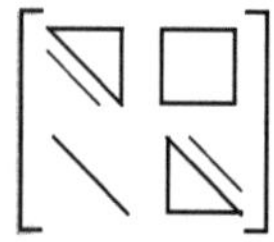

i.e., H_{21} is diagonal and H_{12} is symmetric.

The reduced form (4.35) of a Hamiltonian matrix will be called *PVL form* in the following. An algorithm for computing the transformation given in (4.35) is derived in [82]. It can be implemented using a finite number of similarity transformations and requires $\mathcal{O}(n^3)$ flops. Unfortunately, the PVL form is not preserved under the Hamiltonian *QR* iteration and can therefore not serve for the initial reduction step of the Hamiltonian *QR* algorithm. In the following, we will see that the PVL form can be used in what can be considered as a Hamiltonian *multishift QR* algorithm.

First, we need some more theory. As before, we denote $J = \left[\begin{smallmatrix} 0_n & I_n \\ -I_n & 0_n \end{smallmatrix}\right]$.

Definition 4 A real subspace $\mathcal{L} \subset \mathbb{R}^{2n}$ is *isotropic* iff $x^T J y = 0$ for all $x, y \in \mathcal{L}$. If $\mathcal{L}$ is maximal, i.e., not contained in an isotropic subspace of larger dimension, then $\mathcal{L}$ is a *Lagrangian* subspace (cf. Definition 1).

By the definition of symplectic matrices, we obtain immediately the following lemma.

Lemma 3 *Let $S \in \mathbb{R}^{2n \times 2n}$ be symplectic. Then the first r columns of S, $1 \le r \le n$, span an isotropic subspace of $\mathbb{R}^{2n}$. For $r = n$, this subspace is Lagrangian.*

The basis for the multishift algorithm is contained in the following result.

Proposition 2 ([3]) *Let $H \in \mathbb{R}^{2n \times 2n}$ be a Hamiltonian matrix with spectrum $\Lambda(H) = \Delta_n \cup (-\Delta_n)$, $\Delta_n \cap (-\Delta_n) = \emptyset$, and $\Delta_n = \overline{\Delta_n} = \{\lambda_1, \ldots, \lambda_n\}$. Then the* multishift vector

$$x = \alpha(H - \lambda_1 I_{2n}) \cdots (H - \lambda_n I_{2n}) e_1, \quad \alpha \in \mathbb{R}, \tag{4.36}$$

where $e_1 \in \mathbb{R}^{2n}$ is the first unit vector, is contained in the n-dimensional H-invariant subspace corresponding to $-\Delta_n$. Moreover, this subspace is Lagrangian. In particular, if $\Delta_n \subset \mathbb{C}^+ := \{z \in \mathbb{C} \,|\, \Re(z) > 0\}$, then this Lagrangian subspace is the stable H-invariant subspace.

So, once we know the spectrum of H, we can compute *one* vector that is contained in the subspace required for solving the corresponding CARE. This observation can be combined with the computation of the PVL form in order to derive a *multishift* step as follows – assuming for simplicity that H has no eigenvalues on the imaginary axis.

Using this approach, it is possible to get the whole stable H-invariant subspace. The following theorem will indicate how Algorithm 1 can be used to achieve this.

Algorithm 1 One step of the multishift algorithm for Hamiltonian matrices

---

Multishift step

1. Compute the multishift vector as in (4.36) with $\lambda_j \in \mathbb{C}^+$, $j = 1, \ldots, n$. Choose α in (4.36) such that $\|x\|_2 = 1$. (If this is not possible, i.e., $x = 0$, then exit.)
2. Compute $U_1 \in \mathcal{US}_{2n}$ such that $U_1^T x = \pm e_1$.
3. Set $H_1 = U_1^T H U_1$.
4. Compute the PVL form of H_1, i.e., compute $U_2 \in \mathcal{US}_{2n}$ such that $H_2 = U_2^T H_1 U_2 = (U_1U_2)^T H(U_1U_2)$ is in PVL form.

Theorem 5 *Let $H \in \mathbb{R}^{2n\times 2n}$ be Hamiltonian and let $\mathcal{V}_n$ be an n-dimensional H-invariant Lagrangian subspace corresponding to $\Delta_n \subset \Lambda(H)$ with Δ_n as in Proposition 2. Further, let the multishift vector x from* (4.36) *be computed using $-\Delta_n = \{\lambda_1, \ldots, \lambda_n\}$ as the shifts. If $1 \leq p \leq n$ is the dimension of the minimal isotropic H-invariant subspace $\mathcal{V}_p$ containing x, then after Step 4 of the multishift step, H_2 has the form*

$$H_2 = \left[\begin{array}{cc|cc} A_{11} & A_{12} & G_{11} & G_{21} \\ 0 & A_{22} & G_{21}^T & G_{22} \\ \hline 0 & 0 & -A_{11}^T & 0 \\ 0 & F_{22} & -A_{12}^T & -A_{22}^T \end{array}\right] \begin{array}{l} \}p \\ \}n-p \\ \}p \\ \}n-p \end{array}, \tag{4.37}$$

where $A_{11} \in \mathbb{R}^{p\times p}$, $\Lambda(A_{11}) \subset \Delta_n$, and the Hamiltonian submatrix

$$H_{22} := \begin{bmatrix} A_{22} & G_{22} \\ F_{22} & -A_{22}^T \end{bmatrix} \in \mathbb{R}^{2(n-p)\times 2(n-p)}$$

is in PVL form.

Furthermore, for $U_1, U_2 \in \mathcal{US}_{2n}$ from the multishift step we have

$$U := U_1U_2 = [u_1, \ldots, u_p, u_{p+1}, \ldots, u_{2n}] \in \mathcal{US}_{2n}, \quad u_j \in \mathbb{R}^{2n} \text{ for } j = 1, \ldots, 2n,$$

and $\operatorname{span}\{u_1, \ldots, u_p\} = \mathcal{V}_p \subset \mathcal{V}_n$.

The detailed proof of this result is contained in [10].

The theorem shows that if the multishift vector x from (4.36) has components in *all* directions of a Lagrangian H-invariant subspace, then after one multishift step, a basis for this invariant subspace is given by the first n columns of U_1U_2. Otherwise, the first p columns of U_1U_2 span a p-dimensional H-invariant subspace contained in this subspace and the problem decouples into two subproblems. Algorithm 1 can then repeatedly be applied to the resulting Hamiltonian submatrix $H_{22} \in \mathbb{R}^{2(n-p)\times 2(n-p)}$ until $p = n$. The implementation of this algorithm is described in detail in [2].

As only orthogonal symplectic similarity transformations are used, a multishift step is strongly backward stable. The computational cost of one multishift step for $p = 0$ is around 15 % of the Schur vector method [71]. The complete computational cost depends on the number of iteration steps necessary. In a worst case scenario, i.e., in each step only one basis vector of $\mathscr{V}_n$ is found, the complexity of this algorithm becomes basically $\mathscr{O}(n^4)$. This is rarely observed in praxis, though. On the other hand, rounding errors during the computation, in particular while forming the multishift vector, and the fact that the eigenvalues are usually only known approximately, make it practically impossible that deflation occurs exactly. Often, some iteration steps are necessary to detect deflation when using finite precision arithmetic. Generally speaking, as long as the size of the problem is modest ($n \leq 100$), the method is feasible and the number of required iterations is acceptable.

When solving CAREs, usually the stable H-invariant subspace is required. In that case, Δ_n in Proposition 2 has to be chosen such that $\Re(\lambda_j) < 0$ for all $j = 1, \ldots, n$. Note that the stable H-invariant subspace is Lagrangian; see, e.g., [3, 76]. But observe that in principle, the multishift algorithm can be used to compute the CARE solution corresponding to any Lagrangian H-invariant subspace. This is of particular importance in some applications, e.g., in some $\mathscr{H}_\infty$-control problems, ARE solutions exist and have to be computed if H has eigenvalues on the imaginary axis. As long as these eigenvalues permit a Lagrangian invariant subspace, the corresponding ARE solutions can be computed by the multishift algorithm.

The computation of the multishift vector in (4.36) requires the knowledge of the spectrum of H. Hence, what remains to show is how to obtain the eigenvalues of a Hamiltonian matrix H. One possibility is to run the QR algorithm without accumulating the transformations. But then the problems with eigenvalues close to the imaginary axis as mentioned above have to be expected. A different approach, which costs only one third of the QR algorithm and takes the symmetry of $\Lambda(H)$ into account, was suggested by Van Loan [87]. Consider $K := H^2$. Obviously, if $\lambda \in \Lambda(H)$, then $\lambda_K := \lambda^2 \in \Lambda(K)$. If $\Re(\lambda) \neq 0$, then λ_K is a double eigenvalue of K due to the symmetry of $\Lambda(H)$. Squared Hamiltonian matrices are *skew-Hamiltonian*, that is, they satisfy $KJ = -(KJ)^T$ and therefore have the explicit block structure

$$K = \begin{bmatrix} K_1 & K_2 \\ K_3 & K_1^T \end{bmatrix}, \qquad K_2 = -K_2^T, \quad K_3 = -K_3^T. \tag{4.38}$$

The skew-Hamiltonian structure is preserved under symplectic similarity transformations [87]. Hence, computing the PVL form (4.33) for skew-Hamiltonian matrices yields

$$U^T H^2 U = U^T K U = \begin{bmatrix} \tilde{K}_1 & \tilde{K}_2 \\ 0 & \tilde{K}_1^T \end{bmatrix} = \tag{4.39}$$

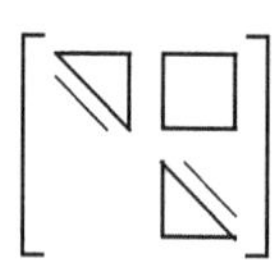

Hence, $\Lambda(K)$ can be obtained by computing the eigenvalues of the upper Hessenberg matrix $\tilde{K}_1$, e.g., by applying the QR iteration to $\tilde{K}_1$. Let $\Lambda(\tilde{K}_1) = \{\mu_1, \ldots, \mu_n\}$, then $\Lambda(H) = \{\pm\sqrt{\mu_1}, \ldots, \pm\sqrt{\mu_n}\}$. Note that no information about eigenvectors or invariant subspaces of H is obtained.

The resulting method is strongly backward stable for K and preserves the symmetry structures of $\Lambda(K)$ and $\Lambda(H)$. An implicit version of this algorithm is also suggested in [87]; U from (4.39) is applied directly to the Hamiltonian matrix such that $\tilde{H} := U^T H U$ is *square-reduced*, i.e., $\tilde{H}^2$ has the form given in (4.39). The disadvantage of Van Loan's method is that a loss of accuracy up to half the number of significant digits of the computed eigenvalues of H is possible. An error analysis in [87] shows that for a computed simple eigenvalue $\tilde{\lambda}$ corresponding to $\lambda \in \Lambda(H)$ we have

$$|\lambda - \tilde{\lambda}| \approx \min\left\{\frac{\varepsilon\|H\|_2^2}{s(\lambda)|\lambda|}, \frac{\sqrt{\varepsilon}\|H\|_2}{s(\lambda)}\right\} = \varepsilon\frac{\|H\|_2}{s(\lambda)} \times \min\left\{\frac{\|H\|_2}{|\lambda|}, \frac{1}{\sqrt{\varepsilon}}\right\}, \tag{4.40}$$

where $s(\lambda)$, the reciprocal condition number of λ, is the cosine of the acute angle between the left and right eigenvectors of H corresponding to λ. Basically, this error estimate indicates that eigenvalues computed by Van Loan's method are as accurate as those computed by a numerically backward stable method provided that $\lambda \approx \|H\|_2$ while for $\lambda \ll \|H\|_2$, the error grows with the ratio $\|H\|_2/|\lambda|$.

Usually, eigenvalues computed by Van Loan's method are satisfactory as shifts for the multishift algorithm and in most other practical circumstances. On the other hand, removing the possible $1/\sqrt{\varepsilon}$ loss of accuracy provides the motivation of the algorithm presented in the next section.

4.4.1.4 A Method Based on the Symplectic URV Decomposition

The method described in this subsection was the key to breaking Van Loan's curse as already described in Chap. 1. As we investigate it here in the context of solving CAREs, we will also need some details and therefore repeat the essential steps.[2]

The central problem of Van Loan's method is that squaring the Hamiltonian matrix leads to a possible loss of half of the accuracy. For products of general matrices, this possible loss of accuracy caused by forming the product can be circumvented by employing the *periodic* or *cyclic QR algorithm* [37, 59, 60].

---

[2]*A personal remark: We derived this method over several months in 1995/1996, regularly meeting with Hongguo Xu, then a Humboldt fellow at TU Chemnitz-Zwickau in Volker's group, on Thursdays in Volker's office. The quest was to avoid the loss of accuracy due to the explicit squaring of H in Van Loan's method. When Hongguo wrote the key step, the symplectic URV decomposition, on Volker's blackboard one Thursday morning, this was one of the most beautiful and enlightening moments in my career as a mathematician.*

If $A = A_1 \cdot A_2 \cdots A_p$, where $A_j \in \mathbb{R}^{n\times n}$, $j = 1, \ldots, p$, then this algorithm computes the real Schur form of A without forming A explicitly. This is achieved by cyclically reducing the factors A_j to (quasi-)upper triangular form:

$$U^T A U = (U_1^T A_1 U_2)(U_2^T A_2 U_3) \cdots (U_p^T A_p U_1) = \left[\diagdown\right] \cdot \left[\diagdown\right] \cdots \left[\diagdown\right]. \tag{4.41}$$

Here, $U_1^T A_1 U_2$ is in real Schur form while $U_j^T A_j U_{(j+1) \bmod p}$, $j = 2, \ldots, p$, are upper triangular such that the product is in real Schur form. The eigenvalues are then obtained from computing the eigenvalues of the 1×1 and 2×2 blocks on the diagonal of the product in (4.41). This method is numerically backward stable and avoids the loss of accuracy in the eigenvalues as the product A is never formed explicitly.

The idea is now to employ this approach to H^2 by replacing the reduction of H^2 to PVL form by $U^T H^2 U = (U^T H V)(V^T H U)$, where $U, V \in \mathcal{US}_{2n}$. This can be achieved by the *symplectic URV-like decomposition* given in [28].

Proposition 3 *For $H \in \mathbb{R}^{2n\times 2n}$ there exist $U, V \in \mathcal{US}_{2n}$ such that*

$$V^T H U = \begin{bmatrix} H_1 & H_3 \\ 0 & -H_2^T \end{bmatrix} = \begin{bmatrix} \diagdown & \square \\ 0 & \diagdown \end{bmatrix}, \tag{4.42}$$

i.e., H_1 is upper triangular and H_2 is upper Hessenberg. If, in addition, H is Hamiltonian, then

$$U^T H^2 U = \begin{bmatrix} H_2 H_1 & H_2 H_3 - (H_2 H_3)^T \\ 0 & (H_2 H_1)^T \end{bmatrix} = \begin{bmatrix} \diagdown & \square \\ 0 & \diagdown \end{bmatrix} \tag{4.43}$$

and the eigenvalues of H are the positive and negative square roots of the eigenvalues of the upper Hessenberg matrix $H_2 H_1$.

That is, using the decomposition given in (4.42) we obtain the PVL form of H^2 without explicitly squaring H. In order to obtain the eigenvalues of H we then apply the periodic QR algorithm to $H_2 H_1$.

In [28] an algorithm for computing the decomposition given in (4.42) is presented. It requires a finite number of transformations. The combined cost of computing the decomposition (4.42) and applying the periodic QR algorithm to $H_2 H_1$ is about $48n^3$ flops – this is $1.5 \times$ the computational cost of Van Loan's method and about 60 % of the cost of the QR algorithm applied to a non-symmetric $2n \times 2n$ matrix. The method is numerically backward stable as only orthogonal transformations are used. The symmetry property of $\Lambda(H)$ is preserved and in this sense the method can be considered to be strongly backward stable.

A detailed error analysis of the above method yields the following result [28]. Essentially (under mild assumptions), for a nonzero and simple eigenvalue λ of a Hamiltonian matrix $H \in \mathbb{R}^{2n\times 2n}$, the algorithm based on the symplectic URV-like decomposition followed by applying the periodic *QR* algorithm to H_2H_1 from (4.42) yields a computed eigenvalue $\tilde{\lambda}$ satisfying

$$|\tilde{\lambda} - \lambda| \leq \frac{2\|H\|\varepsilon}{s(\lambda)} + \mathcal{O}(\varepsilon^2).$$

This is the accuracy to be expected from any backward stable method like the *QR* algorithm and shows that by avoiding to square H, we get the full possible accuracy.

Nevertheless, as Van Loan's method, the approach presented above does not provide the H-invariant subspaces. But based on (4.42) it is possible to derive an algorithm that can be used to compute the stable H-invariant subspace and the solution of the CARE (4.23) [27]. The basis for this algorithm is the following theorem.

Theorem 6 ([27]) *Let $A \in \mathbb{R}^{n\times n}$ and define $B = \left[\begin{smallmatrix} 0 & A \\ A & 0 \end{smallmatrix}\right]$. Then the spectra of A and B are related via $\Lambda(B) = \Lambda(A) \cup (-\Lambda(A))$. Further, let $\Lambda(A) \cap \imath\mathbb{R} = \emptyset$. If the columns of $\left[U_1^T, U_2^T\right]^T \in \mathbb{R}^{2n\times n}$ span an orthogonal basis for a B-invariant subspace such that*

$$B\begin{bmatrix} U_1 \\ U_2 \end{bmatrix} = \begin{bmatrix} U_1 \\ U_2 \end{bmatrix} R, \qquad \Lambda(R) \subset \mathbb{C}^+ \cap \Lambda(B),$$

then range $(U_1 + U_2)$ *is the A-invariant subspace corresponding to $\Lambda(A) \cap \mathbb{C}^+$ and* range $(U_1 - U_2)$ *is the stable A-invariant subspace.*

An orthogonal basis for the subspace defined by range $(U_1 - U_2)$ can be obtained, e.g., from a rank-revealing *QR* decomposition of $U_1 - U_2$; see, e.g., [56].

In general it is of course not advisable to use the above result in order to obtain the stable invariant subspace of a matrix A as one would have to double the dimension and thereby increase the computational cost and required workspace significantly as compared to applying the *QR* algorithm to A. But we will see that for Hamiltonian matrices, the given structure makes this approach very attractive.

Let $H \in \mathbb{R}^{2n\times 2n}$ be Hamiltonian with $\Lambda(H) \cap \imath\mathbb{R} = \emptyset$. Define a permutation matrix $P \in \mathbb{R}^{4n\times 4n}$ by

$$P = \begin{bmatrix} I_n & 0 & 0 & 0 \\ 0 & 0 & I_n & 0 \\ 0 & I_n & 0 & 0 \\ 0 & 0 & 0 & I_n \end{bmatrix}.$$

Then $P^T \left[\begin{smallmatrix} 0 & H \\ H & 0 \end{smallmatrix}\right] P$ is a Hamiltonian matrix in $\mathbb{R}^{4n\times 4n}$. The basic idea is now to employ the decomposition (4.42) in order to make $P^T HP$ block-upper triangular.

Therefor, let $\hat{U}, \hat{V} \in \mathscr{US}_{2n}$ be as in Proposition 3 such that $\hat{V}^T H \hat{U}$ has the form given in (4.42). Then we apply the periodic QR algorithm to $H_2 H_1$. From this we obtain orthogonal matrices $V_1, V_2 \in \mathbb{R}^{n\times n}$ such that both, the product

$$(V_1^T H_2 V_2)(V_2^T H_1 V_1) =: \hat{H}_2 \hat{H}_1$$

and $\hat{H}_2$, are in upper real Schur form while $\hat{H}_1$ is upper triangular. Define

$$U_1 := \hat{U} \begin{bmatrix} V_1 & 0 \\ 0 & V_1 \end{bmatrix}, \quad U_2 := \hat{V} \begin{bmatrix} V_2 & 0 \\ 0 & V_2 \end{bmatrix}, \quad \text{and} \quad U := \begin{bmatrix} U_1 & 0 \\ 0 & U_2 \end{bmatrix}.$$

Then

$$B := P^T U^T \begin{bmatrix} 0 & H \\ H & 0 \end{bmatrix} UP = \begin{bmatrix} 0 & \hat{H}_2 & 0 & \hat{H}_3^T \\ \hat{H}_1 & 0 & \hat{H}_3 & 0 \\ 0 & 0 & 0 & -\hat{H}_1^T \\ 0 & 0 & -\hat{H}_2^T & 0 \end{bmatrix}$$

is Hamiltonian and block upper triangular with $\hat{H}_1$ upper triangular, $\hat{H}_2$ in real Schur form, and $\hat{H}_3 = V_2^T (H_2 H_3 - H_3^T H_2^T) V_1$.

Now let U_3 be orthogonal such that

$$U_3^T \begin{bmatrix} 0 & \hat{H}_2 \\ \hat{H}_1 & 0 \end{bmatrix} U_3 = \begin{bmatrix} T_1 & T_3 \\ 0 & -T_2 \end{bmatrix} \tag{4.44}$$

is in upper real Schur form with $T_j \in \mathbb{R}^{n\times n}$, $j = 1, 2, 3$, and $\Lambda(T_1) = \Lambda(T_2) \subset \mathbb{C}^+$. Note that this is possible as the eigenvalues of $\left[\begin{smallmatrix} 0 & \hat{H}_2 \\ \hat{H}_1 & 0 \end{smallmatrix}\right]$ are exactly those of H^2 and $\Lambda(H) \cap \imath\mathbb{R} = \emptyset$. Hence,

$$\tilde{B} := \begin{bmatrix} U_3^T & 0 \\ 0 & U_3^T \end{bmatrix} B \begin{bmatrix} U_3 & 0 \\ 0 & U_3 \end{bmatrix} = \begin{bmatrix} T_1 & T_3 & R_1 & R_2 \\ 0 & -T_2 & R_2^T & R_3 \\ 0 & 0 & -T_1^T & 0 \\ 0 & 0 & -T_3^T & T_2^T \end{bmatrix}$$

is in Hamiltonian Schur form. In order to apply Theorem 6, we need to reorder the eigenvalues in the Hamiltonian Schur form such that all eigenvalues in the upper left $2n \times 2n$ block are in the open right half plane. This can be achieved, e.g., by the symplectic re-ordering algorithm due to Byers [43, 44]. With this algorithm it is possible to determine $\tilde{U} \in \mathscr{US}_{2n}$ such that

$$\tilde{U}^T \tilde{B} \tilde{U} = \begin{bmatrix} T_1 & \tilde{T}_3 & R_1 & \tilde{R}_2 \\ 0 & \tilde{T}_2 & \tilde{R}_2^T & R_3 \\ 0 & 0 & -T_1^T & 0 \\ 0 & 0 & -\tilde{T}_3^T & -\tilde{T}_2^T \end{bmatrix}, \qquad \Lambda(\tilde{T}_2) = \Lambda(T_2).$$

Now define

$$S := P^T \begin{bmatrix} U_1 & 0 \\ 0 & U_2 \end{bmatrix} P \begin{bmatrix} U_3 & 0 \\ 0 & U_3 \end{bmatrix} \tilde{U}. \tag{4.45}$$

Then $S \in \mathcal{US}_{4n}$ and

$$T := S^T P^T \begin{bmatrix} 0 & H \\ H & 0 \end{bmatrix} PS =: \begin{bmatrix} T_{11} & T_{12} \\ 0 & -T_{11}^T \end{bmatrix} \tag{4.46}$$

is in Hamiltonian Schur form with $\Lambda(T_{11}) \subset \mathbb{C}^+$. Now we can apply Theorem 6 with A replaced by H and $R := T_{11}$.

Corollary 1 *Let $H \in \mathbb{R}^{2n\times 2n}$ be Hamiltonian with $\Lambda(H) \cap \imath\mathbb{R} = \emptyset$ and let S be as in (4.45) such that (4.46) holds. If $PS := \left[\begin{smallmatrix} S_{11} & S_{12} \\ S_{21} & S_{22} \end{smallmatrix}\right]$, with $S_{ij} \in \mathbb{R}^{2n\times 2n}$, then the n-dimensional, stable H-invariant subspace is given by* range $(S_{11} - S_{21})$.

The above transformations yielding S are described in more detail in [27]. The solution of the CARE can be obtained from an orthogonal basis of range $(S_{11} - S_{21})$ computed by a rank-revealing QR decomposition or directly from $S_{11} - S_{21}$; for details see [27]. The latter approach saves a significant amount of work such that the cost of the algorithm described above for computing the stabilizing solution of the CARE (4.23) is approximately 60 % of the cost of the Schur vector method.

Remark 3 The transformation of $\left[\begin{smallmatrix} 0 & \hat{H}_2 \\ \hat{H}_1 & 0 \end{smallmatrix}\right]$ to real Schur form and the computation of the matrix U_3 in (4.44) can be efficiently implemented employing the available structure. An algorithm for this is given in [27].

It is shown in [27] that the algorithm presented above is strongly backward stable in $\mathbb{R}^{4n\times 4n}$. That is, if $\tilde{S}$ is the analogue to S from (4.45) computed in finite precision arithmetic, then

$$\tilde{S}^T P^T \begin{bmatrix} 0 & H \\ H & 0 \end{bmatrix} P\tilde{S} = T + E,$$

with T as in (4.46), $\|E\|_2 \le c\varepsilon\|H\|_2$ for a small constant c and $E \in \mathbb{R}^{4n\times 4n}$ is Hamiltonian. Moreover it is shown in [27] that the computed invariant subspace is as accurate as the maximum of its condition number and the condition number of its complimentary (*antistable*) H-invariant subspace permit. This is to be expected from the fact that at the same time we compute the stable H-invariant subspace, by Theorem 6 we also compute the antistable H-invariant subspace. In that sense the algorithm is not optimal as we would like the accuracy of the computed subspace to be limited only by its own condition number.

The algorithms described in this section are implemented in Fortran 77, see [19], while an implementation of Van Loan's method with scaling is provided in [12]. All these methods are also integrated in SLICOT,[3] the **S**ubroutine **LI**brary in **CO**ntrol **T**heory [26]. This subroutine library is based on a joint initiative of European researchers, in which Volker also was a driving force.

The invariant subspace computation as described above turned out to be not completely satisfactory. For some examples with eigenvalues close to the imaginary axis, unexpectedly large errors are encountered in the computed CARE solutions. Almost 10 years later, Volker and co-workers came up with the observation that the orthogonal and symplectic matrix U that puts H^2 into skew-Hamiltonian Schur form contains information that can be used to construct the stable H-invariant subspace in a recursive fashion [49]. This matrix U is obtained from the symplectic URV decomposition (4.42) with the additional step of accumulating the matrices obtained from the periodic QR algorithm applied to $H_2 H_1$ into U, V. This approach is described in some detail in Sect. 1.4, we therefore refrain from recapitulating it here. An intuitive explanation of the method was given by Watkins in [89], and a block version of the algorithm that can better deal with clusters of eigenvalues is described in [79]. It can be concluded that Volker and co-workers have eventually found a method to solve CAREs via the approach based on invariant subspaces of the Hamiltonian eigenproblem that satisfies all desired properties regarding numerical stability, structure-preservation, and efficiency.

Analogous procedures for solving the DARE via the symplectic pencil approach have not yet been derived. On the other hand, by employing the generalized Cayley transformation approach described in the previous section, one may use the generalizations of this approach to skew-Hamiltonain/Hamiltonian or even/odd matrix pencils described by Volker and co-workers in [15, 24]. Nevertheless, there is still room for further improvement: the methods discussed in [15, 24] extend the approach from [27] rather than the more robust methods from [49, 79]. Also, the available software for structured matrix pencils [33, 34] is still based on the skew-Hamiltonian/Hamiltonian pencil structure rather than the more general structure of even or alternating matrix pencils.

4.4.2 Defect Correction

In this subsection, we re-visit a topic that is very important for obtaining solutions to CAREs and DAREs at highest possible accuracy. If a numerical solution $\tilde{X}$ of the CARE (4.23) is computed, this process is prone to roundoff errors, thus $\tilde{X}$ can only be an approximation to the solution X of (4.23). In particular for algorithms that are not numerically backward stable, like the SR algorithm discussed in Chap. 1 and the previous subsection, methods based on sign and disk functions, but also symplectic QR algorithms in case the final step of obtaining X via $X = -U_2 U_1^{-1}$ as in (4.25) is ill-conditioned, these roundoff errors may lead to deteriorating accuracy. Improving

[3] Available at www.slicot.org

this approximation may be achieved using Newton's method, but alternatively also by *defect correction*. The necessary theory was derived in [80] and [76, Chapter 10]. We will provide this here in some detail, as the defect correction principle has, despite its importance for practice, received little attention since, and in particular new computing platforms including hardware accelerators may use this principle for obtaining fast and reliable algorithms, as has been already suggested for Lyapunov equation in [18].

The defect correction principle is most easily explained for linear systems of equations. Given an approximate solution $\tilde{x}$ of the underlying problem

$$Ax = b,$$

given $A \in \mathbb{R}^{n \times n}$ nonsingular, $b \in \mathbb{R}^n$, we determine the *residual* $r = b - A\tilde{x}$, then compute a solution δ of the linear system

$$A\delta = r,$$

and set

$$x^+ = \tilde{x} + \delta.$$

This process can be iterated until a given stopping criterion is satisfied. Defect correction is successful as long as the residual can be computed accurately enough, i.e., with higher precision than $\tilde{x}$. Such higher precision may be obtained, e.g., using double precision in case $\tilde{x}$ was computed in single precision. There is no need to compute $\tilde{x}$ and δ by the same algorithm, providing great flexibility to the defect correction principle.

The basis for applying defect correction to CAREs is provided by the following theorem due to Mehrmann/Tan [80]

Theorem 7 *Let $X = X^T$ be a solution of* (4.23), *and $\tilde{X}$ a symmetric approximation to this solution. Define $P = X - \tilde{X}$, $\tilde{A} = A - G\tilde{X}$, and let*

$$\tilde{F} = F + A^T \tilde{X} + \tilde{X} A - \tilde{X} G \tilde{X}$$

be the residual of (4.23) *with respect to $\tilde{X}$. Then P satisfies the CARE*

$$0 = \tilde{F} + P\tilde{A} + \tilde{A}^T P - PGP. \tag{4.47}$$

The theorem is proved by simple algebraic manipulations after inserting $X = \tilde{X} + P$ in (4.23).

Due to the non-uniqueness of CARE solutions, it is important to guarantee that the updated approximate solution $\tilde{X} + P$ is still the one of interest, usually the stabilizing solution in control applications. In order to show that this is indeed the case, first it is necessary to check whether the stabilizability property related to (4.23) carries over to (4.47).

Lemma 4 *Given* (4.23) *with* (A, G) *stabilizable, i.e.,*

$$\operatorname{rank}[\lambda I - A,\ G] = n \quad \forall\, \lambda \in \mathbb{C} : \Re(\lambda) \geq 0.$$

Then $\tilde{A} = A - G\tilde{X}$ *is stabilizable as well.*

Proof

$$\begin{aligned}\operatorname{rank}[\lambda I_n - \tilde{A}, G] &= \operatorname{rank}[\lambda I_n - A + G\tilde{X}, G] \\ &= \operatorname{rank}\left[[\lambda I_n - A, G]\begin{bmatrix} I_n & 0 \\ \tilde{X} & I_n \end{bmatrix}\right] = \operatorname{rank}[\lambda I_n - A, G].\end{aligned}$$

□

As we have seen before, the positive semidefinite solution of the CARE (4.23) can be obtained from the invariant subspace of the Hamiltonian matrix

$$H = \begin{bmatrix} A & G \\ F & -A^T \end{bmatrix}$$

corresponding to the eigenvalues in $\mathbb{C}^-$. Forming the Hamiltonian matrix corresponding to the defect CARE (4.47),

$$\tilde{H} = \begin{bmatrix} \tilde{A} & G \\ \tilde{F} & -\tilde{A}^T \end{bmatrix},$$

we see that $\tilde{H} = Q^{-1}HQ$ using

$$Q = \begin{bmatrix} I & 0 \\ -\tilde{X} & I \end{bmatrix}.$$

This immediately implies that $\Lambda(H) = \Lambda(\tilde{H})$, so that also $\tilde{H}$ has exactly n eigenvalues with negative real parts. Now assume the $\tilde{H}$–invariant subspace corresponding to these eigenvalues is spanned by $\begin{bmatrix} Z_1 \\ Z_2 \end{bmatrix} \in \mathbb{R}^{2n\times n}$, that is,

$$\begin{bmatrix} \tilde{A} & G \\ \tilde{F} & -\tilde{A}^T \end{bmatrix}\begin{bmatrix} Z_1 \\ Z_2 \end{bmatrix} = \begin{bmatrix} Z_1 \\ Z_2 \end{bmatrix} Z, \tag{4.48}$$

where all eigenvalues of $Z \in \mathbb{R}^{n\times n}$ have negative real parts.

By Lemma 4, $(\tilde{A}, G)$ is stabilizable. Using standard arguments of control theory [52], it can be shown then that Z_1 is invertible and that $Z_2 Z_1^{-1}$ is symmetric negative semidefinite. Inserting $P := -Z_2 Z_1^{-1}$ in (4.48), we obtain

$$\begin{bmatrix} \tilde{A} & G \\ \tilde{F} & -\tilde{A}^T \end{bmatrix} \begin{bmatrix} I \\ -P \end{bmatrix} = \begin{bmatrix} I \\ -P \end{bmatrix} \tilde{Z}, \tag{4.49}$$

where $\tilde{Z} = Z_1 Z Z_1^{-1}$ has only eigenvalues with negative real parts, too.

Expanding terms in (4.49), the first row yields

$$\tilde{A} - GP = \tilde{Z}, \tag{4.50}$$

implying that P is stabilizing, and from the second row,

$$\tilde{F} + \tilde{A}^T P = -P\tilde{Z} = -P\tilde{A} + PGP,$$

we see that P is a solution of (4.47). In summary, P must be the unique symmetric positive semidefinite solution of (4.47).

Now with $\tilde{A} = A - G\tilde{X}$ it follows from (4.50), that

$$A - G(\tilde{X} + P) = \tilde{Z}$$

is stable, so that in summary we obtain the following result:

Theorem 8 *Suppose the invariant subspace of the Hamiltonian matrix*

$$\tilde{H} = \begin{bmatrix} \tilde{A} & G \\ \tilde{F} & -\tilde{A}^T \end{bmatrix}$$

corresponding to (4.47) *is spanned by the columns of* $\left[\begin{smallmatrix} Z_1 \\ Z_2 \end{smallmatrix}\right] \in \mathbb{R}^{2n\times n}$, *and* $P = -Z_2 Z_1^{-1}$, *then* $\tilde{X} + P$ *is the unique symmetric positive semidefinite solution of* (4.47).

Based on Theorems 7 and 8, we may now formulate a defect correction algorithm for CAREs.

As there are no specifications given for the methods employed in Steps 1 and 3 of Algorithm 2, one could use any numerical method to solve the CAREs, and possibly different ones in Steps 1 and 3. For instance, one could simply use a fast, but potentially unstable, method as the SR algorithm, as there is no need to have an $\tilde{X}$ of high accuracy. As $\|P\|$ will be very small in general, the quadratic term in the defect CARE is basically negligible, Newton's method is a natural choice in Step 3. Another option is to employ an algorithm like the orthogonal symplectic multishift algorithm in which part of the computations from Step 1 can be re-used

Algorithm 2 Defect correction for the CARE (4.23)

Algorithm DC_CARE

Input. $A, G, F \in \mathbb{R}^{n \times n}$ as in (4.23); tolerance $\varepsilon > 0$ for the stopping criterion.

Output. Approximation $\tilde{X}$ to the unique stabilizing solution of (4.23) and an error estimator $P \in \mathbb{R}^{n \times n}$ with $\|P\| \leq \varepsilon$.

Step 1. Compute stabilizing approximate solution $\tilde{X}$ of (4.23) with the favorite CARE solver.

Step 2. Set $P := \tilde{X}, \quad \tilde{X} := 0.$

Step 3.

while $\|P\| > \varepsilon$ **do**

Set

$$\begin{aligned} \tilde{X} &:= \tilde{X} + P \\ \tilde{F} &:= F + A^T \tilde{X} + \tilde{X} A - \tilde{X} G \tilde{X} \\ \tilde{A} &:= A - G \tilde{X} \end{aligned}$$

Compute a stabilizing solution of the defect CARE

$$0 = \tilde{F} + \tilde{A}^T P + P \tilde{A} - PGP$$

end

in Step 3, so that the cost in Step 3 can be reduced. Such a variant is discussed in [2, 8].

The Algorithm DC_CARE is numerically backward stable in the sense that $\tilde{X}$ is the exact solution of the following CARE with perturbed data:

$$0 = F - \tilde{F} + A^T X + XA - XGX,$$

where $\tilde{F}$ is the residual from Theorem 7. If the leading significant digits of the defect P in Step 3 are computed correctly, the approximate solution $\tilde{X}$ converges to the exact stabilizing solution X of (4.23). In practice, this will lead to reduced accuracy of the residual $\tilde{F}$ due to cancelation which leads to a limitation of the obtainable accuracy. Nevertheless, the accuracy of the CARE is often greatly improved by 1–2 steps of defect correction, as several examples in [8] indicate.

A defect correction procedure for the DARE (4.24) can be derived in a completely analogous fashion; see [80] and [76, Chapter 10]. An interesting aspect of further research would be to derive a mixed precision CPU-GPU variant of Algorithm DC_CARE in the fashion of [18]. Also, in case of large-scale sparse solvers for CAREs as recently reviewed in [32], where low-rank approximations to the stabilizing solution are computed, it would be necessary to be able to represent $\tilde{F}$ in low-rank format to use this concept. Whether this is possible or not is an open problem.

4.4.3 *Other Contributions to the Numerical Solution of AREs*

Volker's many contributions to the numerical solution of AREs go well beyond the described methods based on structured eigenproblems. A classical solution method of AREs is Newton's method, given their nature as nonlinear systems of equations. Volker has contributed to the convergence theory of Newton's method applied to the generalized CAREs (4.7) and DAREs (4.8), see [76, § 11]. He attributes these contributions mainly to Ludwig Elsner, but they had not been published before.

A notable contribution to understanding why and when the sign function method for CAREs [45] can be expected to yield accurate solutions was the perturbation analysis derived in [46].

Recently, Volker together with Federico Poloni has explored the use of permuted graph matrices to represent invariant and deflating subspaces of Hamiltonian matrices and pencils. This has often a tremendously positive effect on the numerical accuracy of iterative methods based on the inverse-free iteration/disk function method [5, 11, 31] and doubling-like algorithms [50, 51]. See the Chap. 5 for further details and references.

An important aspect in deriving numerical methods for AREs is to test them on challenging examples and to compare the performance to existing methods using well-defined benchmarks. Volker was the driving force in establishing benchmark collections for CAREs [20] and DAREs [21], see [22] for an overview. This had a very positive effect on the publication attitudes of new numerical methods for AREs as non competitive methods can be identified easily since these benchmark collections became available. Later on, he also inspired a number of other benchmark collections for computational control problems that became part of the SLICOT project, see also [25].

Together with Petko Petkov and Mihail Konstantinov, Volker also contributed and tested mathematical software for solving CAREs [84, 85].

Although Volker had not published on solving large-scale AREs until recently [78], he inspired much of the work in this area by putting his Ph.D. student Thilo Penzl on this track. His thesis [83] is now considered the starting point for many of the currently used methods for large-scale matrix equations, see also the later paper [23] that was published only 8 years after Thilo's unfortunate passing in 1999 due to a tragic accident. A survey on the developments of this prospering field was recently given in [32].

4.5 Avoiding AREs

Already in his early work related to his habilitation thesis and the book [76], Volker often made the point that in solving LQR problems, it may not be necessary to solve AREs explicitly. The basic idea of this can be presented using the continuous-time LQR problem. Borrowing the most recent representation, we can associate an even matrix pencil to the LQR problem (4.1) for the descriptor or LTI system (4.2):

$$\lambda K - L := \lambda \begin{bmatrix} 0 & E & 0 \\ -E^T & 0 & 0 \\ 0 & 0 & 0 \end{bmatrix} - \begin{bmatrix} 0 & A & B \\ A^T & C^T Q C & -C^T S \\ B^T & S^T C & R \end{bmatrix}. \tag{4.51}$$

This matrix pencil considered as a matrix polynomial has the leading coefficient skew-symmetric, the constant term is symmetric, and therefore is skew-symmetric/symmetric, or more general, an alternating matrix polynomial, see the Chaps. 2, 6, and 12 for more on these structures. Assuming for simplicity E, R nonsingular, the key observation now is that if the feedback solution $u(t) = F_c x(t)$ is sought, and an n-dimensional stable deflating subspace of the even pencil $\lambda K - L$ from (4.51) exists, spanned by the columns of $\left[V^T, U^T, W^T\right]^T$ with $U \in \mathbb{R}^{n\times n}$ invertible and V, W of size according to (4.51), then, following [76, § 6], we get

$$F_c = -R^{-1}(B^T X_c E + S^T C) = WU^{-1}.$$

Hence, the optimal feedback control law can be determined without computing the CARE solution! The latter is determined via $X_c E = VU^{-1}$ in this setting. Working with the even matrix pencil in (4.1) has the additional advantage that no linear systems with R need to be solved in forming the coefficients and thus, rounding errors in forming these coefficients are avoided [14]. This principle also carries over to the discrete-time case, see again [76, § 6] for the unstructured setting.

Avoiding the solution of AREs is even more desirable in H_∞ control. We will not go into the details of this problem, that has become a paradigm in robust control. It suffices to understand that the H_∞-optimal controller is usually determined using the so called *γ-iteration* and subsequent controller formulas based on the output of this iteration. The γ-iteration is classically formulated in such a way that (in the continuous-time case) two CAREs need to be solved in each iteration step, and a spectral condition of their product is checked, see, e.g., [57, 90]. The crux is that these CAREs usually have indefinite quadratic terms which does not allow to solve them with Newton's method, and the eigenvalues of the associated Hamiltonian matrices are often close to the imaginary axis (in particular close to the "optimal" γ) which makes them difficult to solve by methods based on invariant or deflating subspaces of the associated Hamiltonian matrices. Moreover, often their norms tend to infinity which yields poorly scaled problems, implying additional numerical difficulties. For all these reasons, it was desirable to avoid the CAREs in the process. Volker and co-workers were able to derive a numerically robust method that achieves this, see [16, 17] for the LTI case and [73] for the descriptor case. Later, also controller formula based on this approach were presented in [13]. For details, we refer to [24], where also other applications of using extended pencil formulations in even or alternating form are discussed.

4.6 Concluding Remarks

Over the last 20 years, Volker and I have collaborated on methods for algebraic Riccati equations, and on how to avoid them. We often had differing opinions on whether or not the ARE is the right concept to use for solving certain control problems. These were always inspiring and fruitful discussions, and they certainly have been driving both of us to improve methods and ideas more and more. Today, we basically agree on which problems should be solved using AREs, and where one should avoid them by using structured matrix pencil methods as discussed, e.g., in Chaps. 1–3, and 5. Therefore, I foresee interesting research tasks in both directions for certainly another decade and more, as with increasing computer power, model complexity in control engineering problems is increasing, and will require new ideas and further development of the methods for DREs, AREs, and the related pencil problems at hand. I truly hope Volker and I will also be part of this development. In writing this chapter, a number of open problems and ideas already evolved, and I hope some of these can lead to improved methods and further inside in the theoretical and numerical treatment of DREs and AREs.

References

1. Abou-Kandil, H., Freiling, G., Ionescu, V., Jank, G.: Matrix Riccati Equations in Control and Systems Theory. Birkhäuser, Basel (2003)
2. Ammar, G.S., Benner, P., Mehrmann, V.: A multishift algorithm for the numerical solution of algebraic Riccati equations. Electron. Trans. Numer. Anal. **1**, 33–48 (1993)
3. Ammar, G.S., Mehrmann, V.: On Hamiltonian and symplectic Hessenberg forms. Linear Algebra Appl. **149**, 55–72 (1991)
4. Arnold, W.F., Laub, A.J.: Generalized eigenproblem algorithms and software for algebraic Riccati equations. Proc. IEEE **72**, 1746–1754 (1984)
5. Bai, Z., Demmel, J., Gu, M.: An inverse free parallel spectral divide and conquer algorithm for nonsymmetric eigenproblems. Numer. Math. **76**(3), 279–308 (1997)
6. Bender, D.J., Laub, A.J.: The linear-quadratic optimal regulator for descriptor systems. IEEE Trans. Automat. Control **32**, 672–688 (1987)
7. Bender, D.J., Laub, A.J.: The linear-quadratic optimal regulator for descriptor systems: discrete-time case. Automatica **23**, 71–85 (1987)
8. Benner, P.: Ein orthogonal symplektischer Multishift Algorithmus zur Lösung der algebraischen Riccatigleichung. Diplomarbeit, RWTH Aachen, Institut für Geometrie und Praktische Mathematik, Aachen (1993) (in German)
9. Benner, P.: Contributions to the Numerical Solution of Algebraic Riccati Equations and Related Eigenvalue Problems. Logos–Verlag, Berlin (1997). *Also:* Dissertation, Fakultät für Mathematik, TU Chemnitz–Zwickau (1997)
10. Benner, P.: Computational methods for linear-quadratic optimization. Supplemento ai Rendiconti del Circolo Matematico di Palermo, Serie II **58**, 21–56 (1999)
11. Benner, P., Byers, R.: Disk functions and their relationship to the matrix sign function. In: Proceedings of the European Control Conference (ECC 97), Brussels, Paper 936. BELWARE Information Technology, Waterloo (1997). CD-ROM

12. Benner, P., Byers, R., Barth, E.: Algorithm 800. fortran 77 subroutines for computing the eigenvalues of Hamiltonian matrices I: the square-reduced method. ACM Trans. Math. Softw. **26**(1), 49–77 (2000)
13. Benner, P., Byers, R., Losse, P., Mehrmann, V., Xu, H.: Robust formulas for H_∞ optimal controllers. Automatica **47**(12), 2639–2646 (2011)
14. Benner, P., Byers, R., Mehrmann, V., Xu, H.: Numerical methods for linear-quadratic and H_∞ control problems. In: Picci, G., Gilliam, D.S. (eds.) Dynamical Systems, Control, Coding, Computer Vision: New Trends, Interfaces, and Interplay. Progress in Systems and Control Theory, vol. 25, pp. 203–222. Birkhäuser, Basel (1999)
15. Benner, P., Byers, R., Mehrmann, V., Xu, H.: Numerical computation of deflating subspaces of skew-Hamiltonian/Hamiltonian pencils. SIAM J. Matrix Anal. Appl. **24**(1), 165–190 (2002)
16. Benner, P., Byers, R., Mehrmann, V., Xu, H.: Robust numerical methods for robust control. Preprint 2004–06, Institut für Mathematik, TU Berlin, D-10623 Berlin (2004). http://www.math.tu-berlin.de/preprints
17. Benner, P., Byers, R., Mehrmann, V., Xu, H.: A robust numerical method for the γ-iteration in $\mathscr{H}_\infty$ control. Linear Algebra Appl. **425**(2–3), 548–570 (2007)
18. Benner, P., Ezzatti, P., Kressner, D., Quintana-Ortí, E.S., Remón, A.: A mixed-precision algorithm for the solution of Lyapunov equations on hybrid CPU-GPU platforms. Parallel Comput. **37**(8), 439–450 (2011). doi:10.1016/j.parco.2010.12.002
19. Benner, P., Kressner, D.: Fortran 77 subroutines for computing the eigenvalues of Hamiltonian matrices II. ACM Trans. Math. Softw. **32**(2), 352–373 (2006)
20. Benner, P., Laub, A.J., Mehrmann, V.: A collection of benchmark examples for the numerical solution of algebraic Riccati equations I: continuous-time case. Technical report SPC 95_22, Fakultät für Mathematik, TU Chemnitz–Zwickau, 09107 Chemnitz (1995). Available from http://www.tu-chemnitz.de/sfb393/spc95pr.html
21. Benner, P., Laub, A.J., Mehrmann, V.: A collection of benchmark examples for the numerical solution of algebraic Riccati equations II: discrete-time case. Technical report SPC 95_23, Fakultät für Mathematik, TU Chemnitz–Zwickau, 09107 Chemnitz (1995). Available from http://www.tu-chemnitz.de/sfb393/spc95pr.html
22. Benner, P., Laub, A.J., Mehrmann, V.: Benchmarks for the numerical solution of algebraic Riccati equations. IEEE Control Syst. Mag. **7**(5), 18–28 (1997)
23. Benner, P., Li, J.R., Penzl, T.: Numerical solution of large Lyapunov equations, Riccati equations, and linear-quadratic control problems. Numer. Linear Algebra Appl. **15**(9), 755–777 (2008)
24. Benner, P., Losse, P., Mehrmann, V., Voigt, M.: Numerical solution of eigenvalue problems for alternating matrix polynomials and their application in control problems for descriptor systems. Preprint MPIMD/13-24, Max Planck Institute Magdeburg (2013). http://www2.mpi-magdeburg.mpg.de/preprints/2013/MPIMD13-24.pdf
25. Benner, P., Mehrmann, V.: Collections of benchmark examples in control. WGS Newsl. **10**, 7–8 (1996)
26. Benner, P., Mehrmann, V., Sima, V., Huffel, S.V., Varga, A.: SLICOT – a subroutine library in systems and control theory. In: Datta, B.N. (ed.) Applied and Computational Control, Signals, and Circuits, vol. 1, chap. 10, pp. 499–539. Birkhäuser, Boston (1999)
27. Benner, P., Mehrmann, V., Xu, H.: A new method for computing the stable invariant subspace of a real Hamiltonian matrix. J. Comput. Appl. Math. **86**, 17–43 (1997)
28. Benner, P., Mehrmann, V., Xu, H.: A numerically stable, structure preserving method for computing the eigenvalues of real Hamiltonian or symplectic pencils. Numer. Math. **78**(3), 329–358 (1998)
29. Benner, P., Mena, H.: BDF methods for large-scale differential Riccati equations. In: De Moor, B., Motmans, B., Willems, J., Van Dooren, P., Blondel, V. (eds.) Proceedings of 16th International Symposium on Mathematical Theory of Network and Systems (MTNS 2004), Leuven, 12p. (2004)
30. Benner, P., Mena, H.: Rosenbrock methods for solving Riccati differential equations. IEEE Trans. Autom. Control **58**(11), 2950–2957 (2013)

31. Benner, P., Quintana-Ortí, E.S., Quintana-Ortí, G.: Solving linear-quadratic optimal control problems on parallel computers. Optim. Methods Softw. **23**(6), 879–909 (2008). doi:10.1080/10556780802058721
32. Benner, P., Saak, J.: Numerical solution of large and sparse continuous time algebraic matrix Riccati and Lyapunov equations: a state of the art survey. GAMM Mitt. **36**(1), 32–52 (2013). doi:10.1002/gamm.201310003
33. Benner, P., Sima, V., Voigt, M.: FORTRAN 77 subroutines for the solution of skew-Hamiltonian/Hamiltonian eigenproblems – part I: algorithms and applications. Preprint MPIMD/13-11, Max Planck Institute Magdeburg (2013). Available from http://www.mpi-magdeburg.mpg.de/preprints/2013/11/
34. Benner, P., Sima, V., Voigt, M.: FORTRAN 77 subroutines for the solution of skew-Hamiltonian/Hamiltonian eigenproblems – part II: implementation and numerical results. Preprint MPIMD/13-12, Max Planck Institute Magdeburg (2013). Available from http://www.mpi-magdeburg.mpg.de/preprints/2013/12/
35. Benner, P., Stykel, T.: Numerical solution of projected algebraic Riccati equations. SIAM J. Numer. Anal **52**(2), 581–600 (2014). doi:10.1137/130923993
36. Bini, D.A., Iannazzo, B., Meini, B.: Numerical Solution of Algebraic Riccati Equations. No. 9 in Fundamentals of Algorithms. SIAM, Philadelphia (2012)
37. Bojanczyk, A., Golub, G.H., Van Dooren, P.: The periodic Schur decomposition. Algorithms and applications. In: Luk, F.T. (ed.) Advanced Signal Processing Algorithms, Architectures, and Implementations III, San Diego. Proceeding SPIE, vol. 1770, pp. 31–42 (1992)
38. Bunch, J.R.: The weak and strong stability of algorithms in numerical algebra. Linear Algebra Appl. **88**, 49–66 (1987)
39. Bunse-Gerstner, A.: Matrix factorization for symplectic QR-like methods. Linear Algebra Appl. **83**, 49–77 (1986)
40. Bunse-Gerstner, A., Byers, R., Mehrmann, V.: Numerical methods for algebraic Riccati equations. In: Bittanti, S. (ed.) Proceeding Workshop on the Riccati Equation in Control, Systems, and Signals, Como, pp. 107–116 (1989)
41. Bunse-Gerstner, A., Mehrmann, V.: A symplectic QR-like algorithm for the solution of the real algebraic Riccati equation. IEEE Trans. Autom. Control **31**, 1104–1113 (1986)
42. Bunse-Gerstner, A., Mehrmann, V., Watkins, D.: An SR algorithm for Hamiltonian matrices based on Gaussian elimination. Methods Oper. Res. **58**, 339–358 (1989)
43. Byers, R.: Hamiltonian and symplectic algorithms for the algebraic Riccati equation. Ph.D. thesis, Cornell University, Department of Computer Science, Ithaca (1983)
44. Byers, R.: A Hamiltonian QR-algorithm. SIAM J. Sci. Stat. Comput. **7**, 212–229 (1986)
45. Byers, R.: Solving the algebraic Riccati equation with the matrix sign function. Linear Algebra Appl. **85**, 267–279 (1987)
46. Byers, R., He, C., Mehrmann, V.: The matrix sign function method and the computation of invariant subspaces. SIAM J. Matrix Anal. Appl. **18**(3), 615–632 (1997)
47. Byers, R., Mehrmann, V.: Symmetric updating of the solution of the algebraic Riccati equation. Methods Oper. Res. **54**, 117–125 (1985)
48. Choi, C., Laub, A.J.: Efficient matrix-valued algorithms for solving stiff Riccati differential equations. IEEE Trans. Autom. Control **35**, 770–776 (1990)
49. Chu, D., Liu, X., Mehrmann, V.: A numerical method for computing the Hamiltonian Schur form. Numer. Math. **105**(3), 375–412 (2006)
50. Chu, E.K.W., Fan, H.Y., Lin, W.W.: A structure-preserving doubling algorithm for continuous-time algebraic Riccati equations. Linear Algebra Appl. **396**, 55–80 (2005)
51. Chu, E.K.W., Fan, H.Y., Lin, W.W., Wang, C.S.: Structure-preserving algorithms for periodic discrete-time algebraic Riccati equations. Int. J. Control **77**(8), 767–788 (2004)
52. Datta, B.N.: Numerical Methods for Linear Control Systems. Elsevier Academic, Amsterdam/Boston (2004)
53. Dieci, L.: Numerical integration of the differential Riccati equation and some related issues. SIAM J. Numer. Anal. **29**(3), 781–815 (1992)
54. Dorf, R.C.: Modern Control Systems, 2nd edn. Addison Wesley, Reading (1974)

55. Flaschka, U., Mehrmann, V., Zywietz, D.: An analysis of structure preserving methods for symplectic eigenvalue problems. RAIRO Autom. Prod. Inform. Ind. **25**, 165–190 (1991)
56. Golub, G.H., Van Loan, C.F.: Matrix Computations, 3rd edn. Johns Hopkins University Press, Baltimore (1996)
57. Green, M., Limebeer, D.J.N.: Linear Robust Control. Prentice-Hall, Englewood Cliffs (1995)
58. Heiland, J.: Decoupling and optimization of differential-algebraic equations with application in flow control. Ph.D. thesis, Fakultät II – Mathematik und Naturwissenschaften, TU Berlin (2014)
59. Hench, J.J., Laub, A.J.: An extension of the QR algorithm for a sequence of matrices. Technical report CCEC-92-0829, ECE Department, University of California, Santa Barbara (1992)
60. Hench, J.J., Laub, A.J.: Numerical solution of the discrete-time periodic Riccati equation. IEEE Trans. Autom. Control **39**, 1197–1210 (1994)
61. Horn, R.A., Johnson, C.R.: Topics in Matrix Analysis. Cambridge University Press, Cambridge (1991)
62. Kunkel, P., Mehrmann, V.: Numerical solution of Riccati differential algebraic equations. Linear Algebra Appl. **137/138**, 39–66 (1990)
63. Kunkel, P., Mehrmann, V.: The linear quadratic control problem for linear descriptor systems with variable coefficients. Math. Control Signals Syst. **10**, 247–264 (1997)
64. Kunkel, P., Mehrmann, V.: Analysis of over- and underdetermined nonlinear differential-algebraic systems with application to nonlinear control problems. Math. Control Signals Syst. **14**, 233–256 (2001)
65. Kunkel, P., Mehrmann, V.: Differential-Algebraic Equations: Analysis and Numerical Solution. Textbooks in Mathematics. EMS Publishing House, Zürich (2006)
66. Kunkel, P., Mehrmann, V.: Optimal control for unstructured nonlinear differential-algebraic equations of arbitrary index. Math. Control Signals Syst. **20**(3), 227–269 (2008)
67. Kunkel, P., Mehrmann, V., Rath, W.: Analysis and numerical solution of control problems in descriptor form. Math. Control Signals Syst. **14**, 29–61 (2001)
68. Lancaster, P., Rodman, L.: The Algebraic Riccati Equation. Oxford University Press, Oxford (1995)
69. Lancaster, P., Tismenetsky, M.: The Theory of Matrices, 2nd edn. Academic, Orlando (1985)
70. Lang, N., Mena, H., Saak, J.: On the benefits of the LDL^T factorization for large-scale differential matrix equation solvers. Preprint MPIMD/14-14, Max Planck Institute Magdeburg (2014). Available from http://www.mpi-magdeburg.mpg.de/preprints/
71. Laub, A.J.: A Schur method for solving algebraic Riccati equations. IEEE Trans. Autom. Control **24**, 913–921 (1979)
72. Lin, W.W., Mehrmann, V., Xu, H.: Canonical forms for Hamiltonian and symplectic matrices and pencils. Linear Algebra Appl. **301–303**, 469–533 (1999)
73. Losse, P., Mehrmann, V., Poppe, L.K., Reis, T.: The modified optimal H_∞ control problem for descriptor systems. SIAM J. Control Optim. **47**(6), 2795–2811 (2008)
74. Mehrmann, V.: The linear quadratic control problem: theory and numerical algorithms. Habilitationsschrift, Universität Bielefeld, Bielefeld (1987)
75. Mehrmann, V.: A symplectic orthogonal method for single input or single output discrete time optimal linear quadratic control problems. SIAM J. Matrix Anal. Appl. **9**, 221–248 (1988)
76. Mehrmann, V.: The Autonomous Linear Quadratic Control Problem, Theory and Numerical Solution. Lecture Notes in Control and Information Sciences, vol. 163. Springer, Heidelberg (1991)
77. Mehrmann, V.: A step toward a unified treatment of continuous and discrete time control problems. Linear Algebra Appl. **241–243**, 749–779 (1996)
78. Mehrmann, V., Poloni, F.: An inverse-free ADI algorithm for computing Lagrangian invariant subspaces. Preprint 14-2014, Institut für Mathematik, TU Berlin (2014)
79. Mehrmann, V., Schröder, C., Watkins, D.: A new block method for computing the Hamiltonian Schur form. Linear Algebra Appl. **431**(3–4), 350–368 (2009)
80. Mehrmann, V., Tan, E.: Defect correction methods for the solution of algebraic Riccati equations. IEEE Trans. Autom. Control **33**, 695–698 (1988)

81. Mehrmann, V., Xu, H.: Explicit solutions for a Riccati equation from transport theory. SIAM J. Matrix Anal. Appl. **30**(4), 1339–1357 (2008)
82. Paige, C.C., Van Loan, C.F.: A Schur decomposition for Hamiltonian matrices. Linear Algebra Appl. **41**, 11–32 (1981)
83. Penzl, T.: Numerische Lösung großer Lyapunov-Gleichungen. Logos–Verlag, Berlin, Germany (1998). Dissertation, Fakultät für Mathematik, TU Chemnitz (1998)
84. Petkov, P.H., Konstantinov, M.M., Gu, D.W., Mehrmann, V.: Numerical solution of matrix Riccati equations: a comparison of six solvers. SLICOT Working Note SLWN1999–10, (1999). Available from www.slicot.org
85. Petkov, P.H., Konstantinov, M.M., Mehrmann, V.: DGRSVX and DMSRIC: Fortran 77 subroutines for solving continuous-time matrix algebraic Riccati equations with condition and accuracy estimates. Technical report SFB393/98-16, Fakultät für Mathematik, TU Chemnitz, 09107 Chemnitz (1998)
86. Reid, W.T.: Riccati Differential Equations. Academic, New York (1972)
87. Van Loan, C.F.: A symplectic method for approximating all the eigenvalues of a Hamiltonian matrix. Linear Algebra Appl. **61**, 233–251 (1984)
88. Voigt, M.: On linear-quadratic optimal control and robustness of differential-algebraic systems. Ph.D. thesis, Faculty for Mathematics, Otto-von-Guericke University, Magdeburg (2015)
89. Watkins, D.S.: On the reduction of a Hamiltonian matrix to Hamiltonian Schur form. Electron. Trans. Numer. Anal. **23**, 141–157 (2006)
90. Zhou, K., Doyle, J.C., Glover, K.: Robust and Optimal Control. Prentice-Hall, Upper Saddle River (1996)

Chapter 5
Permuted Graph Matrices and Their Applications

Federico Poloni

Abstract A *permuted graph matrix* is a matrix $V \in \mathbb{C}^{(m+n)\times m}$ such that every row of the $m \times m$ identity matrix I_m appears at least once as a row of V. Permuted graph matrices can be used in some contexts in place of orthogonal matrices, for instance when giving a basis for a subspace $\mathcal{U} \subseteq \mathbb{C}^{m+n}$, or to normalize matrix pencils in a suitable sense. In these applications the permuted graph matrix can be chosen with bounded entries, which is useful for stability reasons; several algorithms can be formulated with numerical advantage with permuted graph matrices. We present the basic theory and review some applications from optimization or in control theory.

5.1 Introduction

A *graph matrix* is a matrix of the form

$$\mathcal{G}(X) := \begin{bmatrix} I_m \\ X \end{bmatrix}, \quad X \in \mathbb{C}^{n\times m},$$

where I_m is the $m \times m$ identity matrix. The name comes from the set-theoretical definition of graph of a function f as the set of pairs $(x, f(x))$. The image $\operatorname{im} \mathcal{G}(X)$ of a graph matrix is sometimes called *graph subspace*; however, this is improper, since "graph-ness" is a property of the basis matrix $\mathcal{G}(X)$, not of the subspace. Indeed, almost every subspace is a graph subspace: let

$$U = \begin{bmatrix} E \\ A \end{bmatrix}, \quad E \in \mathbb{C}^{m\times m}, \quad A \in \mathbb{C}^{n\times m} \tag{5.1}$$

be any matrix with full column rank; whenever E is invertible, we have $U = \mathcal{G}(AE^{-1})E$, and hence $\operatorname{im} U = \operatorname{im} \mathcal{G}(AE^{-1})$, because post-multiplying by E^{-1} does not change the column space.

F. Poloni (✉)
Dipartimento di Informatica, Università di Pisa, Largo B. Pontecorvo, 3, I–56127 Pisa, Italy
e-mail: fpoloni@di.unipi.it

P. Benner et al. (eds.), *Numerical Algebra, Matrix Theory, Differential-Algebraic Equations and Control Theory*, DOI 10.1007/978-3-319-15260-8_5

It will be useful to introduce a notation that avoids the dependence on the change of basis matrix E. Given $U, V \in \mathbb{C}^{(m+n)\times m}$ with full column rank, we write $U \sim V$ to mean that there exists an invertible $S \in \mathbb{C}^{m\times m}$ such that $U = VS$. In other words, this is the equivalence relation "U has the same column space as V".

Note that, in the above setting, given a generic matrix U, computing $X = AE^{-1}$ is not a good idea numerically, since its top $m \times m$ block E could be ill-conditioned or even singular. A modification of this approach is the following: instead of requiring an identity submatrix in the top block, we can ask for a subset of the rows that, when taken in some order, forms an identity matrix. More formally, we call a matrix $V \in \mathbb{C}^{(m+n)\times m}$ a *permuted graph matrix* if there exist $X \in \mathbb{C}^{n\times m}$ and a permutation matrix $P \in \mathbb{R}^{(m+n)\times(m+n)}$ such that $V = P\mathcal{G}(X)$. This is equivalent to requiring that every row of I_m occurs at least once as a row of V.

It is easy to prove that every subspace is spanned by the columns of a permuted graph matrix. Indeed, let the columns of $U \in \mathbb{C}^{(m+n)\times m}$ form a basis for a given subspace; U must then have full column rank, that is, it must contain an invertible submatrix $E \in \mathbb{C}^{m\times m}$. We can choose a permutation matrix P such that $U = P\left[\begin{smallmatrix} E \\ A \end{smallmatrix}\right]$, with $A \in \mathbb{C}^{n\times m}$. Then, $U \sim P\mathcal{G}(X)$ with $X = AE^{-1}$. We call $P\mathcal{G}(X)$ a *permuted graph representation* of U, or of its column space.

A more interesting result is the following, which shows that we can always find a basis matrix in the form $P\mathcal{G}(X)$ with the additional property that X is bounded in a suitable sense.

Theorem 1 ([22, 31]) *Let $U \in \mathbb{C}^{(m+n)\times m}$ be a matrix with full column rank. Then, there exist $X \in \mathbb{C}^{n\times m}$ and a permutation matrix $P \in \mathbb{R}^{(m+n)\times(m+n)}$ such that $U \sim P\mathcal{G}(X)$ and $\|X\|_{\max} \leq 1$.*

We have used the notation $\|X\|_{\max} := \max_{i,j} |x_{ij}|$, where x_{ij} are the entries of the matrix X; essentially, the theorem states that all the entries of X are bounded in modulus by 1.

In this chapter, we focus on Theorem 1, its extension to Lagrangian subspaces, and the applications of these two results. There are several contexts in numerical linear algebra and in control theory in which it is useful to work with the pair (P, X) as a representation of the subspace $\operatorname{im} U$; we review briefly these applications and the underlying theory.

The chapter is organized as follows. We describe an efficient algorithm for the computation of a permuted graph matrix $P\mathcal{G}(X) \sim U$ in Sect. 5.2; in Sect. 5.3, we present a result regarding their conditioning and introduce two different applications of these matrices in optimization. Another application, skeleton approximation of large-scale matrices, is discussed in Sect. 5.4. A structured version of this technique is presented in Sect. 5.5; in Sect. 5.6 we show how permuted graph representations can be used to work with matrix pencils. In Sect. 5.7 we introduce briefly numerical methods for a standard problem in control theory, constant-coefficient linear-quadratic control, and in Sects. 5.8 and 5.9 we show how two of these algorithms can be improved with the use of permuted graph matrices. Lastly, Sect. 5.10 discusses open issues and research problems.

5.2 Computing Permuted Graph Bases

We start our review by discussing the computation of permuted graph bases in practice. The idea of the proof of Theorem 1 is the following. Choose as E the $m \times m$ submatrix of U that maximizes $|\det E|$ (*maximum-volume submatrix*). Using Cramer's rule on the linear system $XE = A$, one can write $|x_{ij}| = \frac{|\det E'|}{|\det E|}$, for a suitable $m \times m$ submatrix E' of U (depending on i, j), hence the result follows.

Unfortunately, finding an E with this maximizing property is an NP-hard problem [13], so this construction is not useful computationally. We can use instead an iterative algorithm that resembles a lot the so-called "canonical tableaux" implementation of the simplex algorithm [14], in that we update at each step an active set of rows containing an identity submatrix. This procedure is described in [22, 31, 35]; we present it as Algorithm 1. The method produces a permuted graph representation in which each entry of X is bounded in magnitude by a parameter τ. It is advised to choose $\tau > 1$ (for instance $\tau = 2$), to avoid numerical troubles with entries that are exactly 1 and to get faster convergence.

Algorithm 1: Obtaining a permuted graph representation with $\|X\|_{\max} \leq \tau$ [22, 31, 35]

Input: $U \in \mathbb{C}^{(m+n)\times m}$ with full column rank; a threshold value $\tau \geq 1$; an initial permutation P_0 such that the top m rows of $P_0^T U$ form an invertible matrix

Output: A permutation matrix $P \in \mathbb{C}^{(m+n)\times(m+n)}$ and $X \in \mathbb{C}^{n\times m}$ such that $U \sim P\mathcal{G}(X)$ and $\|X\|_{\max} \leq \tau$

Let $P = P_0$, $\left[\begin{smallmatrix} E \\ A \end{smallmatrix}\right] = P^T U$, and $X = AE^{-1}$;

while $\|X\|_{\max} > \tau$ **do**

  take a pair (i, j) such that $|x_{ij}| > \tau$;

  let $P' = P\Pi$, where $\Pi \in \mathbb{C}^{(m+n)\times(m+n)}$ is the permutation that exchanges j and $m + i$;

  find $X' \in \mathbb{C}^{n\times m}$ such that $P\mathcal{G}(X) \sim P'\mathcal{G}(X')$;

  replace (X, P) with (X', P') and continue;

end

In practice, the permutation P can be stored as a sequence of $m+n$ integers, and all the needed operations on it can be performed on a computer in $O(m + n)$ time and space.

The computation of X' in Algorithm 1 can be performed efficiently as well. Here and in the following we use the notation $X_{\mathcal{I}\mathcal{J}}$ to denote the submatrix of $X \in \mathbb{C}^{n\times m}$ containing the rows with indices $\mathcal{I}$ and columns with indices $\mathcal{J}$, where $\mathcal{I}$ (resp. $\mathcal{J}$) is a tuple of distinct indices in $\{1, 2, \ldots, n\}$ (resp. $\{1, 2, \ldots, m\}$). Moreover, we denote by $\mathcal{I}^c$ a tuple composed of all the row (or column) indices that do not belong to $\mathcal{I}$, and with a colon ':' (as in many computer languages) the whole set of admissible row/column indices.

Lemma 1 ([35, Lemma 4.1]) *Let a permutation matrix* $P \in \mathbb{C}^{(m+n)\times(m+n)}$ *and* $X \in \mathbb{C}^{n\times m}$ *be given. Let* $\mathcal{I} = (i_1, i_2, \dots, i_\ell)$ *be distinct elements of* $\{1, 2, \dots, n\}$ *and* $\mathcal{J} = (j_1, j_2, \dots, j_\ell)$ *be distinct elements of* $\{1, 2, \dots, m\}$. *Let* $P' = P\Pi$, *where* Π *is the permutation that swaps* j_k *with* $m + i_k$, *for each* $k = 1, 2, \dots, \ell$, *and leaves everything else unchanged. A matrix* $X' \in \mathbb{C}^{n\times m}$ *such that* $P\mathcal{G}(X) \sim P'\mathcal{G}(X')$ *exists if and only if* $X_{\mathcal{I}\mathcal{J}}$ *is invertible, and in that case it is given by*

$$X' = \begin{bmatrix} X'_{\mathcal{I}\mathcal{J}} & X'_{\mathcal{I}\mathcal{J}^c} \\ X'_{\mathcal{I}^c\mathcal{J}} & X'_{\mathcal{I}^c\mathcal{J}^c} \end{bmatrix} = \begin{bmatrix} (X_{\mathcal{I}\mathcal{J}})^{-1} & -(X_{\mathcal{I}\mathcal{J}})^{-1}X_{\mathcal{I}\mathcal{J}^c} \\ X_{\mathcal{I}^c\mathcal{J}}(X_{\mathcal{I}\mathcal{J}})^{-1} & X_{\mathcal{I}^c\mathcal{J}^c} - X_{\mathcal{I}^c\mathcal{J}}(X_{\mathcal{I}\mathcal{J}})^{-1}X_{\mathcal{I}\mathcal{J}^c} \end{bmatrix}.$$

Lemma 1 shows how to update a permuted graph representation $P\mathcal{G}(X)$ when we change the set of rows where the identity submatrix is located. The operation needed in Algorithm 1 corresponds to the case in which $\mathcal{I} = \{i\}$ and $\mathcal{J} = \{j\}$ have one element.

The map $X \mapsto X'$ appears in other applications as well and is known as *principal pivot transform* [47].

As an initial permutation, in absence of better guesses, one can take the permutation P produced by a rank-revealing QR factorization $U^H = QRP$ [21, Section 5.4.1]. With this choice, one can prove (when $\tau > 1$) that the algorithm terminates in $O(n \log_\tau n)$ steps, with a total cost of $O(n^3 \log_\tau n)$ floating point operations, and converges to a *local* maximizer of $|\det E|$ (that is, a submatrix E such that $|\det E| \geq |\det E'|$ for each other submatrix E' differing from E only by a single row). Moreover, the determinant of the top $m \times m$ submatrix of $P^T U$ increases by a factor greater than τ at each step. In practice, the number of steps is often much lower than the bound, and in many small-scale cases the P coming from rank-revealing QR already gives an X with $\|X\|_{\max} \leq 1$. Indeed, finding the submatrix E with maximum volume $|\det E|$ is a problem that can be explicitly related to the computation of rank-revealing factorizations [42].

Another area of mathematics where these submatrix determinants appear is algebraic geometry: given $U \in \mathbb{C}^{(m+n)\times n}$ with full column rank, the determinants of all possible $\binom{m+n}{m}$ subset of rows (each subset ordered, for instance, in increasing order) are called *Plücker (projective) coordinates* of the subspace $\operatorname{im} U$. Indeed, if we have two matrices spanning the same subspace, U and $V = UE \sim U$, their Plücker coordinates differ only by a common factor $\det E$, and one can show that the converse holds, that is, matrices with the same Plücker coordinates up to a common multiple are equivalent according to $\sim$ and span the same subspace.

5.3 Conditioning of Subspaces and Applications in Optimization

Given a matrix U, its condition number $\kappa(U) := \frac{\sigma_{\max}(U)}{\sigma_{\min}(U)}$ (where $\sigma_{\min}(U)$ and $\sigma_{\max}(U)$ are its minimum and maximum singular value, respectively) measures the sensitivity of its column space $\operatorname{im} U$ with respect to perturbations [46, Page 154].

Hence, if we wish to perform computations involving a subspace, a natural way to operate on it is through an orthogonal basis U_O, for which $\kappa(U_O) = 1$. Suppose that we decide instead to use a basis $U_G = P\mathscr{G}(X)$ which is a permuted graph matrix. How well does it fare with respect to this measure? The answer is given by the following result.

Theorem 2 ([35]) *Let $U_G = P\mathscr{G}(X) \in \mathbb{C}^{(m+n)\times m}$, where the elements x_{ij} of X satisfy the inequality $|x_{ij}| \leq \tau$ for a certain $\tau \geq 1$. Then, $\kappa(U_G) \leq \sqrt{1 + mn\tau^2}$.*

Proof Because of the identity submatrix, $\|U_G v\|_2 \geq \|v\|_2$ for each $v \in \mathbb{C}^m$, hence $\sigma_{\min}(U_G) \geq 1$. On the other hand,

$$\sigma_{\max}(P\mathscr{G}(X)) = \|P\mathscr{G}(X)\|_2 = \|\mathscr{G}(X)\|_2 \leq \sqrt{\|I_m\|_2^2 + \|X\|_2^2} \leq \sqrt{1 + mn\tau^2}. \quad \square \tag{5.2}$$

The condition number $\kappa(U_G)$ is not as small as the perfect $\kappa(U_O) = 1$ of an orthogonal basis, but still it can be explicitly bounded by a linear function of the dimensions and of the chosen threshold τ. A permuted graph basis can hence be used to represent a subspace in a stable way and to operate on it.

Are there contexts in which there is an advantage in using a permuted graph basis U_G rather than an orthogonal one U_O? We sketch two applications here, taken from [48] and [3], respectively. More examples will appear in the next sections.

A problem encountered in optimization is the maximization (or minimization) of functions on the *Grassmann manifold* [1, 48], i.e., the set of m-dimensional subspaces of $\mathbb{C}^{m+n}$. In practice, this means maximizing a given function $f : \mathbb{C}^{(m+n)\times m} \mapsto \mathbb{R}$ such that $f(U) = f(V)$ whenever $U \sim V$. Working with orthogonal bases may lead to some difficulties. First, the parametrization of the Grassmann manifold via orthogonal matrices is ambiguous, since the relation between an orthogonal matrix $U \in \mathbb{C}^{(m+n)\times m}$ and its spanned subspace is not one-to-one. As a consequence, the gradient ∇f is always zero in some directions, and the optimization problem in this formulation is never strictly convex. Moreover, in most iterative algorithms, it is difficult to enforce orthogonality of the next iterate explicitly, so a typical algorithm will make an update in a general direction in $\mathbb{C}^{(m+n)\times m}$ and then restore orthogonality at a later time via projection.

Neither of these problems is unsolvable, and there are now mature algorithms for optimization on matrix manifolds [1]. Nevertheless, using permuted graph bases rather than orthogonal bases allows for a simplification of the problem. The maps $g_P(X) := X \mapsto P\mathscr{G}(X)$ are one-to-one local charts and together constitute an atlas of the Grassmann manifold, so they can be used to reduce the problem to a standard multivariate optimization problem on the space $\mathbb{C}^{nm}$. In practice, one defines for each permutation matrix the auxiliary map $f_P : \mathbb{C}^{n\times m} \to \mathbb{R}$ as $f_P(X) := f(P\mathscr{G}(X))$, and uses a traditional multivariate optimization algorithm on it. We sketch a method, originally from [48], in Algorithm 2: at each step, we check if the entries of the current iterate X have magnitude greater than τ, and if so, we update the permutation. Changing the permutation P with Algorithm 1 is needed in few iterations only, and the previous value of P is a good initial guess, so its cost is typically much less than the generic $O(n^3 \log_\tau n)$.

Algorithm 2: Optimization on the Grassmann manifold [48]

Input: A function $f : \mathbb{C}^{(m+n)\times m} \to \mathbb{R}$ such that $f(U) = f(V)$ whenever $U \sim V$; an initial value $U \in \mathbb{C}^{(m+n)\times m}$; a threshold $\tau > 1$
Output: A (possibly local) minimum of f
Find a permuted graph representation $U \sim P\mathscr{G}(X)$ (with threshold τ);
while *X is not a local minimum of f_P* **do**
  apply one step of a multivariate optimization algorithm (gradient descent, Newton, BFGS...) to f_P, starting from X, obtaining a new point X';
  if $\|X'\|_{\max} > \tau$ **then**
    Use Algorithm 1 to find a permuted graph representation $P''\mathscr{G}(X'')$ of $P\mathscr{G}(X')$, with threshold τ;
    replace (X, P) with (X'', P'') and continue;
  else
    replace X with X' and continue;
  end
end

A different context in optimization in which suitable permutations and graph forms have appeared recently is the preconditioning and solution of large-scale saddle-point problems [3, 15, 16]. We present here the preconditioner for least-squares problems appearing in [3]. A least-squares problem $\min_{x\in\mathbb{C}^m} \|Ux - b\|$, for $U \in \mathbb{C}^{(m+n)\times m}$, can be reformulated as the augmented system

$$\begin{bmatrix} I_{m+n} & U \\ U^H & 0 \end{bmatrix} \begin{bmatrix} r \\ x \end{bmatrix} = \begin{bmatrix} b \\ 0 \end{bmatrix}.$$

Let us take a permuted graph basis $U = P\begin{bmatrix} E \\ A \end{bmatrix} \sim P\mathscr{G}(X)$, with $X = AE^{-1}$; permuting the first $m+n$ entries and partitioning $Pr = \begin{bmatrix} r_E \\ r_A \end{bmatrix}$ and $Pb = \begin{bmatrix} b_E \\ b_A \end{bmatrix}$ conformably with $\begin{bmatrix} E \\ A \end{bmatrix}$, we get the equivalent system

$$\begin{bmatrix} I_m & 0 & E \\ 0 & I_n & A \\ E^H & A^H & 0 \end{bmatrix} \begin{bmatrix} r_E \\ r_A \\ x \end{bmatrix} = \begin{bmatrix} b_E \\ b_A \\ 0 \end{bmatrix}.$$

Finally, eliminating the variables r_E from this system and multiplying by $Q = \begin{bmatrix} I_n & 0 \\ 0 & E^{-H} \end{bmatrix}$ and Q^H on the two sides, one gets the equivalent reduced system

$$\begin{bmatrix} I_n & X \\ X^H & -I_m \end{bmatrix} \begin{bmatrix} r_A \\ Ex \end{bmatrix} = \begin{bmatrix} b_A \\ -b_E \end{bmatrix}.$$

The condition number of this linear system equals $\kappa(P\mathscr{G}(X))$ (see [3, p. 4]), and thus it can be bounded using Theorem 2. In practice, the matrix Q above is applied as a preconditioner; hence, to get faster convergence of preconditioned iterative methods, it is useful to choose a permuted graph basis with a small $\kappa(P\mathscr{G}(X))$. The authors in [3] suggest useful heuristic methods to find one for a large and sparse U.

5.4 Skeleton Approximation

In this section, we consider the problem of finding or approximating $\|M\|_{\max}$ for a large-scale matrix M, not necessarily sparse.

Let $M \in \mathbb{C}^{n\times m}$, and $\mathcal{I}, \mathcal{J}$ be two tuples of ℓ pairwise distinct row and column indices respectively. If $M_{\mathcal{I}\mathcal{J}}$ is invertible, the matrix

$$M_S = M_{:\mathcal{J}} M_{\mathcal{I}\mathcal{J}}^{-1} M_{\mathcal{I}:} \tag{5.3}$$

is called *skeleton approximation* of M along $(\mathcal{I}, \mathcal{J})$ [23], and has the same entries as M on the rows belonging to $\mathcal{I}$ and the columns belonging to $\mathcal{J}$. Moreover, whenever $\operatorname{rank} M \leq \ell$ we have $M_S = M$. If $\mathcal{I}$ and $\mathcal{J}$ are chosen so that $|\det M_{\mathcal{I}\mathcal{J}}|$ is the maximum over all $\ell \times \ell$ submatrices, then one can prove specific approximation properties for the extremal values of M.

Theorem 3 ([22, 23]) *Let $M \in \mathbb{C}^{n\times m}$ and $\ell \leq \min(m, n)$ be given; let $\mathcal{I}, \mathcal{J}$ be ℓ-tuples of pairwise distinct indices chosen so that $|\det M_{\mathcal{I}\mathcal{J}}|$ is maximal and M_S be the skeleton approximation* (5.3). *Then,*

1. $\|M - M_S\|_{\max} \leq (\ell+1)\sigma_{\ell+1}$, *where $\sigma_{\ell+1}$ is the $(\ell+1)$st singular value of M;*
2. $\|M_{\mathcal{I}\mathcal{J}}\|_{\max} \geq \|M\|_{\max}/(2\ell^2+\ell)$.

As stated above, finding a maximum volume submatrix is an NP complete problem already in the case $\ell = m$, so in practice one must resort to heuristics and approximations. A possible implementation, using alternating optimization on $\mathcal{I}$ and $\mathcal{J}$, is given in Algorithm 3. As in Algorithm 1, termination is ensured by the fact

---

Algorithm 3: Alternating optimization algorithm for skeleton approximation.

Input: A matrix $M \in \mathbb{C}^{n\times m}$, possibly sparse or given implicitly as a procedure that returns single entries, rows of columns; an initial guess $\mathcal{J}$; a threshold $\tau \geq 1$

Output: ℓ-tuples of row and column indices $\mathcal{I}, \mathcal{J}$ such that $\|M_{:\mathcal{J}}(M_{\mathcal{I}\mathcal{J}})^{-1}\|_{\max} \leq \tau$ and $\|(M_{\mathcal{I}\mathcal{J}})^{-1}M_{\mathcal{I}:}\|_{\max} \leq \tau$

repeat
- apply Algorithm 1 to $M_{:\mathcal{J}}$, producing a new index set $\mathcal{I}$ that maximizes $|\det M_{\mathcal{I}\mathcal{J}}|$;
- apply Algorithm 1 to $M_{\mathcal{I}:}^H$, producing a new index set $\mathcal{J}'$ that maximizes $|\det M_{\mathcal{I}\mathcal{J}'}|$;
- replace $\mathcal{J}$ with $\mathcal{J}'$ and continue;

until *$\mathcal{I}$ and $\mathcal{J}$ stop changing*;

---

that $|\det M_{\mathcal{I}\mathcal{J}}|$ increases monotonically by a factor larger than τ. As initial guess, one can take for instance a random $\mathcal{J}$, or start with a permuted graph representation of MV for a suitably chosen random full-rank $V \in \mathbb{C}^{m\times\ell}$.

Note that the procedure works on few rows and columns of M at a time, and in fact typically it will not even access many of its entries. Nevertheless, in practice $\|M_{\mathcal{I}\mathcal{J}}\|_{\max}$ is a good approximation of $\|M\|_{\max}$ in many real-world cases where the singular values of M decay sufficiently fast [22]. This method has been used in cases in which the entries of M can be efficiently generated one-by-one, or one row/column at a time; for instance, they might be the values of a bivariate function $f(x, y)$ on a huge grid. Generalizations to problems in more than two variables and tensor approximations can be devised using the same ideas; see, e.g., [41, 45]. This method, in combination with efficient tensor storage techniques, allows for the treatment of massively large maximization/minimization problems, with applications to many computationally challenging problems in quantum physics, computational chemistry and biology.

5.5 Permuted Graph Bases for Lagrangian Subspaces

An n-dimensional subspace $\mathscr{U}$ of $\mathbb{C}^{2n}$ is called *Lagrangian* if $u^H J_{2n} v = 0$ for every $u, v \in \mathscr{U}$, where $J_{2n} := \left[\begin{smallmatrix} 0 & I_n \\ -I_n & 0 \end{smallmatrix}\right]$. Lagrangian subspaces appear naturally in systems and control theory, as we discuss later in Sect. 5.7.

Given $U \in \mathbb{C}^{2n\times n}$ of full column rank, $\operatorname{im} U$ is Lagrangian if and only if $U^H J_{2n} U = 0$. When $U = \mathscr{G}(X)$ is a graph basis, this expands to $X = X^H$, that is, $\operatorname{im} \mathscr{G}(X)$ is Lagrangian if and only if X is Hermitian. The same property does not hold for *permuted* graph bases, though; to recover it, we have to alter the definition to adapt it to this structured case. For $i = 1, 2, \ldots, n$, let

$$S_i = \left[\begin{array}{ccc|ccc} I_{i-1} & & & 0_{i-1} & & \\ & 0 & & & -1 & \\ & & I_{n-i} & & & 0_{n-i} \\ \hline 0_{i-1} & & & I_{i-1} & & \\ & 1 & & & 0 & \\ & & 0_{n-i} & & & I_{n-i} \end{array}\right],$$

where 0_k denotes the zero matrix of size $k \times k$; that is, S_i is the $2n \times 2n$ matrix that acts as $-J_2$ on the i-th and $n+i$-th component of a vector (swapping them and changing sign to one of them) and as the identity matrix on all other components. Clearly the S_i all commute. Let us consider the set of all 2^n possible products that we can build by taking a (possibly empty) subset of them,

$$\mathscr{S}_{2n} := \{S_{i_1} S_{i_2} \cdots S_{i_\ell} \mid 1 \le i_1 < i_2 < \ldots < i_\ell \le n\}, \quad \ell \in \{0, 1, \ldots, n\}.$$

The identity matrix I_{2n} and $-J_{2n}$ are contained in the set, corresponding to the trivial subsets. All the matrices in $\mathscr{S}_{2n}$ are orthogonal and *symplectic* (i.e., they satisfy $S^H J_{2n} S = J_{2n}$), and they are up to sign changes a subgroup of the permutation matrices (the one generated by the transpositions $(i, n+i)$). We call these matrices *symplectic swaps*.

A symplectic swap can be stored as the subset $\{i_1, i_2, \dots, i_\ell\}$, memorized for instance as a length-n binary vector; all the operations that we need can be easily performed on a computer in $O(n)$ space and time.

Using these matrices in place of the permutations, we can build an analogue of the theory of permuted graph bases for symplectic subspaces. Given $U \in \mathbb{C}^{2n\times n}$ and a symplectic swap $S \in \mathscr{S}_{2n}$, whenever the top $n \times n$ submatrix E of $S^T U = \left[\begin{smallmatrix} E \\ A \end{smallmatrix}\right]$ is nonsingular, we can form $X = AE^{-1}$ so that $U \sim S\mathscr{G}(X)$. Using the symplecticity of S, it is easy to check that $\operatorname{im} U$ is Lagrangian if and only if $X = X^H$; hence, if $X = X^H$ for some choice of $S \in \mathscr{S}_{2n}$, then the same property holds for all possible choices.

Since there are only 2^n symplectic swaps, less than the number of essentially different $n\times n$ submatrices of U, it is already nontrivial to see that for any $U \in \mathbb{C}^{2n\times n}$ with full column rank there exists at least one choice of S that gives an invertible E, let alone one with bounded X. Nevertheless, the following result holds.

Theorem 4 ([17, 35]) *Let $U \in \mathbb{C}^{2n\times n}$ have full column rank and satisfy $U^H J_{2n} U = 0$ (i.e., $\operatorname{im} U$ is Lagrangian). Then,*

1. *There exists $S \in \mathscr{S}_{2n}$ so that the top $n \times n$ submatrix E of $S^T U = \left[\begin{smallmatrix} E \\ A \end{smallmatrix}\right]$ is nonsingular, and hence $U \sim S\mathscr{G}(X)$ with $X = X^H = AE^{-1}$.*
2. *There exists $S \in \mathscr{S}_{2n}$ so that the above property holds, and $\|X\|_{\max} \leq \sqrt{2}$.*

Item 1 appeared in [17], and Item 2 in [35]; indeed, one can find X with $|x_{ij}| \leq 1$ when $i = j$ and $|x_{ij}| \leq \sqrt{2}$ otherwise, which is a slightly stronger condition.

The proof of Item 2 is similar to the one for unstructured case: one looks for the symplectic swap S that maximizes $|\det E|$, where $S^T U = \left[\begin{smallmatrix} E \\ A \end{smallmatrix}\right]$. Similarly, for each $\tau \geq \sqrt{2}$ there is an iterative optimization algorithm with complexity $O(n^3 \log_{(\tau^2-1)^{1/2}} n)$ flops which produces a permuted Lagrangian graph representation $U \sim S\mathscr{G}(X)$ with $\|X\|_{\max} = \tau$. As a starting permutation, one can take the S originating from a variant of the rank-revealing QR factorization in which the third term is a symplectic swap rather than a permutation. The proof and the algorithm use ideas similar to the ones in the unstructured case; we refer the reader to [35] for more detail. Here we report only the analogue of Lemma 1, which gives an interesting symmetric variant of the principal pivot transform.

Lemma 2 *Let $S \in \mathscr{S}_{2n}$ be a symplectic swap and $X = X^H \in \mathbb{C}^{n\times n}$. Let $\mathcal{I} = (i_1, i_2, \dots, i_\ell)$ be given, where the i_k are distinct elements of $\{1, 2, \dots, n\}$. Define $S' = SS_{i_1} S_{i_2} \cdots S_{i_\ell}$. Then there exists a matrix $X' = (X')^H \in \mathbb{C}^{n\times n}$ such that $S\mathscr{G}(X) \sim S'\mathscr{G}(X')$ if and only if $X_{\mathcal{I}\mathcal{I}}$ is invertible, and in that case it is given by*

$$X' = \begin{bmatrix} X'_{\mathcal{I}\mathcal{I}} & X'_{\mathcal{I}\mathcal{I}^c} \\ X'_{\mathcal{I}^c\mathcal{I}} & X'_{\mathcal{I}^c\mathcal{I}^c} \end{bmatrix} = \begin{bmatrix} -(X_{\mathcal{I}\mathcal{I}})^{-1} & (X_{\mathcal{I}\mathcal{I}})^{-1} X_{\mathcal{I}\mathcal{I}^c} \\ X_{\mathcal{I}^c\mathcal{I}}(X_{\mathcal{I}\mathcal{I}})^{-1} & X_{\mathcal{I}^c\mathcal{I}^c} - X_{\mathcal{I}^c\mathcal{I}}(X_{\mathcal{I}\mathcal{I}})^{-1} X_{\mathcal{I}\mathcal{I}^c} \end{bmatrix}.$$

Some additional sign book-keeping is needed in addition to the above formula if we wish to get a representation with a symplectic swap as in Theorem 4: indeed, if for a $k \in \{1, 2, \ldots, \ell\}$ the symplectic swap S already contains the factor S_{i_k}, then the product S' includes $S_{i_k}^2$, which acts as $-I_2$ on the i_kth and $(n + i_k)$th entry of a vector. Hence $S' \notin \mathscr{S}_{2n}$; to get back a symplectic swap we need to correct some signs in S' and X'. This is just a technical issue; a MATLAB function that deals with it and produces $S' \in \mathscr{S}_{2n}$ is in [43, file `private/updateSymBasis.m`].

The statement and proof of Theorem 2 hold for permuted Lagrangian graph bases as well, by simply changing P to S. Hence, permuted Lagrangian graph bases $U \sim S\mathscr{G}(X)$ provide a reasonably well-conditioned way to represent a Lagrangian subspace on a computer and perform computational work with it. This time, we have a distinct advantage with respect to orthogonal bases: the fact that the subspace is Lagrangian is equivalent to $X = X^H$, a property which is easy to enforce and deal with computationally. On the other hand, when working with orthogonal bases, it is well possible that a subspace "drifts away" from the manifold of Lagrangian subspaces due to the accumulation of numerical errors. Structure preservation in permuted Lagrangian graph bases will be crucial in Sect. 5.9.

5.6 Representation of Pencils

A *matrix pencil* is a degree-1 matrix polynomial, i.e., an expression of the form $L(z) = L_1 z + L_0$, with $L_0, L_1 \in \mathbb{C}^{n\times m}$ and z an indeterminate. We call a pencil *row-reduced* if $\begin{bmatrix} L_1 & L_0 \end{bmatrix}$ has full row rank, i.e., if there exists no nonzero $v \in \mathbb{C}^n$ such that $v^H(L_1\lambda + L_0) = 0$ for all $\lambda \in \mathbb{C}$. We call a pencil *regular* if $m = n$ and $\det L(z)$ is not the zero polynomial. For a regular $L(z)$, the roots of $\det L(z)$ are called *eigenvalues*, and a vector $v \neq 0$ such that $L(\lambda)v = 0$ is called *(right) eigenvector* relative to the eigenvalue λ. We say that ∞ is an eigenvalue of $L(z)$ (with eigenvector v) whenever 0 is an eigenvalue of $L_0 z + L_1$ (with eigenvector v). A full theory of eigenvalues and eigenvectors of (non necessarily regular) matrix pencils, including an extension of the Jordan canonical form, can be found in the classical book [19].

In this section and the next ones, we focus on eigenvalue and eigenvector problems for pencils; therefore, we are free to replace a pencil with another one having the same eigenvalues and eigenvectors. We say that two matrix pencils $L(z), M(z) \in \mathbb{C}[z]^{n\times n}$ are *left equivalent* (and we write $L(z) \sim M(z)$) if there is an invertible matrix $N \in \mathbb{C}^{n\times n}$ (not depending on z) such that $L(z) = NM(z)$. When this property holds and $L(z)$ and $M(z)$ are regular, clearly they have the same eigenvalues and right eigenvectors. The symbol $\sim$ is the same that we have used for matrices spanning the same subspace, and indeed these two equivalence relations are intimately connected: given $L(z) = L_1 z + L_0$ and $M(z) = M_1 z + M_0$, we have $L(z) \sim M(z)$ if and only if $\begin{bmatrix} L_1 & L_0 \end{bmatrix}^H \sim \begin{bmatrix} M_1 & M_0 \end{bmatrix}^H$.

In the previous sections, we have focused our efforts on finding P and a bounded X so that $U \sim P\mathscr{G}(X)$, for a given matrix U. In view of the above connection, this translates immediately to a result on pencils.

Theorem 5 ([35]) *Let $L(z) = L_1 z + L_0 \in \mathbb{C}[z]^{n\times n}$ be a row-reduced matrix pencil. Then, there exists another matrix pencil $M(z) = M_1 z + M_0 \in \mathbb{C}[z]^{n\times n}$ such that $L(z) \sim M(z)$ and $\begin{bmatrix} M_1 & M_0 \end{bmatrix}^H = P\mathscr{G}(X)$, for a suitable permutation matrix $P \in \mathbb{R}^{2n\times 2n}$ and $X \in \mathbb{C}^{n\times n}$ with $\|X\|_{\max} \le 1$.*

In other words, each column of I_n appears at least once among the columns of M_1 and M_0, and all the entries of these two matrices are bounded by 1.

Similarly, the results of Sect. 5.5 can be used to obtain pencils that are left equivalent to some with special structures. A row-reduced pencil $L(z) = L_1 z + L_0 \in \mathbb{C}^{2n\times 2n}$ is called *Hamiltonian* if $L_1 J_{2n} L_0^H + L_0 J_{2n} L_1^H = 0$; see [32, 40]. Simple manipulations show that this holds if and only if $U = \begin{bmatrix} L_1 & J_{2n} L_0 \end{bmatrix}^H \in \mathbb{C}^{4n\times 2n}$ satisfies $U^H J_{2n} U = 0$, i.e., $\operatorname{im} U$ is Lagrangian. Hence we can reduce to the setting of Theorem 4, obtaining the following result.

Theorem 6 ([35]) *Let $L(z) = L_1 z + L_0 \in \mathbb{C}[z]^{2n\times 2n}$ be a row-reduced Hamiltonian pencil. Then, there exist $S \in \mathscr{S}_{4n}$ and $X = X^H \in \mathbb{C}^{2n\times 2n}$ with $\|X\|_{\max} \le \sqrt{2}$ so that $L(z) \sim M(z)$, with $M(z)$ defined by $\begin{bmatrix} M_1 & M_0 J_{2n} \end{bmatrix}^H = S\mathscr{G}(X)$.*

Notice the structure of $M(z)$: for each $i = 1, 2, \ldots, 2n$, either the ith column of M_1 or the $n \pm i$th column of M_0 is (modulo signs) equal to the ith column of the identity matrix I_{2n}.

It is common in the literature to represent a Hamiltonian pencil with no infinite eigenvalues as $L(z) \sim I_{2n} z - H$, where H is a *Hamiltonian matrix*, i.e., a matrix such that HJ_{2n} is Hermitian: this corresponds to the case $S = I_{4n}$ of Theorem 6. Introducing column swaps in the picture allows us to find a representation that has bounded entries and works without constraints on the eigenvalues.

Another structure that we can deal with is the following. A row-reduced pencil $L(z) = L_1 z + L_0 \in \mathbb{C}[z]^{2n\times 2n}$ is called *symplectic* if $L_1 J_{2n} L_1^H - L_0 J_{2n} L_0^H = 0$; see [32, 40]. If one partitions $L_1 = \begin{bmatrix} L_{10} & L_{11} \end{bmatrix}$, $L_0 = \begin{bmatrix} L_{00} & L_{01} \end{bmatrix}$, with all blocks $2n \times n$, this is equivalent to $U = \begin{bmatrix} L_{10} & L_{01} & L_{11} & L_{00} \end{bmatrix}^H$ spanning a Lagrangian subspace. Note that symplectic swaps act separately on the two blocks composing L_1 and on the two composing L_0. Keeping track of this, one can decompose $S \in \mathscr{S}_{4n}$ into two smaller symplectic swaps, and obtain a simpler statement for the analogue of Theorem 6 for symplectic pencils.

Theorem 7 ([35]) *Let $L(z) = L_1 z + L_0 \in \mathbb{C}[z]^{2n\times 2n}$ be a row-reduced symplectic pencil. Then, there exist two symplectic swaps $S_1, S_2 \in \mathscr{S}_{2n}$ and $X = X^H \in \mathbb{C}^{2n\times 2n}$ with $\|X\|_{\max} \le \sqrt{2}$ so that $L(z) \sim M(z)$, with $M(z)$ defined by*

$$M(z) = \begin{bmatrix} I_n & X_{11} \\ 0 & X_{12}^H \end{bmatrix} S_1 z - \begin{bmatrix} X_{12} & 0 \\ X_{22} & I_n \end{bmatrix} S_2, \quad X = \begin{bmatrix} X_{11} & X_{12} \\ X_{12}^H & X_{22} \end{bmatrix}.$$

Again, the representation with $S_1 = S_2 = I_{2n}$ is widely used [11, 18, 34].

The main advantage of these forms is that we can represent on a computer pencils that are symplectic or Lagrangian, not up to numerical errors but exactly, and at the same time we do not have to deal with the numerical troubles of unduly large entries.

Lemma 1 bounds the quantity $\kappa(\begin{bmatrix} M_1 & M_0 \end{bmatrix}^H)$ for these "permuted graph pencils". Using standard properties of the singular values, one can see that the inverse of this quantity is the relative distance (in the Euclidean norm) to the closest non-row-reduced pencil, i.e.,

$$\kappa(\begin{bmatrix} M_1 & M_0 \end{bmatrix})^{-1} = \frac{\min_{\tilde{M}_1, \tilde{M}_0} \left\| \begin{bmatrix} \tilde{M}_1 & \tilde{M}_0 \end{bmatrix} - \begin{bmatrix} M_1 & M_0 \end{bmatrix} \right\|_2}{\left\| \begin{bmatrix} M_1 & M_0 \end{bmatrix} \right\|_2},$$

where the minimum is taken over all the pencils $\tilde{M}_1 z + \tilde{M}_0$ that are not row-reduced. While having a small $\kappa(\begin{bmatrix} M_1 & M_0 \end{bmatrix})$ seems desirable, because it means that $M(z)$ is far away from a variety of ill-posed problems, it is not clear what exactly this quantity represents in terms of perturbation theory. It is not a condition number for the eigenvalues, nor the distance from the closest singular (i.e., non-regular) pencil. Indeed, all non-row-reduced pencils are singular, but the converse does not hold (see for instance (5.7) in the following for a counterexample).

Hence, from the point of view of perturbation theory and numerical stability, the effectiveness of these special forms can currently only be justified by heuristic reasons.

5.7 Numerical Methods for Linear-Quadratic Optimal Control

Systems and control theory is a branch of engineering and mathematics that leads to an abundance of linear algebra applications. Here we focus on a simple version of the linear-quadratic optimal control problem [34]. The reader will find several chapters in this book dedicated to control theory, but we give a quick introduction to the numerical methods directly here to keep this chapter self-contained and introduce a notation consistent with our exposition.

Given matrices $A \in \mathbb{R}^{n \times n}$, $B, S \in \mathbb{R}^{n \times m}$, $Q = Q^T \in \mathbb{R}^{n \times n}$, $R = R^T \in \mathbb{R}^{m \times m}$, one looks for vector-valued functions $x, \mu : \mathbb{R}_+ \to \mathbb{R}^n$, $u : \mathbb{R}_+ \to \mathbb{R}^m$ such that

$$\begin{bmatrix} 0 & I_n & 0 \\ -I_n & 0 & 0 \\ 0 & 0 & 0 \end{bmatrix} \frac{d}{dt} \begin{bmatrix} \mu(t) \\ x(t) \\ u(t) \end{bmatrix} = \begin{bmatrix} 0 & A & B \\ A^T & Q & S \\ B^T & S^T & R \end{bmatrix} \begin{bmatrix} \mu(t) \\ x(t) \\ u(t) \end{bmatrix}, \quad x(0) = x_0, \lim_{t \to \infty} \begin{bmatrix} \mu(t) \\ x(t) \\ u(t) \end{bmatrix} = 0. \tag{5.4}$$

The textbook solution to this problem goes as follows. First, assuming $R > 0$, one eliminates $u(t)$ and swaps the two remaining equations, obtaining

$$\frac{d}{dt}\begin{bmatrix} x(t) \\ \mu(t) \end{bmatrix} = H \begin{bmatrix} x(t) \\ \mu(t) \end{bmatrix}, \quad H = -J_{2n}M, \quad M = \begin{bmatrix} Q & A^T \\ A & 0 \end{bmatrix} - \begin{bmatrix} S \\ B \end{bmatrix} R^{-1} \begin{bmatrix} S^T & B^T \end{bmatrix}. \tag{5.5}$$

One can prove under mild assumptions that H has n eigenvalues with negative real part and n with positive real part (counted with multiplicities); hence there exists a unique n-dimensional subspace $\mathscr{U}$ such that $H\mathscr{U} \subseteq \mathscr{U}$, and the restriction of H to $\mathscr{U}$ has only eigenvalues with negative real part. A stable solution to the system of ordinary differential equations (5.5) is obtained if and only if $\begin{bmatrix} x(0) \\ \mu(0) \end{bmatrix} \in \mathscr{U}$, and if this happens then $\begin{bmatrix} x(t) \\ \mu(t) \end{bmatrix} \in \mathscr{U}$ for all $t > 0$. How does one determine $\mathscr{U}$? The traditional approach is looking for a graph basis $U = \mathscr{G}(X)$, with $X \in \mathbb{R}^{n\times n}$, which exists under additional assumptions on the problem (typically satisfied). Then the condition $H\mathscr{U} \subseteq \mathscr{U}$ becomes $HU = UT$, for some matrix $T \in \mathbb{C}^{n\times n}$ with all its eigenvalues in the left half-plane; expanding out the products gives

$$\begin{cases} M_{11} + M_{12}X + XM_{21} + XM_{22}X = 0, \\ T = M_{21} + M_{22}X, \end{cases} \qquad \text{with } M = \begin{bmatrix} M_{11} & M_{12} \\ M_{21} & M_{22} \end{bmatrix}. \tag{5.6}$$

The first equation in X alone is called *algebraic Riccati equation*; several solution methods exist. Once X is determined, thanks to the previous observation, we have $\mu(t) = Xx(t)$ for each t, and hence some manipulations give $u(t) = Kx(t)$ with $K = -R^{-1}(B^T X + S^T)$, and $x(t) = \exp((A + BK)t)x_0$.

Although one can prove that $\mathscr{U}$ admits a graph basis, this does not mean that it is a good idea to compute it numerically. The corresponding X might have very large elements. An alternative strategy is computing an orthogonal basis instead. Given any basis $U = \begin{bmatrix} E \\ A \end{bmatrix}$ for $\mathscr{U}$, we can reduce the problem to solving an initial-value ODE problem for $w(t) : \mathbb{R}_+ \to \mathbb{R}^n$ such that $\begin{bmatrix} x(t) \\ \mu(t) \end{bmatrix} = Uw(t)$. A necessary step is computing $w(0) = w_0$, which might still be troublesome numerically if E is ill-conditioned, but all the other steps, notably the eigenvalue computation, benefit from the additional stability associated with working with orthogonal matrices. If needed, the solution X of the Riccati equation can be obtained as well as AE^{-1}.

One can apply this approach of computing an invariant subspace directly to (5.4) as well. Let us call $\mathcal{E}$ and $\mathcal{A}$ the two block-3×3 matrices appearing in the left- and right-hand side of the leftmost equation in (5.4), respectively. This time, $\mathcal{E}$ is singular, but one can generalize the concept of invariant subspaces to matrix pencils. Given a regular pencil $\mathcal{E}z - \mathcal{A} \in \mathbb{C}[z]^{k\times k}$, we say that the image of a $U \in \mathbb{C}^{k\times \ell}$ with full column rank is a *deflating subspace* if there are $V \in \mathbb{C}^{k\times \ell}$, $E, A \in \mathbb{C}^{\ell\times \ell}$ such that $(\mathcal{E}z - \mathcal{A})U = V(Ez - A)$. The eigenvalues of $Ez - A$ are a subset of those of $\mathcal{E}z - \mathcal{A}$, and are called *associated* with the deflating subspace $\operatorname{im} U$. Under the same assumptions that we have made above, $\mathcal{E}z - \mathcal{A}$ has m eigenvalues equal to

∞, n with positive real part and n with negative real part; the finite eigenvalues coincide with those of H. One can solve a generalized eigenvalue problem [21, Section 7.7] to determine the invariant subspace associated with the last ones, and proceed similarly.

Several more general settings exist, most notably *finite-horizon problems* in which the boundary condition at ∞ in (5.4) is replaced by one at a time $t_f > 0$, or problems in which R is not invertible and the assumptions that we made on the location of eigenvalues are not respected. Large-scale problems with pencils exhibiting the same structure appear for instance in model reduction.

In the numerical solution of linear-quadratic control problems, matrix structures play a crucial role. The pencil $J_{2n}z - M$ is Hamiltonian, as well as the matrix H, and the pencil $\mathcal{E}z - \mathcal{A}$ is *even*, i.e., $\mathcal{E} = -\mathcal{E}^H$ and $\mathcal{A} = \mathcal{A}^H$. These pencils and matrices have a distinguishing pairing of eigenvalues; namely, for each eigenvalue λ, one has that $-\bar{\lambda}$ is an eigenvalue as well. For Hamiltonian problems, moreover, $\mathscr{U}$ is Lagrangian, and hence $X = X^T$. For even problems in general this latter property does not hold (although a similar property does hold for the pencil $\mathcal{E}z - \mathcal{A}$ defined in (5.4)).

Numerical methods that exploit these structures are preferable, not only for speed, but especially for accuracy: in an ill-conditioned problem, for instance, an unstructured eigensolver might detect numerically $n + 1$ eigenvalues with negative real part and $n - 1$ with positive real part, a situation which is impossible under the structural constraints, and hence fail to identify correctly the unique n-dimensional invariant subspace. Even when this does not happen, it is a sounder theoretical guarantee to have a low *structured* backward error, that is, to be able to guarantee that the computed solution is the exact solution of a nearby problem respecting the same structure.

There has been extensive numerical research on how to accurately solve Hamiltonian and even eigenvalue problems; countless methods have been suggested [9, 34]: for instance, focusing only on the small-case dense case, there are the Newton method for algebraic Riccati equations [5, 24, 30], QR-type algorithms based on reduction to Hamiltonian or symplectic and Schur forms [10, 18], and structure-preserving versions of matrix iterations [2, 11, 20]. In the next sections, we describe two improvements that can be obtained by using permuted graph bases.

5.8 Permuted Graph Bases for the Deflation of Control Problems

A first area where we can see an improvement by judiciously using permuted graph bases is transforming (5.4) into the form (5.5). Indeed, consider the following formulation of this deflation process. We premultiply $\mathcal{E}z - \mathcal{A}$ by a suitable matrix to obtain an identity submatrix I_{2n+m} in the first $2n$ columns of $\mathcal{E}$ and the last m columns of $\mathcal{A}$, that is,

$$\mathcal{E}z - \mathcal{A} \sim \begin{bmatrix} 0 & I_n & B \\ -I_n & 0 & S \\ 0 & 0 & R \end{bmatrix}^{-1} (\mathcal{E}z - \mathcal{A}) = \begin{bmatrix} I_n & 0 & 0 \\ 0 & I_n & 0 \\ 0 & 0 & 0 \end{bmatrix} z - \begin{bmatrix} H_{22} & H_{21} & 0 \\ H_{11} & H_{12} & 0 \\ R^{-1}B^T & R^{-1}S^T & I \end{bmatrix},$$

where one can see that the H_{ii} are exactly the blocks of H defined in (5.5), albeit swapped. The rightmost pencil is block-triangular with a leading diagonal block of size $2n \times 2n$ and a trailing one of size $m \times m$; its eigenvalues are given by the union of the eigenvalues of these two diagonal blocks. The trailing $m \times m$ block contains the m infinite eigenvalues, and the leading $2n \times 2n$ block contains the $2n$ finite eigenvalues that coincide with the eigenvalues of H. The eigenvectors and deflating subspaces can be related as well; we do not go through the details. This construction shows that the process of reducing (5.4) to (5.5) can be interpreted as performing an equivalence transformation of $\mathcal{E}z - \mathcal{A}$ that enforces an identity submatrix and then deflating the resulting block-triangular pencil. In view of our previous discussion, it looks natural to try to enforce an identity submatrix in a different choice of columns. A special choice of swap matrices is needed to ensure that the deflated pencil is Hamiltonian. The following result can be obtained extending the previous theory to this particular problem.

Theorem 8 ([36]) *Let $\mathcal{E}z - \mathcal{A}$ be a row-reduced pencil with $\mathcal{E}$, $\mathcal{A}$ the two matrices in* (5.4). *There exist matrices M_{ij} such that*

$$\mathcal{E}z - \mathcal{A} \sim \begin{bmatrix} M_{11} & M_{12} & 0 \\ M_{12} & M_{22} & 0 \\ M_{31} & M_{32} & 0 \end{bmatrix} z + \begin{bmatrix} M_{13} & M_{14} & 0 \\ M_{23} & M_{24} & 0 \\ M_{33} & M_{34} & I_m \end{bmatrix},$$

where

$$\begin{bmatrix} 0 & I_n & A & 0 \\ -I_n & 0 & Q & -A^T \\ 0 & 0 & S^T & -B^T \end{bmatrix}^H \sim \begin{bmatrix} M_{11} & M_{12} & -M_{14} & M_{13} \\ M_{21} & M_{22} & -M_{24} & M_{23} \\ M_{31} & M_{32} & -M_{34} & M_{33} \end{bmatrix}^H = S \begin{bmatrix} I_n & 0 & X_{11} & X_{12} \\ 0 & I_n & X_{21} & X_{22} \\ 0 & 0 & X_{31} & X_{32} \end{bmatrix}^H,$$

for some $S \in \mathscr{S}_{2n}$, and $X = \begin{bmatrix} X_{11} & X_{12} \\ X_{21} & X_{22} \end{bmatrix}$ symmetric and such that $\|X\|_{\max} \leq 1$.

An explicit algorithm to obtain X with $\|X\|_{\max} \leq \tau$ for each $\tau \geq 1$ and an initial permutation heuristic inspired by the rank-revealing QRP factorization are provided in [36].

If one performs deflation in this form, $\begin{bmatrix} M_{11} & M_{12} \\ M_{21} & M_{22} \end{bmatrix} z - \begin{bmatrix} M_{13} & M_{14} \\ M_{23} & M_{24} \end{bmatrix}$ is a Hamiltonian pencil left equivalent to $I_{2n}z - H$, already in the format given by Theorem 6.

We report an example with a pencil that is particularly troublesome for most numerical methods. Let $m = n = 1$, and

$$\mathcal{E}z - \mathcal{A} = \begin{bmatrix} 0 & 1 & 0 \\ -1 & 0 & 0 \\ 0 & 0 & 0 \end{bmatrix} z - \begin{bmatrix} 0 & 0 & 1 \\ 0 & 0 & 0 \\ 1 & 0 & \varepsilon \end{bmatrix}. \tag{5.7}$$

Then,

$$\mathcal{E}z - \mathcal{A} \sim \begin{bmatrix} 0 & 0 & -1 \\ -1 & 0 & 0 \\ 0 & -1 & -\varepsilon \end{bmatrix}^{-1} (\mathcal{E}z - \mathcal{A}) = \begin{bmatrix} 1 & 0 & 0 \\ 0 & \varepsilon & 0 \\ 0 & -1 & 0 \end{bmatrix} z + \begin{bmatrix} 0 & 0 & 0 \\ 1 & 0 & 0 \\ 0 & 0 & 1 \end{bmatrix}.$$

The deflated pencil is $\left[\begin{smallmatrix} 1 & 0 \\ 0 & \varepsilon \end{smallmatrix}\right] z + \left[\begin{smallmatrix} 0 & 0 \\ 1 & 0 \end{smallmatrix}\right]$, which is in the form of Theorem 6 with $S = S_2$ and $X = \left[\begin{smallmatrix} 0 & 0 \\ 0 & -\varepsilon \end{smallmatrix}\right]$. Note that the procedure can be performed without trouble even if ε is very small or zero. Several methods for the deflation of an even problem (5.4) to a Hamiltonian one have appeared in literature [27, 28, 34, 44, 49]; in most of them, it is required either that R is nonsingular, or that the kernel of R (and possibly further kernels) are determined accurately. Rank decisions on R have often been considered a crucial part of the deflation procedure; the method outlined here shows instead that it is not the case, and that a Hamiltonian problem can be produced in a stable way without worrying about its singularity (or closeness to singularity).

If $\varepsilon = 0$, then the pencil (5.7) is singular, although both $\mathcal{E}z - \mathcal{A}$ and its transpose are row-reduced; correspondingly, the deflated Hamiltonian pencil is singular, too. So, from a computational point of view, we did not eliminate the problem of singularity, but simply push it to a later stage. Numerical methods for Hamiltonian eigenproblems that do not break down for singular (and close-to-singular) pencils are then required; as far as we know they have not yet appeared in the literature.

5.9 Hamiltonian Pencils and the Doubling Algorithm

A second application of permuted graph bases comes from solving Hamiltonian invariant subspace problems. The starting point for our algorithm is the following result.

Theorem 9 ([6]) *Let* $L(z) = L_1 z + L_0 \in \mathbb{C}[z]^{n\times n}$ *be a regular pencil, and let* $M_0, M_1 \in \mathbb{C}^{n\times n}$ *be such that* $\begin{bmatrix} -M_0 & M_1 \end{bmatrix}$ *has full row rank and* $\begin{bmatrix} -M_0 & M_1 \end{bmatrix} \begin{bmatrix} L_1 \\ L_0 \end{bmatrix} = 0$. *If* $v \in \mathbb{C}^n$ *is an eigenvector of* $L(z)$ *with eigenvalue* λ, *then it is also an eigenvector of the pencil*

$$N(z) = N_1 z + N_0 = M_0 L_1 z + \frac{1}{2}(M_1 L_1 + M_0 L_0) \tag{5.8}$$

with eigenvalue $f(\lambda)$, *where* $f(z) = \frac{1}{2}(z + z^{-1})$.

If we denote by $f^{(k)}$ the composition of f with itself k times and by $\Re z$ the real part of z, we have

$$\lim_{k\to\infty} f^{(k)}(z) = \begin{cases} 1 & \text{if } \Re z > 0, \\ -1 & \text{if } \Re z < 0 \end{cases}$$

(the iteration in this form breaks down if $\Re z = 0$, but as we see in the following this will not be a concern). Hence, if we start from a pencil $L(z)$ with no eigenvalues on the imaginary axis, repeating the transformation $L(z) \mapsto N(z)$, we converge (in a suitable sense) to a pencil $L_\infty(z)$ with eigenvalues 1 and -1 only, from which one can recover the invariant subspace associated with the eigenvalues having negative real part. This iteration is essentially a pencil version of the matrix sign iteration [26, Chapter 5].

Note that one can replace $M(z) = M_1 z + M_0$ with any pencil $M'(z) \sim M(z)$, obtaining then a different $N'(z) \sim N(z)$; so there is some arbitrariness in how to perform the iteration. Some form of normalization needs to be enforced, otherwise N_1 and N_0 could both converge to zero, or diverge, or (even worse) converge to matrices with the same left kernel, giving a non-row-reduced $L_\infty(z)$. Hence one can see a role for permuted graph representations in this setting. A second point in which this technique helps is in computing the kernel $\begin{bmatrix}-M_0 & M_1\end{bmatrix}$. Indeed, the following result is easy to verify.

Lemma 3 *Let $U \sim P\mathscr{G}(X)$ be the permuted graph representation of a matrix $U \in \mathbb{C}^{(m+n)\times m}$ with full column rank. Then, $W = P\begin{bmatrix}-X^H \\ I_n\end{bmatrix} \in \mathbb{C}^{(m+n)\times n}$ is such that $W^H U = 0$. Moreover, the matrix W has full column rank and spans the kernel of U^H.*

Hence, given a permuted graph basis for a subspace, we can determine a permuted graph basis for its left kernel with basically no computational effort.

Another important observation is that in Theorem 9 whenever $L(z)$ is Hamiltonian, then $N(z)$ is Hamiltonian, too, so we can compute at each step a permuted Lagrangian graph representation as normalization. Putting everything together, we get Algorithm 4.

The bulk of the computational cost consists in computing permuted graph bases for $\begin{bmatrix}L_1 \\ L_0\end{bmatrix}$ and permuted Lagrangian graph bases for $\begin{bmatrix}N_1 & N_0 J_{2n}\end{bmatrix}^H$, alternately, together with the matrix products that appear in (5.8). At each step after the first, P and S from the previous steps typically work well as initial guesses; recomputing X and Y from the permutation at the end of Algorithm 1 might be needed for better accuracy.

The algorithm converges quadratically; one can relax the assumptions, allowing for eigenvalues on the imaginary axis; in this case, the algorithm can be proved to converge in every problem for which there exists a Lagrangian deflating subspace [35, 36], but the convergence rate turns to linear. (Actually, Theorem 9 and our analysis above are slightly incomplete even in the case with no eigenvalues on the

Algorithm 4: Inverse-free sign algorithm with permuted graph bases [36]

Input: A Hamiltonian pencil $L(z) = L_1 z + L_0 \in \mathbb{C}[z]^{2n\times 2n}$ without eigenvalues on the imaginary axis; a threshold $\tau > \sqrt{2}$

Output: A basis for the invariant subspace $\mathcal{U}$ of $L(z)$ associated with the eigenvalues in the left half-plane

repeat

compute a permutation matrix $P \in \mathbb{C}^{4n\times 4n}$ and $X \in \mathbb{C}^{2n\times 2n}$ such that $P\mathcal{G}(X) \sim \left[\begin{smallmatrix} L_1 \\ L_0 \end{smallmatrix}\right]$ and $\|X\|_{\max} \leq \tau$, using Algorithm 1;

let $\begin{bmatrix}-M_0 & M_1\end{bmatrix} = \begin{bmatrix}-X & I\end{bmatrix} P^H$;

compute $N(z)$ as in (5.8);

compute $S \in \mathcal{S}_{4n}$, $Y = Y^H \in \mathbb{C}^{2n\times 2n}$ such that $\begin{bmatrix}N_1 & N_0 J_{2n}\end{bmatrix}^H \sim S\mathcal{G}(Y)$ and $\|Y\|_{\max} \leq \tau$, using the symplectic analogue of Algorithm 1 (see Theorem 6 and Sect. 5.5);

replace $L(z)$ with $N(z)$ and continue;

until Y *converges*;

Find the kernel of $L_1 + L_0$, which is $\mathcal{U}$;

imaginary axis, because we do not consider what happens to multiple eigenvalues; we refer the reader to [35, 36] for full detail.)

There are essentially two versions of this algorithm; one is as described above; the other one works by first converting $L_1 z + L_0$ to a symplectic pencil via the transformation $L(z) \mapsto (L_1 + L_0)z + (L_0 - L_1)$ (known as *Cayley transform*), and then applying a transformation analogous to (5.8), that is,

$$N(z) = M_1 L_1 z - M_0 L_0, \tag{5.9}$$

which transforms the eigenvalues according to $g(z) = z^2$, and for which

$$\lim_{k\to\infty} g^{(k)}(z) = \begin{cases} \infty & \text{if } |z| > 1, \\ 0 & \text{if } |z| < 1. \end{cases}$$

In this second case, the last line of Algorithm 4 changes to computing the kernel of L_0. The work [6] contains a general theory of operations with matrix pencils that describes how to produce "matrix pencil versions" of rational functions, such as (5.8) and (5.9) for $f(z)$ and $g(z)$.

This modified version is called *doubling algorithm*; it was introduced (with a different derivation) for unstructured invariant subspace problems without the use of permuted graph bases in [4, 33], and for symplectic problems with the special choice $S = I$ (graph basis without permutation) in [2, 11, 12, 29], and then generalized to make full use of permuted graph bases in [35]. The algorithm that we described first is known as *inverse-free sign method*; it appeared without the use of permuted graph bases in [6], then with $S = I$ in [20], and with permuted graph bases in [36]. Permuted graph bases are important here because they ensure that the iterative procedure produces an exactly Hamiltonian (or symplectic) pencil at each step and steers clear of numerically singular pencils.

A basic implementation in the MATLAB language of Algorithm 4 and its doubling variant is available on [43]; the library also includes Algorithm 1 and several functions to compute permuted graph representations of subspaces and pencils, both in the unstructured and the Lagrangian case.

How well do these algorithms fare in practice, compared to their many competitors and variants that do not make use of permuted graph bases? The work [35] reports computational results on a set of 33 small-scale problems (the same test problems used in [10]) obtained from the benchmark set [7]. This is a benchmark set containing examples of linear-quadratic control problems; it contains both examples from real-life applications and challenging problems created ad-hoc to be difficult to solve. As far as we know, the algorithm in [35] (Algorithm 4 in the variant with transformation (5.9)) is the first numerical algorithm to obtain completely satisfying results in all 33 problems on both these grounds:

- Small subspace residual, that is, $\frac{\|(I-UU^T)HU\|}{\|H\|}$ of the order of machine precision for the computed subspace U;
- Exact preservation of the Lagrangian structure, that is, $U^T J_{2n} U$ either zero or of the order of machine precision.

Algorithm 4 in the variant presented here was tested on another challenging application ($\mathscr{H}_\infty$ control, an optimization procedure which requires solving one after another a set of close-to-unsolvable Riccati equations) in [36]; the results suggest that the variant (5.8) is more stable than (5.9), because it avoids the initial Cayley transform. This is why we chose to highlight (5.8) in this presentation.

Explicit theoretical results proving stability of the algorithm are still an open issue, though. For methods based on orthogonal transformations and reduction to Schur form, the standard technique is a Wilkinson-style backward stability proof ([50, Chapter 3] and [25, Section 19.3]); however, a counterexample in [35, p. 798] shows that the simplest version of a backward stability proof using this technique would not work for doubling-type algorithms. As far as we know, the only stability proof for an algorithm of this family is given in [4], for a doubling algorithm for unstructured pencils based on orthogonal bases. It can be adapted to the other variants; however, it is not completely satisfying as the error growth coefficient is bounded by $1/d^2$, where d is the distance to the closest ill-posed problem, instead of the more natural $1/d$ which constitutes the condition number of the problem. We are not aware of an example in which this larger factor shows up in practice. Another related result is the work in [38, 39], which shows that for Hermitian matrices a carefully chosen variant of doubling achieves a mixed variant of backward and forward stability. Nevertheless, for nonsymmetric problems, proving the stability of doubling-type algorithms is still an open problem, both in the structured and non-structured case.

5.10 Research Directions

There are many possible improvements and open problems related to these topics; some of them have already been presented in our exposition, and we collect here a few more.

An interesting issue is the significance of $\kappa(\left[\begin{smallmatrix} L_1 \\ L_0 \end{smallmatrix}\right])$ for a pencil $L_1 z + L_0$: what is its role in the stability and perturbation theory of the stable invariant subspace problem?

Another research direction is extending the methods presented in Sects. 5.8 and 5.9 to deal with so-called *descriptor systems*, a common variant of the control theory setting described above. The matrix pencils appearing in such problems are similar to the one in (5.4), but the leading submatrix J_{2n} of $\mathcal{E}$ is replaced by $\left[\begin{smallmatrix} 0 & E \\ -E^T & 0 \end{smallmatrix}\right]$, for a matrix $E \in \mathbb{C}^{n \times n}$ that may be singular. With this modification, only part of the structure is preserved: the resulting pencil $\mathcal{E}z - \mathcal{A}$ is still even, but the deflated problem is not Hamiltonian and its stable deflating subspace $\mathscr{U}$ is not Lagrangian. The algorithms that we have presented rely in an essential way on these structures, so modifying them to work with descriptor systems will probably require major changes.

A first attempt to use permuted graph bases in large-scale control problems is in the recent preprint [37]; the underlying algorithm is not a doubling algorithm but low-rank ADI [8], and inverse-free techniques and permuted graph bases are used to adapt it to an invariant subspace formulation. One of the most interesting observations in that work is that Lemma 3, which we have used here only in the case $m = n$, is noteworthy also when $m \ll n$, since it allows one to compute an explicit basis for a large n-dimensional subspace of $\mathbb{C}^{m+n}$ defined as a kernel with basically no effort.

Overall, in control theory there are many possibilities for further applications; many problems can be reduced to the computation of some Lagrangian invariant subspace, and permuted graph bases are a natural choice for this task.

5.11 Conclusions

In this chapter we have presented the basic theory of permuted graph matrices and bases, and shown how techniques based on them are useful in a variety of applications. We believe that they are an interesting alternative, in selected problems, to the ubiquitous orthogonal matrices. They have led to the construction of several efficient algorithms for various tasks, and, as always in research, there is plenty of open problems and possibilities for improvement.

Acknowledgements The author is grateful to C. Mehl, B. Meini and N. Strabič for their useful comments on an early version of this chapter.

References

1. Absil, P.-A., Mahony, R., Sepulchre, R.: Optimization Algorithms on Matrix Manifolds. Princeton University Press, Princeton (2008)
2. Anderson, B.: Second-order convergent algorithms for the steady-state Riccati equation. Int. J. Control **28**(2), 295–306 (1978)
3. Arioli, M., Duff, I.S.: Preconditioning of linear least-squares problems by identifying basic variables. Technical report RAL-P-2014-007s, Rutherford Appleton Laboratory (2014)
4. Bai, Z., Demmel, J., Gu, M.: An inverse free parallel spectral divide and conquer algorithm for nonsymmetric eigenproblems. Numer. Math. **76**(3), 279–308 (1997)
5. Benner, P., Byers, R.: An exact line search method for solving generalized continuous-time algebraic Riccati equations. IEEE Trans. Autom. Control **43**(1), 101–107 (1998)
6. Benner, P., Byers, R.: An arithmetic for matrix pencils: theory and new algorithms. Numer. Math. **103**(4), 539–573 (2006)
7. Benner, P., Laub, A., Mehrmann, V.: A collection of benchmark examples for the numerical solution of algebraic Riccati equations I: the continuous-time case. Technical report SPC 95-22, Forschergruppe 'Scientific Parallel Computing', Fakultät für Mathematik, TU Chemnitz-Zwickau, 1995. Version dated 28 Feb 1996
8. Benner, P., Li, J.-R., Penzl, T.: Numerical solution of large-scale Lyapunov equations, Riccati equations, and linear-quadratic optimal control problems. Numer. Linear Algebra Appl. **15**(9), 755–777 (2008)
9. Bini, D.A., Iannazzo, B., Meini, B.: Numerical Solution of Algebraic Riccati Equations. Volume 9 of Fundamentals of Algorithms. Society for Industrial and Applied Mathematics (SIAM), Philadelphia (2012)
10. Chu, D., Liu, X., Mehrmann, V.: A numerical method for computing the Hamiltonian Schur form. Numer. Math. **105**(3), 375–412 (2007)
11. Chu, E.K.-W., Fan, H.-Y., Lin, W.-W.: A structure-preserving doubling algorithm for continuous-time algebraic Riccati equations. Linear Algebra Appl. **396**, 55–80 (2005)
12. Chu, E.K.-W., Fan, H.-Y., Lin, W.-W., Wang, C.-S.: Structure-preserving algorithms for periodic discrete-time algebraic Riccati equations. Int. J. Control **77**(8), 767–788 (2004)
13. Çivril, A., Magdon-Ismail, M.: On selecting a maximum volume sub-matrix of a matrix and related problems. Theor. Comput. Sci. **410**(47–49), 4801–4811 (2009)
14. Dantzig, G., Thapa, M.: Linear Programming. 1. Springer Series in Operations Research. Springer, New York (1997)
15. Dollar, H.S.: Constraint-style preconditioners for regularized saddle point problems. SIAM J. Matrix Anal. Appl. **29**(2), 672–684 (2007)
16. Dollar, H.S., Wathen, A.J.: Approximate factorization constraint preconditioners for saddle-point matrices. SIAM J. Sci. Comput. **27**(5), 1555–1572 (2006)
17. Dopico, F., Johnson, C.: Complementary bases in symplectic matrices and a proof that their determinant is one. Linear Algebra Appl. **419**(2–3), 772–778 (2006)
18. Fassbender, H.: Symplectic Methods for the Symplectic Eigenproblem. Kluwer Academic/Plenum Publishers, New York (2000)
19. Gantmacher, F.R.: The Theory of Matrices. Vols. 1, 2 (Trans. by K.A. Hirsch). Chelsea Publishing, New York (1959)
20. Gardiner, J.D., Laub, A.J.: A generalization of the matrix-sign-function solution for algebraic Riccati equations. Int. J. Control **44**(3), 823–832 (1986)
21. Golub, G., Van Loan C.: Matrix Computations. Johns Hopkins Studies in the Mathematical Sciences, 3rd edn. Johns Hopkins University Press, Baltimore (1996)
22. Goreinov, S.A., Oseledets, I.V., Savostyanov, D.V., Tyrtyshnikov, E.E., Zamarashkin, N.L.: How to find a good submatrix. In: Matrix Methods: Theory, Algorithms and Applications, pp 247–256. World Scientific, Hackensack (2010)
23. Goreinov, S.A., Tyrtyshnikov, E.E., Zamarashkin, N.L.: A theory of pseudoskeleton approximations. Linear Algebra Appl. **261**, 1–21 (1997)

24. Guo, C.-H., Lancaster, P.: Analysis and modification of Newton's method for algebraic Riccati equations. Math. Comput. **67**(223), 1089–1105 (1998)
25. Higham, N.: Accuracy and Stability of Numerical Algorithms, 2nd edn. Society for Industrial and Applied Mathematics (SIAM), Philadelphia (2002)
26. Higham, N.J.: Functions of Matrices. Theory and Computation. Society for Industrial and Applied Mathematics (SIAM), Philadelphia (2008)
27. Ionescu, V., Oară, C., Weiss, M.: Generalized Riccati Theory and Robust Control. A Popov Function Approach. Wiley, Chichester (1999)
28. Jacobson, D., Speyer, J.: Necessary and sufficient conditions for optimality for singular control problems; a transformation approach. J. Math. Anal. Appl. **33**, 163–187 (1971)
29. Kimura, M.: Convergence of the doubling algorithm for the discrete-time algebraic Riccati equation. Int. J. Syst. Sci. **19**(5), 701–711 (1988)
30. Kleinman, D.: On an iterative technique for Riccati equation computations. IEEE Trans. Autom. Control **13**(1), 114–115 (1968)
31. Knuth, D.E.: Semioptimal bases for linear dependencies. Linear Multilinear Algebra **17**(1), 1–4 (1985)
32. Lin, W.-W., Mehrmann, V., Xu, H.: Canonical forms for Hamiltonian and symplectic matrices and pencils. Linear Algebra Appl. **302/303**, 469–533 (1999) Special issue dedicated to Hans Schneider (Madison, 1998)
33. Malyshev, A.N.: Calculation of invariant subspaces of a regular linear matrix pencil. Sibirsk. Mat. Zh. **30**(4), 76–86, 217 (1989)
34. Mehrmann, V.: The Autonomous Linear Quadratic Control Problem: Theory and Numerical Solution. Volume 163 of Lecture Notes in Control and Information Sciences. Springer, Berlin (1991)
35. Mehrmann, V., Poloni, F.: Doubling algorithms with permuted Lagrangian graph bases. SIAM J. Matrix Anal. Appl. **33**(3), 780–805 (2012)
36. Mehrmann, V., Poloni, F.: Using permuted graph bases in $\mathcal{H}_\infty$ control. Autom. J. IFAC **49**(6), 1790–1797 (2013)
37. Mehrmann, V., Poloni, F.: An inverse-free ADI algorithm for computing Lagrangian invariant subspaces. Technical report 14-2014, Institute of Mathematics, Technische Universität Berlin (2014)
38. Nakatsukasa, Y., Higham, N.J.: Backward stability of iterations for computing the polar decomposition. SIAM J. Matrix Anal. Appl. **33**(2), 460–479 (2012)
39. Nakatsukasa, Y., Higham, N.J.: Stable and efficient spectral divide and conquer algorithms for the symmetric eigenvalue decomposition and the SVD. SIAM J. Sci. Comput. **35**(3), A1325–A1349 (2013)
40. Noferini, V., Poloni, F.: Duality of matrix pencils, Wong chains and linearizations. MIMS E-prints 2013.17, The University of Manchester, 2014. Version deposited on 7 August 2014
41. Oseledets, I.V., Savostyanov, D.V., Tyrtyshnikov, E.E.: Cross approximation in tensor electron density computations. Numer. Linear Algebra Appl. **17**(6), 935–952 (2010)
42. Pan, C.-T.: On the existence and computation of rank-revealing LU factorizations. Linear Algebra Appl. **316**(1–3), 199–222 (2000). Conference Celebrating the 60th Birthday of Robert J. Plemmons (Winston-Salem, 1999)
43. Poloni, F.: Pgdoubling (2012) MATLAB library. Available online on https://bitbucket.org/fph/pgdoubling.
44. Poloni, F., Reis, T.: A deflation approach for large-scale Lur'e equations. SIAM J. Matrix Anal. Appl. **33**(4), 1339–1368 (2012)
45. Savostyanov, D.V.: Quasioptimality of maximum-volume cross interpolation of tensors. Technical report arXiv:1305.1818, arXiv.org (2013)
46. Stewart, G.W., Sun, J.G.: Matrix Perturbation Theory. Computer Science and Scientific Computing. Academic, Boston (1990)
47. Tsatsomeros, M.: Principal pivot transforms: properties and applications. Linear Algebra Appl. **307**(1–3), 151–165 (2000)

48. Usevich, K., Markovsky, I.: Optimization on a Grassmann manifold with application to system identification. Autom. J. IFAC **50**(6), 1656–1662 (2014)
49. Weiss, H., Wang, Q., Speyer, J.: System characterization of positive real conditions. IEEE Trans. Autom. Control **39**(3), 540–544 (1994)
50. Wilkinson, J.H.: The Algebraic Eigenvalue Problem. Monographs on Numerical Analysis. The Clarendon Press/Oxford University Press, New York (1988). Oxford Science Publications.

Chapter 6
Canonical Forms of Structured Matrices and Pencils

Christian Mehl and Hongguo Xu*

Abstract This chapter provides a survey on the development of canonical forms for matrices and matrix pencils with symmetry structures and on their impact in the investigation of application problems. The survey mainly focuses on the results from three topics that have been developed during the past 15 years: structured canonical forms for Hamiltonian and related matrices, structured canonical forms for doubly structured matrices and pencils, and singular value-like decompositions for matrices associated with two sesquilinear forms.

6.1 Introduction

Eigenvalue problems frequently arise in several applications from natural sciences and industry and therefore the corresponding theory is a fundamental topic in Linear Algebra, Matrix Theory, and Numerical Analysis. The practical applications typically lead to matrices, matrix pencils, or matrix polynomials with additional symmetry structures that reflect symmetries in the underlying physics. As a consequence also the *eigenstructures* (i.e., eigenvalues, eigenvectors, root vectors, Jordan blocks, singular blocks and other invariants as, e.g., algebraic, geometric, and partial multiplicities) of such matrices, matrix pencils, and matrix polynomials inherit certain symmetries or patterns. As these reflect the nature and characteristics of the original application problems, they play critical roles both in theory and practice.

*Partially supported by *Alexander von Humboldt Foundation* and by *Deutsche Forschungsgemeinschaft*, through the DFG Research Center MATHEON *Mathematics for key technologies* in Berlin.

C. Mehl (✉)
Institut für Mathematik, Technische Universität Berlin, Sekretariat MA 4-5, Straße des 17. Juni 136, D-10623 Berlin, Germany
e-mail: mehl@math.tu-berlin.de

H. Xu
Department of Mathematics, University of Kansas, 603 Snow Hall, Lawrence, KS 66045, USA
e-mail: feng@ku.edu

P. Benner et al. (eds.), *Numerical Algebra, Matrix Theory, Differential-Algebraic Equations and Control Theory*, DOI 10.1007/978-3-319-15260-8_6

Typically, the solution of structured eigenvalue problems is a challenge, because there is demand for the design of new algorithms that are structure-preserving in each step, so that the corresponding symmetry in the spectrum is maintained in finite precision arithmetic and the obtained results are physically meaningful [48]. Simple variations of the QR algorithm or methods based on standard Krylov subspaces may not be sufficient to achieve this goal so that new ideas and concepts need to be developed. This requires a deeper understanding of the corresponding eigenstructures and therefore the derivation of structured canonical forms is essential. It is the aim of this chapter to review such forms for some particular classes of structured matrices or matrix pencils.

The most important and well-known matrices with symmetry structures are probably real or complex Hermitian, skew-Hermitian, and unitary matrices. Still, there are many other kinds of important structured matrices like complex symmetric, skew-symmetric, and orthogonal matrices as well as nonnegative matrices all of which are discussed in the classical books [13, 14]. In this chapter, we focus on structured matrices that are self-adjoint, skew-adjoint, or unitary with respect to an inner product associated with a possibly indefinite Hermitian or skew-Hermitian matrix and give a brief review on their theory, also including the corresponding matrix pencils that generalize those structures. We do not consider the corresponding structured matrix polynomials in this chapter, but refer the reader to Chap. 12of this book instead.

Let $\mathbb{F}$ be either the real field $\mathbb{R}$ or the complex field $\mathbb{C}$. Suppose $M \in \mathbb{F}^{m\times m}$ is an invertible Hermitian or skew-Hermitain matrix, and define the bilinear or sesquilinear form

$$[x, y]_M = x^* My =: [x, y], \qquad x, y \in \mathbb{F}^m, \tag{6.1}$$

where $*$ is the conjugate transpose, which reduces to just T, the transpose, if $\mathbb{F} = \mathbb{R}$. Then three sets of structured matrices can be defined:

1. The set of M-Hermitian matrices or M-selfadjoint matrices:
$\mathbb{H}_M = \{A \mid A^*M = MA\} = \{A \mid [Ax, y]_M = [x, Ay]_M \text{ for all } x, y \in \mathbb{F}^m\}$.
2. The set of M-skew-Hermitian matrices or M-skew-adjoint matrices:
$\mathbb{S}_M = \{K \mid K^*M = -MK\} = \{K \mid [Kx, y]_M = -[x, Ky]_M \text{ for all } x, y \in \mathbb{F}^m\}$.
3. The set of M-unitary matrices:
$\mathbb{U}_M = \{U \mid U^*MU = M\} = \{U \mid [Ux, Uy]_M = [x, y]_M \text{ for all } x, y \in \mathbb{F}^m\}$.

The concept of M-Hermitian and M-skew-Hermitian matrices can be generalized to matrix pencils via

$$\lambda M - B; \quad \text{with} \quad M = \pm M^*, \quad B = \pm B^*. \tag{6.2}$$

In fact, if M is invertible, the generalized eigenvalue problem with underlying matrix pencil as in (6.2) is equivalent to the eigenvalue problem for the matrix $A = M^{-1}B$, which is M-Hermitian or M-skew-Hermitian, depending on whether

M and B are Hermitian or skew-Hermitian. M-unitary matrices may be related to structured pencils indirectly by using a Cayley-transformation [24, 28, 38].

Another structured matrix pencil of the form

$$\lambda A^* - A,$$

which is called *palindromic* [22, 29, 43, 44], can also be transformed to a Hermitian/skew-Hermitian pencil with a Cayley-transformation and can therefore be considered a generalization of M-unitary matrices as well.

The study of matrices and matrix pencils with the symmetry structures outlined above started about one and a half centuries ago (we refer to the review article [25] and the references therein for more details) and continues to be of strong interest as there are many important applications in several areas of science and engineering, see, e.g., [16, 24, 38, 41, 45, 48, 49, 55].

A particular example is given by *Hamiltonian matrices* that arise, e.g., in systems and control theory [27, 38, 45, 55] and in the theory of dynamical and Hamiltonian systems [19–21]. These matrices are J-skew-Hermitian, where the skew-symmetric matrix J is given by

$$\mathrm{J} := \mathrm{J}_n := \begin{bmatrix} 0 & I_n \\ -I_n & 0 \end{bmatrix} \in \mathbb{R}^{2n \times 2n}. \tag{6.3}$$

(We drop the subscript n whenever it is clear from the context.) Due to their many applications, in particular those in system and control theory, the investigation of Hamiltonian matrices has been an important part of Volker Mehrmann's research interest and he and his coauthors have contributed many results to their theory, like discovering the reason for the difficulty in computing Hamiltonian Hessenberg forms [1], finding necessary and sufficient conditions for the existence of the Hamiltonian Schur form [28], and developing several algorithms for the Hamiltonian eigenvalue problem [4, 8, 38, 39]. For the understanding of the underlying theory, it was crucial to be aware of the presence of additional invariants besides the eigenvalues, eigenvectors, and root vectors of Hamiltonian matrices, the so called *signs* in the *sign characteristic* of purely imaginary eigenvalues. The classical Jordan canonical form cannot display these additional invariants, because it is obtained under general similarity transformations that ignore the special structure of Hamiltonian matrices. Therefore, it was important to develop a canonical form that is obtained under structure-preserving transformations, so that additional information like the sign characteristic is preserved and can be read off.

The phenomenon of presence of a sign characteristic not only occurs for the special case of Hamiltonian matrices, but for all three types of matrices structured with respect to the inner product (6.1) induced by M. To be more precise, it occurs for real eigenvalues of M-Hermitian, purely imaginary eigenvalues of M-skew-Hermitian, and unimodular eigenvalues of M-unitary matrices, as well as for the classes of related matrix pencils as in (6.2). In all cases, the sign characteristic has proven to play a fundamental role in theory and applications, like in the analysis of structured dynamic systems [19–21], in perturbation analysis of structured matrices [31–33, 40], and in the investigation of solutions of Riccati equations [12, 24, 28], to name a few examples.

After introducing the well-known canonical forms for Hermitian pencils and M-Hermitian matrices in the next section, we will give a survey on three related topics in the following sections:

(a) Structured canonical forms for Hamiltonian and related matrices.
(b) Canonical forms for doubly structured matrices.
(c) Singular value-like decompositions for matrices associated with two sesquilinear forms.

Throughout the chapter, we will use the following notation. $A_1 \oplus \cdots \oplus A_m$ is the block diagonal matrix $\mathrm{diag}(A_1, \ldots, A_m)$. The $n \times n$ identity matrix is denoted by I_n and $0_{m\times n}$ (0_n) stand for the $m \times n$ ($n \times n$) zero matrix. If the size is clear from the context, we may use I and 0 instead for convenience. We denote by e_j the jth unit vector, i.e., the jth column of I.

The $n \times n$ reverse identity will be denoted by R_n while $J_n(\alpha)$ stands for the upper triangular $n \times n$ Jordan block with eigenvalue α, that is

$$R_n = \begin{bmatrix} & & 1 \\ & \iddots & \\ 1 & & \end{bmatrix} \in \mathbb{F}^{n\times n}, \quad J_n(\alpha) = \begin{bmatrix} \alpha & 1 & & \\ & \ddots & \ddots & \\ & & \ddots & 1 \\ & & & \alpha \end{bmatrix} \in \mathbb{F}^{n\times n}$$

Finally, the $m \times (m+1)$ singular block in the Kronecker canonical form of matrix pencils is denoted by

$$L_m(\lambda) = \begin{bmatrix} \lambda & 1 & & \\ & \ddots & \ddots & \\ & & \lambda & 1 \end{bmatrix}.$$

6.2 Canonical Forms for Hermitian Pencils and M-Hermitian Matrices

For all the structured pencils of one of the forms in (6.2), the theory of structured Kronecker canonical forms is well-established, see, e.g., [9, 25, 26, 46, 47], following the work from the second half of the nineteenth century [23, 51, 52]. These forms are obtained under congruence transformations $(\lambda M - B) \mapsto X^*(\lambda M - B)X$ with X invertible, because those preserve both the Hermitian and the skew-Hermitian structure of matrices and thus the structure of pencils $\lambda M - B$ of the forms in (6.2). For instance, a *Hermitian pencil* $\lambda M - B$, i.e., a pencil such that both M and

B are Hermitian, has the following structured Kronecker canonical form under congruence.

Theorem 1 *Let $\lambda M - B$ be a complex $n \times n$ Hermitian pencil. Then there exists an invertible matrix X such that*

$$X^*(\lambda M - B)X = \mathcal{J}_C(\lambda) \oplus \mathcal{J}_R(\lambda) \oplus \mathcal{J}_\infty(\lambda) \oplus \mathcal{L}(\lambda), \tag{6.4}$$

where

$$\begin{aligned}
\mathcal{J}_C(\lambda) &= \left(\lambda \begin{bmatrix} 0 & R_{m_1} \\ R_{m_1} & 0 \end{bmatrix} - \begin{bmatrix} 0 & R_{m_1} J_{m_1}(\lambda_1) \\ R_{m_1} J_{m_1}(\bar{\lambda}_1) & 0 \end{bmatrix}\right) \oplus \cdots \\
&\quad \oplus \left(\lambda \begin{bmatrix} 0 & R_{m_p} \\ R_{m_p} & 0 \end{bmatrix} - \begin{bmatrix} 0 & R_{m_p} J_{m_p}(\lambda_p) \\ R_{m_p} J_{m_p}(\bar{\lambda}_p) & 0 \end{bmatrix}\right), \\
\mathcal{J}_R(\lambda) &= s_1 R_{n_1}\big(\lambda I_{n_1} - J_{n_1}(\alpha_1)\big) \oplus \cdots \oplus s_q R_{n_q}\big(\lambda I_{n_q} - J_{n_q}(\alpha_q)\big), \\
\mathcal{J}_\infty(\lambda) &= s_{q+1} R_{k_1}\big(\lambda J_{k_1}(0) - I_{k_1}\big) \oplus \cdots \oplus s_{q+r} R_{k_r}\big(\lambda J_{k_r}(0) - I_{k_r}\big), \\
\mathcal{L}(\lambda) &= \begin{bmatrix} 0 & L_{\ell_1}(\lambda) \\ L_{\ell_1}(\lambda)^T & 0 \end{bmatrix} \oplus \cdots \oplus \begin{bmatrix} 0 & L_{\ell_t}(\lambda) \\ L_{\ell_t}(\lambda)^T & 0 \end{bmatrix} \oplus 0_{\nu \times \nu},
\end{aligned}$$

with $\operatorname{Im} \lambda_j > 0,\ j = 1, \ldots, p;\ \alpha_j \in \mathbb{R},\ j = 1, \ldots, q;\ s_j = \pm 1,\ j = 1, \ldots, q+r$ *and* $p, q, r, t \in \mathbb{N}$.

If two pencils $C(\lambda)$ and $D(\lambda)$ are equivalent, i.e., $X_1 C(\lambda) X_2 = D(\lambda)$ for some invertible matrices X_1, X_2 independent of λ, we use the notation $C(\lambda) \sim D(\lambda)$. It is easy to show that the blocks in (6.4) satisfy

$$\begin{aligned}
\mathcal{J}_C(\lambda) &\sim \big(\lambda I_{m_1} - J_{m_1}(\lambda_1)\big) \oplus \big(\lambda I_{m_1} - J_{m_1}(\bar{\lambda}_1)\big) \oplus \\
&\qquad \cdots \oplus \big(\lambda I_{m_p} - J_{m_p}(\lambda_p)\big) \oplus \big(\lambda I_{m_p} - J_{m_p}(\bar{\lambda}_p)\big) \\
\mathcal{J}_R(\lambda) &\sim \big(\lambda I_{n_1} - J_{n_1}(\alpha_1)\big) \oplus \big(\lambda I_{n_q} - J_{n_q}(\alpha_q)\big) \\
\mathcal{J}_\infty(\lambda) &\sim \big(\lambda J_{k_1}(0) - I_{k_1}\big) \oplus \cdots \oplus \big(\lambda J_{k_r}(0) - I_{k_r}\big) \\
\mathcal{L}(\lambda) &\sim L_{\ell_1}(\lambda) \oplus L_{\ell_1}(\lambda)^T \oplus \cdots \oplus L_{\ell_t}(\lambda) \oplus L_{\ell_t}(\lambda)^T \oplus 0_{\nu \times \nu}.
\end{aligned}$$

Therefore, the classical Kronecker canonical form of the pencil $\lambda M - B$ can easily be read off from the structured version (6.4). In particular, the pairing of blocks elegantly displays the corresponding symmetry in the spectrum: the block $\mathcal{J}_C(\lambda)$ contains the nonreal eigenvalues that occur in complex conjugate pairs $\lambda_j, \bar{\lambda}_j$, both having exactly the same Jordan structures. If the pencil is singular, then the singular blocks – contained in $\mathcal{L}(\lambda)$ – are also paired: each *right singular block* $L_{\ell_j}(\lambda)$ has a corresponding *left singular block* $L_{\ell_j}(\lambda)^T$.

However, the structured canonical form (6.4) has an important advantage over the classical Kronecker canonical form of a Hermitian pencil. It displays additional invariants that are present under congruence transformations, the *signs* $s_1, \ldots, s_{q+r}$ attached to each Jordan block of a real eigenvalue and each Jordan block of the eigenvalue infinity. The collection of these signs is referred to as the *sign characteristic* of the Hermitian pencil [25], see also [15].

As a corollary of Theorem 1, one obtains a canonical form for M-Hermitian matrices, also known as M-selfadjoint matrices, see [15, 25].

Corollary 1 *Let $M \in \mathbb{C}^{n\times n}$ be Hermitian and invertible and let $A \in \mathbb{C}^{n\times n}$ be M-Hermitian. Then there exists an invertible matrix $X \in \mathbb{C}^{n\times n}$ such that*

$$X^{-1}AX = \mathscr{J}_R \oplus \mathscr{J}_C, \quad X^*MX = \mathscr{M}_R \oplus \mathscr{M}_C,$$

where

$$\mathscr{J}_R = J_{n_1}(\alpha_1) \oplus \cdots \oplus J_{n_q}(\alpha_q), \quad \mathscr{M}_R = s_1 R_{n_1} \oplus \cdots \oplus s_q R_{n_q}$$

$$\mathscr{J}_C = \begin{bmatrix} J_{m_1}(\lambda_1) & 0 \\ 0 & J_{m_1}(\bar{\lambda}_1) \end{bmatrix} \oplus \cdots \oplus \begin{bmatrix} J_{m_p}(\lambda_p) & 0 \\ 0 & J_{m_p}(\bar{\lambda}_p) \end{bmatrix}, \ \mathscr{M}_C = R_{2m_1} \oplus \cdots \oplus R_{2m_p},$$

where $\alpha_j \in \mathbb{R}$, $s_j = \pm 1$, $j = 1, \ldots, q$, $\operatorname{Im}\lambda_j > 0$, $j = 1, \ldots, p$, and $p, q \in \mathbb{N}$.

Indeed, the form is easily obtained by recalling that a matrix A is M-Hermitian if and only if the pencil $\lambda M - MA$ is a Hermitian pencil and applying Theorem 1 to this pencil. By convention, we will call s_j in Corollary 1 the *sign* of the Jordan block $J_{n_j}(\alpha_j)$.

For the other three types of matrix pencils in (6.2), structured canonical forms can be derived directly from (6.4). If $\lambda M - B$ is Hermitian/skew-Hermitian, skew-Hermitian/Hermitian, or skew-Hermitian/skew-Hermitian, then Theorem 1 can be applied to the Hermitian pencils $\lambda M - (-iB)$, $\lambda(-iM) - B$ or $\lambda(-iM) - (-iB)$, respectively, to obtain (6.4). As a consequence, these pencils also have a sign characteristic. In the case of pencils of "mixed" structure, i.e., one matrix being Hermitian and the other skew-Hermitian, now the purely imaginary eigenvalues (including the eigenvalue infinity) have signs.

For the case of real pencils of the form (6.2), also real structured Kronecker canonical forms under real congruence transformations are known. We refer the reader to [26, 47] for details.

6.3 Structured Canonical Forms for Hamiltonian Matrices

When the matrix defining the inner product (6.1) is the skew-symmetric matrix J from (6.3), then a J-Hermitian matrix is called *skew-Hamiltonian*, a J-skew-Hermitian is called *Hamiltonian*, and a J-unitary matrix is called *symplectic*.

In many applications, in particular in systems and control, *invariant Lagrangian subspaces* are of interest. A *Lagrangian subspace* is an n-dimensional subspace $\mathscr{L} \subseteq \mathbb{F}^{2n}$ that is J-neutral, i.e., $[x, y]_{\mathrm{J}} = 0$ for all $x, y \in \mathscr{L}$. Suppose the columns of the matrix $W_1 \in \mathbb{F}^{2n\times n}$ span an invariant Lagrangian subspace $\mathscr{L}$ of a Hamiltonian matrix $H \in \mathbb{F}^{n\times n}$. Then there exists a $2n \times n$ matrix W_2 such that $W = [W_1, W_2]$ is symplectic. Indeed, one may choose $W_2 = \mathrm{J}^T W_1(W_1^* W_1)^{-1}$. Then

$$W^*\mathrm{J}W = \begin{bmatrix} W_1^*\mathrm{J}W_1 & W_1^*\mathrm{J}W_2 \\ W_2^*\mathrm{J}W_1 & W_2^*\mathrm{J}W_2 \end{bmatrix} = \begin{bmatrix} 0 & I_n \\ -I_n & 0 \end{bmatrix} = \mathrm{J},$$

because $W_1^*\mathrm{J}W_1 = 0 = (W_1^* W_1)^{-1} W_1^*\mathrm{J}\mathrm{J}\mathrm{J}^T W_1(W_1^* W_1)^{-1} = W_2^*\mathrm{J}W_2$ as $\mathscr{L}$ is J-neutral. Since $\mathscr{L}$ is also an invariant subspace of H, we obtain

$$W^{-1}HW = \begin{bmatrix} T & D \\ 0 & -T^* \end{bmatrix}, \qquad D = D^*. \tag{6.5}$$

From the decomposition (6.5), we can easily see that a necessary condition for the existence of an invariant Lagrangian subspace is that the algebraic multiplicities of all purely imaginary eigenvalues must be even, because any purely imaginary eigenvalue $i\alpha$ of T is also an eigenvalue of $-T^*$. This condition, however, is not sufficient as the following example shows.

Example 1 Consider the Hamiltonian matrices

$$H_1 = \begin{bmatrix} 0 & 0 & 1 & 0 \\ 0 & 0 & 0 & 1 \\ -1 & 0 & 0 & 0 \\ 0 & -1 & 0 & 0 \end{bmatrix} = \mathrm{J}_2, \quad H_2 = \begin{bmatrix} 0 & 1 & 0 & 0 \\ -1 & 0 & 0 & 0 \\ 0 & 0 & 0 & 1 \\ 0 & 0 & -1 & 0 \end{bmatrix} = \begin{bmatrix} \mathrm{J}_1 & 0 \\ 0 & -\mathrm{J}_1^T \end{bmatrix}$$

Then the matrix H_1 does not have a decomposition (6.5) since for any symplectic matrix W, by definition, $W^* H_1 W = W^*\mathrm{J}_2 W = \mathrm{J}_2$. The matrix H_2 on the other hand already is of the form (6.5). Surprisingly, the matrices H_1 and H_2 are similar. It is easy to check that they both have the semi-simple eigenvalues i and $-i$, which both have the algebraic multiplicity two.

To explain this surprising behavior, a closer look at a structured canonical form of Hamiltonian matrices is necessary. One way is to consider instead of a Hamiltonian matrix H the iJ-Hermitian matrix iH and to apply Corollary 1. This yields the existence of an invertible matrix X such that

$$X^{-1}HX = \mathscr{H}_I \oplus \mathscr{H}_C, \quad X^*JX = \mathscr{M}_I \oplus \mathscr{M}_C$$

where

$$\mathscr{H}_I = iJ_{n_1}(\alpha_1) \oplus \cdots \oplus iJ_{n_q}(\alpha_q), \quad \mathscr{M}_I = s_1 iR_{n_1} \oplus \cdots \oplus s_q iR_{n_q}$$

$$\mathscr{H}_C = i\begin{bmatrix} J_{m_1}(\lambda_1) & 0 \\ 0 & J_{m_1}(\bar{\lambda}_1) \end{bmatrix} \oplus \cdots \oplus i\begin{bmatrix} J_{m_p}(\lambda_p) & 0 \\ 0 & J_{m_p}(\bar{\lambda}_p) \end{bmatrix},$$

$$\mathscr{M}_C = iR_{2m_1} \oplus \cdots \oplus iR_{2m_p}.$$

Here, s_j is the sign of the Jordan block $J_{n_j}(i\alpha_j)$ of H, for $j = 1, \ldots, q$, i.e., the purely imaginary eigenvalues come with a sign characteristic. Although this canonical form reveals these additional invariants, one cannot tell immediately whether a decomposition as (6.5) exists. One possible way to proceed is to apply further transformations to transform X^*JX back to J, say by constructing an invertible matrix Y such that

$$\mathscr{H} := (XY)^{-1}H(XY), \quad (XY)^*J(XY) = J$$

Then, the matrix XY is symplectic, because $(XY)^*J(XY) = J$. Clearly, there are many such transformations and then the task is to choose among all these transformations a particular one so that $\mathscr{H}$ is as close to a block upper triangular form (6.5) as possible. In [28] such an *optimal* canonical form is presented in the sense that the (2,1) block of $\mathscr{H}$ has the lowest possible rank. The result is given in the following theorem.

Theorem 2 ([28]) *Let $H \in \mathbb{C}^{2n\times 2n}$ be a Hamiltonian matrix. Then there exists a symplectic matrix $W \in \mathbb{C}^{2n\times 2n}$ such that*

$$W^{-1}HW = \left[\begin{array}{cccc|cccc} T_c & & & & 0 & & & \\ & T_{ie} & & & & D_{ie} & & \\ & & T_{io} & & & & D_{io} & \\ & & & T_{ior} & & & & D_{ior} \\ \hline & & & & -T_c^* & & & \\ & & & & & -T_{ie}^* & & \\ & & & & & & -T_{io}^* & \\ & & & M_{ior} & & & & -T_{ior}^* \end{array}\right],$$

where the blocks have the following properties:

(i)

$$T_c = J_{m_1}(\lambda_1) \oplus J_{m_2}(\lambda_2) \oplus \cdots \oplus J_{m_p}(\lambda_p),$$

where $\lambda_1, \ldots, \lambda_p \in \mathbb{C}$ *with* $\operatorname{Re}\lambda_1, \ldots, \operatorname{Re}\lambda_p > 0$.

(ii)

$$T_{ie} = J_{n_1}(i\alpha_1) \oplus \cdots \oplus J_{n_q}(i\alpha_q), \quad D_{ie} = s_1 e_{n_1} e_{n_1}^* \oplus \cdots \oplus s_q e_{n_q} e_{n_q}^*,$$

where $\alpha_1, \ldots, \alpha_q \in \mathbb{R}$ *and* $s_1, \ldots, s_q = \pm 1$. *Each sub-matrix*

$$\begin{bmatrix} J_{n_j}(i\alpha_j) & s_j e_{n_j} e_{n_j}^* \\ 0 & -(J_{n_j}(i\alpha_j))^* \end{bmatrix}$$

corresponds to an even-sized Jordan block $J_{2n_j}(i\alpha_j)$ *of* H *with sign* s_j.

(iii)

$$T_{io} = T_{io}^{(1)} \oplus \cdots \oplus T_{io}^{(r)}, \quad D_{io} = D_{io}^{(1)} \oplus \cdots \oplus D_{io}^{(r)},$$

and

$$T_{io}^{(j)} = \begin{bmatrix} J_{\ell_j}(i\beta_j) & 0 & -\frac{\sqrt{2}}{2} e_{\ell_j} \\ 0 & J_{k_j}(i\beta_j) & -\frac{\sqrt{2}}{2} e_{k_j} \\ 0 & 0 & i\beta_j \end{bmatrix}, \quad D_{io}^{(j)} = \frac{\sqrt{2}i}{2}\sigma_j \begin{bmatrix} 0 & 0 & e_{\ell_j} \\ 0 & 0 & -e_{k_j} \\ -e_{\ell_j}^* & e_{k_j}^* & 0 \end{bmatrix},$$

where $\beta_1, \ldots, \beta_r \in \mathbb{R}$ *and* $\sigma_1, \ldots, \sigma_r = \pm 1$. *For each* $j = 1, \ldots, r$, *the sub-matrix*

$$\begin{bmatrix} T_{io}^{(j)} & D_{io}^{(j)} \\ 0 & -(T_{io}^{(j)})^* \end{bmatrix}$$

corresponds to two odd-sized Jordan blocks of H *associated with the same purely imaginary eigenvalue* $i\beta_j$. *The first is* $J_{2\ell_j+1}(i\beta_j)$ *with sign* σ_j *and the second is* $J_{2k_j+1}(i\beta_j)$ *with sign* $-\sigma_j$.

(iv)

$$T_{ior} = T_{ior}^{(1)} \oplus \cdots \oplus T_{ior}^{(t)}, \quad M_{ior} = M_{ior}^{(1)} \oplus \cdots \oplus M_{ior}^{(t)}, \quad D_{ior} = D_{ior}^{(1)} \oplus \cdots \oplus D_{ior}^{(t)},$$

where

$$T_{ior}^{(j)} = \begin{bmatrix} J_{\eta_j}(i\gamma_j) & 0 & -\frac{\sqrt{2}}{2} e_{\eta_j} \\ 0 & J_{\nu_j}(i\delta_j) & -\frac{\sqrt{2}}{2} e_{\nu_j} \\ 0 & 0 & \frac{i}{2}(\gamma_j + \delta_j) \end{bmatrix}, \quad M_{ior}^{(j)} = \sigma_{r+j} \begin{bmatrix} 0\,0 & 0 \\ 0\,0 & 0 \\ 0\,0 & -\frac{1}{2}(\gamma_j - \delta_j) \end{bmatrix},$$

$$D_{ior}^{(j)} = \frac{\sqrt{2}i}{2}\sigma_{r+j} \begin{bmatrix} 0 & 0 & e_{\eta_j} \\ 0 & 0 & -e_{\nu_j} \\ -e_{\eta_j}^* & e_{\nu_j}^* & -\frac{\sqrt{2}i}{2}(\gamma_j - \delta_j) \end{bmatrix}$$

and $\gamma_j, \delta_j \in \mathbb{R}$, $\gamma_j \neq \delta_j$, $\sigma_{r+j} = \pm 1$ *for* $j = 1, \ldots, t$. *The submatrix*

$$\begin{bmatrix} T_{ior}^{(j)} & D_{ior}^{(j)} \\ M_{ior}^{(j)} & -(T_{ior}^{(j)})^* \end{bmatrix}$$

corresponds to two odd-sized Jordan blocks of H *associated with two distinct purely imaginary eigenvalues* $i\gamma_j$ *and* $i\delta_j$. *The first one is* $J_{2\eta_j+1}(i\gamma_j)$ *with sign* σ_{r+j} *and the second one is* $J_{2\nu_j+1}(i\delta_j)$ *with sign* $-\sigma_{r+j}$.

Thus, the spectrum of H can be read off from the Hamiltonian submatrices

$$H_c := \begin{bmatrix} T_c & 0 \\ 0 & -T_c^* \end{bmatrix}, \; H_{ie} := \begin{bmatrix} T_{ie} & D_{ie} \\ 0 & -T_{ie}^* \end{bmatrix},$$

$$H_{io} := \begin{bmatrix} T_{io} & D_{io} \\ 0 & -T_{io}^* \end{bmatrix}, \; H_{ior} := \begin{bmatrix} T_{ior} & D_{ior} \\ M_{ior} & -T_{ior}^* \end{bmatrix}$$

The submatrix H_c contains all Jordan blocks associated with eigenvalues that are not purely imaginary. To be more precise, T_c contains all the Jordan blocks of eigenvalues of H with positive real parts, and $-T_c^*$ contains all the Jordan blocks of eigenvalues of H with negative real parts. The submatrix H_{ie} contains all even-sized Jordan blocks associated with purely imaginary eigenvalues of H, whereas H_{io} and H_{ior} contain all Jordan blocks associated with purely imaginary eigenvalues of H that have odd sizes. Here, H_{io} consists of pairs of Jordan blocks of (possibly different) odd sizes that are associated with the same purely imaginary eigenvalue, but have opposite signs. On the other hand, H_{ior} consists of the remaining Jordan blocks that do not allow such a pairing. In particular, if H_{ior} contains more than one Jordan block associated to a particular purely imaginary eigenvalue, then all such blocks must have the same sign in the sign characteristic.

While the canonical form in Theorem 2 looks quite complicated at first sight, its advantage is that the conditions for the existence of a decomposition of the form (6.5) can now be trivially derived by requesting the submatrix H_{ior} being void. Thus, with the interpretation of H_{ior} in terms of the sign characteristic, we immediately obtain the following result that is in accordance with a corresponding result in [42] in terms of M-selfadjoint matrices.

Theorem 3 ([28]) *A Hamiltonian matrix* H *has a decomposition (6.5) if and only if for each purely imaginary eigenvalue of* H, *it has an even number of odd-sized Jordan blocks half of which have sign* $+1$ *and half of which have sign* -1.

The theorem also gives necessary and sufficient conditions for the existence of the *Hamiltonian Schur form*. A Hamiltonian matrix H is said to have a Hamiltonian Schur form, if it allows a decomposition of the form (6.5) with T being upper triangular and W being both symplectic and unitary, i.e., satisfying $W^*JW = J$ and $W^*W = I$. Under the same conditions as in Theorem 3, we obtain the existence of a symplectic matrix W such that $W^{-1}HW$ is in the Hamiltonian canonical form of

Theorem 2 without the blocks from H_{ior}. Since the blocks T_c, T_{ie}, and T_{io} are upper triangular, we find that $W^{-1}HW$ has the form (6.5) with T being upper triangular. A Hamiltonian Schur form can then be derived by performing a symplectic QR-like decomposition to the symplectic matrix W, see [7, 28].

Corollary 2 ([28]) *Let H be a Hamiltonian matrix. Then there exists a unitary and symplectic matrix W such that $W^{-1}HW$ has the form (6.5) with T being upper triangular if and only if for each purely imaginary eigenvalue, H has an even number of odd-sized Jordan blocks half of which have sign $+1$ and half of which have sign -1.*

The following example, borrowed from [30], shows that the two Jordan blocks that are paired in one of the particular submatrices of H_{io} in Theorem 2 may indeed have different sizes.

Example 2 Consider the two matrices

$$H = \begin{bmatrix} i & 1 & 1 & 0 \\ 0 & i & 0 & 0 \\ 0 & 0 & i & 0 \\ 0 & 0 & -1 & i \end{bmatrix} \quad \text{and} \quad X = \frac{1}{\sqrt{2}} \begin{bmatrix} 2i & 0 & i & 2i \\ 0 & 1 & -i & -i \\ 0 & 0 & 1 & 0 \\ 0 & -i & 1 & 1 \end{bmatrix}.$$

Then H is a Hamiltonian matrix in Hamiltonian Schur form and X is the transformation matrix that brings the pair $(iH, i\mathrm{J})$ into the canonical form of Corollary 1:

$$X^{-1}(iH)X = \begin{bmatrix} -1 & 1 & 0 & 0 \\ 0 & -1 & 1 & 0 \\ 0 & 0 & -1 & 0 \\ 0 & 0 & 0 & -1 \end{bmatrix}, \quad X^*(iJ)X = \begin{bmatrix} 0 & 0 & 1 & 0 \\ 0 & 1 & 0 & 0 \\ 1 & 0 & 0 & 0 \\ 0 & 0 & 0 & -1 \end{bmatrix},$$

Thus, H has the eigenvalue i and two corresponding Jordan blocks with sizes 3 and 1. The Jordan block of size 3 has the sign $+1$ and the Jordan block of size 1 has the sign -1 thus satisfying the condition of Theorem 3.

Example 3 Revisiting the matrices H_1 and H_2 from Example 1, one can easily check that the eigenvalues i and $-i$ of H_2 have one Jordan block with sign $+1$ and one Jordan block with sign -1 each. In fact, H_2 is a matrix of the form H_{io} as in Theorem 2. On the other hand, for the matrix H_1 the signs corresponding to i are both $+1$ and the signs corresponding to $-i$ are both -1. In fact, H_1 is in the canonical form of Theorem 2 corresponding exactly to a matrix in the form H_{ior}.

However, for the matrix $X = [e_1, e_3, e_2, e_4]$, which is not symplectic, we obtain that $X^{-1}H_1X = H_2$ is in the form (6.5). Although the transformation with X maps the Hamiltonian matrix H_1 to the Hamiltonian matrix H_2, it is not a structure-preserving transformation in the sense that for small Hamiltonian perturbations $H_1 + \Delta H$ the transformed matrix $H_2 + X^{-1}\Delta HX$ is in general not Hamiltonian. This fact in a sense allows the similarity transformation with X to take a bypass by ignoring the sign constraints shown in Theorem 3. It was shown in [28] that the existence of the decomposition (6.5) with a non-symplectic similarity transformation only requires the algebraic multiplicities of all purely imaginary eigenvalues of H to be even.

In the case that the Hamiltonian matrix under consideration is real, there is also a canonical form under real symplectic similarity, see [28, Theorem 22]. In this case, the eigenvalues of a Hamiltonian matrix are not only symmetric with respect to the imaginary axis, but also with respect to the real axis. Thus, in particular the Jordan blocks associated with purely imaginary eigenvalues $i\alpha$, $\alpha > 0$ occur in complex conjugate pairs and it turns out that their signs in the sign characteristic are related. It can be shown that if $J_{m_1}(i\alpha), \ldots, J_{m_p}(i\alpha)$ are the Jordan blocks of a Hamiltonian matrix H associated with the eigenvalue $i\alpha$ and having the signs $s_1, \ldots, s_p$, then the signs of the corresponding Jordan blocks $J_{m_1}(-i\alpha), \ldots, J_{m_p}(-i\alpha)$ are $-s_1, \ldots, -s_p$, respectively. Another key difference between the real and the complex case is the behavior of the eigenvalue 0 when H is singular. While in the complex case this eigenvalue can be treated as any other purely imaginary eigenvalue, it has a special Jordan structure in the real case: each odd-sized Jordan block associated with zero must have an even number of copies and in the corresponding sign characteristic, half of the signs must be $+1$ and half of the signs must be -1. In contrast, there is no such pairing for Jordan blocks associated with zero that have even sizes. This extraordinary behavior of the eigenvalue zero leads to a real version of Theorem 3 that yields slightly different conditions in comparison with the complex case.

Theorem 4 ([28]) *A real Hamiltonian matrix H has a decomposition (6.5) with a real symplectic transformation matrix W if and only if for each nonzero purely imaginary eigenvalue, H has an even number of odd-sized Jordan blocks half of which have sign $+1$ and half of which have sign -1.*

For most of the problems arising from systems and control, one actually is interested in special invariant Lagrangian subspaces of Hamiltonian matrices. For instance, for the existence of solutions of algebraic Riccati equations [24, 38, 55] one is interested in the invariant Lagrangian subspaces of a Hamiltonian matrix corresponding to the eigenvalues in the closed or open left half complex plane. A more general question is the following: if H is a $2n \times 2n$ Hamiltonian matrix and a list Λ of n of its eigenvalues (counted with multiplicities) is prescribed, does there exists an invariant Lagrangian subspace associated with the eigenvalues in Λ, and if so, is this subspace unique? This question can be answered with the help of Theorem 3 or its corresponding real version. As we already know, the existence of an invariant Lagrangian subspace for a Hamiltonian matrix H is equivalent to the existence of a decomposition of the form (6.5). From (6.5), the spectrum of H is the union of the spectra of both T and $-T^*$. So one may assume that H has pairwise distinct non purely imaginary eigenvalues

$$\lambda_1, -\bar{\lambda}_1, \ldots, \lambda_p, -\bar{\lambda}_p, \quad \text{with algebraic multiplicities} \quad \nu_1, \nu_1, \ldots, \nu_p, \nu_p$$

and pairwise distinct purely imaginary eigenvalues

$$i\alpha_1, \ldots, i\alpha_q \quad \text{with algebraic multiplicities} \quad 2\mu_1, \ldots, 2\mu_q.$$

In order to have an invariant Lagrangian subspace, or, equivalently, a decomposition (6.5), it is necessary that the spectrum of T contains λ_j and $-\bar{\lambda}_j$ with algebraic multiplicities k_j and $\nu_j - k_j$, respectively, for each $j = 1, \ldots, p$, and μ_j copies of $i\alpha_j$ for each $j = 1, \ldots, q$. Let $\Omega(H)$ denote the set of all possible spectra for T in a decomposition of the form (6.5) of H. Then this set contains $\prod_{j=1}^{p}(\nu_j + 1)$ different selections, because k_j can be any number from 0 to ν_j for each j. Among them there are 2^p selections that contain either λ_j or $-\bar{\lambda}_j$, but not both, for all j. This subset of $\Omega(H)$ is denoted by $\tilde{\Omega}(H)$.

The answer to the question of existence of invariant Lagrangian subspaces with a prescribed spectrum is then given in the following theorem.

Theorem 5 ([12]) *A Hamiltonian matrix H has an invariant Lagrangian subspace corresponding to every $\omega \in \Omega(H)$ if and only if the conditions Theorem 3 (or Theorem 4 in real case) hold. Concerning uniqueness, we have the following conditions.*

(i) For every $\omega \in \Omega(H)$, H has a unique corresponding invariant Lagrangian subspace if and only if for every non purely imaginary eigenvalue λ_j (and $-\bar{\lambda}_j$) H has only a single Jordan block, and for every purely imaginary eigenvalue $i\alpha_j$, H only has even-sized Jordan blocks all of them having the same sign.

(ii) For every $\omega \in \tilde{\Omega}(H)$, H has a unique corresponding invariant Lagrangian subspace if and only if for every purely imaginary eigenvalue $i\alpha_j$, H has only even-sized Jordan blocks all of them having the same sign.

When the Lagrangian invariant subspaces corresponding to the eigenvalues in $\Omega(H)$ or $\tilde{\Omega}(H)$ are not unique, then it is possible to parameterize their bases. Moreover, the results in Theorem 5 can be used to study Hermitian solutions to algebraic Riccati equations, see [12].

We will now turn to skew-Hamiltonian matrices. Analogously to the case of Hamiltonian matrices, it can be shown that if the columns of W_1 span an invariant Lagrangian subspace of a skew-Hamiltonian matrix K, then there exists a symplectic matrix $W = [W_1, W_2]$ such that

$$W^{-1}KW = \begin{bmatrix} T & D \\ 0 & T^* \end{bmatrix}, \quad D = -D^*. \tag{6.6}$$

Structured canonical forms for complex skew-Hamiltonian matrices can be constructed in the same way as for complex Hamiltonian matrices, using the fact that K is skew-Hamiltonian if and only if iK is Hamiltonian. Thus, the conditions for the existence of invariant Lagrangian subspaces are the same as in Theorem 3 replacing "purely imaginary eigenvalues" with "real eigenvalues".

Interestingly, for any real skew-Hamiltonian matrix the real version of the decomposition (6.6) always exists, see [50]. Also, it is proved in [10] that for a real skew-Hamiltonian matrix K, there always exists a real symplectic matrix W such that

$$W^{-1}KW = N \oplus N^T,$$

where N is in real Jordan canonical form. The result shows clearly that every Jordan block of K has an even number of copies.

Finally, if $S \in \mathbb{F}^{n\times n}$ is a symplectic matrix and if the columns of W_1 span an invariant Lagrangian invariant subspace of S, then similar to the Hamiltonian case one can show that there exists a symplectic matrix $W = [W_1, W_2]$ such that

$$W^{-1}SW = \begin{bmatrix} T & D \\ 0 & T^{-*} \end{bmatrix}, \quad DT^* = (DT^*)^*. \tag{6.7}$$

The case of symplectic matrices can be reduced to the case of Hamiltonian matrices with the help of the *Cayley transformation*, see Chap. 2 in this book for details on the Cayley transformation. Therefore, structured Jordan canonical forms for symplectic matrices can be derived using the structured canonical forms for Hamiltonian matrices in Theorem 2 and its real version. Then, conditions for the existence of a decomposition of the form (6.7) can be obtained which are essentially the same as in Theorems 3 and 4, with purely imaginary eigenvalues replaced by unimodular eigenvalues in the symplectic case, see [28].

6.4 Doubly Structured Matrices and Pencils

In this section we discuss canonical forms of doubly structured matrices and pencils. This research was mainly motivated by applications from quantum chemistry [2, 3, 16, 41, 49]. In linear response theory, one has to solve a generalized eigenvalue problem with a pencil of the form

$$\lambda \begin{bmatrix} C & Z \\ -Z & -C \end{bmatrix} - \begin{bmatrix} E & F \\ F & E \end{bmatrix}, \tag{6.8}$$

where C, E, F are $n \times n$ Hermitian matrices and Z is skew-Hermitian. The simplest response function model is the time-dependent Hartree-Fock model (also called random phase approximation) in which the pencil (6.8) takes the simpler structure $C = I$ and $Z = 0$ so that the corresponding eigenvalue problem can be reduced to a standard eigenvalue problem with a matrix of the form

$$A = \begin{bmatrix} E & F \\ -F & -E \end{bmatrix}$$

with E and F being Hermitian. It is straightforward to check that A is Hamiltonian (or J-skew-Hermitian) and M-Hermitian, where

$$M = \begin{bmatrix} I_n & 0 \\ 0 & -I_n \end{bmatrix}.$$

In the general setting, we consider matrices that are structured with respect to two invertible Hermitian or skew-Hermitian matrices K and M. Because any skew-Hermitian matrix K can be transformed to the Hermitian matrix iK and any K-skew-Hermitian matrix A can be transformed to the K-Hermitian matrix iA, we may assume that both K and M are invertible and Hermitian and consider two cases only:

(a) A is K-Hermitian and M-Hermitian, i.e., $KA = A^*K,\ MA = A^*M$,
(b) A is K-Hermitian and M-skew-Hermitian, i.e., $KA = A^*K,\ MA = -A^*M$.

The task is now to find an invertible matrix X to perform a transformation

$$\mathscr{A} = X^{-1}AX, \quad \mathscr{K} = X^*KX, \quad \mathscr{M} = X^*MX$$

so that the canonical form of Corollary 1 (or the corresponding version for M-skew-Hermitian matrices) for both pairs (A, K) and (A, M) can simultaneously be recovered. As shown in [34], this is not always possible, because the situation is too general. So it is reasonable to restrict oneself to the situation where the pencil $\lambda K - M$ is *nondefective*, meaning that all the eigenvalues of the Hermitian pencil $\lambda K - M$ are semisimple. (This assumption is satisfied in the case $K = i\,\mathrm{J}_n$ and $M = \mathrm{diag}(I_n, -I_n)$ which is relevant for the applications in quantum chemistry.) Then by (6.4), there exists an invertible matrix Q such that

$$Q^*(\lambda K - M)Q = (\lambda K_1 - M_1) \oplus \cdots \oplus (\lambda K_p - M_p), \tag{6.9}$$

where, for each $j = 1, \ldots, p$, either

$$\lambda K_j - M_j = \lambda \begin{bmatrix} 0 & 1 \\ 1 & 0 \end{bmatrix} - \begin{bmatrix} 0 & \lambda_j \\ \bar{\lambda}_j & 0 \end{bmatrix}$$

containing a pair of nonreal eigenvalues $\lambda_j, \bar{\lambda}_j$, or

$$\lambda K_j - M_j = s_j\big(\lambda \left[\,1\,\right] - \left[\,\alpha_j\,\right]\big), \quad s_j = \pm 1,$$

containing a single real eigenvalue α_j with a sign s_j. (We highlight that the same eigenvalues $\lambda_j, \bar{\lambda}_j$ or α_j, respectively, may appear multiple times among the blocks $\lambda K_1 - M_1, \ldots, \lambda K_p - M_p$.)

Under this assumption, the following structured canonical form can be obtained for a matrix that is doubly structured in the sense of case (a) above.

Theorem 6 ([34]) *Suppose K, M are Hermitian and invertible, such that the pencil $\lambda K - M$ is nondefective. Suppose A is both K-Hermitian and M-Hermitian. Then there exists an invertible matrix X such that*

$$\begin{aligned}\mathscr{A} &:= X^{-1}AX = A_1 \oplus A_2 \oplus \cdots \oplus A_p \\ \mathscr{K} &:= X^*KX \;= K_1 \oplus K_2 \oplus \cdots \oplus K_p \\ \mathscr{M} &:= X^*MX = M_1 \oplus M_2 \oplus \cdots \oplus M_p,\end{aligned}$$

where for each $j = 1, 2, \ldots, p$ the blocks A_j, K_j, M_j are in one of the following forms.

(i) Blocks associated with a pair of conjugate complex eigenvalues of A:

$$A_j = \begin{bmatrix} J_{m_j}(\lambda_j) & 0 \\ 0 & J_{m_j}(\bar{\lambda}_j) \end{bmatrix}, \quad K_j = \begin{bmatrix} 0 & R_{m_j} \\ R_{m_j} & 0 \end{bmatrix}, \quad M_j = \begin{bmatrix} 0 & \gamma_j R_{m_j} \\ \bar{\gamma}_j R_{m_j} & 0 \end{bmatrix},$$

where $\lambda_j \in \mathbb{C} \setminus \mathbb{R}$, $\gamma_j = c_j + id_j \neq 0$ with $c_j, d_j \in \mathbb{R}$ and $d_j \geq 0$.

(ii) Blocks associated with real eigenvalues of A and real eigenvalues of $\lambda K - M$:

$$A_j = J_{n_j}(\alpha_j), \quad K_j = s_j R_{n_j}, \quad M_j = s_j \eta_j R_{n_j},$$

where $s_j = \pm 1$, $0 \neq \eta_j \in \mathbb{R}$, and $\alpha_j \in \mathbb{R}$. The sign of the block A_j with respect to K is s_j and the sign with respect to M is $\operatorname{sign}(s_j \eta_j)$.

(iii) Blocks associated with real eigenvalues of A and a pair of conjugate complex eigenvalues of $\lambda K - M$:

$$A_j = \begin{bmatrix} J_{n_j}(\alpha_j) & 0 \\ 0 & J_{n_j}(\alpha_j) \end{bmatrix}, \quad K_j = \begin{bmatrix} 0 & R_{n_j} \\ R_{n_j} & 0 \end{bmatrix}, \quad M_j = \begin{bmatrix} 0 & \gamma_j R_{n_j} \\ \bar{\gamma}_j R_{n_j} & 0 \end{bmatrix},$$

where $\alpha_j \in \mathbb{R}$ and $\gamma_j = c_j + id_j$ with $c_j, d_j \in \mathbb{R}$ and $d_j > 0$. Thus A_j contains a pair of two $n_j \times n_j$ Jordan blocks of A associated with the same real eigenvalue α_j. The pairs of corresponding signs are $(+1, -1)$ with respect to both K and M.

It is easily seen that the structured canonical forms for A with respect to K and M, respectively, can immediately be read off from the canonical form in Theorem 6. In addition, the structured canonical form of $\lambda K - M$ as in (6.9) can easily be derived from $\lambda \mathscr{K} - \mathscr{M}$. Therefore, Theorem 6 combines three different structured canonical forms into one.

On the other hand, Theorem 6 shows that the presence of two structures in A leads to additional restrictions in the Jordan structure of A which can be seen from the blocks of type (iii) of Theorem 6. This block is *indecomposable* in a sense that there does not exist any transformation of the form $(A_j, K_j, M_j) \mapsto (Y^{-1}A_jY, Y^*K_jY, Y^*M_jY)$ that simultaneously block-diagonalizes all three matrices. As a consequence, the Jordan structure of a matrix A that is both K-Hermitian and M-Hermitian is rather restricted if the pencil $\lambda K - M$ (is defective and) has only nonreal eigenvalues. In that case, each Jordan block associated with a real eigenvalue of A must occur an even number of times in the Jordan canonical form of A. In particular, all real eigenvalues of A must have even algebraic multiplicity.

In case (b), i.e., when A is K-Hermitian and M-skew-Hermitian, then the eigenstructure of A has even richer symmetry than in case (a), because now the spectrum has to be symmetric to both the real and the imaginary axes. Also, the Jordan blocks associated with real eigenvalues of A will have signs with respect to K while the ones associated with purely imaginary eigenvalues will have signs with respect to M. Thus, the eigenvalue zero will play a special role, because it will have signs both with respect to K and to M. A structured canonical form for this case will be given in the next theorem, for which we need extra notation. By Σ_n, we denote the $n \times n$ anti-diagonal matrix alternating sign matrix, i.e.,

$$\Sigma_n = \begin{bmatrix} & & & (-1)^0 \\ & & (-1)^1 & \\ & \iddots & & \\ (-1)^{n-1} & & & \end{bmatrix}.$$

Theorem 7 ([34]) *Suppose K, M are Hermitian and invertible, and $\lambda K - M$ is nondefective. Suppose A is both K-Hermitian and M-skew-Hermitian. Then there exists an invertible matrix X such that*

$$\begin{aligned} \mathscr{A} &:= X^{-1}AX = A_1 \oplus A_2 \oplus \cdots \oplus A_p \\ \mathscr{K} &:= X^*KX \;= K_1 \oplus K_2 \oplus \cdots \oplus K_p \\ \mathscr{M} &:= X^*MX = M_1 \oplus M_2 \oplus \cdots \oplus M_p, \end{aligned}$$

where for $j = 1, 2, \ldots, p$ the blocks A_j, K_j, M_j are in one of the following forms.

(i) Blocks associated with nonreal, non purely imaginary eigenvalues of A:

$$A_j = \begin{bmatrix} J_{m_j}(\lambda_j) & 0 & 0 & 0 \\ 0 & -J_{m_j}(\lambda_j) & 0 & 0 \\ 0 & 0 & J_{m_j}(\bar{\lambda}_j) & 0 \\ 0 & 0 & 0 & -J_{m_j}(\bar{\lambda}_j) \end{bmatrix},$$

$$K_j = \begin{bmatrix} 0 & 0 & R_{m_j} & 0 \\ 0 & 0 & 0 & R_{m_j} \\ R_{m_j} & 0 & 0 & 0 \\ 0 & R_{m_j} & 0 & 0 \end{bmatrix}, \quad M_j = \begin{bmatrix} 0 & 0 & 0 & \gamma_j R_{m_j} \\ 0 & 0 & \gamma_j R_{m_j} & 0 \\ 0 & \bar{\gamma}_j R_{m_j} & 0 & 0 \\ \bar{\gamma}_j R_{m_j} & 0 & 0 & 0 \end{bmatrix},$$

where $\lambda_j = a_j + ib_j$ with $a_j, b_j \in \mathbb{R}$ and $a_j b_j > 0$, and the parameter γ_j satisfies one of the following three mutually exclusive conditions: (a) $\gamma_j = \beta_j$ with $\beta_j > 0$, (b) $\gamma_j = i\beta_j$ with $\beta_j > 0$, or (c) $\gamma_j = c_j + id_j$ with $c_j, d_j \in \mathbb{R}$ and $c_j d_j > 0$.

(ii) *Blocks associated with a pair of real eigenvalues* $\pm\alpha_j$ *of* A *and nonreal non purely imaginary eigenvalues of* $\lambda K - M$:

$$A_j = \begin{bmatrix} J_{n_j}(\alpha_j) & 0 & 0 & 0 \\ 0 & -J_{n_j}(\alpha_j) & 0 & 0 \\ 0 & 0 & J_{n_j}(\alpha_j) & 0 \\ 0 & 0 & 0 & -J_{n_j}(\alpha_j) \end{bmatrix},$$

$$K_j = \begin{bmatrix} 0 & 0 & R_{n_j} & 0 \\ 0 & 0 & 0 & R_{n_j} \\ R_{n_j} & 0 & 0 & 0 \\ 0 & R_{n_j} & 0 & 0 \end{bmatrix}, \quad M_j = \begin{bmatrix} 0 & 0 & 0 & \gamma_j R_{n_j} \\ 0 & 0 & \gamma_j R_{n_j} & 0 \\ 0 & \bar{\gamma}_j R_{n_j} & 0 & 0 \\ \bar{\gamma}_j R_{n_j} & 0 & 0 & 0 \end{bmatrix},$$

where $0 < \alpha_j \in \mathbb{R}$ *and* $\gamma_j = c_j + i d_j$ *with* $c_j, d_j \in \mathbb{R}$ *and* $c_j d_j > 0$. *The two Jordan blocks associated with* α_j *have the signs* 1 *and* -1 *with respect to* K *and the two Jordan blocks associated with* $-\alpha_j$ *also have the signs* 1 *and* -1 *with respect to* K.

(iii) *Blocks associated with a pair of real eigenvalues* $\pm\alpha_j$ *of* A *and real or purely imaginary eigenvalues of* $\lambda K - M$:

$$A_j = \begin{bmatrix} J_{n_j}(\alpha_j) & 0 \\ 0 & -J_{n_j}(\alpha_j) \end{bmatrix},$$

$$K_j = s_j \begin{bmatrix} R_{n_j} & 0 \\ 0 & \left(\frac{\gamma_j}{|\gamma_j|}\right)^2 R_{n_j} \end{bmatrix}, \quad M_j = \begin{bmatrix} 0 & \gamma_j R_{n_j} \\ \bar{\gamma}_j R_{n_j} & 0 \end{bmatrix},$$

where $0 < \alpha_j \in \mathbb{R}$, $s_j = \pm 1$, $\gamma_j = \beta_j$ *or* $\gamma_j = i\beta_j$ *with* $0 < \beta_j \in \mathbb{R}$. *The Jordan block of* A *associated with* α_j *has the sign* s_j *with respect to* K *and the one associated with* $-\alpha_j$ *has the sign* $(-1)^{n_j+1} s_j (\gamma_j/|\gamma_j|)^2$ *with respect to* K.

(iv) *Blocks associated with a pair of purely imaginary eigenvalues* $\pm i\alpha_j$ *of* A *and nonreal non purely imaginary eigenvalues of* $\lambda K - M$:

$$A_j = \begin{bmatrix} iJ_{n_j}(\alpha_j) & 0 & 0 & 0 \\ 0 & -iJ_{n_j}(\alpha_j) & 0 & 0 \\ 0 & 0 & iJ_{n_j}(\alpha_j) & 0 \\ 0 & 0 & 0 & -iJ_{n_j}(\alpha_j) \end{bmatrix},$$

$$K_j = \begin{bmatrix} 0 & 0 & 0 & R_{n_j} \\ 0 & 0 & R_{n_j} & 0 \\ 0 & R_{n_j} & 0 & 0 \\ R_{n_j} & 0 & 0 & 0 \end{bmatrix}, \quad M_j = \begin{bmatrix} 0 & 0 & \gamma_j R_{n_j} & 0 \\ 0 & 0 & 0 & \gamma_j R_{n_j} \\ \bar{\gamma}_j R_{n_j} & 0 & 0 & 0 \\ 0 & \bar{\gamma}_j R_{n_j} & 0 & 0 \end{bmatrix},$$

where $0 < \alpha_j \in \mathbb{R}$ *and* $\gamma_j = c_j + i d_j$ *with* $c_j, d_j \in \mathbb{R}$ *and* $c_j d_j > 0$. *The two Jordan blocks associated with* $i\alpha_j$ *have the signs* 1 *and* -1 *with respect to* M *and the two Jordan blocks associated with* $-i\alpha_j$ *also have the signs* 1 *and* -1 *with respect to* M.

(v) *Blocks associated with a pair of purely imaginary eigenvalues* $\pm i\alpha_j$ *of* A *and real or purely imaginary eigenvalues of* $\lambda K - M$:

$$A_j = \begin{bmatrix} iJ_{n_j}(\alpha_j) & 0 \\ 0 & -iJ_{n_j}(\alpha_j) \end{bmatrix},$$

$$K_j = \begin{bmatrix} 0 & R_{n_j} \\ R_{n_j} & 0 \end{bmatrix}, \quad M_j = s_j|\gamma_j| \begin{bmatrix} R_{n_j} & 0 \\ 0 & \left(\frac{|\gamma_j|}{\gamma_j}\right)^2 R_{n_j} \end{bmatrix},$$

where $0 < \alpha_j$ *in* $\mathbb{R}$, $s_j = \pm 1$, $\gamma_j = \beta_j$ *or* $\gamma_j = i\beta_j$ *with* $0 < \beta_j \in \mathbb{R}$. *The Jordan block of* A *associated with* $i\alpha_j$ *has sign* s_j *with respect to* M *and the one associated with* $-\alpha_j$ *has sign* $(-1)^{n_j+1} s_j (\gamma_j/|\gamma_j|)^2$ *with respect to* M.

(vi) *A pair of blocks associated with the eigenvalue zero of* A *and nonreal, non purely imaginary eigenvalues of* $\lambda K - M$:

$$A_j = \begin{bmatrix} J_{n_j}(0) & 0 \\ 0 & J_{n_j}(0) \end{bmatrix}, \quad K_j = \begin{bmatrix} 0 & R_{n_j} \\ R_{n_j} & 0 \end{bmatrix},$$

$$M_j = s_j \begin{bmatrix} 0 & \gamma_j \Sigma_{n_j} \\ (-1)^{n_j+1}\bar{\gamma}_j \Sigma_{n_j} & 0 \end{bmatrix},$$

where $s_j = \pm 1$, $\gamma_j = c_j + i d_j$ *with* $c_j, d_j \in \mathbb{R}$ *and* $c_j d_j > 0$. *The two Jordan blocks of* A *associated with the eigenvalue* 0 *have the signs* 1 *and* -1 *with respect to both* K *and* M.

(vii) *A pair of blocks associated with the eigenvalue zero of* A *and real or purely imaginary eigenvalues of* $\lambda K - M$:

$$A_j = \begin{bmatrix} J_{n_j}(0) & 0 \\ 0 & J_{n_j}(0) \end{bmatrix}, \quad K_j = \begin{bmatrix} 0 & R_{n_j} \\ R_{n_j} & 0 \end{bmatrix}, \quad M_j - \begin{bmatrix} 0 & \gamma_j \Sigma_{n_j} \\ -\gamma_j \Sigma_{n_j} & 0 \end{bmatrix},$$

where $\gamma_j = \beta_j$ *if* n_j *is even and* $\gamma_j = i\beta_j$ *if* n_j *is odd for some* $0 < \beta_j \in \mathbb{R}$. *The two Jordan blocks of* A *associated with the eigenvalue zero of* A *have the signs* 1 *and* -1 *with respect to both* K *and* M.

(viii) *A single block associated with the eigenvalue zero of* A *and real or purely imaginary eigenvalues of* $\lambda K - M$:

$$A_j = J_{n_j}(0), \quad K_j = s_j R_{n_j}, \quad M_j = \sigma_j \gamma_j \Sigma_{n_j},$$

where $s_j, \sigma_j = \pm 1$; *and* $\gamma_j = \beta_j$ *if* n_j *is odd and* $\gamma_j = i\beta_j$ *if* n_j *is even for some* $0 < \beta_j \in \mathbb{R}$. *The Jordan block of* A *associated with the eigenvalue zero has the sign* s_j *with respect to* K *and the sign* $\frac{\gamma_j}{|\gamma_j|}\sigma_j i^{n_j-1}$ *with respect to* M.

Theorem 7 shows the intertwined connection of the three different structures: the double structure of A with respect to K and M and the structure of the Hermitian pencil $\lambda K - M$. The property of being K-Hermitian forces the spectrum of A to be symmetric with respect to the real axis and the property of being M-skew-Hermitian forces the spectrum to be symmetric with respect to the imaginary axis. The particular structure of blocks, however, depends in addition on the eigenvalues of the Hermitian pencil $\lambda K - M$. Interestingly, there is not only a distinction between real and nonreal eigenvalues of $\lambda K - M$, but also the purely imaginary eigenvalues of $\lambda K - M$ play a special role. This effect can in particular be seen in the blocks associated with the eigenvalue zero, the only point in the complex plane that is both real and purely imaginary. Depending on the type of the corresponding eigenvalues of $\lambda K - M$, we have the following cases:

(a) Real eigenvalues of $\lambda K - M$: in this case, *even*-sized Jordan blocks of A associated with zero must occur in pairs (vii), but *odd*-sized Jordan blocks need not (viii);
(b) Purely imaginary eigenvalues of $\lambda K - M$: in this case, *odd*-sized Jordan blocks of A associated with zero must occur in pairs (vii), but *even*-sized Jordan blocks need not (viii);
(c) Nonreal, non purely imaginary eigenvalues of $\lambda K - M$: in this case, all Jordan blocks of A associated with the eigenvalue zero must occur in pairs (vi).

Structured canonical forms for A as a K-Hermitian matrix, for A as an M-Hermitian matrix and for the Hermitian pencil $\lambda K - M$ can be easily derived from the canonical form in Theorem 7, so again the result combines three different canonical forms into one.

As the particular application from quantum chemistry shows, there is also interest in doubly structured generalized eigenvalue problems. In general, we can consider a matrix pencil $\lambda A - B$ with both A, B being doubly structured with respect to two invertible Hermitian or skew-Hermitian matrices K and M. It turns out that a structured Weierstraß canonical form for a regular doubly structured pencil can easily be derived by using the results of the matrix case as in Theorems 6 and 7.

Theorem 8 ([34]) *Suppose K, M are both invertible and each is either Hermitian or skew-Hermitian, i.e.,*

$$K^* = \sigma_K K, \quad M^* = \sigma_M M, \qquad \sigma_K, \sigma_M = \pm 1.$$

Let $\lambda A - B$ be a regular pencil (that is $\det(\lambda A - B) \not\equiv 0$*) with A, B satisfying*

$$A^* K = \varepsilon_A KA, \quad A^* M = \delta_A MA, \qquad \varepsilon_A, \delta_A = \pm 1$$

$$B^* K = \varepsilon_B KB, \quad B^* M = \delta_B MB, \qquad \varepsilon_B, \delta_B = \pm 1.$$

Then there exist invertible matrices X, Y such that

$$Y^{-1}(\lambda A - B)X = \lambda \begin{bmatrix} I & 0 \\ 0 & E \end{bmatrix} - \begin{bmatrix} H & 0 \\ 0 & I \end{bmatrix}$$

$$X^*KY = \begin{bmatrix} K_1 & 0 \\ 0 & K_2 \end{bmatrix}, \qquad X^*MY = \begin{bmatrix} M_1 & 0 \\ 0 & M_2 \end{bmatrix},$$

where E is nilpotent and all three matrices in (H, K_1, M_1) and (E, K_2, M_2), respectively, have the same sizes. Furthermore, we have that

$$K_1^* = \sigma_K \varepsilon_A K_1, \;\; M_1^* = \sigma_M \delta_A M_1; \;\; H^* K_1 = (\varepsilon_A \varepsilon_B) K_1 H, \;\; H^* M_1 = (\delta_A \delta_B) M_1 H,$$
$$K_2^* = \sigma_K \varepsilon_B K_2, \;\; M_2^* = \sigma_M \delta_B M_2; \;\; E^* K_2 = (\varepsilon_A \varepsilon_B) K_2 E, \;\; E^* M_2 = (\delta_A \delta_B) M_2 E.$$

Clearly, H is a doubly structured matrix associated with the Hermitian or skew-Hermitian matrices K_1, M_1, and E is a doubly structured matrix associated with the Hermitian or skew Hermitian matrices K_2, M_2. Thus, the pencil $\lambda A - B$ is decoupled and becomes $(\lambda I - H) \oplus (\lambda E - I)$. Hence a structured Weierstraß canonical form of $\lambda A - B$ can be derived by applying the results in Theorems 6 and 7 to H and E separately.

Note that in Theorem 8 one does not require $\lambda K - M$ to be nondefective. However, in order to apply Theorems 6 or 7 to obtain structured Jordan canonical forms for the matrices H and E, the condition that both $\lambda K_1 - M_1$ and $\lambda K_2 - M_2$ are nondefective is necessary.

Finally, we point out that for the special type of doubly structured matrices and matrix pencils from linear response theory [16, 41, 49], necessary and sufficient conditions for the existence of structured Schur-like forms (obtained under unitary transformations) were provided in [36].

6.5 Structured Singular Value Decompositions

The singular value decomposition (SVD) is an important tool in Matrix Theory and Numerical Linear Algebra. For a given matrix $A \in \mathbb{C}^{m \times n}$ it computes unitary matrices X, Y such that Y^*AX is diagonal with nonnegative diagonal entries. The condition that X and Y are unitary can be interpreted in such a way that the standard Euclidean inner product is preserved by the transformation with X and Y. Thus, to be more precise, we have a transformation on the matrix triple (A, I_n, I_m) that yields the canonical form

$$Y^*AX = \begin{bmatrix} \Delta & 0 \\ 0 & 0 \end{bmatrix}, \quad X^* I_n X = I_n, \quad Y^* I_m Y = I_m,$$

where Δ is diagonal with positive diagonal entries. But the singular value decomposition even yields more information as the nonzero singular values are the square roots of the positive eigenvalues of the matrices AA^* and A^*A. Thus, in addition to a canonical form for A under unitary equivalence, the SVD simultaneously provides two spectral decompositions

$$Y^*(AA^*)Y = \begin{bmatrix} \Delta^2 & 0 \\ 0 & 0 \end{bmatrix}, \quad X^*(A^*A)X = \begin{bmatrix} \Delta^2 & 0 \\ 0 & 0 \end{bmatrix},$$

for the Hermitian (positive semi-definite) matrices AA^* and A^*A.

This concept can be generalized to the case of possibly indefinite inner products. Suppose that the two spaces $\mathbb{C}^n$ and $\mathbb{C}^m$ are equipped with inner products given by the Hermitian invertible matrices $K \in \mathbb{C}^{n\times n}$ and $M \in \mathbb{C}^{m\times m}$, respectively. Then the task is to find invertible matrices X and Y such that

$$\mathscr{A} = Y^*AX, \quad \mathscr{K} = X^*KX, \quad \mathscr{M} = Y^*MY, \tag{6.10}$$

are in a canonical form, so that also the canonical forms of the K-Hermitian matrix T and the M-Hermitian matrix Z can easily be derived, where

$$T = K^{-1}A^*M^{-1}A, \qquad Z = M^{-1}AK^{-1}A^*. \tag{6.11}$$

Equivalently, we obtain structured canonical forms for the two Hermitian pencils $\lambda K - A^*M^{-1}A$ and $\lambda M - AK^{-1}A^*$

The transformation (6.10) has several mathematical applications. For instance, the existence of a generalization of a polar decompositions for a matrix A in a space equipped with an indefinite inner product as in (6.1) given by the invertible Hermitian matrix $\widetilde{M} \in \mathbb{C}^{n\times n}$ is related to the matrices $AA^{[*]}$ and $A^{[*]}A$, where $A^{[*]} := \widetilde{M}^{-1}A^*\widetilde{M}$. By definition, a matrix $A \in \mathbb{C}^{n\times n}$ is said to have an $\widetilde{M}$*-polar decomposition*, if there exists an $\widetilde{M}$-Hermitian matrix H and an $\widetilde{M}$-unitary matrix U, such that $A = UH$, see [5, 6]. In contrast to the classical polar decomposition in the case of the Euclidean inner product, an $\widetilde{M}$-polar decomposition need not exist for a given matrix $A \in \mathbb{C}^{n\times n}$. In [37], it was proved that a matrix $A \in \mathbb{C}^{n\times n}$ allows an $\widetilde{M}$-polar decomposition if and only if the two $\widetilde{M}$-Hermitian matrices $AA^{[*]}$ and $A^{[*]}A$ have the same canonical forms (as in Corollary 1) – a fact that was already conjectured in [18]. If a canonical form under a transformation as in (6.10) is given with the matrices T and Z as in (6.11), then we have that $AA^{[*]} = MZM^{-1}$ and $A^{[*]}A = T$ with $K = \widetilde{M}$ and $M = \widetilde{M}^{-1}$. Thus, structured canonical forms can easily be derived from the canonical form under the transformation (6.10).

On the other hand, the simultaneous transformation (6.10) provides more flexibility in solving the eigenvalue problem of a structured matrix as $B = A^*M^{-1}A$ from a numerical point of view. That is, instead of performing similarity transformations on B, one may use two-sided transformations on A. For example, when $K =$ J and $M = I$, a structured condensed form for a matrix A was proposed in [53] and a numerical method was given in [54].

In the case when A is invertible (hence square), the following theorem provides the desired canonical form for the transformation in (6.10). Here, we use the notation $J_m^2(\alpha)$ for the square of a Jordan block $J_m(\alpha)$ of size m associated with the eigenvalue α.

Theorem 9 ([35]) *Let $A \in \mathbb{C}^{n\times n}$ be nonsingular and let $K, M \in \mathbb{C}^{n\times n}$ be Hermitian and invertible. Then there exist invertible matrices $X, Y \in \mathbb{C}^{n\times n}$ such that*

$$Y^*AX = A_c \oplus A_r, \qquad X^*KX = K_c \oplus K_r, \qquad Y^*MY = M_c \oplus M_r. \tag{6.12}$$

Consequently, for the K-Hermitian matrix $T = K^{-1}A^*M^{-1}A$ and the M-Hermitian matrix $Z = M^{-1}AK^{-1}A^*$, one has

$$X^{-1}TX = T_c \oplus T_r, \qquad Y^{-1}ZY = Z_c \oplus Z_r. \tag{6.13}$$

The diagonal blocks in (6.12) and (6.13) have the following forms.

(i)

$$\begin{aligned}
A_c &= \begin{bmatrix} J_{m_1}(\mu_1) & 0 \\ 0 & J_{m_1}(\bar{\mu}_1) \end{bmatrix} \oplus \cdots \oplus \begin{bmatrix} J_{m_p}(\mu_p) & 0 \\ 0 & J_{m_p}(\bar{\mu}_p) \end{bmatrix},\\
K_c &= \begin{bmatrix} 0 & R_{m_1} \\ R_{m_1} & 0 \end{bmatrix} \oplus \cdots \oplus \begin{bmatrix} 0 & R_{m_p} \\ R_{m_p} & 0 \end{bmatrix},\\
M_c &= \begin{bmatrix} 0 & R_{m_1} \\ R_{m_1} & 0 \end{bmatrix} \oplus \cdots \oplus \begin{bmatrix} 0 & R_{m_p} \\ R_{m_p} & 0 \end{bmatrix},\\
T_c &= \begin{bmatrix} J_{m_1}^2(\mu_1) & 0 \\ 0 & J_{m_1}^2(\bar{\mu}_1) \end{bmatrix} \oplus \cdots \oplus \begin{bmatrix} J_{m_p}^2(\mu_p) & 0 \\ 0 & J_{m_p}^2(\bar{\mu}_p) \end{bmatrix},\\
Z_c &= \begin{bmatrix} J_{m_1}^2(\mu_1) & 0 \\ 0 & J_{m_1}^2(\bar{\mu}_1) \end{bmatrix}^* \oplus \cdots \oplus \begin{bmatrix} J_{m_p}^2(\mu_p) & 0 \\ 0 & J_{m_p}^2(\bar{\mu}_p) \end{bmatrix}^*,
\end{aligned}$$

where $\mu_j = a_j + ib_j$ with $0 < a_j, b_j \in \mathbb{R}$ for $j = 1, \ldots, p$. For each j, both the diagonal block $\operatorname{diag}\left(J_{m_j}^2(\mu_j), J_{m_j}^2(\bar{\mu}_j)\right)$ of T_c as well as the diagonal block $\operatorname{diag}\left(J_{m_j}^2(\mu_j), J_{m_j}^2(\bar{\mu}_j)\right)^*$ of Z_c are similar to a matrix consisting of two $m_j \times m_j$ Jordan blocks, one of them associated with the nonreal and non purely imaginary eigenvalue μ_j^2 and the other one with $\bar{\mu}_j^2$.

(ii)

$$\begin{aligned}
A_r &= J_{n_1}(\beta_1) \oplus \cdots \oplus J_{n_q}(\beta_q),\\
K_r &= s_1 R_{n_1} \oplus \cdots \oplus s_q R_{n_q},\\
M_r &= \sigma_1 R_{n_1} \oplus \cdots \oplus \sigma_q R_{n_q},\\
T_r &= s_1\sigma_1 J_{n_1}^2(\beta_1) \oplus \cdots \oplus s_q\sigma_q J_{n_q}^2(\beta_q),\\
Z_r &= s_1\sigma_1 \left(J_{n_1}^2(\beta_1)\right)^* \oplus \cdots \oplus s_q\sigma_q \left(J_{n_q}^2(\beta_q)\right)^*,
\end{aligned}$$

where $\beta_j > 0$, *and* $s_j, \sigma_j = \pm 1$ *for* $j = 1, \ldots, q$. *For each* j, *the block* $s_j\sigma_j J_{n_j}^2(\beta_j)$ *of* T_r *is similar to an* $n_j \times n_j$ *Jordan block associated with a real eigenvalue* $s_j\sigma_j\beta_j^2$ *of* T *with the sign with respect to* K *being*

$$\begin{cases} s_j & \text{if } n_j \text{ is odd, or if } n_j \text{ is even and } s_j\sigma_j = 1, \\ \sigma_j & \text{if } n_j \text{ is even and } s_j\sigma_j = -1, \end{cases}$$

and the block $s_j\sigma_j\left(J_{n_j}^2(\beta_j)\right)^*$ *of* Z_r *is similar to an* $n_j \times n_j$ *Jordan block associated with a real eigenvalue* $s_j\sigma_j\beta_j^2$ *of* Z *with the sign with respect to* M *being*

$$\begin{cases} \sigma_j & \text{if } n_j \text{ is odd, or if } n_j \text{ is even and } s_j\sigma_j = 1, \\ s_j & \text{if } n_j \text{ is even and } s_j\sigma_j = -1. \end{cases}$$

For a general rectangular matrix $A \in \mathbb{C}^{m\times n}$, the situation is more complicated because of (a) the rectangular form of A and (b) the presence of the eigenvalue 0 in T or Z. Indeed, note that these two matrices T and Z can be represented as products of the same two factors, but with different order, i.e., $T = BC$ and $Z = CB$, where $B = K^{-1}A^*$ and $C = M^{-1}A$. By a well-known result [11], the Jordan structures of the nonzero eigenvalues of the two matrix products BC and CB are identical, while this is not the case for the eigenvalue zero. Despite this additional complexity in the problem of finding a canonical form under the transformation (6.10), a complete answer is still possible as shown in the next theorem.

Theorem 10 ([35]) *Let* $A \in \mathbb{C}^{m\times n}$, *and let* $K \in \mathbb{C}^{n\times n}$ *and* $M \in \mathbb{C}^{m\times m}$ *be Hermitian and invertible. Then there exist invertible matrices* $Y \in \mathbb{C}^{m\times m}$ *and* $X \in \mathbb{C}^{n\times n}$ *such that*

$$\begin{aligned} Y^*AX &= A_c \oplus A_r \oplus A_1 \oplus A_2 \oplus A_3 \oplus A_4, \\ X^*KX &= K_c \oplus K_r \oplus K_1 \oplus K_2 \oplus K_3 \oplus K_4, \\ Y^*MY &= M_c \oplus M_r \oplus M_1 \oplus M_2 \oplus M_3 \oplus M_4. \end{aligned} \tag{6.14}$$

Moreover, for the K*-Hermitian matrix* $T = K^{-1}A^*M^{-1}A \in \mathbb{C}^{n\times n}$ *and for the* M*-Hermitian matrix* $Z = M^{-1}AK^{-1}A^* \in \mathbb{C}^{m\times m}$ *we have that*

$$\begin{aligned} X^{-1}TX &= T_c \oplus T_r \oplus T_1 \oplus T_2 \oplus T_3 \oplus T_4, \\ Y^{-1}ZY &= Z_c \oplus Z_r \oplus Z_1 \oplus Z_2 \oplus Z_3 \oplus Z_4. \end{aligned}$$

The blocks $A_c, A_r, K_c, K_r, M_c, M_r$ *have the same forms as in* (6.12). *Therefore, the blocks* T_c, T_r *and* Z_c, Z_r *have the same forms as in* (6.13). *The remaining blocks are associated with the eigenvalue* 0 *of* T *and* Z *and have the following forms.*

(i)

$$A_1 = 0_{\ell\times k},\ K_1 = \text{diag}(I_{k_1}, -I_{k_2}),\ M_1 = \text{diag}(I_{\ell_1}, -I_{\ell_2}),\ T_1 = 0_k,\ Z_1 = 0_\ell,$$

where $k_1 + k_2 = k$ and $\ell_1 + \ell_2 = \ell$. So there are k copies of 1×1 Jordan blocks associated with the eigenvalue 0 of T such that k_1 of them have the sign $+1$ and k_2 of them the sign -1 with respect to K, and there are ℓ copies of 1×1 Jordan blocks associated with the eigenvalue 0 of Z such that ℓ_1 of them have the sign $+1$ and ℓ_2 of them the sign -1.

(ii)

$$\begin{aligned}
A_2 &= J_{2r_1}(0) \oplus J_{2r_2}(0) \oplus\cdots\oplus J_{2r_u}(0)\,,\\
K_2 &= R_{2r_1} \oplus R_{2r_2} \oplus\cdots\oplus R_{2r_u}\,,\\
M_2 &= R_{2r_1} \oplus R_{2r_2} \oplus\cdots\oplus R_{2r_u}\,,\\
T_2 &= J^2_{2r_1}(0) \oplus J^2_{2r_2}(0) \oplus\cdots\oplus J^2_{2r_u}(0)\,,\\
Z_2 &= \left(J^2_{2r_1}(0)\right)^T \oplus \left(J^2_{2r_2}(0)\right)^T \oplus\cdots\oplus \left(J^2_{2r_u}(0)\right)^T.
\end{aligned}$$

For each $j = 1,\ldots,u$, the block $J^2_{2r_j}(0)$ of T_2 is similar to a matrix consisting of two copies of the Jordan block $J_{r_j}(0)$ of T with one of them having the sign $+1$ and the other having the sign -1 with respect to K, and the block $\left(J^2_{2r_j}(0)\right)^T$ is similar to a matrix consisting of two copies of the Jordan block $J_{r_j}(0)$ of Z with one of them having the sign $+1$ and the other having the sign -1 with respect to M.

(iii)

$$\begin{aligned}
A_3 &= \begin{bmatrix} I_{s_1} \\ 0 \end{bmatrix}_{(s_1+1)\times s_1} \oplus \begin{bmatrix} I_{s_2} \\ 0 \end{bmatrix}_{(s_2+1)\times s_2} \oplus\cdots\oplus \begin{bmatrix} I_{s_v} \\ 0 \end{bmatrix}_{(s_v+1)\times s_v},\\
K_3 &= \phi_1 R_{s_1} \oplus \phi_2 R_{s_2} \oplus\cdots\oplus \phi_v R_{s_v}\,,\\
M_3 &= \psi_1 R_{s_1+1} \oplus \psi_2 R_{s_2+1} \oplus\cdots\oplus \psi_v R_{s_v+1}\,,\\
T_3 &= \phi_1\psi_1 J_{s_1}(0) \oplus \phi_2\psi_2 J_{s_2}(0) \oplus\cdots\oplus \phi_v\psi_v J_{s_v}(0)\,,\\
Z_3 &= \phi_1\psi_1 J^T_{s_1+1}(0) \oplus \phi_2\psi_2 J^T_{s_2+1}(0) \oplus\cdots\oplus \phi_v\psi_v J^T_{s_v+1}(0),
\end{aligned}$$

where for $j = 1,\ldots,v$, $\phi_j = 1$ and $\psi_j = \pm 1$ if s_j is even, and $\phi_j = \pm 1$ and $\psi_j = 1$ if s_j is odd. Hence, for each j, the block $\phi_j\psi_j J_{s_j}(0)$ of T_3 is a modified $s_j \times s_j$ Jordan block associated with the eigenvalue 0 of T with sign ϕ_j if s_j is odd and ψ_j if s_j is even; the block $\phi_j\psi_j J^T_{s_j+1}(0)$ of Z_3 is a modified $(s_j+1)\times(s_j+1)$ Jordan block associated with the eigenvalue 0 of Z with sign ϕ_j if s_j is odd and ψ_j if s_j is even.

(iv)

$$
\begin{aligned}
A_4 &= \begin{bmatrix} 0 & I_{t_1} \end{bmatrix}_{t_1\times(t_1+1)} \oplus \begin{bmatrix} 0 & I_{t_2} \end{bmatrix}_{t_2\times(t_2+1)} \oplus\cdots\oplus \begin{bmatrix} 0 & I_{t_w} \end{bmatrix}_{t_w\times(t_w+1)},\\
K_4 &= \theta_1 R_{t_1+1} \oplus \theta_2 R_{t_2+1} \oplus\cdots\oplus \theta_w R_{t_w+1},\\
M_4 &= \rho_1 R_{t_1} \oplus \rho_2 R_{t_2} \oplus\cdots\oplus \rho_w R_{t_w},\\
T_4 &= \theta_1\rho_1 J_{t_1+1}(0) \oplus \theta_2\rho_2 J_{t_2+1}(0) \oplus\cdots\oplus \theta_w\rho_w J_{t_w+1}(0),\\
Z_4 &= \theta_1\rho_1 J_{t_1}^T(0) \oplus \theta_2\rho_2 J_{t_2}^T(0) \oplus\cdots\oplus \theta_w\rho_w J_{t_w}^T(0),
\end{aligned}
$$

where for $j = 1,\dots,w$, $\theta_j = 1$ and $\rho_j = \pm 1$ if t_j is odd, and $\theta_j = \pm 1$ and $\rho_j = 1$ if t_j is even. Hence, for each j, the block $\theta_j\rho_j J_{t_j+1}(0)$ of T_4 is a modified $(t_j+1)\times(t_j+1)$ Jordan block associated with the eigenvalue 0 of T with sign ρ_j if t_j is odd and θ_j if t_j is even; the block $\theta_j\rho_j J_{t_j}^T(0)$ of Z_4 is a modified $t_j\times t_j$ Jordan block associated with the eigenvalue 0 of Z with sign ρ_j if t_j is odd and θ_j if t_j is even.

Theorem 10 shows that the sizes of the Jordan blocks and signs associated with the eigenvalue zero may be different for the matrices T and Z. Still, they are related and the canonical form for A exactly explains in which way.

As mentioned earlier, the investigation of the canonical forms of the matrices $AA^{[*]}$ and $A^{[*]}A$ is crucial if one wants to check if $A \in \mathbb{C}^{n\times n}$ has an M-polar decomposition with respect to the invertible Hermitian matrix M. Therefore, the possible difference in the canonical forms of $AA^{[*]}$ and $A^{[*]}A$ has been analyzed in [17]. With the canonical form from Theorem 10 there is now a complete classification of all possible canonical forms for the matrices $AA^{[*]}$ and $A^{[*]}A$ for a general matrix A.

A real version of (6.14) for real A, K, M can be derived essentially in the same way. In the case that all A, K, and M are real and at least one of K and M is skew-symmetric, the real canonical forms of the simultaneous transformation (6.10) can be derived too, but with some additional techniques. The details can be found in [35].

6.6 Conclusion

Applications in different areas provide a variety of eigenvalue problems with different symmetry structures that lead to symmetries in the spectra of the corresponding matrices or matrix pencils. It is crucial to use structure-preserving algorithms so that the symmetry in the spectra is not lost due to roundoff errors in the numerical computation and that the computed results are physically meaningful. For the understanding of the behavior of these algorithms and the effect in the corresponding perturbation theory, structured canonical forms are an essential tool.

In this survey, we have presented three particular structured canonical forms with respect to matrices that carry one or two structures with respect to possible indefinite inner products. Moreover, we have highlighted the important role that the sign characteristic plays in the understanding of the behavior of Hamiltonian matrices under structure-preserving transformations.

References

1. Ammar, G.S., Mehrmann, V.: On Hamiltonian and symplectic Hessenberg forms. Linear Algebra Appl. **149**, 55–72 (1991)
2. Bai, Z., Li, R.-C.: Minimization principles of the linear response eigenvalue problem I: theory. SIAM J. Matrix Anal. Appl. **33**(4), 1075–1100 (2012)
3. Bai, Z., Li, R.-C.: Minimization principles of the linear response eigenvalue problem II: computation. SIAM J. Matrix Anal. Appl. **34**(2), 392–416 (2013)
4. Benner, P., Byers, R., Mehrmann, V., Xu, H.: Numerical methods for linear quadratic and H_∞ control problems. In: Picci, G., Gillian, D.S. (eds.) Dynamical Systems, Control, Coding, Computer Vision. Progress in Systems and Control Theory, vol. 25, pp. 203–222. Birkhäuser Verlag, Basel (1999)
5. Bolschakov, Y., Reichstein, B.: Unitary equivalence in an indefinite scalar product: an analogue of singular-value decomposition. Linear Algebra Appl. **222**, 155–226 (1995)
6. Bolschakov, Y., van der Mee, C.V.M., Ran, A.C.M., Reichstein, B., Rodman, L.: Polar decompositions in finite-dimensional indefinite scalar product spaces: general theory. Linear Algebra Appl. **261**, 91–141 (1997)
7. Bunse-Gerstner, A.: Matrix factorization for symplectic methods. Linear Algebra Appl. **83**, 49–77 (1986)
8. Chu, D., Liu, X., Mehrmann, V.: A numerical method for computing the Hamiltonian Schur form. Numer. Math. **105**, 375–412 (2007)
9. Djokovic, D.Z., Patera, J., Winternitz, P., Zassenhaus, H.: Normal forms of elements of classical real and complex Lie and Jordan algebras. J. Math. Phys. **24**, 1363–1374 (1983)
10. Faßbender, H., Mackey, D.S., Mackey, N., Xu, H.: Hamiltonian square roots of skew–Hamiltonian matrices. Linear Algebra Appl. **287**, 125–159 (1999)
11. Flanders, H.: Elementary divisors of AB and BA. Proc. Am. Math. Soc. **2**, 871–874 (1951)
12. Freiling, G., Mehrmann, V., Xu, H.: Existence, uniqueness and parametrization of Lagrangian invariant subspaces. SIAM J. Matrix Anal. Appl. **23**, 1045–1069 (2002)
13. Gantmacher, F.R.: Theory of Matrices. Vol. 1. Chelsea, New York (1959)
14. Gantmacher, F.R.: Theory of Matrices. Vol. 2. Chelsea, New York (1959)
15. Gohberg, I., Lancaster, P., Rodman, L.: Indefinite Linear Algebra and Applications. Birkhäuser, Basel (2005)
16. Hansen, A., Voigt, B., Rettrup, S.: Large-scale RPA calculations of chiroptical properties of organic molecules: program RPAC. Int. J. Quantum Chem. **XXIII**, 595–611 (1983)
17. Kes, J., Ran, A.C.M.: On the relation between $XX^{[*]}$ and $X^{[*]}X$ in an indefinite inner product space. Oper. Matrices **1**, 181–197 (2007)
18. Kintzel, U.: Polar decompositions and procrustes problems in finite dimenionsional indefinite scalar product spaces. PhD thesis, TU Berlin, Institute of Mathematics (2005)
19. Krein, M.G.: The basic propositions in the theory of λ-zones of stability of a canonical system of linear differential equations with periodic coefficients. Oper. Theory Adv. Appl, vol. 7. Birkhäuser-Verlag, Basel (1988)
20. Krein, M.G., Langer, H.: On some mathematical principles in the linear theory of damped oscillations of continua, I. Integral Equ. Oper. Theory **1**, 364–399 (1978)

21. Krein, M.G., Langer, H.: On some mathematical principles in the linear theory of damped oscillations of continua, II. Integral Equ. Oper. Theory **1**, 539–566 (1978)
22. Kressner, D., Schröder, C., Watkins, D.S.: Implicit QR algorithms for palindromic and even eigenvalue problems. Numer. Algorithms **51**, 209–238 (2009)
23. Kronecker, L.: Algebraische Reduction der Schaaren bilineare Formen. S. B. Akad., Berlin, pp 1225–1237 (1890)
24. Lancaster, P., Rodman, L.: The Algebraic Riccati Equation. Oxford University Press, Oxford (1995)
25. Lancaster, P., Rodman, L.: Canonical forms for Hermitian matrix pairs under strict equivalence and congruence. SIAM Rev. **47**, 407–443 (2005)
26. Lancaster, P., Rodman, L.: Canonical forms for symmetric/skew symmetric real pairs under strict equivalence and congruence. Linear Algebra Appl. **406**, 1–76 (2005)
27. Laub, A.J.: A Schur method for solving algebraic Riccati equations. IEEE Trans. Autom. Control **AC-24**, 913–921 (1979)
28. Lin, W.-W., Mehrmann, V., Xu, H.: Canonical forms for Hamiltonian and symplectic matrices and pencils. Linear Algebra Appl. **301–303**, 469–533 (1999)
29. Mackey, D.S., Mackey, N., Mehl, C., Mehrmann, V.: Numerical methods for palindromic eigenvalue problems: computing the anti-triangular Schur form. Numer. Linear Algebra Appl. **16**, 63–68 (2009)
30. Mehl, C.: Finite dimensional inner product spaces and applications in numerical analysis. In: Alpay, D. (ed.) Operator Theory. Springer, Basel (2015)
31. Mehl, C., Mehrmann, V., Ran, A.C.M., Rodman, L.: Perturbation analysis of Lagrangian invariant subspaces of symplectic matrices. Linear Multilinear Algebra **57**, 141–184 (2009)
32. Mehl, C., Mehrmann, V., Ran, A.C.M., Rodman, L.: Eigenvalue perturbation theory of structured matrices under generic structured rank one perturbations. Linear Algebra Appl. **435**, 687–716 (2011)
33. Mehl, C., Mehrmann, V., Ran, A.C.M., Rodman, L.: Perturbation theory of selfadjoint matrices and sign characteristics under generic structured rank one perturbations. Linear Algebra Appl. **436**, 4027–4042 (2012)
34. Mehl, C., Mehrmann, V., Xu, H.: Canonical forms for doubly structured matrices and pencils. Electron. J. Linear Algebra **7**, 112–151 (2000)
35. Mehl, C., Mehrmann, V., Xu, H.: Structured decompositions for matrix triples: SVD-like concepts for structured matrices. Oper. Matrices **3**, 303–356 (2009)
36. Mehl, C., Mehrmann, V., Xu, H.: Singular-value-like decompositions for complex matrix triples. J. Comput. Appl. Math. **233**, 1245–1276 (2010)
37. Mehl, C., Ran, A.C.M., Rodman, L.: Polar decompositions of normal operators in indefinite inner product spaces. Oper. Theory Adv. Appl. **162**, 277–292 (2006)
38. Mehrmann, V.: The Autonomous Linear Quadratic Control Problem, Theory and Numerical Solution. Number 163 in Lecture Notes in Control and Information Sciences. Springer, Heidelberg (1991)
39. Mehrmann, V., Watkins, D.: Structure-preserving methods for computing eigenpairs of large sparse skew-Hamiltonian/Hamiltonian pencils. SIAM J. Sci. Comput. **22**, 1905–1925 (2000)
40. Mehrmann, V., Xu, H.: Perturbation of purely imaginary eigenvalues of Hamiltonian matrices under structured perturbations. Electron J. Linear Algebra **17**, 234–257 (2008)
41. Olson, J., Jensen, H.J.A., Jørgensen, P.: Solution of large matrix equations which occur in response theory. J. Comput. Phys. **74**, 265–282 (1988)
42. Ran, A.C.M., Rodman, L.: Stability of invariant maximal semidefinite subspaces I. Linear Algebra Appl. **62**, 51–86 (1984)
43. Schröder, C.: A structured Kronecker form for the palindromic eigenvalue problem. Proc. Appl. Math. Mech. **6**, 721–722 (2006). GAMM Annual Meeting, Berlin, 2006
44. Schröder, C.: Palindromic and even eigenvalue problems – analysis and numerical methods. PhD thesis, TU Berlin, Institut für Mathematik, Str. des 17. Juni 136, D-10623 Berlin, Germany (2008)

45. Stoorvogel, A.: The H_∞ Control Problem: A State-Space Approach. Prentice-Hall, Englewood Cliffs (1992)
46. Thompson, R.C.: The characteristic polynomial of a principal submatrix of a Hermitian pencil. Linear Algebra Appl. **14**, 135–177 (1976)
47. Thompson, R.C.: Pencils of complex and real symmetric and skew matrices. Linear Algebra Appl. **147**, 323–371 (1991)
48. Tisseur, F., Meerbergen, K.: The quadratic eigenvalue problem. SIAM Rev. **43**, 235–286 (2001)
49. Tsiper, E.: A classical mechanics technique for quantum linear response. J. Phys. B (Lett.) **34**, L401–L407 (2001)
50. Van Loan, C.F.: A symplectic method for approximating all the eigenvalues of a Hamiltonian matrix. Linear Algebra Appl. **61**, 233–251 (1984)
51. Wedderburn, J.H.M.: Lectures on Matrices. Dover, New York (1964)
52. Weierstraß, K.: Zur Theorie der quadratischen und bilinearen Formen. Monatsber. Akad. Wiss., Berlin, pp 310–338 (1868)
53. Xu, H.: An SVD-like matrix decomposition and its applications. Linear Algebra Appl. **368**, 1–24 (2003)
54. Xu, H.: A numerical method for computing an SVD-like decomposition. SIAM J. Matrix Anal. Appl. **26**, 1058–1082 (2005)
55. Zhou, K., Doyle, J.C., Glover, K.: Robust and Optimal Control. Prentice-Hall, Englewood Cliffs, New Jersey (1995)

Chapter 7
Perturbation Analysis of Matrix Equations and Decompositions

Mihail M. Konstantinov and Petko H. Petkov

Abstract A matrix computational problem is a function which maps a set of data (usually in the form of a collection of matrices) into a matrix space whose elements are the desired solutions. If a particular data is perturbed then the corresponding solution is also perturbed. The goal of the norm-wise perturbation analysis is to estimate the norm of the perturbation in the solution as a function of the norms of the perturbations in the data. In turn, in the component-wise perturbation analysis the modules of the elements of the solution are estimated as functions of the modules of the perturbations in the data.

The perturbation analysis can be local and nonlocal. In the local analysis it is supposed that the perturbations in the data are asymptotically small and a local bound for the perturbation in the solution is constructed which is valid for first order (small) perturbations. A disadvantage of the local analysis is that normally it does not have a priori measure on how 'small' the data perturbations must be in order to guarantee the results from the local estimates being correct. A desirable property of the local bounds is formulated as follows: a perturbation bound is asymptotically exact when for some perturbations it is arbitrarily close to the actual perturbed quantity.

On the other hand the nonlocal perturbation analysis produces perturbation estimates which are rigorously valid in a certain set of data perturbations. The price of this advantage is that the nonlocal perturbation bounds may be too pessimistic in certain cases and/or the domain of validity of these bounds may be relatively small. However, a desirable property of the nonlocal bounds is that within first order perturbations they coincide with the improved local bounds.

In this chapter we consider the basic methods for perturbation analysis of matrix algebraic equations and unitary (orthogonal in particular) matrix decompositions.

The nonlocal perturbation analysis of matrix equations includes several steps: (a) reformulation of the perturbed problem as an equivalent operator equation with

M.M. Konstantinov
University of Architecture, Civil Engineering and Geodesy, 1046 Sofia, Bulgaria
e-mail: mmk_fte@uacg.bg

P.H. Petkov (✉)
Technical University of Sofia, 1000 Sofia, Bulgaria
e-mail: php@tu-sofia.bg

P. Benner et al. (eds.), *Numerical Algebra, Matrix Theory, Differential-Algebraic Equations and Control Theory*, DOI 10.1007/978-3-319-15260-8_7

respect to the perturbation in the solution; (b) construction of a Lyapunov majorant for the corresponding operator; (c) application of fixed point principles in order to prove that the perturbed equation has a solution; (d) estimation of the solution of the associated majorant equation. The latter estimate gives the desired nonlocal perturbation bound.

The nonlocal perturbation analysis of unitary matrix decompositions is based on a systematic use of the method of splitting operators and vector Lyapunov majorants. In this way nonlocal perturbation bounds are derived for the basic unitary decompositions of matrices (QR decomposition and Schur decomposition in particular) and for important problems in the theory of linear time-invariant control systems: transformation into unitary canonical form and synthesis of closed-loop systems with desired equivalent form.

7.1 Introduction

When solving a matrix computational problem in finite machine arithmetic obeying the IEEE Standard 754-2008 [29] there are three main factors determining the accuracy of the computed solution: (a) the parameters of the machine arithmetic and in particular the rounding unit, (b) the sensitivity of the problem and in particular its condition number, and (c) the properties of the numerical algorithm and in particular the parameters specifying its numerical stability. Only taking into account these factors it is possible to derive an error estimate for the computed solution, see e.g. [27, 28, 38, 75]. Without such an estimate a computational procedure cannot be accepted as reliable.

The sensitivity of matrix computational problems can be revealed by the methods and techniques of perturbation (or sensitivity) analysis. In turn, the perturbation analysis may be norm-wise and componentwise [38, 87] (below we present results on norm-wise analysis). But the necessity of such an analysis is motivated by at least two other reasons. First, it enlightens the very nature of the problem independently on its practical applicability. And second, the mathematical models of real systems and processes are subject to parametric and measurement uncertainties [95]. Within these uncertainties we actually have a family of models. Such a family can be characterized by the methods of perturbation analysis.

In this chapter we consider three classes of matrix perturbation problems: matrix equations, unitary decompositions of matrices and modal control of controllable systems. Besides numerous articles, there are many dissertations and books devoted to these and related problems, see e.g. [2, 4, 16–18, 31, 32, 86] and [38, 45–47]. Related matrix and control problems are considered in [5, 23, 60, 66, 75].

A considerable contribution to the perturbation theory of matrix equations and decompositions as well as to the corresponding numerical methods is made by Volker Mehrmann and many of his coauthors since 1991, see [1, 10, 12, 38–41, 64, 68–70, 72] for theoretical considerations and [6–9, 25, 28, 65–67, 71, 76, 78] for numerical methods, algorithms and software.

7.2 Notation

In what follows we denote by $\mathbb{R}^{p\times q}$ (resp. $\mathbb{C}^{p\times q}$) the space of $p \times q$ matrices over $\mathbb{R}$ (resp. $\mathbb{C}$); we write $\mathbb{R}^p$ for $\mathbb{R}^{p\times 1}$ and $\mathbf{A}^\top$, $\overline{\mathbf{A}}$ and $\mathbf{A}^{\mathrm{H}} = \overline{\mathbf{A}}^\top$ for the transpose, complex conjugate and complex conjugate transpose of the matrix $\mathbf{A}$ with elements $a_{k,l}$, respectively (we use bold for matrices and vectors). For block matrices we use MATLAB[1]-like notation, e.g. $\mathbf{A} = \begin{bmatrix} \mathbf{A}_{1,1} & \mathbf{A}_{1,2} \\ \mathbf{A}_{2,1} & \mathbf{A}_{2,2} \end{bmatrix} = [\mathbf{A}_{1,1}, \mathbf{A}_{1,2}; \mathbf{A}_{2,1}, \mathbf{A}_{2,2}]$. In particular the vectorized column-wise form of the $(p \times q)$-matrix $\mathbf{A} = [\mathbf{a}_1, \mathbf{a}_2, \ldots, \mathbf{a}_q]$ with columns $\mathbf{a}_k$ is the column pq-vector $\mathrm{vec}(\mathbf{A}) = [\mathbf{a}_1; \mathbf{a}_2; \ldots; \mathbf{a}_q]$.

The size of the involved matrices (in particular the size of the identity matrix $\mathbf{I}$) shall be clear from the context but we also use the notation $\mathbf{I}_n$ for the identity $(n \times n)$-matrix and $\mathbf{0}_{p\times q}$ for the zero $(p \times q)$-matrix.

The Frobenius and the 2-norm of a matrix $\mathbf{A}$ are denoted as $\|\mathbf{A}\|$ and $\|\mathbf{A}\|_2$. We recall that $\|\mathbf{A}\|^2 = \sum_{k,l} |a_{k,l}|^2$ and $\|\mathbf{A}\|_2$ is the square root of the maximum eigenvalue of the matrix $\mathbf{A}^{\mathrm{H}}\mathbf{A}$ (or the maximum singular value of $\mathbf{A}$). The Kronecker product of the matrices $\mathbf{A} = [a_{k,l}]$ and $\mathbf{B}$ is $\mathbf{A} \otimes \mathbf{B} = [a_{k,l}\mathbf{B}]$.

The Frobenius norm is very useful in matrix perturbation problems. Indeed, if $\mathbf{Y} = \mathbf{AXB}$ then $\mathrm{vec}(\mathbf{Y}) = (\mathbf{B}^\top \otimes \mathbf{A})\mathrm{vec}(\mathbf{X})$ and $\|\mathbf{Y}\| \le \|(\mathbf{B}^\top \otimes \mathbf{A})\|_2 \|\mathbf{X}\|$, where the equality $\|\mathbf{Y}\| = \|(\mathbf{B}^\top \otimes \mathbf{A})\|_2 \|\mathbf{X}\|$ is reachable. In addition, if the matrix $\mathbf{A}$ is perturbed to $\mathbf{A} + \mathbf{E}$ then we usually have a bound $\|\mathbf{E}\| \le \delta$ on the perturbation $\mathbf{E}$ in terms its Frobenius norm $\|\mathbf{E}\|$ rather than a bound $\|\mathbf{E}\|_2 \le \delta$ in terms of its 2-norm $\|\mathbf{E}\|_2$.

Finally, the notation ':=' means 'equal by definition'.

7.3 Problem Statement

Explicit matrix problems may be written as

$$\mathbf{X} = \Phi(\mathbf{A}),$$

where the data $\mathbf{A}$ and the result $\mathbf{X}$ are nonzero real or complex matrices (or collection of matrices), while implicit problems are defined via matrix equations

$$F(\mathbf{A}, \mathbf{X}) = 0.$$

The function Φ satisfies the Lipschitz condition

$$\|\Phi(\mathbf{A} + \mathbf{E}) - \Phi(\mathbf{A})\| \le L\|\mathbf{E}\|,$$

---

[1]MATLAB® is a trademark of MathWorks, Inc.

where $\mathbf{E}$ is a certain perturbation in the data $\mathbf{A}$. Here for some $\eta > 0$ the Lipschitz constant $L = L(\mathbf{A}, \eta)$ is the supremum of the quantity

$$\frac{\|\Phi(\mathbf{A} + \mathbf{E}) - \Phi(\mathbf{A})\|}{\|\mathbf{E}\|}$$

over all $\mathbf{E}$ satisfying $0 < \|\mathbf{E}\| \leq \eta$. Denoting by

$$\mathbf{Y} = \mathbf{Y}(\mathbf{A}, \mathbf{E}) := \Phi(\mathbf{A} + \mathbf{E}) - \Phi(\mathbf{A})$$

the perturbation in the solution we obtain

$$\frac{\|\mathbf{Y}\|}{\|\mathbf{X}\|} \leq K \frac{\|\mathbf{E}\|}{\|\mathbf{A}\|}, \ \mathbf{X} \neq 0, \ \mathbf{A} \neq 0, \tag{7.1}$$

where the quantity

$$K = K(\mathbf{A}, \eta) := L(\mathbf{A}, \eta) \frac{\|\mathbf{A}\|}{\|\mathbf{X}\|} \tag{7.2}$$

is the relative condition number of the problem. We stress that the estimate (7.1), (7.2) is nonlocal since it holds true for all $\mathbf{E}$ with $\|\mathbf{E}\| \leq \eta$.

Let the problem be solved by a numerically stable algorithm in floating-point arithmetic with rounding unit u (in double precision mode $\mathrm{u} = 2^{-53} \simeq 1.1 \times 10^{-16}$). Then the computed solution $\tilde{\mathbf{X}}$ may be represented as

$$\tilde{\mathbf{X}} = \tilde{\Phi}(\mathbf{A}),$$

where $\tilde{\Phi}(\mathbf{A})$ is close to $\Phi(\tilde{\mathbf{A}})$ for some data $\tilde{\mathbf{A}}$ which in turn is close to $\mathbf{A}$ in the sense that [28, 75]

$$\|\tilde{\Phi}(\mathbf{A}) - \Phi(\tilde{\mathbf{A}})\| \leq b\mathrm{u}\|\mathbf{X}\| \ \text{ and } \ \|\tilde{\mathbf{A}} - \mathbf{A}\| \leq a\mathrm{u}\|\mathbf{A}\| \tag{7.3}$$

for some positive constants a, b depending on the algorithm (and, eventually, on the data).

It may be shown [28] that within first order terms in u we have the relative accuracy estimate

$$\frac{\|\tilde{\mathbf{X}} - \mathbf{X}\|}{\|\mathbf{X}\|} \leq \mathrm{u}(aK + b). \tag{7.4}$$

For $a = 0$ the algorithm is *forwardly numerically stable*, while for $b = 0$ it is *backwardly numerically stable* according to the definitions given in [94], see also [28].

The inequality (7.4) in view of (7.3) is one of the most useful estimates in matrix perturbation analysis. It reveals the influence of the three main factors determining the accuracy of the solution: the machine arithmetic (via u), the problem (via K) and the algorithm (via a and b). But there are some difficulties in applying this approach.

First, it is hard to estimate the constants a and b. That is why heuristically it is often assumed that $a = 1$ and $b = 0$. This is the case when the algorithm is backwardly stable (the computed solution is the exact solution of a close problem) and the only errors in the computational process are introduced when rounding the data $\mathbf{A}$ to some machine matrix $\mathrm{fl}(\mathbf{A})$ with $\|\mathrm{fl}(\mathbf{A}) - \mathbf{A}\| \le \mathrm{u}\|\mathbf{A}\|$. This is a very successful approach as the computational practice suggests. Under the heuristic assumption the relative error in the computed solution is bounded by $K\mathrm{u}$,

$$\frac{\|\tilde{\mathbf{X}} - \mathbf{X}\|}{\|\mathbf{X}\|} \le K\mathrm{u},$$

and for $K\mathrm{u} \ll 1$ (which is most often the case) we may expect about $-\log_{10}(K\mathrm{u})$ true decimal digits in the solution.

Second, to estimate the conditioning of the problem may also be difficult. Various aspects of the conditioning of computational problems are considered in [13, 21, 22, 34, 80]. It may be hard to compute $L(\mathbf{A}, \eta)$ and, in addition, it is not clear how to determine η a priori. That is why it is often assumed that

$$L \approx L(\mathbf{A}, 0) = \|\Phi'(\mathbf{A})\|,$$

where $\Phi'(\mathbf{A})$ is the Fréchet derivative of Φ computed at the data $\mathbf{A}$. This results in a local perturbation analysis when the norm of the perturbation in the solution is assumed to be bounded by a linear function of the perturbation in the data. Unfortunately this assumption may be severely violated as the next simple example shows.

Consider the linear scalar equation

$$AX = 1.$$

For $A = 1$ the solution is $X = 1$. Let the data $A = 1$ be perturbed to $1 + E$, where $E > -1$. Then the solution $X = 1$ is perturbed to $1 + Y$, where

$$Y = \frac{-E}{1 + E}.$$

For any $\eta \in (0, 1)$ the Lipschitz constant is

$$L(1, \eta) = \frac{1}{1 - \eta}$$

and the correct perturbation bound (7.1) is

$$|Y| \le L(1, \eta)|E|, \ |E| \le \eta.$$

If we use $L(1,0) = 1$ instead of $L(1,\eta)$, then the local analysis gives the approximate estimate $|Y| \leq |E|$ (with no restrictions on E) while at the same time $|Y| \to \infty$ for $E \to -1$! Moreover, the local bound "works" even for $E = -1$ when there is no solution at all. Of course, we know that E should be small but in a real problem we do not know what "small" means.

The drawbacks of the local analysis may be overcome by the techniques of nonlocal perturbation analysis. In this case a quantity $r > 0$ and a non-decreasing function

$$f : [0,r] \to \mathbb{R}_+$$

are defined such that $f(0) = 0$ and

$$\|\mathbf{Y}\| \leq f(\|\mathbf{E}\|), \ \|\mathbf{E}\| \leq r.$$

This is the desired nonlocal (and in general nonlinear) perturbation estimate.

In many cases $\mathbf{A}$ is not a single matrix but a collection

$$\mathbf{A} = (\mathbf{A}_1, \mathbf{A}_2, \ldots, \mathbf{A}_m) \in \mathscr{A}$$

of m matrices $\mathbf{A}_k$. If the data matrices are perturbed as $\mathbf{A}_k \to \mathbf{A}_k + \mathbf{E}_k$ with

$$\|\mathbf{E}_k\| \leq \delta_k \ (k = 1, 2, \ldots, m)$$

the problem is to estimate (locally or nonlocally) the norm $\|\mathbf{Y}\|$ of the perturbation $\mathbf{Y}$ in the solution $\mathbf{X}$ as a function of the perturbation vector

$$\delta = [\delta_1; \delta_2; \ldots; \delta_m] \in \mathbb{R}^m_+.$$

7.4 Matrix Equations

7.4.1 Introductory Remarks

Consider the matrix equation

$$F(\mathbf{A}, \mathbf{X}) = \mathbf{0}, \tag{7.5}$$

where $\mathbf{A}$ is a collection of matrices as above, $\mathbf{X} \in \mathscr{X}$ is the solution and the function

$$F(\mathbf{A}, \cdot) : \mathscr{X} \to \mathscr{X}$$

is Fréchet differentiable ($\mathscr{X}$ is a certain space of matrices), while the function

$$F(\cdot, \mathbf{X}) : \mathscr{A} \to \mathscr{X}$$

is at least Fréchet pseudo-differentiable [38]. The latter case occurs in complex equations when the data includes both the matrix $\mathbf{A}_k$ and its complex conjugate $\overline{\mathbf{A}}_k$ for some k. The correct treatment of this case was firstly given in [43], see also [38, 47] and [92].

Denoting by

$$\mathbf{E} = (\mathbf{E}_1, \mathbf{E}_2, \ldots, \mathbf{E}_m) \in \mathscr{A}$$

the collection of perturbations, the perturbed equation is written as

$$F(\mathbf{A} + \mathbf{E}, \mathbf{X} + \mathbf{Y}) = \mathbf{0}. \tag{7.6}$$

If the Fréchet derivative $F_{\mathbf{X}}(\mathbf{A}, \mathbf{X})$ is invertible we may rewrite (7.6) as an equivalent operator equation

$$\mathbf{Y} = \Pi(\mathbf{A}, \mathbf{X}, \mathbf{E}, \mathbf{Y}), \tag{7.7}$$

where

$$\begin{aligned}
\Pi(\mathbf{A}, \mathbf{X}, \mathbf{E}, \mathbf{Y}) &:= -F_{\mathbf{X}}^{-1}(\mathbf{A}, \mathbf{X})(F_{\mathbf{A}}(\mathbf{A}, \mathbf{X})(\mathbf{E}) + G(\mathbf{A}, \mathbf{X}, \mathbf{E}, \mathbf{Y})), \\
G(\mathbf{A}, \mathbf{X}, \mathbf{E}, \mathbf{Y}) &:= F(\mathbf{A} + \mathbf{E}, \mathbf{X} + \mathbf{Y}) - F(\mathbf{A}, \mathbf{X}) - F_{\mathbf{A}}(\mathbf{A}, \mathbf{X})(\mathbf{E}) - F_{\mathbf{X}}(\mathbf{A}, \mathbf{X})(\mathbf{Y})
\end{aligned}$$

(we must have in mind that $F(\mathbf{A}, \mathbf{X}) = \mathbf{0}$ for the particular solution $\mathbf{X}$).

If for example

$$F(\mathbf{A}, \mathbf{X}) := \mathbf{A}_1 + \mathbf{A}_2\mathbf{X}\mathbf{A}_3 + \mathbf{A}_4\mathbf{X}\mathbf{A}_5\mathbf{X}\mathbf{A}_6 + \mathbf{A}_7\mathbf{X}\mathbf{A}_8\mathbf{X}\mathbf{A}_9\mathbf{X}\mathbf{A}_{10} + \mathbf{A}_{11}\mathbf{X}^{-1}\mathbf{A}_{12}$$

then

$$\begin{aligned}
F_{\mathbf{X}}(\mathbf{A}, \mathbf{X})(\mathbf{Y}) = {} & \mathbf{A}_2\mathbf{Y}\mathbf{A}_3 + \mathbf{A}_4\mathbf{X}\mathbf{A}_5\mathbf{Y}\mathbf{A}_6 + \mathbf{A}_4\mathbf{Y}\mathbf{A}_5\mathbf{X}\mathbf{A}_6 + \mathbf{A}_7\mathbf{X}\mathbf{A}_8\mathbf{X}\mathbf{A}_9\mathbf{Y}\mathbf{A}_{10} \\
& + \mathbf{A}_7\mathbf{X}\mathbf{A}_8\mathbf{Y}\mathbf{A}_9\mathbf{X}\mathbf{A}_{10} + \mathbf{A}_7\mathbf{Y}\mathbf{A}_8\mathbf{X}\mathbf{A}_9\mathbf{X}\mathbf{A}_{10} - \mathbf{A}_{11}\mathbf{X}^{-1}\mathbf{Y}\mathbf{X}^{-1}\mathbf{A}_{12}.
\end{aligned}$$

The perturbation analysis of algebraic matrix equation is subject to numerous studies, e.g. [19, 20, 26, 33, 37, 49, 50, 79, 81] and [2, 3, 35, 38, 51, 92, 93], see also the bibliography in the monograph [38]. A general framework for such an analysis is given in [38, 41]. Perturbation analysis of general coupled matrix quadratic equations is given in [36]. Such an analysis for the H_∞ problem involving two Riccati equations and other relations is done in [59], while perturbation analysis of differential matrix quadratic equations is presented in [33, 42].

7.4.2 Local Estimates

Neglecting second order terms in δ in (7.7) it is possible to derive an expression

$$\mathbf{y} \approx \mathbf{z} := \sum_{k=1}^{m} \mathbf{L}_k \mathbf{e}_k,$$

where

$$\mathbf{y} := \mathrm{vec}(\mathbf{Y}), \ \mathbf{e}_k := \mathrm{vec}(\mathbf{E}_k)$$

and $\mathbf{L}_k$ are certain matrices. Since

$$\|\mathbf{Y}\| = \|\mathbf{y}\| \approx \|\mathbf{z}\|$$

the problem is to find a tight bound on $\|\mathbf{z}\|$ as a function of δ. Such an improved norm-wise estimate is given in [38, 55]

$$\|\mathbf{z}\| \le \mathrm{est}(\mathbf{L};\delta) := \min\left\{ \|\mathbf{L}\|_2\|\delta\|, \sqrt{\delta^\top \Lambda(\mathbf{L})\delta} \right\}, \tag{7.8}$$

where $\mathbf{L} := [\mathbf{L}_1, \mathbf{L}_2, \ldots, \mathbf{L}_m]$ and $\Lambda = \Lambda(\mathbf{L}) \in \mathbb{R}_+^{m\times m}$ is a matrix with elements

$$\lambda_{k,l} := \left\| \mathbf{L}_k^{\mathrm{H}} \mathbf{L}_l \right\|_2 \ (k, l = 1, 2, \ldots, m).$$

Note that $\mathrm{est}(\mathbf{L};\cdot)$ is a continuous first order non-differentiable function $\mathbb{R}_+^m \to \mathbb{R}_+$.

The estimate (7.8) is in general better than the linear estimate

$$\|\mathbf{z}\| \le \sum_{k=1}^{m} \|\mathbf{L}_k\|_2 \delta_k$$

based on the individual absolute condition numbers $\|\mathbf{L}_k\|_2$.

Componentwise local estimates for various classes of matrix equations are also known, see [38] and the bibliography therein.

7.4.3 Nonlocal Estimates

Nonlocal perturbation estimates for matrix equations may be derived by the technique of Lyapunov majorants [24, 30, 38, 63] and using fixed point principles.

The exact Lyapunov majorant for Eq. (7.7) is

$$l(\delta, \rho) := \max\{\|\Pi(\mathbf{A}, \mathbf{X}, \mathbf{E}, \mathbf{Y})\| : \|\mathbf{E}_k\| \le \delta_k, \ \|\mathbf{Y}\| \le \rho\}$$

(the dependence of l on $\mathbf{A}$ and $\mathbf{X}$ is not marked since the latter two quantities are considered fixed in the framework of the perturbation problem).

The exact Lyapunov majorant for nonlinear algebraic matrix equations is nonlinear and strictly convex in ρ. However, with rare exceptions, exact Lyapunov majorants cannot be constructed explicitly. That is why we use Lyapunov majorants

$$h(\delta, \rho) \ge l(\delta, \rho)$$

which are not exact but can be constructed in an explicit form.

The technique of Lyapunov majorants uses the majorant equation

$$\rho = h(\delta, \rho) \tag{7.9}$$

for determining of ρ as a function of δ. Complete analysis of different types of majorant equations is presented in [38, 47].

Let $\Delta \subset \mathbb{R}^m_+$ be the set of all $\delta \in \mathbb{R}^m_+$ such that Eq. (7.9) has a nonnegative solution. The following facts are established in [38] under certain general assumptions (see also [24, 63]).

- The interior Δ^o of Δ is nonempty.
- For a part of the boundary of Δ the majorant equation has double solution.
- For $\delta \in \Delta^o$ the majorant equation has two positive solutions $f(\delta)$, $g(\delta)$ such that $f(\delta) < g(\delta)$.

Moreover, the function f is increasing in its arguments and $f(0) = 0$. This function is referred to as the *small solution* to the majorant equation.

Since the operator $\Pi(\mathbf{A}, \mathbf{X}, \mathbf{E}, \cdot)$ transforms the central ball of radius $f(\delta)$ into itself, then according to the Schauder fixed point principle equation (7.7) has a solution Y such that

$$\|\mathbf{Y}\| \le f(\delta), \ \delta \in \Delta. \tag{7.10}$$

This is the desired nonlocal nonlinear perturbation estimate.

Explicit expressions for $f(\delta)$ can be derived for polynomial and fractional-affine matrix equations [38]. If for example the matrix equation is quadratic then

$$h(\delta, \rho) = a_0(\delta) + a_1(\delta)\rho + a_2(\delta)\rho^2, \tag{7.11}$$

where $a_0(\delta), a_1(\delta)$ are expressions of type $\mathrm{est}(\mathbf{L}, \delta)$. In particular $a_0(0) = a_1(0) = 0$ and hence

$$\|\mathbf{Y}\| \le f(\delta) := \frac{2a_0(\delta)}{1 - a_1(\delta) + \sqrt{(1 - a_1(\delta))^2 - 4a_0(\delta)a_2(\delta)}} \tag{7.12}$$

for

$$\delta \in \Delta := \left\{\delta : \mathbb{R}_+^m : a_1(\delta) + 2\sqrt{a_0(\delta)a_2(\delta)} \leq 1\right\}.$$

Relations (7.11), (7.12) constitute the nonlocal estimate in this case.

We note finally that local, nonlocal, norm-wise and componentwise perturbation estimates for linear, polynomial and fractional-affine algebraic matrix equations are given in the monograph [38]. Such estimates are used in modern reliable computational methods and algorithms [66, 71, 75].

7.4.4 Linear Equations

As a first example consider the Lyapunov matrix equation

$$F(\mathbf{A}, \mathbf{X}) := \mathbf{A}_1 + \mathbf{A}_2\mathbf{X} + \mathbf{X}\mathbf{A}_2^{\mathrm{H}} = \mathbf{0}, \tag{7.13}$$

where $\mathbf{A}_1, \mathbf{A}_2, \mathbf{X} \in \mathbb{C}^{n\times n}$. Suppose that

$$\lambda_p(\mathbf{A}_2) + \overline{\lambda}_q(\mathbf{A}_2) \neq 0 \ \ (p, q = 1, 2, \ldots, n),$$

where $\lambda_p(\mathbf{A}_2)$ are the eigenvalues of the matrix $\mathbf{A}_2$ counted according to their algebraic multiplicities. Under this assumption the matrix

$$\mathbf{A}_0 := \mathbf{I}_n \otimes \mathbf{A}_2 + \overline{\mathbf{A}_2} \otimes \mathbf{I}_n \in \mathbb{C}^{n^2\times n^2}$$

of the Lyapunov operator $\mathbf{X} \mapsto \mathbf{A}_2\mathbf{X}+\mathbf{X}\mathbf{A}_2^{\mathrm{H}}$ is invertible and Eq. (7.13) has a unique solution $\mathbf{X}$. Moreover, if $\mathbf{A}_1^{\mathrm{H}} = \mathbf{A}_1$ then $\mathbf{X}^{\mathrm{H}} = \mathbf{X}$ as well.

The perturbed Lyapunov equation is

$$F(\mathbf{A}+\mathbf{E}, \mathbf{X}+\mathbf{Y}) = \mathbf{A}_1 + \mathbf{E}_1 + (\mathbf{A}_2+\mathbf{E}_2)(\mathbf{X}+\mathbf{Y}) + (\mathbf{X}+\mathbf{Y})(\mathbf{A}_2+\mathbf{E}_2)^{\mathrm{H}} = \mathbf{0}, \tag{7.14}$$

where the perturbations in the data are bounded as

$$\|\mathbf{E}_k\| \leq \delta_k \ \ (k = 1, 2).$$

The condition

$$\delta_2 < \delta_2^0 := \frac{1}{2l_1}, \ \ l_1 := \left\|\mathbf{A}_0^{-1}\right\|_2$$

is sufficient for Eq. (7.14) to have a unique solution. At the same time when $\delta_2 \geq \delta_2^0$ this equation may have no solution or may have a variety of solutions.

We have

$$\mathbf{y} := \mathrm{vec}(\mathbf{Y}) = \mathbf{z} + \mathrm{O}(\|\delta\|^2),\ \delta \to 0,$$

where

$$\mathbf{z} := \mathbf{L}_1\mathbf{e}_1 + \mathbf{L}_2\mathbf{e}_2 + \mathbf{L}_3\overline{\mathbf{e}}_2,\ \mathbf{e}_k = \mathrm{vec}(\mathbf{E}_k)$$

and

$$\mathbf{L}_1 := -\mathbf{A}_0^{-1},\ \mathbf{L}_2 := \mathbf{L}_1(\mathbf{X}^\top \otimes \mathbf{I}_n),\ \mathbf{L}_3 := \mathbf{L}_1(\mathbf{I}_n \otimes \mathbf{X})\mathscr{V}_n.$$

Here $\mathscr{V}_n \in \mathbb{R}^{n^2\times n^2}$ is the vec-permutation matrix such that $\mathrm{vec}(\mathbf{Z}^\top) = \mathscr{V}_n\mathrm{vec}(\mathbf{Z})$.

According to [38, 47] we have the local perturbation estimate

$$\|\mathbf{z}\| \le \mathrm{est}(\mathbf{M}_1, \mathbf{M}_2; \delta_1, \delta_2),$$

where

$$\mathbf{M}_1 := \begin{bmatrix} \mathbf{L}_{10} & -\mathbf{L}_{11} \\ \mathbf{L}_{11} & \mathbf{L}_{10} \end{bmatrix},\ \mathbf{M}_2 := \begin{bmatrix} \mathbf{L}_{20} + \mathbf{L}_{30} & \mathbf{L}_{31} - \mathbf{L}_{21} \\ \mathbf{L}_{21} + \mathbf{L}_{31} & \mathbf{L}_{20} - \mathbf{L}_{30} \end{bmatrix}$$

and $\mathbf{L}_k = \mathbf{L}_{k0} + \imath\mathbf{L}_{k1}$, $\imath^2 = -1$.

To obtain a nonlocal estimate we rewrite the equivalent operator equation for $\mathbf{Y}$ in a vector form for $\mathbf{y} = \mathrm{vec}(\mathbf{Y})$ as

$$\mathbf{y} = \pi(\mathbf{A}, \mathbf{E}, \mathbf{y}) := \mathbf{L}_1\mathbf{e}_1 + \mathbf{L}_2\mathbf{e}_2 + \mathbf{L}_3\overline{\mathbf{e}}_2 + \mathbf{L}_1\mathrm{vec}(\mathbf{E}_2\mathbf{Y} + \mathbf{Y}\mathbf{E}_2^{\mathrm{H}}).$$

Therefore the Lyapunov majorant h is defined by

$$\|\pi(\mathbf{A}, \mathbf{E}, \mathbf{y})\| \le h(\delta, \rho) := \mathrm{est}(\mathbf{M}_1, \mathbf{M}_2; \delta_1, \delta_2) + 2l_1\delta_2\rho,\ \|\mathbf{y}\| \le \rho.$$

Hence for $\delta_2 < \delta_2^0$ we have the nonlocal estimate

$$\|\mathbf{Y}\| = \|\mathbf{y}\| \le \frac{\mathrm{est}(\mathbf{M}_1, \mathbf{M}_2; \delta_1, \delta_2)}{1 - 2l_1\delta_2}.$$

Similar norm-wise bounds as well as component-wise bounds for more general linear matrix equations

$$\mathbf{B}_0 + \sum_{k=1}^{r} \mathbf{B}_k\mathbf{X}\mathbf{C}_k = \mathbf{0} \tag{7.15}$$

with $m = 2r + 1$ matrix coefficients $\mathbf{A} = (\mathbf{B}_0, \mathbf{B}_1, \mathbf{C}_1, \ldots, \mathbf{B}_r, \mathbf{C}_r)$ are analyzed in a similar way, see e.g. [38, 56]. Here the coefficients $\mathbf{B}_k, \mathbf{C}_k$ may not be independent and relations such as $\mathbf{B}_k = \mathbf{C}_k^{\mathrm{H}}$ are possible for some indices k.

7.4.5 Quadratic Equations

Consider the matrix quadratic equation

$$F(\mathbf{A}, \mathbf{X}) := \mathbf{A}_1 + \mathbf{A}_2\mathbf{X} + \mathbf{X}\mathbf{A}_3 + \mathbf{X}\mathbf{A}_4\mathbf{X} = \mathbf{0}, \tag{7.16}$$

where the coefficients $\mathbf{A}_k$ and the solution $\mathbf{X}$ are matrices of appropriate size (we shall use the same notations $\mathbf{A}_k$, $\mathbf{L}_k$, a_k, h etc. for different quantities in each subsection of this paper). Important particular case of a quadratic equation is the Riccati equation

$$\mathbf{A}_1 + \mathbf{A}_2\mathbf{X} + \mathbf{X}\mathbf{A}_2^{\top} - \mathbf{X}\mathbf{A}_3\mathbf{X} = \mathbf{0}; \ \ \mathbf{A}_1^{\top} = \mathbf{A}_1, \ \mathbf{A}_3^{\top} = \mathbf{A}_3,$$

arising in the theory of control and filtering of linear continuous time-invariant systems.

Let the matrices $\mathbf{A}_k$ in (7.16) be perturbed as $\mathbf{A}_k + \mathbf{E}_k$ with $\|\mathbf{E}_k\| \leq \delta_k$ $(k = 1, 2, 3, 4)$ and let $\mathbf{X} + \mathbf{Y}$ be the solution to the perturbed equation

$$F(\mathbf{A} + \mathbf{E}, \mathbf{X} + \mathbf{Y}) = 0.$$

Suppose that the matrix

$$\mathbf{A}_0 := \mathbf{I} \otimes (\mathbf{A}_2 + \mathbf{X}\mathbf{A}_4) + (\mathbf{A}_3 + \mathbf{A}_4\mathbf{X})^{\top} \otimes \mathbf{I}$$

of the linear matrix operator $\mathbf{Y} \mapsto (\mathbf{A}_2 + \mathbf{X}\mathbf{A}_4)\mathbf{Y} + \mathbf{Y}(\mathbf{A}_3 + \mathbf{A}_4\mathbf{X})$ is invertible and denote

$$\mathbf{L}_1 := -\mathbf{A}_0^{-1}, \ \mathbf{L}_2 := \mathbf{L}_1(\mathbf{X}^{\top} \otimes \mathbf{I}), \ \mathbf{L}_3 := \mathbf{L}_1(\mathbf{I} \otimes \mathbf{X}), \ \mathbf{L}_4 := \mathbf{L}_1(\mathbf{X}^{\top} \otimes \mathbf{X}).$$

Then we have the local bound

$$\|\mathbf{Y}\| = a_0(\delta) + \mathrm{O}(\|\delta\|^2), \ \delta \to 0,$$

and the nonlocal bound (7.12), where (see [47])

$$\begin{aligned} a_0(\delta) &:= \operatorname{est}(\mathbf{L}_1, \mathbf{L}_2, \mathbf{L}_3, \mathbf{L}_4; \delta_1, \delta_2, \delta_3, \delta_4), \\ a_1(\delta) &:= \operatorname{est}(\mathbf{L}_1, \mathbf{L}_2, \mathbf{L}_3; \delta_2 + \delta_3, \delta_4, \delta_4), \\ a_2(\delta) &:= l_1(\|\mathbf{A}_4\|_2 + \delta_4), \ l_1 := \|\mathbf{L}_1\|_2. \end{aligned}$$

Matrix quadratic equations involving more general expressions of the form $\mathbf{B}_k\mathbf{X}\mathbf{C}_k\mathbf{X}\mathbf{D}_k$ with matrix coefficients $\mathbf{B}_k, \mathbf{C}_k, \mathbf{D}_k$ are analyzed similarly [50].

7.4.6 Polynomial Equations of Degree $d > 2$

Matrix polynomial equations of degree $d > 2$ in X and with m matrix coefficients give rise to Lyapunov majorants

$$h(\delta, \rho) := \sum_{k=0}^{d} a_k(\delta)\rho^k, \ \delta = [\delta_1; \delta_2; \ldots; \delta_m] \in \mathbb{R}_+^m.$$

Here a_k are continuous nonnegative non-decreasing functions in δ of type est or polynomials in δ satisfying

$$a_0(0) = 0, \ \ a_1(0) < 1, \ \ a_d(0) > 0$$

(usually we even have $a_1(0) = 0$). An example of a simple third degree matrix equation is

$$\mathbf{A}_1 + \mathbf{A}_2\mathbf{X} + \mathbf{A}_3\mathbf{X}^2 + \mathbf{A}_4\mathbf{X}^3 = 0. \tag{7.17}$$

For δ sufficiently small (in particular we must at least guarantee that $a_1(\delta) < 1$) the ME

$$\rho = h(\delta, \rho)$$

in ρ has small positive solution $\rho = f(\delta)$ vanishing together with δ and such that the Frobenius norm $\|\mathbf{Y}\|$ of the perturbation $\mathbf{Y}$ in the solution $\mathbf{X}$ satisfies

$$\|\mathbf{Y}\| \le f(\delta), \ \delta \in \Delta \subset \mathbb{R}_+^m.$$

Here Δ is the domain of all δ for which the majorant equation has nonnegative roots.

The boundary $\partial\Delta$ of Δ is defined by the pair of equations

$$\rho = h(\delta, \rho), \quad \frac{\partial h(\delta, \rho)}{\partial \rho} = 1 \tag{7.18}$$

and the inequalities $\delta_k \ge 0$ $(k = 1, 2, \ldots, m)$. Hence for $\delta \in \partial\Delta$ either the discriminant of the algebraic equation

$$a_0(\delta) - (1 - a_1(\delta))\rho + \sum_{k=2}^{d} a_k(\delta)\rho^k = 0$$

in ρ is zero or $\delta_k = 0$ for some k.

In general there is no convenient explicit expression for $f(\delta)$ when $d > 2$. Therefore the problem is to find a tight easily computable upper bound $\hat{f}(\delta)$ for $f(\delta)$. For this purpose the dth degree Lyapunov majorant $h(\delta, \rho)$ is replaced by a second degree Lyapunov majorant

$$\hat{h}(\delta, \rho) := a_0(\delta) + a_1(\delta)\rho + \hat{a}_2(\delta)\rho^2 \geq h(\delta, \rho), \ \rho \in [0, \tau(\delta)].$$

Here the positive quantity $\tau(\rho)$ satisfies $h(\delta, \tau(\delta)) \leq \tau(\delta)$.

Denoting by $\hat{f}(\delta)$ the small solution of the new ME

$$\rho = \hat{h}(\delta, \rho) \tag{7.19}$$

we get the perturbation estimate

$$\|\mathbf{Y}\| \leq \hat{f}(\delta) := \frac{2a_0(\delta)}{1 - a_1(\delta) + \sqrt{(1 - a_1(\delta))^2 - 4a_0(\delta)\hat{a}_2(\delta)}} \tag{7.20}$$

provided

$$a_1(\delta) + 2\sqrt{a_0(\delta)\hat{a}_2(\delta)} \leq 1 \ \text{ and } \ h(\delta, \tau(\delta)) \leq \tau(\delta).$$

We note that both $f(\delta)$ and $\hat{f}(\delta)$ have asymptotic order

$$\frac{a_0(\delta)}{1 - a_1(0)} + \mathrm{O}(\|\delta\|^2), \ \delta \to 0.$$

To find $\hat{h}(\delta, \rho)$ and $\tau(\delta)$ we proceed as follows. For any $\tau > 0$ and $\rho \leq \tau$ we have

$$h(\delta, \rho) \leq a_0(\delta) + a_1(\delta)\rho + \beta(\delta, \tau)\rho^2,$$

where

$$\beta(\delta, \tau) := a_2(\delta) + \sum_{k=2}^{d-1} a_{k+1}(\delta)\tau^{k-1}. \tag{7.21}$$

Let now $\tau(\delta)$ be a positive nondecreasing expression in δ and $\rho \leq \tau(\delta)$. Then we may find an upper bound $\hat{a}_2(\delta)$ for $\beta(\delta, \tau(\delta))$ and use it in the estimate (7.20). Choosing different expressions for $\tau(\delta)$ we obtain different upper bounds $\hat{a}_2(\delta)$ for $\beta(\delta, \tau(\delta))$ and different Lyapunov majorants $\hat{h}(\delta, \rho)$. As a result we get different estimates $\|\mathbf{Y}\| \leq \hat{f}(\delta)$.

It must be stressed that if the ME (7.19) has positive roots then its small root $\hat{f}(\delta)$ does not exceed the value of ρ for which the second equation

$$1 = \omega(\delta, \rho) := \sum_{k=0}^{d-1} (k+1) a_{k+1}(\delta) \rho^k \tag{7.22}$$

in (7.18) is fulfilled. For sufficiently small δ it is fulfilled

$$\omega(\delta, 0) = a_1(\delta) < 1 \ \text{ and } \ \omega(\delta, r) > 1, \ \ r := (d a_d(\delta))^{1/(1-d)}.$$

Hence there is a unique positive solution $\rho = \tau(\delta)$ of Eq. (7.22). This solution may exist even when the ME (7.19) has no positive solutions. But if Eq. (7.19) has positive solutions $\hat{f}(\delta) \le \hat{g}(\delta)$ then

$$f(\delta) \le \tau(\delta) \le g(\delta) \text{ and } h(\delta, \tau(\delta)) \le \tau(\delta)$$

by necessity. Using this approach we distinguish two cases.

The case $d = 3$**.** Here $\tau(\delta)$ may be computed directly from (7.22) which in this case is a quadratic equation

$$3a_3(\delta)\tau^2 + 2a_2(\delta)\tau - (1 - a_1(\delta)) = 0.$$

We have

$$\tau(\delta) = \frac{1 - a_1(\delta)}{a_2(\delta) + \sqrt{a_2^2(\delta) + 3a_3(\delta)(1 - a_1(\delta))}}. \tag{7.23}$$

For $\rho \le \tau(\delta)$ we have

$$h(\delta, \rho) \le \hat{h}(\delta, \rho) := a_0(\delta) + a_1(\delta)\rho + \hat{a}_2(\delta)\rho^2, \ \hat{a}_2(\delta) := a_2(\delta) + a_3(\delta)\tau(\delta). \tag{7.24}$$

Hence $\hat{h}(\delta, \rho)$ is a new quadratic Lyapunov majorant. As a result we get the nonlocal perturbation estimate (7.20) with $\hat{a}_2(\delta)$ defined in (7.24).

Consider again the matrix cubic equation (7.17). Suppose that the matrix

$$\mathbf{A}_0 := \mathbf{I} \otimes (\mathbf{A}_2 + \mathbf{A}_3\mathbf{X} + \mathbf{A}_4\mathbf{X}^2) + \mathbf{X}^\top \otimes \mathbf{A}_3 + (\mathbf{X}^\top \otimes \mathbf{A}_4)(\mathbf{I} \otimes \mathbf{X} + \mathbf{X}^\top \otimes \mathbf{I})$$

of the linear operator

$$\mathbf{Y} \mapsto (\mathbf{A}_2 + \mathbf{A}_3\mathbf{X} + \mathbf{A}_4\mathbf{X}^2)\mathbf{Y} + \mathbf{A}_3\mathbf{Y}\mathbf{X} + \mathbf{A}_4(\mathbf{X}\mathbf{Y} + \mathbf{Y}\mathbf{X})\mathbf{X}$$

is invertible. Denote

$$\mathbf{L}_1 := -\mathbf{A}_0^{-1},\ \mathbf{L}_k := \mathbf{L}_1\left((\mathbf{X}^{k-1})^\top \otimes \mathbf{I}\right),\ l_k := \|\mathbf{L}_k\|_2\ (k=1,2,3,4),\ x := \|\mathbf{X}\|_2.$$

Then the Lyapunov majorant for Eq. (7.17) is

$$a_0(\delta) + a_1(\delta)\rho + a_2(\delta)\rho^2 + a_3(\delta)\rho^3,$$

where

$$\begin{aligned}
a_0(\delta) &:= \mathrm{est}(\mathbf{L}_1, \mathbf{L}_2, \mathbf{L}_3, \mathbf{L}_4; \delta_1, \delta_2, \delta_3, \delta_4),\\
a_1(\delta) &:= l_1\delta_2 + (l_1 x + l_2)\delta_3 + (l_1 x^2 + l_2 x + l_3)\delta_4,\\
a_2(\delta) &:= \|\mathbf{L}_1(\mathbf{I}\otimes \mathbf{A}_3)\|_2 + (1+x)\left\|\mathbf{L}_1\left(\mathbf{X}^\top \otimes \mathbf{A}_4\right)\right\|_2 + \|\mathbf{L}_1(\mathbf{I}\otimes(\mathbf{A}_4\mathbf{X}))\|_2\\
&\quad + l_1\delta_3 + (l_2 + 2l_1 x)\delta_4,\\
a_3(\delta) &:= \|\mathbf{L}_1(\mathbf{I}\otimes \mathbf{A}_4)\|_2 + l_1\delta_4.
\end{aligned}$$

Now we may apply the estimate (7.20) in view of (7.24).

The case $d > 3$**.** Here the estimation of the quantity $\tau(\delta)$ is more subtle. We shall work with certain easily computable quantities

$$\gamma_{k+1}(\delta) \geq a_{k+1}(\delta)\tau^{k-1}(\delta)$$

in (7.21), see [38].

Consider again Eq. (7.22) in ρ for a given small δ which guarantees that (7.22) has a (unique) root $\rho = \tau(\delta)$. For $k = 2, 3, \ldots, d-1$ we have

$$(k+1)a_{k+1}(\delta)\tau^k(\delta) \leq 1 - a_1(\delta)$$

and

$$\tau(\delta) \leq \left(\frac{1 - a_1(\delta)}{(k+1)a_{k+1}(\delta)}\right)^{1/k},\ a_{k+1}(\delta) > 0.$$

Hence

$$a_{k+1}(\delta)\tau^{k-1}(\delta) \leq \gamma_{k+1}(\delta) := a_{k+1}^{1/k}(\delta)\left(\frac{1 - a_1(\delta)}{k+1}\right)^{1-1/k}$$

and

$$\beta(\delta, \tau(\delta)) \leq \hat{a}_2(\delta) := a_2(\delta) + \sum_{k=2}^{d-1}\gamma_{k+1}(\delta). \tag{7.25}$$

Hence we get again the nonlocal perturbation estimate (7.20) with $\hat{a}_2(\delta)$ defined in (7.25).

7.4.7 Fractional-Affine Equations

Fractional-affine matrix equations (FAME) involve inversions of affine expressions in $\mathbf{X}$ such as the left-hand side of Eq. (7.15). A famous example of a FAME is the equation

$$\mathbf{A}_1 - \mathbf{X} + \mathbf{A}_2^{\mathrm{H}}(\mathbf{I} + \mathbf{A}_3\mathbf{X})^{-1}\mathbf{A}_2 = 0$$

arising in linear-quadratic optimization and filtering of discrete-time dynamic systems [66, 75]. Here the matrices $\mathbf{A}_1 = \mathbf{A}_1^{\mathrm{H}}$ and $\mathbf{A}_3 = \mathbf{A}_3^{\mathrm{H}}$ are nonnegative definite, the pair $[\mathbf{A}_2, \mathbf{A}_3)$ is stabilizable and the pair $(\mathbf{A}_1, \mathbf{A}_2]$ is detectable (we denote matrix pairs so that the state matrix $\mathbf{A}_2$ is near the square bracket).

The Lyapunov majorant $h(\delta, \rho)$ for such equations may be chosen as a rational function of ρ with coefficients depending on the perturbation vector δ (see [47] for more details)

$$h(\delta, \rho) := b_0(\delta) + b_1(\delta)\rho + \frac{b_2(\delta) + b_3(\delta)\rho + b_4(\delta)\rho^2}{b_5(\rho) - b_6(\delta)\rho}. \tag{7.26}$$

Here the following conditions are fulfilled (some of them for δ sufficiently small).

1. The functions $b_1, b_2, \ldots, b_6$ are nonnegative and continuous in δ.
2. The functions b_k are nondecreasing for $k \neq 5$, while the function b_5 is positive and non-increasing in δ.
3. The relations

$$b_0(0) = b_2(0) = 0,\ b_1(0) < 1,\ b_5(0) > 0,\ b_1(0) + \frac{b_3(0)}{b_5(0)} < 1$$

take place.

Denote

$$\begin{aligned}
c_0(\delta) &:= b_2(\delta) + b_0(\delta)b_5(\delta), \\
c_1(\delta) &:= b_5(\delta)(1 - b_1(\delta)) + b_0(\delta)b_6(\delta) - b_3(\delta), \\
c_2(\delta) &:= b_4(\delta) + b_6(\delta)(1 - b_1(\delta)).
\end{aligned}$$

Then the majorant equation $\rho = h(\delta, \rho)$ takes the form

$$c_2(\delta)\rho^2 - c_1(\delta)\rho + c_0(\delta) = 0.$$

On the other hand we have $c_0(0) = 0$ and $c_1(0) > 0$. Hence for small δ it is fulfilled that $c_1(\delta) > 0$ and $c_1^2(\delta) > 4c_0(\delta)c_2(\delta)$. Set

$$\Delta := \left\{\delta \in \mathbb{R}_+^m : c_1(\delta) > 0,\ c_1^2(\delta) \geq 4c_0(\delta)c_2(\delta)\right\}.$$

It may be shown that the set Δ has nonempty interior. Hence the perturbation bound corresponding to the Lyapunov majorant (7.26) is

$$f(\delta) = \frac{2c_0(\delta)}{c_1(\delta) + \sqrt{c_1^2(\delta) - 4c_0(\delta)c_2(\delta)}},\ \delta \in \Delta. \tag{7.27}$$

As a particular example consider the FAME

$$F(\mathbf{A}, \mathbf{X}) := \mathbf{A}_1 + \mathbf{A}_2\mathbf{X} + \mathbf{X}\mathbf{A}_3 + \mathbf{A}_4\mathbf{X}^{-1}\mathbf{A}_5 = \mathbf{0}. \tag{7.28}$$

The perturbation analysis of such equations uses the technique of Lyapunov majorants combined with certain useful matrix identities.

The solution $\mathbf{X}$ of (7.28) is invertible and so is the matrix $\mathbf{Z} := \mathbf{X} + \mathbf{Y}$ for small matrix perturbations $\mathbf{Y}$. If in particular $\|\mathbf{Y}\| \leq \rho < \sigma$ then $\mathbf{Z}$ is invertible and

$$\|\mathbf{Z}^{-1}\|_2 \leq \frac{1}{\sigma - \rho},$$

where $\sigma = \|\mathbf{X}^{-1}\|^{-1}$ is the minimum singular value of $\mathbf{X}$. We also have the identities

$$\begin{aligned}\mathbf{Z}^{-1} &= \mathbf{X}^{-1} - \mathbf{X}^{-1}\mathbf{Y}\mathbf{Z}^{-1} = \mathbf{X}^{-1} - \mathbf{Z}^{-1}\mathbf{Y}\mathbf{X}^{-1} \\ &= \mathbf{X}^{-1} - \mathbf{X}^{-1}\mathbf{Y}\mathbf{X}^{-1} + \mathbf{X}^{-1}\mathbf{Y}\mathbf{Z}^{-1}\mathbf{Y}\mathbf{X}^{-1}.\end{aligned}$$

As a result we get the following matrix identity typical for the proper perturbation analysis of FAME [47, 54]

$$\begin{aligned}F(\mathbf{A} + \mathbf{E}, \mathbf{X} + \mathbf{Y}) = &\ F(\mathbf{A}, \mathbf{X}) + F_{\mathbf{X}}(\mathbf{A}, \mathbf{X})(\mathbf{Y}) \\ &+ F_0(\mathbf{A}, \mathbf{X}, \mathbf{E}) + F_1(\mathbf{A}, \mathbf{X}, \mathbf{E}, \mathbf{Y}) + F_2(\mathbf{A}, \mathbf{X}, \mathbf{Y}),\end{aligned}$$

where the Fréchet derivative $F_{\mathbf{X}}(\mathbf{A}, \mathbf{X})$ is determined by

$$F_{\mathbf{X}}(\mathbf{A}, \mathbf{X})(\mathbf{Y}) = \mathbf{A}_2\mathbf{Y} + \mathbf{Y}\mathbf{A}_3 - \mathbf{A}_4\mathbf{X}^{-1}\mathbf{Y}\mathbf{X}^{-1}\mathbf{A}_5$$

and

$$\begin{aligned}F_0(\mathbf{A}, \mathbf{X}, \mathbf{E}) &:= \mathbf{E}_1 + \mathbf{E}_2\mathbf{X} + \mathbf{X}\mathbf{E}_3 + \mathbf{A}_4\mathbf{X}^{-1}\mathbf{E}_5 + \mathbf{E}_4\mathbf{X}^{-1}\mathbf{A}_5 + \mathbf{E}_4\mathbf{Z}^{-1}\mathbf{E}_5, \\ F_1(\mathbf{A}, \mathbf{X}, \mathbf{E}, \mathbf{Y}) &:= \mathbf{E}_2\mathbf{Y} + \mathbf{Y}\mathbf{E}_3 - \mathbf{A}_4\mathbf{X}^{-1}\mathbf{Y}\mathbf{Z}^{-1}\mathbf{E}_5 - \mathbf{E}_4\mathbf{Z}^{-1}\mathbf{Y}\mathbf{X}^{-1}\mathbf{A}_5, \\ F_2(\mathbf{A}, \mathbf{X}, \mathbf{Y}) &:= \mathbf{A}_4\mathbf{X}^{-1}\mathbf{Y}\mathbf{Z}^{-1}\mathbf{Y}\mathbf{X}^{-1}\mathbf{A}_5.\end{aligned}$$

Suppose that the matrix

$$\mathbf{A}_0 := \mathbf{I} \otimes \mathbf{A}_2 + \mathbf{A}_3^\top \otimes \mathbf{I} - \mathbf{B}$$

of the linear matrix operator $F_{\mathbf{X}}(\mathbf{A}, \mathbf{X})$ is invertible, where

$$\mathbf{B} := \left(\mathbf{X}^{-1}\mathbf{A}_5\right)^\top \otimes \left(\mathbf{A}_4\mathbf{X}^{-1}\right),$$

and denote

$$\mathbf{L}_1 := -\mathbf{A}_0^{-1},\ \mathbf{L}_2 := \mathbf{L}_1(\mathbf{I} \otimes \mathbf{X}),\ \mathbf{L}_3 := \mathbf{L}_1(\mathbf{I} \otimes \mathbf{X}),$$
$$\mathbf{L}_4 := \mathbf{L}_1\left(\left(\mathbf{X}^{-1}\mathbf{A}_5\right)^\top \otimes \mathbf{I}\right),\ \mathbf{L}_5 := \mathbf{L}_1\left(\mathbf{I} \otimes \left(\mathbf{A}_4\mathbf{X}^{-1}\right)\right).$$

As it is shown in [47] the Lyapunov majorant of type (7.26) for Eq. (7.28) is determined by the relations

$$b_0(\delta) := \mathrm{est}(\mathbf{L}_1, \mathbf{L}_2, \mathbf{L}_3, \mathbf{L}_4; \delta_1, \delta_2, \delta_3, \delta_4),\ b_1(\delta) := l_1(\delta_2 + \delta_3),\ b_2(\delta) := l_1\delta_4\delta_5,$$
$$b_3(\delta) := \mathrm{est}(\mathbf{L}_4, \mathbf{L}_5; \delta_4, \delta_5),\ b_4 := \|\mathbf{L}_1\mathbf{B}\|_2,\ b_5 = \sigma,\ b_6 := 1,\ l_1 := \|\mathbf{L}_1\|_2. \quad (7.29)$$

Therefore the perturbation bound for Eq. (7.28) is given by (7.27), where the quantities $b_1, b_2, \ldots, b_6$ are defined in (7.29).

7.5 Matrix Decompositions

7.5.1 Introductory Remarks

A complex or real square matrix $\mathbf{U}$ is said to be *unitary* if $\mathbf{U}^\mathrm{H}\mathbf{U} = \mathbf{I}$, where $\mathbf{I}$ is the identity matrix. The group of unitary $(n \times n)$-matrices is denoted as $\mathscr{U}(n)$. Real unitary matrices satisfy $\mathbf{U}^\top\mathbf{U} = \mathbf{I}$ and are also called *orthogonal* – a term which is not very suitable since orthogonality is a relation between two objects. The columns $\mathbf{u}_k$ of an unitary matrix $\mathbf{U}$ satisfy $\mathbf{u}_k^\mathrm{H}\mathbf{u}_l = \delta_{k,l}$ (the Kronecker delta).

Unitary matrices and transformations play a major role in theoretical and numerical matrix algebra [23, 27]. Moreover, a computational matrix algorithm for implementation in finite machine arithmetic can hardly be recognized as reliable unless it is based on unitary transformations (algorithms like Gaussian elimination are among the rare exceptions from this rule).

It suffices to mention the *QR decomposition* and *singular value decomposition (SVD)* of rectangular matrices, the *Schur decomposition* and *anti-triangular Schur decomposition* of square matrices and the *Hamiltonian-Schur* and *block-Schur decomposition* of Hamiltonian matrices [1, 62].

Unitary $(n \times n)$-matrices $\mathbf{U}$ have excellent numerical properties since they are easily inverted (without rounding errors) and have small norm:

$$\mathbf{U}^{-1} = \mathbf{U}^{\mathrm{H}} \text{ and } \|\mathbf{U}\|_2 = 1, \; \|\mathbf{U}\| = \sqrt{n}\,.$$

Perturbation bounds for these and other matrix decompositions are presented in many articles and monographs [11, 14, 15, 40, 52, 57, 58, 84–90].

A unified effective approach to the perturbation analysis of matrix problems involving unitary matrices was firstly proposed in [52, 74]. It is called the *Method of Splitting Operators and Lyapunov Majorants (MSOLM)*. This method and its main applications are presented in [44, 45]. The main applications of MSOLM include the unitary matrix decompositions mentioned above as well as certain control problems. Among the latter we shall mention the synthesis of linear systems with desired equivalent form (pole assignment synthesis in particular) and the transformation into canonical or condensed unitary (orthogonal in particular) form.

7.5.2 Splitting Operators and Lyapunov Majorants

MSOLM is based on splitting of matrix operators $\mathscr{P} : \mathbb{C}^{m\times n} \to \mathbb{C}^{m\times n}$ and their matrix arguments $\mathbf{X}$ into strictly lower, diagonal and strictly upper parts

$$\mathbf{X}_1 = \mathrm{Low}(\mathbf{X}), \; \mathbf{X}_2 = \mathrm{Diag}(\mathbf{X}) \text{ and } \mathbf{X}_3 = \mathrm{Up}(\mathbf{X}),$$

namely

$$\mathbf{X} = \mathbf{X}_1 + \mathbf{X}_2 + \mathbf{X}_3 \text{ and } \mathscr{P} = \mathscr{P}_1 + \mathscr{P}_2 + \mathscr{P}_3,$$

where

$$\mathscr{P}_1 := \mathrm{Low} \circ \mathscr{P}, \; \mathscr{P}_2 := \mathrm{Diag} \circ \mathscr{P} \text{ and } \mathscr{P}_3 := \mathrm{Up} \circ \mathscr{P}.$$

If for example $\mathbf{X} = [x_{k,l}] \in \mathbb{C}^{3\times 3}$ then

$$\mathbf{X}_1 = \begin{bmatrix} 0 & 0 & 0 \\ x_{2,1} & 0 & 0 \\ x_{3,1} & x_{3,2} & 0 \end{bmatrix}, \; \mathbf{X}_2 = \begin{bmatrix} x_{1,1} & 0 & 0 \\ 0 & x_{2,2} & 0 \\ 0 & 0 & x_{3,3} \end{bmatrix} \text{ and } \mathbf{X}_3 = \begin{bmatrix} 0 & x_{1,2} & x_{1,3} \\ 0 & 0 & x_{2,3} \\ 0 & 0 & 0 \end{bmatrix}.$$

The operators Low, Diag and Up are projectors of the matrix space $\mathbb{C}^{m\times n}$ on the subspaces of strictly lower triangular, diagonal and strictly upper triangular matrices. The properties of splitting operators are studied in detail in [44, 45].

Let for simplicity $m = n$ and denote by $\mathscr{T}_n \subset \mathbb{C}^{n\times n}$ the set of upper triangular matrices. Then we have

$$\mathrm{Up}(\mathbf{X}) = \mathrm{Low}(\mathbf{X}^{\top})^{\top}, \; \mathscr{T}_n = (\mathrm{Diag} + \mathrm{Up})(\mathbb{C}^{n\times n}).$$

If $\mathbf{e}_k \in \mathbb{C}^n$ is the unit vector with $\mathbf{e}_k(l) = \delta_{k,l}$ (the Kronecker delta) then

$$\mathrm{Low}(\mathbf{X}) = \sum_{l=1}^{n-1} \sum_{k=l+1}^{n} \mathbf{e}_k \mathbf{e}_k^\top \mathbf{X} \mathbf{e}_l \mathbf{e}_l^\top, \ \mathrm{Diag}(\mathbf{X}) = \sum_{k=1}^{n} \mathbf{e}_k \mathbf{e}_k^\top \mathbf{X} \mathbf{e}_k \mathbf{e}_k^\top.$$

The vectorizations of the splitting operators contain many zeros. That is why we prefer to work with the *compressed vectorizations* of the splitting operators Low and Diag, namely

$$\mathrm{lvec} : \mathbb{C}^{n\times n} \to \mathbb{C}^{\nu}, \ \mathrm{dvec} : \mathbb{C}^{n\times n} \to \mathbb{C}^{n},$$

where $\nu := n(n-1)/2$ and

$$\mathrm{lvec}(\mathbf{X}) := [x_{2,1}, x_{3,1}, \ldots, x_{n,1}, x_{3,2}, x_{4,2}, \ldots, x_{n,2}, \ldots, x_{n-1,n-2}, x_{n,n-2}, x_{n,n-1}]^\top \in \mathbb{C}^{\nu},$$
$$\mathrm{dvec}(\mathbf{X}) := [x_{1,1}, x_{2,2}, \ldots, x_{n,n}]^\top \in \mathbb{C}^{n}.$$

Thus the vector lvec($\mathbf{X}$) contains the strictly lower part of the matrix Low($\mathbf{X}$) spanned column-wise and dvec($\mathbf{X}$) is the vector of the diagonal elements of the matrix Diag($\mathbf{X}$).

We have

$$\mathrm{lvec}(\mathbf{X}) := \mathbf{M}_{\mathrm{lvec}} \mathrm{vec}(X), \ \mathrm{dvec}(\mathbf{X}) := \mathbf{M}_{\mathrm{dvec}} \mathrm{vec}(X),$$

where the matrices $\mathbf{M}_{\mathrm{lvec}}$, $\mathbf{M}_{\mathrm{dvec}}$ of the operators lvec, dvec are given by

$$\mathbf{M}_{\mathrm{lvec}} := [\mathrm{diag}(\mathbf{N}_1, \mathbf{N}_2, \ldots, \mathbf{N}_{n-1}), \mathbf{0}_{\nu\times n}] \in \mathbb{R}^{\nu\times n^2},$$
$$\mathbf{M}_{\mathrm{dvec}} := \mathrm{diag}\left(\mathbf{e}_1^\top, \mathbf{e}_2^\top, \ldots, \mathbf{e}_n^\top\right) \in \mathbb{R}^{n\times n^2},$$

where

$$\mathbf{N}_k := [\mathbf{0}_{(n-k)\times k}, \mathbf{I}_{n-k}] \in \mathbb{R}^{(n-k)\times n} \ (k = 1, 2, \ldots, n-1).$$

Let

$$\mathrm{lvec}^\dagger : \mathbb{C}^{\nu} \to \mathbb{C}^{n\times n}$$

be the right inverse of the operator lvec such that

$$\mathrm{lvec} \circ \mathrm{lvec}^\dagger = \mathbf{I}_{\nu\times\nu} \ \text{ and } \ \mathrm{lvec}^\dagger \circ \mathrm{lvec} = \mathrm{Low}.$$

Then the matrix of $\mathrm{lvec}^\dagger$ is

$$\mathbf{M}_{\mathrm{lvec}^\dagger} = \mathbf{M}_{\mathrm{lvec}}^\top.$$

An important property of the above splittings is that if $\mathbf{T} \in \mathscr{T}_n$ and $\mathbf{X} \in \mathbb{C}^{n\times n}$ then the matrices $\mathrm{Low}(\mathbf{XT})$ and $\mathrm{Low}(\mathbf{TX})$ depend *only* on the strictly lower part $\mathrm{Low}(\mathbf{X})$ of $\mathbf{X}$ rather than on the whole matrix $\mathbf{X}$.

In a more general setting we have the following result [44]. Let the matrix operator $\mathscr{L} : \mathbb{C}^{n\times n} \to \mathbb{C}^{n\times n}$ be determined from

$$\mathscr{L}(\mathbf{X}) = \mathbf{AXB},$$

where $\mathbf{A}, \mathbf{B} \in \mathscr{T}_n$ are given matrices. Then

$$\mathrm{Low} \circ \mathscr{L} = \mathrm{Low} \circ \mathscr{L} \circ \mathrm{Low}.$$

The matrix problems considered in this and next sections may be formulated as follows. We have a collection $\mathbf{A} = (\mathbf{A}_1, \mathbf{A}_2, \ldots, \mathbf{A}_m) \in \mathscr{A}$ of data matrices $\mathbf{A}_k$. The resulting matrix (or the solution)

$$\mathbf{R} = \Psi(\mathbf{A}, \mathbf{U})$$

is an upper triangular (or anti-triangular) matrix, where $\mathbf{U} = (\mathbf{U}_1, \mathbf{U}_2, \ldots, \mathbf{U}_k)$ is a collection of matrices $\mathbf{U}_p \in \mathscr{U}(n)$. At the same time the matrix arguments $\mathbf{A}$ and $\mathbf{U}$ of Ψ are not independent. Moreover, in many problems the matrix collection $\mathbf{U}$ is determined by the data $\mathbf{A}$ "almost uniquely" via the requirement

$$\mathrm{Low}(\Psi(\mathbf{A}, \mathbf{U})) = \mathbf{0}.$$

(we stress that the result $\mathbf{R}$ may also be a collection rather than a single matrix).

In certain problems with $k = 1$ the matrix $\mathbf{U} = \mathbf{U}_1$ is uniquely determined by the data. This may be the case when $\mathbf{R}$ is the canonical form of $\mathbf{A}$ for certain multiplicative action of the group $\mathscr{U}(n)$ on the set of data [45, 48]. Let for example the matrix $\mathbf{A} \in \mathbb{C}^{n\times n}$ be invertible and

$$\mathbf{A} = \mathbf{UR}$$

be its QR decomposition, where $\mathbf{U} \in \mathscr{U}(n)$ and $\mathrm{Low}(\mathbf{R}) = \mathbf{0}$ (for unity of notations we denote this decomposition as $\mathbf{A} = \mathbf{UR}$ instead of the widely used $\mathbf{A} = \mathbf{QR}$). Under certain additional requirements we may consider

$$\mathbf{R} = \Psi(\mathbf{A}, \mathbf{U}) := \mathbf{U}^{\mathrm{H}}\mathbf{A}$$

as the canonical form of $\mathbf{A}$ under the left multiplicative action of the group $\mathscr{U}(n)$. Since the diagonal elements of $\mathbf{R}$ are nonzero we may force them to be real and positive. In this case $\mathbf{U}$ is determined uniquely and $\mathbf{R}$ is the canonical form of $\mathbf{A}$ for this action [45].

In other problems the transformation matrix $\mathbf{U} \in \mathscr{U}(n)$ cannot be chosen uniquely. For example, in the Schur decomposition $\mathbf{A} = \mathbf{URU}^{\mathrm{H}}$ of $\mathbf{A}$, the Schur form $\mathbf{R}$ (either canonical or only condensed) satisfies $\mathrm{Low}(\mathbf{R}) = \mathbf{0}$. At the same time any other matrix $\nu\mathbf{U}$ is also unitary and transforms $\mathbf{A}$ into $\mathbf{R}$ provided that $\nu = \exp(\imath\omega)$ and $\omega \in \mathbb{R}$.

However, in practice condensed rather than canonical forms are used. In this case, due to the non-uniqueness of the condensed forms and their perturbations, the perturbation bounds are valid only for *some* (but not for all) of the solutions of the perturbed problem.

Let $\mathbf{E} = (\mathbf{E}_1, \mathbf{E}_2, \ldots, \mathbf{E}_m)$ be a perturbation in the collection $\mathbf{A}$ such that

$$\|\mathbf{E}_k\| \le \delta_k \ \ (k = 1, 2, \ldots, m).$$

Suppose that the perturbed problem with data $\mathbf{A} + \mathbf{E}$ has a solution

$$\mathbf{R} + \mathbf{Z} = \Psi(\mathbf{A} + \mathbf{E}, \mathbf{U} + \mathbf{V}) \subset \mathscr{T}_n,$$

such that the perturbation $\mathbf{Z}$ in $\mathbf{R}$ satisfies $\mathrm{Low}(\mathbf{Z}) = \mathbf{0}$ and let

$$\mathbf{U} + \mathbf{V} = \mathbf{U}(\mathbf{I} + \mathbf{X}) \in \mathscr{U}(n), \ \mathbf{X} := \mathbf{U}^{\mathrm{H}}\mathbf{V}$$

be the perturbed transformation matrix.

The norm-wise perturbation problem is to estimate the norm of the perturbation

$$\mathbf{Z} = Z(\mathbf{A}, \mathbf{E}, \mathbf{X}) = \Psi(\mathbf{A} + \mathbf{E}, \mathbf{U}(\mathbf{I} + \mathbf{X})) - \Psi(\mathbf{A}, \mathbf{U})$$

in the solution $\mathbf{R}$ as well as the norm of the perturbation $\mathbf{V}$ in the transformation matrix $\mathbf{U}$ as functions of the perturbation vector

$$\delta = [\delta_1; \delta_2; \ldots; \delta_m] \in \mathbb{R}^m_+,$$

e.g.

$$\|\mathbf{Z}\| \le f_{\mathbf{R}}(\delta), \ \|\mathbf{V}\| = \|\mathbf{X}\| \le f_{\mathbf{U}}(\delta), \tag{7.30}$$

where the nonnegative valued functions $f_{\mathbf{R}}$ and $f_{\mathbf{U}}$ are continuous and nondecreasing in δ, and $f_{\mathbf{R}}(0) = f_{\mathbf{U}}(0) = 0$.

We stress again that the perturbed problem may have solutions in which $\mathbf{V}$ is not small even when $\mathbf{E}$ is small or even zero (in the latter case the problem is actually not perturbed). This may occur when we deal with condensed forms or with canonical forms for which $\mathbf{U}$ is not uniquely determined.

To illustrate this phenomenon consider again the Schur decomposition $\mathbf{A} = \mathbf{U}\mathbf{R}\mathbf{U}^{\mathrm{H}}$ of the matrix $\mathbf{A}$ and let $\mathbf{E} = \mathbf{0}$. Choosing $\mathbf{V} = -2\mathbf{U}$ we see that the matrix $\mathbf{U} + \mathbf{V} = -\mathbf{U} \in \mathscr{U}(n)$ also solves the problem, i.e. transforms $\mathbf{A}$ into $\mathbf{R}$. However, in this case the norm $\|\mathbf{V}\| = 2\sqrt{n}$ of the perturbation $\mathbf{V}$ is the maximum possible and does not satisfy any estimate of type (7.30). Similar effects may arise in some perturbation problems in control theory for which the solution set is not even bounded! So what is way out of this situation?

We can only assert that the perturbation estimates (7.30) are valid for *some* perturbations $\mathbf{Z}$ and $\mathbf{V}$. At the same time the perturbed problem may have other solutions which are not small and for which the inequalities (7.30) does not hold true. We may formalize these considerations as follows.

For a given fixed perturbation $\mathbf{E}$ let $\mathbb{V}_{\mathbf{E}} \subset \mathbb{C}^{n\times n}$ be the set of all $\mathbf{V}$ satisfying the perturbed problem

$$\mathrm{Low}(\Psi(\mathbf{A}+\mathbf{E},\mathbf{U}+\mathbf{V})-\Psi(\mathbf{A},\mathbf{U}))=\mathbf{0},\ \mathbf{U}+\mathbf{V}\in\mathscr{U}(n).$$

Since the set $\mathbb{V}_E$ is defined by a system of polynomial equations it is compact and the infimum

$$\inf\{\|\mathbf{V}\|:\mathbf{V}\in\mathbb{V}_{\mathbf{E}}\}=\|\mathbf{V}_0\|$$

is reached for some $\mathbf{V}_0 \in \mathbb{V}_{\mathbf{E}}$. Now we may choose $\mathbf{V}=\mathbf{V}_0$ and claim that the estimates (7.30) will be valid for this particular value of $\mathbf{V}$.

Since the matrix $\mathbf{I}+\mathbf{X}$ is unitary we have $(\mathbf{I}+\mathbf{X})^{\mathrm{H}}(\mathbf{I}+\mathbf{X})=\mathbf{I}$ and

$$\mathbf{X}^{\mathrm{H}}+\mathbf{X}+\mathbf{X}^{\mathrm{H}}\mathbf{X}=\mathbf{0}.$$

Hence

$$\mathbf{X}^{\mathrm{H}}=-\mathbf{X}+\mathrm{O}(\|\mathbf{X}\|^2),\ \mathbf{X}\to\mathbf{0}.$$

Now the problem is to estimate the norm of $\mathbf{X}$. Splitting $\mathbf{X}$ as $\mathbf{X}_1+\mathbf{X}_2+\mathbf{X}_3$ above, we rewrite the perturbed problem as an operator equation

$$\mathbf{X}=\Pi(\mathbf{A},\mathbf{E},\mathbf{X}),$$

or as a system of three operator equations

$$\begin{aligned}\mathbf{X}_1&=\Pi_1(\mathbf{A},\mathbf{E},\mathbf{X}),\\ \mathbf{X}_2&=\Pi_2(\mathbf{X}):=-0.5\,\mathrm{Diag}(\mathbf{X}^{\mathrm{H}}\mathbf{X}),\\ \mathbf{X}_3&=\Pi_3(\mathbf{X}):=-\mathrm{Up}(\mathbf{X}^{\mathrm{H}})-\mathrm{Up}(\mathbf{X}^{\mathrm{H}}\mathbf{X}).\end{aligned} \tag{7.31}$$

The right-hand side $\Pi_1(\mathbf{A},\mathbf{E},\mathbf{X})$ of the first equation in (7.31) depends on the particular problem, while the second and the third equalities are *universal equations*. The only information that we need about the universal equations is that

$$\|\Pi_2(\mathbf{X})\|\le 0.5\|\mathbf{X}\|^2,\ \|\Pi_3(\mathbf{X})\|\le\|\mathbf{X}_1\|+\mu_n\|\mathbf{X}\|^2,$$

where

$$\mu_n := \sqrt{\frac{n-1}{2n}}. \tag{7.32}$$

Further on we introduce a generalized norm $|\cdot| : \mathbb{C}^{n\times n} \to \mathbb{R}^3_+$ from

$$|\mathbf{X}| := [\|\mathbf{X}_1\|; \|\mathbf{X}_2\|; \|\mathbf{X}_3\|] \in \mathbb{R}^3_+ .$$

In certain problems the splitting of $\mathbf{X}$ and Π is done in $p > 3$ parts $\mathbf{X}_1, \mathbf{X}_2, \ldots, \mathbf{X}_p$ and $\Pi_1, \Pi_2, \ldots, \Pi_p$. Here the generalized norm $|\mathbf{X}|$ of $\mathbf{X}$ is a nonnegative p-vector (see [40]),

$$|\mathbf{X}| := [\|\mathbf{X}_1\|; \|\mathbf{X}_2\|; \ldots; \|\mathbf{X}_p\|] \in \mathbb{R}^p_+ .$$

Here the *vector Lyapunov majorant*

$$\mathbf{h} = [h_1; h_2; \ldots; h_p] : \mathbb{R}^m_+ \times \mathbb{R}^p_+ \to \mathbb{R}^p_+$$

is a continuous function which satisfies the following conditions.

1. For all perturbations $\mathbf{E}$ and matrices $\mathbf{X} \in \mathbb{C}^{n\times n}$, and for all vectors $\delta \in \mathbb{R}^m_+$, $\xi \in \mathbb{R}^p_+$ with

$$|\mathbf{E}| \preceq \delta,\ |\mathbf{X}| \preceq \xi$$

   it is fulfilled

$$|\Pi(\mathbf{E}, \mathbf{X})| \preceq \mathbf{h}(\delta, \xi)$$

   where $\preceq$ is the component-wise partial order relation.
2. For each $\delta \in \mathbb{R}^m_+$ the function $\mathbf{h}(\delta, \cdot) : \mathbb{R}^p_+ \to \mathbb{R}^p_+$ is differentiable in the domain $\mathbb{R}^p_+ \backslash \{\mathbf{0}\}$.
3. The elements h_k of $\mathbf{h}$ are non-decreasing strictly convex functions of all their arguments and $\mathbf{h}(\mathbf{0}, \mathbf{0}) = \mathbf{0}$.
4. There is a continuous matrix function $J : \mathbb{R}^m_+ \times \mathbb{R}^p_+ \to \mathbb{R}^{p\times p}$ such that

$$\frac{\partial \mathbf{h}(\delta, \xi)}{\partial \xi} \preceq J(\delta, \xi);\ \mathbf{0} \preceq \delta,\ \mathbf{0} \prec \boldsymbol{\xi}$$

   and the spectral radius of $J(\mathbf{0}, \mathbf{0})$ is less than 1 (here $\mathbf{0} \prec \xi$ means that all elements of ξ are positive).

Applying the Schauder fixed point principle it may be shown [38] that under these conditions and for some $\delta_0 \succ 0$ the vector majorant equation

$$\xi = \mathbf{h}(\delta, \xi),\ \delta \preceq \delta_0$$

has a solution $\xi = \mathbf{f}(\delta) \succeq \mathbf{0}$ which tends to $\mathbf{0}$ together with δ. Finally the desired perturbation estimate for $\mathbf{X}$ is

$$|\mathbf{X}| \preceq \mathbf{f}(\delta)$$

which also yields

$$\|\mathbf{V}\| = \|\mathbf{X}\| = \sqrt{\|\mathbf{X}_1\|^2 + \|\mathbf{X}_2\|^2 + \|\mathbf{X}_3\|^3} \leq \|\mathbf{f}(\delta)\|.$$

To construct the operator $\Pi_1(\mathbf{A}, \mathbf{E}, \cdot)$ in (7.31) we represent $\mathbf{Z}$ as

$$\mathbf{Z} = Z(\mathbf{A}, \mathbf{E}, \mathbf{X}) = \mathscr{L}(\mathbf{A})(\mathbf{X}) - \Omega(\mathbf{A}, \mathbf{E}, \mathbf{X}),$$

where $\mathscr{L}(\mathbf{A})(\mathbf{X})$ is the main part of $Z(\mathbf{A}, \mathbf{E}, \mathbf{X})$ with respect to $\mathbf{X}$ (in particular $\mathscr{L}(\mathbf{A})$ can be the Fréchet derivative of $\mathbf{Z}$ in $\mathbf{X}$ computed for $\mathbf{E} = \mathbf{0}, \mathbf{X} = \mathbf{0}$). In turn, the expression

$$\Omega(\mathbf{A}, \mathbf{E}, \mathbf{X}) := \mathscr{L}(\mathbf{A})(\mathbf{X}) - Z(\mathbf{A}, \mathbf{E}, \mathbf{X})$$

contains first order terms in $\|\mathbf{E}\|$ and higher order terms in $\|\mathbf{E}\| + \|\mathbf{X}\|$.

Since $\mathrm{Low}(\mathbf{Z}) = \mathbf{0}$ we have

$$\mathscr{L}_{\mathrm{low}}(\mathbf{A})(\mathrm{lvec}(\mathbf{X})) = \mathrm{lvec}(\Omega(\mathbf{A}, \mathbf{E}, \mathbf{X})).$$

Here $\mathscr{L}_{\mathrm{low}}(\mathbf{A}) : \mathbb{C}^{\nu} \to \mathbb{C}^{\nu}$ is the lower compression of $\mathscr{L}(\mathbf{A})$ with matrix

$$\mathbf{L}_{\mathrm{low}}(\mathbf{A}) := \mathbf{M}_{\mathrm{lvec}} \mathbf{L}(\mathbf{A}) \mathbf{M}_{\mathrm{lvec}}^{\top}, \tag{7.33}$$

where $\mathbf{L}(\mathbf{A}) \in \mathbb{C}^{\nu \times \nu}$ is the matrix of the operator $\mathscr{L}(\mathbf{A})$.

Under certain generic conditions the operator $\mathscr{L}_{\mathrm{low}}(\mathbf{A})$ is invertible although $\mathscr{L}(\mathbf{A})$ is not. Thus we have

$$\mathbf{X}_1 = \Pi_1(\mathbf{A}, \mathbf{E}, \mathbf{X}) := \mathrm{lvec}^{\dagger} \circ L_{\mathrm{low}}^{-1}(\mathbf{A}) \circ \mathrm{lvec}(\Omega(\mathbf{A}, \mathbf{E}, \mathbf{X})).$$

The explicit expression for $\Pi_1(\mathbf{A}, \mathbf{E}, \mathbf{X})$ may not be constructed. Instead, to apply the technique of vector Lyapunov majorants it suffices to use the estimate

$$\|\Pi_1(\mathbf{A}, \mathbf{E}, \mathbf{X})\| \leq \lambda \|\Omega(\mathbf{A}, \mathbf{E}, \mathbf{X})\|,$$

where

$$\lambda := \left\| L_{\mathrm{low}}^{-1}(\mathbf{A}) \right\|_2 . \tag{7.34}$$

Fortunately, a Lyapunov majorant for $\Omega(\mathbf{A}, \mathbf{E}, \mathbf{X})$ is usually constructed relatively easy.

This is in brief the general scheme for perturbation analysis of matrix problems involving unitary transformations. Applications of this scheme to particular problems are outlines in the next subsections.

7.5.3 QR Decomposition

Perturbation analysis of the QR decomposition

$$\mathbf{A} = \mathbf{U}\mathbf{R},\ \mathrm{Low}(\mathbf{R}) = 0,\ \mathbf{U} \in \mathscr{U}(n)$$

of the matrix $\mathbf{A} \in \mathbb{C}^{n\times n}$ is done in [88]. Later on such an analysis has been performed by the MSOLM [44] thus getting tighter perturbation bounds.

Here the matrix $\mathbf{L}_{\mathrm{low}}(\mathbf{A})$ from (7.33) is

$$\mathbf{L}_{\mathrm{low}}(\mathbf{A}) := \mathbf{M}_{\mathrm{lvec}}(\mathbf{R}^\top \otimes \mathbf{I})\mathbf{M}_{\mathrm{lvec}}^\top = \mathbf{M}_{\mathrm{lvec}}((\mathbf{A}^\top\overline{\mathbf{U}}) \otimes \mathbf{I})\mathbf{M}_{\mathrm{lvec}}^\top,$$

where

$$\mathbf{R} = \mathbf{U}^{\mathrm{H}}\mathbf{A} = [r_{k,l}]\ (k, l = 1, 2, \dots, n).$$

The eigenvalues of the matrix $\mathbf{L}_{\mathrm{low}}(\mathbf{A})$ are $r_{1,1}$ with multiplicity $n-1$, $r_{2,2}$ with multiplicity $n-2,\dots$, and $r_{n-1,n-1}$ with multiplicity 1. Let either rank($\mathbf{A}$) $= n$, or rank($\mathbf{A}$) $= n-1$. In the latter case let us rearrange the columns of $\mathbf{A}$ so that the first $n-1$ columns of $\mathbf{A}$ are linearly independent. Then the matrix $\mathbf{L}_{\mathrm{low}}(\mathbf{A})$ is invertible.

The perturbed QR decomposition is

$$\mathbf{A} + \mathbf{E} = \mathbf{U}(\mathbf{I} + \mathbf{X})(\mathbf{R} + \mathbf{Z}),\ \mathbf{I} + \mathbf{X} \in \mathscr{U}(n),\ \mathrm{Low}(\mathbf{Z}) = \mathbf{0}$$

and the problem is to estimate $\|\mathbf{Z}\|$ and $\|\mathbf{X}\|$ as functions of $\delta := \|\mathbf{E}\|$.

In this case the vector majorant equation is equivalent to a quadratic equation which yields the estimates [44, 45]

$$\begin{aligned} \|\mathbf{Z}\| &\le f_{\mathbf{R}}(\delta) := \|\mathbf{A}\|_2 f_{\mathbf{U}}(\delta) + \delta, \\ \|\mathbf{X}\| &\le f_{\mathbf{U}}(\delta) := \frac{2\lambda\delta}{\sqrt{1 - 2\lambda\mu_n\delta + \sqrt{w(\delta)}}} \end{aligned} \tag{7.35}$$

provided

$$\delta \le \frac{1}{\lambda\left(2\mu_n + \sqrt{2 + 8\mu_n^2}\right)}.$$

Here

$$w(\delta) := (1 - 2\lambda\mu_n\delta)^2 - 2\lambda^2(1 + 4\mu_n^2)\delta^2$$

and the quantities μ_n and λ are given in (7.32) and (7.34) respectively.

7.5.4 Singular Value Decomposition

Consider the singular value decomposition (SVD)

$$\mathbf{A} = \mathbf{U}_1\mathbf{R}\mathbf{U}_2^{\mathrm{H}}, \ \ \mathbf{U}_1, \mathbf{U}_2 \in \mathscr{U}(n), \ \mathrm{Low}(\mathbf{R}) = \mathrm{Low}(\mathbf{R}^\top) = \mathbf{0},$$

of the invertible matrix $\mathbf{A} \in \mathbb{C}^{n\times n}$ (the case of general matrices $\mathbf{A}$ is more subtle but is treated similarly). We have $\mathbf{R} = \mathrm{diag}(r_{1,1}, r_{2,2}, \ldots, r_{n,n})$, where $r_{1,1} \geq r_{2,2} \geq \cdots \geq r_{n,n} > 0$. The singular values $\sigma_k = r_{k,k}$ of $\mathbf{A}$ are the square roots of the eigenvalues of the matrix $\mathbf{A}^{\mathrm{H}}\mathbf{A}$. Thus $\mathbf{R}$ is the canonical form of $\mathbf{A}$ for the multiplicative action

$$\mathbf{A} \mapsto \mathbf{U}_1^{\mathrm{H}}\mathbf{A}\mathbf{U}_2$$

of the group $\mathscr{U}(n) \times \mathscr{U}(n)$ on the set of invertible matrices.

Using splittings we may introduce a condensed form $\hat{\mathbf{R}}$ for this action from

$$\mathrm{Low}(\hat{\mathbf{R}}) = \mathrm{Low}(\hat{\mathbf{R}}^\top) = \mathbf{0}$$

without ordering the elements of $\hat{\mathbf{R}}$.

Let $\mathbf{A}$ be perturbed to $\mathbf{A} + \mathbf{E}$ with $\varepsilon := \|\mathbf{E}\|_2 < \sigma_n$ and let

$$\mathbf{A} + \mathbf{E} = \mathbf{U}_1(\mathbf{I} + \mathbf{X}_1)(\mathbf{R} + \mathbf{Z})(\mathbf{I} + \mathbf{X}_2^{\mathrm{H}})\mathbf{U}_2^{\mathrm{H}},$$

where

$$\mathbf{I} + \mathbf{X}_k \in \mathscr{U}(n), \ \mathrm{Low}(\mathbf{Z}) = \mathrm{Low}(\mathbf{Z}^\top) = \mathbf{0},$$

be the SVD of the matrix $\mathbf{A} + \mathbf{E}$. Here

$$\mathbf{R} + \mathbf{Z} = \mathrm{diag}(\tau_1, \tau_2, \ldots, \tau_n)$$

and

$$0 < \sigma_k - \varepsilon \leq \tau_k \leq \sigma_k + \varepsilon \ \ (k = 1, 2, \ldots, n).$$

Thus we have the perturbation estimates

$$\|\mathbf{Z}\|_2 \le \varepsilon \ \text{and} \ \|\mathbf{Z}\| \le \varepsilon\sqrt{n}\,.$$

This reflects the well known fact that the SVD is well conditioned. In this case a variant of MSOLM may be used to estimate the norms of the perturbations $\mathbf{X}_1$, $\mathbf{X}_2$ as well.

7.5.5 Schur Decomposition

The sensitivity of subspaces connected to certain eigenvalue problems are considered in [82, 83]. Nonlocal and local perturbation analysis of the Schur decomposition

$$\mathbf{A} = \mathbf{U}\mathbf{R}\mathbf{U}^{\mathrm{H}}, \ \ \mathbf{U} \in \mathscr{U}(n), \ \mathrm{Low}(\mathbf{R}) = \mathbf{0},$$

of the matrix $\mathbf{A} \in \mathbb{C}^{n\times n}$ is first done in [52]. Here the matrix $\mathbf{L}_{\mathrm{low}}(\mathbf{A})$ from (7.33) is

$$\mathbf{L}_{\mathrm{low}}(\mathbf{A}) := \mathbf{M}_{\mathrm{lvec}}(\mathbf{I} \otimes \mathbf{R} - \mathbf{R}^{\top} \otimes \mathbf{I})\mathbf{M}_{\mathrm{lvec}}^{\top}, \ \ \mathbf{R} = [r_{p,q}] = \mathbf{U}^{\mathrm{H}}\mathbf{A}\mathbf{U}.$$

The eigenvalues of the matrix $\mathbf{L}_{\mathrm{low}}(A)$ are

$$r_{p,p} - r_{q,q} = \lambda_p(\mathbf{A}) - \lambda_q(\mathbf{A}) \ \ (q = 1, 2, \ldots, n-1, \ p = q+1, q+2, \ldots, n).$$

Hence it is invertible if and only if $\mathbf{A}$ has distinct eigenvalues $\lambda_1(\mathbf{A}), \lambda_2(\mathbf{A}), \ldots,$ $\lambda_n(\mathbf{A})$ (note that if $\mathbf{A}$ has multiple eigenvalues the Schur form of $\mathbf{A} + \mathbf{E}$ may even be discontinuous as a function of $\mathbf{E}$!).

The perturbed Schur decomposition is

$$\mathbf{A} + \mathbf{E} = \mathbf{U}(\mathbf{I} + \mathbf{X})(\mathbf{R} + \mathbf{Z})(\mathbf{I} + \mathbf{X}^{\mathrm{H}})\mathbf{U}^{\mathrm{H}},$$

where

$$\mathbf{I} + \mathbf{X} \in \mathscr{U}(n), \ \mathrm{Low}(\mathbf{Z}) = \mathbf{0}.$$

The corresponding vector majorant equation is equivalent to a 6th degree algebraic equation. After certain manipulations it is replaced by a quadratic equation which yields explicit nonlocal perturbation estimates of type (7.35), see [47, 52].

7.5.6 Anti-triangular Schur Decomposition

The *anti-triangular Schur decomposition* of the matrix $\mathbf{A} \in \mathbb{C}^{n\times n}$ is described in [65]

$$\mathbf{A} = \overline{\mathbf{U}}\mathbf{R}\mathbf{U}^{\mathrm{H}}, \ \mathbf{U} \in \mathscr{U}(n),$$

where the matrix $\mathbf{R}$ is anti-triangular,

$$\mathbf{R} = \begin{bmatrix} 0 & 0 & \dots & 0 & r_{1,n} \\ 0 & 0 & \dots & r_{2,n-1} & r_{2,n} \\ \vdots & \vdots & \ddots & \vdots & \vdots \\ 0 & r_{n-1,2} & \dots & r_{n-1,n-1} & r_{n-1,n} \\ r_{n,1} & r_{n,2} & \dots & r_{n,n-1} & r_{n,n} \end{bmatrix}$$

This decomposition arises in solving palindromic eigenvalue problems [65].

Special matrix splittings are derived for the perturbation analysis of this decomposition and a variant of MSOLM for this purpose is presented recently in [15].

7.5.7 Hamiltonian-Schur Decomposition

The Hamiltonian-Schur decomposition of a Hamiltonian matrix

$$\mathbf{A} = \begin{bmatrix} \mathbf{A}_1 & \mathbf{A}_2 \\ \mathbf{A}_3 & -\mathbf{A}_1^{\mathrm{H}} \end{bmatrix} \in \mathbb{C}^{2n\times 2n}, \ \mathbf{A}_2^{\mathrm{H}} = \mathbf{A}_2, \ \mathbf{A}_3^{\mathrm{H}} = \mathbf{A}_3$$

is considered in [64]. When $\mathbf{A}$ has no imaginary eigenvalues there exist a matrix

$$\mathbf{U} = \begin{bmatrix} \mathbf{U}_1 & \mathbf{U}_2 \\ -\mathbf{U}_2 & \mathbf{U}_1 \end{bmatrix} \in \mathscr{U}\mathscr{S}(2n),$$

where $\mathscr{U}\mathscr{S}(2n) \subset \mathscr{U}(2n)$ is the group of unitary symplectic matrices, such that

$$\mathbf{R} := \mathbf{U}^{\mathrm{H}}\mathbf{A}\mathbf{U} = \begin{bmatrix} \mathbf{R}_1 & \mathbf{R}_2 \\ \mathbf{0} & -\mathbf{R}_1^{\mathrm{H}} \end{bmatrix}, \ \mathrm{Low}(\mathbf{R}_1) = \mathbf{0}, \ \mathbf{R}_2^{\mathrm{H}} = \mathbf{R}_2.$$

Less condensed forms

$$\hat{\mathbf{R}} = \begin{bmatrix} \hat{\mathbf{R}}_{1,1} & \hat{\mathbf{R}}_{1,2} \\ \mathbf{0} & \hat{\mathbf{R}}_{2,2} \end{bmatrix}$$

(block Hamiltonian-Schur form relative to $\mathscr{U}\mathscr{S}(2n)$ and block Schur form relative $\mathscr{U}(2n)$) of Hamiltonian matrices are introduced in [40]. The (1,1) and (2,2) blocks in $\hat{\mathbf{R}}$ are less structured in comparison with the corresponding blocks of $\mathbf{R}$.

Local and nonlocal perturbation analysis of the Hamiltonian-Schur and block-Schur forms of Hamiltonian matrices using MSOLM is presented in [40, 45]. For Hamiltonian-Schur (resp. block-Shur) forms the vector Lyapunov majorant $\mathbf{h}$ has 4 components and the vector majorant equation is equivalent to a 8th degree (resp. 6th degree) algebraic equation. After certain manipulations it is replaced by a bi-quadratic equation which yields explicit nonlocal perturbation bounds.

7.6 Control Problems

Perturbation analysis of control problems is subject to many investigations [45, 58, 73, 90]. Here we briefly mention two such major problems.

7.6.1 Unitary Canonical Forms

Unitary canonical forms (UCF) of linear time-invariant control systems

$$\mathbf{x}'(t) = \mathbf{A}\mathbf{x}(t) + \mathbf{B}\mathbf{u}(t),$$

where $\mathbf{x}(t) \in \mathbb{C}^n$ is the state vector, $\mathbf{u}(t) \in \mathbb{C}^m$ is the control vector and

$$\mathbf{A} \in \mathbb{C}^{n\times n},\ \mathbf{B} \in \mathbb{C}^{n\times m},\ \operatorname{rank}(\mathbf{B}) = m < n,$$

have been introduced in [48, 61]. The rigorous definition of these forms is given in [48]. The perturbation analysis of UCF is done in [57, 58, 74, 90] by MSOLM. We stress that UCF now play a major role in the analysis and synthesis of linear time-invariant systems.

The action of the group $\mathscr{U}(n)$ on the set of controllable matrix pairs $[\mathbf{A}, \mathbf{B}) \in \mathbb{C}^{n\times n} \times \mathbb{C}^{n\times m}$ is given by

$$[\mathbf{A}, \mathbf{B}) \mapsto [\mathbf{R}, \mathbf{T}) := [\mathbf{U}^{\mathrm{H}}\mathbf{A}\mathbf{U}, \mathbf{U}^{\mathrm{H}}\mathbf{B}),\ \mathbf{U} \in \mathscr{U}(n).$$

(we prefer to denote by $\mathbf{B}$ and $\mathbf{A}$ the system matrices instead of the more 'consistent' notation $\mathbf{A}_1$ and $\mathbf{A}_2$).

The canonical pair $[\mathbf{R}, \mathbf{T})$ has a very involved structure depending on the controllability indexes of $[\mathbf{A}, \mathbf{B})$, see [48]. In particular the matrix $[\mathbf{T}, \mathbf{R}] \in \mathbb{C}^{n\times(n+m)}$

is block upper triangular. Hence $\mathbf{R}$ is a block Hessenberg matrix as shown below

$$[\mathbf{T}, \mathbf{R}] = \begin{bmatrix} \mathbf{T}_{1,0} & \mathbf{R}_{1,1} & \mathbf{R}_{1,2} & \dots & \mathbf{R}_{1,p-2} & \mathbf{R}_{1,p-1} & \mathbf{R}_{1,p} \\ \mathbf{0} & \mathbf{R}_{2,1} & \mathbf{R}_{2,2} & \dots & \mathbf{R}_{2,p-2} & \mathbf{R}_{2,p-1} & \mathbf{R}_{2,p} \\ \mathbf{0} & \mathbf{0} & \mathbf{R}_{3,2} & \dots & \mathbf{R}_{3,p-2} & \mathbf{R}_{3,p-1} & \mathbf{R}_{3,p} \\ \vdots & \vdots & \vdots & \ddots & \vdots & \vdots & \vdots \\ \mathbf{0} & \mathbf{0} & \mathbf{0} & \dots & \mathbf{R}_{p-1,p-2} & \mathbf{R}_{p-1,p-1} & \mathbf{R}_{p-1,p} \\ \mathbf{0} & \mathbf{0} & \mathbf{0} & \dots & \mathbf{0} & \mathbf{R}_{p,p-1} & \mathbf{R}_{p,p} \end{bmatrix} =$$

where $\mathbf{T}_{1,0} \in \mathbb{R}^{m_1 \times m_0}$, $\mathbf{R}_{k,k-1} \in \mathbb{R}^{m_k \times m_{k-1}}$, $m_1 = m$ and

$$\mathrm{Low}(\mathbf{T}_{1,0}) = \mathbf{0}, \;\; \mathrm{Low}(\mathbf{R}_{k,k-1}) = \mathbf{0} \; (k = 1, 2, \dots, p).$$

Here $p > 1$ is the controllability index and $m_1 \geq m_2 \geq \dots \geq m_p \geq 1$ are the conjugate Kronecker indexes of the pair $[\mathbf{A}, \mathbf{B})$. Note that in the generic case $m = m_1 = m_2 = \dots = m_{p-1}$.

The complete set of arithmetic and algebraic invariants relative to the action of various unitary (orthogonal in particular) matrix groups is described in [45, 48].

Consider the perturbed pair $[\mathbf{A} + \mathbf{E}_2, \mathbf{B} + \mathbf{E}_1)$ with

$$\|\mathbf{E}_1\| \leq \delta_1, \;\; \|\mathbf{E}_2\|_2 \leq \delta_2,$$

which is reduced to UCF $[\mathbf{R} + \mathbf{Z}, \mathbf{T} + \mathbf{Y})$ by the transformation matrix $\mathbf{U}(\mathbf{I} + \mathbf{X}) \in \mathscr{U}(n)$. The perturbation problem here is to estimate the norms of $\mathbf{X}$, $\mathbf{Y}$ and $\mathbf{Z}$ as functions of the perturbation vector $\delta = [\delta_1; \delta_2] \in \mathbb{R}^2_+$. Local and nonlocal perturbation estimates for this problem are presented in [45, 57, 58, 73, 90]. The most general and tight results are those given in [58].

7.6.2 *Modal Control*

Consider the linear time-invariant system

$$\mathbf{x}'(t) = \mathbf{A}\mathbf{x}(t) + \mathbf{B}\mathbf{u}(t), \;\; \mathbf{y}(t) = \mathbf{C}\mathbf{x}(t),$$

where $(\mathbf{C}, \mathbf{A}, \mathbf{B}) \in \mathbb{C}^{r \times n} \times \mathbb{C}^{n \times n} \times \mathbb{C}^{n \times m}$ and $mr \geq n$, $r \leq n$, $m < n$. We suppose that the triple $(\mathbf{C}, \mathbf{A}, \mathbf{B})$ is complete, i.e. that the pair $[\mathbf{A}, \mathbf{B})$ is controllable and the pair $(\mathbf{C}, \mathbf{A}]$ is observable.

The static feedback $\mathbf{u}(t) = \mathbf{K}\mathbf{y}(t)$ results in the closed-loop system

$$\mathbf{x}'(t) = (\mathbf{A} + \mathbf{BKC})\mathbf{x}(t).$$

The purpose of the modal control is to find an output feedback matrix $\mathbf{K} \in \mathbb{C}^{m\times r}$ so as the closed-loop system matrix $\mathbf{A} + \mathbf{BKC}$ to have certain desirable properties. In particular it should have a prescribed set $\{\lambda_1, \lambda_2, \ldots, \lambda_n\}$ of eigenvalues (pole assignment synthesis), or, more generally, should be similar to a given matrix $\mathbf{D} \in \mathbb{C}^{n\times n}$, i.e.

$$\mathbf{U}^{-1}(\mathbf{A} + \mathbf{BKC})\mathbf{U} = \mathbf{D} \tag{7.36}$$

for some $\mathbf{U} \in \Gamma$, where the matrix group $\Gamma \subset \mathbb{C}^{n\times n}$ is either the unitary group $\mathscr{U}(n)$ or the group $\mathscr{GL}(n)$ of invertible matrices. In particular we may choose a desired form $\mathbf{D}$ with $\mathrm{Low}(\mathbf{D}) = \mathbf{0}$ and with diagonal elements $d_{k,k} = \lambda_k$ ($k = 1, 2, \ldots, n$). Suppose that $\Gamma = \mathscr{U}(n)$ in order to achieve reliability of the numerical procedure for feedback synthesis as proposed in [77].

Conditions for solvability of the problem of determining $\mathbf{U}$ and $\mathbf{K}$ from (7.36) are given in [45, 77]. When the problem is solvable its solution for $\mathbf{K}$ is an $(mr - n)$-parametric algebraic variety $\mathscr{K}(\mathbf{C}, \mathbf{A}, \mathbf{B}, \mathbf{D}) \subset \mathbb{C}^{m\times r}$.

If the data $(\mathbf{C}, \mathbf{A}, \mathbf{B})$ and $\mathbf{D}$ is perturbed to $(\mathbf{C} + \mathbf{E}_3, \mathbf{A} + \mathbf{E}_2, \mathbf{B} + \mathbf{E}_1)$ and $\mathbf{D} + \mathbf{E}_4$ with $\|\mathbf{E}_k\| \le \delta_k$ ($k = 1, 2, 3, 4$), then under certain conditions the perturbed problem

$$(\mathbf{I} + \mathbf{X}^{\mathrm{H}})\mathbf{U}^{\mathrm{H}}(\mathbf{A} + \mathbf{E}_2 + (\mathbf{B} + \mathbf{E}_1)(\mathbf{K} + \mathbf{Z})(\mathbf{C} + \mathbf{E}_3))\mathbf{U}(\mathbf{I} + \mathbf{X}) = \mathbf{D} + \mathbf{E}_4$$

has a solution $\mathbf{Z}, \mathbf{X}$ with $\mathbf{I} + \mathbf{X} \in \mathscr{U}(n)$. The task now is to estimate the quantities $\|\mathbf{Z}\|$ and $\|\mathbf{X}\|$ as functions of the perturbation vector $\delta = [\delta_1; \delta_2; \delta_3; \delta_4] \in \mathbb{R}^4_+$.

Perturbation analysis for the pole assignment synthesis problem is presented in [91] for the particular case when $r = n$ and the desired poles λ_k are pairwise distinct, using specific matrix techniques. However, this restriction on λ_k is not necessary and more general and tighter perturbation bounds may be derived. This is done in [53] using MSOLM.

An important feature of the feedback synthesis of linear systems is the possibility to use the freedom in the solution $\mathbf{K} \in \mathscr{K}(\mathbf{C}, \mathbf{A}, \mathbf{B}, \mathbf{D})$ when $mr > n$ for other design purposes. A reliable algorithm for this purpose is proposed in [77].

Acknowledgements The authors would like to thank an anonymous referee for the helpful comments.

References

1. Ammar, G., Mehrmann, V.: On Hamiltonian and symplectic Hessenberg forms. Linear Algebra Appl. **149**, 55–72 (1991)
2. Angelova, V.: Perturbation analysis of linear multivariable control systems. PhD thesis, Bulgarian Academy of Sciences, Sofia (1994)

3. Angelova, V.: Perturbation analysis for the complex linear matrix equation $\alpha X + \sigma A^{H} XA = I$, $\alpha, \sigma = \pm 1$. C. R. Acad. Bulg. Sci. Tech. Theor. des Syst. **56**(12), 47–52 (2003). ISSN:1310-1331
4. Baumgärtel, H.: Analytic Perturbation Theory for Matrices and Operators. Birkhäuser Verlag, Basel (1985). ISBN:3764316640
5. Benner, P.: Contributions to the Numerical Solution of Algebraic Riccati Equations and Related Eigenvalue Problems. Logos-Verlag, Berlin (1997). ISBN:9783931216702
6. Benner, P., Byers, R., Mehrmann, V., Xu, H.: Numerical computation of deflating subspaces of skew-Hamiltonian/Hamiltonian pencils. SIAM J. Matrix Anal. Appl. **24**, 165–190 (2002)
7. Benner, P., Kressner, D., Mehrmann, V.: Skew-Hamiltonian and Hamiltonian eigenvalue problems: theory, algorithms, and applications. In: Drmac, Z., Marusic, M., Tutek, Z. (eds.) Proceedings of the Conference on Applied Mathematics and Scientific Computing, Brijuni, 23–27 June 2003, pp. 3–39. Springer, Berlin (2005). ISBN:1-4020-3196-3
8. Benner, P., Laub, A.J., Mehrmann, V.: Benchmarks for the numerical solution of algebraic Riccati equations. IEEE Control Syst. Mag. **17**, 18–28 (1997)
9. Benner, P., Mehrmann, V., Xu, H.: A note on the numerical solution of complex Hamiltonian and skew-Hamiltonian eigenvalue problems. Electron. Trans. Numer. Anal. **8**, 115–126 (1999)
10. Benner, P., Mehrmann, V., Xu, H.: Perturbation analysis of the eigenvalue problem of a formal product of matrices. BIT **42**, 1–43 (2002)
11. Bhatia, R.: Matrix factorizations and their perturbations. Linear Algebra Appl. **197–198**, 245–276 (1994)
12. Bora, S., Mehrmann, V.: Linear perturbation theory for structured matrix pencils arising in control theory. SIAM J. Matrix Anal. Appl. **28**, 148–169 (2006)
13. Demmel, J.: On condition numbers and the distance to the nearest ill-posed problem. Numer. Math. **51**, 251–289 (1987)
14. Chang, X., Paige, C., Stewart, G.: Perturbation analysis for the QR factorization. SIAM J. Matrix Anal. Appl. **18**, 775–791 (1997)
15. Chen, X., Li, W., Ng, M.: Perturbation analysis for antitriangular Schur decomposition. SIAM J. Matrix Anal. Appl. **33**, 325–335 (2012)
16. Dontchev, A.: Perturbations, Approximations and Sensitivity Analysis of Optimal Control Systems. Springer, Berlin (1983). ISBN:0387124632
17. Eslami, M.: Theory of Sensitivity in Dynamic Systems. Springer, Berlin (1994). ISBN:0387547614
18. Gahinet, P.: Perturbational and topological aspects of sensitivity in control theory. PhD thesis, University of California at Santa Barbara, Santa Barbara (1989)
19. Gahinet, P., Laub, A.: Computable bounds for the sensitivity of the algebraic Riccati equation. SIAM J. Control Optim. **28**, 1461–1480 (1990)
20. Gahinet, P., Laub, A., Kenney, C., Hewer, G.: Sensitivity of the algebraic Riccati equation. IEEE Trans. Autom. Control **35**, 1209–1217 (1990)
21. Gautschi, W.: On the condition of algebraic equations. Numer. Math. **21**, 405–424 (1973)
22. Gohberg, I., Koltracht, I.: Mixed, componentwise and structural condition numbers. SIAM J. Matrix Anal. Appl. **14**, 688–704 (1993)
23. Golub, G., Van Loan, C.: Matrix Computations, 3rd edn. John Hopkins University Press, Baltimore (1996). ISBN:0801854148
24. Grebenikov, E., Ryabov, Y.: Constructive Methods for Analysis of Nonlinear Systems (in Russian). Nauka, Moscow (1979)
25. Gu, D., Petkov, P., Konstantinov, M., Mehrmann, V.: Condition and error estimates in the solution of Lyapunov and Riccati equations. SLICOT working note 2000-1, Department of Electrical Engineering, University of Leuven (2000). http://www.icm.tu-bs.de/NICONET/
26. Hewer, G., Kenney, C.: The sensitivity of the stable Lyapunov equation. SIAM J. Control Optim. **26**, 321–344 (1988)
27. Higham, N.: Accuracy and Stability of Numerical Algorithms, 2nd edn. SIAM, Philadelphia (2002). ISBN:0898715210

28. Higham, N., Konstantinov, M., Merhmann, V., Petkov, P.: The sensitivity of computational control problems. IEEE Control Syst. Mag. **24**, 28–43 (2004)
29. IEEE: IEEE Standard for Floating-Point Arithmetic 754–2008. IEEE, New York (2008). ISBN:9780738157535
30. Kantorovich, L.: Principle of majorants and Newton's method (in Russian). Proc. Acad. Sci. USSR **76**, 17–20 (1951)
31. Kato, T.: Perturbation Theory for Linear Operators. Springer, Berlin (1980) (Reprint 1995). ISBN:354058661X
32. Kawelke, J.: Perturbation and error analysis considerations in robust control. PhD thesis, University of Leicester, Leicester (1997)
33. Kenney, C., Hewer, G.: The sensitivity of the algebraic and differential Riccati equations. SIAM J. Control Optim. **28**, 50–69 (1990)
34. Kenney, C., Laub, A.: Condition estimates for matrix functions. SIAM J. Matrix Anal. Appl. **10**, 191–209 (1989)
35. Konstantinov, M., Angelova, V.: Sensitivity analysis of the differential matrix Riccati equation based on the associated linear differential system. Adv. Comput. Math. **7**, 295–301 (1997)
36. Konstantinov, M., Angelova, V., Petkov, P., Gu, D., Tsachouridis, V.: Perturbation analysis of coupled matrix Riccati equations. Linear Algebra Appl. **359**, 197–218 (2002)
37. Konstantinov, M., Christov, N., Petkov, P.: Perturbation analysis of linear control problems. Prepr. 10th IFAC Congress 9, 16–21, Munich (1987)
38. Konstantinov, M., Gu, D., Merhmann, V., Petkov, P.: Perturbation Theory for Matrix Equations. North-Holland, Amsterdam (2003). ISBN:0444513159
39. Konstantinov, M., Mehrmann, V., Petkov, P.: On properties of Sylvester and Lyapunov operators. Linear Algebra Appl. **32**, 35–71 (2000)
40. Konstantinov, M., Merhmann, V., Petkov, P.: Perturbation analysis of Hamiltonian Schur and block-Schur forms. SIAM J. Matrix Anal. Appl. **23**, 387–424 (2001)
41. Konstantinov, M., Mehrmann, V., Petkov, P., Gu, D.: A general framework for the perturbation theory of matrix equations. Prepr. 760-02, Institute for Mathematics, Technical University of Berlin, Berlin (2003)
42. Konstantinov, M., Pelova, G.: Sensitivity of the solutions to differential matrix Riccati equations. IEEE Trans. Autom. Control **36**, 213–215 (1991)
43. Konstantinov, M., Petkov, P.: A note on Perturbation theory for algebraic Riccati equations. SIAM J. Matrix Anal. Appl. **21**, 327 (1999)
44. Konstantinov, M., Petkov, P.: The method of splitting operators and Lyapunov majorants in perturbation linear algebra and control. Numer. Func. Anal. Opt. **23**, 529–572 (2002)
45. Konstantinov, M., Petkov, P.: Perturbation analysis in linear algebra and control theory (in Bulgarian). Stud. BIAR Math. Sci. **6**, Sofia (2003). ISBN:9549526186
46. Konstantinov, M., Petkov, P.: Perturbation methods in linear algebra and control. Appl. Comput. Math. **7**, 141–161 (2008)
47. Konstantinov, M., Petkov, P.: Lyapunov majorants for perturbation analysis of matrix equations. In: Proceedings of 38-th Spring Conference on UBM, Borovetz, pp. 70–80 (2009). ISSN:13133330
48. Konstantinov, M., Petkov, P., Christov, N.: Invariants and canonical forms for linear multivariable systems under the action of orthogonal transformation groups. Kybernetika **17**, 413–421 (1981)
49. Konstantinov, M., Petkov, P., Christov, N.: Perturbation analysis of the continuous and discrete matrix Riccati equations. In: Proceedings of 1986 American Control Conference, Seattle, pp. 636–639 (1986)
50. Konstantinov, M., Petkov, P., Christov, N.: Perturbation analysis of matrix quadratic equations. SIAM J. Sci. Stat. Comput. **11**, 1159–1163 (1990)
51. Konstantinov, M., Petkov, P., Christov, N.: Perturbation analysis of the discrete matrix Riccati equation. Kybernetika **29**, 18–29 (1993)
52. Konstantinov, M., Petkov, P., Christov, N.: Nonlocal perturbation analysis of the Schur system of a matrix. SIAM J. Matrix Anal. Appl. **15**, 383–392 (1994)

53. Konstantinov, M., Petkov, P., Christov, N.: Sensitivity analysis of the feedback synthesis problem. IEEE Trans. Autom. Control **42**, 568–573 (1997)
54. Konstantinov, M., Petkov, P., Christov, N.: New results in the perturbation analysis of matrix algebraic equations. Appl. Comput. Math. **9**, 153–161 (2010)
55. Konstantinov, M., Petkov, P., Gu, G.: Improved perturbation bounds for general quadratic matrix equations. Numer. Func. Anal. Opt. **20**, 717–736 (1999)
56. Konstantinov, M., Petkov, P., Gu, D., Mehrmann, V.: Sensitivity of general Lyapunov equations. Report 98-15, Department of Engineering, Leicester University (1998)
57. Konstantinov, M., Petkov, P., Gu, D., Postlethwaite, I.: Perturbation analysis of orthogonal canonical forms. Technical report 95-4, Department of Engineering, Leicester University, Leicester (1995)
58. Konstantinov, M., Petkov, P., Gu, D., Postlethwaite, I.: Perturbation analysis of orthogonal canonical forms. Linear Algebra Appl. **251**, 267–291 (1997)
59. Konstantinov, M., Petkov, P., Kawelke, J., Gu, D., Postlethwaite, I.: Perturbation analysis of the H_∞ control problems. In: Proceedings of 2nd Asian Control Conference, Seoul, pp. 427–430 (1997)
60. Lancaster, P., Rodman, L.: The Algebraic Riccati Equation. Oxford University Press, Oxford (1995). ISBN:0198537956
61. Laub, A., Linnemann, A.: Hessenberg and Hessenberg/triangular forms in linear system theory. Int. J. Control **44**, 1523–1547 (1986)
62. Laub, A., Meyer, K.: Canonical forms for symplectic and Hamiltonian matrices. Celest. Mech. **9**, 213–238 (1974)
63. Lika, D., Ryabov, Y.: Iterative Methods and Lyapunov Majorant Equations in Nonlinear Oscillation Theory (in Russian). Shtiinca, Kishinev (1974)
64. Lin, W., Mehrmann., V., Xu, H.: Canonical forms for Hamiltonian and symplectic matrices and pencils. Linear Algebra Appl. **302–303**, 469–533 (1999)
65. Mackey, D., Mackey, N., Mehl, C., Mehrmann, V.: Numerical methods for palindromic eigenvalue problems: computing the anti-triangular Schur form. Numer. Linear Algebra **16**, 63–86 (2009)
66. Mehrmann, V.: The Autonomous Linear Quadratic Control Problem. Lecture Notes in Control and Information Sciences, vol. 163. Springer, Heidelberg (1991). ISBN:3540541705
67. Mehrmann, V.: Numerical methods for eigenvalue and control problems. In: Blowey, J.F., Craig, A.W., Shardlow, T. (eds.) Frontiers in Numerical Analysis, pp. 303–349. Springer, Berlin (2003). ISBN:978-3-540-44319-3
68. Mehrmann, V., Xu, H.: An analysis of the pole placement problem. I. The single-input case. Electron. Trans. Numer. Anal. **4**, 89–105 (1996)
69. Mehrmann, V., Xu, H.: An analysis of the pole placement problem. II. The multi-input case. Electron. Trans. Numer. Anal. **5**, 77–97 (1997)
70. Mehrmann, V., Xu, H.: Choosing poles so that the single-input pole placement problem is well conditioned. SIAM J. Matrix Anal. Appl. **19**, 664–681 (1998)
71. Mehrmann, V., Xu, H.: Numerical methods in control. J. Comput. Appl. Math. **123**, 371–394 (2000)
72. Mehrmann, V., Xu, H.: Perturbation of purely imaginary eigenvalues of Hamiltonian matrices under structured perturbations. Electron. J. Linear Algebra **17**, 234–257 (2008)
73. Petkov, P.: Perturbation bounds for orthogonal canonical forms and numerical stability analysis. IEEE Trans. Autom. Control **38**, 639–643 (1993)
74. Petkov, P., Christov, N., Konstantinov, M.: A new approach to the perturbation analysis of linear control problems. Prepr. 11th IFAC Congress, Tallin, pp. 311–316 (1990)
75. Petkov, P., Christov, N., Konstantinov, M.: Computational Methods for Linear Control Systems. Prentice Hall, Hemel Hempstead (1991). ISBN:0131618032
76. Petkov, P., Konstantinov, M., Gu, D., Mehrmann, V.: Numerical solution of matrix Riccati equations – a comparison of six solvers. NICONET report 99–10, Department of Electrical Engineering, University of Leuven (1999). http://www.icm.tu-bs.de/NICONET/

77. Petkov, P., Konstantinov, M., Gu, D., Postlethwaite, I.: Optimal eigenstructure assignment of linear systems. In: Proceedings of 13th IFAC Congress, San Francisco, vol. C, pp. 109–114 (1996)
78. Petkov, P., Konstantinov, M., Mehrmann, V.: DGRSVX and DMSRIC: Fortran 77 subroutines for solving continuous-time matrix algebraic Riccati equations with condition and accuracy estimates. Prepr. SFB393/98-16, Fak. für Math., Technische Universität Chemnitz (1998)
79. Popchev, I., Angelova, V.: Residual bound of the matrix equation $X + A^{\mathrm{H}} X^{-1} A + B^{\mathrm{H}} X^{-1} B = I$, C.R. Acad. Bulg. Sci. **66**(10), 1379–1384 (2013). ISSN:1310–1331
80. Rice, J.: A theory of condition. SIAM J. Numer. Anal. **3**, 287–310 (1966)
81. Rohn, J.: New condition numbers for matrices and linear systems. Computing **41**, 167–169 (1989)
82. Stewart, G.: Error bounds for approximate invariant subspaces of closed linear operators. SIAM J. Numer. Anal. **8**, 796–808 (1971)
83. Stewart, G.: Error and perturbation bounds for subspaces associated with certain eigenvalue problems. SIAM Rev. **15**, 727–764 (1973)
84. Stewart, G.: Perturbation bounds for the QR factorization of a matrix. SIAM J. Numer. Anal. **14**, 509–518 (1977)
85. Stewart, G.: On the perturbation of LU, Cholesky and QR factorizations. SIAM J. Matrix Anal. Appl. **14**, 1141–1145 (1993)
86. Stewart, G., Sun, J.: Matrix Perturbation Theory. Academic, New York (1990). ISBN:0126702306
87. Sun, J.: Componentwise perturbation bounds for some matrix decompositions. BIT **32**, 702–714 (1992)
88. Sun, J.: On perturbation bounds for the QR factorization. Linear Algebra Appl. **215**, 95–111 (1995)
89. Sun, J.: Perturbation bounds for the generalized Schur decomposition. SIAM J. Matrix Anal. Appl. **16**, 1328–1340 (1995)
90. Sun, J.: Perturbation analysis of system Hessenberg and Hessenberg-triangular forms. Linear Algebra Appl. **241**, 811–849 (1996)
91. Sun, J.: Perturbation analysis of the pole assignment problem. SIAM J. Matrix Anal. A. **17**, 313–331 (1996)
92. Sun, J.: Perturbation theory for algebraic Riccati equations. SIAM J. Matrix Anal. Appl. **19**, 39–65 (1998)
93. Sun, J.: Sensitivity analysis of the discrete-time algebraic Riccati equation. Linear Algebra Appl. **275–276**, 595–615 (1998)
94. Vulchanov, N., Konstantinov, M.: Modern Mathematical Methods for Computer Calculations, Part 1 (2nd edn., in Bulgarian). Studies of BIAR in Math. Sci., **1**, Demetra Publishing House, Sofia (2005). ISBN:9548949016
95. Weinmann, A.: Uncertain Models and Robust Control. Springer, Wien (1991). ISBN:3211822992

Chapter 8
Structured Eigenvalue Perturbation Theory

Shreemayee Bora and Michael Karow

Abstract We give an overview of Volker Mehrmann's work on structured perturbation theory of eigenvalues. In particular, we review his contributions on perturbations of structured pencils arising in control theory and of Hamiltonian matrices. We also give a brief outline of his work on structured rank one perturbations.

8.1 Introduction

The core research interests of Volker Mehrmann include mathematical modelling of real world processes and the design of numerically stable solutions of associated problems. He has almost four decades of research experience in these areas and perturbation analysis of the associated challenging eigenvalue problems form an integral part of his research. Often the challenges provided by these eigenvalue problems are due to the fact that the associated matrices have a special structure leading to symmetries in the distribution of their eigenvalues. For example the solution of continuous time linear quadratic optimal control problems and the vibration analysis of machines, buildings and vehicles lead to generalized eigenvalue problems where the coefficient matrices of the matrix pencil have a structure that alternates between Hermitian and skew-Hermitian. Due to this, their eigenvalues occur in pairs $(\lambda, -\bar{\lambda})$ when the matrices are complex and quadruples $(\lambda, \bar{\lambda}, -\lambda, -\bar{\lambda})$ when they are real. In either case, the eigenvalues are symmetrically placed with respect to the imaginary axis. This is referred to as Hamiltonian spectral symmetry as it is also typically displayed by eigenvalues of Hamiltonian matrices. Note that a matrix A of even size, say $2n$, is said to be Hamiltonian if $(JA)^* = JA$ where $J = \begin{bmatrix} 0 & I_n \\ -I_n & 0 \end{bmatrix}$.

S. Bora (✉)
Department of Mathematics, Indian Institute of Technology Guwahati, Guwahati 781039, Assam, India
e-mail: shbora@iitg.ernet.in

M. Karow
Institut für Mathematik, Technische Universität Berlin, Sekretariat MA 4-5, Straße des 17. Juni 136, D-10623 Berlin, Germany
e-mail: karow@math.tu-berlin.de

P. Benner et al. (eds.), *Numerical Algebra, Matrix Theory, Differential-Algebraic Equations and Control Theory*, DOI 10.1007/978-3-319-15260-8_8

On the other hand, discrete time linear quadratic optimal control problems lead to generalized eigenvalue problems associated with matrix pencils of the form $A - zA^*$ where A^* denotes the complex conjugate transpose of the matrix A [15, 46, 51, 63]. For these pencils, the eigenvalues occur in pairs $(\lambda, 1/\bar{\lambda})$ in the complex case and in quadruples $(\lambda, 1/\lambda, \bar{\lambda}, 1/\bar{\lambda})$ in the real case which imply that the eigenvalues are symmetrically placed with respect to the unit circle. This symmetry is referred to as symplectic spectral symmetry as it is typically possessed by symplectic matrices. Note that a matrix S of size $2n \times 2n$ is said to be symplectic if JS is a unitary matrix.

It is now well established that solutions of structured eigenvalue problems by algorithms that preserve the structure are more efficient because often they need less storage space and time than other algorithms that do not do so. Moreover, computed eigenvalues of stable structure preserving algorithms also reflect the spectral symmetry associated with the structure which is important in applications. For more on this, as well as Volker Mehrmann's contributions to structure preserving algorithms, we refer to the Chap. 1 by A. Bunse-Gerstner and H. Faßbender.

Perturbation analysis of eigenvalue problems involves finding the sensitivity of eigenvalues and eigenvectors of matrices, matrix pencils and polynomials with respect to perturbations and is essential for many applications, important among which is the design of stable and accurate algorithms. Typically, this is measured by the condition number which gives the rate of change of these quantities under perturbations to the data. For example, the condition number of a simple eigenvalue λ of an $n \times n$ matrix A, is defined by

$$\kappa(\lambda) := \limsup_{\epsilon \to 0} \left\{ \frac{|\tilde{\lambda} - \lambda|}{\epsilon} : \tilde{\lambda} \in \Lambda(A + \Delta),\ \Delta \in \mathbb{C}^{n \times n},\ \epsilon > 0 \text{ and } \|\Delta\| < \epsilon \|A\| \right\},$$

where $\Lambda(A + \Delta)$ denotes the spectrum of $A + \Delta$. Given any $\mu \in \mathbb{C}$ and $x \in \mathbb{C}^n \setminus \{0\}$, perturbation analysis is also concerned with computing backward errors $\eta(\mu)$ and $\eta(\mu, x)$. They measure minimal perturbations $\Delta \in \mathbb{C}^{n \times n}$ (with respect to a chosen norm) to A such that μ is an eigenvalue of $A + \Delta$ in the first instance and an eigenvalue of $A + \Delta$ with corresponding eigenvector x in the second instance. In particular, if $\eta(\mu, x)$ is sufficiently small for eigenvalue-eigenvector pairs computed by an algorithm, then the algorithm is (backward) stable. Moreover, $\eta(\mu, x) \times \kappa(\mu)$ is an approximate upper bound on the (forward) error in the computed pair (μ, x).

In the case of structured eigenvalue problems, the perturbation analysis of eigenvalues and eigenvectors with respect to structure preserving perturbations is important for the stability analysis of structure preserving algorithms and other applications like understanding the behaviour of dynamical systems associated with such problems. This involves finding the structured condition numbers of eigenvalues and eigenvectors and structured backward errors of approximate eigenvalues and eigenvalue-eigenvector pairs. For instance, if $A \in \mathbf{S} \subset \mathbb{C}^{n \times n}$, then the structured condition number of any simple eigenvalue λ of A with respect to structure preserving perturbations is defined by

$$\kappa^{\mathbf{S}}(\lambda) := \limsup_{\epsilon \to 0} \left\{ \frac{|\tilde{\lambda} - \lambda|}{\epsilon} : \tilde{\lambda} \in \Lambda(A + \Delta),\ A + \Delta \in \mathbf{S}, \epsilon > 0 \text{ and } \|\Delta\| < \epsilon \|A\| \right\}.$$

Similarly, the structured backward error $\eta^{\mathbf{S}}(\mu)$ (resp., $\eta^{\mathbf{S}}(\mu, x)$) is a measure of the minimum structure preserving perturbation to $A \in \mathbf{S}$ such that $\mu \in \mathbb{C}$ (resp. $(\mu, x) \in \mathbb{C} \times \mathbb{C}^n$) is an eigenvalue (resp. eigenvalue-eigenvector pair) for the perturbed problem. Evidently, $\kappa^{\mathbf{S}}(\lambda) \leq \kappa(\lambda)$ and $\eta^{\mathbf{S}}(\mu) \geq \eta(\mu)$. Very often, the difference between condition numbers with respect to arbitrary and structure preserving perturbations is small for most eigenvalues of structured problems. In fact, they are equal for purely imaginary eigenvalues of problems with Hamiltonian spectral symmetry and eigenvalues on the unit circle of problems with symplectic spectral symmetry [2, 14, 33, 36, 55]. Such eigenvalues are called *critical* eigenvalues as they result in a breakdown in the eigenvalue pairing. The same also holds for structured and conventional backward errors of approximate critical eigenvalues of many structured eigenvalue problems [1, 14].

However the conventional perturbation analysis via condition numbers and backward errors does not always capture the full effect of structure preserving perturbations on the movement of the eigenvalues. This is particularly true of certain critical eigenvalues whose structured condition numbers do not always indicate the significant difference in their directions of motion under structure preserving and arbitrary perturbations [5, 48]. It is important to understand these differences in many applications. For instance, appearance of critical eigenvalues may lead to a breakdown in the spectral symmetry resulting in loss of uniqueness of deflating subspaces associated with the non-critical eigenvalues and leading to challenges in numerical computation [24, 46, 52–54]. They also result in undesirable physical phenomena like loss of passivity [5, 8, 27]. Therefore, given a structured eigenvalue problem without any critical eigenvalues, it is important to find the distance to a nearest problem with critical eigenvalues. Similarly, if the structured eigenvalue problem already has critical eigenvalues, then it is important to investigate the distance to a nearest problem with the same structure which has no such eigenvalues. Finding these distances pose significant challenges due to the fact that critical eigenvalues are often associated with an additional attributes called sign characteristics which restrict there movement under structure preserving perturbations. These specific 'distance problems' come within the purview of structured perturbation analysis and are highly relevant to practical problems. Volker Mehrmann is one of the early researchers to realise the significance of structured perturbation analysis to tackle these issues.

It is difficult to give a complete overview of Volker Mehrmann's wide body of work in eigenvalue perturbation theory [3–5, 10, 11, 13, 15, 31, 34, 35, 41–45, 48]. For instance one of his early papers in this area is [10] where along with co-authors Benner and Xu, he extends some of the classical results of the perturbation theory for eigenvalues, eigenvectors and deflating subspaces of matrices and matrix pencils to a formal product $A_1^{s_1} A_2^{s_2} \cdot A_p^{s_p}$ of p square matrices $A_1, A_2, \ldots, A_p$ where $s_1, s_2, \ldots, s_p \in \{-1, 1\}$. With co-authors Konstantinov and Petkov, he investigates the effect of perturbations on defective matrices in [35]. In [13] he looks into structure preserving algorithms for solving Hamiltonian and skew-Hamiltonian eigenvalue problems and compares the conditioning of eigenvalues and invariant subspaces with respect to structure preserving and arbitrary perturbations. In fact, the structure preserving perturbation theory of the Hamiltonian and the skew-Hamiltonian eigenvalue problem has been a recurrent theme of his research [5, 34, 48]. He has also been deeply interested in structure preserving perturbation analysis of the structured eigenvalue problems that arise in control theory and their role in the design of efficient, robust and accurate methods for solving problems of computational control [11, 15, 31]. The computation of Lagrangian invariant subspaces of symplectic matrices arises in many applications and this can be difficult specially when the matrix has eigenvalues very close to the unit circle. With co-authors, Mehl, Ran and Rodman, he investigates the perturbation theory of such subspaces in [41]. With the same co-authors, he has also investigated the effect of low rank structure preserving perturbations on different structured eigenvalue problems in a series of papers [42–45]. Volker Mehrmann has also undertaken the sensitivity and backward error analysis of several structured polynomial eigenvalue problems with co-author Ahmad in [3, 4].

In this article we give a brief overview of Volker Mehrmann's contributions in three specific topics of structured perturbation theory. In Sect. 8.2, we describe his work with co-author Bora on structure preserving linear perturbation of some structured matrix pencils that occur in several problems of control theory. In Sect. 8.3, we describe his work with Xu, Alam, Bora, Karow and Moro on the effect of structure preserving perturbations on purely imaginary eigenvalues of Hamiltonian and skew-Hamiltonian matrices. Finally, in Sect. 8.4 we give a brief overview of Volker Mehrmann's research on structure preserving rank one perturbations of several structured matrices and matrix pencils with co-authors Mehl, Ran and Rodman.

8.2 Structured Perturbation Analysis of Eigenvalue Problems Arising in Control Theory

One of the early papers of Volker Mehrmann in the area of structured perturbation analysis is [15] where he investigates the effect of structure preserving linear perturbations on matrix pencils that typically arise in robust and optimal control theory. Such control problems typically involve constant co-efficient dynamical systems of the form

$$E\dot{x} = Ax + Bu, \qquad x(\tau_0) = x^0, \tag{8.1}$$

where $x(\tau) \in \mathbb{C}^n$ is the state, x^0 is an initial vector, $u(\tau) \in \mathbb{C}^m$ is the control input of the system and the matrices $E, A \in \mathbb{C}^{n,n}$, $B \in \mathbb{C}^{n,m}$ are constant. The objective of linear quadratic optimal control is to find a control law $u(\tau)$ such that the closed loop system is asymptotically stable and the performance criterion

$$\mathscr{S}(x,u) = \int_{\tau_0}^{\infty} \begin{bmatrix} x(\tau) \\ u(\tau) \end{bmatrix}^T \begin{bmatrix} Q & S \\ S^* & R \end{bmatrix} \begin{bmatrix} x(\tau) \\ u(\tau) \end{bmatrix} d\tau \tag{8.2}$$

is minimized, where $Q = Q^* \in \mathbb{C}^{n,n}$, $R = R^* \in \mathbb{C}^{m,m}$ and $S \in \mathbb{C}^{n,m}$. Application of the maximum principle [46, 51] leads to the problem of finding a stable solution to the two-point boundary value problem of Euler-Lagrange equations

$$N_c \begin{bmatrix} \dot{\mu} \\ \dot{x} \\ \dot{u} \end{bmatrix} = H_c \begin{bmatrix} \mu \\ x \\ u \end{bmatrix}, \qquad x(\tau_0) = x^0, \qquad \lim_{\tau \to \infty} \mu(\tau) = 0, \tag{8.3}$$

leading to the structured matrix pencil

$$H_c - \lambda N_c := \begin{bmatrix} 0 & A & B \\ A^* & Q & S \\ B^* & S^* & R \end{bmatrix} - \lambda \begin{bmatrix} 0 & E & 0 \\ -E^* & 0 & 0 \\ 0 & 0 & 0 \end{bmatrix} \tag{8.4}$$

in the continuous time case. Note that H_c and N_c are Hermitian and skew-Hermitian respectively, due to which the eigenvalues of the pencil occur in pairs $(\lambda, -\bar{\lambda})$ if the matrices are complex and in quadruples $(\lambda, \bar{\lambda}, -\lambda, -\bar{\lambda})$ when the matrices are real. In fact, for the given pencil it is well known that if E is invertible, then under the usual control theoretic assumptions [46, 66, 67], it has exactly n eigenvalues on the left half plane, n eigenvalues on the right half plane and m infinite eigenvalues. One of the main concerns in the perturbation analysis of these pencils is to find structure preserving perturbations that result in eigenvalues on the imaginary axis. In [15], the authors considered the pencils $H_c - \lambda N_c$ which had no purely imaginary eigenvalues or infinite eigenvalues (i.e. the block E is invertible) and investigated the effect of structure preserving linear perturbations of the form $H_c - \lambda N_c + t(\Delta H_c - \lambda \Delta N_c)$ where

$$\Delta H_c := \begin{bmatrix} 0 & \Delta A & \Delta B \\ (\Delta A)^* & \Delta Q & \Delta S \\ (\Delta B)^* & (\Delta S)^* & \Delta R \end{bmatrix} \text{ and } \Delta N_c := \begin{bmatrix} 0 & \Delta E & 0 \\ -(\Delta E)^* & 0 & 0 \\ 0 & 0 & 0 \end{bmatrix} \tag{8.5}$$

are fixed matrices such that ΔQ and ΔR are symmetric, $E + \Delta E$ is invertible and t is a parameter that varies over $\mathbb{R}$. The aim of the analysis was to find the smallest value(s) of the parameter t such that the perturbed pencil has an eigenvalue on the imaginary axis in which case their is loss of spectral symmetry and uniqueness of the deflating subspace associated with the eigenvalues on the left half plane.

The discrete-time analogue to the linear quadratic control problem leads to slightly different matrix pencils of the form [46, 47]

$$H_d - \lambda N_d = \begin{bmatrix} 0 & A & B \\ -E^* & Q & S \\ 0 & S^* & R \end{bmatrix} - \lambda \begin{bmatrix} 0 & E & 0 \\ -A^* & 0 & 0 \\ -B^* & 0 & 0 \end{bmatrix}. \tag{8.6}$$

The eigenvalues of $H_d - \lambda N_d$ occur in pairs $(\lambda, 1/\bar{\lambda})$ when the pencil is complex and in quadruples $(\lambda, 1/\bar{\lambda}, \bar{\lambda}, 1/\lambda)$ when the pencil is real. For such pencils, the critical eigenvalues are the ones on the unit circle. With the assumption that $H_d - \lambda N_d$ has no critical eigenvalues, [15] investigates the smallest value of the parameter t such that the perturbed pencils $H_d + t\Delta H_d - \lambda(N_d + t\Delta N_d)$ have an eigenvalue on the unit circle, where

$$\Delta H_d = \begin{bmatrix} 0 & \Delta A & \Delta B \\ -(\Delta E)^* & \Delta Q & \Delta S \\ 0 & (\Delta S)^* & \Delta R \end{bmatrix} \text{ and } \Delta N_d = \begin{bmatrix} 0 & \Delta E & 0 \\ -(\Delta A)^* & 0 & 0 \\ -(\Delta B)^* & 0 & 0 \end{bmatrix} \tag{8.7}$$

are fixed matrices that preserve the structure of H_d and N_d respectively. Note that in this case, the loss of spectral symmetry can lead to non-uniqueness of the deflating subspace associated with eigenvalues inside the unit circle.

Analogous investigations were also made in [15] for a slightly different set of structured matrix pencils that were motivated by problems of H_∞ control. The method of γ-iteration suggested in [12] for robust control problems arising in frequency domain [26, 68] results in pencils of the form

$$\hat{H}_c(t) - \lambda \hat{N}_c := \begin{bmatrix} 0 & A & B \\ A^* & 0 & S \\ B^* & S^* & R(t) \end{bmatrix} - \lambda \begin{bmatrix} 0 & E & 0 \\ -E^* & 0 & 0 \\ 0 & 0 & 0 \end{bmatrix} \tag{8.8}$$

in the continuous time case and in pencils of the form

$$\hat{H}_d(t) - \lambda \hat{N}_d = \begin{bmatrix} 0 & A & B \\ -E^* & 0 & S \\ 0 & S^* & R(t) \end{bmatrix} - \lambda \begin{bmatrix} 0 & E & 0 \\ -A^* & 0 & 0 \\ -B^* & 0 & 0 \end{bmatrix} \tag{8.9}$$

in the discrete time case where in each case,

$$R(t) = \begin{bmatrix} R_{11} - tI & R_{12} \\ R_{12}^* & R_{22} \end{bmatrix}$$

is an indefinite Hermitian matrix. Each of the structured pencils vary with the positive parameter t (playing the role of the parameter γ in the γ-iteration), while the other coefficients are constant in t. Here too, the key question investigated was the smallest value(s) of the parameter t for which the pencils $\hat{H}_c(t) - \lambda \hat{N}_c$ and $\hat{H}_d(t) - \lambda \hat{N}_d$ have critical eigenvalues.

The authors developed a general framework for dealing with the structured perturbation problems under consideration in [15] and derived necessary and sufficient conditions for the perturbed matrix pencils to have critical eigenvalues. The following was one of the key results.

Theorem 1 *Consider a matrix pencil* $H_c - \lambda N_c$ *as given in* (8.4). *Let the matrices* ΔH_c *and* ΔN_c *be as in* (8.5) *and*

$$P(t, \gamma) := [A - i\gamma E + t(\Delta A - i\gamma \Delta E) \;\; B + t\Delta B],$$

$$Z(t) := \begin{bmatrix} Q + t\Delta Q & S + t\Delta S \\ (S + t\Delta S)^* & R + t\Delta R \end{bmatrix}.$$

Let $V(t, \gamma)$ *be any orthonormal basis of the kernel of* $P(t, \gamma)$, *and let* $W(t, \gamma)$ *be the range of* $P(t, \gamma)^*$.

Then for given real numbers $t \neq 0$ *and* γ, *the purely imaginary number* $i\gamma$ *is an eigenvalue of the matrix pencil* $(H_c + t\Delta H_c, N_c + t\Delta N_c)$ *if and only if*

$$Z(t)(V(t, \gamma)) \cap W(t, \gamma) \neq \emptyset.$$

In particular, it was observed that if the cost function is chosen in such a way that the matrices Q, R and S are free of perturbation, then the pencil $H_c + t\Delta H_c - \lambda(N_c + t\Delta N_c)$ has a purely imaginary eigenvalue $i\gamma$ if and only if

$$\begin{bmatrix} Q & S \\ S^* & R \end{bmatrix} V(t, \gamma) \cap W(t, \gamma) \neq \emptyset.$$

In particular, if the matrix $\begin{bmatrix} Q & S \\ S^* & R \end{bmatrix}$ associated with the cost function is nonsingular and the kernel of $P(t, \gamma)$ is an invariant subspace of the matrix, then $i\gamma$ is not an eigenvalue of $H_c + t\Delta H_c - \lambda(N_c + t\Delta N_c)$. Similar results were also obtained for the other structures.

8.3 Perturbation Theory for Hamiltonian Matrices

Eigenvalue problems associated with Hamiltonian matrices play a central role in computing various important quantities that arise in diverse control theory problems like robust control, gyroscopic systems and passivation of linear systems. For example, optimal H_∞ control problems involve Hamiltonian matrices of the form

$$\mathscr{H}(\gamma) = \begin{bmatrix} F & G_1 - \gamma^{-2} G_2 \\ H & -F^T \end{bmatrix}$$

where $F, G_1, G_2, H \in \mathbb{R}^{n,n}$ are such that G_1, G_2 and H are symmetric positive semi-definite and γ is a positive parameter [26, 38, 68]. In such cases, it is important to identify the smallest value of γ for which all the eigenvalues of $\mathcal{H}(\gamma)$ are purely imaginary. The stabilization of linear second order gyroscopic systems [37, 63] requires computing the smallest real value of δ such that all the eigenvalues of the quadratic eigenvalue problem $(\lambda^2 I + \lambda(2\delta G) - K)x = 0$ are purely imaginary. Here G is a non-singular skew-Hermitian matrix and K is a Hermitian positive definite matrix. This is equivalent to finding the smallest value of δ such that all the eigenvalues of the Hamiltonian matrix $\mathcal{H}(\delta) = \begin{bmatrix} -\delta G & K + \delta^2 G^2 \\ I_n & -\delta G \end{bmatrix}$ are purely imaginary. Finally Hamiltonian matrices also arise in the context of passivation of non-passive dynamical systems. Consider a linear time invariant control system described by

$$\dot{x} = Ax + Bu,$$
$$y = Cx + Du,$$

where $A \in \mathbb{F}^{n,n}, B \in \mathbb{F}^{n,m}, C \in \mathbb{F}^{m,n}$, and $D \in \mathbb{F}^{m,m}$ are real or complex matrices such that all the eigenvalues of A are on the open left half plane, i.e., the system is asymptotically stable and D has full column rank. The system is said to be passive if there exists a non-negative scalar valued function Θ such that the dissipation inequality

$$\Theta(x(t_1)) - \Theta(x(t_0)) \leq \int_{t_0}^{t_1} u^* y + y^* u \, dt$$

holds for all $t_1 \geq t_0$, i.e., the system absorbs supply energy. This is equivalent to checking whether the Hamiltonian matrix

$$\mathcal{H} = \begin{bmatrix} A - B(D + D^*)^{-1} C & -B(D + D^*)^{-1} B^* \\ C^*(D + D^*)C & -(A - B(D + D^*)^{-1} C)^* \end{bmatrix}$$

has any purely imaginary eigenvalues. A non-passive system is converted to a passive one by making small perturbations to the matrices A, B, C, D [19, 25, 27] such that the eigenvalues of the corresponding perturbed Hamiltonian matrix move off the imaginary axis.

Many linear quadratic optimal control problems require the computation of the invariant subspace of a Hamiltonian matrix $H \in \mathbb{F}^{n,n}$ associated with eigenvalues on the left half plane. Structure preserving algorithms are required for the efficient computation of such subspaces and the aim of such algorithms is to transform H via eigenvalue and structure preserving transformations to the Hamiltonian Schur form $\Sigma := \begin{bmatrix} T & R \\ 0 & -T^* \end{bmatrix}$ where $R \in \mathbb{F}^{n,n}$ is Hermitian and $T \in \mathbb{F}^{n,n}$ is upper triangular if $\mathbb{F} = \mathbb{C}$ and quasi-upper triangular if $\mathbb{F} = \mathbb{R}$. It is well known that there exists a unitary symplectic matrix U (which is orthogonal if $\mathbb{F} = \mathbb{R}$)

such that $U^*HU = \Sigma$ if H has no purely imaginary eigenvalue [50] although Mehrmann along with co-authors Lin and Xu have shown in [40] that Hamiltonian matrices with purely imaginary eigenvalues also have a Hamiltonian Schur form under special circumstances.

Due to these motivations, a significant part of Volker Mehrmann's research in structured perturbation theory focuses on Hamiltonian matrices and especially on the effect of Hamiltonian perturbations on purely imaginary eigenvalues. One of his early contributions in this area is [34] co-authored with Konstantinov and Petkov, in which the importance of the Hamiltonian Schur form motivates the authors to perform a perturbation analysis of the form under Hamiltonian perturbations. In this work, the authors also introduce a Hamiltonian block Schur form which is seen to be relatively less sensitive to perturbations.

Volker Mehrmann's first significant contribution to the perturbation analysis of purely imaginary eigenvalues of Hamiltonian matrices is a joint work [48] with Hongguo Xu. Here Xu and Mehrmann describe the perturbation theory for purely imaginary eigenvalues of Hamiltonian matrices with respect to Hamiltonian and non-Hamiltonian perturbations. It was observed that when a Hamiltonian matrix $\mathscr{H}$ is perturbed by a small Hamiltonian matrix, whether a purely imaginary eigenvalue of $\mathscr{H}$ will stay on the imaginary axis or move away from it is determined by the inertia of a Hermitian matrix associated with that eigenvalue.

More precisely, let $i\alpha, \alpha \in \mathbb{R}$, be a purely imaginary eigenvalue of $\mathscr{H}$ and let $X \in \mathbb{C}^{2n,p}$ be a full column rank matrix. Suppose that the columns of X span the right invariant subspace $\ker(\mathscr{H} - i\alpha I)^{2n}$ associated with $i\alpha$ so that

$$\mathscr{H}X = X\mathscr{R} \text{ and } \Lambda(\mathscr{R}) = \{i\alpha\} \tag{8.10}$$

for some square matrix $\mathscr{R}$. Here and in the sequel $\Lambda(M)$ denotes the spectrum of the matrix M. Since $\mathscr{H}$ is Hamiltonian, relation (8.10) implies that

$$X^*J\mathscr{H} = -\mathscr{R}^*X^*J. \tag{8.11}$$

Since $\Lambda(\mathscr{R}^*) = \{i\alpha\}$, it follows that the columns of the full column rank matrix J^*X span the left invariant subspace associated with $i\alpha$. Hence, $(J^*X)^*X = X^*JX$ is nonsingular and the matrix

$$Z_\alpha = iX^*JX \tag{8.12}$$

is Hermitian and nonsingular. The inertia of Z_α plays the central role in the main structured perturbation theory result of [48] for the spectral norm $\|\cdot\|_2$.

Theorem 2 ([48]) *Consider a Hamiltonian matrix $\mathscr{H} \in \mathbb{F}^{2n,2n}$ with a purely imaginary eigenvalue $i\alpha$ of algebraic multiplicity p. Suppose that $X \in \mathbb{F}^{2n,p}$ satisfies* Rank $X = p$ *and (8.10), and that Z_α as defined in (8.12) is congruent to* $\begin{bmatrix} I_\pi & 0 \\ 0 & -I_\mu \end{bmatrix}$ *(with $\pi + \mu = p$).*

If $\mathscr{E}$ is Hamiltonian and $\|\mathscr{E}\|_2$ is sufficiently small, then $\mathscr{H}+\mathscr{E}$ has p eigenvalues $\lambda_1, \ldots, \lambda_p$ (counting multiplicity) in the neighborhood of $i\alpha$, among which at least $|\pi - \mu|$ eigenvalues are purely imaginary. In particular, we have the following cases.

1. *If Z_α is definite, i.e. either $\pi = 0$ or $\mu = 0$, then all $\lambda_1, \ldots, \lambda_p$ are purely imaginary with equal algebraic and geometric multiplicity, and satisfy*

$$\lambda_j = i(\alpha + \delta_j) + O(\|\mathscr{E}\|_2^2),$$

   where $\delta_1, \ldots, \delta_p$ are the real eigenvalues of the pencil $\lambda Z_\alpha - X^*(J\mathscr{E})X$.
2. *If there exists a Jordan block associated with $i\alpha$ of size larger than 2, then generically for a given $\mathscr{E}$ some eigenvalues among $\lambda_1, \ldots, \lambda_p$ will no longer be purely imaginary.*

   If there exists a Jordan block associated with $i\alpha$ of size 2, then for any $\epsilon > 0$, there always exists a Hamiltonian perturbation matrix $\mathscr{E}$ with $\|\mathscr{E}\|_2 = \epsilon$ such that some eigenvalues among $\lambda_1, \ldots, \lambda_p$ will have nonzero real part.
3. *If $i\alpha$ has equal algebraic and geometric multiplicity and Z_α is indefinite, then for any $\epsilon > 0$, there always exists a Hamiltonian perturbation matrix $\mathscr{E}$ with $\|\mathscr{E}\|_2 = \epsilon$ such that some eigenvalues among $\lambda_1, \ldots, \lambda_p$ have nonzero real part.*

The above theorem has implications for the problem of passivation of dynamical systems as mentioned before, where the goal is to find a smallest Hamiltonian perturbation to a certain Hamiltonian matrix associated with the system such that the perturbed matrix has no purely imaginary eigenvalues. Indeed, Theorem 2 states that a purely imaginary eigenvalue $i\alpha$ can be (partly) removed from the imaginary axis by an arbitrarily small Hamiltonian perturbation if and only if the associated matrix Z_α in (8.12) is indefinite. The latter implies that $i\alpha$ has algebraic multiplicity at least 2. This results in the following Theorem in [48].

Theorem 3 ([48]) *Suppose that $\mathscr{H} \in \mathbb{C}^{2n,2n}$ is Hamiltonian and all its eigenvalues are purely imaginary. Let $\mathbb{H}_{2n}$ be the set of all $2n \times 2n$ Hamiltonian matrices and let $\mathbb{S}$ be the set of Hamiltonian matrices defined by*

$$\mathbb{S} = \left\{ \begin{array}{l} \mathscr{E} \in \mathbb{H}_{2n} \mid \mathscr{H} + \mathscr{E} \text{ has an imaginary eigenvalue with algebraic} \\ \text{multiplicity} > 1 \text{ and the corresponding } Z_\alpha \text{ in (8.12) is indefinite} \end{array} \right\}.$$

Let

$$\epsilon_0 := \min_{\mathscr{E} \in \mathbb{S}} \|\mathscr{E}\|_2. \tag{8.13}$$

If every eigenvalue of $\mathscr{H}$ has equal algebraic and geometric multiplicity and the corresponding matrix Z_α as in (8.12) is definite, then for any Hamiltonian matrix $\mathscr{E}$ with $\|\mathscr{E}\|_2 \le \epsilon_0$, $\mathscr{H} + \mathscr{E}$ has only purely imaginary eigenvalues. For any $\epsilon > \epsilon_0$, there always exists a Hamiltonian matrix $\mathscr{E}$ with $\|\mathscr{E}\|_2 = \epsilon$ such that $\mathscr{H} + \mathscr{E}$ has an eigenvalue with non zero real part.

Mehrmann and Xu also consider skew-Hamiltonian matrices in [48]. These are $2n \times 2n$ matrices $\mathscr{K}$ such that $J\mathscr{K}$ is skew-Hermitian. Clearly, if $\mathscr{K}$ is skew-Hamiltonian, then $i\mathscr{K}$ is Hamiltonian so that the critical eigenvalues of $\mathscr{K}$ are its real eigenvalues. Therefore the effect of skew-Hamiltonian perturbations on the real eigenvalues of a complex skew-Hamiltonian matrix is expected to be the same as the effect of Hamiltonian perturbations on the purely imaginary eigenvalues of a complex Hamiltonian matrix. However, the situation is different if the skew-Hamiltonian matrix $\mathscr{K}$ is real. In such cases, the canonical form of $\mathscr{K}$ is given by the following result.

Theorem 4 ([23, 62]) *For any skew- Hamiltonian matrix $\mathscr{K} \in \mathbb{R}^{2n,2n}$, there exists a real symplectic matrix $\mathscr{S}$ such that*

$$\mathscr{S}^{-1}\mathscr{K}\mathscr{S} = \begin{bmatrix} K & 0 \\ 0 & K^T \end{bmatrix}$$

where $\mathscr{K}$ is in real Jordan canonical form.

The above result shows that each Jordan block associated with a real eigenvalue of $\mathscr{K}$ occurs twice and consequently every real eigenvalue is of even algebraic multiplicity. The following result in [48] summarizes the effect of structure preserving perturbations on the real eigenvalues of a real skew-Hamiltonian matrix.

Theorem 5 ([48]) *Consider the skew-Hamiltonian matrix $\mathscr{K} \in \mathbb{R}^{2n,2n}$ with a real eigenvalue α of algebraic multiplicity $2p$.*

1. *If $p = 1$, then for any skew-Hamiltonian matrix $\mathscr{E} \in \mathbb{R}^{2n,2n}$ with sufficiently small $\|\mathscr{E}\|_2$, $\mathscr{K} + \mathscr{E}$ has a real eigenvalue λ close to α with algebraic and geometric multiplicity 2, which has the form*

$$\lambda = \alpha + \eta + O(\|\mathscr{E}\|_2^2),$$

where η is the real double eigenvalue of the 2×2 matrix pencil $\lambda X^T JX - X^T(J\mathscr{E})X$, and X is a full column rank matrix so that the columns of X span the right eigenspace $\ker(\mathscr{K} - \alpha I)$ associated with α.

2. *If α is associated with a Jordan block of size larger than 2, then generically for a given $\mathscr{E}$ some eigenvalues of $\mathscr{K} + \mathscr{E}$ will no longer be real. If there exists a Jordan block of size 2 associated with α, then for every $\epsilon > 0$, there always exists $\mathscr{E}$ with $\|\mathscr{E}\|_2 = \epsilon$ such that some eigenvalues of $\mathscr{K} + \mathscr{E}$ are not real.*
3. *If the algebraic and geometric multiplicities of α are equal and are greater than 2, then for any $\epsilon > 0$, there always exists $\mathscr{E}$ with $\|\mathscr{E}\|_2 = \epsilon$ such that some eigenvalues of $\mathscr{K} + \mathscr{E}$ are not real.*

Mehrmann and Xu use Theorems 2 and 4 in [48] to analyse the properties of the symplectic URV algorithm that computes the eigenvalues of a Hamiltonian matrix in a structure preserving way. One of the main conclusions of this analysis is that if a Hamiltonian matrix $\mathscr{H}$ has a simple non-zero purely imaginary eigenvalue say

$i\alpha$, then the URV algorithm computes a purely imaginary eigenvalue say, $i\hat{\alpha}$ close to $i\alpha$. In such a case, they also find a relationship between $i\alpha$ and $i\hat{\alpha}$ that holds asymptotically. However, they show that if $i\alpha$ is multiple or zero, then the computed eigenvalue obtained from the URV algorithm may not be purely imaginary.

The structured perturbation analysis of purely imaginary eigenvalues of Hamiltonian matrices in [48] was further extended by Volker Mehrmann along with co-authors Alam, Bora, Karow and Moro in [5]. This analysis was used to find explicit Hamiltonian perturbations to Hamiltonian matrices that move eigenvalues away from the imaginary axis. In the same work, the authors also provide a numerical algorithm for finding an upper bound to the minimal Hamiltonian perturbation that moves all eigenvalues of a Hamiltonian matrix outside a vertical strip containing the imaginary axis. Alternative algorithms for this task have been given in [17, 27] and [28]. These works were motivated by the problem of passivation of dynamical systems mentioned previously.

One of the main aims of the analysis in [5] is to identify situations under which arbitrarily small Hamiltonian perturbations to a Hamiltonian matrix move its purely imaginary eigenvalues away from the imaginary axis with further restriction that the perturbations are real if the original matrix is real. Motivated by the importance of the Hermitian matrix Z_α introduced via (8.12) in [48], the authors make certain definitions in [5] to set the background for the analysis. Accordingly, any two vectors x and y from $\mathbb{F}^{2n}$ are said to be *J-orthogonal* if $x^*Jy = 0$. Subspaces $\mathscr{X}, \mathscr{Y} \subseteq \mathbb{F}^{2n}$ are said to be *J-orthogonal* if $x^*Jy = 0$ for all $x \in \mathscr{X}$, $y \in \mathscr{Y}$. A subspace $\mathscr{X} \subseteq \mathbb{F}^{2n}$ is said to be *J-neutral* if $x^*Jx = 0$ for all $x \in \mathscr{X}$. The subspace $\mathscr{X}$ is said to be *J-nondegenerate* if for any $x \in \mathscr{X} \setminus \{0\}$ there exists $y \in \mathscr{X}$ such that $x^*Jy \neq 0$. The subspaces of $\mathbb{F}^{2n}$ invariant with respect to $\mathscr{H}$ were then investigated with respect to these properties and the following was one of the major results in this respect.

Theorem 6 *Let $\mathscr{H} \in \mathbb{F}^{2n,2n}$ be Hamiltonian. Let $i\alpha_1, \ldots, i\alpha_p \in i\mathbb{R}$ be the purely imaginary eigenvalues of $\mathscr{H}$ and let $\lambda_1, \ldots, \lambda_q \in \mathbb{C}$ be the eigenvalues of $\mathscr{H}$ with negative real part. Then the $\mathscr{H}$-invariant subspaces* $\ker(\mathscr{H} - i\alpha_k I)^{2n}$ *and* $\ker(\mathscr{H} - \lambda_j I)^{2n} \oplus \ker(\mathscr{H} + \overline{\lambda}_j I)^{2n}$ *are pairwise J-orthogonal. All these subspaces are J-nondegenerate. The subspaces*

$$\mathscr{X}_-(\mathscr{H}) := \bigoplus\nolimits_{j=1}^{q} \ker(\mathscr{H} - \lambda_j I)^{2n},$$

$$\mathscr{X}_+(\mathscr{H}) := \bigoplus\nolimits_{j=1}^{q} \ker(\mathscr{H} + \overline{\lambda_j} I)^{2n}$$

are J-neutral.

An important result of [5] is the following.

Theorem 7 *Suppose that $\mathscr{H} \in \mathbb{F}^{2n,2n}$ is Hamiltonian and $\lambda \in \mathbb{C}$ is an eigenvalue of $\mathscr{H}$ such that* $\ker(\mathscr{H} - \lambda I)^{2n}$ *contains a J-neutral invariant subspace of dimension d. Let $\delta_1, \ldots, \delta_d$ be arbitrary complex numbers such that* $\max_k |\delta_k| < \epsilon$. *Then there exists a Hamiltonian perturbation $\mathscr{E}$ such that $\|\mathscr{E}\|_2 = O(\epsilon)$ and $\mathscr{H} + \mathscr{E}$*

has eigenvalues $\lambda + \delta_k, k = 1, \ldots, d$. *The matrix* $\mathscr{E}$ *can be chosen to be real if* $\mathscr{H}$ *is real.*

This together with Theorem 6 implies that eigenvalues of $\mathscr{H}$ with non zero real parts can be moved in any direction in the complex plane by an arbitrarily small Hamiltonian perturbation. However such a result does not hold for the purely imaginary eigenvalues as the following argument shows. Suppose that a purely imaginary eigenvalue $i\alpha$ is perturbed to an eigenvalue $i\alpha + \delta$ with $\operatorname{Re}\delta \neq 0$ by a Hamiltonian perturbation $\mathscr{E}_\delta$. Then the associated eigenvector v_δ of the Hamiltonian matrix $\mathscr{H} + \mathscr{E}_\delta$ is J-neutral by Theorem 6. By continuity it follows that $i\alpha$ must have an associated J-neutral eigenvector $v \in \ker(\mathscr{H} - i\alpha I)$. This in turn implies that the associated matrix Z_α is indefinite and so in particular, the eigenvalue $i\alpha$ must have multiplicity at least 2. In view of this, the following definition was introduced in [5].

Definition 1 Let $\mathscr{H} \in \mathbb{F}^{2n,2n}$ be Hamiltonian and $i\alpha$, $\alpha \in \mathbb{R}$ (with the additional assumption that $\alpha \neq 0$ if $\mathscr{H}$ is real) be an eigenvalue of $\mathscr{H}$. Then $i\alpha$ is of positive, negative or mixed sign characteristic if the matrix Z_α given by (8.12) is positive definite, negative definite or indefinite respectively.

This definition clearly implies that only purely imaginary eigenvalues of $\mathscr{H}$ that are of mixed sign characteristic possess J-neutral eigenvectors. So, the key to removing purely imaginary eigenvalues of $\mathscr{H}$ away from the imaginary axis is to initially generate an eigenvalue of mixed sign characteristic via a Hamiltonian perturbation. It was established in [5] that this can be achieved only by merging two imaginary eigenvalues of opposite sign characteristic and the analysis and investigations involved in the process utilized the concepts of Hamiltonian backward error and Hamiltonian pseudospectra.

The *Hamiltonian backward error* associated with a complex number $\lambda \in \mathbb{C}$ is defined by

$$\eta^{\mathrm{Ham}}(\lambda, \mathscr{H}) := \inf\{\, \|\mathscr{E}\| : \ \mathscr{E} \in \mathbb{F}^{2n,2n} \text{ Hamiltonian}, \ \lambda \in \Lambda(\mathscr{H} + \mathscr{E})\}. \qquad (8.14)$$

Note that in general $\eta^{\mathrm{Ham}}(\lambda, \mathscr{H})$ is different for $\mathbb{F} = \mathbb{C}$ and for $\mathbb{F} = \mathbb{R}$. We use the notation $\eta_F^{\mathrm{Ham}}(\lambda, \mathscr{H})$ and $\eta_2^{\mathrm{Ham}}(\lambda, \mathscr{H})$, when the norm in (8.14) is the Frobenius norm and the spectral norm, respectively. The complex Hamiltonian backward error for nonimaginary λ is discussed in the theorem below, see [5, 32].

Theorem 8 *Let* $\mathscr{H} \in \mathbb{C}^{2n,2n}$ *be a Hamiltonian matrix, and let* $\lambda \in \mathbb{C}$ *be such that* $\operatorname{Re}\lambda \neq 0$. *Then we have*

$$\eta_F^{\mathrm{Ham}}(\lambda, \mathscr{H}) = \min_{\|x\|_2=1} \left\{ \sqrt{2\|(\mathscr{H} - \lambda I)x\|_2^2 - |x^* J \mathscr{H} x|^2} \ : x \in \mathbb{C}^{2n},\ x^*Jx = 0 \right\}, \qquad (8.15)$$

$$\eta_2^{\mathrm{Ham}}(\lambda, \mathscr{H}) = \min_{\|x\|_2=1} \{\|(\mathscr{H} - \lambda I)x\|_2 : x \in \mathbb{C}^{2n},\ x^*Jx = 0\}. \qquad (8.16)$$

In particular, we have $\eta_2^{\mathrm{Ham}}(\lambda, \mathscr{H}) \leq \eta_F^{\mathrm{Ham}}(\lambda, \mathscr{H}) \leq \sqrt{2}\, \eta_2^{\mathrm{Ham}}(\lambda, \mathscr{H})$.

Suppose that the minima in (8.15), and (8.16) are attained for $u \in \mathbb{C}^{2n}$ and $v \in \mathbb{C}^{2n}$, respectively. Let

$$\mathscr{E}_1 := \frac{(\lambda I - \mathscr{H})uu^* + Juu^*(\lambda I - \mathscr{H})^* J}{\|u\|_2^2} + \frac{u^* J(\lambda I - \mathscr{H})uJuu^*}{\|u\|_2^4}$$

and

$$\mathscr{E}_2 := \|(\mathscr{H} - \lambda\, I)v\|_2\, J\, \begin{bmatrix} w & v \end{bmatrix} \begin{bmatrix} w^* v & 1 \\ 1 & v^* w \end{bmatrix}^{-1} \begin{bmatrix} w^* \\ v^* \end{bmatrix},$$

where $w := J(\mathscr{H} - \lambda\, I)v / \|(\mathscr{H} - \lambda\, I)v\|_2$. *Then*

$$\|\mathscr{E}_1\|_F = \eta_F^{\mathrm{Ham}}(\lambda, \mathscr{H}) \text{ and } (\mathscr{H} + \mathscr{E}_1)u = \lambda u,$$
$$\|\mathscr{E}_2\|_2 = \eta_2^{\mathrm{Ham}}(\lambda, \mathscr{H}) \text{ and } (\mathscr{H} + \mathscr{E}_2)v = \lambda v.$$

A minimizer v of the right hand side of (8.16) can be found via the following method. For $t \in \mathbb{R}$ let $F(t) = (\mathscr{H} - \lambda\, I)^*(\mathscr{H} - \lambda\, I) + i\, t\, J$. Let $t_0 = \arg\max_{t \in \mathbb{R}} \lambda_{\min}(F(t))$. Then there exists a normalized eigenvector v to the minimum eigenvalue of $F(t_0)$ such that $v^* J v = 0$. Thus,

$$\eta_2^{\mathrm{Ham}}(\lambda, \mathscr{H}) = \|(\mathscr{H} - \lambda\, I)v\|_2 = \max_{t \in \mathbb{R}} \sqrt{\lambda_{\min}(F(t))}.$$

The proposition below from [5] deals with the Hamiltonian backward error for the case that $\lambda = i\omega$ is purely imaginary. In this case an optimal perturbation can also be constructed as a real matrix if $\mathscr{H}$ is real. In the sequel M^+ denotes the Moore-Penrose generalized inverse of M.

Proposition 1 *Let $\mathscr{H} \in \mathbb{F}^{2n,2n}$ be Hamiltonian and $\omega \in \mathbb{R}$. Let v be a normalized eigenvector of the Hermitian matrix $J(\mathscr{H} - i\omega I)$ corresponding to an eigenvalue $\lambda \in \mathbb{R}$. Then $|\lambda|$ is a singular value of the Hamiltonian matrix $\mathscr{H} - i\omega I$ and v is an associated right singular vector.*

Further, the matrices

$$\mathscr{E} = \lambda J v v^*, \tag{8.17}$$
$$\mathscr{K} = \lambda J [v\ \bar{v}]\, [v\ \bar{v}]^+ \tag{8.18}$$

are Hamiltonian, $\mathscr{K}$ is real and we have $(\mathscr{H} + \mathscr{E})v = (\mathscr{H} + \mathscr{K})v = i\omega v$. Furthermore, $\|\mathscr{E}\|_F = \|\mathscr{E}\|_2 = \|\mathscr{K}\|_2 = |\lambda|$ and $\|\mathscr{K}\|_F \le \sqrt{2}\,|\lambda|$.

Moreover, suppose that λ is an eigenvalue of $J(\mathscr{H} - i\omega I)$ of smallest absolute value and let $\sigma_{\min}(\mathscr{H} - i\omega I)$ be the smallest singular value of $\mathscr{H} - i\omega I$. Then

$|\lambda| = \sigma_{\min}(\mathcal{H} - i\omega I)$ *and we have*

$$\eta_F^{\mathrm{Ham}}(i\omega, \mathcal{H}) = \eta_2^{\mathrm{Ham}}(i\omega, \mathcal{H}) = |\lambda| = \|\mathcal{E}\|_2, \text{ when } \mathbb{F} = \mathbb{C},$$
$$\eta_F^{\mathrm{Ham}}(i\omega, \mathcal{H}) = \sqrt{2}\,\eta_2^{\mathrm{Ham}}(i\omega, \mathcal{H}) = \sqrt{2}|\lambda| = \|\mathcal{K}\|_F, \text{ when } \mathbb{F} = \mathbb{R} \text{ and } \omega \neq 0.$$

The above result shows that the real and complex Hamiltonian backward errors for purely imaginary numbers $i\omega$, $\omega \in \mathbb{R}$ can be easily computed and in fact they are equal with respect to the 2-norm. Moreover, they depend on the eigenvalue of smallest magnitude of the Hermitian matrix $J(\mathcal{H} - i\omega I)$ and the corresponding eigenvector v is such that v and Jv are respectively the right and left eigenvectors corresponding to $i\omega$ as an eigenvalue of the minimally perturbed Hamiltonian matrix $\mathcal{H} + \Delta\mathcal{H}$ where $\Delta\mathcal{H} = \mathcal{E}$ if $\mathcal{H}$ is complex and $\Delta\mathcal{H} = \mathcal{K}$ when $\mathcal{H}$ is real. The authors also introduce the eigenvalue curves $\lambda_{\min}(\omega), \omega \in \mathbb{R}$ in [5] satisfying

$$|\lambda_{\min}(\omega)| = \min\{\,|\lambda| : \lambda \text{ is an eigenvalue of } J(\mathcal{H} - i\omega I)\} \tag{8.19}$$

and show that the pair $(\lambda_{\min}(\omega), v(\omega))$ where $J(\mathcal{H} - i\omega I)v(\omega) = \lambda_{\min}(\omega)v(\omega)$ is a piecewise analytic function of ω. Moreover,

$$\frac{d}{d\omega}\lambda_{\min}(\omega) = v(\omega)^* J v(\omega)$$

for all but finitely many points $\omega_1, \omega_2, \ldots, \omega_p \in \mathbb{R}$ at which there is loss of analyticity. These facts are used to show that $i\omega$ is an eigenvalue of $\mathcal{H} + \Delta\mathcal{H}$ with a J-neutral eigenvector if and only if it is a local extremum of $\lambda_{\min}(\omega)$. Hamiltonian ϵ-pseudospectra of $\mathcal{H}$ are then introduced into the analysis in [5] to show that the local extrema of $\lambda_{\min}(\omega)$ are the points of coalescence of certain components of the Hamiltonian pseudospectra.

Given any $A \in \mathbb{C}^{n,n}$ and $\epsilon \geq 0$, the *ϵ-pseudospectrum* of A is defined as

$$\Lambda_\epsilon(A; \mathbb{F}) = \bigcup_{\|E\|_2 \leq \epsilon} \{\Lambda(A + E) : E \in \mathbb{F}^{n,n}\}.$$

It is well-known [64] that in the complex case when $\mathbb{F} = \mathbb{C}$, we have

$$\Lambda_\epsilon(A; \mathbb{C}) = \{z \in \mathbb{C} : \ \sigma_{\min}(A - zI) \leq \epsilon\},$$

where, $\sigma_{\min}(\cdot)$ denotes the minimum singular value. The *Hamiltonian ϵ-pseudospectrum* is defined by

$$\Lambda_\epsilon^{\mathrm{Ham}}(\mathcal{H}; \mathbb{F}) = \bigcup_{\|\mathcal{E}\|_2 \leq \epsilon} \{\Lambda(\mathcal{H} + \mathcal{E}) : \mathcal{E} \in \mathbb{F}^{2n,2n} \text{ and } (J\mathcal{E})^* = J\mathcal{E}\}.$$

It is obvious that

$$\Lambda_\epsilon^{\mathrm{Ham}}(\mathscr{H};\mathbb{C}) = \{\, z \in \mathbb{C} : \ \eta_2^{\mathrm{Ham}}(z,\mathscr{H}) \le \epsilon \,\},$$

where $\eta_2^{\mathrm{Ham}}(z,\mathscr{H})$ is the Hamiltonian backward error as defined in (8.14) with respect to the spectral norm. The Hamiltonian pseudospectrum is in general different for $\mathbb{F} = \mathbb{C}$ and for $\mathbb{F} = \mathbb{R}$. However, the real and complex Hamiltonian pseudospectra coincide on the imaginary axis due to Proposition 1.

Corollary 1 *Let $\mathscr{H} \in \mathbb{C}^{2n,2n}$ be Hamiltonian. Consider the pseudospectra $\Lambda_\epsilon(\mathscr{H};\mathbb{F})$ and $\Lambda_\epsilon^{\mathrm{Ham}}(\mathscr{H};\mathbb{F})$. Then,*

$$\begin{aligned}\Lambda_\epsilon^{\mathrm{Ham}}(\mathscr{H};\mathbb{C}) \cap i\mathbb{R} = \Lambda_\epsilon^{\mathrm{Ham}}(\mathscr{H};\mathbb{R}) \cap i\mathbb{R} &= \Lambda_\epsilon(\mathscr{H};\mathbb{C}) \cap i\mathbb{R} = \Lambda_\epsilon(\mathscr{H};\mathbb{R}) \cap i\mathbb{R} \\ &= \{i\omega \ : \ \omega \in \mathbb{R},\ \sigma_{\min}(\mathscr{H} - i\omega I) \le \epsilon\} \\ &= \{i\omega \ : \ \omega \in \mathbb{R},\ |\lambda_{\min}(J(\mathscr{H} - i\omega I))| \le \epsilon\}.\end{aligned}$$

Analogous results for other perturbation structures have been obtained in [55].

Definition 1 introduced in [5] for purely imaginary eigenvalues of a Hamiltonian matrix is then extended to the components of the Hamiltonian pseudospectra $\Lambda_\epsilon^{\mathrm{Ham}}(\mathscr{H},\mathbb{F})$ as follows.

Definition 2 Let $\mathscr{H} \in \mathbb{F}^{2n,2n}$. A connected component $\mathscr{C}_\epsilon(\mathscr{H})$ of $\Lambda_\epsilon^{\mathrm{Ham}}(\mathscr{H},\mathbb{F})$ is said to have positive (resp., negative) sign characteristic if for all Hamiltonian perturbations $\mathscr{E}$ with $\|\mathscr{E}\|_2 \le \epsilon$ each eigenvalue of $\mathscr{H} + \mathscr{E}$ that is contained in $\mathscr{C}_\epsilon(\mathscr{H})$ has positive (resp., negative) sign characteristic.

In view of the above definition, if a component $\mathscr{C}_\epsilon(\mathscr{H})$ of $\Lambda_\epsilon^{\mathrm{Ham}}(\mathscr{H},\mathbb{F})$ has positive (resp., negative) sign characteristic then $\mathscr{C}_\epsilon(\mathscr{H}) \subset i\mathbb{R}$ and all eigenvalues of $\mathscr{H}$ that are contained in $\mathscr{C}_\epsilon(\mathscr{H})$ have positive (resp., negative) sign characteristic. Consequently, such components are necessarily subsets of the imaginary axis. In fact an important result in [5] is that sign characteristic of $\mathscr{C}_\epsilon(\mathscr{H})$ is completely determined by the sign characteristic of the eigenvalues of $\mathscr{H}$ that are contained in $\mathscr{C}_\epsilon(\mathscr{H})$.

Theorem 9 *Let $\mathscr{H} \in \mathbb{F}^{2n,2n}$ and let $\mathscr{C}_\epsilon(\mathscr{H})$ be a connected component of $\Lambda_\epsilon^{\mathrm{Ham}}(\mathscr{H},\mathbb{F})$. For a Hamiltonian matrix $\mathscr{E} \in \mathbb{F}^{2n,2n}$ with $\|\mathscr{E}\|_2 \le \epsilon$, let $X_{\mathscr{E}}$ be a full column rank matrix whose columns form a basis of the direct sum of the generalized eigenspaces* $\ker(\mathscr{H} + \mathscr{E} - \lambda I)^{2n}$, $\lambda \in \mathscr{C}_\epsilon(\mathscr{H}) \cap \Lambda(\mathscr{H} + \mathscr{E})$. *Set $Z_{\mathscr{E}} := -iX_{\mathscr{E}}^* J X_{\mathscr{E}}$. Then the following conditions are equivalent.*

(a) The component $\mathscr{C}_\epsilon(\mathscr{H})$ has positive (resp., negative) sign characteristic.
(b) All eigenvalues of $\mathscr{H}$ that are contained in $\mathscr{C}_\epsilon(\mathscr{H})$ have positive (resp., negative) sign characteristic.
(c) The matrix Z_0 associated with $\mathscr{E} = 0$ is positive (resp., negative) definite.
(d) The matrix $Z_{\mathscr{E}}$ is positive (resp., negative) definite for all Hamiltonian matrix $\mathscr{E}$ with $\|\mathscr{E}\|_2 \le \epsilon$.

Apart from characterising the sign characteristic of components of $\Lambda_\epsilon^{\mathrm{Ham}}(\mathscr{H})$ in terms of the sign characteristic of the eigenvalues of $\mathscr{H}$ contained in them, Theorem 9 implies that the only way to produce minimal Hamiltonian perturbations to $\mathscr{H}$ such that the perturbed matrices have purely imaginary eigenvalues with J-neutral eigenvectors is to allow components of $\Lambda_\epsilon^{\mathrm{Ham}}(\mathscr{H})$ of different sign characteristic to coalesce. Also in view of the analysis of the eigenvalue curves $w \to \lambda_{\min}(J(\mathscr{H} - iwI))$, it follows that the corresponding points of coalescence are precisely the local extrema of these curves.

If all the eigenvalues of a Hamiltonian matrix $\mathscr{H}$ are purely imaginary and of either positive or negative sign characteristic, [5] provides a procedure for constructing a minimal 2-norm Hamiltonian perturbation $\Delta\mathscr{H}$ based on Theorem 9, that causes at least one component of the Hamiltonian ϵ-pseudospectrum of $\mathscr{H}$ to be of mixed sign characteristic. This component is formed from the coalescence of components of positive and negative sign characteristic of the pseudospectrum and the perturbation $\Delta\mathscr{H}$ induces a point of coalescence of the components (which is purely imaginary) as an eigenvalue of $\mathscr{H} + \Delta\mathscr{H}$ with a J-neutral eigenvector. Therefore, any further arbitrarily small Hamiltonian perturbation $\mathscr{E}$ results in a non-imaginary eigenvalue for the matrix $\mathscr{H} + \Delta\mathscr{H} + \mathscr{E}$. The details of this process are given below. The following theorem from [5] is a refinement of Theorem 3.

Theorem 10 *Let $\mathscr{H} \in \mathbb{F}^{2n,2n}$ be a Hamiltonian matrix whose eigenvalues are all purely imaginary, and let $f(\omega) = \sigma_{\min}(\mathscr{H} - i\omega I)$, $\omega \in \mathbb{R}$. Define*

$$\rho_{\mathbb{F}}(\mathscr{H}) := \inf\{\, \|\mathscr{E}\|_2 \,:\, \mathscr{E} \in \mathbb{F}^{2n,2n},\ (J\mathscr{E})^* = J\mathscr{E},\ \mathscr{H} + \mathscr{E} \text{ has a non-imaginary eigenvalue}\},$$

$$R_{\mathbb{F}}(\mathscr{H}) := \inf\{\, \|\mathscr{E}\|_2 \,:\, \mathscr{E} \in \mathbb{F}^{2n,2n},\ (J\mathscr{E})^* = J\mathscr{E},\ \mathscr{H} + \mathscr{E} \text{ has a } J\text{-neutral eigenvector}\}$$

Furthermore, let ϵ_0 be defined as in (8.13). Then the following assertions hold.

(i) If at least one eigenvalue of $\mathscr{H}$ has mixed sign characteristic then $\epsilon_0 = R_{\mathbb{F}}(\mathscr{H}) = \rho_{\mathbb{F}}(\mathscr{H}) = 0$.

(ii) Suppose that each eigenvalue of $\mathscr{H}$ has either positive or negative sign characteristic. Let

$$i\mathscr{I}_1, \ldots, i\mathscr{I}_q \subset i\mathbb{R}$$

denote the closed intervals on the imaginary axis whose end points are adjacent eigenvalues of $\mathscr{H}$ with opposite sign characteristics. Then we have

$$\epsilon_0 = R_{\mathbb{F}}(\mathscr{H}) = \rho_{\mathbb{F}}(\mathscr{H}) = \min_{1 \le k \le q} \max_{\omega \in \mathscr{I}_k} f(\omega).$$

(iii) *Consider an interval* $\mathscr{I} \in \{\mathscr{I}_1, \ldots, \mathscr{I}_q\}$ *satisfying*

$$\min_{1\le k\le q}\max_{\omega\in\mathscr{I}_k} f(\omega) = \max_{\omega\in\mathscr{I}} f(\omega) = f(\omega_0), \quad \omega_0 \in \mathscr{I}.$$

Suppose that $i\mathscr{I}$ *is given by* $i\mathscr{I} = [i\alpha, i\beta]$. *Then the function* f *is strictly increasing in* $[\alpha, \omega_0]$ *and strictly decreasing in* $[\omega_0, \beta]$. *For* $\epsilon < \epsilon_0$, *we have* $i\omega_0 \notin \Lambda_\epsilon^{\mathrm{Ham}}(\mathscr{H}, \mathbb{F})$, $\mathscr{C}_\epsilon(\mathscr{H}, i\alpha) \cap \mathscr{C}_\epsilon(\mathscr{H}, i\beta) = \emptyset$ *and* $i\omega_0 \in \mathscr{C}_{\epsilon_0}(\mathscr{H}, i\alpha) = \mathscr{C}_{\epsilon_0}(\mathscr{H}, i\beta) = \mathscr{C}_{\epsilon_0}(\mathscr{H}, i\alpha) \cup \mathscr{C}_{\epsilon_0}(\mathscr{H}, i\beta)$ *where* $\mathscr{C}_{\epsilon_0}(\mathscr{H}, i\alpha)$ *and* $\mathscr{C}_{\epsilon_0}(\mathscr{H}, i\beta)$ *are components of* $\mathscr{H}$ *containing* $i\alpha$ *and* $i\beta$ *respectively. Moreover, if* $i\alpha$ *has positive sign characteristic and* $i\beta$ *has negative sign characteristic, then the eigenvalue curves* $\lambda_{\min}(w)$ *(satisfying (8.19)) are such that* $\lambda_{\min}(\omega) = f(\omega)$ *for all* $\omega \in [\alpha, \beta]$. *On the other hand, if* $i\alpha$ *has negative sign characteristic and* $i\beta$ *has positive sign characteristic then* $\lambda_{\min}(\omega) = -f(\omega)$ *for all* $\omega \in [\alpha, \beta]$. *In both cases there exists a* J*-neutral normalized eigenvector* v_0 *of* $J(\mathscr{H} - i\omega_0 I)$ *corresponding to the eigenvalue* $\lambda_{\min}(\omega_0)$.

(iv) *For the* J*-neutral normalized eigenvector vector* v_0 *mentioned in part (iii), consider the matrices*

$$\begin{aligned}
\mathscr{E}^0 &:= \lambda_{\min}(\omega_0) J v_0 v_0^*,\\
\mathscr{K}^0 &:= \lambda_{\min}(\omega_0) J [v_0\ \overline{v_0}][v_0, \overline{v_0}]^+,\\
\mathscr{E}_\mu &:= \mu v_0 v_0^* + \bar{\mu} J v_0 v_0^* J,\\
\mathscr{K}_\mu &:= [\mu v_0, \overline{\mu v_0}][v_0, \overline{v_0}]^+ + J([v_0, \overline{v_0}]^+)^*[\mu v, \overline{\mu v_0}]^* J [\mu v_0, \overline{\mu v_0}]\\
&\quad + J[v_0, \overline{v_0}][v_0, \overline{v_0}]^+ J [v_0, \overline{v_0}]^+, \qquad \mu \in \mathbb{C}.
\end{aligned}$$

Then $\mathscr{E}^0$ *is Hamiltonian,* $\mathscr{K}^0$ *is real and Hamiltonian,* $(\mathscr{H} + \mathscr{E}^0)v_0 = (\mathscr{H} + \mathscr{K}^0)v_0 = i\omega_0 v_0$ *and* $\|\mathscr{E}^0\|_2 = \|\mathscr{K}^0\|_2 = f(\omega_0)$. *For any* $\mu \in \mathbb{C}$ *the matrix* $\mathscr{E}_\mu$ *is Hamiltonian, and* $(\mathscr{H} + \mathscr{E}^0 + \mathscr{E}_\mu)v_0 = (i\omega_0 + \mu)v_0$. *If* $\omega_0 = 0$ *and* $\mathscr{H}$ *is real then* v_0 *can be chosen as a real vector. Then* $\mathscr{E}^0 + \mathscr{E}_\mu$ *is a real matrix for all* $\mu \in \mathbb{R}$. *If* $\omega_0 \neq 0$ *and* $\mathscr{H}$ *is real then for any* $\mu \in \mathbb{C}$, $\mathscr{K}_\mu$ *is a real Hamiltonian matrix satisfying* $(\mathscr{H} + \mathscr{K}^0 + \mathscr{K}_\mu)v_0 = (i\omega_0 + \mu)v_0$.

Theorem 10 is the basis for the construction an algorithm in [5] which produces a Hamiltonian perturbation to any Hamiltonian matrix $\mathscr{H}$ such that either all the eigenvalues of the perturbed Hamiltonian matrix lie outside an infinite strip containing the imaginary axis or any further arbitrarily small Hamiltonian perturbation results in a matrix with no purely imaginary eigenvalues. This is done by repeated application of the perturbations specified by the Theorem 10 to the portion of the Hamiltonian Schur form of $\mathscr{H}$ that corresponds to the purely imaginary eigenvalues of positive and negative sign characteristic. Each application, brings at least one pair of purely imaginary eigenvalues together on the imaginary axis to form eigenvalue(s) of mixed sign characteristic of the perturbed Hamiltonian matrix. Once this happens, the Hamiltonian Schur form of the perturbed matrix can again be utilised to construct subsequent perturbations that affect only the portion of the matrix corresponding to purely imaginary eigenvalues of positive and negative sign characteristic. When the sum of all such perturbations is considered, all the purely imaginary eigenvalues of the resulting Hamiltonian matrix are of mixed sign characteristic and there exist Hamiltonian perturbations of arbitrarily small norm that can remove all of them from the imaginary axis. These final perturbations can also be designed in a way that all the eigenvalues of the perturbed Hamiltonian matrix are outside a pre-specified infinite strip containing the imaginary axis. The procedure leads to an upper bound on the minimal Hamiltonian perturbations that achieve the desired objective.

As mentioned earlier, critical eigenvalues of other structured matrices and matrix pencils are also associated with sign characteristics and many applications require the removal of such eigenvalues. For example, critical eigenvalues of Hermitian pencils $L(z) = A - zB$ are the ones on the real line or at infinity. If such pencils are also definite, then minimal Hermitian perturbations that result in at least one eigenvalue with a non zero purely imaginary part with respect to the norm $|||L||| := \sqrt{\|A\|_2^2 + \|B\|_2^2}$, (where $\|\cdot\|_2$ denotes the spectral norm), is the Crawford number of the pencil [30, 39, 58]. The Crawford number was first introduced in [20] although the name was first coined only in [57]. However, it has been considered as early as in the 1960s in the work of Olga Taussky [61] and its theory and computation has generated a lot of interest since then [9, 18, 21, 22, 29, 30, 57–60, 65]. Recently, definite pencils have been characterised in terms of the distribution of their real eigenvalues with respect to their sign characteristic [6, 7]. This has allowed the extension of the techniques used in [5] to construct a minimal Hermitian perturbation $\Delta L(z) = \Delta_1 - z\Delta_2$ (with respect to the norm $|||\cdot|||$) to a definite pencil $L(z)$ that results in a real or infinite eigenvalue of mixed sign characteristic for the perturbed pencil $(L + \Delta L)(z)$. The Crawford number of $L(z)$ is equal to $|||\Delta L|||$ as it can be shown that there exists a further Hermitian perturbation $\tilde{L}(z) = \tilde{\Delta}_1 - z\tilde{\Delta}_2$ that can be chosen to be arbitrarily small such that the Hermitian pencil $(L+\Delta L+\tilde{L})(z)$ has a pair of eigenvalues with non zero imaginary parts. The challenge in these cases is to formulate suitable definitions of the sign characteristic of eigenvalues at infinity that consider the effect of continuously changing Hermitian perturbations on these very important attributes of real or infinite eigenvalues. This work has been done in [56] and [16]. In fact, the work done in [56] also provides answers to analogous problems for Hermitian matrix polynomials with real eigenvalues and also extend to the case of matrix polynomials with co-efficient matrices are all skew-Hermitian or alternately Hermitian and skew-Hermitian.

8.4 Structured Rank One Perturbations

Motivated by applications in control theory Volker Mehrmann along with co-authors Mehl, Ran and Rodman investigated the effect of structured rank one perturbations on the Jordan form of a matrix [42–45]. We mention here two of the main results. In the following, a matrix $A \in \mathbb{C}^{n,n}$ is said to be selfadjoint with respect to the

Hermitian matrix $H \in \mathbb{C}^{n,n}$ if $A^* H = HA$. Also a subset of a vector space V over $\mathbb{R}$ is said to be generic, if its complement is contained in a proper algebraic subset of V. A Jordan block of size n to the eigenvalue λ is denoted by $\mathcal{J}_n(\lambda)$. The symbol $\oplus$ denotes the direct sum.

Theorem 11 *Let $H \in \mathbb{C}^{n\times n}$ be Hermitian and invertible, let $A \in \mathbb{C}^{n\times n}$ be H-selfadjoint, and let $\lambda \in \mathbb{C}$. If A has the Jordan canonical form*

$$\left(\mathcal{J}_{n_1}(\lambda)^{\oplus \ell_1}\right) \oplus \cdots \oplus \left(\mathcal{J}_{n_m}(\lambda)^{\oplus \ell_m}\right) \oplus \tilde{A},$$

where $n_1 > \cdots > n_m$ and where $\Lambda(\tilde{A}) \subseteq \mathbb{C} \setminus \{\lambda\}$ and if $B \in \mathbb{C}^{n\times n}$ is a rank one perturbation of the form $B = uu^* H$, then generically (with respect to $2n$ independent real variables that represent the real and imaginary components of u) the matrix $A + B$ has the Jordan canonical form

$$\left(\mathcal{J}_{n_1}(\lambda)^{\oplus \ell_1 - 1}\right) \oplus \left(\mathcal{J}_{n_2}(\lambda)^{\oplus \ell_2}\right) \oplus \cdots \oplus \left(\mathcal{J}_{n_m}(\lambda)^{\oplus \ell_m}\right) \oplus \tilde{\mathcal{J}},$$

where $\tilde{\mathcal{J}}$ contains all the Jordan blocks of $A + B$ associated with eigenvalues different from λ.

The theorem states that under a generic H-selfadjoint rank one perturbation precisely one of the largest Jordan blocks to each eigenvalue of the perturbed H-selfadjoint matrix A splits into distinct eigenvalues while the other Jordan blocks remain unchanged. An analogous result holds for unstructured perturbations. In view of these facts the following result on Hamiltonian perturbations is surprising.

Theorem 12 *Let $J \in \mathbb{C}^{n\times n}$ be skew-symmetric and invertible, let $A \in \mathbb{C}^{n\times n}$ be J-Hamiltonian (with respect to transposition, i.e. $A^T J = -JA$) with pairwise distinct eigenvalues $\lambda_1, \lambda_2, \cdots, \lambda_p, \lambda_{p+1} = 0$ and let B be a rank one perturbation of the form $B = uu^T J \in \mathbb{C}^{n\times n}$.*

For every λ_j, $j = 1, 2, \ldots, p+1$, let $n_{1,j} > n_{2,j} > \ldots > n_{m_j,j}$ be the sizes of Jordan blocks in the Jordan form of A associated with the eigenvalue λ_j, and let there be exactly $\ell_{k,j}$ Jordan blocks of size $n_{k,j}$ associated with λ_j in the Jordan form of A, for $k = 1, 2, \ldots, m_j$.

(i) If $n_{1,p+1}$ is even (in particular, if A is invertible), then generically with respect to the components of u, the matrix $A + B$ has the Jordan canonical form

$$\bigoplus_{j=1}^{p+1} \left(\left(\mathcal{J}_{n_{1,j}}(\lambda_j)^{\oplus \ell_{1,j}-1}\right) \oplus \left(\mathcal{J}_{n_{2,j}}(\lambda_j)^{\oplus \ell_{2,j}}\right) \oplus \cdots \oplus \left(\mathcal{J}_{n_{m_j,j}}(\lambda_j)^{\oplus \ell_{m_j,j}}\right)\right) \oplus \tilde{\mathcal{J}},$$

where $\tilde{\mathcal{J}}$ contains all the Jordan blocks of $A + B$ associated with eigenvalues different from any of $\lambda_1, \ldots, \lambda_{p+1}$.

(ii) If $n_{1,p+1}$ is odd (in this case $\ell_{1,p+1}$ is even), then generically with respect to the components of u, the matrix $A + B$ has the Jordan canonical form

$$\bigoplus_{j=1}^{p} \left(\left(\mathcal{J}_{n_{1,j}}(\lambda_j)^{\oplus \ell_{1,j}-1} \right) \oplus \left(\mathcal{J}_{n_{2,j}}(\lambda_j)^{\oplus \ell_{2,j}} \right) \oplus \cdots \oplus \left(\mathcal{J}_{n_{m_j,j}}(\lambda_j)^{\oplus \ell_{m_j,j}} \right) \right)$$
$$\oplus \left(\mathcal{J}_{n_{1,p+1}}(0)^{\oplus \ell_{1,p+1}-2} \right) \oplus \left(\mathcal{J}_{n_{2,p+1}}(0)^{\oplus \ell_{2,p+1}} \right)$$
$$\oplus \cdots \oplus \left(\mathcal{J}_{n_{m_{p+1},p+1}}(0)^{\oplus \ell_{m_{p+1},p+1}} \right) \oplus \mathcal{J}_{n_{1,p+1}+1}(0) \oplus \tilde{\mathcal{J}},$$

(iii) In either case (1) *or* (2)*, generically the part $\tilde{\mathcal{J}}$ has simple eigenvalues.*

The surprising fact here is that concerning the change of Jordan structure, the largest Jordan blocks of odd size, say n, corresponding to the eigenvalue 0 of a J-Hamiltonian matrix are exceptional. Under a generic structured rank one perturbation two of them are replaced by one block corresponding to the eigenvalue 0 of size $n + 1$ and some 1×1 Jordan blocks belonging to other eigenvalues.

In the paper series [42–45] the authors inspect changes of the Jordan canonical form under structured rank one perturbations for various other classes of matrices with symmetries also.

8.5 Concluding Remarks

In this article we have given a brief overview of Volker Mehrmann's contributions on eigenvalue perturbations for pencils occurring in control theory, Hamiltonian matrices and for structured rank one matrix perturbations. We believe that Theorem 10 can be extended to cover perturbations of the pencil (8.4) or, more generally, to some other structured matrix pencils and polynomials.

References

1. Adhikari, B., Alam, R.: Structured backward errors and pseudospectra of structured matrix pencils. SIAM J. Matrix Anal. Appl. **31**, 331–359 (2009)
2. Adhikari, B., Alam, R., Kressner, D.: Structured eigenvalue condition numbers and linearizations for matrix polynomials. Linear Algebra. Appl. **435**, 2193–2221 (2011)
3. Ahmad, Sk.S., Mehrmann, V.: Perturbation analysis for complex symmetric, skew symmetric even and odd matrix polynomials. Electron. Trans. Numer. Anal. **38**, 275–302 (2011)
4. Ahmad, Sk.S., Mehrmann, V.: Backward errors for Hermitian, skew-Hermitian, H-even and H-odd matrix polynomials. Linear Multilinear Algebra **61**, 1244–1266 (2013)
5. Alam, R., Bora, S., Karow, M., Mehrmann, V., Moro, J.: Perturbation theory for Hamiltonian matrices and the distance to bounded realness. SIAM J. Matrix Anal. Appl. **32**, 484–514 (2011)
6. Al-Ammari, M.: Analysis of structured polynomial eigenvalue problems. PhD thesis, School of Mathematics, University of Manchester, Manchester (2011)

7. Al-Ammari, M., Tisseur, F.: Hermitian matrix polynomials with real eigenvalues of definite type. Part 1 Class. Linear Algebra Appl. **436**, 3954–3973 (2012)
8. Antoulas, T: Approximation of Large-Scale Dynamical Systems. SIAM, Philadelphia (2005)
9. Ballantine, C.S.: Numerical range of a matrix: some effective criteria. Linear Algebra Appl. **19**, 117–188 (1978)
10. Benner, P., Mehrmann, V., Xu, H.: Perturbation analysis for the eigenvalue problem of a formal product of matrices. BIT Numer. Math. **42**, 1–43 (2002)
11. Benner, P., Kressner, D., Mehrmann, V.: Structure preservation: a challenge in computational control. Future Gener. Comput. Syst. **19**, 1243–1252 (2003)
12. Benner, P., Byers, R., Mehrmann, V., Xu, H.: Robust method for robust control. Preprint 2004–2006, Institut für Mathematik, TU Berlin (2004)
13. Benner, P., Kressner, D., Mehrmann, V.: Skew-Hamiltonian and Hamiltonian eigenvalue problems: theory, algorithms and applications. In: Drmac, Z., Marusic, M., Tutek, Z. (eds.) Proceedings of the Conference on Applied Mathematics and Scientific Computing, Brijuni, pp. 3–39. Springer (2005). [ISBN:1-4020-3196-3]
14. Bora, S.: Structured eigenvalue condition number and backward error of a class of polynomial eigenvalue problems. SIAM J. Matrix Anal. Appl. **31**, 900–917 (2009)
15. Bora, S., Mehrmann, V.: Linear perturbation theory for structured matrix pencils arising in control theory. SIAM J. Matrix Anal. Appl. **28**, 148–191 (2006)
16. Bora, S., Srivastava, R.: Distance problems for Hermitian matrix pencils with eigenvalues of definite type (submitted)
17. Brüll, T., Schröder, C.: Dissipativity enforcement via perturbation of para-Hermitian pencils. IEEE Trans. Circuit. Syst. I Regular Paper **60**, 164–177 (2012)
18. Cheng, S.H., Higham, N.J.: The nearest definite pair for the Hermitian generalized eigenvalue problem. Linear Algebra Appl. **302/303**, 63–76 (1999). Special issue dedicated to Hans Schneider (Madison, 1998)
19. Coelho, C.P., Phillips, J.R., Silveira, L.M.: Robust rational function approximation algorithm for model generation. In: Proceedings of the 36th ACM/IEEE Design Automation Conference (DAC), New Orleans, pp. 207–212 (1999)
20. Crawford, C.R.: The numerical solution of the generalised eigenvalue problem. Ph.D. thesis, University of Michigan, Ann Arbor (1970)
21. Crawford, C.R.: Algorithm 646: PDFIND: a routine to find a positive definite linear combination of two real symmetric matrices. ACM Trans. Math. Softw. **12**, 278–282 (1986)
22. Crawford, C.R., Moon, Y.S.: Finding a positive definite linear combination of two Hermitian matrices. Linear Algebra Appl. **51**, pp. 37–48 (1983)
23. Faßbender, H., Mackey, D.S., Mackey, N., Xu, H.: Hamiltonian square roots of skew-Hamiltonian matrices. Linear Algebra Appl. **287**, 125–159 (1999)
24. Freiling, G., Mehrmann, V., Xu, H.: Existence, uniqueness and parametrization of Lagrangian invariant subspaces. SIAM J. Matrix Anal. Appl. **23**, 1045–1069 (2002)
25. Freund, R.W., Jarre, F.: An extension of the positive real lemma to descriptor systems. Optim. Methods Softw. **19**, 69–87 (2004)
26. Green, M., Limebeer, D.J.N.: Linear Robust Control. Prentice-Hall, Englewood Cliffs (1995)
27. Grivet-Talocia, S.: Passivity enforcement via perturbation of Hamiltonian matrices. IEEE Trans. Circuits Syst. **51**, pp. 1755–1769 (2004)
28. Guglielmi, N., Kressner, D., Lubich, C.: Low rank differential equations for Hamiltonian matrix nearness problems. Oberwolfach preprints, OWP 2013-01 (2013)
29. Guo, C.-H., Higham, N.J., Tisseur, F.: An improved arc algorithm for detecting definite Hermitian pairs. SIAM J. Matrix Anal. Appl. **31**, 1131–1151 (2009)
30. Higham, N.J., Tisseur, F., Van Dooren, P.: Detecting a definite Hermitian pair and a hyperbolic or elliptic quadratic eigenvalue problem, and associated nearness problems. Linear Algebra Appl. **351/352**, 455–474 (2002). Fourth special issue on linear systems and control
31. Higham, N.J., Konstantinov, M.M., Mehrmann, V., Petkov, P.Hr.: The sensitivity of computational control problems. Control Syst. Mag. **24**, 28–43 (2004)

32. Karow, M.: μ-values and spectral value sets for linear perturbation classes defined by a scalar product. SIAM J. Matrix Anal. Appl. **32**, 845–865 (2011)
33. Karow, M., Kressner, D., Tisseur, F.: Structured eigenvalue condition numbers. SIAM J. Matrix Anal. Appl. **28**, 1052–1068 (2006)
34. Konstantinov, M.M., Mehrmann, V., Petkov, P.Hr.: Perturbation analysis for the Hamiltonian Schur form. SIAM J. Matrix Anal. Appl. **23**, 387–424 (2002)
35. Konstantinov, M.M., Mehrmann, V., Petkov, P.Hr.: Perturbed spectra of defective matrices. J. Appl. Math. **3**, pp. 115–140 (2003)
36. Kressner, D., Peláez, M.J., Moro, J.: Structured Hölder condition numbers for multiple eigenvalues. SIAM J. Matrix Anal. Appl. **31**, 175–201 (2009)
37. Lancaster, P.: Strongly stable gyroscopic systems. Electron. J. Linear Algebra **5**, 53–67 (1999)
38. Lancaster, P., Rodman, L.: The Algebraic Riccati Equation. Oxford University Press, Oxford (1995)
39. Li, C.-K., Mathias, R.: Distances from a Hermitian pair to diagonalizable and nondiagonalizable Hermitian pairs. SIAM J. Matrix Anal. Appl. **28**, 301–305 (2006) (electronic)
40. Lin, W.-W., Mehrmann, V., Xu, H.: Canonical forms for Hamiltonian and symplectic matrices and pencils. Linear Algebra Appl. **301–303**, 469–533 (1999)
41. Mehl, C., Mehrmann, V., Ran, A., Rodman, L.: Perturbation analysis of Lagrangian invariant subspaces of symplectic matrices. Linear Multilinear Algebra **57**, 141–185 (2009)
42. Mehl, C., Mehrmann, V., Ran, A.C.M., Rodman, L.: Eigenvalue perturbation theory of classes of structured matrices under generic structured rank one perturbations. Linear Algebra Appl. **435**, 687–716 (2011)
43. Mehl, C., Mehrmann, V., Ran, A.C.M., Rodman, L.: Perturbation theory of selfadjoint matrices and sign characteristics under generic structured rank one perturbations. Linear Algebra Appl. **436**, 4027–4042 (2012)
44. Mehl, C., Mehrmann, V., Ran, A.C.M., Rodman, L.: Jordan forms of real and complex matrices under rank one perturbations. Oper. Matrices **7**, 381–398 (2013)
45. Mehl, C., Mehrmann, V., Ran, A.C.M., Rodman, L.: Eigenvalue perturbation theory of symplectic, orthogonal, and unitary matrices under generic structured rank one perturbations. BIT **54**, 219–255 (2014)
46. Mehrmann, V.: The Autonomous Linear Quadratic Control Problem, Theory and Numerical Solution. Lecture Notes in Control and Information Science, vol. 163. Springer, Heidelberg (1991)
47. Mehrmann, V.: A step towards a unified treatment of continuous and discrete time control problems. Linear Algebra Appl. **241–243**, 749–779 (1996)
48. Mehrmann, V., Xu, H.: Perturbation of purely imaginary eigenvalues of Hamiltonian matrices under structured perturbations. Electron. J. Linear Algebra **17**, 234–257 (2008)
49. Overton, M., Van Dooren, P.: On computing the complex passivity radius. In: Proceedings of the 4th IEEE Conference on Decision and Control, Seville, pp. 760–7964 (2005)
50. Paige, C., Van Loan, C.: A Schur decomposition for Hamiltonian matrices. Linear Algebra Appl. **14**, 11–32 (1981)
51. Pontryagin, L.S., Boltyanskii, V., Gamkrelidze, R., Mishenko, E.: The Mathematical Theory of Optimal Processes. Interscience, New York (1962)
52. Ran, A.C.M., Rodman, L.: Stability of invariant maximal semidefinite subspaces. Linear Algebra Appl. **62**, 51–86 (1984)
53. Ran, A.C.M., Rodman, L.: Stability of invariant Lagrangian subspaces I. In: Gohberg, I. (ed.) Operator Theory: Advances and Applications, vol. 32, pp. 181–218. Birkhäuser, Basel (1988)
54. Ran, A.C.M., Rodman, L.: Stability of invariant Lagrangian subspaces II. In: Dym, H., Goldberg, S., Kaashoek, M.A., Lancaster, P. (eds.) Operator Theory: Advances and Applications, vol. 40, pp. 391–425. Birkhäuser, Basel (1989)
55. Rump, S.M.: Eigenvalues, pseudospectrum and structured perturbations. Linear Algebra Appl. **413**, 567–593 (2006)
56. Srivastava, R.: Distance problems for Hermitian matrix polynomials – an ϵ-pseudospectra based approach. PhD thesis, Department of Mathematics, IIT Guwahati (2012)

57. Stewart, G.W.: Perturbation bounds for the definite generalized eigenvalue problem. Linear Algebra Appl. **23**, 69–85 (1979)
58. Stewart, G.W., Sun, J.-G.: Matrix perturbation theory. Computer Science and Scientific Computing. Academic, Boston (1990)
59. Sun, J.-G.: A note on Stewart's theorem for definite matrix pairs. Linear Algebra Appl. **48**, 331–339 (1982)
60. Sun, J.-G.: The perturbation bounds for eigenspaces of a definite matrix-pair. Numer. Math. **41**, 321–343 (1983)
61. Taussky, O.: Positive-definite matrices. In: Inequalities (Proceeding Symposium Wright-Patterson Air Force Base, Ohio, 1965), pp. 309–319. Academic, New York (1967)
62. Thompson, R.C.: Pencils of complex and real symmetric and skew matrices. Linear Algebra Appl. **147**, pp. 323–371 (1991)
63. Tisseur, F., Meerbergen, K.: The quadratic eigenvalue problem. SIAM Rev. **43**, 235–286 (2001)
64. Trefethen, L.N., Embree, M.: Spectra and Pseudospectra: The Behavior of Nonnormal Matrices and Operators. Princeton University Press, Princeton (2005)
65. Uhlig, F.: On computing the generalized Crawford number of a matrix. Linear Algebra Appl. **438**, 1923–1935 (2013)
66. Van Dooren, P.: The generalized eigenstructure problem in linear system theory. IEEE Trans. Autom. Control **AC-26**, 111–129 (1981)
67. Van Dooren, P.: A generalized eigenvalue approach for solving Riccati equations. SIAM J. Sci. Stat. Comput. **2**, 121–135 (1981)
68. Zhou, K., Doyle, J.C., Glover, K.: Robust and Optimal Control. Prentice-Hall, Upper Saddle River (1995)

Chapter 9
A Story on Adaptive Finite Element Computations for Elliptic Eigenvalue Problems

Agnieszka Międlar

What we have done for ourselves alone dies with us; what we have done for others and the world remains and is immortal

Albert Pike

Abstract We briefly survey the recent developments in adaptive finite element approximations of eigenvalue problems arising from elliptic, second-order, selfadjoint partial differential equations (PDEs). The main goal of this paper is to present the variety of subjects and corresponding results contributing to this very complex and broad area of research, and to provide a reader with a relevant sources of information for further investigations.

9.1 Introduction

The PDE eigenvalue problems can be divided into several categories depending on different criterion. This article intends to introduce the state-of-art results and new advancements for elliptic, second-order, selfadjoint and compact partial differential operators. A well-known example of this class is the *Laplace* eigenvalue problem: *Find* $\lambda \in \mathbb{R}$ *and* $u \subset H_0^1(\Omega)$ *such that*

$$\begin{aligned} \Delta u &= \lambda u \qquad \text{in } \Omega \\ u &= 0 \qquad \text{on } \partial\Omega, \end{aligned} \tag{9.1}$$

where $\Omega \in \mathbb{R}^d, d = 1, 2, \ldots$ *is a bounded, polyhedral Lipschitz domain and* $\partial\Omega$ *is its boundary.* This simple but how important model problem can, however, illustrate many interesting phenomena of eigenvalues and eigenfunctions of general elliptic

A. Międlar (✉)
ANCHP – MATHICSE, École Polytechnique Fédérale de Lausanne, Bâtiment MA, Station 8, CH-1015 Lausanne, Switzerland

Institut für Mathematik, Technische Universität Berlin, Sekretariat MA 4-5, Straße des 17. Juni 136, D-10623 Berlin, Germany
e-mail: agnieszka.miedlar@epfl.ch; miedlar@math.tu-berlin.de

P. Benner et al. (eds.), *Numerical Algebra, Matrix Theory, Differential-Algebraic Equations and Control Theory*, DOI 10.1007/978-3-319-15260-8_9

selfadjoint partial differential equations. We refer the reader to a wonderful survey article [71] on Laplace eigenvalue problem which contains many references to the original papers.

Throughout this survey we are going to concentrate on selfadjoint problems, however, we mention some results and further reading for the non-selfadjoint case when relevant.

The paper is organized as follows. We present a variationally stated PDE eigenvalue problem, discuss its properties and present its Galerkin approximation in Sects. 9.1.1 and 9.1.2. In Sect. 9.1.3 we introduce the main ingredients of adaptive FEM. The whole Sect. 9.2 is dedicated to error analysis of AFEM. We discuss an a priori as well as an a posteriori error estimators in Sects. 9.2.1 and 9.2.2, respectively. The state-of-art eigensolvers used in the context of adaptive FEM are presented in Sect. 9.3, whereas issues like convergence and optimality are the subject of Sect. 9.4. Last but not least, Sect. 9.5 sheds some light on the fundamental role of linear algebra not only in eigenvalue, but in all adaptive FE computations.

9.1.1 An Eigenvalue Problem for Partial Differential Equation

Let V and U be real Hilbert spaces with $V \subset U$ densely and continuously embedded in H, e.g. $V := H_0^1(\Omega)$ and $U := L^2(\Omega)$, and $\|\cdot\|_V$ and $\|\cdot\|_U$ the associated norms, respectively. Let $a(\cdot,\cdot) : V \times V \to \mathbb{R}$ and $(\cdot,\cdot)_U : U \times U \to \mathbb{R}$ be symmetric and continuous bilinear forms. We consider the following variationally stated eigenvalue problem: *Find $\lambda \in \mathbb{R}$ and $u \in V$ such that*

$$a(u,v) = \lambda(u,v)_U \quad \text{for all} \quad v \in V. \tag{9.2}$$

We assume that $a(\cdot,\cdot)$ is continuous, i.e., there exists $\beta := \|a\|_V < \infty$ such that

$$|a(w,v)| \leq \beta\|w\|_V\|v\|_V, \tag{9.3}$$

and V-elliptic (coercive), i.e., there exists $\alpha > 0$ such that

$$a(v,v) \geq \alpha\|v\|_V^2 \quad \text{for all} \ \ v \in V. \tag{9.4}$$

Remark 1 The bilinear form $a(\cdot,\cdot)$ satisfies all properties of the scalar product on V. The norm induced by $a(\cdot,\cdot)$ is the so-called *energy norm*

$$|||\cdot||| := a(\cdot,\cdot)^{1/2}, \tag{9.5}$$

which is equivalent to the standard norm $\|\cdot\|_V$ in V. Namely

$$\alpha\|v\|_V^2 \leq |||v|||^2 \leq \beta\|v\|_V^2. \tag{9.6}$$

Due to conditions (9.3) and (9.4) the existence and uniqueness of a weak eigenpair (λ, u) is a simple consequence of the classical *Lax-Milgram Lemma* [95], see, e.g., [66, Theorem 2.12, p. 52], [77, Theorem 6.5.9, p. 140], [118, Theorem 5.5.1, p. 133] or [122, §5.5, Theorem 13]. For the selfadjoint eigenvalue problem the existence of a unique eigenpair (λ, u) can also be proved using the *Riesz Representation Theorem* [122, §5.4, Theorem 11].

Let us consider, for any $f \in V$, the following (variational) boundary value problem

$$\textit{Find } w \in V \quad \textit{such that} \quad a(w, v) = (f, v)_U \quad \text{for all } v \in V.$$

Following the results in classical spectral approximation theory [44, 83, 132, 133] we introduce a linear, compact solution operator $\mathfrak{T} : V \to V$, such that for any given $f \in V$

$$a(\mathfrak{T} f, v) = (f, v)_U \quad \text{for all } v \in V.$$

Now, using the definition of the operator $\mathfrak{T}$ with $f = u$ and Eq. (9.2) yields

$$a(\mathfrak{T} u, v) = (u, v)_U = \frac{1}{\lambda} a(u, v) = a(\frac{1}{\lambda} u, v) \quad \text{for all } v \in V. \tag{9.7}$$

Hence $(\lambda^{(i)}, u^{(i)})$ is a solution of the eigenvalue problem (9.2) if and only if $(\mu^{(i)}, u^{(i)})$, $\mu^{(i)} = \frac{1}{\lambda^{(i)}}$, is an eigenpair of the operator $\mathfrak{T}$. The assumptions on $a(\cdot, \cdot)$ guarantees that the operator $\mathfrak{T}$ has countably many real and positive eigenvalues whose inverses are denoted by

$$0 < \lambda^{(1)} \le \lambda^{(2)} \le \ldots \le \lambda^{(i)} \le \ldots ,$$

according to their multiplicity, and the corresponding (normalized) eigenfunctions $u^{(i)}$ form the orthogonal basis for the space U. Obviously, the eigenfunctions $u^{(i)}$ are orthogonal with respect to $a(\cdot, \cdot)$ and $(\cdot, \cdot)_U$, i.e.,

$$a(u^{(i)}, u^{(j)}) = \lambda^{(i)} (u^{(i)}, u^{(j)})_U = 0, \quad \text{for } i \neq j.$$

For further details see, e.g. [14, Chapter 2, pp. 692–714], [44, Chapter 4, pp. 203–204] or [25, 90, 127].

9.1.2 The Galerkin Approximation and the Finite Element Method (FEM)

In order to find an approximation (λ_h, u_h) to the exact solution of a variational problem (9.2), we consider the idea of approximating the exact solution by an element from a given finite dimensional subspace V_h, known as the *Galerkin* method (also known as *Bubnov-Galerkin* or *Ritz-Galerkin* method in the selfadjoint case) [48].

For $V_h \subseteq V$ the variationally stated eigenvalue problem (9.2) is approximated by the *discrete eigenvalue problem*: *Find* $\lambda_h \in \mathbb{R}$ *and* $u_h \in V_h$ *such that*

$$a(u_h, v_h) = \lambda_h (u_h, v_h)_U \quad \text{for all} \quad v_h \in V_h. \tag{9.8}$$

Remark 2 One should mention that there exist a number of other approximate methods, i.e., the Petrov-Galerkin method, the generalized Galerkin method, the method of weighted residuals, collocation methods etc., see [118, 122]. In general, the trial space U_h where the solution u_h lives and test space V_h are not related to each other. Intuitively, the trial space is responsible for the approximability of the solution, whereas the test space for stability (or quasi-optimality) of the discretization method, see [9, 131]. The Galerkin method is simple to analyze, since both spaces are taken to be the same, i.e., $U_h = V_h$, however, in many cases one should consider them to be distinct [82].

Since $V_h \subseteq V$, the bilinear form $a(\cdot,\cdot)$ is also bounded and coercive on V_h. Therefore, the existence of a unique Galerkin solution (λ_h, u_h) is inherited from the well-posedness of the original problem [122]. Analogously, there exists a discrete compact solution operator $\mathfrak{T}_h : V \to V$ such that for any given $f \in V$

$$a(\mathfrak{T}_h f, v) = (f, v)_U \text{ for all} \quad v \in V_h,$$

and therefore the eigenpairs of $\mathfrak{T}_h$ can be identified with those of (9.8), accordingly.

At this point, let us discuss some of the possible choices for space V_h, namely, basics of the Finite Element Method (FEM) [48]. For simplicity, we restrict ourself to consider only polygonal domains in $\mathbb{R}^2$.

Let $\mathcal{T}_h$ be a partition (triangulation) of a domain Ω into elements (triangles) T, such that

$$\mathcal{T}_h := \bigcup_{T \in \mathcal{T}_h} = \overline{\Omega},$$

and any two distinct elements in $\mathcal{T}_h$ share at most a common edge E or a common vertex υ. For each element $T \in \mathcal{T}_h$ by $\mathcal{E}(T)$ and $\mathcal{N}(T)$ we denote the set of corresponding edges and vertices, respectively, where $\mathcal{E}_h$ and $\mathcal{N}_h$ denote all edges and vertices in $\mathcal{T}_h$. Likewise, we define a *diameter* (the length of the longest edge) of an element as h_T. For each edge E we denote its length by h_E and the unit normal vector by $\mathbf{n}_E$. The label h associated with the triangulation $\mathcal{T}_h$ denotes its mesh size and is given as $h := \max_{T \in \mathcal{T}_h} h_T$. We say that the triangulation is *regular* in the sense of Ciarlet [48] if there exist a positive constant ρ such that

$$\frac{h_T}{d_T} < \rho,$$

with d_T being the diameter of the largest ball that may be inscribed in element T, i.e., the minimal angle of all triangles in $\mathscr{T}_h$ is bounded away from zero. Of course the choice of triangle elements is not a restriction of the finite element method and is made only in order to clarify the notation.

Consider a regular triangulation $\mathscr{T}_h$ of Ω and the set of polynomials $\mathbb{P}_p$ of total degree $p \geq 1$ on $\mathscr{T}_h$, which vanish on the boundary of Ω, see, e.g., [29]. Then the Galerkin discretization (9.8) with $V_h^p \subset V$, $\dim V_h^p = n_h$, taken as

$$V_h^p(\Omega) := \left\{v_h \in C^0(\overline{\Omega}) : v_h|_T \in \mathbb{P}_p \text{ for all } T \in \mathscr{T}_h \text{ and } v_h = 0 \text{ on } \partial\Omega\right\},$$

is called a *finite element discretization*. The Finite Element Method (FEM) [48] is a Galerkin method with a special choice of the approximating subspace, namely, the subspace of piecewise polynomial functions, i.e., continuous in Ω and polynomials on each $T \in \mathscr{T}_h$. For the sake of simplicity we consider only $\mathbb{P}_1$ finite elements, i.e., $p = 1$, and use $V_h := V_h^1$. The last condition, which is crucial from the practical point of view, states that the space V_h should have a canonical basis of functions with small supports over $\mathscr{T}_h$. It is easily seen that the simplest set satisfying this condition is the set $\left\{\varphi_h^{(1)}, \ldots, \varphi_h^{(n_h)}\right\}$ of the *Lagrange basis* (also known as *nodal* or *hat functions*) [48]. With this special choice of basis, the solution u_h is determined by its values at the n_h grid points of $\mathscr{T}_h$ and it can be written as

$$u_h = \sum_{i=1}^{n_h} u_{h,i}\varphi_h^{(i)}.$$

Then the discretized problem (9.8) reduces to a *generalized algebraic eigenvalue problem* of the form

$$A_h \mathbf{u}_h = \lambda_h B_h \mathbf{u}_h, \tag{9.9}$$

where the matrices

$$A_h := [a(\varphi_h^{(i)}, \varphi_h^{(j)})]_{1 \leq i,j \leq n_h}, \quad B_h := [(\varphi_h^{(i)}, \varphi_h^{(j)})_U]_{1 \leq i,j \leq n_h}$$

are called *stiffness* and *mass* matrix, respectively. The representation vector $\mathbf{u}_h$ is defined as

$$\mathbf{u}_h := [u_{h,i}]_{1 \leq i \leq n_h}.$$

Since $\left\{\varphi_h^{(1)}, \ldots, \varphi_h^{(n_h)}\right\}$ are chosen to have a small support over $\mathscr{T}_h$, the resulting matrices A_h, B_h are sparse. The symmetry of A_h, B_h and positive definiteness of B_h are inherited from the properties of the bilinear forms $a(\cdot,\cdot)$ and $(\cdot,\cdot)_U$, accordingly.

For conforming approximations, i.e., $V_h \subset V$, the *Courant-Fisher min-max characterization*, see, e.g. [14, Equation (8.42), p. 699], [130, Equation (23), p. 223] or [49, 138], implies that exact eigenvalues are approximated from above, i.e.,

$$\lambda^{(i)} \leq \lambda_h^{(i)}, \quad i = 1, 2, \ldots, n_h.$$

On the contrary, for the nonconforming discretizations, e.g., the Crouzeix-Raviart method [29, 50] or [56, Sections 1.2.6, 3.2.3], the discrete eigenvalues provide lower bounds [8, 25, 40] for the exact eigenvalues. The convergence of discrete eigenvalues/eigenfunctions towards their continuous counterparts preserving the multiplicities and preventing spurious solutions is discussed in details in [25, 26].

9.1.3 The Adaptive Finite Element Method (AFEM)

The standard finite element method would proceed from the selection of a mesh and basis to the generation of a solution. However, it is well-known that the overall accuracy of the numerical approximation in determined by several factors: the regularity of the solution (smoothness of the eigenfunctions), the approximation properties of the finite element spaces, i.e., the search and test space, the accuracy of the eigenvalue solver and its influence on the total error. The most efficient approximations of smooth functions can be obtained using large higher-order finite elements (p-FEM), where the local singularities, arising from re-entrant corners, interior or boundary layers, can be captured by small low-order elements (h-FEM) [47]. Unfortunately, in real-world applications, none of those phenomena are known a priori. Therefore, constructing an optimal finite dimensional space to improve the accuracy of the solution requires refining the mesh and (or) basis and performing the computations again. A more efficient procedure try to decrease the mesh size (*h-adaptivity*) or (and) increase the polynomial degree of the basis (*p-refinement*) automatically such that the accurate approximation can be obtained at a lower computational cost, retaining the overall efficiency. This adaptation is based on the local contributions of the global *error estimates*, the so-called *refinement indicators*, extracted from the numerical approximation. This simple algorithmic idea is called *Adaptive Finite Element Method (AFEM)* and can be described as a following loop

SOLVE ⟶ ESTIMATE ⟶ MARK ⟶ REFINE (⟶ back to SOLVE)

The number of manuscripts addressing adaptive finite element methods is constantly growing, and its importance can not be underestimated. In the following sections we present a small fraction of material presented in [4, 14, 16, 17, 29, 32, 48, 55–57, 66, 67, 73, 77, 82, 83, 93, 94, 113, 114, 117, 118, 121, 123, 128, 130, 136].

The application of the adaptive FEM to the variationally stated eigenvalue problem (9.2) yields to the following scheme: first the eigenvalue problem is solved on some initial mesh $\mathscr{T}_0$ to provide a finite element approximation (λ_h, u_h) of the continuous eigenpair (λ, u). Afterwards, the total error in the computed solution is estimated by some *error estimator* η_h. Unfortunately, even the most accurate global error estimators itself do not guarantee the efficiency of the adaptive algorithm. If the global error is sufficiently small, the adaptive algorithm terminates and returns (λ_h, u_h) as a final approximation, otherwise, the local contributions of the error are estimated on each element. A local *error indicator* (*refinement indicator*) for an element $T \in \mathscr{T}_h$ is usually denoted by η_T and related to a global error estimator η_h through

$$\eta_h = \Big(\sum_{T \in \mathscr{T}_h} \eta_T^2 \Big)^{1/2}.$$

Based on those, the elements for refinement are selected and form the set $\mathscr{M} \subset \mathscr{T}_h$ of *marked elements*. The final step involves the actual refinement of marked elements and creating a new mesh. Since the resulting mesh may possess hanging nodes, an additional closure procedure is applied is order to obtain a new regular (conforming) mesh. As a consequence, the adaptive finite element method (AFEM) generates a sequence of nested triangulations $\mathscr{T}_0, \mathscr{T}_1, \ldots$ with corresponding nested spaces

$$V_0 \subseteq V_1 \subseteq \ldots \subseteq V_{n_h} \subset V.$$

In the upcoming chapters we will concentrate particularly on SOLVE and ESTIMATE steps, however, let us shortly discuss the marking and refinement procedures.

As we already know the set $\mathscr{M}$ of marked elements is determined based on the sizes of refinement indicators η_T. Now, a question arises: How do we decide which elements $T \in \mathscr{T}_h$ should be added to the set $\mathscr{M}$ such that the newly obtained adaptive mesh fulfil the regularity condition. The process of selecting the elements of $\mathscr{M}$ is called the *marking criterion* or the *marking strategy*. Let us keep in mind that by marking an element we actually mean marking all its edges. The simplest marking strategy takes a fixed rate (e.g. 50 %) of elements of $\mathscr{T}_h$ with the largest values of η_T or elements T for which the refinement indicators η_T are larger than some fixed threshold $\tau \in \mathbb{R},\ \tau > 0$, i.e.,

$$\mathscr{M} := \{T \in \mathscr{T}_h : \tau \leq \eta_T\}.$$

Notice that the choice of a threshold τ is essential for the efficiency of the adaptive algorithm since it directly determines the size of the set $\mathscr{M}$. A more sophisticated strategy is the *maximum criterion*, where the elements selection is based on a fixed fraction Θ of the maximal refinement indicator in $\mathscr{T}_h$, i.e.,

$$\tau := \Theta \max_{T \in \mathscr{T}_h} \eta_T,$$

with $0 \leq \Theta \leq 1$. The most interesting, especially in the context of optimality of a standard adaptive finite element method, is a *Dörfler marking* strategy [53], where the set of marked elements is defined as the subset $\mathcal{M} \subseteq \mathcal{T}_h$ of smallest cardinality such that

$$(1-\Theta)^2 \sum_{T \in \mathcal{T}_h} \eta_T^2 \leq \sum_{T \in \mathcal{M}} \eta_T^2, \tag{9.10}$$

where $0 \leq \Theta \leq 1$. i.e., $\Theta = 0$ corresponds to $\mathcal{M} = \mathcal{T}_h$ and $\Theta = 1$ to $\mathcal{M} = \emptyset$. For more details see [31].

The refinement of the finite element space can be performed using various techniques like moving grid points (*r-refinement*), subdividing elements of a fixed grid (*h-refinement*), applying locally higher-order basis functions (*p-refinement*) or any combinations of those [47], see Fig. 9.1 for illustration.

For the sake of exposition, here, we consider only the *h-refinement* of the triangle elements. The most common h-refinement subdivision rules (techniques) based on edge marking are presented in Fig. 9.2.

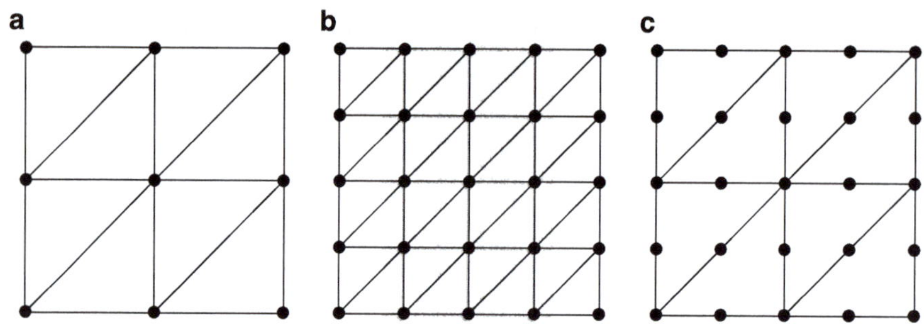

Fig. 9.1 (**a**) Original mesh, (**b**) a uniform *h-refinement* and (**c**) a uniform *p-refinement*

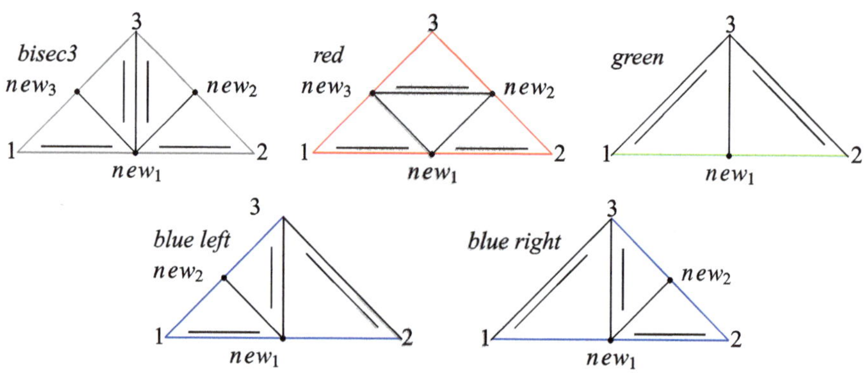

Fig. 9.2 *Bisec3*, *red*, *green* and *blue* refinement. The edges marked by MARK step are colored. The new reference edge is marked through a second line in parallel opposite the new vertices new_1, new_2 or new_3 [38]

As we have mentioned before, applying these refinement procedures may lead to nonconforming meshes with the so-called *hanging nodes*. Therefore, the *closure algorithm* [38] is applied to overcome this drawback and get a regular triangulation. Summarizing, if any edge E of the element T is marked for the refinement, the reference edge (e.g. longest edge) of T should also be marked. Since each element has $k = 0, 1, 2,$ or 3 edges marked for the refinement, if $k \geq 1$, the reference edge belongs to it. Moreover, the choice of a refinement method, see Fig. 9.2, depends on k, for instance, if $k = 1$ the *green* refinement is used etc. For more details about adaptive refinement strategies see, e.g., [4, 47, 135]. In the remainder of this survey we will focus on the h-adaptivity, hence, we will denote the finite dimensional space over the partition $\mathscr{T}_h$ as V_h and the associated Galerkin approximation as (λ_h, u_h).

9.2 Error Estimates

Over the years, research in fundamental spectral approximation by [10–13, 59, 83, 91, 130] resulted in the unified spectral approximation framework, nowadays referred to as a so-called *Babuška-Osborn Theory* [14, Theorem 7.1–7.4, 8.1–8.4].

Theorem 1 ([14, 25]) *Let $u^{(i)}$ be a unit eigenfunction associated with the eigenvalue $\lambda^{(i)}$ of multiplicity m and let $u_h^{(i)}, \dots, u_h^{(i+m-1)}$ denote the eigenfunctions associated with the m discrete eigenvalues converging to $\lambda^{(i)}$, then for each index i*

$$\lambda^{(i)} \leq \lambda_h^{(j)} \leq \lambda^{(i)} + C \sup_{\substack{u \in E_{\lambda^{(i)}} \\ \|u\|=1}} \inf_{v \in V_h} \|u - v\|_V^2, \quad j = i, \dots, i + m - 1,$$

and there exists $w_h^{(i)} \in span\{u_h^{(i)}, \dots, u_h^{(i+m-1)}\}$ such that

$$\|u^{(i)} - w_h^{(i)}\|_V \leq C \sup_{\substack{u \in E_{\lambda^{(i)}} \\ \|u\|=1}} \inf_{v \in V_h} \|u - v\|_V,$$

where $E_{\lambda^{(i)}}$ denotes the eigenspace associated with $\lambda^{(i)}$ and C is a generic constant.

Remark 3 For $\lambda^{(i)}$ being simple, we get the following estimates

$$|\lambda^{(i)} - \lambda_h^{(i)}| \leq C \sup_{\substack{u \in E_{\lambda^{(i)}} \\ \|u\|=1}} \inf_{v \in V_h} \|u - v\|_V^2 \quad \text{and} \quad \|u^{(i)} - u_h^{(i)}\|_V \leq C \sup_{\substack{u \in E_{\lambda^{(i)}} \\ \|u\|=1}} \inf_{v \in V_h} \|u - v\|_V.$$

Theorem 1 have some important consequences. First, it is easy to notice that the convergence rate of the eigenvalue/eigenfunction of interest is directly related to the approximability of the associated eigenfunctions (eigenspace). Namely, the approximation rate for the eigenvalue is double with respect to the approximation rate of the corresponding eigenfunctions, which is a well-known fact in the standard perturbation theory for the matrix eigenvalue problems and explains nicely why the eigenvalues are usually much more accurate than the corresponding eigenfunctions. On the other hand, the approximability of the eigenfunctions depends strongly on their regularity and the approximability conditions of the discrete solution space V_h, e.g. the finite element space. Hence, following [127, Theorem 3.6], we consider the regularity result for the eigenvalue problem (9.2) as

$$\|u\|_{H^2(\Omega)} = C\sqrt{\lambda},$$

where C is a generic constant depending on the data, but not on λ itself. The latter condition, is a well-known phenomena, called the *best approximation property* of the solution space. The choice of V_h being a space of piecewise polynomials of degree p guarantees the best approximation property of the finite element space [25, 48], i.e.,

$$\inf_{v \in V_h} \|u - v\|_{L^2(\Omega)} \leq Ch^{\min\{p+1,r\}} \|u\|_{H^r(\Omega)},$$

$$\inf_{v \in V_h} \|u - v\|_{H^1(\Omega)} \leq Ch^{\min\{p,r-1\}} \|u\|_{H^r(\Omega)},$$

where r is the regularity of the eigenspace (in the case of a simple eigenvalue, the regularity of the corresponding eigenfunction). These two facts immediately explain the common fact of deteriorating rates of convergence in the presence of singularities (singular eigenfunctions). Moreover, the eigenfunctions present more oscillations when the associated eigenvalue increases and the largest eigenvalues are much harder to approximate [38].

The *Babuška-Osborn Theory* provided important basis for further developments in a priori and a posteriori estimates and designing efficient numerical algorithms. In the remainder of this section we will focus in more details on estimating the eigenvalue/eigenfunction errors.

9.2.1 A Priori Error Estimates

In general, a priori error estimators give information about the asymptotic behavior of the error or the stability of the applied solver independent of the actually computed approximation. Likewise, they require particular regularity conditions of the solution, the stability properties of the discrete operator or the continuous solution u itself [4, 135] and approximability conditions of the discrete solution spaces. Except of some simple one-dimensional boundary value problems, where an optimal finite element space can be constructed based on a priori estimates, see [112, Lecture 1] or [113] for details. All these conditions make a priori error estimators not computable and hardly applicable in practice. Of course, if some a priori information about the solution is known it can be relevant for the construction of efficient numerical algorithms, e.g. *a priori adaptivity technique* [117], unfortunately, this is often not the case.

One of the simplest a priori error results obtained in [121] gives estimates to the piecewise linear eigenvalue/eigenfunction error in $L^2(\Omega)$ and energy norm depending on the regularity of the solution space, i.e.,

$$|||u-u_h||| \leq Ch^r, \quad \|u-u_h\|_{L^2(\Omega)} \leq Ch^r\,|||u-u_h|||, \quad |\lambda-\lambda_h| \leq C\,|||u-u_h|||^2 \leq Ch^{2r},$$

for $u \in H^{1+r}(\Omega)$, where $r \in (0,1]$ is a regularity constant and $C > 0$ is a constant depending on the eigenvalue λ, the eigenfunction u and on the triangulation $\mathcal{T}_h$, see, e.g. [77, Corollary 11.2.21, p. 264]. Note that if $r = 1$ (convex domain), the solution u has to fulfil a $H^2(\Omega)$-regularity condition, which is very restrictive and excludes a large class of problems, e.g., the Laplace eigenvalue problem on the L-shape domain. More details can be found in [72] for nonconvex domains and in [1, 2, 65] for higher order polynomials. As mentioned before, also here, the eigenvalue approximation is much more accurate (double) than the corresponding eigenfunction approximation, i.e., the eigenfunctions are first order convergent in $H^1(\Omega)$ and second order convergent in $L^2(\Omega)$. In [14] this result was generalized for multiple eigenvalues. Also, in this case, the error behaves like $\mathcal{O}(h)$ for the energy norm of the eigenfunction and $\mathcal{O}(h^2)$ for the eigenvalue. Several further results include improved eigenvector estimates in [12, 13], refined estimates in $H^1(\Omega)$–norm together with the lower and upper bound for the eigenvalue error in the case of multiple eigenvalues [43, 45], to mention just a few.

In [90, Theorems 3.1, 3.2 and 3.3] some new a priori FEM error estimates for simple, multiple and clustered eigenvalues were proposed. The error estimate for a simple eigenvalue [90, Theorem 2.7] depend on the continuous eigenfunction $u^{(i)}$ and its approximability properties in space V_h, but do not involve any underdetermined constants. Analogously, for the multiple or clustered eigenvalues [90, Theorem 2.11], the a priori error estimates depend only on approximability properties of eigenfunctions in the corresponding eigenspace (invariant subspace). Moreover for clustered eigenvalues the presented estimates are *cluster robust*, i.e., they do not depend on the width of the cluster. This work has improved several previous results involving the approximability properties of all previous eigenvectors and easily explained different convergence rates of Ritz values approximating a multiple eigenvalue [12–14]. To conclude, we present an a priori eigenvalue error estimator for the FEM approximation of a simple eigenvalue introduced in [90, Theorems 3.1 and 3.2], i.e.,

Theorem 2 *Knyazev [90, Theorem 3.2] For a fixed index i such that* $1 \leq i \leq n_h$ *suppose that*

$$d_{i,V_h} := \min_{j=1,\ldots,i-1} |\lambda_h^{(j)} - \lambda^{(i)}| \neq 0,$$

then

$$0 \le \frac{\lambda_h^{(i)} - \lambda^{(i)}}{\lambda_h^{(i)}} \le \left(1 + \max_{j=1,\ldots,i-1} \frac{(\lambda_h^{(j)}\lambda^{(i)})^2}{|\lambda_h^{(j)} - \lambda^{(i)}|^2} |||(I - P_h)\mathfrak{T} P_{h,1,\ldots,i-1}|||^2\right)$$
$$\sin^2 \angle_{|||\cdot|||}(u_i, V_h),$$

where $P_h : V \to V_h$ *is an* $|||\cdot|||$*-orthogonal projector on* V_h*, i.e.,* for all $u \in V$, $a(P_h u - u, v) = 0$, for all $v \in V_h$ *and* $P_{h,1,\ldots,i-1}$ *is the* $|||\cdot|||$*-orthogonal projector onto* $E_{h,1,\ldots,i-1} = span\left\{u_h^{(1)}, \ldots, u_h^{(i-1)}\right\}$.

Remark 4 If $i = j$, Theorem 2 turns into [90, Theorem 3.1], namely

$$0 \le \frac{\lambda_h^{(i)} - \lambda^{(i)}}{\lambda_h^{(i)}} \le |||(I - P_h)P_{1,\ldots,i}|||^2. \tag{9.11}$$

Finally, we would like to mention an a priori error estimate obtained in [125, Theorem 3], [85, Theorem 3.3], [127, Theorem 4.11], i.e.,

Theorem 3 (Saad [125, Theorem 3], [127, Theorem 4.11]) *Let* $(\mu^{(i)}, u^{(i)})$, $1 \le i \le n_h$ *be the* i*-th eigenpair of the operator* $\mathfrak{T}$ *defined in* (9.7) *with normalization* $\|u^{(i)}\|_{H^1(\Omega)} = 1$. *Suppose that*

$$\widehat{d}_{i,V_h} := \min_{j \ne i} |\mu_h^{(j)} - \mu^{(i)}| \ne 0,$$

then there exists $u_h^{(i)}$ *such that*

$$\|u^{(i)} - u_h^{(i)}\|_{H^1(\Omega)} \le \left(1 + \frac{\|(I - P_h)\mathfrak{T} P_h\|^2_{H^{-1}(\Omega)}}{\widehat{d}^2_{i,V_h}}\right)^{1/2} \inf_{v \in V_h} \|u^{(i)} - v\|_{H^1(\Omega)}.$$

The aforementioned theorem is a special case of a more general result for eigenspaces and invariant subspaces proposed in [85, Theorem 2.1, 3.1]. Obviously, the a priori error estimators are not limited to those listed above and include work presented in [75, 87, 92, 115] etc.

9.2.2 A Posteriori Error Estimates

Although a priori error estimates are usually not available for practical problems, we would still like to determine the quality of the numerical approximation, i.e., obtain a reliable estimate of the error $e_h = u - u_h$ in a specified norm $\| \cdot \|$ or quantity of interest (quality measure), e.g., energy norm $|||\cdot|||$ [4, 34, 135], or to terminate the algorithm as soon as a prescribed tolerance ε is reached, e.g. $\|e_h\| \leq \varepsilon$. Therefore, we need some computable quantity η_h (*a posteriori error estimator*) which can estimate the actual error $\|e_h\|$, i.e.,

$$\|u - u_h\| \approx \eta_h.$$

The formal definition, see [34], states as follows:

Definition 1 (A posteriori error estimator) A computable quantity η_h is called *a posteriori error estimator* if it can be extracted from the computed numerical solution u_h and the given data of the problem, e.g. the known domain Ω and its boundary $\partial\Omega$.

There are several important practical requirements on a posteriori error estimators. First, as the definition states, they should be computable. Secondly, in contrast to a priori error estimators, they depend on the stability properties of the continuous operator which are known and use the approximate solution itself to check its quality. Last but not least, calculating the estimator should be cheaper than computing the new numerical approximation (e.g., assembling the matrices). Besides, it is of great importance, especially in the context of the AFEM, to be able to extract the local contribution of the error estimator, i.e., the refinement indicators η_T,

$$\eta_h = \Big(\sum_{T \in \mathcal{T}_h} \eta_T^2 \Big)^{1/2}.$$

As far as a global upper bound is sufficient to assure the accuracy of the solution, an a posteriori error estimator should also provide local lower bound for the true error. These properties of a posteriori error estimator η_h are called *reliability (guaranteed upper bound)*

$$\|\!|e_h|\!\| \leq C_{rel}\eta_h + h.o.t_{rel}$$

and *local efficiency*

$$\eta_T \leq C_{eff} \|e_h\|_{\widehat{T}} + h.o.t._{eff},$$

with constants $C_{rel}, C_{eff} > 0$ independent of the mesh size h or polynomial degree p, $\widehat{T}$ the union of T and neighbouring elements and higher-order terms $h.o.t_{rel}$, $h.o.t_{eff}$ related to data oscillations. Both these bounds are crucial from the point of convergence and optimality, respectively. Namely, it is well-known that a reliable and efficient a posteriori error estimator, i.e.,

$$\frac{1}{C_{eff}}\eta_h \leq \|\!|e_h|\!\| \leq C_{rel}\eta_h.$$

decays with the same rate as the actual computational error up to higher-order terms. We discuss this issue in detail in Sect. 9.4. The aforementioned definition of the reliability, where the constant C_{rel} is present, is called a weak form of reliability. In the ideal situation we would like the estimator to satisfy

$$|||e_h||| \leq \eta_h,$$

which is very rarely the case.

In order to conclude, the main goal is to obtain an accurate solution with an optimal use of resources and guarantee that the a posteriori error estimator captures the behavior of the actual error as $h \to 0$. In practice, we are often interested in the asymptotical exactness or efficiency of the a posteriori error estimator. Following [4], we call the error estimator η_h *asymptotically exact* if

$$\lim_{h\to 0} \theta = 1,$$

where $\theta := \frac{\eta_h}{|||e_h|||}$ is called a global *efficiency index*. An error estimator is called *efficient* if its efficiency index θ and its inverse are bounded independent on the mesh size [34].

The pioneering work of Babuška and Rheinboldt [15] initiated decades of research devoted to a posteriori error estimates. We strongly encourage the reader to further explore the vast literature on the a posteriori error estimates, see e.g., [4, 16, 32, 135, 136]. Following [47, 135], the a posteriori error estimators can be classified as *residual error estimators (explicit error estimators)*, *local problem-based estimators (implicit error estimators)*, *averaging estimators (recovery-based estimators)*, *hierarchical estimators (multilevel estimators)* and *goal-oriented error estimators*. For the sake of exposition, let us concentrate on the residual type estimators and provide only some general information for the other classes.

9.2.2.1 The Residual Error Estimators (Explicit Error Estimators)

Whereas a priori error estimators relate the error $\|u - u_h\|_V$ to the regularity of the exact solution, residual a posteriori error estimators consider the connection of the error to the residual of the computed finite element solution u_h.
Let us consider the residual

$$Res_h(\cdot) := \lambda_h(u_h, \cdot)_U - a(u_h, \cdot)_U \in V^*$$

and the residual equation

$$Res_h(v) = a(u - u_h, v) - (\lambda u - \lambda_h u_h, v)_U \quad \text{for all } v \in V, \tag{9.12}$$

where V^* denotes the dual of the space V.

First of all, notice that the Galerkin orthogonality property does not hold for variationally stated eigenvalue problems, namely

$$a(u-u_h, v_h) = (\lambda u - \lambda_h u_h, v_h)_U \neq 0 \quad \text{for some } v_h \in V_h. \tag{9.13}$$

Secondly, since $e_h \in V$, Eq. (9.12) combined with the higher-order term [54]

$$(\lambda u - \lambda_h u_h, e_h)_U = \frac{\lambda + \lambda_h}{2} \|e_h\|_U^2 \tag{9.14}$$

imply

$$\|\!|u-u_h|\!\|^2 = \frac{\lambda + \lambda_h}{2} \|u-u_h\|_U^2 + Res_h(u-u_h), \tag{9.15}$$

which provides the crucial relation in the residual type error analysis, namely, the equivalence between the energy norm of the error and the residual, which, up to the higher-order terms was proved for the selfadjoint eigenvalue problems in [38].

Theorem 4 *Let $e_h = u - u_h$ and $Res_h(\cdot) := \lambda_h(u_h, \cdot)_U - a(u_h, \cdot)$. Then the following holds*

$$\alpha \|u-u_h\|_V \lesssim \|Res_h\|_{V^*} \lesssim \beta \|u-u_h\|_V, \tag{9.16}$$

where $0 < \alpha \leq \beta < \infty$ are the coercivity and continuity constants introduced in (9.4) and (9.3), respectively.

Proof The coercivity (9.4), the residual equation (9.12) and (9.14) imply

$$\begin{aligned}
\alpha \|u-u_h\|_V &\leq \frac{a(e_h, e_h)}{\|e_h\|_V} = \frac{Res_h(e_h)}{\|e_h\|_V} + \frac{(\lambda + \lambda_h)}{2} \frac{\|e_h\|_U^2}{\|e_h\|_V} \\
&\leq \sup_{v \in V} \frac{Res_h(v)}{\|v\|_V} + \frac{(\lambda + \lambda_h)}{2} \frac{\|e_h\|_U^2}{\|e_h\|_V} \\
&= \|Res_h\|_{V^*} + \frac{(\lambda + \lambda_h)}{2} \frac{\|e_h\|_U^2}{\|e_h\|_V}.
\end{aligned}$$

Since $\frac{\lambda+\lambda_h}{2} \|e_h\|_U^2$ was proved to be of higher order, see e.g. [54], it can be neglected and the left inequality holds. Furthermore the continuity (9.3) implies

$$\begin{aligned}
Res_h(v) &= \lambda_h(u_h, v)_U - a(u_h, v) \\
&= \lambda_h(u_h, v)_U - a(u_h, v) + a(u, v) - \lambda(u, v)_U \\
&= a(u-u_h, v) + (\lambda_u u_h - \lambda u, v)_U
\end{aligned}$$

$$= a(u - u_h, v) + \frac{(\lambda + \lambda_h)}{2}\|v\|_U^2$$

$$\leq \beta\|u - u_h\|_V\|v\|_V + \frac{(\lambda + \lambda_h)}{2}\|v\|_U^2 \tag{9.17}$$

and therefore

$$\|Res_h\|_{V*} = \frac{Res_h(v)}{\|v\|_V} \leq \beta\|u - u_h\|_V + \frac{(\lambda + \lambda_h)}{2}\frac{\|v\|_U^2}{\|v\|_V},$$

which completes the proof.

Theorem 4 proves that the dual norm of the residual, i.e., $\|Res_h\|_{V*}$ is equivalent to the error $\|u - u_h\|_V$. Nevertheless, it is still a challenge to estimate the dual norm of the residual $Res_h(v)$ in the most reliable and efficient way.

Remark 5 Since the standard norm $\|\cdot\|_V$ in V is equivalent to the energy norm $|||\cdot|||$, see Remark 1, the dual norm of the residual $\|Res_h\|_{V*}$ is also equivalent to the energy norm of the error, i.e., $|||u - u_h|||$.

Now, exploiting the variational eigenvalue problem (9.2) and its Galerkin discretization (9.8) it is easy to derive a simple residual type a posteriori estimator

$$\eta_{res,h} \equiv \left(\sum_{T\in\mathcal{T}_h} h_T^2 \|\Delta u_h + \lambda_h u_h\|_{L^2(T)}^2 + \sum_{E\in\mathcal{E}_h} h_E \| [\nabla u_h \cdot \mathbf{n}_E] \|_{L^2(E)}^2 \right)^{1/2},$$

$h_T := diam(T)$ and $h_E := length(E)$, such that

$$|||u - u_h|||^2 \leq C\,\eta_{res,h}|||u - u_h||| + \frac{\lambda + \lambda_h}{2}\|u - u_h\|_{L^2(\Omega)}^2, \tag{9.18}$$

see [27, Section 6.3], [54, Theorem 3.1], [137, Section 4], or the earlier work of Larson [92]. Here, the constant $C > 0$ depends on the minimal angle allowed in the mesh elements, on the Poincaré-Friedrichs inequality constant (which is a function of the volume of Ω and the area of the portion of $\partial\Omega$ corresponding to the Dirichlet condition, see [135, p. 11]) and on the constants α, β from the Lax-Milgram conditions (9.3) and (9.4), see [95]. However, the possibly large value of C can produce a significant overestimate of the error, see e.g. [36, 46].

In (9.18) the energy norm of the error is bounded by the sum of local contributions of the interior (volumetric) element residuals $\Delta u_h + \lambda_h u_h$, measuring how good the finite element approximations λ_h, u_h satisfy the original PDE in its strong form on the interior of the domain, and of the edge residuals, the jumps of the gradient of u_h over the element edges E, reflecting the accuracy of the approximation [31, 135]. Here h_T, h_E denote the mesh-depending weights and $\|\cdot\|_{L^2(T)}$, $\|\cdot\|_{L^2(E)}$ the problem dependent, local norms.

As it was shown in [54] that the L^2-norm of the error is of higher order than the energy norm of the error (9.18) represents an a posteriori estimate for the energy norm of the error. This residual a posteriori error estimator is reliable in a weaker form with the constant C in, e.g. [36, 37, 134], and it is locally efficient, see e.g. [4, 135]. The asymptotic exactness of the estimator usually does not hold in practical computations. Many interesting results on residual type a posteriori error estimates for eigenvalue/eigenvector approximations were proposed in the last two decades, see e.g. [74, 92, 108, 115, 137], to mention only few.

The residual a posteriori estimators, though well-understood and well-established in practice, may significantly overestimate the actual error due to the possibly very large constant C present in the bound. Therefore, several other techniques, which we now briefly review, were introduced over the last years, see, e.g. [4, 16, 47, 69, 136].

9.2.2.2 Local Problem-Based Estimators (Implicit Estimators)

In the case of explicit error estimators all information about the total error is obtained only from the computed approximation. The main idea behind implicit estimators is to enrich these information by solving some supplementary problem, e.g., local analogues of the residual equations. In order to capture the local behavior of the solution and to get accurate information about the error, the local problems usually involve only small subdomains of Ω (subdomain or element residual method) and more accurate finite element spaces, see [69] for more details. In terms of complexity the solution of all local problems should cost less than assembling the stiffness matrix of the original discrete problem. Implicit error estimators for boundary value problems, e.g. partition of unity or equilibration estimators, are discussed in [4, Chapter 3], [47, Section 6.3.2], [66, Section 15.3], [135, Section 1.3] and [3, 31, 42]. A proof of the local efficiency for this type of estimator can be found, e.g., in [4], whereas reliability and asymptotic exactness are usually satisfied in practical computations. To the best of the author's knowledge there are no local problem-based error estimators designed specifically for eigenvalue problems.

9.2.2.3 Averaging Estimators (Recovery-Based Estimators)

These error estimators, also known as ZZ-estimators [145], exploit a local extrapolation or averaging techniques. Due to the high accuracy, practical effectiveness and robustness they are widely used in engineering applications. In general, the error of the approximation is controlled by a difference of a low-order approximation (e.g., a piecewise constant function) and a finite element solution obtained in the space of higher-order elements (e.g., globally continuous piecewise linear functions) satisfying more restrictive continuity conditions than the approximation itself e.g. [4, Chapter 4], [47, Section 6.3.3] or [135, Section 1.5]. For example, if a quantity to be recovered is a gradient ∇u_h, the main idea is to compare the smoothed and unsmoothed gradients to estimate the actual error. Reference [34] gives a nice overview of averaging techniques in a posteriori finite element error analysis in general, whereas [16, 143, 144, 146, 147] discuss the gradient recovery-based estimators in details. A local averaging technique for eigenvalue problems was proposed in [97]. Here, we present an improved averaging a posteriori error estimator neglecting the volumetric contributions introduced in [38].

Let $\mathscr{A}_h : L^2(\Omega) \to \mathscr{S}^1(\mathscr{T}_h)$, $\mathscr{S}^1(\mathscr{T}_h) := \mathbb{P}_1 \cap H_0^1(\Omega)$ be a local averaging operator, i.e.,

$$\mathscr{A}_h(v) := \sum_{z \in \mathscr{N}_h} \frac{1}{|\omega_z|} \Big(\int_{\omega_z} v \, dx \Big) \varphi_z,$$

with a nodal hat function φ_z and a nodal patch ω_z associated with node z. Then the averaging error estimator for problem (9.2) reads

$$\eta_{avg,h}^2 := \sum_{T \in \mathscr{T}_h} \| \mathscr{A}_h(\nabla u_h) - \nabla u_h \|_{L^2(T)}^2. \tag{9.19}$$

The reliability of the averaging error estimator (9.19) is proved in [38], whereas the efficiency follows from the fact that the averaging estimator is locally equivalent to the residual estimator [33, 34, 135]. The proof of the asymptotic exactness can be found, e.g., in [4, 135]. More details on recovery-based a posteriori error estimators for higher-order polynomials can be found in [35, Section 9.4]. Recovery type a posteriori error estimates for the eigenvalues and eigenfunctions of selfadjoint elliptic equations by the projection method are derived in [104, 139] and [96] for conforming and nonconforming finite elements, respectively.

9.2.2.4 Hierarchical Estimators (Multilevel Estimators)

The main idea of a hierarchical error estimator is to evaluate the residual obtained for the finite element solution $u_h \in V_h$ with respect to another finite element space $V_{h'}$ satisfying $V_h \subset V_{h'} \subset V$. Then the error $|||u - u_h|||$ can be bounded by

$$\eta_{hie,h} := |||u_{h'} - u_h|||,$$

where $u_{h'} \in V_{h'}$, see [20, 52, 58], [47, Chapter 6], or [135, Section 1.4] for details. The finite element space $V_{h'}$ corresponds usually to a refinement $\mathscr{T}_{h'}$ of $\mathscr{T}_h$ or consists of higher-order finite elements. The idea behind goes back to a so-called *saturation assumption* [21] stating that the error of a fine discrete solution $u_{h'}$ is supposed to be smaller than the error of the coarse solution u_h in the sense of an error reduction property, i.e.,

$$|||u_{h'} - u||| \leq \sigma |||u_h - u|||,$$

where $\sigma \in (0,1)$. Good general references concerning hierarchical estimators are [20, 21, 52, 58]. Hierarchical error estimators for eigenvalue problems are discussed in [98, 108].

9.2.2.5 Goal-Oriented Estimators

The objective in goal-oriented error estimation is to determine the accuracy of the finite element solution u_h with respect to some physically relevant scalar quantity of interest given as a linear functional $J(\cdot) : V \to \mathbb{R}$ of the solution u, e.g. velocity, flow rates, deformations, stresses or lifts and drags in the case of Navier-Stokes problems etc. The error in the quantity of interest is then related to the residual, i.e.,

$$\eta_h := |J(u) - J(u_h)| \approx \sum_{T \in \mathscr{T}_h} \rho_T(u_h)\omega_T,$$

where $\rho_T(u_h)$ denotes the so-called "cell residuals" of the approximate solution, and ω_T a corresponding "cell weights". The latter are obtained from the solution u^* of the so-called *dual problem*, which, in practice, is replaced by its locally postprocessed discrete approximation u_h^*. In order to make this abstract concept a little more explicit, for a simple boundary value problem $\mathscr{L}u = f$ the cell residuals read

$$\|\mathscr{L}u_h - f\|_{L^2(T)} + h_T^{1/2}\|[\nabla u_h \cdot \mathbf{n}_E]\|_{L^2(\partial T)},$$

with ∂T being the boundary of an element $T \in \mathscr{T}_h$. Probably, one of the most well-known techniques of goal-oriented error estimation is the *Dual Weighted Residual (DWR)* method introduced in [119]. The reliability, efficiency and asymptotic exactness of goal-oriented estimators are typically hard to prove, however, they are very successful in many challenging practical applications. For eigenvalue problems, the full potential of a goal-oriented a posteriori error estimation was demonstrated in [81] as a successful application of the DWR method to non-selfadjoint operators. For eigenvalues λ and λ_h being simple, the DWR a posteriori error estimator of the following form is proposed

$$|\lambda - \lambda_h| \leq c \sum_{T \in \mathscr{T}_h} h_T^2 \Big(\rho_T(\lambda_h, u_h) + \rho_T^*(\lambda_h^*, u_h^*)\Big),$$

where $\rho_T(\lambda_h, u_h)$ and $\rho_T(\lambda_h^*, u_h^*)$ denote cell residuals of the primal and dual problem, respectively. See [18, Chapter 7], [81, 116] for more details.

9.3 Adaptive Finite Element Eigenvalue Solver

The choice of a proper iterative eigenvalue solver is an integral part of the successful adaptive finite element scheme. We present some well-established iterative methods which admit the quasi-optimal computational complexity on uniform meshes. However, since generated meshes are refined adaptively, there is an increasing demand for designing efficient and reliable matrix-free eigensolvers with mesh size independent convergence rates. We will discuss this issue, in more details, in the following sections.

Let us consider the generalized algebraic eigenvalue problem

$$Ax = \lambda Bx \tag{9.20}$$

resulting from the finite element discretization of (9.8), namely,

$$A = A_h, \quad B = B_h \quad \text{and} \quad x = \mathbf{u}_h,$$

defined as in (9.9).

9.3.1 *PINVIT*

The *preconditioned inverse iteration* (PINVIT), introduced and analyzed in series of papers [89, 105, 106, 109], is an iterative method for solving the generalized eigenvalue problem (9.20) written as a system of linear equations $Ax_{k+1} = \lambda(x_k)Bx_k$, where the new iterate x_{k+1} is determined as

$$x_{k+1} = x_k - M^{-1}\big(Ax_k - \lambda(x_k)Bx_k\big),$$

with a symmetric and positive definite optimally scaled preconditioner (e.g., multigrid preconditioner) M^{-1} of the matrix A such that

$$\|I - M^{-1}A\|_A \leq \gamma, \quad \gamma \in [0, 1).$$

The corresponding error propagation equation

$$x_{k+1} - \lambda(x_k)A^{-1}Bx_k = (I - M^{-1}A)\big(x_k - \lambda(x_k)A^{-1}Bx_k\big),$$

not only illustrates the dependence between the initial error $x_k - \lambda(x_k)A^{-1}Bx_k$, the new iterate error $x_{k+1} - \lambda(x_k)A^{-1}Bx_k$ and the error propagation matrix (reducer) $I - M^{-1}A$, but presents a more general relation between preconditioned gradient eigensolvers [68, 126] and preconditioned inverse iteration. PINVIT can be viewed as a counterpart of multigrid algorithms for the solution of boundary value problems.

As a simple mesh-free eigensolver, with convergence independent on the largest eigenvalue and the mesh size, it is perfectly suitable for grid-dependent eigenvalue problems. More details about the subspace version of the method can be found in [30, 107]. A continuous counterpart of the preconditioned inverse iteration was proposed and analyzed in [142].

9.3.2 LO(B)PCG

Let us assume that we are given the smallest eigenvalue λ_1 of our problem. Then obtaining the corresponding eigenvector requires solving the homogeneous linear system $(A - \lambda_1 B)x_1 = 0$. The method of choice in this case would be a *(preconditioned) conjugate gradient* ((P)CG) method [28]. Though, in practice, the exact eigenvalue is not known, the underlying idea is still useful and can be combined with the standard *preconditioned steepest descent* (PSD) method [78, 126]. A sharp convergence estimate and a subspace variant of PSD combined with AFEM are discussed in [110, 111].

The *Locally Optimal (Block) Preconditioned Conjugate Gradient* (LO(B) PSCG) [86] method combines a three-term recurrence method with the robust and simple Rayleigh-Ritz minimization procedure which allows (allowing) efficient solutions of large and sparse eigenvalue problems. The main idea of the method is to determine a new eigenvalue/eigenvector approximation as the smallest Ritz value/vector with respect to the three-dimensional space $\mathrm{span}\{x_k,\ M^{-1}(Ax_k - \lambda(x_k)Bx_k),\ x_{k-1}\}$. The new iterate x_{k+1} is now determined as

$$x_{k+1} = x_k - \vartheta_k x_{k-1} - \xi_k M^{-1}(Ax_k - \lambda(x_k)Bx_k),$$

where

$$(\vartheta_k, \xi_k) = \arg\min_{(\vartheta,\xi)} \lambda(x_k - \vartheta x_{k-1} - \xi M^{-1}(Ax_k - \lambda(x_k)Bx_k)).$$

It is important to notice that the preconditioner is not used in inner iterations to solve the linear system, but it is directly integrated into a Krylov-based iteration. LO(B)PCG is broadly used eigensolver within the AFEM due to its low memory requirements (only one additional vector has to be stored), reasonable complexity (few additional inner products to determine the Rayleigh-Ritz projection) and its convergence. On every step, the LO(B)PSCG is not slower than the preconditioned steepest descent in terms of the maximizing the Rayleigh quotient [84], however, in practice the convergence is much faster and robust than PSD or Jacobi-Davidson. A commonly used implementation of the method, released by its developer under GNU Lesser GPL, together with several benchmark model problems, is available as *Block Locally Optimal Preconditioned Eigenvalue Xolver* (BLOPEX) [88].

9.3.3 *Two-Grid Discretization*

Already in 1979, the idea of using the multigrid-method for solving mesh eigenproblems was introduced in [76]. A simple *one-stage method* requires computations of one eigenpair on the coarse grid and approximates further fine grid eigenpairs in a recursive way. Its computational effort is proportional to the dimension of the finite dimensional space and convergence is proved also for the approximate eigenpairs. A well-known example of the class of multigrid approaches is the *two-grid discretization* method introduced in [140, 141]. The idea of the method is quite simple and uses the underlying expertise from the study of boundary value problems. In particular, we consider to linear finite element spaces $V_H(\Omega) \subset V_h(\Omega) \subset H_0^1(\Omega)$, e.g., coarse and fine space, respectively. The solution of an eigenvalue problem on a fine grid is reduced to the solution of an eigenvalue problem on a much coarser grid (mesh size H) followed by the solution of a boundary value problem on the fine grid (mesh size h), whereas the resulting solution maintains an asymptotically optimal accuracy for $H = \mathcal{O}(\sqrt{h})$. We can summarize the method within three steps:

Step 1 Find $(\lambda_H, u_H) \in (\mathbb{R}, V_H)$ s.t. $a(u_H, v_H) = \lambda_H (u_H, v_H)_U$, for all $v_H \in V_H$.
Step 2 Find $u_h \in V_h$ s.t. $a(u_h, v_h) = \lambda_H (u_H, v_h)_U$, for all $v_h \in V_h$.
Step 3 Compute λ_h as Rayleigh quotient of u_h.

In other words the method can be reformulated as finding a correction e_h in the fine space, such that

$$a(e_h, v_h) = \lambda_H (u_H, v_h)_U - a(u_H, v_h) \text{ for all } v_h \in V_h$$

and setting

$$u_h = u_H + e_h.$$

9.4 Convergence and Optimality Results

In the classical sense the convergence of the FEM requires that for each value $h \to 0$ the approximation error is of required order or accuracy. When dealing with AFEM the goal is to show that the method is a contraction between two consecutive loops. The algebraic convergence rates for adaptive FEM, under the assumption of the exact solution of the algebraic eigenvalue problem, were first proved in [103] and later on improved in [129]. A first convergence results for adaptive finite element methods for eigenvalue problems have been obtained in [64]. Assuming a sufficiently fine initial mesh, Dörfler's strategy for marking separately error and oscillation indicators, and enforcing the interior node property, the authors proved an error reduction result for consecutive iterates, which is essential for proving quasi-optimality, but very hard to satisfy in practice. Uniform convergence and optimal complexity, relaxing the assumptions of [64], was introduced in [51]. In order to prove convergence, marking of oscillation terms is not required. Moreover, the optimal complexity was shown without any additional assumptions on the data. At the same time, an independent adaptive finite element eigenvalue solver (AFEMES) enabling a contraction property up to higher-order terms (also known as Q-linear convergence) and global strong convergence, was proposed in [38]. Also this result requires no assumptions on the inner node property and small enough mesh size. Furthermore, the same authors provided the first adaptive finite element method combined with an iterative algebraic eigenvalue solver of asymptotic quasi-optimal computational complexity [39]. Another important contribution to be mentioned is the convergence result given in [62]. Here, despite less restrictive initial assumptions (any initial triangulation and marking strategy is allowed) and only minimal refinement of marked elements, the convergence was proved for simple as well as for the multiple eigenvalues. A recent article [35] presents a general framework on optimality of adaptive schemes covering linear as well as nonlinear problems, which embeds the previous results of [38, 51]. The authors consider optimality and convergence of the adaptive algorithm with an optimal convergence rate guaranteed by the efficiency of the error estimator η_h, see [35, Theorem 4.5]. In particular, in the case of determining a simple eigenvalue, following [35, Lemma 3.4 and Proposition 10.5], [39, Lemma 4.2] and the convergence of the conforming finite element discretization [130], one can prove that the following four properties,

(A1) Stability on non-refined elements,
(A2) Reduction property on refined elements,
(A3) General quasi-orthogonality,
(A4) Discrete reliability of the error estimator;

together with sufficiently small initial mesh size h_0, are sufficient for optimal convergence of an adaptive scheme. Finally, in conclusion, we point the reader to [39, 60, 61] for more results on clustered eigenvalues, nonconforming AFEM, inexact solves and algorithms of optimal computational complexity.

9.5 The Role of Linear Algebra in AFEM for PDE Eigenvalue Problems

In this section we would like to point the attention of the reader to a very important, though commonly neglected, aspect of practical realization of adaptive FEM. The majority of the AFEM publications consider *exact* solutions of the algebraic problems (linear systems or eigenvalue problems). When the cost for solving these problems is small and the problems itself are well conditioned independently of the mesh refinement, see [19] and [32, Section 9.5], this assumption is acceptable.

However, in real-world applications, adaptive finite element methods are used for challenging, very large and often ill-conditioned problems, for which an exact (up to machine precision) solution is not available. Notice that even a small algebraic residual does not guarantee a good accuracy of the resulting solution, neither for linear systems nor eigenvalue problems. We refer to [79], [5, Section 4] for more details. Moreover, solving the linear algebraic problems to a (much) higher accuracy than the order of the discretization error not only does not improve the overall accuracy but also significantly increases the computational cost [66, Section 13.4.1].

Because of these reasons, in the following, we will advocate for considering the algebraic error as an integral part of the adaptive FEM, especially, in practical applications. Hence, when estimating the total error we will aim at estimates of the form

$$\|u - u_h^{(n)}\| \approx \eta_{h,n}, \tag{9.21}$$

where $\eta_{h,n}$ is a function of the approximate solution $u_h^{(n)}$ (or $\lambda_h^{(n)}$ and $u_h^{(n)}$) of the linear algebraic problem. Moreover, the fact that the algebraic problems are not solved exactly (and the Galerkin orthogonality does not hold when u_h is replaced by $u_h^{(n)}$) should be also taken into account in the derivation of *all* a posteriori error estimators discussed in Sect. 9.2.2.

A pioneering work in this direction was published in 1995 by Becker, Johnson, and Rannacher [22]. Although dedicated to boundary value problems, it proposes a posteriori error estimates in the $H^1(\Omega)$- and $L^2(\Omega)$-norms that incorporate algebraic errors and design of the adaptive algorithm, and they suggest stopping criterion for the multigrid computations. Several aspects concerning the interplay between discretization and algebraic computation in adaptive FEM are discussed in a recent survey [6].

Now, at step SOLVE of the adaptive FEM applied to problem (9.2), the generalized eigenvalue problem is solved inexactly and we obtain an eigenvector approximation $\mathbf{u}_h^{(n)}$ and a corresponding eigenvalue approximation $\lambda_h^{(n)}$, associated with the following *algebraic errors*

$$\mathbf{u}_h - \mathbf{u}_h^{(n)} \in \mathbb{R}^{n_h} \quad \text{or} \quad u_h - u_h^{(n)} \in V_h, \quad \text{and} \quad \lambda_h - \lambda_h^{(n)}.$$

The *total errors* are then given as a sum of the discretization and the algebraic error, i.e.,

$$u - u_h^{(n)} = (u - u_h) + (u_h - u_h^{(n)}) \quad \text{and} \tag{9.22}$$

$$\lambda - \lambda_h^{(n)} = (\lambda - \lambda_h) + (\lambda_h - \lambda_h^{(n)}). \tag{9.23}$$

For boundary value problems minimizing the total error can be achieved by applying the CG method, which naturally minimizes the algebraic energy norm of the error. However, the same task is much more complicated in the case of eigenvalue problems which, by their nature, are nonlinear. Even the definition of an appropriate (in the physical sense) norm to measure the error for the eigenvalue problem is not trivial and still under intensive consideration, see [80].

In [98], exploiting backward error analysis and saturation assumption, the authors introduce a residual a posteriori error estimators for total errors (9.22) and (9.23) and develop an adaptive FEM, called AFEMLA (LA standing for linear algebra), which incorporates the inexact iterative eigensolver, i.e., the Lanczos method. In particular, this new approach allows for mesh-free adaptation, which is of great interest in the context of the *discrete finite element modeling* [99], being known in engineering practice for decades.

The concept of a *functional backward error and condition number* introduced in [7] for boundary value problems is used again in [101] for selfadjoint eigenvalue problems in order to analyze the continuous dependence of the inexact solution on the data, in particular to analyze the approximation error and the backward stability of the algebraic eigenvalue problem. This resulted in a combined residual a posteriori error estimator and a balanced AFEM algorithm, where the stopping criteria are based on the variant of the shift-invert Lanczos method introduced in [80]. A similar direction was considered in [70] in the context of bound-constrained optimization; the ideas introduced there can be applied to the minimization of the Rayleigh-quotient in the case of eigenvalue computations.

When dealing with inexact AFEM, issues such as convergence and optimality are of even greater interest. The convergence of the perturbed preconditioned inverse iteration (PPINVIT), see Sect. 9.3.1, i.e., an algorithm in which the application of the operator is performed approximately, was proved in [24] with bounds for the convergence rate depending on the eigenvalue gap and the quality of the preconditioner. Regarding the optimality of AFEM for eigenvalue problems, in [124] the authors exploited the theory of *best N-term approximation*. Namely, the number of degrees of freedom needed to obtain the AFEM solution of a given accuracy should be proportional to the number of degrees of freedom needed to approximate the exact solution up to the same accuracy. Under the assumption that the iteration error $|||u_h - u_h^{(n)}|||^2 + |\lambda_h - \lambda_h^{(n)}|$ for two consecutive AFEM steps is small in comparison with the size of the residual a posteriori error estimate quasi-optimality of the inexact inverse iteration coupled with adaptive finite element method (AFEM) for a class of elliptic eigenvalue problems was proved in [39]. Moreover, the proposed method admits also a quasi-optimal complexity. A similar analysis of convergence and a quasi-optimality of the inexact inverse iteration coupled with adaptive finite element methods was presented in [142] for operator eigenvalue problem.

The aforementioned results have been derived in the context of selfadjoint eigenvalue problems. To deal with non-selfadjoint problems, one can follow results in [23, 100] and their DWR approach. Here duality techniques are used to estimate the error in the target quantities in terms of the weighted primal and dual residuals, i.e.,

$$Res_h(u_h, \lambda_h)(v) \equiv \lambda_h (u_h, v)_U - a(u_h, v), \tag{9.24}$$

$$Res_h^*(u_h^*, \lambda_h^*)(v) \equiv \lambda_h^* (v, u_h^*)_U - a(v, u_h^*), \tag{9.25}$$

respectively. The resulting estimates, based on a perturbation argument, can be written as

$$\lambda - \lambda_h^{(n)} \lesssim \left(\eta_{h,n} + \eta_{h,n}^* + \eta_{h,n}^{(it)} \right), \tag{9.26}$$

with the primal and the dual eigenvalue residual estimators

$$\eta_{h,n} \equiv \frac{1}{2} Res_h(u_h^{(n)}, \lambda_h^{(n)})(I_{2h}^{(2)} u_h^{*(n+1)} - u_h^{*(n)}), \tag{9.27}$$

$$\eta_{h,n}^* \equiv \frac{1}{2} Res_h^*(u_h^{*(n)}, \lambda_h^{*(n)})(I_{2h}^{(2)} u_h^{(n+1)} - u_h^{(n)}), \tag{9.28}$$

the iteration error indicator

$$\eta_{h,n}^{(it)} = Res_h(u_h^{(n)}, \lambda_h^{(n)})(u_h^{*(n)}), \tag{9.29}$$

and the interpolation operator $I_{2h}^{(2)}$. For more details we refer to [120]. Another approach, based on a *homotopy method* which allows adaptivity in space, in the homotopy step-size as well as in the stopping criteria for the iterative algebraic eigenvalue solvers has been derived in [41], see also [63, 102].

9.6 Concluding Remarks

This short survey gives a very brief introduction to the adaptive approximation of PDE eigenvalue problems, but it is far away from being complete in any sense. At this point, we excuse for any missing contributions about whose existence we were not aware in the time of preparation of this manuscript. Due to the lack in space, we mentioned only shortly some results on non-selfadjoint eigenvalue problems, and did not consider at all a very important class of nonlinear eigenvalue problems. As our study on adaptive FEM has no end, we will leave the reader with their own thoughts, questions and ideas to contemplate. There are still many doors to be open and we encourage researchers from many fields such as mathematical and numerical PDE analysis and discretization, functional analysis and matrix computations to write further chapters of this wonderful story.

Acknowledgements The author would like to thank Federico Poloni for careful reading of the paper and providing valuable comments. The work of the author has been supported by the DFG Research Fellowship under the DFG GEPRIS Project *Adaptive methods for nonlinear eigenvalue problems with parameters* and Chair of Numerical Algorithms and High-Performance Computing, Mathematics Institute of Computational Science and Engineering (MATHICSE), École Polytechnique Fédérale de Lausanne, Switzerland.

References

1. Agmon, S., Douglis, A., Nirenberg, L.: Estimates near the boundary for solutions of elliptic partial differential equations satisfying general boundary conditions. I. Commun. Pure Appl. Math. **12**, 623–727 (1959)
2. Agmon, S., Douglis, A., Nirenberg, L.: Estimates near the boundary for solutions of elliptic partial differential equations satisfying general boundary conditions. II. Commun. Pure Appl. Math. **17**, 35–92 (1964)
3. Ainsworth, M., Oden, J.T.: A posteriori error estimators for second order elliptic systems. II. An optimal order process for calculating self-equilibrating fluxes. Comput. Math. Appl. **26**(9), 75–87 (1993)
4. Ainsworth, M., Oden, J.T.: A Posteriori Error Estimation in Finite Element Analysis. Pure and Applied Mathematics. Wiley-Interscience [Wiley], New York (2000)
5. Arioli, M.: A stopping criterion for the conjugate gradient algorithm in a finite element method framework. Numer. Math. **97**(1), 1–24 (2004)
6. Arioli, M., Liesen, J., Międlar, A., Strakoš, Z.: Interplay between discretization and algebraic computation in adaptive numerical solution of elliptic PDE problems. GAMM-Mitt. **36**(1), 102–129 (2013)
7. Arioli, M., Noulard, E., Russo, A.: Stopping criteria for iterative methods: applications to PDE's. Calcolo **38**(2), 97–112 (2001)
8. Armentano, M.G., Durán, R.G.: Asymptotic lower bounds for eigenvalues by nonconforming finite element methods. Electron. Trans. Numer. Anal. **17**, 93–101 (electronic) (2004)
9. Arnold, D.N., Babuška, I., Osborn, J.: Selection of finite element methods. In: Atluri, S.N., Gallagher, R.H., Zienkiewicz, O.C. (eds.) Hybrid and Mixed Finite Element Methods (Atlanta, 1981), chapter 22, pp. 433–451. Wiley-Interscience [Wiley], New York (1983)
10. Babuška, I.: Error-bounds for finite element method. Numer. Math. **16**, 322–333 (1970/1971)
11. Babuška, I., Aziz, A.K.: Survey lectures on the mathematical foundations of the finite element method. In: The Mathematical Foundations of the Finite Element Method with Applications to Partial Differential Equations (Proceedings of a Symposium Held at the University of Maryland, Baltimore, 1972), pp. 1–359. Academic, New York (1972). With the collaboration of G. Fix and R. B. Kellogg
12. Babuška, I., Osborn, J.E.: Estimates for the errors in eigenvalue and eigenvector approximation by Galerkin methods, with particular attention to the case of multiple eigenvalues. SIAM J. Numer. Anal. **24**(6), 1249–1276 (1987)
13. Babuška, I., Osborn, J.E.: Finite element-Galerkin approximation of the eigenvalues and eigenvectors of selfadjoint problems. Math. Comput. **52**(186), 275–297 (1989)
14. Babuška, I., Osborn, J.E.: Eigenvalue Problems. Volume II of Handbook of Numerical Analysis. North Holland, Amsterdam (1991)
15. Babuška, I., Rheinboldt, W.C.: Error estimates for adaptive finite element computations. SIAM J. Numer. Anal. **15**(4), 736–754 (1978)
16. Babuška, I., Strouboulis, T.: The Finite Element Method and Its Reliability. Numerical Mathematics and Scientific Computation. Oxford University Press, New York (2001)
17. Babuška, I., Whiteman, J.R., Strouboulis, T.: Finite Elements – An Introduction to the Method and Error Estimation. Oxford Press, New York (2011)

18. Bangerth, W., Rannacher, R.: Adaptive Finite Element Methods for Differential Equations. Lectures in Mathematics ETH Zürich. Birkhäuser Verlag, Basel (2003)
19. Bank, R.E., Scott, L.R.: On the conditioning of finite element equations with highly refined meshes. SIAM J. Numer. Anal. **26**(6), 1383–1394 (1989)
20. Bank, R.E., Smith, R.K.: A posteriori error estimates based on hierarchical bases. SIAM J. Numer. Anal. **30**(4), 921–935 (1993)
21. Bank, R.E., Weiser, A.: Some a posteriori error estimators for elliptic partial differential equations. Math. Comput. **44**(170), 283–301 (1985)
22. Becker, R., Johnson, C., Rannacher, R.: Adaptive error control for multigrid finite element methods. Computing **55**(4), 271–288 (1995)
23. Becker, R., Rannacher, R.: An optimal control approach to a posteriori error estimation in finite element methods. Acta Numer. **10**, 1–102 (2001)
24. Binev, P., Dahmen, W., DeVore, R.: Adaptive finite element methods with convergence rates. Numer. Math. **97**(2), 219–268 (2004)
25. Boffi, D.: Finite element approximation of eigenvalue problems. Acta Numer. **19**, 1–120 (2010)
26. Boffi, D., Brezzi, F., Gastaldi, L.: On the problem of spurious eigenvalues in the approximation of linear elliptic problems in mixed form. Math. Comput. **69**(229), 121–140 (2000)
27. Boffi, D., Gardini, F., Gastaldi, L.: Some remarks on eigenvalue approximation by finite elements. In: Blowey, J., Jensen, M. (eds.) Frontiers in Numerical Analysis – Durham 2010. Volume 85 of Springer Lecture Notes in Computational Science and Engineering, pp. 1–77. Springer, Heidelberg (2012)
28. Bradbury, W.W., Fletcher, R.: New iterative methods for solution of the eigenproblem. Numer. Math. **9**, 259–267 (1966)
29. Braess, D.: Finite Elements, 3rd edn. Cambridge University Press, Cambridge (2007). Theory, Fast Solvers, and Applications in Elasticity Theory. Translated from the German by Larry L. Schumaker
30. Bramble, J.H., Pasciak, J.E., Knyazev, A.V.: A subspace preconditioning algorithm for eigenvector/eigenvalue computation. Adv. Comput. Math. **6**(2), 159–189 (1997) (1996)
31. Brenner, S.C., Carstensen, C.: Finite element methods. In: Stein, E., de Borst, R., Huges, T.J.R. (eds.) Encyclopedia of Computational Mechanics, vol. I, pp. 73–114. Wiley, New York (2004)
32. Brenner, S.C., Scott, L.R.: The Mathematical Theory of Finite Element Methods. Volume 15 of Texts in Applied Mathematics, 3rd edn. Springer, New York (2008)
33. Carstensen, C.: All first-order averaging techniques for a posteriori finite element error control on unstructured grids are efficient and reliable. Math. Comput. **73**(247), 1153–1165 (electronic) (2004)
34. Carstensen, C.: Some remarks on the history and future of averaging techniques in a posteriori finite element error analysis. ZAMM Z. Angew. Math. Mech. **84**(1), 3–21 (2004)
35. Carstensen, C., Feischl, M., Page, M., Praetorius, D.: Axioms of adaptivity. Comput. Math. Appl. **67**(6), 1195–1253 (2014)
36. Carstensen, C., Funken, S.A.: Constants in Clément-interpolation error and residual based a posteriori error estimates in finite element methods. East-West J. Numer. Math. **8**(3), 153–175 (2000)
37. Carstensen, C., Funken, S.A.: Fully reliable localized error control in the FEM. SIAM J. Sci. Comput. **21**(4), 1465–1484 (2000)
38. Carstensen, C., Gedicke, J.: An oscillation-free adaptive FEM for symmetric eigenvalue problems. Numer. Math. **118**(3), 401–427 (2011)
39. Carstensen, C., Gedicke, J.: An adaptive finite element eigenvalue solver of asymptotic quasi-optimal computational complexity. SIAM J. Numer. Anal. **50**(3), 1029–1057 (2012)
40. Carstensen, C., Gedicke, J.: Guaranteed lower bounds for eigenvalues. Math. Comput. **83**(290), 2605–2629 (2014)
41. Carstensen, C., Gedicke, J., Mehrmann, V., Międlar, A.: An adaptive homotopy approach for non-selfadjoint eigenvalue problems. Numer. Math. **119**(3), 557–583 (2011)

42. Carstensen, C., Merdon, C.: Estimator competition for Poisson problems. J. Comput. Math. **28**(3), 309–330 (2010)
43. Chatelin, F.: Spectral Approximation of Linear Operators. Computer Science and Applied Mathematics. Academic [Harcourt Brace Jovanovich Publishers], New York (1983). With a foreword by P. Henrici, With solutions to exercises by Mario Ahués
44. Chatelin, F.: Spectral Approximation of Linear Operators. Volume 65 of Classics in Applied Mathematics. Society for Industrial and Applied Mathematics (SIAM), Philadelphia (2011). Reprint of the 1983 original (Academic, New York)
45. Chatelin, F., Lemordant, M.J.: La méthode de Rayleigh-Ritz appliquée à des opérateurs différentiels elliptiques—ordres de convergence des éléments propres. Numer. Math. **23**, 215–222 (1974/1975)
46. Cheddadi, I., Fučík, R., Prieto, M.I., Vohralík, M.: Computable a posteriori error estimates in the finite element method based on its local conservativity: improvements using local minimization. In: Dobrzynski, C., Frey, P., Pebay, Ph. (eds.) Pre and Post Processing in Scientific Computing (CEMRACS 2007), Luminy, 23rd July–31st August, 2007 (2007)
47. Chen, Z.: Finite Element Methods and Their Applications. Scientific Computation. Springer, Berlin (2005)
48. Ciarlet, P.G.: The Finite Element Method for Elliptic Problems. Volume 40 of Classics in Applied Mathematics. Society for Industrial and Applied Mathematics (SIAM), Philadelphia (2002). Reprint of the 1978 original (North-Holland, Amsterdam)
49. Courant, R., Hilbert, D.: Methods of Mathematical Physics, vol. I. Interscience Publishers, New York (1953)
50. Crouzeix, M., Raviart, P.-A.: Conforming and nonconforming finite element methods for solving the stationary Stokes equations. I. Rev. Française Automat. Informat. Recherche Opérationnelle Sér. Rouge **7**(R-3), 33–75 (1973)
51. Dai, X., Xu, J., Zhou, A.: Convergence and optimal complexity of adaptive finite element eigenvalue computations. Numer. Math. **110**(3), 313–355 (2008)
52. Deuflhard, P., Leinen, P., Yserentant, H.: Concepts of an adaptive hierarchical finite element code. IMPACT Comput. Sci. Eng. **1**(1), 3–35 (1989)
53. Dörfler, W.: A convergent adaptive algorithm for Poisson's equation. SIAM J. Numer. Anal. **33**(3), 1106–1124 (1996)
54. Durán, R.G., Padra, C., Rodríguez, R.: A posteriori error estimates for the finite element approximation of eigenvalue problems. Math. Models Methods Appl. Sci. **13**(8), 1219–1229 (2003)
55. Elman, H., Silvester, D., Wathen, A.: Finite Elements and Fast Iterative Solvers with Applications in Incompressible Fluid Dynamics. Numerical Mathematics and Scientific Computation, 2nd edn. Oxford University Press, Oxford (2014)
56. Ern, A.A., Guermond, J.-L.: Theory and practice of finite elements. Volume 159 of Applied Mathematical Sciences. Springer, New York (2004)
57. Evans, L.C.: Partial Differential Equations. Volume 19 of Graduate Studies in Mathematics, 2nd edn. American Mathematical Society, Providence (2010)
58. Ferraz-Leite, S., Ortner, C., Praetorius, D.: Convergence of simple adaptive Galerkin schemes based on h-h/2 error estimators. Numer. Math. **116**(2), 291–316 (2010)
59. Fix, G.J.: Eigenvalue approximation by the finite element method. Adv. Math. **10**, 300–316 (1973)
60. Gallistl, D.: Adaptive nonconforming finite element approximation of eigenvalue clusters. Comput. Methods Appl. Math. **14**(4), 509–535 (2014)
61. Gallistl, D.: An optimal adaptive FEM for eigenvalue clusters. Numer. Math. (2014). Accepted for publication
62. Garau, E.M., Morin, P., Zuppa, C.: Convergence of adaptive finite element methods for eigenvalue problems. Math. Models Methods Appl. Sci. **19**(5), 721–747 (2009)
63. Gedicke, J., Carstensen, C.: A posteriori error estimators for convection-diffusion eigenvalue problems. Comput. Methods Appl. Mech. Eng. **268**, 160–177 (2014)

64. Giani, S., Graham, I.G.: A convergent adaptive method for elliptic eigenvalue problems. SIAM J. Numer. Anal. **47**(2), 1067–1091 (2009)
65. Gilbarg, D., Trudinger, N.S.: Elliptic partial differential equations of second order. Classics in Mathematics. Springer, Berlin (2001). Reprint of the 1998 edition
66. Gockenbach, M.S.: Understanding and Implementing the Finite Element Method. Society for Industrial and Applied Mathematics (SIAM), Philadelphia (2006)
67. Gockenbach, M.S.: Partial Differential Equations, 2nd edn. Society for Industrial and Applied Mathematics (SIAM), Philadelphia (2011). Analytical and Numerical Methods
68. Godunov, S.K., Ogneva, V.V., Prokopov, G.P.: On the convergence of the modified method of steepest descent in the calculation of eigenvalues. Am. Math. Soc. Transl. Ser. 2 **105**, 111–116 (1976)
69. Grätsch, T., Bathe, K.-J.: A posteriori error estimation techniques in practical finite element analysis. Comput. Struct. **83**(4–5), 235–265 (2005)
70. Gratton, S., Mouffe, M., Toint, P.L.: Stopping rules and backward error analysis for bound-constrained optimization. Numer. Math. **119**(1), 163–187 (2011)
71. Grebenkov, D.S., Nguyen, B.-T.: Geometrical structure of Laplacian eigenfunctions. SIAM Rev. **55**(4), 601–667 (2013)
72. Grisvard, P.: Elliptic Problems in Nonsmooth Domains. Volume 24 of Monographs and Studies in Mathematics. Pitman (Advanced Publishing Program), Boston (1985)
73. Grossmann, C., Roos, H.-G.: Numerical treatment of partial differential equations. Universitext. Springer, Berlin (2007). Translated and revised from the 3rd (2005) German edition by Martin Stynes
74. Grubišić, L., Ovall, J.S.: On estimators for eigenvalue/eigenvector approximations. Math. Comput. **78**(266), 739–770 (2009)
75. Grubišić, L., Veselić, K.: On Ritz approximations for positive definite operators. I. Theory. Linear Algebra Appl. **417**(2–3), 397–422 (2006)
76. Hackbusch, W.: On the computation of approximate eigenvalues and eigenfunctions of elliptic operators by means of a multi-grid method. SIAM J. Numer. Anal. **16**(2), 201–215 (1979)
77. Hackbusch, W.: Elliptic Differential Equations. Volume 18 of Springer Series in Computational Mathematics. Springer, Berlin (1992). Translated from the author's revision of the 1986 German original by Regine Fadiman and Patrick D. F. Ion
78. Hestenes, M.R., Karush, W.: Solutions of $Ax = \lambda Bx$. J. Res. Nat. Bur. Stand. **47**, 471–478 (1951)
79. Hestenes, M.R., Stiefel, E.: Methods of conjugate gradients for solving linear systems. J. Research Nat. Bur. Stand. **49**, 409–436 (1952)
80. Hetmaniuk, U.L., Lehoucq, R.B.: Uniform accuracy of eigenpairs from a shift-invert Lanczos method. SIAM J. Matrix Anal. Appl. **28**(4), 927–948 (2006)
81. Heuveline, V., Rannacher, R.: A posteriori error control for finite element approximations of elliptic eigenvalue problems. Adv. Comput. Math. **15**(1–4), 107–138 (2001)
82. Johnson, C.: Numerical solution of partial differential equations by the finite element method. Dover, Mineola (2009). Reprint of the 1987 edition
83. Kato, T.: Perturbation Theory for Linear Operators. Classics in Mathematics. Springer, Berlin (1995). Reprint of the 1980 edition
84. Knyazev, A.V.: A preconditioned conjugate gradient method for eigenvalue problems and its implementation in a subspace. In: Eigenwertaufgaben in Natur- und Ingenieurwissenschaften und ihre numerische Behandlung, Oberwolfach 1990. International Series of Numerical Mathematics, pp. 143–154. Birkhäuser, Basel (1991)
85. Knyazev, A.V.: New estimates for Ritz vectors. Math. Comput. **66**(219), 985–995 (1997)
86. Knyazev, A.V.: Toward the optimal preconditioned eigensolver: locally optimal (block) preconditioned conjugate gradient method. SIAM J. Sci. Comput. **23**(2), 517–541 (2001)
87. Knyazev, A.V., Argentati, M.E.: Rayleigh-Ritz majorization error bounds with applications to FEM. SIAM J. Matrix Anal. Appl. **31**(3), 1521–1537 (2009)

88. Knyazev, A.V., Lashuk, I., Argentati, M.E., Ovchinnikov, E.: Block locally optimal preconditioned eigenvalue xolvers (BLOPEX) in hypre and PETSc. SIAM J. Sci. Comput. **25**(5), 2224–2239 (2007)
89. Knyazev, A.V., Neymeyr, K.: A geometric theory for preconditioned inverse iteration. III: A short and sharp convergence estimate for generalized eigenvalue problems. Linear Algebra Appl. **358**, 95–114 (2003)
90. Knyazev, A.V., Osborn, J.E.: New a priori FEM error estimates for eigenvalues. SIAM J. Numer. Anal. **43**(6), 2647–2667 (2006)
91. Kolata, W.G.: Approximation in variationally posed eigenvalue problems. Numer. Math. **29**(2), 159–171 (1977/1978)
92. Larson, M.G.: A posteriori and a priori error analysis for finite element approximations of self-adjoint elliptic eigenvalue problems. SIAM J. Numer. Anal. **38**(2), 608–625 (2000)
93. Larson, M.G., Bengzon, F.: The Finite Element Method: Theory, Implementation, and Applications. Volume 10 of Texts in Computational Science and Engineering. Springer, Heidelberg (2013)
94. Larsson, S., Thomée, V.: Partial Differential Equations with Numerical Methods. Volume 45 of Texts in Applied Mathematics. Springer, Berlin (2003)
95. Lax, P.D., Milgram, A.N.: Parabolic equations. In: Bers, L., Bochner, S., John, F. (eds.) Contributions to the Theory of Partial Differential Equations. Annals of Mathematics Studies, no. 33, pp. 167–190. Princeton University Press, Princeton (1954)
96. Liu, H., Sun, J.: Recovery type a posteriori estimates and superconvergence for nonconforming FEM of eigenvalue problems. Appl. Math. Model. **33**(8), 3488–3497 (2009)
97. Mao, D., Shen, L., Zhou, A.: Adaptive finite element algorithms for eigenvalue problems based on local averaging type a posteriori error estimates. Adv. Comput. Math. **25**(1–3), 135–160 (2006)
98. Mehrmann, V., Międlar, A.: Adaptive computation of smallest eigenvalues of self-adjoint elliptic partial differential equations. Numer. Linear Algebra Appl. **18**(3), 387–409 (2011)
99. Mehrmann, V., Schröder, C.: Nonlinear eigenvalue and frequency response problems in industrial practice. J. Math. Ind. **1**, Art. 7, 18 (2011)
100. Meidner, D., Rannacher, R., Vihharev, J.: Goal-oriented error control of the iterative solution of finite element equations. J. Numer. Math. **17**(2), 143–172 (2009)
101. Międlar, A.: Functional perturbation results and the balanced AFEM algorithm for self-adjoint PDE eigenvalue problems. Preprint 817, DFG Research Center MATHEON, Berlin (2011)
102. Międlar, A.: Inexact Adaptive Finite Element Methods for Elliptic PDE Eigenvalue Problems. PhD thesis, Technische Universität Berlin (2011)
103. Morin, P., Nochetto, R.H., Siebert, K.G.: Convergence of adaptive finite element methods. SIAM Rev. **44**(4), 631–658 (2002). Revised reprint of "Data oscillation and convergence of adaptive FEM" [SIAM J. Numer. Anal. **38**(2), 466–488 (2000)]
104. Naga, A., Zhang, Z., Zhou, A.: Enhancing eigenvalue approximation by gradient recovery. SIAM J. Sci. Comput. **28**(4), 1289–1300 (electronic) (2006)
105. Neymeyr, K.: A geometric theory for preconditioned inverse iteration. I: extrema of the rayleigh quotient. Linear Algebra Appl. **322**, 61–85 (2001)
106. Neymeyr, K.: A geometric theory for preconditioned inverse iteration. II: convergence estimates. Linear Algebra Appl. **331**, 87–104 (2001)
107. Neymeyr, K.: A geometric theory for preconditioned inverse iteration applied to a subspace. Math. Comput. **71**(237), 197–216 (electronic) (2002)
108. Neymeyr, K.: A posteriori error estimation for elliptic eigenproblems. Numer. Linear Algebra Appl. **9**(4), 263–279 (2002)
109. Neymeyr, K.: A geometric theory for preconditioned inverse iteration. IV: on the fastest converegence cases. Linear Algebra Appl. **415**, 114–139 (2006)
110. Neymeyr, K.: A geometric convergence theory for the preconditioned steepest descent iteration. SIAM J. Numer. Anal. **50**(6), 3188–3207 (2012)
111. Neymeyr, K., Zhou, M.: The block preconditioned steepest descent iteration for elliptic operator eigenvalue problems. Electron. Trans. Numer. Anal. **41**, 93–108 (2014)

112. Nochetto, R.H.: Adaptive finite element methods for elliptic PDE. 2006 CNA Summer School, Probabilistic and Analytical Perpectives in Contemporary PDE (2006)
113. Nochetto, R.H., Siebert, K.G., Veeser, A.: Theory of adaptive finite element methods: an introduction. In: DeVore, R., Kunoth, A. (eds.) Multiscale, Nonlinear and Adaptive Approximation, pp. 409–542. Springer, Berlin (2009)
114. Nochetto, R.H., Veeser, A.: Primer of adaptive finite element methods. In: Naldi, G., Russo, G. (eds.) Multiscale and adaptivity: modeling, numerics and applications. Volume 2040 of Lecture Notes in Mathematics, pp. 125–225. Springer, Heidelberg (2012)
115. Nystedt, C.: A priori and a posteriori error estimates and adaptive finite element methods for a model eigenvalue problem. Technical Report 1995-05, Department of Mathematics, Chalmers University of Technology (1995)
116. Oden, J.T., Prudhomme, S.: Goal-oriented error estimation and adaptivity for the finite element method. Comput. Math. Appl. **41**(5–6), 735–756 (2001)
117. Quarteroni, A.: Numerical Models for Differential Problems. Volume 8 of MS&A. Modeling, Simulation and Applications, 2nd edn. Springer, Milan (2014). Translated from the fifth (2012) Italian edition by Silvia Quarteroni
118. Quarteroni, A., Valli, A.: Numerical Approximation of Partial Differential Equations. Springer Series in Computational Mathematics. Springer, Berlin (2008)
119. Rannacher, R.: Error control in finite element computations. An introduction to error estimation and mesh-size adaptation. In: Bulgak, H., Zenger, Ch. (eds.) Error Control and Adaptivity in Scientific Computing (Antalya, 1998). Volume 536 of NATO Science, Series C, Mathematical and Physical Sciences, pp. 247–278. Kluwer Academic, Dordrecht (1999)
120. Rannacher, R., Westenberger, A., Wollner, W.: Adaptive finite element solution of eigenvalue problems: balancing of discretization and iteration error. J. Numer. Math. **18**(4), 303–327 (2010)
121. Raviart, P.-A., Thomas, J.-M.: Introduction à l'Analyse Numérique des Équations aux Dérivées Partielles. Collection Mathématiques Appliquées pour la Maîtrise. Masson, Paris (1983)
122. Reddy, B.D.: Introductory Functional Analysis. Volume 27 of Texts in Applied Mathematics. Springer, New York (1998). With applications to boundary value problems and finite elements
123. Repin, S.I.: A Posteriori Estimates for Partial Differential Equations. Volume 4 of Radon Series on Computational and Applied Mathematics. Walter de Gruyter GmbH & Co. KG, Berlin (2008)
124. Rohwedder, T., Schneider, R., Zeiser, A.: Perturbed preconditioned inverse iteration for operator eigenvalue problems with applications to adaptive wavelet discretization. Adv. Comput. Math. **34**(1), 43–66 (2011)
125. Saad, Y.: On the rates of convergence of the Lanczos and the block-Lanczos methods. SIAM J. Numer. Anal. **17**(5), 687–706 (1980)
126. Samokish, A.: The steepest descent method for an eigen value problem with semi-bounded operators. Izv. Vyssh. Uchebn. Zaved. Mat. **5**, 105–114 (1958, in Russian)
127. Sauter, S.: Finite Elements for Elliptic Eigenvalue Problems: Lecture Notes for the Zürich Summerschool 2008. Preprint 12–08, Institut für Mathematik, Universität Zürich (2008). http://www.math.uzh.ch/compmath/fileadmin/math/preprints/12_08.pdf
128. Šolín, P.: Partial Differential Equations and the Finite Element Method. Pure and Applied Mathematics (New York). Wiley-Interscience [Wiley], Hoboken (2006)
129. Stevenson, R.: Optimality of a standard adaptive finite element method. Found. Comput. Math. **7**(2), 245–269 (2007)
130. Strang, G., Fix, G.J.: An Analysis of the Finite Element Method. Prentice-Hall Series in Automatic Computation. Prentice-Hall, Englewood Cliffs (1973)
131. Bui-Thanh, T., Ghattas, O., Demkowicz, L.: A relation between the discontinuous Petrov–Galerkin methods and the discontinuous galerkin method. ICES Report 11–45, The Institute for Computational Engineering and Sciences, The University of Texas at Austin, Austin, Texas 78712

132. Vaĭnikko, G.M.: Asymptotic error bounds for projection methods in the eigenvalue problem. Ž. Vyčisl. Mat. i Mat. Fiz. **4**, 405–425 (1964)
133. Vaĭnikko, G.M.: On the rate of convergence of certain approximation methods of galerkin type in eigenvalue problems. Izv. Vysš. Učebn. Zaved. Matematika **2**, 37–45 (1966)
134. Veeser, A., Verfürth, R.: Explicit upper bounds for dual norms of residuals. SIAM J. Numer. Anal. **47**(3), 2387–2405 (2009)
135. Verfürth, R.: A Review of A Posteriori Error Estimation and Adaptive Mesh-Refinement Techniques. Wiley/Teubner, New York/Stuttgart (1996)
136. Verfürth, R.: A Posteriori Error Estimation Techniques for Finite Element Methods. Numerical Mathematics and Scientific Computation. Oxford University Press, Oxford (2013)
137. Walsh, T.F., Reese, G.M., Hetmaniuk, U.L.: Explicit a posteriori error estimates for eigenvalue analysis of heterogeneous elastic structures. Comput. Methods Appl. Mech. Eng. **196**(37–40), 3614–3623 (2007)
138. Weinberger, H.F.: Variational methods for eigenvalue approximation. Society for Industrial and Applied Mathematics, Philadelphia (1974). Based on a series of lectures presented at the NSF-CBMS Regional Conference on Approximation of Eigenvalues of Differential Operators, Vanderbilt University, Nashville, Tenn., 26–30 June 1972, Conference Board of the Mathematical Sciences Regional Conference Series in Applied Mathematics, No. 15
139. Wu, H., Zhang, Z.: Enhancing eigenvalue approximation by gradient recovery on adaptive meshes. IMA J. Numer. Anal. **29**(4), 1008–1022 (2009)
140. Xu, J., Zhou, A.: A two-grid discretization scheme for eigenvalue problem. Math. Comput. **70**, 17–25 (1999)
141. Xu, J., Zhou, A.: Local and parallel finite element algorithms based on two-grid disretizations. Math. Comput. **69**, 881–909 (2000)
142. Zeiser, A.: On the optimality of the inexact inverse iteration coupled with adaptive finite element methods. Preprint 57, DFG-SPP 1324 (2010)
143. Zhang, Z., Naga, A.: A new finite element gradient recovery method: superconvergence property. SIAM J. Sci. Comput. **26**(4), 1192–1213 (electronic) (2005)
144. Zhang, Z., Yan, N.: Recovery type a posteriori error estimates in finite element methods. Korean J. Comput. Appl. Math. **8**(2), 235–251 (2001)
145. Zienkiewicz, O.C., Zhu, J.: A simple error estimator and adaptive procedure for practical engineering analysis. Int. J. Numer. Methods Eng. **24**(2), 337–357 (1987)
146. Zienkiewicz, O.C., Zhu, J.Z.: The superconvergent patch recovery and a posteriori error estimates. I. The recovery technique. Int. J. Numer. Methods Eng. **33**(7), 1331–1364 (1992)
147. Zienkiewicz, O.C., Zhu, J.Z.: The superconvergent patch recovery and a posteriori error estimates. II. Error estimates and adaptivity. Int. J. Numer. Methods Eng. **33**(7), 1365–1382 (1992)

Chapter 10
Algebraic Preconditioning Approaches and Their Applications

Matthias Bollhöfer

Abstract We will review approaches to numerically treat large-scale systems of equations including preconditioning, in particular those methods which are suitable for solving linear systems in parallel. We will also demonstrate how completion techniques can serve as a useful tool to prevent ill-conditioned systems. Beside parallel aspects for preconditioning, multilevel factorization methods will be investigated and finally we will demonstrate how these methods can be combined for approximate matrix inversion methods.

10.1 Introduction

Solving linear systems of the form $Ax = b$, where $A \in \mathbb{R}^{n\times n}$ is nonsingular, $x, b \in \mathbb{R}^n$ efficiently is an ubiquitous problem in many scientific applications such as solving partial differential equations, inverting matrices or parts of matrices or computing eigenstates in computational physics and many other application areas. For specific application problems, methods that are tailored to the underlying problem often serve best as problem-dependent solver, e.g. multigrid methods [36, 69, 71] are among the best methods for solving large classes of partial differential equations efficiently. However, when the underlying application problems do not posses enough problem-dependent structure information to allow for specific solution methods, more general methods are needed. Often enough, sparse direct solution methods (e.g. [22, 23, 64]) are very efficient and even if their efficiency with respect to computation time and memory is not quite satisfactory, their robustness is a strong argument to prefer these kind of methods, in particular, because only a small number of parameters needs to be adapted, if any. In contrast to that, preconditioned Krylov subspace solvers [33, 34, 61] are a frequently used alternative whenever an efficient preconditioner is available to solve the system in a reasonable amount of time. Nowadays as multicore and manycore architectures become standard even for desktop computers, parallel approaches to raise efficiency have gained attraction

M. Bollhöfer (✉)
TU Braunschweig, Institute for Computational Mathematics, D-38106 Braunschweig, Germany
e-mail: m.bollhoefer@tu-bs.de

P. Benner et al. (eds.), *Numerical Algebra, Matrix Theory, Differential-Algebraic Equations and Control Theory*, DOI 10.1007/978-3-319-15260-8_10

and are not anymore restricted to supercomputers. Many parallelization strategies are based on divide & conquer principles which decompose the whole problem into a sequence of smaller problems to be treated independently plus an additional coupling system to reveal the original problem [3, 11, 15, 35, 51, 64]. Among many parallelization approaches to solve linear systems, general black-box approaches are based on splitting the system or, partitioning the system appropriately into one part that is easily treated in parallel and a remaining part. Due to the rapidly increasing number of cores available for parallel solution techniques, direct solvers are often replaced by hybrid solvers in order to solve some part of the system directly while the additional coupling system is solved iteratively (see e.g. [32, 49]). With respect to their core part, these methods are based on a similar parallelization principle. To describe the breadth of parallel preconditioning approaches for efficiently solving linear systems would be too much to be covered by this article. Here we will focus on selected aspects which can also be used for efficient multilevel incomplete factorization techniques and for inverting parts of a matrix.

We will start in Sect. 10.2 reviewing splitting and partitioning methods for solving block-tridiagonal systems in parallel, in particular parallel direct and hybrid methods are often based on this kind of approach. After that we will display in Sect. 10.3, how similar methods can be set up even when the system is not block-tridiagonal. Section 10.4 will state how splitting-type methods can be improved to avoid ill-conditioned systems. Next we will demonstrate in Sect. 10.5 how algebraic multilevel preconditioners can be easily analyzed and improved and finally Sect. 10.6 demonstrates how multilevel methods on the one hand and parallel partitioning methods on the other hand can be employed for approximate matrix inversion.

10.2 Hybrid Solution Methods

With ever increasing size, large-scale systems are getting harder to be solved by direct methods and often enough, out-of-core techniques are required in order to solve systems, even in a parallel environment, since the memory consumption may exceed the available main memory. As a compromise between direct methods and preconditioned Krylov subspace methods, hybrid solvers that mix both ideas can be used. We briefly describe the two most common approaches that allow for efficient parallel treatment as well as for hybrid solution methods. Suppose that

$$A = C - EF^T, \tag{10.1}$$

where $C \in \mathbb{R}^{n\times n}$ is nonsingular and $E, F^T \in \mathbb{R}^{n\times q}$ are of lower rank $q \ll n$. The Sherman-Morrison-Woodbury formula

$$A^{-1} = C^{-1} + C^{-1}E(I - F^TC^{-1}E)^{-1}F^TC^{-1} \tag{10.2}$$

yields that solving $Ax = b$ is equivalent to

$$\text{solve } Cy = b, \text{ set } r := F^T y, \text{ solve } Rz = r, \text{ set } c = b + Ez, \text{ solve } Cx = c.$$

Here one has to solve two systems $Cy = b$, $Cx = c$ with C directly and a further small system $Rz = r$ with

$$R = I - F^T C^{-1} E \in \mathbb{R}^{q \times q}. \tag{10.3}$$

One can easily verify that R is nonsingular as well. The bottleneck of this splitting approach consists of computing the small system R explicitly which is most time-consuming. Usually having a small rank q, solving $CU = E$ can be performed efficiently using direct methods. The matrix U is sometimes [11] also called "spike matrix", since it refers to the non-trivial block columns of $C^{-1}A$. If it pays off, one could avoid solving the system $Cx = c$ by using the relation $x = Uz + y$ instead. However, when the rank is increasing, significantly more time is consumed. Thus, alternatively to solving $Rz = r$ directly, iterative solution methods that only require matrix-vector products are a favorable alternative and this finally yields a hybrid solution method [11, 49, 51]. A natural way of obtaining a splitting (10.1) for large-scale sparse matrices consists of partitioning the matrix A into two diagonal blocks plus a few nonzero off-diagonal entries outside the block-diagonal pattern which are then obviously of lower rank, i.e.,

$$\begin{aligned} A &= \begin{pmatrix} C_1 & 0 \\ 0 & C_2 \end{pmatrix} - \begin{pmatrix} 0 & E_1 F_2^T \\ E_2 F_1^T & 0 \end{pmatrix} \\ &= \begin{pmatrix} C_1 & 0 \\ 0 & C_2 \end{pmatrix} - \begin{pmatrix} E_1 & 0 \\ 0 & E_2 \end{pmatrix} \begin{pmatrix} 0 & F_1 \\ F_2 & 0 \end{pmatrix}^T \equiv C - EF^T. \end{aligned} \tag{10.4}$$

This procedure can be recursively applied to C_1 and C_2 to obtain a nested sequence of splittings and solving the systems via the Sherman–Morrison–Woodbury formula (10.2) can be performed recursively as well [67]. Although being elegant, splitting (10.4) has the drawback that the recursive application of splittings may also lead to higher complexity [50, 51]. More efficiently, an immediate parallelization approach with p processors prefers to substitute (10.4) by

$$A = \begin{pmatrix} C_1 & & & 0 \\ & C_2 & & \\ & & \ddots & \\ 0 & & & C_p \end{pmatrix} - EF^T \tag{10.5}$$

for suitably chosen EF^T. For block-tridiagonal systems having $m \geqslant p$ or significantly more diagonal blocks, EF^T is easily constructed. Suppose for simplicity that $m = l \cdot p$ for some $l \in \{1, 2, 3, \dots, \}$. Then we have

$$A = \begin{pmatrix} A_{11} & A_{12} & & & 0 \\ A_{21} & A_{22} & A_{23} & & \\ & \ddots & \ddots & \ddots & \\ & & A_{m-1,m-2} & A_{m-1,m-1} & A_{m-1,m} \\ 0 & & & A_{m,m-1} & A_{mm} \end{pmatrix}, \quad C_i = (A_{rs})_{r,s=(i-1)l+1,\dots,il} \tag{10.6}$$

and

$$EF^T = \begin{pmatrix} 0 & E_{12}F_{12}^T & & & & 0 \\ E_{21}F_{21}^T & 0 & E_{23}F_{23}^T & & & \\ & \ddots & \ddots & & \ddots & \\ & & E_{p-1,p-2}F_{p-1,p-2}^T & & 0 & E_{p-1,p}F_{p-1,p}^T \\ 0 & & & & E_{p,p-1}F_{p,p-1}^T & 0 \end{pmatrix},$$

where

$$E_{i,i+1}F_{i,i+1}^T = \begin{pmatrix} 0 & 0 \\ -A_{il,il+1} & 0 \end{pmatrix}, \quad E_{i+1,i}F_{i+1,i}^T = \begin{pmatrix} 0 & -A_{il+1,il} \\ 0 & 0 \end{pmatrix}$$

and one could even employ a low rank factorization of $A_{il,il+1}$ and $A_{il+1,il}$ to decrease the rank further. We can take advantage of instantaneously splitting the initial system into p parts since we only obtain a single coupling system R, which is usually small but hard to solve in parallel. Besides, computing R now only requires solving $C_i U_i = (E_{i,i-1}, E_{i,i+1})$, $i = 1, \dots, p-1$ simultaneously without further recursion. Here $E_{0,1}$ and $E_{p,p+1}$ are void. Because of its ease, this variant may be preferred to the recursive approach.

Another approach for solving systems $Ax = b$ in parallel consists of partitioning the initial system A into subsystems rather than splitting the matrix A. This approach is favorable in particular in cases where the diagonal blocks of A can be assumed to be safely nonsingular (i.e., the case of positive definite matrices or diagonal dominant matrices). In this case we partition A as

$$A = \left(\begin{array}{ccc|c} C_1 & & 0 & E_{1,p+1} \\ & \ddots & & \vdots \\ 0 & & C_p & E_{p,p+1} \\ \hline F_{1,p+1}^T & \cdots & F_{p,p+1}^T & C_{p+1} \end{array}\right) \equiv \left(\begin{array}{c|c} C & E \\ \hline F^T & C_{p+1} \end{array}\right) \tag{10.7}$$

and solving $Ax = b$ is easily obtained from the block LU decomposition of the system. I.e., partition

$$x^T = \left(x_1^T \cdots x_p^T \middle| x_{p+1}^T\right) \equiv \left(\hat{x}^T \middle| x_{p+1}^T\right),$$
$$b^T = \left(b_1^T \cdots b_p^T \middle| b_{p+1}^T\right) \equiv \left(\hat{b}^T \middle| b_{p+1}^T\right).$$

Then x is obtained as follows.

$$\text{solve } Cy = \hat{b}, \text{ set } r := b_{p+1} - F^T y, \text{ solve } Sx_{p+1} = r,$$
$$\text{set } c = b - Ex_{p+1}, \text{ solve } C\hat{x} = c.$$

Here we set $S := C_{p+1} - F^T C^{-1} E$ as the Schur complement. Similar to the case of splitting A as in (10.5) the major amount of work here is spent in computing S, i.e., computing $C_i U_i = E_{i,p+1}, i = 1, \ldots, p$. A natural alternative would also be in this case to solve $Sx_{p+1} = r$ using iterative solution methods which again leads to a hybrid solution method [1, 32, 51]. We like to point out that within the context of solving partial differential equations, these kind of methods are usually called domain decomposition methods, see e.g. [65, 74], which will definitely be beyond the scope of this paper. Instead we will focus on several algebraic aspects.

Example 1 We demonstrate the difference of the splitting approach (10.5) and the partitioning approach (10.7) as direct and hybrid solvers when the block diagonal system is factored using LU decomposition. The alternatives are either generating and solving the coupling systems R and S directly or to avoid explicit computation and use an iterative solver instead. For simplicity we choose the problem $-\Delta u = f$ in $\Omega = [0, 1]^2$, with Dirichlet boundary conditions and 5-point-star discretization. We display a simplified parallel model, where we measure only the maximum amount of computation time over all blocks p whenever a system with C_i is treated. In Fig. 10.1 we compare the direct method versus the hybrid method for (10.5) and (10.7) based on the initial block-tridiagonal structure of the underlying system with natural ordering. We use MATLAB for these experiments. As we can see from Fig. 10.1, the direct approach is only feasible for small p, since otherwise R and S become too big as confirmed by theoretical estimates in [51]. Moreover, the computation of the "spike-matrix" U requires solving $2N$ systems with each diagonal block C_i. We can also see that there is no great difference between the splitting approach and the partitioning approach, although in the splitting approach the system is roughly twice as big and nonsymmetric which is the reason for using Bi-CGSTAB [25] as iterative solver. For the partitioning approach CG [38] can be used. Both iterative solvers use a relative residual of 10^{-8} for termination.We also remark at this point that the number of iteration steps significantly increases as the number of blocks p increases (as expected by the domain decomposition theory).

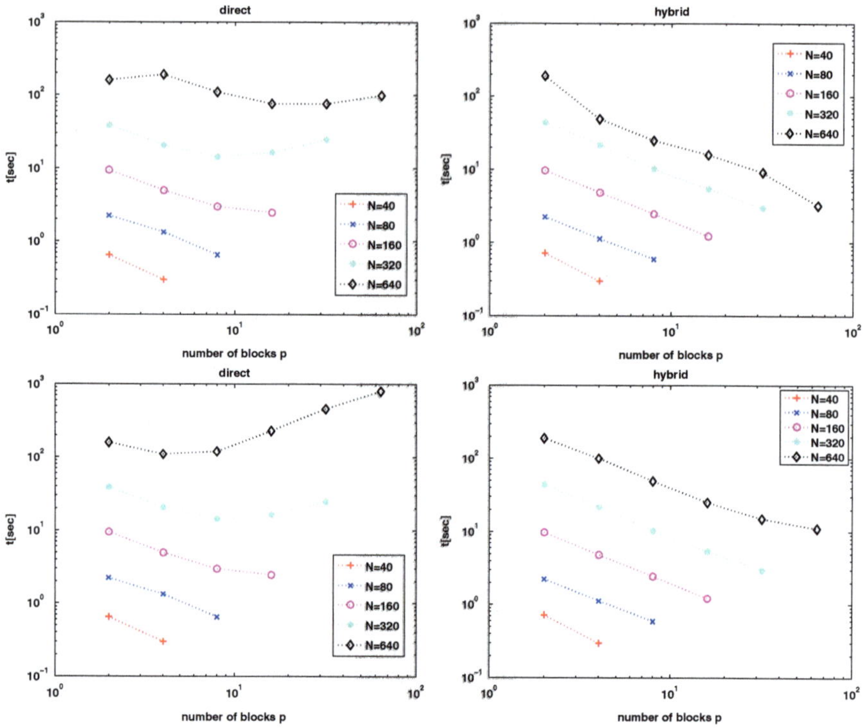

Fig. 10.1 Direct methods (*left*), hybrid methods (*right*), splitting approaches (*top*), partitioning approaches (*bottom*)

Both approaches based on splittings as in (10.5) or based on partitionings (10.7) are relatively similar with respect to parallelization and computational amount of work. The splitting-based approach allows to modify the blocks if necessary, the partitioning-based approach is simpler since it does not rely on especially constructed splittings which is advantageous when the diagonal blocks are safely nonsingular. In Sect. 10.3 we will compare both approaches and further generalize them in particular for systems that are not necessarily block-tridiagonal.

10.3 Reordering and Partitioning the System

We will now generalize how to split A as in (10.1) or to partition A as in (10.7). First of all we discuss the situation when the (block-)diagonal part of A is far away from having large entries, e.g. in the sense of some diagonal dominance measure [62] such as

$$r_i = \frac{|a_{ii}|}{\sum_{j=1}^{n} |a_{ij}|} \in [0, 1], \; i = 1, \ldots, n. \tag{10.8}$$

Note that a value of r_i larger than $\frac{1}{2}$ refers to a diagonal dominant row. The use of maximum weight matchings [8, 26, 27] is often very helpful to improve the diagonal dominance and to hopefully obtain diagonal blocks that are better conditioned. Maximum weight matchings replace A by

$$A^{(1)} = D_l A D_r \Sigma \tag{10.9}$$

where $D_l, D_r \in \mathbb{R}^{n \times n}$ are nonsingular, nonnegative diagonal matrices and $\Sigma \in \mathbb{R}^{n \times n}$ is a permutation matrix such that

$$|a_{ij}^{(1)}| \leqslant 1, \ |a_{ii}^{(1)}| = 1, \ \text{ for all } i, j = 1, \ldots, n.$$

Algorithms for computing maximum weight matchings for sparse matrices [26, 27] are experimentally often very fast of complexity $\mathcal{O}(n + nz)$, where nz refers to the number of nonzero elements of A. Note that theoretical bounds are much worse and also that maximum weight matchings are known to be strongly sequential. We illustrate the effect of maximum weight matchings in the following example. For details we refer to [26].

Example 2 We consider the sample matrix "west0479" (available from the University of Florida collection) of size $n = 479$ and number of nonzeros $nz = 1887$. In Fig. 10.2 we illustrate the effect of maximum weight matching for this particular matrix. The diagonal dominance measure r_i from (10.8) changes on the average from $\frac{1}{n}\sum_i r_i^{(old)} \approx 5.7 \cdot 10^{-3}$ initially to $\frac{1}{n}\sum_i r_i^{(new)} \approx 0.49$ after maximum weight matching is applied.

Even if the system is well-suited with respect to its diagonal blocks, partitioning the matrix into p blocks remains to be done prior to solving the system in a hybrid

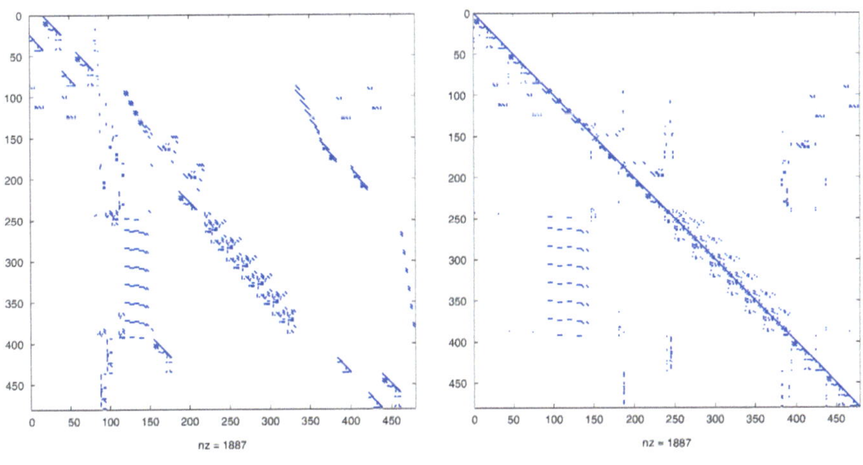

Fig. 10.2 Sample matrix before reordering and rescaling (*left*) and afterwards (*right*)

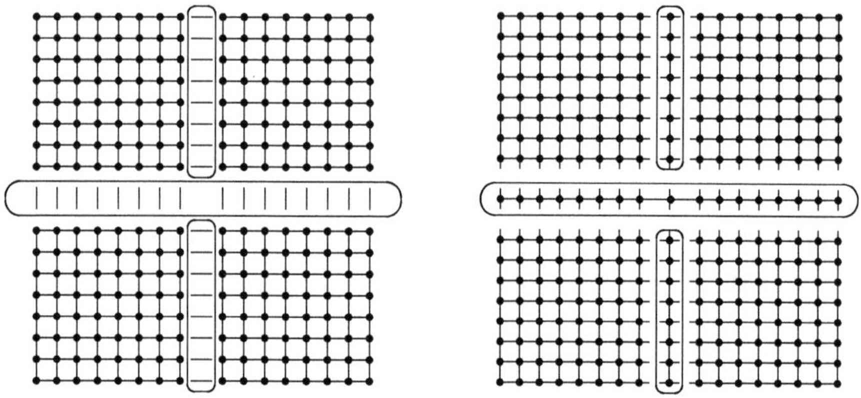

Fig. 10.3 Nested dissection by edges (*left*) and nested dissection by nodes (*right*)

fashion or to invert parts of the system. To do so, multilevel nested dissection [40, 41] can be used. Formally $A^{(1)}$ is replaced by

$$A^{(2)} = \Pi^T A^{(1)} \Pi$$

for some permutation matrix $\Pi \in \mathbb{R}^{n \times n}$. When targeting a splitting of A such as in (10.5), nested dissection by edges is the natural partitioning of the system whereas reordering the system matrix A as in (10.7) requires nested dissection by nodes. We illustrate the difference between both permutation strategies using the following simple undirected graph of a matrix in Example 3. Note that $G(A)$ is called (undirected) graph of A, if it consists of nodes $\mathscr{V} = \{1, \ldots, n\}$ and edges $\mathscr{E} = \{\{i, j\} : a_{ij} \neq 0 \text{ or } a_{ji} \neq 0, \forall i \neq j\}$.

Example 3 We consider an example that frequently applies in solving partial differential equations for a model problem. The graph we use is simply a grid (see Fig. 10.3).

To reorder the system with respect to the nested dissection approach there exist fast reordering tools, e.g., the MeTis software package [39].

Up to now rescaling and reordering the system matrix can be considered as relatively cheap compared to solving $Ax = b$ or inverting parts of A [8, 26, 27].

10.3.1 Reordering the System for a Splitting-Type Approach

Now we describe how the preprocessing step can in particular advance splitting or partitioning the system compared with only using a block-tridiagonal structure as in (10.6). Here we may assume that the underlying matrix is not just block-tridiagonal but sparse. We will start with partitioning the graph with respect to the edges.

Definition 1 Suppose that $A \in \mathbb{R}^{n\times n}$, $\mathscr{V} = \{1,\ldots,n\}$. Let $\mathscr{C}_1\dot\cup\cdots\dot\cup\mathscr{C}_p = \mathscr{V}$ be a disjoint union of $\mathscr{V}$, partitioning $\mathscr{V}$ into p disjoint subsets. We define $G_M(A) := (\mathscr{V}_M, \mathscr{E}_M)$, where $\mathscr{V}_M = \{1,\ldots,p\}$,

$$\mathscr{E}_M = \{\{r,s\} \subset \mathscr{V}_M \times \mathscr{V}_M : r \neq s, \text{ there exist } i \in \mathscr{C}_r, j \in \mathscr{C}_s, \text{ such that } a_{ij} \neq 0\}.$$

We call $G_M(A)$ block or modified graph of A with respect to $\mathscr{C}_1,\ldots,\mathscr{C}_p$.

$G_M(A)$ can be regarded as block graph of A after reordering A such that the entries of $\mathscr{C}_1,\ldots,\mathscr{C}_p$ are taken in order of appearance and using the associated block matrix shape, i.e., given a suitable permutation matrix $\Pi \in \mathbb{R}^{n\times n}$ we obtain

$$\Pi^T A \Pi = \begin{pmatrix} A_{11} & \cdots & A_{1p} \\ \vdots & & \vdots \\ A_{p1} & \cdots & A_{pp} \end{pmatrix}$$

and many blocks A_{ij} are expected to be zero or of low rank.

Let $e_1,\ldots,e_n$ be the standard unit vector basis of $\mathbb{R}^n$. We denote by I_r the matrix of column unit vectors from $\mathscr{C}_r$, i.e.,

$$I_r = (e_j)_{j\in\mathscr{C}_r}, r = 1,\ldots,p.$$

Then after reordering A with respect to $\mathscr{C}_1,\ldots,\mathscr{C}_p$ we obtain

$$P^T A P = C - EF^T$$

where

$$C = \begin{pmatrix} A_{11} & & 0 \\ & \ddots & \\ 0 & & A_{pp} \end{pmatrix}, \; EF^T = \sum_{\{r,s\}\in\mathscr{E}_M} (I_r, I_s) \begin{pmatrix} 0 & -A_{rs} \\ -A_{sr} & 0 \end{pmatrix} (I_r, I_s)^T.$$

If we compute some low rank factorization $-A_{rs} = E_{rs}F_{rs}^T, -A_{sr} = E_{sr}F_{sr}^T$, then we obtain E and F in a similar way compared with the block tridiagonal case. Suppose that $m = \#\mathscr{E}_M$ and the edges $\{r,s\}$ of $\mathscr{E}_M$ are taken in a suitable order $\{r_1,s_1\},\ldots,\{r_m,s_m\}$. Then we define E, F via

$$E = (E_1,\ldots,E_m), \; F = (F_1,\ldots,F_m), \tag{10.10}$$

where

$$E_i = (I_{r_i}, I_{s_i}) \cdot \begin{pmatrix} E_{r_i,s_i} & 0 \\ 0 & E_{s_i,r_i} \end{pmatrix}, \; F_i = (I_{r_i}, I_{s_i}) \cdot \begin{pmatrix} 0 & F_{r_i,s_i} \\ F_{s_i,r_i} & 0 \end{pmatrix}. \tag{10.11}$$

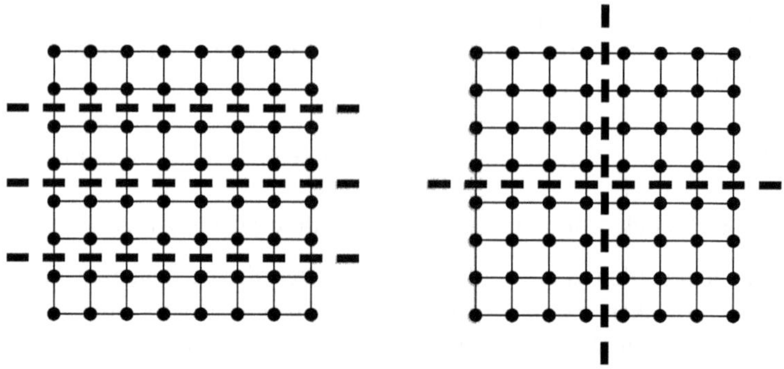

Fig. 10.4 Partitioning the grid into 4 sub-grids horizontally (*left*) and in checker board fashion (*right*)

We note that if A_{r_i,s_i} and A_{s_i,r_i} have rank q_{r_i,s_i}, q_{s_i,r_i}, then the total rank of E, F is

$$q = \sum_{\{r,s\}\in\mathscr{E}_M} (q_{rs} + q_{sr}) \,. \tag{10.12}$$

For general sparse matrices this might lead to a significantly smaller q compared with the case where A is reordered into a block-tridiagonal shape as in Sect. 10.2. We will illustrate this effect in the following example.

Example 4 Consider a matrix A such that its graph is a grid with $M \times M$ grid points, i.e., $n = M^2$. Suppose further that the number of processors p can be written $p = P^2$ and that M is a multiple of P.

For $p = 4$ we illustrate in Fig. 10.4 two different canonical ways of partitioning the underlying graph. A graph like in Fig. 10.4 may serve as a toy problem for some class of partial differential equations. In the simplest case for the elliptic boundary value problem $-\Delta u = f$ in $[0, 1]^2$ with Dirichlet boundary conditions and 5-point-star difference stencil a graph similar to Fig. 10.4 is obtained. The edges would refer to numerical values -1, the cross points would refer to diagonal entries with value 4. In this case the left partitioning of the domain in Fig. 10.4 would lead to

$$E = \left(\begin{array}{cc|cc|cc|c}
0 & & & & & & \\
I & 0 & & & & & \\ \hline
0 & I & & & & & \\
 & 0 & 0 & & & & \\
 & & I & 0 & & & \\ \hline
 & & 0 & I & & & \\
 & & & 0 & 0 & & \\
 & & & & I & 0 & \\ \hline
 & & & & 0 & I & \\
 & & & & & 0 &
\end{array}\right), \quad
F = \left(\begin{array}{cc|cc|cc|c}
0 & & & & & & \\
0 & -I & & & & & \\ \hline
-I & 0 & & & & & \\
 & 0 & 0 & & & & \\
 & & 0 & -I & & & \\ \hline
 & & -I & 0 & & & \\
 & & & 0 & 0 & & \\
 & & & & 0 & -I & \\ \hline
 & & & & -I & 0 & \\
 & & & & & 0 &
\end{array}\right).$$

Each of the identity matrices has size M. The generalization to p sub-blocks is straightforward and would lead to E, F of size $n \times (2(p-1)M)$.

In contrast to this, the checker board partitioning in Fig. 10.4 would lead to E and F which look almost as follows

$$E = \left(\begin{array}{cc|cc|cc|cc}
 & & & & & & 0 & I \\
0 & & & & & & & 0 \\
I & 0 & & & & & & \\
\hline
0 & I & & & & & & \\
 & 0 & 0 & & & & & \\
 & & I & 0 & & & & \\
\hline
 & & 0 & I & & & & \\
 & & & 0 & 0 & & & \\
 & & & & I & 0 & & \\
\hline
 & & & & 0 & I & & \\
 & & & & & 0 & 0 & \\
 & & & & & & I & 0
\end{array}\right), \quad F = \left(\begin{array}{cc|cc|cc|cc}
 & & & & & & -I & 0 \\
0 & & & & & & & 0 \\
0 & -I & & & & & & \\
\hline
-I & 0 & & & & & & \\
 & 0 & 0 & & & & & \\
 & & 0 & -I & & & & \\
\hline
 & & -I & 0 & & & & \\
 & & & 0 & 0 & & & \\
 & & & & 0 & -I & & \\
\hline
 & & & & -I & 0 & & \\
 & & & & & 0 & 0 & \\
 & & & & & & 0 & -I
\end{array}\right).$$

Here, the identity matrices are only of size $M/2$. Strictly speaking, the identity matrices overlap at the center of the grid. We skip this detail for ease of description. For the checker board partitioning the generalization to $p = P^2$ blocks would lead to a rank proportional to $\sqrt{p}M$ which is significantly less compared with the first case as it grows slower with respect to p.

10.3.2 Reordering the System for a Partitioning-Type Approach

In contrast to splitting the initial system we now partition it, which means that rather than using nested dissection by edges, we now require nested dissection by nodes as illustrated in Example 3. In this case partitioning the system with respect to the underlying graph can also be advantageous compared to the strategy where A is simply permuted to block-tridiagonal form. We will illustrate this in Example 5.

Example 5 We consider again a graph of a matrix A that can be represented as a grid in two spatial dimensions. Suppose that the number of processors p can be written $p = P^2$. We assume that the number n of grid points can be written as $n = (M + P - 1)^2$ and that M is a multiple of P. For $p = 4$ we illustrate in Fig. 10.5 two obvious ways of partitioning the graph. If we again consider the 5-point-star difference stencil for discretizing the problem $-\Delta u = f$ in $[0, 1]^2$ we still end up with a matrix partitioning

$$A = \left(\begin{array}{c|c} C & E \\ \hline F^T & C_{p+1} \end{array}\right).$$

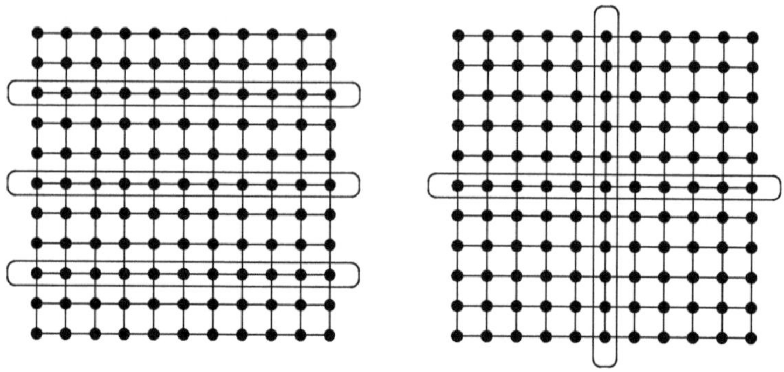

Fig. 10.5 Partitioning the grid into 4 sub-grids horizontally (*left*) and in checker board fashion (*right*)

For the horizontal partitioning approach in general the identity matrices have size $M + p - 1$. The size of the Schur-complement $S = C_{p+1} - F^T C^{-1} E$ is identical to the number of nodes we removed, i.e., its size is $(p - 1)(M + p - 1)$.

In the checker board partitioning case the Schur-complement will have size $2(P - 1)(M + P - 1) - (P - 1)^2$ which is roughly of order $2\sqrt{p}M$ for $p \ll M$. Therefore the checker board partitioning leads to a significantly smaller Schur-complement with respect to p compared with the horizontal approach.

10.3.3 Splitting-Type Approach Versus Partitioning-Type Approach

After we have illustrated how to preprocess a system $A \rightarrow A^{(2)}$ such that the system is either suitable for a splitting-type approach (10.5) or a partitioning-type approach (10.7), we will now highlight the common properties, the major differences and which approach should be preferred depending on the situation.

First of all, with respect to parallelization, one has to distinguish whether reordering the system is a suitable option. This is important since depending on the application, the original system matrix A may not be available in total, but it could be distributed over different machines. This is in particular the case for distributed memory machines, where the problem is already generated in parallel. In this case partitioning the system by permutation refers to re-distributing the system in order to obtain a better load balance. This in turn can become quite expensive. When using finite element application for partial differential equations, domain decomposition methods partition the physical domain and the nodes on the interfaces between the domains share the neighbouring subdomains. Algebraically this refers to the partitioning-type approach (10.7). Otherwise, if there is no natural background why a specific node should share two or more different parts of the system, a more natural distribution in practical applications would be the splitting-type approach (10.5). For shared memory systems like modern multicore or upcoming manycore architectures we would usually have the whole system matrix A available and we are free to decide which approach should be our method of choice.

A major difference between the splitting-type method (10.5) and the partitioning-type approach is the size of the systems R and S in similar circumstances like Examples 4 and 5, where the size of R is approximately twice as big as that of S for the block tridiagonal case and for the checker board case the difference is even larger. This is because in the partitioning–type approach the size of S is exactly the number of nodes to be taken out by nested dissection (by nodes), while in the splitting case the size of R is bounded by twice the number of edges (or the number of off-diagonal entries) taken out from graph using nested dissection by edges. The number of edges is usually bigger than the number of nodes and one even obtains a factor 2. On the other hand the rank $q_{rs} + q_{sr}$ of the matrices A_{rs} and A_{sr} that are taken out is the local contribution to the size of R and certainly the rank could be also less than the number of edges. However, there is one improvement that can be obtained for free in the splitting case, which is referred to as minimum rank decoupling [51, 68]. Suppose for simplicity that $q_{rs} = q_{sr}$. If these numbers differ, we could enlarge the factorization $E_{rs}F_{rs}^T$ or $E_{sr}F_{sr}^T$ of smaller size by zeros. Alternatively to (10.5) we could use the splitting

$$A = \begin{pmatrix} C_1(X) & & & 0 \\ & C_2(X) & & \\ & & \ddots & \\ 0 & & & C_p(X) \end{pmatrix} - E(X)X^{-1}F(X)^T, \tag{10.13}$$

where we replace locally for any $r < s$

$$\begin{pmatrix} 0 & E_{rs}F_{rs}^T \\ E_{sr}F_{sr}^T & 0 \end{pmatrix}$$

by

$$\begin{pmatrix} E_{rs}X_{rs}F_{sr}^T & E_{rs}F_{rs}^T \\ E_{sr}F_{sr}^T & E_{sr}X_{rs}^{-1}F_{rs}^T \end{pmatrix}$$

for some nonsingular $X_{rs} \in \mathbb{R}^{q_{rs}\times q_{rs}}$ and modify the diagonal blocks $C_1, \ldots, C_p$ appropriately to compensate the changes in the block diagonal position. The advantage of this modification consists of reducing the local rank by a factor 2 since

$$\begin{pmatrix} E_{rs}X_{rs}F_{sr}^T & E_{rs}F_{rs}^T \\ E_{sr}F_{sr}^T & E_{sr}X_{rs}^{-1}F_{rs}^T \end{pmatrix} = \begin{pmatrix} E_{rs}X_{rs} \\ E_{sr} \end{pmatrix} X_{rs}^{-1} \begin{pmatrix} X_{rs}F_{sr}^T & F_{rs}^T \end{pmatrix}. \tag{10.14}$$

In the simplest case we could choose $X_{rs} = I$. The associated diagonal matrices C_r are changed to

$$C_r(X) := C_r + \sum_{\substack{s:s>r \\ \{r,s\}\in\mathscr{E}_M}} E_{rs}X_{rs}F_{rs}^T + \sum_{\substack{s:s<r \\ \{r,s\}\in\mathscr{E}_M}} E_{rs}X_{rs}^{-1}F_{rs}^T$$

adding only low-rank contributions to C_r. For sparse matrices these modifications only change entries of C_r that are connected to neighbouring blocks. Thus, if $p \ll n$, only a lower-rank part of small size is changed in C_r. For partial differential equations one could read this modification as imposing some kind of inner boundary condition and a natural question will be how to suitably choose

$$X = \operatorname{diag}(X_{rs})_{\{r,s\}\in\mathscr{E}_M}\,. \tag{10.15}$$

This will be subject of the next section.

To end this section we will demonstrate the benefits of minimum rank decoupling ($X = I$) and using graph partitioning rather than working with a block-tridiagonal shape.

Example 6 We continue with the problem $-\Delta u = f$ on the unit square in two spatial dimensions and N grid points in each spatial dimension. Here we obtain that $F = E$ and we also have that R is symmetric positive definite. This allows to fully exploit symmetry not only for each C_i, but also R using the Cholesky decomposition, resp. the conjugate gradient method. We use the same settings as in Example 1, except that we perform the numerical experiments for the splitting-type approach (10.5) only.

In contrast to Example 1, the size of the "spike-matrix" U now only requires solving $4 \cdot N/p$ systems in parallel rather than $2N$ systems. With increasing size of processors this reduces the overhead for computing the "spike-matrix" significantly. Moreover, as illustrated in Example 5, the size of R also grows much slower than in the block-tridiagonal case and the number of CG steps also increases more slowly. In total this makes the direct approach much more competitive for larger p and explains the remarkable improvement in Fig. 10.6 compared to Fig. 10.1 with respect to the computation time and the scalability.

10.4 Minimum Rank Decoupling and Completion

We will now discuss the problem of choosing X in (10.15) in the minimum rank decoupling case. This problem is connected to the problem of matrix completion [28, 29]. For the problem of completion one is interested in determining a suitable X such that $W(X) = \begin{pmatrix} W_{11} & W_{12} \\ W_{21} & X \end{pmatrix}$ has certain desired properties, e.g. a small norm, an inverse with small norm or a small condition number. For details we refer to [28, 29].

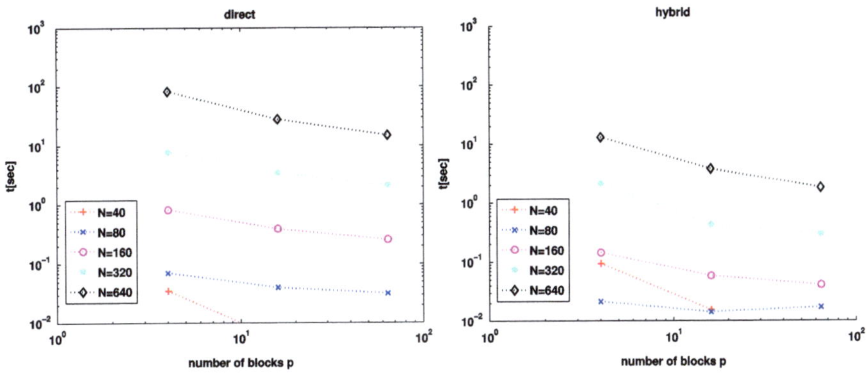

Fig. 10.6 Splitting-based direct method (*left*), splitting-based hybrid method (*right*)

Here the completion problem comes along with the choice of X from (10.15). We will follow the arguments in [28, 29]. Suppose that $\mathscr{E}_M = \{\{r_1, s_1\}, \ldots, \{r_m, s_m\}\}$ with the convention that we use $r_i < s_i$. Given the splitting (10.13) depending on X we set

$$E^{(1)} = \left(E_1^{(1)}, \ldots, E_m^{(1)}\right), \quad F^{(1)} = \left(F_1^{(1)}, \ldots, F_m^{(1)}\right),$$
$$E^{(2)} = \left(E_1^{(2)}, \ldots, E_m^{(2)}\right), \quad F^{(2)} = \left(F_1^{(2)}, \ldots, F_m^{(2)}\right),$$

where for any $\{r_i, s_i\} \in \mathscr{E}_M$ such that $r_i < s_i$ we define

$$E_i^{(1)} = I_{r_i} E_{r_i,s_i}, \; E_i^{(2)} = I_{s_i} E_{s_i,r_i}, \quad F_i^{(1)} = I_{r_i} F_{s_i,r_i}, \; F_i^{(2)} = I_{s_i} F_{r_i,s_i}.$$

Then the minimum rank decoupling (10.14) can be written as

$$E(X)X^{-1}F(X)^T = \underbrace{(E^{(1)}X + E^{(2)})}_{E(X)} X^{-1} \underbrace{(F^{(1)}X^T + F^{(2)})^T}_{F(X)^T}$$

and the block diagonal part $C(X)$ is analogously characterized by

$$C(X) = C + E^{(1)}X(F^{(1)})^T + E^{(2)}X^{-1}(F^{(2)})^T.$$

Here, as before, C refers to the unmodified block diagonal part and $A = C(X) - E(X)X^{-1}F(X)^T$. If our matrix A is block-tridiagonal, then $E^{(1)}, F^{(1)}$ refer to modifications in the lower right block of a diagonal block C_i, whereas $E^{(2)}$, $F^{(2)}$ refer to the upper left corners. Using the Sherman-Morrison-Woodbury formula we find that

$$A^{-1} = C(X)^{-1} + C(X)^{-1}E(X)\underbrace{\left(X - F(X)^T C(X)^{-1}E(X)\right)^{-1}}_{\equiv R(X)^{-1}} F(X)^T C(X)^{-1}$$

and a natural objective is to improve the properties of $C(X)$ or of the coupling system

$$R(X) = X - F(X)^T C(X)^{-1} E(X).$$

Rewriting $R(X)^{-1}$ (again using (10.2)) we can see that

$$R(X)^{-1} = X^{-1} - X^{-1}(F^{(1)}X^T + F^{(2)})^T A^{-1}(E^{(1)}X + E^{(2)})X^{-1}.$$

Taking into account that usually we only have two factors EF^T instead of three factors $E(X)X^{-1}F(X)^T$, we would factorize $X = X_L X_U$ and replace $E(X)X^{-1}F(X)^T$ by $(E^{(1)}X_L + E^{(2)}X_U^{-1}) \cdot (F^{(1)}X_U^T + F^{(2)}X_L^{-T})^T$. This in turn means that $R(X)$ should approximate X rather than I and similarly, $R(X)^{-1}$ has to approximate X^{-1}. If we wish approximate a multiple αX^{-1} of X^{-1} we conclude that using $Y = X^{-1}$ we obtain in the optimal case

$$0 = \alpha Y - R(X)^{-1} = (\alpha - 1)Y - (F^{(1)} + F^{(2)}Y^T)^T A^{-1}(E^{(1)} + E^{(2)}Y). \tag{10.16}$$

Note that (10.16) is called algebraic Riccati equation with respect to Y. For the application of numerical methods for solving Riccati equations we refer to [4, 19, 20, 43, 44, 53]. Here we mention a simple criterion when this quadratic equation simplifies. Since we will not follow this direction in detail we leave the proof to the reader.

Proposition 1 *Suppose that $(F^{(1)})^T C^{-1} E^{(2)} = 0$ and $(F^{(2)})^T C^{-1} E^{(1)} = 0$. Then (10.16) is equivalent to*

$$0 = \alpha Y - (D + Y) - (D + Y)B(D + Y), \tag{10.17}$$

where $B = (F^{(2)})^T A^{-1} E^{(2)}$ and $D = (F^{(1)})^T C^{-1} E^{(1)}$.

Example 7 We continue Examples 3, 4 for the case of a block-tridiagonal partitioning. Note that in the case of minimum rank decoupling we will obviously have $E = F$ and

$$E = \left(\begin{array}{c|c|c} 0 & & \\ I & & \\ \hline I & & \\ 0 & 0 & \\ & I & \\ \hline & I & \\ & 0 & 0 \\ & & I \\ \hline & & I \\ & & 0 \end{array}\right),$$

since the trivial choice $X = I$ modifies the original off-diagonal blocks of type

$$\begin{pmatrix} 0 & I \\ I & 0 \end{pmatrix}$$

in the minimum rank case to blocks of type

$$\begin{pmatrix} I & I \\ I & I \end{pmatrix} = \begin{pmatrix} I \\ I \end{pmatrix} \begin{pmatrix} I & I \end{pmatrix}.$$

In this case we will have

$$E = F = E^{(1)} + E^{(2)} = F^{(1)} + F^{(2)} \equiv \left(\begin{array}{c|c|c} 0 & & \\ I & & \\ \hline 0 & & \\ & 0 & \\ & I & \\ \hline & 0 & \\ & & 0 \\ & & I \\ \hline & & 0 \\ & & \end{array}\right) + \left(\begin{array}{c|c|c} & & \\ 0 & & \\ \hline I & & \\ 0 & & \\ & 0 & \\ \hline & I & \\ & 0 & \\ & & 0 \\ \hline & & I \\ & & 0 \end{array}\right).$$

For modifying $C(X)$ we aim to reduce $\|C(X)\|$ or $\|C(X)^{-1}\|$. Moreover, with respect to the sparsity of C, we cannot afford much more than a diagonal matrix X. As long as $\|X\|, \|X^{-1}\| \leqslant \kappa$ for some constant $\kappa > 0$, the norm of $C(X)$ is suitably bounded. In contrast to that, $\|C(X)^{-1}\|$ might still be large. Completion can be directly used to bound the norm of the inverse of

$$W(X) = \left(\begin{array}{c|cc} C & E^{(1)} & E^{(2)} \\ \hline (F^{(1)})^T & -X^{-1} & 0 \\ (F^{(2)})^T & 0 & -X \end{array}\right)$$

since the associated Schur complement in the top left corner satisfies

$$C(X) = C - \left(E^{(1)}\ E^{(2)}\right) \begin{pmatrix} -X^{-1} & 0 \\ 0 & -X \end{pmatrix}^{-1} \left(F^{(1)}\ F^{(2)}\right)^T.$$

Since $C(X)^{-1}$ is the leading top left block of $W(X)^{-1}$, a bound for the norm of $W(X)$ also leads to a bound for $\|C(X)^{-1}\|$. Following [29] we define α_0 via

$$\alpha_0 = \min\left\{\sigma_{\min}\left[C, E^{(1)}, E^{(2)}\right], \sigma_{\min}\left[C^T, F^{(1)}, F^{(2)}\right]\right\}, \tag{10.18}$$

where $\sigma_{\min}$ denotes the associated smallest singular value.

Lemma 1 *We define for any $r = 1,\dots,p$,*

$$C_{E,r} := C_r C_r^T + \sum_{s:s\neq r} E_{rs}E_{rs}^T,\ C_{F,r} := C_r^T C_r + \sum_{s:s\neq r} F_{rs}^T F_{rs}.$$

Then we have that

$$\alpha_0^2 = \min_r\{\lambda_{\min}(C_{E,r}),\lambda_{\min}(C_{F,r})\}.$$

Proof It is clear that α_0^2 can be obtained from the smallest eigenvalue of $CC^T + E^{(1)}(E^{(1)})^T + E^{(2)}(E^{(2)})^T$ and $C^TC + (F^{(1)})^T F^{(1)} + (F^{(2)})^T F^{(2)}$.

By definition we have

$$CC^T + E^{(1)}(E^{(1)})^T + E^{(2)}(E^{(2)})^T = CC^T + \sum_{i=1}^m E_i^{(1)}(E_i^{(1)})^T + \sum_{i=1}^m E_i^{(2)}(E_i^{(2)})^T,$$

which is block-diagonal by construction and precisely reduces to the block-diagonal matrix $\mathrm{diag}(C_{E,1},\dots,C_{E,p})$. Note that since we always assume that $r_i < s_i$, the local sum over all $s : s \neq r$ covers both sums with $E_i^{(1)}$ and $E_i^{(2)}$. Similar arguments apply to $C^TC + (F^{(1)})^T F^{(1)} + (F^{(2)})^T F^{(2)}$.

As consequence of Lemma 1 we can compute α_0 for each diagonal block separately in parallel. This simplifies the overall complexity.

We note that given $\alpha < \alpha_0$, the general unconstrained solution X rather than $\begin{pmatrix} -X^{-1} & 0 \\ 0 & -X \end{pmatrix}$ of $\|W(X)^{-1}\|_2 \leqslant \frac{1}{\alpha}$ is is stated explicitly in [29]. In addition we would like to point out that the singular values $\sigma_1,\dots,\sigma_n$ [33] of any matrix C can be determined by

$$\sigma_l \equiv \sigma_l(C) = \max_{\substack{\dim U=l \\ \dim V=l}} \min_{\substack{u\in U\setminus\{0\} \\ v\in V\setminus\{0\}}} \frac{v^T C u}{\|v\|_2\,\|u\|_2}.$$

Furthermore since C and $C(X)$ are block-diagonal, we can compute the singular values of each $C_r(X)$ independently. Having

$$C_r(X) = C_r + \sum_{s:s>r} E_{rs}X_{rs}F_{sr}^T + \sum_{s:s<r} E_{rs}X_{rs}^{-1}F_{sr}^T,$$

for some neighbouring diagonal blocks $s \in \{s_1,\dots,s_t\}$ of C_r, we can locally choose $U_r^* \perp (F_{s_1,r},\dots,F_{s_t,r})$ and $V_r^* \perp (E_{r,s_1},\dots,E_{r,s_t})$. We define $q_r = \sum_{s:s\neq r} q_{rs}$, where q_{rs} refers to the number of columns (i.e., the rank) of E_{rs} and F_{rs} and define

$$\sigma^*_{n_r-q_r} := \min_{\substack{u\in U_r^*\setminus\{0\}\\ v\in V_r^*\setminus\{0\}}} \frac{v^T C_r u}{\|v\|_2\ \|u\|_2}. \tag{10.19}$$

Then we immediately obtain

$$\sigma_l(C_r) \geqslant \sigma^*_{n_r-q_r} \geqslant \sigma_{n_r}(C_r),\ \ \sigma_l(C_r(X)) \geqslant \sigma^*_{n_r-q_r} \geqslant \sigma_{n_r}(C_r(X)) \text{ and } \sigma^*_{n_r-q_r} \geqslant \alpha_0$$

for any $l \leqslant n_r - q_r$, $r = 1, \ldots, p$. This also shows that $\sigma^*_{n_r-q_r}$ gives an upper bound for α_0 which cannot be improved. This is also reasonable since the remaining rows and columns of $C_r(X)$ coincide with those of A up to some zeros. Due to the sparsity of the off-diagonal blocks A_{rs}, A_{sr}, our matrices U_r^* and V_r^* would cover many unit vectors associated with unknowns that are not connected with neighbouring blocks in the sense of the underlying graph $G_M(A)$.

Example 8 We will discuss again the equation $-\Delta u = f$ on the unit square $\Omega = [0,1]^2$ in two spatial dimensions. We give a simplified model of different boundary conditions, namely Dirichlet boundary conditions $u = g$ on Γ_D and some kind of Neumann-type boundary conditions $\partial u/\partial \nu = 0$ on Γ_N.

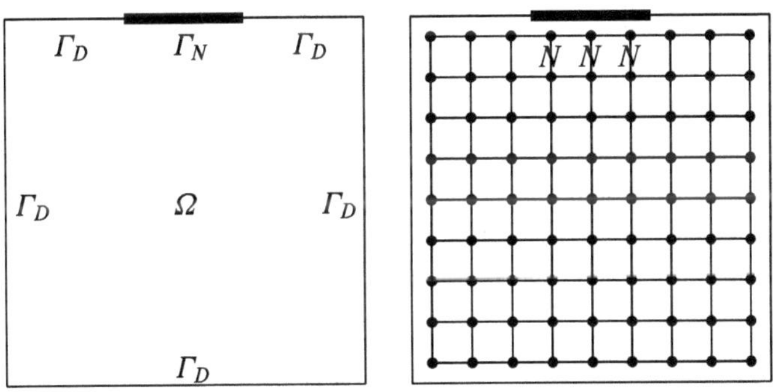

To simplify the discussion we use the 5-point-star difference stencil which leads to a matrix with 4 on the main diagonal and -1 in the off-diagonal positions as described in Example 4. At the positions associated with Γ_N we reduce the diagonal entry from 4 to 3 which refers to first order Neumann boundary conditions. We divide the domain into a checker board of 9 subdomains which corresponds to a block-diagonal splitting with 9 diagonal blocks. For each of the diagonal block we sketch the associated relevant singular values. We will choose a grid size of total size 150×150. This means if we have $p = 9 = 3 \times 3$ diagonal blocks, then each diagonal block is of size $n_r = 2500$.The rank q_r is between 100 and 200 depending on the diagonal block. We will compare each local $\sigma^*_{n_r-q_r}$ with

1. $\sigma_{n_r-q_r}(C_r)$ and $\sigma_{n_r}(C_r)$ of the original block-diagonal matrix and with
2. $\sigma_{n_r-q_r}(C_r(I))$ and $\sigma_{n_r}(C_r(I))$ for minimum-rank decoupling using $X = I$.

$\sigma^*_{n_r-q_r}$

$2.5 \cdot 10^{-3}$	$8.1 \cdot 10^{-3}$	$2.5 \cdot 10^{-3}$
$5.1 \cdot 10^{-3}$	$8.2 \cdot 10^{-3}$	$5.1 \cdot 10^{-3}$
$2.5 \cdot 10^{-3}$	$5.1 \cdot 10^{-3}$	$2.5 \cdot 10^{-3}$

$\frac{\sigma_{n_r-q_r}(C_r)}{\sigma_{n_r}(C_r)}$

$4.9 \cdot 10^{-1}$ $2.4 \cdot 10^{-3}$	$7.4 \cdot 10^{-1}$ $7.6 \cdot 10^{-3}$	$4.9 \cdot 10^{-1}$ $2.4 \cdot 10^{-3}$
$7.2 \cdot 10^{-1}$ $4.8 \cdot 10^{-3}$	$9.6 \cdot 10^{-1}$ $7.6 \cdot 10^{-3}$	$7.2 \cdot 10^{-1}$ $4.8 \cdot 10^{-3}$
$4.9 \cdot 10^{-1}$ $2.4 \cdot 10^{-3}$	$7.2 \cdot 10^{-1}$ $4.8 \cdot 10^{-3}$	$4.9 \cdot 10^{-1}$ $2.4 \cdot 10^{-3}$

$\frac{\sigma_{n_r-q_r}(C_r(I))}{\sigma_{n_r}(C_r(I))}$

$5.0 \cdot 10^{-1}$ $2.5 \cdot 10^{-3}$	$7.6 \cdot 10^{-1}$ $7.8 \cdot 10^{-3}$	$5.0 \cdot 10^{-1}$ $2.5 \cdot 10^{-3}$
$7.4 \cdot 10^{-1}$ $4.9 \cdot 10^{-3}$	$1.0 \cdot 10^{0}$ $7.9 \cdot 10^{-3}$	$7.4 \cdot 10^{-1}$ $4.9 \cdot 10^{-3}$
$5.0 \cdot 10^{-1}$ $2.5 \cdot 10^{-3}$	$7.4 \cdot 10^{-1}$ $4.9 \cdot 10^{-3}$	$5.0 \cdot 10^{-1}$ $2.5 \cdot 10^{-3}$

We can see in this specific example that $\sigma^*_{n_r-q_r}$ serves as a fairly well upper bound for $\sigma_{n_r}(C)$ and $\sigma_{n_r}(C(I))$. This is easily explained by the nature of the partial differential equation, since $\sigma^*_{n_r-q_r}$ refers to the smallest singular value of the subsystem which leaves out the nodes at the interfaces. This system is in general only slightly smaller but with similar properties as each C_r and $C_r(I)$, except that one can read omitting the nodes near the interfaces as choosing Dirichlet boundary conditions everywhere.

We can now easily apply the analytic solution of the completion problem from [29] but we like to note that the constraint with X and X^{-1} is in general not satisfied. We will focus on each local problem involving the blocks $(r, r), (r, s), (s, r), (s, s)$. This simplifies the completion problem dramatically and also allows to treat it for each pair of neighbouring diagonal blocks separately.

Lemma 2 *Let $\{r, s\} \in \mathscr{E}_M$ such that $r < s$. Let*

$$\alpha_0 = \min\left\{\sigma_{\min}\left(\begin{array}{cc|cc} C_r & 0 & E_{rs} & 0 \\ 0 & C_s & 0 & E_{sr} \end{array}\right), \sigma_{\min}\left(\begin{array}{cc|cc} C_r^T & 0 & F_{rs} & 0 \\ 0 & C_s^T & 0 & F_{sr} \end{array}\right)\right\}.$$

Given $\alpha < \alpha_0$ such that α is not a singular value of C_r or C_s, then the general solution X of

$$\left\| \left(\begin{array}{cc|cc} C_r & 0 & E_{rs} & 0 \\ 0 & C_s & 0 & E_{sr} \\ \hline F_{sr}^T & 0 & X_{rr} & X_{rs} \\ 0 & F_{rs}^T & X_{sr} & X_{ss} \end{array}\right)^{-1} \right\|_2 \leqslant \frac{1}{\alpha}$$

satisfies $X_{rs} = 0 = X_{sr}^T$ *and*

$$X_{rr} = F_{sr}^T C_r^T \left(C_r C_r^T - \alpha^2 I\right)^{-1} E_{rs} + \alpha Y_{rr}, \\ X_{ss} = F_{rs}^T C_s^T \left(C_s C_s^T - \alpha^2 I\right)^{-1} E_{sr} + \alpha Y_{ss}, \tag{10.20}$$

where Y_{rr}, Y_{ss} *may be any matrices such that*

$$Y_{rr}\left(I - E_{rs}^T (C_r C_r^T + E_{rs} E_{rs}^T - \alpha^2 I)^{-1} E_{rs}\right) Y_{rr}^T \geqslant I + F_{sr}^T (C_r^T C_r - \alpha^2 I)^{-1} F_{sr}$$
$$Y_{ss}\left(I - E_{sr}^T (C_s C_s^T + E_{sr} E_{sr}^T - \alpha^2 I)^{-1} E_{sr}\right) Y_{ss}^T \geqslant I + F_{rs}^T (C_s^T C_s - \alpha^2 I)^{-1} F_{rs}$$

in the sense of quadratic forms.

Proof We set

$$\hat{C} = \mathrm{diag}(C_r, C_s), \hat{E} = \mathrm{diag}(E_{rs}, E_{sr}), \hat{F} = \mathrm{diag}(F_{sr}, F_{rs}).$$

Except for the block-diagonal structure of X this lemma exactly reveals Theorem 3.1 from [29], which states that there exists X such that

$$X = \hat{F}^T \hat{C}^T \left(\hat{C}\hat{C}^T - \alpha^2 I\right)^{-1} \hat{E} + \alpha Y, \text{ where}$$
$$Y\left(I - \hat{E}^T (\hat{C}\hat{C}^H + \hat{E}\hat{E}^T - \alpha^2 I)^{-1} \hat{E}\right) Y^T \geqslant I + \hat{F}^T (\hat{C}^T \hat{C} - \alpha^2 I)^{-1} \hat{F}.$$

The underlying block structure of $\hat{C}$, $\hat{E}$ and $\hat{F}$ obviously induce the block structure of X.

We like to mention that often enough (say in applications arising from partial differential equations), the diagonal part of a matrix is well-conditioned enough to be used, i.e., rather than using the complete inverses in Lemma 2, we could work with the diagonal parts before inverting the matrices. In this case, simplified versions of X_{rr}, X_{ss} from Lemma 2 could be used to define $-X_{rs}^{-1}, -X_{rs}$.

We like to mention that the Hermitian case can be treated more easily as stated in Theorem 2.1 in [29]. Even if A and C are symmetric and positive definite, $E = F$ and if X is chosen positive definite as well, the constraint minimization problem

$$\left\| \left(\begin{array}{c|cc} C & E^{(1)} & E^{(2)} \\ \hline (E^{(1)})^T & -X^{-1} & 0 \\ (E^{(2)})^T & 0 & -X \end{array} \right)^{-1} \right\|_2 \leqslant \frac{1}{\alpha}$$

refers to a Hermitian but indefinite problem. In this case we always have

$$\lambda_l(C_r(X)) \equiv \sigma_l(C_r(X)) \geqslant \lambda_l(C_r)$$

since in the sense of quadratic forms we have

$$C_r(X) = C_r + \sum_{s:s>r} E_{rs} X_{rs} E_{rs}^T + \sum_{s:s<r} E_{rs} X_{rs}^{-1} E_{rs}^T \geqslant C_r.$$

This can be observed in Example 8. Thus $\|C_r(X)^{-1}\|$ can only become better than $\|C_r^{-1}\|$ and the same applies to the condition number as long as $\|C_r(X)\| \approx \|C_r\|$.

Example 9 We will continue Example 8, except that the elliptic operator $-u_{xx} - u_{yy}$ is now replaced by $-\varepsilon u_{xx} - \varepsilon u_{yy}$ with varying coefficient ε as illustrated below.

24	4	24
4	1955	4
24	4	24

For simplicity we assume that the larger value is taken on the interfaces. We like to stress that each local interface between two neighbouring diagonal blocks is essentially of the following type for a suitable $\alpha \geqslant \beta$ (e.g. $\alpha = 3 \cdot 1955$, $\beta = 4$)

$$\begin{pmatrix} A_{rr} & A_{rs} \\ A_{sr} & A_{ss} \end{pmatrix} = \left(\begin{array}{ccc|ccc} \alpha+\beta & -\beta & & & & \\ -\beta & \ddots & \ddots & & -\beta I & \\ & \ddots & \ddots & & & \\ \hline & & & 4\beta & -\beta & \\ & -\beta I & & -\beta & \ddots & \ddots \\ & & & & \ddots & \ddots \end{array}\right)$$

Since minimum rank decoupling adds positive semidefinite matrices to the diagonal blocks, we propose to move the interface nodes (which reflect the jumps of ε), to the diagonal blocks with larger ε. In this case the diagonal entries of the blocks with larger ε have relatively small diagonal entries at the nodes connected to the neighbouring blocks, e.g., for the $(2,2)$ block, the diagonal entries are $4\varepsilon = 7820$ for the inner nodes, whereas the diagonal entries of the $(2,2)$ system in the extremal four corners are only half as big. Since we work with splittings rather than with removing nodes to a remaining Schur complement system, this effect cannot be avoided. We illustrate this effect by stating the largest and smallest singular value

$\sigma_{\max}(C_r)$, $\sigma_{\min}(C_r)$ for each diagonal block C_r of the unmodified block diagonal matrix C (we will use $N = 40$).

$$\frac{\sigma_1(C_r)}{\sigma_{n_r}(C_r)}$$

$1.9 \cdot 10^2$ $5.7 \cdot 10^{-2}$	$3.2 \cdot 10^1$ $4.7 \cdot 10^{-2}$	$1.9 \cdot 10^2$ $5.7 \cdot 10^{-2}$
$3.2 \cdot 10^1$ $2.9 \cdot 10^{-2}$	$1.6 \cdot 10^4$ $3.9 \cdot 10^{-1}$	$3.2 \cdot 10^1$ $2.9 \cdot 10^{-2}$
$1.9 \cdot 10^2$ $5.7 \cdot 10^{-2}$	$3.2 \cdot 10^1$ $2.9 \cdot 10^{-2}$	$1.9 \cdot 10^2$ $5.7 \cdot 10^{-2}$

Knowing that for large ε the diagonal entries close to the interfaces are less than in the interior of the diagonal block, one can use this information to increase the diagonal entries, e.g., the $(2,2)$ block. Choosing $X = 40 \cdot I$ for all X_{rs} in the $(2,2)$ block and X_{rs}^{-1} outside improves the condition number dramatically. Similarly, for the four blocks in the corner of the domain we could increase the diagonal entries further using $X = 4 \cdot I$. This improves the condition number of several diagonal blocks significantly while other diagonal blocks are hardly affected.

$$\frac{\sigma_1(C_r(X))}{\sigma_{n_r}(C_r(X))}$$

$1.9 \cdot 10^2$ $7.4 \cdot 10^{-2}$	$3.2 \cdot 10^1$ $4.7 \cdot 10^{-2}$	$1.9 \cdot 10^2$ $7.4 \cdot 10^{-2}$
$3.2 \cdot 10^1$ $3.0 \cdot 10^{-2}$	$1.6 \cdot 10^4$ $2.4 \cdot 10^1$	$3.2 \cdot 10^1$ $3.0 \cdot 10^{-2}$
$1.9 \cdot 10^2$ $7.4 \cdot 10^{-2}$	$3.2 \cdot 10^1$ $3.0 \cdot 10^{-2}$	$1.9 \cdot 10^2$ $7.4 \cdot 10^{-2}$

Finally, with respect to ε, the best condition is obtained in the order 1955/04/24. This example demonstrates that completion is able to improve the condition number up to two orders of magnitude in this example and leading to a lower rank between A and $C(X)$ at the same time. We also reiterate that part of this success is moving the interface nodes to the diagonal blocks with larger ε.

10.5 Algebraic Multilevel Preconditioning

So far we have discussed how to improve the diagonal blocks in block-diagonal splitting and for both approaches, the splitting-type approach and the partitioning-type approach we have assumed that the systems are solved directly. Often enough, in practice we prefer to solve these systems iteratively using preconditioned Krylov subspace solvers. Since many application problems arise from the discretization of partial differential equations, preconditioning methods based on (algebraic) multilevel methods are preferred. Therefore this section will discuss algebraic multilevel methods and we will also give some ideas how splitting or partitioning the original system as in Sect. 10.3 may be used to parallelize the approach. Multilevel methods [36, 69] in general are popular for solving systems arising from partial differential equations. However, when information about some kind of grid hierarchy is not available, one often has to use algebraic approaches to construct multilevel methods which mimic the behaviour of multigrid methods using analogous terminology such as smoothing and coarse grid correction. As long as the system arises from partial differential equations, say using finite element discretization, one has additional information about the underlying physical problem and in this case one may use agglomeration techniques in order to glue together clusters of element matrices to successively build an algebraic coarsening hierarchy (cf. e.g. [17, 21, 37, 70]). Somehow in the opposite direction of this development, recent approaches to finite element aggregation are based on a relatively simple aggregation approach but instead they are supplemented with flexible Krylov subspace solvers at every level (also referred to as K-cycle), see e.g. [54, 56, 57]. In a similar direction, algebraic multilevel Krylov methods use K-cycles as well but shift the coarse grid operator additionally [30, 31]. Further approaches such as [5, 6, 60, 66] strongly focus on the underlying matrix and construct the multilevel hierarchy algebraically. This eventually justifies to employ multilevel incomplete factorization as basis of the coarsening process, either when there is a strong link to an underlying partial differential equation [5, 6, 58, 72, 73] or using purely algebraic methodology such as diagonal dominance, independent sets or related ideas, see e.g., [14, 16, 24, 63]. We will link earlier work on algebraic multilevel methods [12, 13] to illustrate the theoretical and practical performance of the multilevel incomplete factorization method, therefore we will restrict the description to these class of methods. Following [14], we rescale and reorder the initial system $A \in \mathbb{R}^{n,n}$ to obtain

$$\hat{A} = P^T DADP,$$

where $D \in \mathbb{R}^{n,n}$ is a nonsingular diagonal matrix and $P \in \mathbb{R}^{n,n}$ is a permutation matrix. Here D is chosen such that DAD has all diagonal entries equal to 1. Fill-reducing algorithms such as (approximate) minimum degree [2] or multilevel nested dissection [40, 41] can be used afterwards to prevent (incomplete) factorization methods from producing too much fill. Then we perform a partial approximate LDL^T factorization of type

$$\Pi^T \hat{A} \Pi = \begin{pmatrix} B & E^T \\ E & C \end{pmatrix} = \begin{pmatrix} L_B & 0 \\ L_E & I \end{pmatrix} \begin{pmatrix} D_B & 0 \\ 0 & S_C \end{pmatrix} \begin{pmatrix} L_B^T & E_F^T \\ 0 & I \end{pmatrix} + \mathscr{E} \qquad (10.21)$$

where we allow further symmetric permutations $\Pi \in \mathbb{R}^{n,n}$ for stability of the factorization. Here D_B refers to a nonsingular diagonal matrix, L_B is lower triangular with unit diagonal part and $\mathscr{E}$ refers to some appropriate perturbation. Furthermore we have $B \approx L_B D_B L_B^T$, $L_E D_B L_B^T \approx E$. Eventually we obtain a remaining approximate Schur complement $S_C \approx C - EB^{-1}F$ that consists of all delayed pivots which were not suitable to serve as pivots during the approximate factorization. Applying the whole procedure to S_C then leads to a multilevel incomplete factorization, where level-by-level, the size of the remaining Schur complement is reduced until it reaches a size such that it can be easily factorized, say, by a dense Cholesky factorization method. Multilevel incomplete factorization methods as described here are well-established, see e.g. [5, 16, 24, 62, 63]. For ease of notation we collect the permutation matrices Π and P in a single permutation matrix and call it again P. Often enough, L_B is not stored explicitly, but implicitly defined via $L_E := EL_B^{-T} D_B^{-1}$ saving some memory at the cost of solving an additional system. We like to point out that in this case $\mathscr{E}$ from (10.21) has an empty $(1, 2)$ block and $(2, 1)$ block. The same applies if $S_C := C - L_E D_B L_E^T$ is chosen. Having only one block $\mathscr{E}_B$ different from zero does not necessarily mean that the approximate factorization is more accurate, since this $\mathscr{E}_B$ propagates through the factorization and using the approximate factorization for preconditioning requires to apply the inverses in Eq. (10.21) which may lift the influence of $\mathscr{E}$.

Example 10 We illustrate for the model problem $-\Delta u = f$ in two spatial dimensions and $N = 100$ grid points in each direction the skeleton of a multilevel factorization.

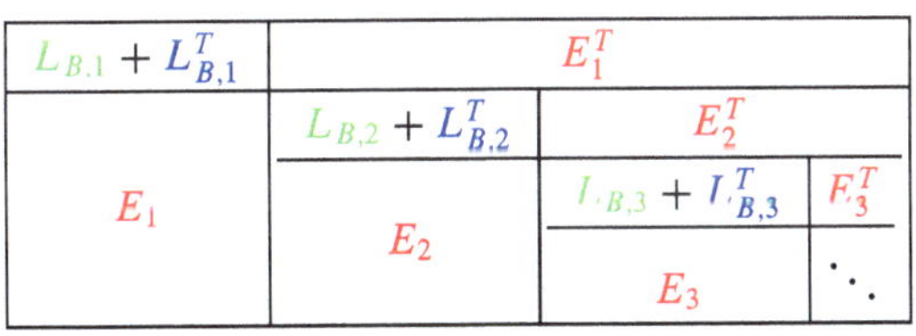

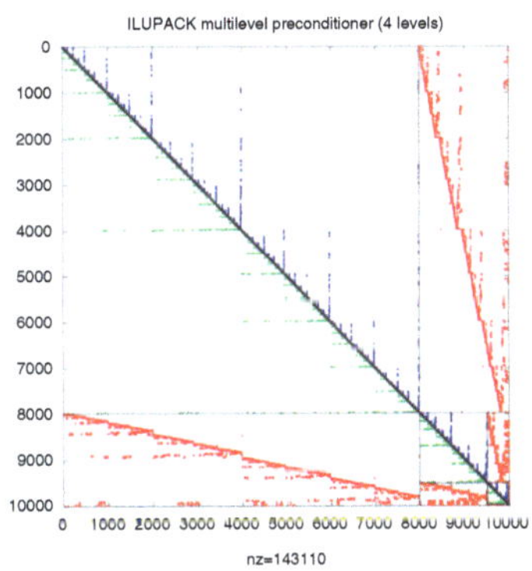

We like to emphasize that at least a single level factorization yields an approximate inverse of type

$$\hat{A}^{-1} \approx \begin{pmatrix} I \\ 0 \end{pmatrix} (L_B D_B L_B^T)^{-1} \begin{pmatrix} I \\ 0 \end{pmatrix}^T + Q S_C^{-1} Q^T \text{ where } Q = \begin{pmatrix} -L_B^{-T} L_E^T \\ I \end{pmatrix}. \tag{10.22}$$

Approximate inverse preconditioners of this type are well-studied in literature and we like to study how a preconditioner of type

$$M^{(1)} = LL^T + Q S_C^{-1} Q^T \tag{10.23}$$

will approximate $\hat{A}$ for some nonsingular $L \in \mathbb{R}^{n,n}$. In the limit, when $LL^T \to \begin{pmatrix} I \\ 0 \end{pmatrix} (L_B D_B L_B^T)^{-1} \begin{pmatrix} I \\ 0 \end{pmatrix}^T$ we will also obtain some information about the (multilevel) incomplete factorization as preconditioner. Here we can imagine that for some positive σ, τ, we could have, e.g.,

$$L = \begin{pmatrix} \sigma L_B^{-T} D_B^{-1/2} & 0 \\ 0 & \tau I \end{pmatrix}.$$

We denote by m the remaining block size of $C \in \mathbb{R}^{m,m}$. It was shown in [12], that the optimal preconditioner for $\hat{A}$ of type $LL^T + QZ^{-1}Q^T$, $Q \in \mathbb{R}^{n,m}$, $Z \in \mathbb{R}^{m,m}$, with respect to the condition number of the preconditioned system is given by choosing Q as the matrix of eigenvectors $Q_{opt} = [q_1, \dots, q_m]$ of $L^T \hat{A} L$ with respect to its m smallest eigenvalues $\lambda_1, \dots, \lambda_m$ and $Z = Q^T A Q$ is almost optimal. It is obvious that for any nonsingular X, $Q_{opt} \to Q_{opt} X$, $Z \to (X^{-T} Z X^{-1})$ is optimal as well, i.e., Q has to approximate the invariant subspace of $L^T \hat{A} L$ associated with its smallest eigenvalues. Taking into account the optimality of Q_{opt} the natural question arises for the preconditioner $M^{(1)}$ from (10.23) how close the specific choice Q matches the optimal Q_{opt}. We like to mention that $L^T \hat{A} L$ must have eigenvalues less than or equal to 1 which is satisfied for sufficiently small σ and τ. Note also that since we have scaled the original system A initially and since $D_B^{-1/2} L_B^{-1} B L_B^{-T} D_B^{-1/2} \approx I$, σ and τ need not be chosen too small. Indeed we may expect that $\sigma, \tau = \mathcal{O}(1)$ is already sufficient. One can also verify easily that in the limit case as $\tau \to 0$, we have

$$(L^T \hat{A} L)^{-1} = \mathcal{O}(1) + L^{-1} \begin{pmatrix} -B^{-1}E^T \\ I \end{pmatrix} (C - EB^{-1}E^T)^{-1} \begin{pmatrix} -B^{-1}E^T \\ I \end{pmatrix}^T L^{-T}.$$

This illustrates that asymptotically as $\tau \to 0$ the largest m eigenvalues of $(L^T \hat{A} L)^{-1}$ and their associated invariant subspace is fairly well approximated by

$$L^{-1} \begin{pmatrix} -B^{-1}E^T \\ I \end{pmatrix}$$

which is therefore close to the optimal rank m choice. This in turn justifies choosing

$$Q = \begin{pmatrix} -L_B^{-T} L_E^T \\ I \end{pmatrix}$$

for the preconditioner $M^{(1)}$ and $S_C \approx Q^T A Q$ as almost optimal choice. We will next illustrate how Theorem 4 from [12] describes the quality of the preconditioner $M^{(1)}$ and we will further use this Theorem in order to improve the multilevel incomplete factorization preconditioner (10.21).

There are two key properties that need to be fulfilled. First of all, we need W such that

$$W^T \hat{A} Q = 0 \text{ and } \Delta W^T \hat{A} W - W^T L^{-T} L^{-1} W \text{ positive semidefinite} \qquad (10.24)$$

for some $\Delta > 0$. Second, the approximate Schur complement S_C has to satisfy

$$\gamma Q^T \hat{A} Q \leqslant S_C \leqslant \Gamma Q^T \hat{A} Q$$

in the sense of quadratic forms for some $0 < \gamma \leqslant \Gamma$. Then

$$\operatorname{cond}\big(\big(M^{(1)}\big)^{-1/2} \hat{A} \big(M^{(1)}\big)^{-1/2}\big) \leqslant \frac{\gamma}{(\gamma+1)\max\{\Gamma, \Delta\}}.$$

While γ and Γ are quite natural bounds, the delicate question is the size of Δ. One can easily verify that using $E = L_E D_B L_B^T$ and $Z = C - L_E D_B L_E^T$ we have

$$\hat{A} Q = \begin{pmatrix} -\mathscr{E}_B \tilde{B}^{-1} E^T \\ Z \end{pmatrix}, \text{ where } \tilde{B} = L_B D_B L_B^T. \qquad (10.25)$$

This allows to define

$$W^T := \begin{pmatrix} I & -\mathscr{E}_B \tilde{B}^{-1} E^T Z^{-1} \end{pmatrix}$$

and to bound Δ. We will not follow this approach in detail and use a different way to examine the multilevel ILU as preconditioner, but even here we can immediately see that $\mathscr{E}_B \tilde{B}^{-1} E^T S_C^{-1}$ plays a central role. Usually the multilevel factorization on each level is set up such that B can be easily approximated by $\tilde{B}$ using some criterion such as diagonal dominance or diagonal blocks of small size whereas S_C is usually more critical. This in turn means that even a small error $\mathscr{E}_B$ may be amplified by S_C^{-1} significantly. This is in line with algebraic multigrid theory (see e.g. [55]), that for approximate inverses (here $\tilde{B}^{-1}$) in multigrid methods it is not enough to approximate the original matrix B^{-1} sufficiently.

We will now give a simple theorem to state the approximation quality of (10.21) directly.

Theorem 1 *Consider the approximate factorization from (10.21) and assume that* $E = L_E D_B L_B^T$. *Furthermore, suppose that* $\tilde{B} = L_B D_B L_B^T$ *satisfies*

$$\lambda B \leqslant \tilde{B} \leqslant \Lambda B$$

for some $0 < \lambda \leqslant \Lambda$ *and that there exist* $0 < \gamma \leqslant \Gamma$ *such that*

$$\gamma Z \leqslant S_C \leqslant \Gamma Z, \text{ where } Z = C - E\tilde{B}^{-1}E^T$$

and we assume that Z is positive definite. Define the preconditioned system T by

$$T = \begin{pmatrix} D_B & 0 \\ 0 & S_C \end{pmatrix}^{-1/2} \begin{pmatrix} L_B & 0 \\ L_E & I \end{pmatrix}^{-1} \hat{A} \begin{pmatrix} L_B & 0 \\ L_E & I \end{pmatrix}^{-T} \begin{pmatrix} D_B & 0 \\ 0 & S_C \end{pmatrix}^{-1/2}.$$

Then

$$\operatorname{cond}(T) \leqslant \frac{\max\{\Lambda, \Gamma\}(\frac{\Lambda}{\lambda} + \sqrt{\Lambda}\|H\|_2)}{\min\{\Lambda, \gamma\}(1 - \sqrt{\Lambda}\|H\|_2)},$$

where

$$H = D_B^{-1/2} L_B^{-1} \mathscr{E}_B \tilde{B}^{-1} E^T Z^{-1/2},$$

provided that $\sqrt{\Lambda}\|H\|_2 < 1$.

Proof We have that

$$C - L_E D_B L_E^T = C - (EL_B^{-T} D_B^{-1}) D_B (EL_B^{-T} D_B^{-1})^T = C - E\tilde{B}^{-1} E^T = Z.$$

From (10.25) we immediately obtain

$$\hat{A} \begin{pmatrix} -L_B^{-T} L_E^T \\ I \end{pmatrix} = \hat{A} Q = \begin{pmatrix} -\mathscr{E}_B \tilde{B}^{-1} E^T \\ Z \end{pmatrix}.$$

We define T_Λ via

$$T_\Lambda := \begin{pmatrix} \frac{1}{\Lambda} D_B & 0 \\ 0 & Z \end{pmatrix}^{-1/2} \begin{pmatrix} L_B & 0 \\ L_E & I \end{pmatrix}^{-1} \hat{A} \begin{pmatrix} L_B & 0 \\ L_E & I \end{pmatrix}^{-T} \begin{pmatrix} \frac{1}{\Lambda} D_B & 0 \\ 0 & Z \end{pmatrix}^{-1/2}.$$

Since we know that Q is the second block column of $\begin{pmatrix} I & 0 \\ L_E L_B^{-1} & I \end{pmatrix}^{-T}$ it follows that

$$\begin{aligned} T_\Lambda &= \begin{pmatrix} \sqrt{\Lambda} D_B^{-1/2} L_B^{-1} & 0 \\ 0 & Z^{-1/2} \end{pmatrix} \begin{pmatrix} B & -\mathscr{E}_B \tilde{B}^{-1} E^T \\ -\mathscr{E}_B \tilde{B}^{-1} E^T & Z \end{pmatrix} \\ &\quad \times \begin{pmatrix} \sqrt{\Lambda} L_B^{-T} D_B^{-1/2} & 0 \\ 0 & Z^{-1/2} \end{pmatrix} \\ &= \begin{pmatrix} \Lambda D_B^{-1/2} L_B^{-1} B L_B^{-T} D_B^{-1/2} & -\sqrt{\Lambda} D_B^{-1/2} L_B^{-1} \mathscr{E}_B \tilde{B}^{-1} E^T Z^{-1/2} \\ -\sqrt{\Lambda} Z^{-1/2} \mathscr{E}_B \tilde{B}^{-1} E^T L_B^{-T} D_B^{-1/2} & I \end{pmatrix}. \end{aligned}$$

We can see that the (1, 2) block exactly refers to $-\sqrt{\Lambda}H$. Since $\tilde{B} \leqslant \Lambda B$ in quadratic forms it follows that $\Lambda D_B^{-1/2} L_B^{-1} B L_B^{-T} D_B^{-1/2} \geqslant I$. This in turn implies that

$$T_\Lambda - \begin{pmatrix} I & -\sqrt{\Lambda}H \\ -\sqrt{\Lambda}H^T & I \end{pmatrix} = \begin{pmatrix} \Lambda D_B^{-1/2} L_B^{-1} B L_B^{-T} D_B^{-1/2} - I & 0 \\ 0 & 0 \end{pmatrix}$$

is positive semidefinite. Thus on one hand we have

$$\lambda_{\min}(T_\Lambda) > 1 - \sqrt{\Lambda}\|H\|_2,$$

provided that $\Lambda\|H\|_2 < 1$. On the other hand we have

$$\begin{aligned}\lambda_{\max}(T_\Lambda) = \|T_\Lambda\|_2 &\leqslant \|\Lambda D_B^{-1/2} L_B^{-1} B L_B^{-T} D_B^{-1/2}\|_2 + \sqrt{\Lambda}\|H\|_2 \\ &\leqslant \frac{\Lambda}{\lambda} + \sqrt{\Lambda}\|H\|_2.\end{aligned}$$

To conclude the proof, we point out that the preconditioned system refers to $T \equiv T_1$ and we obviously have that

$$\min\{\frac{1}{\Lambda}, \frac{1}{\Gamma}\} \begin{pmatrix} \Lambda\tilde{B}^{-1} & 0 \\ 0 & Z^{-1} \end{pmatrix} \leqslant \begin{pmatrix} \tilde{B}^{-1} & 0 \\ 0 & S_C^{-1} \end{pmatrix} \leqslant \max\{\frac{1}{\Lambda}, \frac{1}{\gamma}\} \begin{pmatrix} \Lambda\tilde{B}^{-1} & 0 \\ 0 & Z^{-1} \end{pmatrix}$$

which directly implies

$$\lambda_{\min}(T) \geqslant \min\{\frac{1}{\Lambda}, \frac{1}{\Gamma}\}(1 - \sqrt{\Lambda}\|H\|_2), \ \lambda_{\max}(T) \leqslant \max\{\frac{1}{\Lambda}, \frac{1}{\gamma}\}(\frac{\Lambda}{\lambda} + \sqrt{\Lambda}\|H\|_2).$$

We give an interpretation of the bound obtained by Theorem 1. In practice, λ and Λ are expected to be close to 1, so this effect can be ignored, i.e. we essentially have

$$\text{cond}(T) \lessapprox \frac{\Gamma(1 + \|H\|_2)}{\gamma(1 - \|H\|_2)}.$$

Furthermore we note that

1. $\hat{A}$ is diagonally scaled, thus $\|E\|$ is moderately bounded,
2. $\|\mathscr{E}_B\|$ is considerably small when B is well-suited (say diagonally dominant or close to it).

Thus the main effects are how well S_C approximates Z and how $Z^{-1/2}$ amplifies the remaining small terms in $\|H\|_2$ which is not known in advance.

There are some techniques to keep the influence of $Z^{-1/2}$ smaller and to improve approximating Z by S_C, which we will discuss in the sequel. First of all, we like to point out that similar influence of Z^{-1} or S_C^{-1} is also illustrated by (10.24). We can improve Δ (cf. [13]) by considering

$$\Delta W^T \hat{A}^2 W - W^T \hat{A} W$$

instead. Besides, considering the preconditioner $M^{(2)}$ from [12]

$$M^{(2)} = 2LL^T - LL^T \hat{A} LL^T + (I - LL^T A) \begin{pmatrix} -\tilde{B}^{-1} E^T \\ I \end{pmatrix} S_C^{-1} \times \begin{pmatrix} -\tilde{B}^{-1} E^T \\ I \end{pmatrix}^T (I - ALL^T)$$

will yield improved bounds since in this case, essentially only

$$\Delta W^T (2\hat{A}^2 - \hat{A}^3) W, W^T \hat{A} W$$

are taken into account for the estimates (cf. Theorem 4 in [12]). We will not go into the details of deriving bounds for this case. but mention that this preconditioner one can read as replacing $\tilde{B}$ by more accurate approximations and thus reducing the error $\mathscr{E}_B$. Indeed, $M^{(2)}$ is obtained from the simple 2-level multilevel scheme

$$I - M^{(2)} \hat{A} \equiv (I - \begin{pmatrix} \tilde{B}^{-1} & 0 \\ 0 & 0 \end{pmatrix} \hat{A})(I - \begin{pmatrix} -\tilde{B}^{-1} E^T \\ I \end{pmatrix} S_C^{-1} \begin{pmatrix} -\tilde{B}^{-1} E^T \\ I \end{pmatrix}^T \hat{A})(I - \begin{pmatrix} \tilde{B}^{-1} & 0 \\ 0 & 0 \end{pmatrix} \hat{A}) \quad (10.26)$$

which demonstrates how a multilevel incomplete factorization can be easily upgraded to serve as algebraic multigrid method, see e.g. [59]. In the sense of multigrid methods, the first and the third factor are usually considered as smoothing while the factor in the middle reveals the coarse grid correction. We will demonstrate the difference between the simple multilevel incomplete factorization and its induced algebraic multigrid.

Example 11 Again we will consider the well-known problem $-\Delta u = f$ on a unit square in two spatial dimensions with Dirichlet boundary conditions and N grid points in every direction. We compare

1. The multilevel incomplete factorization from [14] with its default options (in particular a drop tolerance of 10^{-2} and preconditioned conjugate gradient method that stops when the relative error in the energy norm drops below 10^{-6}). This gives a multilevel incomplete Cholesky factorization (MLIC)
2. The associated algebraic multigrid method associated with $M^{(2)}$

Both operators will serve as preconditioners for the conjugate gradient method. Beside the computation time and the number of CG steps we will state the relative fill $\frac{nz(LDL^T)}{nz(A)}$ for the number of nonzero entries.

N	MLIC computation		MLIC-CG		$M^{(2)}$-CG	
	[sec]	fill	[sec]	steps	[sec]	steps
100	$6.6 \cdot 10^{-2}$	2.7	$5.1 \cdot 10^{-2}$	26	$9.1 \cdot 10^{-2}$	23
200	$2.5 \cdot 10^{-1}$	2.8	$3.4 \cdot 10^{-1}$	43	$6.7 \cdot 10^{-1}$	38
400	$1.3 \cdot 10^{0}$	2.8	$3.3 \cdot 10^{0}$	77	$6.7 \cdot 10^{0}$	66
800	$5.8 \cdot 10^{0}$	2.8	$3.0 \cdot 10^{1}$	135	$6.6 \cdot 10^{1}$	119
1600	$2.6 \cdot 10^{1}$	2.9	$2.4 \cdot 10^{2}$	237	$5.7 \cdot 10^{2}$	221

As we can see, although the number of iteration steps is slightly reduced the total computational amount of work even increases. Besides, none of the methods scales linearly.

We can see from Example 11, simply improving the quality of $\tilde{B}$ is not enough which is well-known (see e.g. [55]). Here the source is two-fold. On the one hand one has to ensure that the perturbation is sensitive with respect to H. On the other hand S_C needs to approximate Z sufficiently. Here we attempt to approach these requirement by computing a modified multilevel incomplete factorization that is exact for the vector e with all ones. Besides, when upgrading the multilevel ILU to an algebraic multigrid method, more natural improvements can be achieved borrowing the smoothing and coarse grid methodology from AMG. In the sense of AMG, in (10.26) we replace

$$\begin{pmatrix} \tilde{B}^{-1} & 0 \\ 0 & 0 \end{pmatrix} \longrightarrow \begin{cases} G^{-1} \\ G^{-T} \end{cases}$$

by more general approximations such as the inverse of the lower triangular part of $\hat{A}$ (Gauss-Seidel) or a damped inverse of the diagonal part (Jacobi). To preserve symmetry, one uses G^{-T} in the first factor of (10.26) and G^{-1} in the third factor. Additionally, since the middle factor in (10.26) solves the coarse grid only approximately in the multilevel case, one recursive call refers to the traditional V-cycle while two recursive calls are referred to as W-cycle. We note that since the factorization in (10.21) is exact for e, we have $\mathscr{E}_B e = 0$, $Be = \tilde{B}e$ and $S_C e = Ze$ and e can be regarded as sample vector for the low frequencies. Several algebraic multigrid methods and incomplete factorization methods make use of this improvement, see e.g. [66, 72, 73].

Example 12 We will continue Example 11 and consider the following preconditioners in two spatial dimensions, except that now the multilevel incomplete factorization (10.21) is exact for e. We will compare the following preconditioners.

1. Modified multilevel incomplete Cholesky (MLIC)
2. V-cycle AMG with one Gauss-Seidel forward and one Gauss-Seidel backward smoothing step (AMGV)
3. W-cycle AMG with one Gauss-Seidel forward and one Gauss-Seidel backward smoothing step (AMGW)

N	MLIC comput.		MLIC-CG		AMGV-CG		AMGW-CG	
	[sec]	fill	[sec]	steps	[sec]	steps	[sec]	steps
100	$6.5 \cdot 10^{-2}$	2.9	$3.2 \cdot 10^{-2}$	15	$6.9 \cdot 10^{-2}$	14	$8.7 \cdot 10^{-2}$	8
200	$3.1 \cdot 10^{-1}$	3.0	$1.5 \cdot 10^{-1}$	18	$3.9 \cdot 10^{-1}$	16	$4.4 \cdot 10^{-1}$	8
400	$1.6 \cdot 10^{0}$	3.1	$9.5 \cdot 10^{-1}$	20	$2.6 \cdot 10^{0}$	19	$2.6 \cdot 10^{0}$	8
800	$7.4 \cdot 10^{0}$	3.1	$5.4 \cdot 10^{0}$	23	$1.6 \cdot 10^{1}$	21	$1.5 \cdot 10^{1}$	8
1600	$3.4 \cdot 10^{1}$	3.2	$2.6 \cdot 10^{1}$	25	$8.1 \cdot 10^{1}$	24	$8.3 \cdot 10^{1}$	9

Although the number of CG steps, in particular for W-cycle, is better, the overall complexity is best for the multilevel ILU, because the approach is simpler and the intermediate coarse grid systems are not required. The latter are known to fill-up during the coarsening process.

Example 13 We will conclude this section with another example AF_SHELL3 from sheet metal forming, available at the University of Florida sparse matrix collection, to demonstrate the flexibility of algebraic multilevel ILU preconditioning. The symmetric positive definite system has a size of $n = 504855$ with approximately 35 nonzero entries per row. We will compare the methods without test vector e and with e.

without test vector e

MLIC comput.		MLIC-CG		AMGV-CG		AMGW-CG	
[sec]	fill	[sec]	steps	[sec]	steps	[sec]	steps
$5.1 \cdot 10^{1}$	3.9	$9.7 \cdot 10^{1}$	79	$2.5 \cdot 10^{2}$	82	$3.2 \cdot 10^{2}$	42

with test vector e

MLIC comput.		MLIC-CG		AMGV-CG		AMGW-CG	
[sec]	fill	[sec]	steps	[sec]	steps	[sec]	steps
$6.0 \cdot 10^{1}$	4.2	$5.0 \cdot 10^{1}$	40	$1.2 \cdot 10^{2}$	38	$1.9 \cdot 10^{2}$	20

Similar to Example 12, the ILU performs best, although not in terms of iteration steps. Again, using e to improve the method is beneficial.

We finally like to mention that the partitioning approach as indicated in Sect. 10.3 for nested dissection by nodes may also serve as parallelization approach prior to the incomplete factorization.

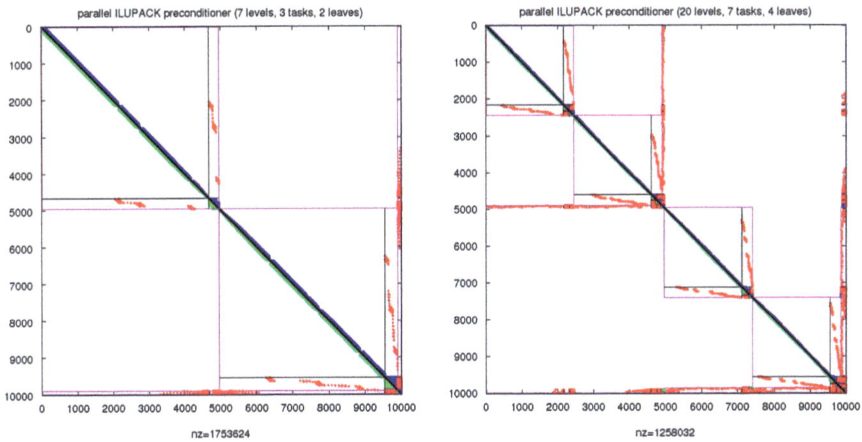

Fig. 10.7 Parallel multilevel incomplete factorization, $p = 2$ (*left*), $p = 4$ (*right*)

Example 14 We consider again the model problem $-\Delta u = f$ and sketch in Fig. 10.7 the parallel multilevel incomplete factorization in the cases $p = 2, 4$ and $N = 100$ grid points in each direction.

10.6 Approximate Inversion Using Multilevel Approximation

In this final section we will illustrate how most of the aspects discussed in the previous sections can be usefully united for the approximate inversion of matrices. Functions of entries of inverses of matrices like all diagonal entries of a sparse matrix inverse or its trace arise in several important computational applications such as density functional theory [42], covariance matrix analysis in uncertainty quantification [7], simulation of quantum field theories [45], vehicle acoustics optimization [52], or when evaluating Green's functions in computational nanolelectronics [48]. Often enough, modern computational methods for matrix inversion are based on reordering or splitting the system into independent parts [11, 51], since in this case the (approximate) inverse triangular factors tend to be relatively sparse which simplifies their computation [46, 47, 67, 68]. Here we will use the following ingredients.

1. We will use the partitioning approach (10.7) from Sect. 10.2 for partitioning the systems. If some of the diagonal blocks were ill-conditioned, one could alternatively fall back to the splitting approach (10.5) and use a completion approach.
2. The multilevel incomplete factorization from Sect. 10.5 will be used as approximate factorization.

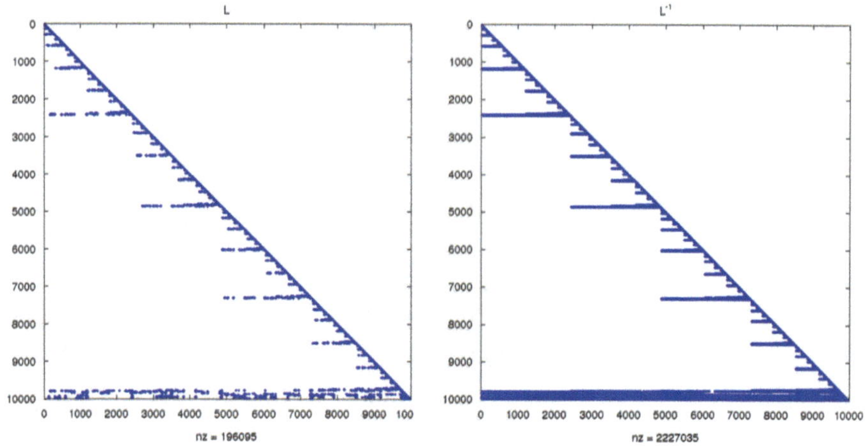

Fig. 10.8 Triangular factor and its (approximate) inverse after nested dissection reordering

For the multilevel incomplete factorization we scale and reorder at each level the system using nested dissection. In principle, an approximate factor

$$L = \left(\begin{array}{ccc|c} L_{11} & & 0 & \\ & \ddots & & 0 \\ 0 & & L_{p-1,p-1} & \\ \hline L_{p1} & \cdots & L_{p,p-1} & L_{pp} \end{array}\right)$$

is easily inverted. This structure keeps the inverse factor sparse and can be applied recursively and is used in approximate inverse methods [18] and is part of several methods for exact inversion [46, 47, 67].

Example 15 Consider the problem $-\Delta u = f$ on the unit square in two spatial dimensions with 5-point-star-stencil. The system will be reordered with nested dissection [40]. Figure 10.8 illustrates the incomplete Cholesky factor and its (approximate) inverse. Although its approximate inverse uses about ten times more memory it still approximately sparse.

Next we like to mention that multilevel incomplete factorizations can be rewritten as a single-level factorization. Consider the incomplete factorization (10.21) and suppose that $P_2^T D_2 S_C D_2 P_2 = L_C D_C L_C^T + \mathscr{E}_C$. One can easily verify that substitution into (10.21) leads to a factorization of the form $\hat{P}^T \hat{D} \hat{A} \hat{D} \hat{P} = \hat{L}\hat{L}^T + \hat{\mathscr{E}}$ with modified permutation matrix $\hat{P}$, new diagonal matrix $\hat{D}$, lower triangular matrix $\hat{L}$ and some perturbation $\hat{\mathscr{E}}$. The triangular factors from Example 15 already refer to a multilevel factorization that was formally rewritten as a single-level factorization.

When inverting the triangular factorization (10.21) we already know that

$$A^{-1} \approx DP\left[\begin{pmatrix} I \\ 0 \end{pmatrix} \tilde{B}^{-1} \begin{pmatrix} 0 & I \end{pmatrix} + \begin{pmatrix} -L_B^{-T} L_E^T \\ I \end{pmatrix} S_C^{-1} \begin{pmatrix} -L_E L_B^{-1} & I \end{pmatrix}\right] P^T D,$$

where equality holds if $\mathscr{E} = 0$ and in particular selected entries such as the diagonal entries of A^{-1} are dominated by S_C^{-1} when $\|\tilde{B}^{-1}\|$ is well-bounded. Here again, as before, we have set $\tilde{B} = L_B D_B L_B^T$. One can compute the diagonal entries of the inverses separately from the sum [67]. Computing the diagonal entries $\mathscr{D}(\tilde{B}^{-1})$ of $\tilde{B}^{-1}$ is easily achieved because of the nested dissection partition and the multilevel approach. It is harder to compute the remaining Schur complement S_C^{-1} in general. But again in a multilevel setting, S_C is substituted until eventually only a system of small size is left over. If we construct the multilevel factorization such that $L_E L_B^{-1}$ is bounded [14], then the influence of the diagonal entries $\mathscr{D}(L_B^{-T} L_E^T S_C^{-1} L_E L_B^{-1})$ in the inversion of $\hat{A}$ remains on the same order as $\|S_C^{-1}\|$. To construct $\tilde{B}$ that is easy to invert and to keep $\|L_E L_B^{-1}\|$ bounded justifies to use a multilevel approach instead of a single level incomplete factorization.

Example 16 We consider the linear operator A that is obtained from $-\Delta u = f$ on the unit square in two spatial dimensions using as before 5-point-difference stencil, Dirichlet boundary conditions and N grid points in each spatial direction. Here $\mathscr{D}(A^{-1})$ is explicitly known which simplifies numerical comparisons. We will use a multilevel incomplete factorization from [14] using different drop tolerances τ. Pivoting is introduced such that successively $\|L_B^{-1}\|$, $\|L_E L_B^{-1}\|$ are approximately kept below a given threshold κ; here we will choose $\kappa = 100$. For details of this strategy we refer to [14].

N	τ	$\frac{\|\hat{A}-\hat{L}D\hat{L}^T\|}{\|\hat{A}\|}$	$\frac{\|\mathscr{D}(A^{-1}-\hat{P}\hat{D}\,L^{-T}D^{-1}L^{-1}\hat{D}\hat{P}^T)\|}{\|\mathscr{D}(A^{-1})\|}$	$\frac{\|\text{trace}(A^{-1}-\hat{P}\hat{D}\,L^{-T}D^{-1}L^{-1}\hat{D}\hat{P}^T)\|}{\|\text{trace}(A^{-1})\|}$
50	10^{-4}	$4.2 \cdot 10^{-5}$	$1.1 \cdot 10^{-4}$	$5.2 \cdot 10^{-6}$
100	10^{-4}	$1.8 \cdot 10^{-5}$	$1.9 \cdot 10^{-5}$	$2.6 \cdot 10^{-6}$
200	10^{-4}	$1.4 \cdot 10^{-5}$	$3.8 \cdot 10^{-5}$	$2.3 \cdot 10^{-6}$
50	10^{-5}	$2.6 \cdot 10^{-6}$	$5.5 \cdot 10^{-6}$	$1.2 \cdot 10^{-7}$
100	10^{-5}	$2.2 \cdot 10^{-6}$	$1.2 \cdot 10^{-6}$	$6.2 \cdot 10^{-8}$
200	10^{-5}	$3.2 \cdot 10^{-4}$	$1.6 \cdot 10^{-3}$	$1.8 \cdot 10^{-4}$
50	10^{-6}	$8.5 \cdot 10^{-16}$	$8.1 \cdot 10^{-15}$	$5.4 \cdot 10^{-16}$
100	10^{-6}	$2.1 \cdot 10^{-8}$	$1.8 \cdot 10^{-8}$	$3.1 \cdot 10^{-10}$
200	10^{-6}	$1.1 \cdot 10^{-5}$	$6.1 \cdot 10^{-5}$	$1.0 \cdot 10^{-5}$

The displayed norm here is always $\| \bullet \|_\infty$. We point out that the multilevel incomplete factorization is not yet fit for approximate inversion. For this reason we do not display the computation time. We can see that the error with respect to the inverse is of the same order as the drop tolerance or at most one order greater which demonstrates the effectiveness of this approach.

Finally we mention that to turn the multilevel approach into an efficient method for approximate inversion, the approach would have to be modified to

$$\begin{pmatrix} W_B & 0 \\ W_E & I \end{pmatrix} \hat{A} \begin{pmatrix} W_B^T & W_E^T \\ 0 & I \end{pmatrix} = \begin{pmatrix} D_B & 0 \\ 0 & S_C \end{pmatrix} + \mathscr{E}$$

which refers to a multilevel approximate inverse-type approach generalizing the AINV method [9, 10]. This will be subject of future research and algorithms.

10.7 Conclusions

In this paper we have demonstrated that several Numerical Linear Algebra methods can be efficiently used in many recent preconditioning techniques and matrix inversion methods. They give deep information about the underlying approximation and help to improve these methods.

References

1. Agullo, E., Giraud, L., Guermouche, A., Haidar, A., Roman, J.: Parallel algebraic domain decomposition solver for the solution of augmented systems. Adv. Eng. Softw. **60–61**, 23–30 (2012)
2. Amestoy, P., Davis, T.A., Duff, I.S.: An approximate minimum degree ordering algorithm. SIAM J. Matrix Anal. Appl. **17**(4), 886–905 (1996)
3. Amestoy, P.R., Duff, I.S., L'Excellent, J.-Y.: Multifrontal parallel distributed symmetric and unsymmetric solvers. Comput. Methods Appl. Mech. Eng. **184**, 501–520 (2000)
4. Ammar, G., Benner, P., Mehrmann, V.: A multishift algorithm for the numerical solution of algebraic Riccati equations. Electron. Trans. Numer. Anal. **1**, 33–48 (1993)
5. Axelsson, O., Vassilevski, P.: Algebraic multilevel preconditioning methods I. Numer. Math. **56**, 157–177 (1989)
6. Axelsson, O., Vassilevski, P.: Algebraic multilevel preconditioning methods II. SIAM J. Numer. Anal. **27**, 1569–1590 (1990)
7. Bekas, C., Curioni, A., Fedulova, I.: Low-cost high performance uncertainty quantification. Concurr. Comput. Pract. Exp. **24**, 908–920 (2011)
8. Benzi, M., Haws, J.C., Tůma, M.: Preconditioning highly indefinite and nonsymmetric matrices. SIAM J. Sci. Comput. **22**(4), 1333–1353 (2000)
9. Benzi, M., Meyer, C.D., Tůma, M.: A sparse approximate inverse preconditioner for the conjugate gradient method. SIAM J. Sci. Comput. **17**, 1135–1149 (1996)
10. Benzi, M., Tůma, M.: A sparse approximate inverse preconditioner for nonsymmetric linear systems. SIAM J. Sci. Comput. **19**(3), 968–994 (1998)
11. Berry, M.W., Sameh, A.: Multiprocessor schemes for solving block tridiagonal linear systems. Int. J. Supercomput. Appl. **1**(3), 37–57 (1988)
12. Bollhöfer, M., Mehrmann, V.: Algebraic multilevel methods and sparse approximate inverses. SIAM J. Matrix Anal. Appl. **24**(1), 191–218 (2002)
13. Bollhöfer, M., Mehrmann, V.: Some convergence estimates for algebraic multilevel preconditioners. In: Olshevsky, V. (ed.) Fast Algorithms for Structured Matrices: Theory and Applications, vol. 323, pp. 293–312. AMS/SIAM, Providence/Philadelphia (2003)

14. Bollhöfer, M., Saad, Y.: Multilevel preconditioners constructed from inverse–based ILUs. SIAM J. Sci. Comput. **27**(5), 1627–1650 (2006)
15. Bondeli, S.: Parallele Algorithmen zur Lösung tridiagonaler Gleichungssysteme. Dissertationsschrift, ETH Zürich, Department Informatik, Institut f. Wissenschaftliches Rechnen (1991)
16. Botta, E., Wubs, F.: Matrix renumbering ILU: an effective algebraic multilevel ILU-decomposition. SIAM J. Matrix Anal. Appl. **20**, 1007–1026 (1999)
17. Brezina, M., Cleary, A.J., Falgout, R.D., Henson, V.E., Jones, J.E., Manteuffel, T.A., McCormick, S.F., Ruge, J.W.: Algebraic multigrid based on element interpolation (AMGe). SIAM J. Sci. Comput. **22**, 1570–1592 (2000)
18. Bridson, R., Tang, W.-P.: Ordering, anisotropy and factored sparse approximate inverses. SIAM J. Sci. Comput. **21**(3), 867–882 (1999)
19. Bunse-Gerstner, A., Mehrmann, V.: A symplectic QR-like algorithm for the solution of the real algebraic Riccati equation. IEEE Trans. Autom. Control **AC-31**, 1104–1113 (1986)
20. Byers, R., Mehrmann, V.: Symmetric updating of the solution of the algebraic Riccati equation. Methods Oper. Res. **54**, 117–125 (1986)
21. Chartier, T., Falgout, R.D., Henson, V.E., Jones, J., Manteuffel, T., McCormick, S., Ruge, J., Vassilevski, P.S.: Spectral AMGe (ρAMGe). SIAM J. Sci. Comput. **25**, 1–26 (2004)
22. Davis, T.A., Duff, I.S.: A combined unifrontal/multifrontal method for unsymmetric sparse matrices. ACM Trans. Math. Softw. **25**(1), 1–19 (1999)
23. Demmel, J.W., Eisenstat, S.C., Gilbert, J.R., Li, X.S., Liu, J.W.H.: A supernodal approach to sparse partial pivoting. SIAM J. Matrix Anal. Appl. **20**(3), 720–755 (1999)
24. Van der Ploeg, A., Botta, E., Wubs, F.: Nested grids ILU–decomposition (NGILU). J. Comput. Appl. Math. **66**, 515–526 (1996)
25. Van der Vorst, H.A.: Bi-CGSTAB: a fast and smoothly converging variant of Bi-CG for the solution of nonsymmetric linear systems. SIAM J. Sci. Stat. Comput. **13**(2), 631–644 (1992)
26. Duff, I.S., Koster, J.: The design and use of algorithms for permuting large entries to the diagonal of sparse matrices. SIAM J. Matrix Anal. Appl. **20**(4), 889–901 (1999)
27. Duff, I.S., Koster, J.: On algorithms for permuting large entries to the diagonal of a sparse matrix. SIAM J. Matrix Anal. Appl. **22**(4), 973–996 (2001)
28. Elsner, L., He, C., Mehrmann, V.: Minimizing the condition number of a positive definite matrix by completion. Numer. Math. **69**, 17–24 (1994)
29. Elsner, L., He, C., Mehrmann, V.: Minimization of the norm, the norm of the inverse and the condition number of a matrix by completion. Numer. Linear Algebra Appl. **2**(2), 155–171 (1995)
30. Erlangga, Y.A., Nabben, R.: Multilevel projection-based nested Krylov iteration for boundary value problems. SIAM J. Sci. Comput. **30**(3), 1572–1595 (2008)
31. Erlangga, Y.A., Nabben, R.: Algebraic multilevel Krylov methods. SIAM J. Sci. Comput. **31**(5), 3417–3437 (2009)
32. Giraud, L., Haidar, A., Saad, Y.: Sparse approximations of the Schur complement for parallel algebraic hybrid linear solvers in 3D. Numer. Math. Theory Methods Appl. **3**(3), 276–294 (2010)
33. Golub, G.H., Van Loan, C.F.: Matrix Computations, 4th edn. The Johns Hopkins University Press, Baltimore (2012)
34. Greenbaum, A.: Iterative Methods for Solving Linear Systems. Frontiers in Applied Mathematics. SIAM, Philadelphia (1997)
35. Gupta, A., Joshi, M., Kumar, V.: WSSMP: a high-performance serial and parallel symmetric sparse linear solver. In: PARA'98 Workshop on Applied Parallel Computing in Large Scale Scientific and Industrial Problems, Umea (1998)
36. Hackbusch, W.: Multigrid Methods and Applications. Springer, Berlin/Heidelberg/New York (1985)
37. Henson, V.E., Vassilevski, P.S.: Element-free AMGe: general algorithms for computing interpolation weights in AMG. SIAM J. Sci. Comput. **23**(2), 629–650 (2001)
38. Hestenes, M., Stiefel, E.: Methods of conjugate gradients for solving linear systems. J. Res. Nat. Bur. Stand. **49**, 409–436 (1952)

39. Karypis, G., Kumar, V.: MeTis: unstrctured graph partitioning and sparse matrix ordering system, version 2.0 (1995)
40. Karypis, G., Kumar, V.: A coarse-grain parallel formulation of multilevel k-way graph-partitioning algorithm. In: Proceedings of the 8th SIAM Conference on Parallel Processing for Scientific Computing, Minneapolis. SIAM (1997)
41. Karypis, G., Kumar, V.: A fast and high quality multilevel scheme for partitioning irregular graphs. SIAM J. Sci. Comput. **20**(1), 359–392 (1998)
42. Kohn, W., Sham, L., et al.: Self-consistent equations including exchange and correlation effects. Phys. Rev. **140**(4A), A1133–A1138 (1965)
43. Kunkel, P., Mehrmann, V.: Numerical solution of Riccati differential algebraic equations. Linear Algebra Appl. **137/138**, 39–66 (1990)
44. Lancaster, P., Rodman, L.: The Algebraic Riccati Equation. Oxford University Press, Oxford, New York (1995)
45. Lee, D., Ipsen, I.: Zone determinant expansions for nuclear lattice simulations. Phys. Rev. C **68**, 064003–1–064003–8 (2003)
46. Li, S., Ahmed, S., Klimeck, G., Darve, E.: Computing entries of the inverse of a sparse matrix using the FIND algorithm. J. Comput. Phys. **227**(22), 9408–9427 (2008)
47. Lin, L., Yang, C., Meza, J.C., Lu, J., Ying, L., Weinan, E.: SelInv – an algorithm for selected inversion of a sparse symmetric matrix. ACM Trans. Math. Softw. **37**(4), 40:1–40:19 (2011)
48. Luisier, M., Boykin, T., Klimeck, G., Fichtner, W.: Atomistic nanoelectronic device engineering with sustained performances up to 1.44 pflop/s. In: 2011 International Conference for High Performance Computing, Networking, Storage and Analysis (SC), Seattle, pp. 1–11. IEEE (2011)
49. Manguoglu, M., Sameh, A.H., Schenk, O.: PSPIKE: a parallel hybrid sparse linear system solver. In: Euro-Par, Delft, pp. 797–808 (2009)
50. Mehrmann, V.: Divide and conquer algorithms for tridiagonal linear systems. In: Hackbusch, W. (ed.) Parallel Algorithms for PDEs, Proceedings of the 6th GAMM-Seminar, Kiel. Notes on Numerical Fluid Mechanics, pp. 188–199. Vieweg, Braunschweig (1990)
51. Mehrmann, V.: Divide & conquer methods for block tridiagonal systems. Parallel Comput. **19**, 257–279 (1993)
52. Mehrmann, V., Schröder, C.: Nonlinear eigenvalue and frequency response problems in industrial practice. J. Math. Ind. **1**, 7 (2011)
53. Mehrmann, V., Tan, E.: Defect correction methods for the solution of algebraic Riccati equations. IEEE Trans. Autom. Control **AC-33**, 695–698 (1988)
54. Muresan, A.C., Notay, Y.: Analysis of aggregation-based multigrid. SIAM J. Sci. Comput. **30**(2), 1082–1103 (2008)
55. Notay, Y.: Using approximate inverses in algebraic multigrid methods. Numer. Math. **80**, 397–417 (1998)
56. Notay, Y.: An aggregation-based algebraic multigrid method. Electron. Trans. Numer. Anal. **37**, 123–146 (2010)
57. Notay, Y., Vassilevski, P.S.: Recursive Krylov-based multigrid cycles. Numer. Linear Algebra Appl. **15**, 473–487 (2008)
58. Reusken, A.: A multigrid method based on incomplete Gaussian elimination. Preprint 95-13, Dept. of Mathematics and Computer Science, Eindhoven University of Technology (1995)
59. Reusken, A.: A multigrid method based on incomplete Gaussian elimination. Numer. Linear Algebra Appl. **3**, 369–390 (1996)
60. Ruge, J., Stüben, K.: Algebraic multigrid. In: McCormick, S. (ed.) Multigrid Methods, pp. 73–130. SIAM, Philadelphia (1987)
61. Saad, Y.: Iterative Methods for Sparse Linear Systems, 2nd edn. SIAM, Philadelphia (2003)
62. Saad, Y.: Multilevel ILU with reorderings for diagonal dominance. SIAM J. Sci. Comput. **27**, 1032–1057 (2005)
63. Saad, Y., Suchomel, B.J.: ARMS: an algebraic recursive multilevel solver for general sparse linear systems. Numer. Linear Algebra Appl. **9**, 359–378 (2002)

64. Schenk, O., Gärtner, K.: Solving unsymmetric sparse systems of linear equations with PARDISO. J. Future Gener. Comput. Syst. **20**(3), 475–487 (2004)
65. Smith, B., Bjorstad, P., Gropp, W.: Domain Decomposition, Parallel Multilevel Methods for Elliptic Partial Differential Equations. Cambridge University Press, Cambridge (1996)
66. Stüben, K.: An Introduction to Algebraic Multigrid. In: Trottenberg, U., Oosterlee, C.W., Schüller, A. (eds.) Multigrid, pp. 413–532. Academic, London (2001). Appendix A
67. Tang, J.M., Saad, Y.: Domain-decomposition-type methods for computing the diagonal of a matrix inverse. SIAM J. Sci. Comput. **33**(5), 2823–2847 (2011)
68. Tang, J.M., Saad, Y.: A probing method for computing the diagonal of a matrix inverse. Numer. Linear Algebra Appl. **19**(3), 485–501 (2012)
69. Trottenberg, U., Oosterlee, C.W., Schüller, A.: Multigrid. Academic, London (2001)
70. Vanek, P., Mandel, J., M. Brezina. Algebraic multigrid by smoothed aggregation for second order and fourth order elliptic problems. Computing **56**, 179–196 (1996)
71. Vassilevski, P.S.: Multilevel Block Factorization Preconditioners. Springer, New York/London (2008)
72. Wagner, C., Wittum, G.: Adaptive filtering. Numer. Math. **78**, 305–328 (1997)
73. Wagner, C., Wittum, G.: Filtering decompositions with respect to adaptive test vectors. In: Hackbusch, W., Wittum, G. (eds.) Multigrid Methods V. Lecture Notes in Computational Science and Engineering, vol. 3, pp. 320–334. Springer (1998)
74. Xu, J., Zou, J.: Some nonoverlapping domain decomposition methods. SIAM Rev. **40**(4), 857–914 (1998)

Part II
Matrix Theory

Chapter 11
Continuous Matrix Factorizations

Erik S. Van Vleck

Abstract Continuous matrix factorizations show great promise in a number of contexts. In this chapter we survey results on continuous matrix factorizations paying particular attention to smooth matrix factorizations of fundamental matrix solutions of linear differential equations and differential-algebraic equations with special emphasis on smooth QR and smooth SVD.

11.1 Introduction

Continuous matrix factorizations are useful in a number of contexts, notably in determining stability spectra such as Lyapunov exponents and Sacker–Sell spectrum, continuation of invariant subspaces, control, etc. Their usefulness becomes apparent when time independent factorizations are no longer applicable or do not provide useful information. Among the continuous matrix factorizations that have been much studied are the smooth singular value decomposition (SVD), smooth QR decomposition, and smooth Schur factorization. Depending on the context these can be decompositions of a given matrix function or of a matrix function such as a fundamental matrix solution of a time dependent linear differential equation that is known to exist but is difficult to obtain explicitly.

Our goal in this chapter is to survey developments in continuous matrix factorizations from Lyapunov and Perron to present day. We discuss results on continuous QR, Schur, and SVD of smooth matrix functions, but will emphasize continuous matrix factorizations of (full rank) fundamental matrix solutions first for time dependent linear differential equations and then for a class of time dependent linear differential algebraic equations (DAEs). Two instructive examples are the smooth QR factorization of general smooth matrix functions and the important case in which the matrix function is a fundamental matrix solution of a linear (time dependent) differential equation. One important difference between the general case and the case of fundamental matrix solutions is that given a full rank initial

---

E.S. Van Vleck (✉)
Department of Mathematics, University of Kansas, Lawrence, KS 66045, USA
e-mail: erikvv@ku.edu

P. Benner et al. (eds.), *Numerical Algebra, Matrix Theory, Differential-Algebraic Equations and Control Theory*, DOI 10.1007/978-3-319-15260-8_11

condition, the fundamental matrix solution is full rank for all time. In addition, if a fundamental matrix solution exists in a classical sense, then it must be at least C^1 smooth. Both general matrix functions and fundamental matrix solutions can have smooth matrix factorizations that fail to exist. For example, in the case of smooth SVD, the crossing of two time dependent singular values can lead to the non-existence of the smooth SVD.

A motivating example in which continuous matrix factorizations are useful is in the computation of stability spectra such as Lyapunov exponents and Sacker–Sell spectrum. These stability spectra reduce in both cases to the real parts of eigenvalues in the time independent setting. We next present background on these stability spectra. Consider the n-dimensional linear system

$$\dot{x} = A(t)x\,, \tag{11.1}$$

where A is continuous and bounded: $\sup_t \|A(t)\| < \infty$ with, for simplicity, real entries. We will focus throughout on $A(t)$ with real entries unless otherwise noted.

In [51], Sacker and Sell introduced a spectrum for (11.1) based upon exponential dichotomy: The Sacker–Sell, or Exponential Dichotomy, spectrum (Σ_{ED}) is given by those values $\lambda \in \mathbb{R}$ such that the shifted system $\dot{x} = [A(t) - \lambda I]x$ does not have exponential dichotomy. Recall that the system (11.1) has *exponential dichotomy* if for a fundamental matrix solution X there exists a projection P and constants $\alpha, \beta > 0$, and $K, L \geq 1$, such that

$$\begin{aligned} \|X(t)PX^{-1}(s)\| &\leq Ke^{-\alpha(t-s)}, \quad t \geq s,\\ \|X(t)(I-P)X^{-1}(s)\| &\leq Le^{\beta(t-s)}, \quad t \leq s. \end{aligned} \tag{11.2}$$

It is shown in [51] that Σ_{ED} is given by the union of at most n closed, disjoint, intervals. Thus, it can be written, for some $1 \leq k \leq n$, as

$$\Sigma_{\mathrm{ED}} := [a_1, b_1] \cup \cdots \cup [a_k, b_k], \tag{11.3}$$

where the intervals are disjoint. The complement of Σ_{ED} is called the *resolvent*: It is given by all values $\lambda \in \mathbb{R}$ for which the shifted system has exponential dichotomy.

To define a spectrum in terms of Lyapunov exponents, let X be a fundamental matrix solution of (11.1) and consider the quantities

$$\lambda_i = \limsup_{t\to\infty} \frac{1}{t} \ln ||X(t)e_i||\,,\ i = 1,\ldots,n, \tag{11.4}$$

where e_i denotes the i-th standard unit vector. When $\sum_{i=1}^{n} \lambda_i$ is minimized with respect to all possible fundamental matrix solutions, then the λ_i's are called the upper Lyapunov exponents, or simply Lyapunov exponents or characteristic numbers, of the system and the corresponding fundamental matrix solution is called *normal*.

The Lyapunov exponents satisfy

$$\sum_{i=1}^{n} \lambda_i \geq \limsup_{t\to\infty} \frac{1}{t} \int_0^t \mathrm{Tr}(A(s))ds \tag{11.5}$$

where $\mathrm{Tr}(A(\cdot))$ is the trace of $A(\cdot)$. Linear systems for which the Lyapunov exponents exist as limits were called regular by Lyapunov.

Definition 1 A system is *regular* (Lyapunov) if the time average of the trace has a finite limit and equality holds in (11.5).

An important and instructive example of a continuous matrix factorizations is the continuous QR factorization of a fundamental matrix solution of (11.1). If we write the fundamental matrix solution $X(t) = Q(t)R(t)$ where $Q(t)$ is orthogonal, $R(t)$ is upper triangular with positive diagonal entries, $X(t_0)$ is invertible for some $t = t_0$, then upon differentiating we obtain

$$\dot{Q}(t)R(t) + Q(t)\dot{R}(t) = \dot{X}(t) = A(t)X(t) = A(t)Q(t)R(t).$$

Upon differentiating $Q^T(t)Q(t) = I$, we note that $Q^T(t)\dot{Q}(t)$ is skew-symmetric and since $R(t)$ is upper triangular and invertible (since $X(t)$ and $Q(t)$ are invertible), then

$$S(Q(t), A(t)) := Q^T(t)\dot{Q}(t) = Q^T(t)A(t)Q(t) - \dot{R}(t)R^{-1}(t)$$

so that the strict lower triangular part of $S(Q(t), A(t))$ may be determined by the corresponding part of $Q^T(t)A(t)Q(t)$ and the rest of $S(Q(t), A(t))$ is determined by skew-symmetry. Thus, we have the following equations to determine $Q(t)$ and the upper triangular coefficient matrix function $B(t)$ for $R(t)$,

$$\dot{Q}(t) = Q(t) \cdot S(Q(t), A(t)), \quad B(t) = Q^T(t)A(t)Q(t) - S(Q(t), A(t)).$$

This provides a means for obtaining fundamental matrix solutions for time dependent linear differential equations since $R(t)$ (assuming an upper triangular initial condition for R) may be obtained using backward substitution and solving linear scalar differential equations. In general $X(t) = \exp(\int_0^t A(s)ds)$ is a fundamental matrix solution for (11.1) if and only if

$$A(t) \cdot \int_0^t A(s)ds = \int_0^t A(s)ds \cdot A(t), \quad \forall t,$$

i.e., $A(t)$ commutes with its integral for all t (see [30]).

We note also the following example (see [11]) that shows that eigenvalues of $A(t)$ do not generally provide stability information Consider $A(t) = U^{-1}(t)A_0U(t)$ where

$$A_0 = \begin{pmatrix} -1 & -5 \\ 0 & -1 \end{pmatrix} \quad \text{and} \quad U(t) = \begin{pmatrix} \cos(t) & \sin(t) \\ -\sin(t) & \cos(t) \end{pmatrix}.$$

But the fundamental matrix solution is

$$\begin{pmatrix} \exp(t)(\cos(t) + \frac{1}{2}\sin(t)) & \exp(-3t)(\cos(t) - \frac{1}{2}\sin(t)) \\ \exp(t)(\sin(t) - \frac{1}{2}\cos(t)) & \exp(-3t)(\sin(t) + \frac{1}{2}\cos(t)) \end{pmatrix}$$

Note that this is an application of classical Floquet theory to transform a periodic coefficient matrix to a constant coefficient matrix. It is also worth emphasizing that whereas for a constant coefficient $A(t) \equiv A$, a similarity transformation of A is a change of variables for (11.1), $Q(t)$ that transforms $A(t)$ to $B(t)$ is a change of variables, but not a similarity transformation. However, if $A(t)$ is symmetric with negative eigenvalues for all t, then a Lyapunov function argument shows that the zero solution is stable. However, this result is far from sharp.

Consider a full rank matrix function $C \in C^k(R, \mathbb{C}^{m\times n})$, $m \geq n$, with $A(0) = Q(0)R(0)$ a given QR factorization where $Q(0)$ is unitary and $R(0)$ is upper triangular with positive diagonal elements. Equations for $Q(t)$ and $R(t)$ can be derived as follows (see [13]), write $A(t) = Q(t)R(t)$, $Q^*(t)Q(t) = I$, and differentiate to obtain

$$\dot{A}(t) = \dot{Q}(t)R(t) + Q(t)\dot{R}(t), \quad \dot{Q}^*(t)Q(t) + Q^*(t)\dot{Q}(t) = 0.$$

Then $H(t) := Q^*(t)\dot{Q}(t)$ is skew-Hermitian and we obtain

$$\begin{cases} \dot{R}(t) = Q^*(t)\dot{A}(t) - H(t)R(t), \\ \dot{Q}(t) = \dot{A}(t)R^{-1}(t) - Q(t)\dot{R}(t)R^{-1}(t) = (I - Q(t)Q^*(t))\dot{A}(t)R^{-1}(t) + Q(t)H(t). \end{cases}$$

A smooth skew-Hermitian $H(t)$ can then be defined based upon the requirement that $\dot{R}(t)$ is upper triangular.

Next, we will provide some additional background and discuss the analytic singular value decomposition and work of Kato, Reinboldt, Bunse-Gerstner, Byers, Mehrmann, and Nichols, and Dieci and Eirola in Sect. 11.2. In Sect. 11.3 we discuss the role of continuous matrix factorizations (QR and SVD) in determining Lyapunov exponents and Sacker–Sell spectrum of time dependent linear differential equations. This includes methods and justification for using such factorizations to extract both Lyapunov exponents and the analogue of eigenvectors, Lyapunov vectors. We will also discuss the perturbation theory for finding Lyapunov exponents and vectors using the smooth QR factorization of a fundamental matrix solution. We turn our attention to differential-algebraic equations in Sect. 11.4 and focus on the use of QR and SVD to determine stability spectra in the case of strangeness-free DAEs.

11.2 Smooth Decompositions of Matrices

A natural and important extension of matrix factorizations is to smooth decompositions of parameter dependent matrix functions. Orthogonal matrices (or unitary matrices in the complex case) serve as an important building block for modern matrix algorithms. They can be computed stably and hence are natural candidates for extending such techniques to the time or parameter dependent case.

The book of Kato [31] provides solid theoretical foundations for decompositions of parameter dependent matrices. In the case in which the matrix function is real analytic and Hermitian there exists an analytic Schur decomposition. Another important work on smooth decompositions is the work of Reinboldt [50] on the smooth QR decomposition of a matrix function. In this work a smoothly varying left null space of the matrix function is used to construct a smooth system of local charts. The numerical method developed in [50] uses a reference decomposition $A(t_0) = Q(t_0)R(t_0)$ with $Q(t_0)$ orthogonal and $R(t_0)$ upper triangular to determine $Q(t)$ by minimizing $\|Q(t_0) - Q(t)\|_F$. An explicit formula for the minimizer is obtained and this is used to determine $Q(t)$ in a neighborhood of $t = t_0$.

In [6] Bunse-Gerstner, Byers, Mehrmann, and Nichols extend the singular value decomposition to a path of matrices $A(t)$. An analytic singular value decomposition of a path of matrices $A(t)$ is an analytic path of factorizations $A(t) = U(t)\Sigma(t)V^T(t)$ where $U(t)$ and $V(t)$ are orthogonal of factorizations $A(t) = U(t)\Sigma(t)V^T(t)$ where $U(t)$ and $V(t)$ are orthogonal and $\Sigma(t)$ is diagonal. To maintain differentiability the diagonal elements of $S(t)$ are not restricted to being positive. Existence, uniqueness, and algorithms are obtained and in particular it is shown that a real analytic $A(t)$ admits a real analytic SVD, and a full-rank, smooth path $A(t)$ with distinct singular values admits a smooth SVD. Differential equations are derived for the left factor based upon minimizing the arc length $\int_a^b \|U'(t)\|dt$.

In [13] smooth orthonormal decompositions of smooth time varying matrices are considered. Emphasis is on the QR, Schur, SVD of matrix function as well as block analogues. Sufficient conditions for the existence of such decompositions are given and differential equations derived. In particular, for a smooth QR factorization of $A(t)$ one needs $A(t)$ full rank, for the smooth Schur or SVD one requires simple eigenvalues or singular values, respectively. Several approaches are considered to weaken these conditions and the generic case is considered.

11.3 Factorization of Fundamental Matrix Solutions and Approximation of Lyapunov Exponents

Techniques for numerical approximation of Lyapunov exponents are based upon smooth matrix factorizations of fundamental matrix solutions X, to bring it into a form from which it is easier to extract the Lyapunov exponents. In practice, two techniques have been studied: based on the QR factorization of X and based on the SVD (singular value decomposition) of X. Although these techniques have been adapted to the case of incomplete decompositions (useful when only a few Lyapunov exponents are needed), or to problems with Hamiltonian structure, we focus on the general case when the entire set of Lyapunov exponents is sought. For extensions, see the references.

11.3.1 SVD Methods

Here one seeks to compute the SVD of X: $X(t) = U(t)\Sigma(t)V^T(t)$, for all t, where U and V are orthogonal and $\Sigma = \text{diag}(\sigma_i,\ i = 1\dots,n)$, with $\sigma_1(t) \geq \sigma_2(t) \geq \dots \geq \sigma_n(t)$. If the singular values are distinct, the following differential equations for U, V and Σ hold. Letting $G = U^T AU$, they are

$$\dot{U} = UH, \quad \dot{V}^T = -KV^T, \quad \dot{\Sigma} = D\Sigma, \tag{11.6}$$

where $D = \text{diag}(G)$, $H^T = -H$, $K^T = -K$, and for $i \neq j$,

$$H_{ij} = \frac{G_{ij}\sigma_j^2 + G_{ji}\sigma_i^2}{\sigma_j^2 - \sigma_i^2}, \quad K_{ij} = \frac{(G_{ij} + G_{ji})\sigma_i\sigma_j}{\sigma_j^2 - \sigma_i^2}. \tag{11.7}$$

From the SVD of X the Lyapunov exponents may be obtained as

$$\limsup_{t\to\infty} \frac{1}{t} \ln \sigma_i(t)\,. \tag{11.8}$$

11.3.2 QR Methods

Triangular time dependent coefficient matrices were employed by Lyapunov [41] to show the existence of so-called normal basis that defines Lyapunov exponents and subsequently by Perron [49]. A time dependent version of the Gram-Schmidt process was derived by Diliberto [29] although the use of matrix equations is algorithmically superior. The use of matrix differential equations in general appears to have several advantages over simply making continuous time versions of matrix algorithms.

The idea of QR methods is to seek the factorization of a fundamental matrix solution as $X(t) = Q(t)R(t)$, for all t, where Q is an orthogonal matrix valued function and R is an upper triangular matrix valued function with positive diagonal entries. The validity of this factorization has been known since Perron [49] and Diliberto [29], and numerical techniques based upon the QR factorization date back at least to [3]. QR techniques come in two flavors, continuous and discrete, and methods for quantifying the error in approximation of Lyapunov exponents have been developed in both cases (see [18, 25, 26, 28, 53]).

11.3.2.1 Continuous QR

Upon differentiating the relation $X = QR$ and using (11.1), we have

$$AQR = Q\dot{R} + \dot{Q}R \quad \text{or} \quad \dot{Q} = AQ - QB\,, \tag{11.9}$$

where $\dot{R} = BR$, hence B must be upper triangular. Now, let us formally set $S = Q^T\dot{Q}$ and note that since Q is orthogonal then S must be skew-symmetric. Now, from $B = Q^TAQ - Q^T\dot{Q}$ it is easy to determine at once the strictly lower triangular part of S (and from this, all of it), and the entries of B.

Given $X(0) = Q_0R_0$, we have

$$\dot{Q} \;=\; QS(Q,A)\,, \quad Q(0) = Q_0\,, \tag{11.10}$$

$$\dot{R} \;=\; B(t)R\,, \quad R(0) = R_0\,,\; B := Q^TAQ - S(Q,A)\,, \tag{11.11}$$

The diagonal entries of R are used to retrieve the exponents:

$$\limsup_{t\to\infty} \frac{1}{t}\int_0^t (Q^T(s)A(s)Q(s))_{ii}ds\,, \quad i = 1,\ldots,n\,. \tag{11.12}$$

since $B_{ii} = (Q^T(s)A(s)Q(s))_{ii}$ the diagonal of the upper triangular coefficient matrix function.

Some variations include the case in which some but not all Lyapunov exponents are required so that $Q(t) \in \mathbb{R}^{m\times n}$ with $m > n$ satisfies (see [22])

$$\dot{Q}(t) = (I - Q(t)Q^T(t))A(t) + Q(t)S(Q(t),A(t))$$

and the case of complex coefficient matrix function $A(t)$ in which $Q(t) \in \mathbb{C}^{n\times n}$ is unitary and (see [20])

$$\dot{Q} = Q\,H(Q,t),$$

with $H(Q,t)$ satisfying

$$(H)_{lm} = \begin{cases} (Q^*AQ)_{lm}, & \text{if } l > m; \\ i\Im(Q^*AQ)_{ll}, & \text{if } l = m; \\ -\overline{(H)_{ml}}, & \text{otherwise.} \end{cases}$$

11.3.2.2 Discrete QR

Here one seeks the QR factorization of the fundamental matrix X at discrete points $0 = t_0 < t_1 < \cdots < t_k < \cdots$, where $t_k = t_{k-1} + h_k$, $h_k > 0$. The method requires the QR factorization of $Y(t_{k+1}) \in R^{m\times n}$, QR factorization of $X(t_{k+1})$, at a matrix solution of (11.1) with linearly independent columns. Let $X_0 = Q_0 R_0$, and suppose we seek the QR factorization of $X(t_{k+1})$. For $j = 0, \ldots, k$, progressively define $Z_{j+1}(t) = X(t, t_j) Q_j$, where $X(t, t_j)$ solves (11.1) for $t \geq t_j$, $X(t_j, t_j) = I$, and Z_{j+1} is the solution of

$$\begin{cases} \dot{Z}_{j+1} = A(t) Z_{j+1}, & t_j \leq t \leq t_{j+1} \\ Z_{j+1}(t_j) = Q_j. \end{cases} \tag{11.13}$$

Update the QR factorization as

$$Z_{j+1}(t_{j+1}) = Q_{j+1} R_{j+1}, \tag{11.14}$$

and finally observe that

$$X(t_{k+1}) = Q_{k+1} [R_{k+1} R_k \cdots R_1 R_0] \tag{11.15}$$

is the QR factorization of $X(t_{k+1})$.The Lyapunov exponents are obtained from the relation

$$\limsup_{k\to\infty} \frac{1}{t_k} \sum_{j=0}^{k} \log(R_j)_{ii}, i = 1, \ldots, n. \tag{11.16}$$

11.3.2.3 Determining Lyapunov Exponents

The key to continuity of Lyapunov exponents with respect to perturbations (see [1]) is integral separation (in the case of distinct Lyapunov exponents) and upper and lower central exponents (in the case of non-distinct Lyapunov exponents). These concepts also play a central role in determining when Lyapunov exponents may be obtained from the diagonal elements of the upper triangular coefficient matrix $B(t)$ as in (11.12). We next summarize the classical theory of continuity of Lyapunov exponents and outline the justification for obtaining Lyapunov exponents from the diagonal elements of the transformed upper triangular coefficient matrix.

Although a convenient and quite reasonable assumption in many practical situations, see [45], regularity is not sufficient to guarantee stability of the Lyapunov exponents, which is what we need to have in order to pursue computational procedures for their approximation. Stability for the LEs means that small perturbations in the function of coefficients, A, produce small changes in the LEs. Millionschikov (see [42, 43]) and Bylov and Izobov (see [7]) gave conditions under which the

LEs are stable, and further proved that these conditions are *generic*, see [46, p. 21], in the class of linear systems with continuous bounded coefficients. The key assumption needed is **integral separation**: "A fundamental matrix solution (written columnwise) $X(t) = [X_1(t), \ldots, X_m(t)]$ is integrally separated if for $j = 1, \ldots, m-1$, there exist $a > 0$ and $d > 0$ such that

$$\frac{||X_j(t)||}{||X_j(s)||} \cdot \frac{||X_{j+1}(s)||}{||X_{j+1}(t)||} \geq de^{a(t-s)}, \tag{11.17}$$

for all $t, s : t \geq s$". In [7] (see also [1] Theorem 5.4.8), it is proved that "If the system (11.1) has different characteristic exponents $\lambda_1 > \cdots > \lambda_m$, then they are stable if and only if there exists a fundamental matrix solution with integrally separated columns". For a good introduction to integral separation and some of its consequences see [1] and the references therein, and the complementary works of Palmer [47, 48]. The following theorem provides conditions (close to the necessary and sufficient conditions for continuity of Lyapunov exponents) under which Lyapunov exponents may be extracted from the diagonal elements of the transformed coefficient matrix B obtained using QR methods.

Theorem 1 (Theorem 5.1 of [27]) *For $t \geq 0$, consider $\dot{x} = B(t)x$ with $B(\cdot)$ bounded, continuous, and upper triangular. If either of the following conditions hold*

(i) The diagonal elements of B are integrally separated, i.e., there exist $a, d > 0$ such that

$$\int_s^t \big(B_{ii}(\tau) - B_{jj}(\tau)\big)d\tau \geq a(t-s) - d, \qquad t \geq s \geq 0\,,\ i < j\,, \tag{11.18}$$

or

(ii) The diagonal elements of B are not all integrally separated and for non-integrally separated diagonal elements, $B_{ii}(t)$ and $B_{jj}(t)$, within an upper triangular block, for every $\epsilon > 0$ there exists $M_{ij}(\epsilon) > 0$ such that

$$\left|\int_s^t (B_{ii}(\tau) - B_{jj}(\tau))d\tau\right| \leq M_{ij}(\epsilon) + \epsilon(t-s),\ t \geq s, \tag{11.19}$$

then the Lyapunov spectrum Σ_{L} is obtained as

$$\Sigma_{\mathrm{L}} := \bigcup_{j=1}^{n}[\lambda_j^i, \lambda_j^s]\,, \quad \lambda_j^i = \liminf_{t\to\infty}\frac{1}{t}\int_0^t B_{jj}(s)ds\,, \quad \lambda_j^s = \limsup_{t\to\infty}\frac{1}{t}\int_0^t B_{jj}(s)ds\,, \tag{11.20}$$

and Σ_{L} is continuous with respect to perturbations.

11.3.2.4 QR Error Analysis

In this section we review the error analysis that has been developed for the QR method for determining stability spectra. This starts from the backward error analysis in [25] where it is shown that uniformly bounded local errors for approximating the orthogonal factor Q with the continuous QR method or uniformly bounded local errors in approximating the fundamental matrix solution X lead to a backward error analysis in which the numerical solution solves a nearby problem (near to B or R). If the problem is sufficiently well-conditioned as measured by the strength of the integral separation in the problem, then the results in [26, 28] (for continuous QR) and [2, 53] (for discrete QR) provide a forward error analysis. This forward error analysis quantifies the error in the Lyapunov exponents (and other stability spectra) for Lyapunov exponents that are continuous with respect to perturbations as well as a long time global error analysis for the orthogonal factor Q in the integrally separated case.

Suppose we want the QR factorization at $X(t_k)$, for some sequence of points t_k, $k = 0, 1, 2, \ldots$, with $t_0 = 0$. At any such point t_k, we can write

$$X(t_k) = \Phi(t_k, t_{k-1}) \ldots \Phi(t_2, t_1)\Phi(t_1, 0)X_0 \,, \tag{11.21}$$

where

$$\dot{\Phi}(t, t_{j-1}) = A(t)\Phi(t, t_{j-1}) \,, \;\; \Phi(t_{j-1}, t_{j-1}) = I \,, \quad t_{j-1} \le t \le t_j \,, j = 1, 2, \ldots, k \,. \tag{11.22}$$

Now, let $X_0 = Q(t_0)R(t_0)$, where $Q(t_0) \in \mathbb{R}^{n\times n}$ is orthogonal and $R(t_0) \in \mathbb{R}^{n\times n}$ is upper triangular with positive diagonal entries. Then, for $j = 1, 2, \ldots, k$, recursively consider

$$\begin{aligned} &\dot{\Psi}(t, t_{j-1}) = A(t)\Psi(t, t_{j-1}) \,, \;\; \Psi(t_{j-1}, t_{j-1}) = Q(t_{j-1}) \\ &\text{and factor } \Psi(t_j, t_{j-1}) = Q(t_j)R(t_j, t_{j-1}) \,, \end{aligned} \tag{11.23}$$

where $Q(t_j)$ are orthogonal and $R(t_j, t_{j-1})$ are upper triangular with positive diagonal. Then, we have the QR factorization of $X(t_k)$

$$X(t_k) = Q(t_k)R(t_k, t_{k-1}) \ldots R(t_2, t_1)R(t_1, t_0)R(t_0) \,. \tag{11.24}$$

In practice, we cannot solve for the transition matrices Φ in (11.21) exactly, and we will actually compute

$$X_k = X(t_k, t_{k-1}) \ldots X(t_2, t_1)X(t_1, t_0)X_0 \,, \tag{11.25}$$

where the matrices $X(t_j, t_{j-1})$ are approximations to $\Phi(t_j, t_{j-1})$, $j = 1, \dots, k$. Letting $Q(t_0) = Q_0$, and progressively setting

$$X(t_j, t_{j-1}) Q_{j-1} = Q_j R_j \ , \quad j = 1, \dots, k \ ,$$

the numerical discrete QR method will read

$$X_k = Q_k R_k R_{k-1} \dots R_2 R_1 R(t_0) \, . \tag{11.26}$$

Theorem 2 (Theorem 3.1 [25]) *For $k = 1, 2, \dots$, and $j = 1, \dots, k$, let $Q(t_j)$ and $R(t_j, t_{j-1})$ be the exact Q and R terms in (11.24), and let X_k be given in (11.25). We have*

$$X_k = Q(t_k)[R(t_k, t_{k-1}) + E_k] \dots [R(t_2, t_1) + E_2][R(t_1, t_0) + E_1] R(t_0) \ , \tag{11.27}$$

where

$$E_j = Q^T(t_j) N_j Q(t_{j-1}) \ , \ j = 1, \dots, k \ , \tag{11.28}$$

and N_j are the local errors obtained when approximating $\Phi(t_j, t_{j-1})$ by $X(t_j, t_{j-1})$: $N_j = X(t_j, t_{j-1}) - \Phi(t_j, t_{j-1})$, $j = 1, \dots, k$.

As a consequence of the above, the numerical realization of the discrete QR method as expressed by (11.26) finds the exact QR factorization of the sequence on the right-hand-side of (11.27).

Using Theorem 2 together with several estimates, the following theorem establishes the backward error analysis in the case of the discrete QR method. In particular, it is shown that under mild assumptions that the computed orthogonal and upper triangular factors correspond to a perturbation of the exact upper triangular coefficient matrix B. The following theorem provides componentwise bounds on the perturbation. We first write the upper triangular $R(t_{j+1}, t_j) = D(t_{j+1}, t_j) + U_{j+1}$ where $D(t_{j+1}, t_j)$ is diagonal and U_{j+1} is strictly upper triangular. Define δ_j by

$$\delta_j := \min_{1 \le p \le n} \frac{1}{\min\left(1, \exp(\int_{t_j}^{t_{j+1}} B_{pp}(t) dt)\right)} , \tag{11.29}$$

and $\nu_j := \|U_{j+1}\|$ as the departure from normality of the exact triangular transition matrix $R(t_{j+1}, t_j)$. We note that in what follows it is not necessary that $\delta_j \nu_j < 1$ for problems of finite dimension n.

Theorem 3 (Theorem 3.12 [25]) *Consider the system (11.1). Let $\{t_j\}$, $j = 0, 1, \dots$, $t_0 = 0 < t_1 < t_2 < \dots$, be the sequence of points (converging to ∞) generated by the numerical realization of the discrete QR method.*

At each t_j, $j = 1, 2, \ldots$, the exact discrete QR method delivers the factorization (11.24), where $R(t_{j+1}, t_j) = e^{h_j B_j}$, $h_j = t_{j+1} - t_j$, $j = 0, 1, \ldots$, is the solution of the upper triangular system:

$$\dot{\tilde{R}} = B_j \tilde{R}\,,\ t_j \le t < t_{j+1}\,,\ \tilde{R}(0) = R(t_0)\,,\ j = 0, \ldots, k-1\,, \tag{11.30}$$

where the matrices $B_j \in \mathbb{R}^{n \times n}$ are upper triangular and satisfy

$$R(t_{j+1}, t_j) = e^{h_j B_j}\,,\ h_j := t_{j+1} - t_j\,,\ j = 0 \ldots, k-1\,. \tag{11.31}$$

The numerical discrete QR method, instead, gives the QR factorization of the matrix in (11.25), that is in (11.27). Assume that $\|E_{j+1}\| = \|\Phi(t_{j+1}, t_j) - X(t_{j+1}, t_j)\| \le \mathtt{TOL}$, for all $j = 0, 1, \ldots$. Finally, assume that

$$\rho_j := \|E_{j+1}\| \cdot [\min_{1 \le i \le n} (1, D_{ii}(t_{j+1}, t_j))]^{-1} \le \rho < 1\,,\quad \forall j = k, \ldots, 0,\quad k = 1, 2, \ldots\,. \tag{11.32}$$

Then, the numerical discrete QR method finds the (exact) QR factorization of the system

$$\dot{\hat{R}} = \hat{B}_j \hat{R}\,,\ t_j \le t < t_{j+1}\,,\ \hat{R}(0) = R_0\,,\ j = 0, \ldots, k-1\,, \tag{11.33}$$

where the matrices $\hat{B}_j \in \mathbb{R}^{n \times n}$ satisfy

$$\hat{R}(t_{j+1}, t_j) = e^{h_j \hat{B}_j}\,,\ h_j := t_{j+1} - t_j\,,\ j = 0, \ldots, k-1\,, \tag{11.34}$$

$$h_j \hat{B}_j = h_j B_j + L_j + O(\|E_{j+1}\|^2)\,, \tag{11.35}$$

and we have the bounds

$$|L_j| \le \left[\frac{1 - (\delta_j \nu_j)^n}{1 - \delta_j \nu_j} \right]^2 |F_{j+1}| \tag{11.36}$$

where the general (p,q)-entry of $|F_{j+1}|$ is bounded as

$$\left(|F_{j+1}|\right)_{(p,q)} \le \mathtt{TOL} / \min_{i=p,q} \left(\exp\left(\int_{t_j}^{t_{j+1}} B_{ii}(t)dt\right)\right)\,.$$

□

The analogue of Theorem 3 for the continuous QR algorithm is Theorem 3.16 in [25].

With the continuous QR algorithm, Theorem 3.16 in [25] together with the nonlinear variation of constants formula and a fixed point argument applied to $\dot{Q} = Q \cdot S(Q, B + E)$ provides the basis for the error analysis for Lyapuonv exponents and also global error bounds for the orthogonal factor (see [26] and [28]). The analysis is based upon showing that if the error $E(t)$ is small enough as compared to the integral separation in the system, then $Q(t) \approx I$ for all t. Both the case of robust, distinct Lyapunov exponents and robust but non-distinct Lyapunov exponents have been investigated.

For the discrete *QR* algorithm one may use Theorem 2 (Theorem 3.1 in [25]) as the starting point to analyze

$$\overline{Q}_{j+1}\overline{R}_j = [R_j + E_j]\overline{Q}_j.$$

Here the approach is based upon application of the Newton-Kantorovich Theorem (see [2, 53]) to again show that with small enough error as compared to the integral separation in the system, there exists a sequence of orthogonal Q_n that remain close to the identity.

11.3.3 Growth/Decay Directions

Whereas Lyapunov exponents and Sacker–Sell spectrum provide time dependent stability information analogous to the real parts of eigenvalues of matrices, here we seek the analogue of eigenvectors, i.e., those initial conditions for (11.1) for which the asymptotic rate of growth/decay given by the Lyapunov exponents is realized. Here we survey results from [24] on using the smooth QR and SVD of fundamental matrix solutions to determine these directions and then turn our attention to the rate (exponential!) at which these directions are approached. That the rate of convergence is exponential means that these directions may be well approximated in finite time.

11.3.3.1 SVD

In general we are only assured that the continuous SVD algorithm is well-posed if the diagonal elements of coefficient matrix function D in (11.6) is integrally separated:

$$\int_0^t (G_{k,k}(\tau) - G_{k+1,k+1}(\tau))d\tau \geq at - d, \quad a > 0,\ d \in R,\ t \geq 0,\ k = 1, \dots, n-1\,. \tag{11.37}$$

It is then a fairly immediate consequence of the definition of the matrix function K in (11.6) that if (11.37) holds, then $K(t) \to 0$ as $t \to \infty$ and the following holds.

Lemma 1 (Lemma 7.1 [24]) *Let (11.37) hold. Then, the orthogonal matrix function $V(t) \to \overline{V}$, as $t \to \infty$, where $\overline{V}$ is a constant orthogonal matrix.*

This provides an orthogonal representation for the growth directions (analogues of eigenvectors) associated with the growth rates given by the Lyapunov exponents may be determined under the assumption of integral separation by $\lim_{t\to\infty} V(t)$ (see [14–16, 24]).

11.3.3.2 QR

The analogous result for the QR decomposition of a fundamental matrix solution is obtained by writing $X(t) = Q(t)R(t) = Q(t)\text{diag}(R(t))Z(t)$ and we have the following.

Lemma 2 (Lemma 7.3 [24]) *Consider the upper triangular system $\dot{R} = BR$ where B is bounded and continuous, and assume that the diagonal of B is integrally separated, as in (11.39). Then, $R \to \text{diag}(R)\overline{Z}$ as $t \to \infty$, where $\overline{Z}$ is a constant upper triangular matrix with 1's on the diagonal.*

Proof Write $R = DZ$, where $Z = D^{-1}R$, and $D = \text{diag}(R)$. Then D satisfies $\dot{D} = \text{diag}(B)D$ and Z satisfies $\dot{Z} = EZ$ where $E = D^{-1}(B - \text{diag}(B))D$. Then, $E_{ij} = B_{ij} \cdot \frac{R_{jj}}{R_{ii}}$ for $i < j$ and $E_{ij} = 0$ for $i \geq j$. Now,

$$\frac{R_{jj}}{R_{ii}} = \frac{R_{jj}}{R_{ii}}(0)\; e^{\int_0^t (B_{jj}-B_{ii})d\tau}\,.$$

Let $j = i + k$, for some $k = 1, \ldots$. The diagonal of B is integrally separated (see (11.39)), and so

$$\int_0^t (B_{jj} - B_{ii})d\tau \;\leq\; -k(at - d)\,,$$

from which $E_{ij} \to 0$ exponentially fast as $t \to \infty$ and the result follows.

This provides a unit upper triangular representation for the growth directions associated with the growth rates given by the Lyapunov exponents may be determined under certain conditions by $\lim_{t\to\infty} \text{diag}(R^{-1}(t))R(t)$ may be further determined by $\lim_{t\to\infty} \text{diag}(R^{-1}(t))R(t)$ (see [15, 16, 24]).

Next, we turn our attention to the rate of convergence to the growth/decay directions. The results in [14, 15] show that the rate of convergence is exponential. We next briefly state the result in the QR case. Consider the system

$$\dot{x} = B(t)x, \tag{11.38}$$

with B continuous, bounded and upper triangular. In what follows, we will assume that the diagonal elements of B satisfy one of the following assumptions:

(i) B_{ii} and B_{jj} are integrally separated, i.e. there exist $a, d > 0$ such that

$$\int_s^t \big(B_{ii}(\tau) - B_{jj}(\tau)\big)d\tau \geq a(t-s) - d, \qquad t \geq s \geq 0\,,\ i < j\,, \tag{11.39}$$

(ii) B_{ii} and B_{jj} are not integrally separated, but $\forall \epsilon > 0$ there exists $M_{ij}(\epsilon) > 0$ such that

$$\left|\int_s^t \big(B_{ii}(\tau) - B_{jj}(\tau)\big)d\tau\right| \leq M_{ij}(\epsilon) + \epsilon(t-s), \qquad t \geq s \geq 0\,,\ i < j\,. \tag{11.40}$$

For $Z(t) = \text{diag(R(t))}^{-1}R(t)$ write the unique QR decomposition of $Z^T(t)$ as $Z^T(t) = U(t)M(t)$ where $M(t)$ has positive diagonal elements. For $0 \leq \tau < 1$ define $\beta_{ij}(t) = \|P_i(t+\tau)P_j(t)\|$ where $P_i(t)$ denote the orthogonal projection onto the span of the first i columns of $U(t)$ which we denote by $U_i(t)$. Then if $U_i(t) \to \overline{U}_i$ then $P_i(t) \to \overline{P}_i$. If we let $\alpha_{ij}(t) = \|P_i(t)\overline{P}_j\|$, then the following shows that the rate of convergence is exponential with the rate of convergence roughly given as the gap between consecutive Lyapunov exponents.

Theorem 4 (Theorem 20 [15]) *Let the diagonal elements of B satisfy assumption (11.39) or (11.40). Then for all $i, j = 1, \ldots, p$, $i \neq j$,*

$$\chi^s(\alpha_{ij}) \leq A|\lambda_i - \lambda_j|, \tag{11.41}$$

where $A = \max_{k \neq l}(\chi(\beta_{kl})/|\lambda_k - \lambda_l|)$ and $\chi(\beta_{kl}) = \limsup_{t\to\infty} \frac{1}{t}\log(|\beta_{kl(t)}|)$, the Lyapunov exponent of $\beta_{kl}(t)$.

The proof is similar to the result for the SVD (in the integrally separated case), i.e., see Theorem 5.4 in [14].

11.3.4 Computational Techniques

There has been interest recently in techniques for preserving the orthogonality of approximate solutions to matrix differential equations with orthogonal solutions, such as in the case of QR factorization of a fundamental matrix solution of (11.1) and the U and V factors in the SVD. Preserving orthogonality is important when finding Lyapunov exponents since an orthogonal transformation preserves the sum of the Lyapunov exponents. Techniques based on a continuous version of the Gram-Schmidt process are a natural idea, but they have not proven to be reliable numerically due to the loss of orthogonality when the differential equations that describe the continuous Gram-Schmidt process are approximated numerically. Still, a host of successful techniques have emerged. Some parallel the algorithmic development of symplectic techniques for Hamiltonian systems, and maintain automatically orthogonality of the computed solution: among these are Gauss Runge-Kutta methods (see [8, 20]), as well as several others which automatically maintain orthogonality, see [10, 21, 23]. However, extensive practical experience with orthogonality-preserving methods has lead to the adoption of so-called projection techniques, whereby a possibly non-orthogonal solution is projected onto an orthogonal one, without loss of order of accuracy. Codes based upon projected integrators for approximating $Q(t)$ and Lyapunov exponents are described in [19]. The codes are designed for both a given time dependent linear differential equation and for nonlinear differential equations using either an existing code to solve the nonlinear equation and then employing the codes for linear problems or in solving the nonlinear equations simultaneously with the linearized equations.

11.4 Differential-Algebraic Equations

In [36, 38, 40] Lyapunov and Sacker–Sell spectral theory is extended from ordinary differential equations (ODEs) to nonautonomous differential-algebraic equations (DAEs). Using orthogonal changes of variables, the original DAE system is transformed into appropriate condensed forms, for which concepts such as Lyapunov exponents, Bohl exponents, exponential dichotomy and spectral intervals of various kinds can be analyzed via the resulting underlying ODE. The focus is on strangeness-free, linear, time varying DAEs. This means that under weak conditions the DAE and its derivatives may be transformed into a system that is strangeness-free, i.e., the differential and algebraic parts of the equations are easily separated.

Consider the linear time-varying DAE

$$E(t)\dot{x} = A(t)x + f(t), \tag{11.42}$$

on the half-line $I = [0, \infty)$, together with an initial condition $x(0) = x_0$. Stability spectra is associated with the homogenous equation

$$E(t)\dot{x} = A(t)x, \tag{11.43}$$

which allows for the analysis of the asymptotic behavior or the growth rate of solutions to initial value problems. Assume that $E, A \in C(I, \mathbb{R}^{n\times n})$ and $f \in C(I, \mathbb{R}^n)$ are sufficiently smooth functions where the notation $C(I, \mathbb{R}^{n\times n})$ denotes the space of continuous functions from I to $\mathbb{R}^{n\times n}$.

A complete theory as well a detailed analysis of the relationship between different index concepts can be found in [34]. It is assumed that the homogeneous equation is strangeness-free and has the form

$$E(t)\dot{x} = A(t)x, \quad t \in I, \tag{11.44}$$

where

$$E(t) = \begin{bmatrix} E_1(t) \\ 0 \end{bmatrix}, \; A(t) = \begin{bmatrix} A_1(t) \\ A_2(t) \end{bmatrix},$$

$E_1 \in C(I, \mathbb{R}^{d \times n})$ and $A_2 \in C(I, \mathbb{R}^{(n-d) \times n})$ are such that the matrix

$$\bar{E}(t) := \begin{bmatrix} E_1(t) \\ A_2(t) \end{bmatrix} \tag{11.45}$$

is invertible for all t. As a direct consequence, then E_1 and A_2 are of full row-rank. We assume that the matrix functions are sufficiently smooth so that the strangeness-free DAE (11.44) is in fact differentiation-index 1, see [34].

Definition 2 Suppose that $U \in C(I, \mathbb{R}^{n \times n})$ and $V \in C^1(I, \mathbb{R}^{n \times n})$ are nonsingular matrix functions such that V and V^{-1} are bounded. Then the transformed DAE system

$$\tilde{E}(t)\dot{\tilde{x}} = \tilde{A}(t)\tilde{x}, \tag{11.46}$$

with $\tilde{E} = UEV$, $\tilde{A} = UAV - UE\dot{V}$ and $x = V\tilde{x}$ is called *globally kinematically equivalent* to (11.44) and the transformation is called a *global kinematic equivalence transformation.* If $U \in C^1(I, \mathbb{R}^{n \times n})$ and, furthermore, also U and U^{-1} are bounded then we call this a *strong global kinematic equivalence transformation.*

It is clear that the Lyapunov exponents of a DAE system as well as the normality of a basis formed by the columns of a fundamental solution matrix are preserved under global kinematic equivalence transformations. The following lemma is the key to constructing and understanding QR methods and it is in fact a simplified version of [36, Lemma 7].

Lemma 3 (Lemma 4 [40]) *Consider a strangeness-free DAE system of the form (11.44) with continuous coefficients and a minimal fundamental solution matrix X. Then there exist matrix functions $V \in C(I, \mathbb{R}^{n \times d})$ and $U \in C^1(I, \mathbb{R}^{n \times d})$ with orthonormal columns such that in the fundamental matrix equation $E\dot{X} = AX$ associated with (11.44), the change of variables $X = UR$, with $R \in C^1(I, \mathbb{R}^{d \times d})$ upper triangular with positive diagonal elements, and the multiplication of both sides of the system from the left with V^T leads to the system*

$$\mathscr{E}\dot{R} = \mathscr{A}R, \tag{11.47}$$

where $\mathscr{E} := V^T EU$ is nonsingular, $\mathscr{A} := V^T AU - V^T E\dot{U}$, and both of them are upper triangular.

System (11.47) is an implicit ODE, since $\mathscr{E}$ is nonsingular. It is called *essentially underlying implicit ODE system* (EUODE) of (11.44), and it can be turned into an ODE by multiplication with $\mathscr{E}^{-1}$ from the left. Lyapunov exponents may be calculated using the same basic ideas as in the ODE case using a proper definition of fundamental matrix solutions for DAEs. Sacker–Sell spectrum may be calculated using the shifted DAE system

$$E(t)\dot{x} = [A(t) - \lambda E(t)]x.$$

Among the issues that make extension to DAEs difficult is in computing the transformation that brings the DAE to EUODE form and that the sensitivity of the Lyapunov exponent and other stability spectrum relies not only on changes to $A(t)$ but also to changes to $E(t)$ that may be close to singular.

In [40] numerical methods based on QR factorization for computing all or some Lyapunov or Sacker–Sell spectral intervals for linear differential-algebraic equations. In addition, a perturbation and error analysis for these methods is presented. We investigate how errors in the data and in the numerical integration affect the accuracy of the approximate spectral intervals over long time intervals. The paper [38] is concerned with the use of the SVD approach for the numerical approximation of Lyapunov and Sacker–Sell spectrum for linear DAEs. This includes approximation of the spectrum and the associated solution subspaces. Numerical methods based on smooth singular value decompositions are introduced for computing all or only some spectral intervals and their associated leading directions.

References

1. Adrianova, L.Ya.: Introduction to Linear Systems of Differential Equations (Transl. from the Russian by Peter Zhevandrov). Translations of Mathematical Monographs, vol. 146, x+204pp. American Mathematical Society, Providence (1995)
2. Badawy, M., Van Vleck, E.: Perturbation theory for the approximation of stability spectra by QR methods for sequences of linear operators on a Hilbert space. Linear Algebra Appl. **437**(1), 37–59 (2012)
3. Benettin, G., Galgani, L., Giorgilli, A., Strelcyn, J.-M.: Lyapunov exponents for smooth dynamical systems and for Hamiltonian systems; a method for computing all of them. Part 1: theory, and Part 2: numerical applications. Meccanica **15**, 9–20, 21–30 (1980)
4. Bohl, P.: Über Differentialungleichungen. J. F. d. Reine Und Angew. Math. **144**, 284–313 (1913)
5. Breda, D., Van Vleck, E.: Approximating Lyapunov exponents and Sacker-Sell spectrum for retarded functional differential equations. Numer. Math. **126**(2), 225–257 (2014)
6. Bunse-Gerstner, A., Byers, R., Mehrmann, V., Nichols, N.K.: Numerical computation of an analytic singular value decomposition of a matrix valued function. Numer. Math. **60**, 1, 1–39 (1991)
7. Bylov, B.F., Izobov, N.A.: Necessary and sufficient conditions for stability of characteristic exponents of a linear system. Differ. Uravn. **5**, 1794–1903 (1969)
8. Calvo, M.P., Iserles, A., Zanna, A.: Numerical solution of isospectral flows. Math. Comput. **66**(220), 1461–1486 (1997)

9. Champneys, A.R., Kuznetsov, Yu.A., Sandstede, B.: A numerical toolbox for homoclinic bifurcation analysis. Int. J. Bifur. Chaos Appl. Sci. Eng. **6**(5), 867–887 (1996)
10. Chu, M.T.: On the continuous realization of iterative processes. SIAM Rev. **30**, 375–387 (1988)
11. Coppel, W.A.: Dichotomies in Stability Theory. Lecture Notes in Mathematics, vol. 629, ii+98 pp. Springer, Berlin/New York (1978)
12. Demmel, J.W., Dieci, L., Friedman, M.J.: Computing connecting orbits via an improved algorithm for continuing invariant subspaces. SIAM J. Sci. Comput. **22**(1), 81–94 (2000)
13. Dieci, L., Eirola, T.: On smooth decompositions of matrices. SIAM J. Matrix Anal. Appl. **20**(3), 800–819 (1999) (electronic)
14. Dieci, L., Elia, C.: The singular value decomposition to approximate spectra of dynamical systems. Theoretical aspects. J. Differ. Eqn. **230**(2), 502–531 (2006)
15. Dieci, L., Elia, C., Van Vleck, E.: Exponential dichotomy on the real line: SVD and QR methods. J. Differ. Eqn. **248**(2), 287–308 (2010)
16. Dieci, L., Elia, C., Van Vleck, E.: Detecting exponential dichotomy on the real line: SVD and QR algorithms. BIT **51**(3), 555–579 (2011)
17. Dieci, L., Friedman, M.J.: Continuation of invariant subspaces. Numer. Linear Algebra Appl. **8**(5), 317–327 (2001)
18. Dieci, L., Jolly, M., Rosa, R., Van Vleck, E.: Error on approximation of Lyapunov exponents on inertial manifolds: the Kuramoto-Sivashinsky equation. J. Discret. Contin. Dyn. Syst. Ser. B **9**(3–4), 555–580 (2008)
19. Dieci, L., Jolly, M.S., Van Vleck, E.S.: Numerical techniques for approximating Lyapunov exponents and their implementation. ASME J. Comput. Nonlinear Dyn. **6**, 011003–1–7 (2011)
20. Dieci, L., Russell, R.D., Van Vleck, E.S.: Unitary integrators and applications to continuous orthonormalization techniques. SIAM J. Numer. Anal. **31**(1), 261–281 (1994)
21. Dieci, L., Russell, R.D., Van Vleck, E.S.: On the computation of Lyapunov exponents for continuous dynamical systems. SIAM J. Numer. Anal. **34**(1), 402–423 (1997)
22. Dieci, L., Van Vleck, E.S.: Computation of a few Lyapunov exponents for continuous and discrete dynamical systems. Numerical methods for ordinary differential equations (Atlanta, 1994). Appl. Numer. Math. **17**(3), 275–291 (1995)
23. Dieci, L., Van Vleck, E.S.: Computation of orthonormal factors for fundamental solution matrices. Numer. Math. **83**(4), 599–620 (1999)
24. Dieci, L., Van Vleck, E.S.: Lyapunov spectral intervals: theory and computation. SIAM J. Numer. Anal. **40**(2), 516–542 (2002) (electronic)
25. Dieci, L., Van Vleck, E.S.: On the error in computing Lyapunov exponents by QR methods. Numer. Math. **101**(4), 619–642 (2005)
26. Dieci, L., Van Vleck, E.S.: Perturbation theory for approximation of Lyapunov exponents by QR methods. J. Dyn. Differ. Eqn. **18**(3), 815–840 (2006)
27. Dieci, L., Van Vleck, E.S.: Lyapunov and Sacker-Sell spectral intervals. J. Dyn. Differ. Eqn. **19**(2), 265–293 (2007)
28. Dieci, L., Van Vleck, E.S.: On the error in QR integration. SIAM J. Numer. Anal. **46**(3), 1166–1189 (2008)
29. Diliberto, S.P.: On systems of ordinary differential equations. In: Contributions to the Theory of Nonlinear Oscillations, Annals of Mathematical Studies, vol. 20, pp. 1–38. Princeton University Press, Princeton (1950)
30. Holtz, O., Mehrmann, V., Schneider, H.: Matrices that commute with their derivative. On a letter from Schur to Wielandt. Linear Algebra Appl. **438**(5), 2574–2590 (2013)
31. Kato, T.: Perturbation Theory for Linear Operators, 2nd edn. Grundlehren der Mathematischen Wissenschaften. Band 132, xxi+619 pp. Springer, Berlin/New York (1976)
32. Kressner, D.: The periodic QR algorithm is a disguised QR algorithm. Linear Algebra Appl. **417**(2–3), 423–433 (2006)
33. Kressner, D.: A periodic Krylov-Schur algorithm for large matrix products. Numer. Math. **103**(3), 461–483 (2006)

34. Kunkel, P., Mehrmann, V.: Differential-Algebraic Equations. Analysis and Numerical Solution. EMS Textbooks in Mathematics, viii+377 pp. European Mathematical Society (EMS), Zürich (2006)
35. Leimkuhler, B.J., Van Vleck, E.S.: Orthosymplectic integration of linear Hamiltonian systems. Numer. Math. **77**(2), 269–282 (1997)
36. Linh, V.H., Mehrmann, V.: Lyapunov, Bohl and Sacker-Sell spectral intervals for differential-algebraic equations. J. Dyn. Differ. Eqn. **21**(1), 153–194 (2009)
37. Linh, V.H., Mehrmann, V.: Spectral analysis for linear differential-algebraic equations. In: 8th AIMS Conference on Dynamical Systems, Differential Equations and Applications, Dresden, 2011. Discrete and Continuous Dynamical Systems Supplement, vol. II, pp. 991–1000. ISBN:978-1-60133-008-6; 1-60133-008-1
38. Linh, V.H., Mehrmann, V.: Approximation of spectral intervals and leading directions for differential-algebraic equation via smooth singular value decompositions. SIAM J. Numer. Anal. **49**(5), 1810–1835 (2011)
39. Linh, V.H., Mehrmann, V.: Spectra and leading directions for linear DAEs. In: Control and Optimization with Differential-Algebraic Constraints. Advances in Design and Control, vol. 23, pp. 59–78. SIAM, Philadelphia (2012)
40. Linh, V.H.: Mehrmann, V., Van Vleck, E.S.: QR methods and error analysis for computing Lyapunov and Sacker-Sell spectral intervals for linear differential-algebraic equations. Adv. Comput. Math. **35**(2–4), 281–322 (2011)
41. Lyapunov, A.: Problém géneral de la stabilité du mouvement. Int. J. Control **53**, 531–773 (1992)
42. Millionshchikov, V.M.: Structurally stable properties of linear systems of differential equations. Differ. Uravn. **5**, 1775–1784 (1969)
43. Millionshchikov V.M.: Systems with integral division are everywhere dense in the set of all linear systems of differential equations. Differ. Uravn. **5**, 1167–1170 (1969)
44. Oliveira, S., Stewart, D.E.: exponential splittings of products of matrices and accurately computing singular values of long products. In: Proceedings of the International Workshop on Accurate Solution of Eigenvalue Problems, University Park, 1998. Linear Algebra Applications, vol. 3091–3, pp. 175–190 (2000)
45. Oseledec, V.I.: A multiplicative ergodic theorem. Lyapunov characteristic numbers for dynamical systems. Trans. Mosc. Math. Soc. **19**, 197 (1998)
46. Palmer, K.J.: The structurally stable systems on the half-line are those with exponential dichotomy. J. Differ. Eqn. **33**, 16–25 (1979)
47. Palmer, K.J.: Exponential dichotomy, integral separation and diagonalizability of linear sys temsof ordinary differential equations. J. Differ. Eqn. **43**, 184–203 (1982)
48. Palmer, K.J.: Exponential separation, exponential dichotomy and spectral theory for linear s ystems of ordinary differential equations. J. Differ. Eqn. **46**, 324–345 (1982)
49. Perron, O.: Die Ordnungszahlen Linearer Differentialgleichungssysteme. Math. Zeits. **31**, 748–766 (1930)
50. Rheinboldt, W.C.: On the computation of multidimensional solution manifolds of parametrized equations. Numer. Math. **53**(1–2), 165–181 (1988)
51. Sacker, R.J., Sell, G.R.: A spectral theory for linear differential systems. J. Differ. Eqn. **7**, 320–358 (1978)
52. Stewart, D.E.: A new algorithm for the SVD of a long product of matrices and the stability of products. Electron. Trans. Numer. Anal. **5**, 29–47 (1997) (electronic)
53. Van Vleck, E.S. On the error in the product QR decomposition. SIAM J. Matrix Anal. Appl. **31**(4), 1775–1791 (2009/2010)

Chapter 12
Polynomial Eigenvalue Problems: Theory, Computation, and Structure

D. Steven Mackey*, **Niloufer Mackey, and Françoise Tisseur**†

Abstract Matrix polynomial eigenproblems arise in many application areas, both directly and as approximations for more general nonlinear eigenproblems. One of the most common strategies for solving a polynomial eigenproblem is via a linearization, which replaces the matrix polynomial by a matrix pencil with the same spectrum, and then computes with the pencil. Many matrix polynomials arising from applications have additional algebraic structure, leading to symmetries in the spectrum that are important for any computational method to respect. Thus it is useful to employ a structured linearization for a matrix polynomial with structure. This essay surveys the progress over the last decade in our understanding of linearizations and their construction, both with and without structure, and the impact this has had on numerical practice.

*Supported by Ministerio de Economía y Competitividad of Spain through grant MTM2012-32542.

†Supported by Engineering and Physical Sciences Research Council grant EP/I005293.

D.S. Mackey
Department of Mathematics, Western Michigan University, 1903 W Michigan Ave, Kalamazoo, MI 49008, USA
e-mail: steve.mackey@wmich.edu

N. Mackey (✉)
Department of Mathematics, Western Michigan University, 1903 W Michigan Ave, Kalamazoo, MI 49008, USA
e-mail: nil.mackey@wmich.edu

F. Tisseur
School of Mathematics, The University of Manchester, Alan Turing Building, M13 9PL Manchester, UK
e-mail: francoise.tisseur@manchester.ac.uk

P. Benner et al. (eds.), *Numerical Algebra, Matrix Theory, Differential-Algebraic Equations and Control Theory*, DOI 10.1007/978-3-319-15260-8_12

12.1 Introduction

Nonlinear eigenvalue problems of the form

$$P(\lambda)x = 0\,, \quad x \in \mathbb{C}^n, \quad x \neq 0\,,$$

where $P(\lambda)$ is an $m \times n$ matrix-valued function of a scalar variable λ, are playing an increasingly important role in classical and contemporary applications. The simplest, but still most important among these problems are the *polynomial eigenproblems*, where $P(\lambda)$ is an $m \times n$ matrix polynomial

$$P(\lambda) = \sum_{i=0}^{k} \lambda^i A_i\,, \quad A_i \in \mathbb{C}^{m\times n}\,. \tag{12.1}$$

Such problems arise directly from applications, from finite element discretizations of continuous models, or as approximations to more general nonlinear eigenproblems, as detailed in the survey articles [69, 79]. The trend towards extreme designs, such as high speed trains, optoelectronic devices, micro-electromechanical systems, and "superjumbo" jets such as the Airbus 380, presents a challenge for the computation of the resonant frequencies of these structures as these extreme designs often lead to eigenproblems with poor conditioning.

However, the physics that underlies problems arising from applications can lead to algebraic structure in their mathematical formulation. Numerical methods that preserve this structure keep key qualitative features such as eigenvalue symmetries from being obscured by finite precision error.

A recurring theme running through much of the work of Volker Mehrmann has been the preservation of structure – in the pursuit of condensed forms, and in the development of numerical algorithms. To quote from the 2004 paper titled *Nonlinear Eigenvalue Problems: A Challenge for Modern Eigenvalue Methods* by Mehrmann and Voss [69]:

> The task of numerical linear algebra then is to design numerical methods that are accurate and efficient for the given problem. The methods should exploit to a maximal extent the sparsity and structure of the coefficient matrices. Furthermore, they should be as accurate as the approximation of the underlying operator problem permits, and they should include error and condition estimates.

One of the most common strategies for solving a polynomial eigenproblem is via a *linearization*, which replaces the given matrix polynomial $P(\lambda)$ by a *matrix pencil* $L(\lambda) = \lambda X + Y$ with the same eigenvalues as P. The eigenproblem for $L(\lambda)$ is then solved with general pencil algorithms like the *QZ* algorithm, or with methods designed to work effectively on the specific types of pencils produced by the linearization process. If the matrix polynomial has some structure, then the linearization should also have that structure, and the algorithm employed on the linearization should preserve that structure.

The most commonly used linearizations in numerical practice have been the Frobenius companion forms. Although these pencils have many desirable properties, including the extreme ease with which they can be constructed, they have one significant drawback. They *do not preserve* any of the most important and commonly occurring matrix polynomial structures – Hermitian, alternating, or palindromic. Thus in order to implement the structure preservation principle on the linearization strategy, it is necessary to have more linearizations available, in particular ones that preserve the structure of the original polynomial. It is also useful to have a large palette of easily constructible linearizations, even in the absence of any structure to be preserved. For example, it may be possible to improve numerical accuracy by selecting an "optimal" linearization, but only if there are many linearizations available to choose from.

This essay will illustrate the influence that the structure preservation principle has had on the development of linearizations of matrix polynomials, on the impact our improved understanding of linearizations in general has had on numerical practice, and Mehrmann's key contributions to that effort.

12.2 Basic Concepts

We use $\mathbb{N}$ to denote the set of nonnegative integers, $\mathbb{F}$ for an arbitrary field, $\mathbb{F}[\lambda]$ for the ring of polynomials in one variable with coefficients from the field $\mathbb{F}$, and $\mathbb{F}(\lambda)$ for the field of rational functions over $\mathbb{F}$.

A matrix polynomial of grade k has the form

$$P(\lambda) = \sum_{i=0}^{k} \lambda^i A_i, \qquad \text{with} \quad A_i \in \mathbb{F}^{m \times n}. \tag{12.2}$$

Here we allow any of the coefficient matrices, including A_k, to be the zero matrix. The *degree* of a nonzero matrix polynomial retains its usual meaning as the largest integer j such that the coefficient of λ^j in $P(\lambda)$ is nonzero. The *grade* of a nonzero matrix polynomial is a choice of integer k at least as large as its degree [22, 59, 61]. It signals that the polynomial is to be viewed as an element of a particular *vector space* – the $\mathbb{F}$-vector space of all matrix polynomials of degree less than or equal to k. Choosing a grade, in effect, specifies the finite-dimensional vector space of discourse.

If $m = n$ and $\det P(\lambda)$ is not the identically zero polynomial, then $P(\lambda)$ is said to be *regular*; equivalently, $P(\lambda)$ is regular if it is invertible when viewed as a matrix with entries in the field of rational functions $\mathbb{F}(\lambda)$. Otherwise, $P(\lambda)$ is said to be *singular* (note that this includes all rectangular matrix polynomials with $m \neq n$). The rank of $P(\lambda)$ is its rank when viewed as a matrix with entries in the field $\mathbb{F}(\lambda)$, or equivalently, the size of the largest nonzero minor of $P(\lambda)$. For simplicity, in many cases we may suppress the dependence on λ when referring to a matrix polynomial.

An $m \times m$ polynomial $E(\lambda)$ is said to be *unimodular* if $\det E(\lambda)$ is a *nonzero constant*, i.e., $E(\lambda)$ has an inverse that is also a matrix polynomial [32]. The canonical form of a matrix polynomial $P(\lambda)$ under a transformation $E(\lambda)P(\lambda)F(\lambda)$ by unimodular matrix polynomials $E(\lambda)$ and $F(\lambda)$ is referred to as the *Smith form* of $P(\lambda)$. This form was first developed for integer matrices by H.J.S. Smith [76] in the context of solving linear systems of Diophantine equations [51]. It was then extended by Frobenius in [30] to matrix polynomials.

Theorem 1 (Smith form (Frobenius, 1878)[30]) *Let $P(\lambda)$ be an $m \times n$ matrix polynomial over an arbitrary field $\mathbb{F}$. Then there exists $r \in \mathbb{N}$, and unimodular matrix polynomials $E(\lambda)$ and $F(\lambda)$ over $\mathbb{F}$ of size $m \times m$ and $n \times n$, respectively, such that*

$$E(\lambda)P(\lambda)F(\lambda) = \operatorname{diag}(d_1(\lambda), \ldots, d_{\min\{m,n\}}(\lambda)) =: D(\lambda), \tag{12.3}$$

where each $d_i(\lambda)$ is in $\mathbb{F}[\lambda]$, $d_1(\lambda), \ldots, d_r(\lambda)$ are monic, $d_{r+1}(\lambda), \ldots, d_{\min\{m,n\}}(\lambda)$ are identically zero, and $d_1(\lambda), \ldots, d_r(\lambda)$ form a divisibility chain, *that is, $d_j(\lambda)$ is a divisor of $d_{j+1}(\lambda)$ for $j = 1, \ldots, r-1$. Moreover, the $m \times n$ diagonal matrix polynomial $D(\lambda)$ is unique, and the number r is equal to the rank of P.*

The nonzero diagonal elements $d_j(\lambda)$, $j = 1, \ldots, r$ in the Smith form $D(\lambda)$ are called the *invariant factors* or *invariant polynomials* of $P(\lambda)$.

The uniqueness of $D(\lambda)$ in Theorem 1 implies that the Smith form is *insensitive to field extensions*. In other words, the Smith forms of $P(\lambda)$ over $\mathbb{F}$ and over any extension field $\widetilde{\mathbb{F}} \supseteq \mathbb{F}$ are identical. Consequently, the following notions of the partial multiplicity sequences, eigenvalues, and elementary divisors of $P(\lambda)$ are well-defined.

Definition 1 (Partial Multiplicity Sequences and Jordan Characteristic) Let $P(\lambda)$ be an $m \times n$ matrix polynomial of rank r over a field $\mathbb{F}$. For any λ_0 in the algebraic closure $\overline{\mathbb{F}}$, the invariant polynomials $d_i(\lambda)$ of P, for $1 \leq i \leq r$, can each be uniquely factored as

$$d_i(\lambda) = (\lambda - \lambda_0)^{\alpha_i}\, p_i(\lambda) \quad \text{with} \quad \alpha_i \geq 0\,,\; p_i(\lambda_0) \neq 0\,. \tag{12.4}$$

The sequence of exponents $(\alpha_1, \alpha_2, \ldots, \alpha_r)$ for any given $\lambda_0 \in \overline{\mathbb{F}}$ satisfies the condition $0 \leq \alpha_1 \leq \alpha_2 \leq \cdots \leq \alpha_r$ by the divisibility chain property of the Smith form, and is called the *partial multiplicity sequence of P at $\lambda_0 \in \overline{\mathbb{F}}$*, denoted $\mathcal{J}(P, \lambda_0)$. The collection of all the partial multiplicity sequences of P is called the *Jordan characteristic* of P.

Note that we allow any, even all, of the exponents α_i in a partial multiplicity sequence $\mathcal{J}(P, \lambda_0)$ to be zero. Indeed, this occurs for all but a finite number of $\lambda_0 \in \overline{\mathbb{F}}$. These exceptional λ_0 with at least one nonzero entry in $\mathcal{J}(P, \lambda_0)$ are of course just the eigenvalues of $P(\lambda)$.

Definition 2 (Eigenvalues and Elementary Divisors) A scalar $\lambda_0 \in \overline{\mathbb{F}}$ is a (finite) *eigenvalue* of a matrix polynomial P whenever its partial multiplicity sequence $(\alpha_1, \alpha_2, \ldots, \alpha_r)$ is not the zero sequence. The *elementary divisors* for an eigenvalue λ_0 of P are the collection of factors $(\lambda - \lambda_0)^{\alpha_i}$ with $\alpha_i \neq 0$, including repetitions. The *algebraic multiplicity* of an eigenvalue λ_0 is the sum $\alpha_1 + \alpha_2 + \cdots + \alpha_r$ of the terms in its partial multiplicity sequence, while the *geometric multiplicity* is the number of nonzero terms in this sequence. An eigenvalue λ_0 is said to be *simple* if its algebraic multiplicity is one; λ_0 is *semisimple* if its algebraic and geometric multiplicities are equal, equivalently, if all of the nonzero terms in its partial multiplicity sequence are equal to one.

It is worth noting that defining the eigenvalues of a matrix polynomial via the Smith form subsumes the more restrictive notion of the eigenvalues as the roots of $\det P(\lambda)$, which is completely inadequate for singular matrix polynomials. We also stress the importance of viewing the partial multiplicities of a fixed λ_0 as a *sequence*. In a number of situations, especially for matrix polynomials with structure [58–60], it is essential to consider certain *subsequences* of partial multiplicities, which can be subtly constrained by the matrix polynomial structure. Indeed, even the zeroes in the partial multiplicity sequences of structured matrix polynomials can sometimes have nontrivial significance [58–60].

Matrix polynomials may also have infinite eigenvalues, with a corresponding notion of elementary divisors at ∞. In order to define the elementary divisors at ∞ we need one more preliminary concept, that of the reversal of a matrix polynomial.

Definition 3 (j-reversal) Let $P(\lambda)$ be a nonzero matrix polynomial of degree $d \geq 0$. For $j \geq d$, the *j-reversal* of P is the matrix polynomial $\text{rev}_j\, P$ given by

$$(\text{rev}_j\, P)(\lambda) := \lambda^j P(1/\lambda). \tag{12.5}$$

In the special case when $j = d$, the j-reversal of P is called the *reversal* of P and is sometimes denoted by just $\text{rev}\, P$.

Definition 4 (Elementary divisors at ∞) Let $P(\lambda)$ be a nonzero matrix polynomial of grade k and rank r. We say that $\lambda_0 = \infty$ is an eigenvalue of P if and only if 0 is an eigenvalue of $\text{rev}_k\, P$, and the partial multiplicity sequence of P at $\lambda_0 = \infty$ is defined to be the same as that of the eigenvalue 0 for $\text{rev}_k\, P$, that is $\mathcal{J}(P, \infty) := \mathcal{J}(\text{rev}_k\, P, 0)$. If this partial multiplicity sequence is $(\alpha_1, \alpha_2, \ldots, \alpha_r)$, then for each $\alpha_i \neq 0$ we say there is an elementary divisor of degree α_i for the eigenvalue $\lambda_0 = \infty$ of P.

If $P(\lambda) = \sum_{i=0}^{g} \lambda^i A_i$ has grade k and rank r, then P has an eigenvalue at ∞ if and only if the rank of the leading coefficient matrix A_k is strictly less than r. For a regular polynomial P this just means that A_k is singular. Observe that if $k > \deg P$, then $A_k = 0$ and P necessarily has r elementary divisors at ∞.

Definition 5 (Spectral Structure of a Matrix Polynomial) The collection of all the eigenvalues of a matrix polynomial $P(\lambda)$, both finite and infinite, is the *spectrum* of P. The collection of all the elementary divisors of P, both finite and infinite, including repetitions, constitutes the *spectral structure* of P.

The two most frequently used equivalence relations that preserve spectral structure between matrix polynomials are *unimodular equivalence* and *strict equivalence*. They can be used only between matrix polynomials of the same size.

Definition 6 A pair of $m \times n$ matrix polynomials P and Q over a fixed but arbitrary field $\mathbb{F}$ are said to be

(a) *Unimodularly equivalent*, denoted $P \sim Q$, if there exist unimodular matrix polynomials $E(\lambda)$ and $F(\lambda)$ over $\mathbb{F}$ such that $E(\lambda)P(\lambda)F(\lambda) = Q(\lambda)$,
(b) *Strictly equivalent*, denoted $P \cong Q$, if there exist invertible (constant) matrices E and F over $\mathbb{F}$ such that $E \cdot P(\lambda) \cdot F = Q(\lambda)$.

Of these two relations, unimodular equivalence is the more flexible, as it allows the degrees of the two matrix polynomials to differ, while keeping the list of finite elementary divisors invariant. On the other hand, strict equivalence preserves both finite and infinite elementary divisors, but because the degrees of strictly equivalent matrix polynomials have to be identical, this relation can be a bit restrictive.

Recently the relations of *extended unimodular equivalence* and *spectral equivalence* have been introduced [22] to facilitate the comparison of matrix polynomials that are of different sizes, including rectangular, and of different grades. The underlying goal is to investigate the extent to which it is possible for such diverse matrix polynomials to share the same spectral structure and *and* the same singular structure. These extended equivalences now open up the possibility of choosing linearizations that can take on any size that "works." This is in accord with the notion of "trimmed linearizations" studied by Byers, Mehrmann and Xu in [16]. Another important consequence is that one can now easily generalize the notion of (strong) linearization to (strong) quadratification, and indeed to (strong) ℓ-ification [22]!

12.3 Linearizations

For square matrix polynomials, the notion of *linearization* plays a central role for both theory and computation.

Definition 7 (Linearization) An $nk \times nk$ pencil $L(\lambda) = \lambda X + Y$ is said to be a *linearization* for an $n \times n$ matrix polynomial $P(\lambda)$ of grade k if there exist unimodular $nk \times nk$ matrix polynomials $E(\lambda), F(\lambda)$ such that

$$E(\lambda)L(\lambda)F(\lambda) = \left[\begin{array}{c|c} P(\lambda) & 0 \\ \hline 0 & I_{(k-1)n} \end{array}\right]_{nk \times nk}.$$

If in addition, $\operatorname{rev}_1 L(\lambda) := \lambda X + Y$ is a linearization of $\operatorname{rev}_k P(\lambda)$, then L is said to be a *strong linearization* of P.

The key property of any linearization L of P is that L has the same finite elementary divisors as P, while a strong linearization has the same finite *and* infinite elementary divisors as P. Since there are well-known algorithms for solving the linear eigenproblem this immediately suggests working on a matrix pencil L that is a strong linearization for P.

The linearizations most used in practice are the *first* and *second Frobenius companion forms* $C_1(\lambda) = \lambda X_1 + Y_1$, and $C_2(\lambda) = \lambda X_2 + Y_2$, where

$$X_1 = X_2 = \operatorname{diag}(A_k, I_{(k-1)n}), \tag{12.6a}$$

$$Y_1 = \begin{bmatrix} A_{k-1} & A_{k-2} & \cdots & A_0 \\ -I_n & 0 & \cdots & 0 \\ \vdots & \ddots & \ddots & \vdots \\ 0 & \cdots & -I_n & 0 \end{bmatrix}, \quad \text{and} \quad Y_2 = \begin{bmatrix} A_{k-1} & -I_n & \cdots & 0 \\ A_{k-2} & 0 & \ddots & \vdots \\ \vdots & \vdots & \ddots & -I_n \\ A_0 & 0 & \cdots & 0 \end{bmatrix}. \tag{12.6b}$$

They have several attractive properties:

- there is a uniform template for constructing them directly from the data in P, using no matrix operations on the coefficients of P,
- eigenvectors of P are easily recoverable from eigenvectors of the companion forms,
- they are always strong linearizations for P, no matter whether P is regular or singular.

However, they have one significant drawback – they usually do not reflect the structure that may be present in the original polynomial P.

12.3.1 Ansatz Spaces

During an extended visit by the first two authors to Berlin in 2003, Mehrmann proposed searching for alternatives to the companion linearizations $C_1(\lambda)$ and $C_2(\lambda)$ – alternatives that would share the structure of the parent polynomial $P(\lambda)$.

In joint work with Mehrmann and Mehl, two large vector spaces of pencils that generalize the first and second Frobenius companion forms were introduced in [55]. Christened $\mathbb{L}_1(P)$ and $\mathbb{L}_2(P)$, where P is a regular matrix polynomial, these spaces were conceived as the collection of all pencils satisfying a certain ansatz, which we now briefly describe. With $\Lambda := \left[\lambda^{k-1}, \lambda^{k-2}, \ldots, \lambda, 1\right]^T$, where k is the grade of P, define $\mathbb{L}_1(P)$ as the set of all $kn \times kn$ pencils $L(\lambda)$ satisfying the right ansatz

$$L(\lambda) \cdot (\Lambda \otimes I_n) = v \otimes P(\lambda), \quad \text{for some } v \in \mathbb{F}^k, \tag{12.7}$$

and $\mathbb{L}_2(P)$ as the set of all $kn \times kn$ pencils $L(\lambda)$ satisfying the left ansatz

$$(\Lambda^T \otimes I_n) \cdot L(\lambda) = w^T \otimes P(\lambda), \quad \text{for some } w \in \mathbb{F}^k. \tag{12.8}$$

A direct calculation shows that $C_1(\lambda) \in \mathbb{L}_1(P)$ with right ansatz vector $v = e_1$, and $C_2(\lambda) \in \mathbb{L}_2(P)$ with left ansatz vector $w = e_1$. The pencils in these ansatz spaces were shown to have a number of nice properties:

- like $C_1(\lambda)$ and $C_2(\lambda)$, they are all easily constructible from the coefficients of P,
- eigenvectors of P are easily recoverable; pencils in $\mathbb{L}_1(P)$ reveal right eigenvectors of P, while those in $\mathbb{L}_2(P)$ reveal left eigenvectors,
- for regular P, almost all pencils in these spaces are strong linearizations for P.

Furthermore, each of these spaces is of dimension $k(k-1)n^2 + k$. Thus each represents a relatively large subspace of the full pencil space (which has dimension $2k^2n^2$), and hence is a large source of potential linearizations for P. In fact, these spaces are so large, that for any choice of ansatz vector there are many degrees of freedom available for choosing a potential linearization in $\mathbb{L}_1(P)$ or $\mathbb{L}_2(P)$.

The aim of identifying smaller, but interesting subspaces of these ansatz spaces brings the double ansatz subspace $\mathbb{DL}(P) := \mathbb{L}_1(P) \cap \mathbb{L}_2(P)$ into focus. One sees right away that linearizations in $\mathbb{DL}(P)$ enjoy a two-sided eigenvector recovery property. But a $\mathbb{DL}(P)$-pencil also has an unexpected feature: its right and left ansatz vectors are identical, with this common vector uniquely determining the pencil. An isomorphism between $\mathbb{DL}(P)$ and $\mathbb{F}^k$ now follows, which in turn induces a natural basis for $\mathbb{DL}(P)$. Described in [38], a pencil $\lambda X_i + Y_i$ in this basis has special structure. Every X_i and Y_i is block diagonal, with the diagonal blocks being block-Hankel. In a surprising twist, a completely different construction of Lancaster [48] dating back to the 1960s is proved to also generate this natural basis for $\mathbb{DL}(P)$.

The unique vector $v \in \mathbb{F}^k$ associated with $L(\lambda) \in \mathbb{DL}(P)$ gives us a way to test when $L(\lambda)$ is a linearization for P, and show that almost all pencils in $\mathbb{DL}(P)$ are linearizations for P.

Theorem 2 (Eigenvalue Exclusion Theorem [55]) *Let $P(\lambda)$ be a regular matrix polynomial of grade k and let $L(\lambda) \in \mathbb{DL}(P)$ with ansatz vector $v = [v_1, v_2, \ldots, v_k]^T \in \mathbb{F}^k$. Then $L(\lambda)$ is a linearization for $P(\lambda)$ if and only if no root of the grade $k-1$ scalar polynomial*

$$q(\lambda) = v_1\lambda^{k-1} + v_2\lambda^{k-2} + \cdots + v_{k-1}\lambda + v_k \tag{12.9}$$

is an eigenvalue of $P(\lambda)$. We include ∞ as one of the possible roots of $q(\lambda)$, or as one of the possible eigenvalues of $P(\lambda)$.

The systematizing of the construction of linearizations [55] has spurred exciting new research in this area. The ansatz spaces $\mathbb{L}_1(P)$ and $\mathbb{L}_2(P)$ were recently revisited from a new vantage point [80]. By regarding block matrices as a device to record the matrix coefficients of a bivariate matrix polynomial, and by using the concepts of the Bézoutian function and associated Bézout matrix, shorter proofs of the key results in [55] were obtained, while simultaneously generalizing them from regular matrix polynomials expressed in the standard monomial basis to regular polynomials expressed in any degree-graded basis.

What can be said about the pencils in $\mathbb{L}_1(P)$ and $\mathbb{L}_2(P)$ when the $n \times n$ matrix polynomial P is singular? As was shown recently in [18], almost all of them are still linearizations for P that now allow easy recovery of the left and right minimal indices and minimal bases of P.

Are there linearizations for P that are not in $\mathbb{L}_1(P)$ or $\mathbb{L}_2(P)$? Yes! Consider the cubic matrix polynomial $P(\lambda) = \lambda^3 A_3 + \lambda^2 A_2 + \lambda A_1 + A_0$. In [6], the pencil

$$L(\lambda) = \lambda \begin{bmatrix} 0 & A_3 & 0 \\ I & A_2 & 0 \\ 0 & 0 & I \end{bmatrix} + \begin{bmatrix} -I & 0 & 0 \\ 0 & A_1 & A_0 \\ 0 & -I & 0 \end{bmatrix}$$

was shown to be a linearization for P; but $L(\lambda)$ is neither in $\mathbb{L}_1(P)$ nor $\mathbb{L}_2(P)$, as observed in [55]. We turn next to the discussion of these pencils.

12.3.2 Fiedler Pencils

Another source of linearizations for matrix polynomials was inspired by a 2003 paper of Fiedler [29], in which he showed that the usual companion matrix C of a scalar polynomial $p(\lambda) = \Sigma_{i=1}^{k} a_i \lambda^i$ of degree k can be factored into a product of n sparse matrices M_i which differ only slightly from the $n \times n$ identity matrix:

$$C = \begin{bmatrix} -a_{k-1} & -a_{k-2} & \dots & -a_1 & -a_0 \\ 1 & 0 & \dots & 0 & 0 \\ 0 & 1 & \ddots & & \vdots \\ \vdots & & \ddots & \ddots & 0 \\ 0 & \dots & 0 & 1 & 0 \end{bmatrix} = M_{k-1} M_{k-2} \cdots M_0 \,,$$

where

$$M_j := \begin{bmatrix} I_{k-j-1} & & & \\ & -a_j & 1 & \\ & 1 & 0 & \\ & & & I_{j-1} \end{bmatrix} \quad \text{for } j = 1, \dots, k-1\,, \quad \text{and } M_0 := \begin{bmatrix} I_{k-1} & \\ & -a_0 \end{bmatrix}.$$

Fiedler observed that any permutation of the factors M_i produces a matrix that is similar to C, and hence also a companion matrix for $p(\lambda)$. Furthermore, certain permutations produce companion matrices that are of *low bandwidth*, i.e., pentadiagonal.

The first step in extending Fiedler's results to matrix polynomials was taken by Antoniou and Vologiannidis in [6]. The Fiedler factors are now *block matrices*:

$$M_j := \begin{bmatrix} I_{n(k-j-1)} & & & \\ & -A_j & I_n & \\ & I_n & 0 & \\ & & & I_{n(j-1)} \end{bmatrix} \quad \text{for } j = 1, \ldots, k-1, \quad M_0 := \begin{bmatrix} I_{k-1} & \\ & -A_0 \end{bmatrix},$$

and one extra block matrix, $M_k := \text{diag}[A_k, I_{n(k-1)}]$, which is needed because matrix polynomials cannot, without loss of generality, be assumed to be monic. For any permutation $\sigma = (j_0, j_1, \ldots, j_{k-1})$ of the indices $(0, 1, 2, \ldots, k-1)$, one can now define the associated *Fiedler pencil*

$$F_\sigma(\lambda) := \lambda M_k - M_{j_0} M_{j_1} \cdots M_{j_{k-1}}. \tag{12.10}$$

Each member of this family of Fiedler pencils was shown in [6] to be a strong linearization when P is a regular matrix polynomial over $\mathbb{C}$, by demonstrating strict equivalence to the Frobenius companion pencil. The regularity assumption is essential for this proof strategy to work, so to prove that the Fiedler pencils remain strong linearizations when P is singular requires different techniques. This was done in [19], with the restriction on the field lifted. It was also shown that the left and right minimal indices of a singular P are recoverable from any of its Fiedler pencils. Additionally, eigenvectors can be recovered without added computational cost.

Antoniou and Vologiannidis also introduced in [6] a kind of "generalized" Fiedler pencil; exploiting the fact that every M_j for $j = 1, \ldots, k-1$ is invertible, we can "shift" some of the M_j factors to the λ-term. For example, $F_\sigma(\lambda) := \lambda M_k - M_{j_0} M_{j_1} \cdots M_{j_{k-1}}$ is strictly equivalent to

$$\tilde{F}_\sigma(\lambda) = \lambda M_{j_1}^{-1} M_{j_0}^{-1} M_k M_{j_{k-1}}^{-1} - M_{j_2} \cdots M_{j_{k-2}},$$

so $\tilde{F}_\sigma(\lambda)$ is also a strong linearization. These generalized Fiedler pencils can have additional nice properties, as illustrated by the following example for a general square polynomial $P(\lambda)$ of degree $k = 5$.

$$S(\lambda) = \lambda M_5 M_3^{-1} M_1^{-1} - M_4 M_2 M_0$$

$$= \begin{bmatrix} \lambda A_5 + A_4 & -I_n & & & \\ -I_n & 0 & \lambda I_n & & \\ & \lambda I_n & \lambda A_3 + A_2 & -I_n & \\ & & -I_n & 0 & \lambda I_n \\ & & & \lambda I_n & \lambda A_1 + A_0 \end{bmatrix}.$$

This pencil $S(\lambda)$ is not only a strong linearization for $P(\lambda)$, it is also block-*tridiagonal*. The low bandwidth property of certain Fiedler (and generalized Fiedler) pencils thus opens up the possibility of developing fast algorithms to compute the eigenstructure of high degree matrix polynomials. The eigenvector and minimal basis recovery properties of these generalized Fiedler pencils have been studied in [13].

In more recent work [83], Vologiannidis and Antoniou have extended Fiedler pencils even further, showing that repetitions of the Fiedler factors M_i can sometimes be allowed in the construction of $F_\sigma(\lambda)$ in (12.10), and template-like strong linearizations for P will still be produced. These pencils are sometimes referred to as Fiedler pencils with repetition, and have been shown to be yet another source of structured linearizations [14, 83].

Fiedler pencils have also been shown [21] to be adaptable to *rectangular* matrix polynomials P. In this case, however, the product representation in (12.10) is no longer tractable, and other techniques for constructing these pencils are required. Each Fiedler pencil now has its own characteristic *size* as well as block pattern, but each rectangular Fiedler pencil is still a strong linearization for P. This concretely illustrates a distinctive feature of rectangular matrix polynomials, as contrasted with regular (square) matrix polynomials; a rectangular $m \times n$ matrix polynomial with $m \neq n$ always has strong linearizations of many different sizes, while a regular matrix polynomial has strong linearizations of only one possible size. This phenomenon is explored in more detail in [22]. For more on the impact of Fiedler's work on our understanding of linearizations, see [53].

12.4 Matrix Polynomial Structures

There are several kinds of algebraic structure commonly encountered in matrix polynomials arising in the analysis and numerical solution of systems of ordinary, partial, and delay differential equations. To concisely define these structures, we define the $\star$-adjoint of matrix polynomials, where the symbol $\star$ is used to denote transpose T in the real case $\mathbb{F} - \mathbb{R}$, and either transpose T or conjugate transpose $*$ in the complex case $\mathbb{F} = \mathbb{C}$. Note that the structures under consideration apply only to square matrix polynomials.

Definition 8 (Adjoint of Matrix Polynomials) Let $P(\lambda) = \sum_{i=0}^{k} \lambda^i A_i$ where $A_i \in \mathbb{F}^{n \times n}$ with $\mathbb{F} = \mathbb{R}$ or $\mathbb{C}$ be a matrix polynomial of grade k. Then

$$P^\star(\lambda) := \sum_{i-0}^{k} \lambda^i A_i^\star \tag{12.11}$$

defines the $\star$-*adjoint* $P^\star(\lambda)$.

The three most important matrix polynomial structures in applications are

$$\text{Hermitian/symmetric:} \quad P^{\star}(\lambda) = P(\lambda)\,, \tag{12.12}$$

$$\star\text{-alternating:} \quad P^{\star}(-\lambda) = \pm P(\lambda)\,, \tag{12.13}$$

$$\text{and } \star\text{-palindromic:} \; \mathrm{rev}\,P^{\star}(\lambda) = \pm P(\lambda)\,. \tag{12.14}$$

Also of interest are skew-symmetric matrix polynomials, defined by $P^T(\lambda) = -P(\lambda)$, and the following alternative types of alternating and palindromic structure. Letting $\mathscr{R} \in \mathbb{R}^{n \times n}$ denote an arbitrary involution (i.e., $\mathscr{R}^2 = I$), then $P(\lambda)$ is said to be $\mathscr{R}C\mathscr{R}$-palindromic if $\mathscr{R}\,\mathrm{rev}\,\overline{P}(\lambda)\,\mathscr{R} = \pm P(\lambda)$, and $\mathscr{R}C\mathscr{R}$-alternating if $\mathscr{R}\overline{P}(-\lambda)\mathscr{R} = \pm P(\lambda)$. Note that the C in $\mathscr{R}C\mathscr{R}$ refers to the conjugation operation in the definitions; the name $\star$-alternating was suggested by Mehrmann and Watkins in [71], because the matrix coefficients of such polynomials strictly alternate between symmetric and skew-symmetric (or Hermitian and skew-Hermitian) matrices.

Matrix polynomials (especially quadratic polynomials) with Hermitian structure are well known from the classical problem of vibration analysis, and have been extensively studied for many years [31, 32, 48, 79]. The analysis of rail noise caused by high speed trains also leads to a quadratic eigenproblem (QEP), but one with a complex T-palindromic matrix polynomial. Real and complex T-palindromic QEPs also arise in the numerical simulation of the behavior of periodic surface acoustic wave (SAW) filters [43, 85]. Quadratic eigenproblems with T-alternating polynomials arise in the study of corner singularities in anisotropic elastic materials [7, 8, 70]. Gyroscopic systems [25, 48, 49] also lead to quadratic T-alternating matrix polynomials. Higher degree $*$-alternating and $*$-palindromic polynomial eigenproblems arise in the linear-quadratic optimal control problem; the continuous-time case leads to $*$-alternating polynomials, while the discrete-time problem produces $*$-palindromic ones [15]. The stability analysis of delay-differential equations leads to an $\mathscr{R}C\mathscr{R}$-palindromic QEP [28], while a variant of $\mathscr{R}C\mathscr{R}$-alternating structure (without conjugation) arises in linear response theory from quantum chemistry [66]. Further details on these and other applications can be found in [52, 69, 79], Chaps. 2 and 3 of this Festschrift, and the references therein.

An important feature of the structured matrix polynomials described above are the special symmetry properties of their spectra, some of which are described in the following result. The proof of this composite theorem may be found in [52] or [56], together with [28].

Theorem 3 (Eigenvalue Pairings of Structured Matrix Polynomials) *Let $P(\lambda) = \sum_{i=0}^{k} \lambda^i A_i$, $A_k \neq 0$ be a regular matrix polynomial that has one of the palindromic or alternating structures described above. Then the spectrum of $P(\lambda)$ has the pairing depicted in Table 12.1. Moreover, the algebraic, geometric, and partial multiplicities of the two eigenvalues in each such pair are equal. Note*

Table 12.1 Spectral symmetries

Structure of $P(\lambda)$	Eigenvalue pairing
T-palindromic	$(\lambda, 1/\lambda)$
$*$-palindromic	$(\lambda, 1/\overline{\lambda})$
$\mathcal{R}C\mathcal{R}$-palindromic	$(\lambda, 1/\overline{\lambda})$
T-alternating	$(\lambda, -\lambda)$
$*$-alternating	$(\lambda, -\overline{\lambda})$
$\mathcal{R}C\mathcal{R}$-alternating	$(\lambda, -\overline{\lambda})$

that $\lambda = 0$ *is included here as a possible eigenvalue, with the reciprocal partner* $1/\lambda$ *or* $1/\overline{\lambda}$ *to be interpreted as the eigenvalue* ∞.

The eigenvalue pairings seen in this theorem are sometimes referred to as *symplectic spectrum* and *Hamiltonian spectrum*, because they parallel the eigenvalue structure of symplectic and Hamiltonian matrices. Indeed, this is one of several ways in which palindromic and alternating matrix polynomials may be viewed as generalizations of symplectic and Hamiltonian matrices, respectively. For more on this connection see [52].

Although Theorem 3 says quite a lot about the spectral structure of palindromic and alternating matrix polynomials, there are several issues that are not addressed by this result. For example, do these spectral symmetries still hold in the singular case? And what happens when the spectral pairings degenerate, e.g., at $\lambda_0 = \pm 1$ for T-palindromic polynomials, and at $\lambda_0 = 0$ or ∞ for T-alternating polynomials? Are there any additional constraints on the spectra at these degenerate points?

In joint work with Mehrmann [58, 59], these questions were resolved by characterizing the Smith forms for these structure classes using a novel technique based on the properties of compound matrices. This work showed that the eigenvalue pairings found in Theorem 3 do indeed extend to singular polynomials in these classes. Degenerate eigenvalues, however, have some nontrivial fine structure in their admissible Jordan characteristics. The details are somewhat technical, but the main message can be simply stated. For each of these structure classes, the constraints on the admissible spectral structures of odd grade polynomials in a class differ from the constraints on the even grade polynomials in that class. It is interesting to note, though, that this dichotomy between odd and even grade appears only in the fine structure of the partial multiplicities at the degenerate eigenvalues.

Next, the same compound matrix techniques were brought to bear on skew-symmetric matrix polynomials [60]. A characterization of their Smith forms revealed even multiplicity for all elementary divisors, with no odd/even grade dichotomy in the admissible spectral structures.

12.4.1 Möbius Transformations

A useful investigative tool developed by Mehrmann and his co-authors in the last few years is the extension of linear fractional rational transformations (i.e., Möbius transformations) to transformations that act on matrix polynomials [61]. One of the main motivations for this work is understanding the relationships between different classes of structured matrix polynomials. Clearly such a study can be greatly aided by fashioning transformations that allow results about one structured class to be translated into results about another structured class. This inquiry has its origin in particular examples such as the classical Cayley transformation for converting one matrix structure (e.g., skew-Hermitian or Hamiltonian) into another (unitary or symplectic, respectively). This Cayley transformation was extended from matrices to matrix pencils in [50], and in a 1996 paper by Mehrmann [67]. It was then generalized to matrix polynomials in 2006 by Mehrmann and co-authors [56], where it was shown how palindromic and alternating structures are related via a Cayley transformation of matrix polynomials. The definition of general Möbius transformations in [61] completes this development, providing an important and flexible tool for working with matrix polynomials.

Definition 9 (Möbius Transformation) Let V be the vector space of all $m \times n$ matrix polynomials of grade k over the field $\mathbb{F}$, and let $A \in GL(2, \mathbb{F})$. Then the *Möbius transformation on* V *induced by* A is the map $\mathbf{M}_A : V \to V$ defined by

$$\mathbf{M}_A \left(\sum_{i=0}^{k} B_i \lambda^i \right)(\mu) = \sum_{i=0}^{k} B_i \, (a\mu + b)^i (c\mu + d)^{k-i}, \quad \text{where} \quad A = \begin{bmatrix} a & b \\ c & d \end{bmatrix}.$$

It is worth emphasizing that a Möbius transformation acts on graded polynomials, returning polynomials of the same grade (although the degree may increase, decrease, or stay the same, depending on the polynomial). In fact, $\mathbf{M}_A$ is a linear operator on V. Observe that the Möbius transformations induced by the matrices

$$A_{+1} = \begin{bmatrix} 1 & 1 \\ -1 & 1 \end{bmatrix} \quad \text{and} \quad A_{-1} = \begin{bmatrix} 1 & -1 \\ 1 & 1 \end{bmatrix}$$

are exactly the Cayley transformations $\mathscr{C}_{+1}(P)$ and $\mathscr{C}_{-1}(P)$, respectively, introduced in [56]. Also note that the reversal operation described in Definition 3 is the Möbius transformation $\mathbf{M}_R$ corresponding to the matrix

$$R = \begin{bmatrix} 0 & 1 \\ 1 & 0 \end{bmatrix}.$$

Some of the significant properties of general Möbius transformations proved in [61] include the following:

1. Möbius transformations affect the eigenvalues of P and their partial multiplicity sequences in a simple and uniform way. In particular, if $\mathsf{m}_A(\lambda) = \frac{a\lambda+b}{c\lambda+d}$ denotes the scalar Möbius function on $\mathbb{F} \cup \{\infty\}$ corresponding to the matrix $A = \left[\begin{smallmatrix} a & b \\ c & d \end{smallmatrix}\right] \in GL(2, \mathbb{F})$, then we have that

$$\mathscr{J}\big(\mathbf{M}_A(P), \mu_0\big) \equiv \mathscr{J}\big(P, \mathsf{m}_A(\mu_0)\big) \tag{12.15}$$

   for any $\mu_0 \in \mathbb{F} \cup \{\infty\}$.
2. Eigenvectors are preserved by Möbius transformations, but Jordan chains are not. By (12.15), though, the lengths of Jordan chains are preserved.
3. Möbius transformations preserve minimal indices, and transform minimal bases in a simple and uniform way.
4. Möbius transformations preserve the property of being a strong linearization; that is, if $L(\lambda)$ is a strong linearization for $P(\lambda)$, then $\mathbf{M}_A(L)$ is a strong linearization for $\mathbf{M}_A(P)$. More generally, Möbius transformations preserve the spectral equivalence relation.
5. Möbius transformations preserve sparsity patterns; for example, if P is upper triangular, then $\mathbf{M}_A(P)$ is also upper triangular.

For the study of structured matrix polynomials, perhaps the most significant property of all is that Möbius transformations provide a rich source of bijections *between* classes of structured polynomials, that allow us to conveniently transfer intuition and results about one class to another. Important examples include correspondences between

T-palindromic and T-alternating polynomials,

as well as between the three classes of

Hermitian, $*$-palindromic, and $*$-alternating matrix polynomials.

These last correspondences provide an opportunity to transfer over to $*$-palindromic and $*$-alternating polynomials much of the existing wealth of knowledge about Hermitian matrix polynomials, including results about such special subclasses as hyperbolic polynomials, definite polynomials [40], and other types of Hermitian matrix polynomials with all-real spectrum [4].

Finally, it is worth noting that the idea of linear fractional transformations acting on matrix polynomials has been extended even further to more general rational transformations in [73].

12.5 Structured Linearizations

I'm pickin' up good vibrations – The Beach Boys

When a matrix polynomial $P(\lambda)$ has structure, the linearization strategy for solving the associated polynomial eigenproblem has two parts: first find a suitable structured linearization $L(\lambda)$ for $P(\lambda)$, and then compute the eigenvalues of $L(\lambda)$ using a structure-preserving algorithm. Although our focus is on the first part of this strategy, it is important to note that there has also been much work on the development of structure-preserving algorithms for matrix pencils in the last decade. Examples of some of this work can be found in the papers [28, 45, 47, 57, 65, 68, 70, 71, 75], as well as in the Chaps. 2 and 3 of this Festschrift.

We now turn to developments of the last decade concerning structure-preserving linearizations, focusing mainly on the "big three" types of structure – Hermitian, palindromic and alternating.

12.5.1 In Ansatz Spaces

The pencil spaces $\mathbb{L}_1(P)$ and $\mathbb{L}_2(P)$ introduced by Mehrmann and co-authors [55] were shown in a follow-up paper [56] to provide a rich arena in which to look for linearizations with additional properties like structure preservation or improved numerics, thus realizing the original purpose for their development. Subspaces of pencils that inherit the $\star$-palindromic or $\star$-alternating structure of P were identified, a constructive method to generate these structured pencils described, and necessary and sufficient conditions for them to be strong linearizations established.

There is a close connection between the structure of a pencil in $\mathbb{L}_1(P)$ and the structure of its ansatz vectors. Loosely put, if P is palindromic, then a palindromic pencil in $\mathbb{L}_1(P)$ will have a palindromic ansatz vector, while if P is alternating, then an alternating pencil in $\mathbb{L}_1(P)$ will have an alternating ansatz vector. When P is structured, there is also a very close connection between the double ansatz space $\mathbb{DL}(P)$ and pencils in $\mathbb{L}_1(P)$ that reflect the structure of P. More precisely, let R be the reverse identity matrix, and Σ a diagonal matrix of alternating signs,

$$R_k := \begin{bmatrix} & & 1 \\ & \cdot^{\cdot^{\cdot}} & \\ 1 & & \end{bmatrix}_{k\times k} \quad \text{and} \quad \Sigma_k := \begin{bmatrix} (-1)^{k-1} & & \\ & \ddots & \\ & & (-1)^0 \end{bmatrix}_{k\times k} \tag{12.16}$$

and let $L(\lambda) \in \mathbb{L}_1(P)$ with ansatz vector v. If P is a palindromic matrix polynomial, e.g., if $\operatorname{rev} P^T(\lambda) = P(\lambda)$, then

$$\operatorname{rev} L^T(\lambda) = L(\lambda) \iff \Big(Rv = v, \text{ and } (R \otimes I)L(\lambda) \in \mathbb{DL}(P) \text{ with ansatz vector } v \Big).$$

So to find a palindromic pencil in $\mathbb{L}_1(P)$, begin with a palindromic ansatz vector. Now there is a unique pencil in $\mathbb{DL}(P)$ corresponding to that vector. This pencil can be explicitly constructed using the natural basis for $\mathbb{DL}(P)$ mentioned in Sect. 12.3.1, and described in detail in [56]. Then reversing the order of the block rows of that $\mathbb{DL}(P)$-pencil turns it into a palindromic pencil in $\mathbb{L}_1(P)$. Will this pencil be a linearization for P ? The Eigenvalue Exclusion Theorem stated in Sect. 12.3.1 and proved in [55], determines whether the answer is yea or nay. If the answer is yea, and P is regular, this linearization is automatically also a strong linearization [55].

On the other hand, if P is alternating, say $P^T(-\lambda) = P(\lambda)$, then for $L(\lambda) \in \mathbb{L}_1(P)$ we have

$$L^T(-\lambda)=L(\lambda) \Longleftrightarrow \Big(\Sigma v=v, \text{ and } (\Sigma \otimes I)L(\lambda) \in \mathbb{DL}(P) \text{ with ansatz vector } v\Big),$$

which, as in the palindromic case detailed before, can be used mutatis mutandis to construct an alternating linearization for P. Similar results were proved for the other flavors of palindromicity and alternation, and concrete examples given in [56].

An unexpected property of $\mathbb{DL}(P)$ itself was proved in [38]. Consider the block transpose of a block matrix, defined as follows.

Definition 10 The *block transpose* of a block $k \times \ell$ matrix A with $m \times n$ blocks A_{ij} is the block $\ell \times k$ matrix $A^{\mathcal{B}}$ with $m \times n$ blocks $(A^{\mathcal{B}})_{ij} = A_{ji}$.

Now consider the subspace $\mathbb{B}(P)$ of all block symmetric (with respect to $n \times n$ blocks) pencils in $\mathbb{L}_1(P)$, that is,

$$\mathbb{B}(P) := \{\lambda X + Y \in \mathbb{L}_1(P) \;:\; X^{\mathcal{B}} = X,\ Y^{\mathcal{B}} = Y\}.$$

Then for any P, the subspaces $\mathbb{B}(P)$ and $\mathbb{DL}(P)$ are identical! Thus pencils in $\mathbb{DL}(P)$ always have block symmetric coefficients, even when there is no structure in the matrix coefficients of P. What happens when P is structured? As shown in [38], when P is symmetric, the collection of all symmetric pencils in $\mathbb{L}_1(P)$ is exactly $\mathbb{DL}(P)$, while for Hermitian P the Hermitian pencils in $\mathbb{L}_1(P)$ form a proper (but nontrivial) subspace $\mathbb{H}(P) \subset \mathbb{DL}(P)$.

Among Hermitian matrix polynomials, perhaps the most important are those with all-real spectrum [4]. This includes the definite polynomials, a class of Hermitian polynomials introduced in [40] as a common generalization for hyperbolic polynomials and definite pencils. In this setting, the natural structured linearization question is whether every definite Hermitian polynomial has a linearization that is a definite pencil. This is answered affirmatively in [40]; indeed, it is shown that a Hermitian matrix polynomial P is definite if and only if it has a definite linearization in $\mathbb{H}(P)$, the set of Hermitian pencils in $\mathbb{L}_1(P)$. Thus we see that $\mathbb{L}_1(P)$ is rich enough to provide a structured-preserving (strong) linearization for any definite Hermitian polynomial. It is also worth noting that the results in [40] had a significant impact on the later characterization results of [4].

The double ansatz space has also appeared as the star player in other structured settings. The stability analysis of time-delay systems leads to a palindromic polynomial eigenproblem [28] with an involutory twist – the $n \times n$ complex matrix polynomial $P(\lambda)$ in this problem satisfies

$$\mathscr{R} \cdot \mathrm{rev}\overline{P}(\lambda) \cdot \mathscr{R} = P(\lambda)\,, \tag{12.17}$$

where $\mathscr{R}$ is a real involution (i.e., $\mathscr{R}^2 = I_n$), thus making P an $\mathscr{R}C\mathscr{R}$-palindromic matrix polynomial in the sense described in Sect. 12.4. In order to find structured linearizations in this context, the first issue is to specify an appropriate class of structured pencils to search in; in other words, a suitable involution on the space of $nk \times nk$ pencils must be chosen. In [28] it is shown that the block anti-diagonal matrix $\widehat{\mathscr{R}} := R_k \otimes \mathscr{R}$, where R_k is the $k \times k$ backwards identity matrix as in (12.16), gives a compatible choice of involution. With this choice of involution, it now follows that the right ansatz vector $v \in \mathbb{C}^k$ of any $\widehat{\mathscr{R}}C\widehat{\mathscr{R}}$-palindromic pencil in $\mathbb{L}_1(P)$ must satisfy $Rv = \overline{v}$. For any such vector v, there are many $\widehat{\mathscr{R}}C\widehat{\mathscr{R}}$-palindromic pencils in $\mathbb{L}_1(P)$ with this right ansatz vector, exactly one of which will also be in $\mathbb{DL}(P)$. These results, along with a constructive procedure to build these structured $\mathbb{DL}(P)$-pencils, were presented in [28], where they were also extended to the other variants of $\mathscr{R}C\mathscr{R}$-structure mentioned in Sect. 12.4 by using the linearization theory and techniques developed in [55, 56].

The techniques developed in [56] had an impact on eigenvalue computations occurring in the vibration analysis of rail tracks under excitation from high speed trains [42, 46]; see also the Chap. 3 of this Festschrift. This eigenvalue problem has the form

$$\left(\lambda A(\omega) + B(\omega) + \frac{1}{\lambda} A(\omega)^T\right) x = 0, \tag{12.18}$$

where A, B are large, sparse, parameter-dependent, complex square matrices, with B complex symmetric, and A highly singular. Clearly, for any fixed value of ω, multiplying (12.18) by λ leads to a T-palindromic eigenvalue problem. Solving this problem directly with the QZ-algorithm without respecting its structure resulted in erroneous eigenvalues. However, the use of a T-palindromic linearization from [56] allowed structured deflation of the zero and infinite eigenvalues. The computed frequencies were now accurate to within the range of the discretization error. Thus we see that the computation of "good vibrations" is aided by the use of "good linearizations."

12.5.1.1 Problematic Eigenvalues

For regular matrix polynomials P, the pencils in $\mathbb{DL}(P)$ have repeatedly shown themselves to be prolific sources of structured linearizations.[1] However, pencils in $\mathbb{DL}(P)$ have one significant drawback. Because of the eigenvalue exclusion property described in Theorem 2, for any $L(\lambda) \in \mathbb{DL}(P)$ there is always at least one "problematic eigenvalue" that may prevent L from being a linearization for P;

[1]The story is quite different for singular polynomials P. In that case, *none* of the pencils in $\mathbb{DL}(P)$ is ever a linearization for P, even when P has no structure [18].

these problematic eigenvalues are just the roots of the scalar polynomial $q(\lambda)$ in (12.9), associated with the ansatz vector v for L.

In many situations, this obstruction to $L(\lambda) \in \mathbb{DL}(P)$ being a linearization can be easily side-stepped simply by shifting consideration to a different pencil in $\mathbb{DL}(P)$, since almost every pencil in $\mathbb{DL}(P)$ is a linearization. However, in a structured setting, where the goal is to find a structured linearization, this problematic eigenvalue obstruction *sometimes cannot be avoided*, no matter what pencil in $\mathbb{DL}(P)$ is used.

Consider, for example, the case of a T-palindromic matrix polynomial P of any *even* grade $k \geq 2$. As described in Sect. 12.5.1, any T-palindromic pencil in $\mathbb{L}_1(P)$ is strictly equivalent to a pencil in $\mathbb{DL}(P)$ possessing a palindromic ansatz vector v, i.e., a $v \in \mathbb{F}^k$ such that $Rv = v$. But the scalar polynomial $q(\lambda)$ in (12.9) corresponding to any such v is necessarily palindromic of *odd* grade, and thus must always have -1 as a root. Consequently, any T-palindromic polynomial P of even grade that has the eigenvalue $\lambda_0 = -1$ will *never* have any structured linearization in $\mathbb{L}_1(P)$!

This phenomenon of having an unavoidable problematic eigenvalue obstruction to the existence of any structured linearizations in $\mathbb{L}_1(P)$ occurs for other structures in addition to T-palindromic structure (see [56]). However, it is significant to note that this is only known to occur for structured polynomials of *even* grade.

12.5.2 Among Fiedler Pencils

Modified versions of the generalized Fiedler pencils and Fiedler pencils with repetition described in Sect. 12.3.2 have shown themselves to be particularly valuable sources for not just structure-preserving linearizations, but for *structured companion forms*. Here by the term "companion form" we mean a template for producing a pencil associated to each matrix polynomial P of some fixed size and grade that

- is constructed *directly* from the matrix coefficients of P, without any matrix operations on these coefficients, and
- produces a strong linearization for *every* polynomial P of the given size and grade (both regular *and* singular if the polynomials are square).

Every Fiedler and generalized Fiedler pencil is a companion form in this sense; by contrast, none of the pencils in $\mathbb{DL}(P)$ is ever a companion form because of Theorem 2.

A companion form is said to be structured with respect to a class $\mathscr{C}$ of matrix polynomials, if for every $P \in \mathscr{C}$, the associated companion pencil is also in $\mathscr{C}$. Thus we might have Hermitian companion forms, palindromic companion forms, und so weiter. Structured companion forms derived from generalized Fiedler pencils have appeared in a number of papers [6, 14, 20, 58–60, 83], for a variety of structure classes, including Hermitian, T-palindromic, and T-alternating matrix polynomials.

Here are some simple examples from those papers. Suppose

$$P(\lambda) = \lambda^5 A_5 + \lambda^4 A_4 + \cdots + \lambda A_1 + A_0$$

is a general $n \times n$ polynomial of grade 5. Then in [58] it is shown that the block-tridiagonal pencil template

$$\mathscr{S}_P(\lambda) = \begin{bmatrix} \lambda A_1 + A_0 & \lambda I & & & \\ \lambda I & 0 & I & & \\ & I & \lambda A_3 + A_2 & \lambda I & \\ & & \lambda I & 0 & I \\ & & & I & \lambda A_5 + A_4 \end{bmatrix}_{5n \times 5n} \tag{12.19}$$

is a companion form for the set of all $n \times n$ matrix polynomials of grade 5. Note that $\mathscr{S}_P(\lambda)$ in (12.19) is a simplified version of an example that first appeared in [6]. It is clear how $\mathscr{S}_P(\lambda)$ can be extended to a companion form for any other *odd* grade. Also noteworthy is that $\mathscr{S}_P(\lambda)$ is not just a companion form, it is also both a symmetric and a Hermitian companion form; i.e., if P is symmetric (Hermitian), then $\mathscr{S}_P$ will also be symmetric (Hermitian). Many more symmetric and Hermitian companion forms can be constructed by the methods developed in [83].

Pre-multiplying $\mathscr{S}_P$ by a certain diagonal ± 1 matrix (a strict equivalence) now immediately produces a T-even companion form

$$\mathscr{E}_P(\lambda) = \begin{bmatrix} \lambda A_1 + A_0 & \lambda I & & & \\ -\lambda I & 0 & -I & & \\ & -I & -\lambda A_3 - A_2 & -\lambda I & \\ & & \lambda I & 0 & I \\ & & & I & \lambda A_5 + A_4 \end{bmatrix}_{5n \times 5n},$$

as shown in [58]. Pre-multiplication of $\mathscr{S}_P$ by $R_k \otimes I_n$ (another strict equivalence) reverses the order of the block rows, giving

$$\mathscr{P}_P(\lambda) = \begin{bmatrix} & & & I & \lambda A_5 + A_4 \\ & & \lambda I & 0 & I \\ & I & \lambda A_3 + A_2 & \lambda I & \\ \lambda I & 0 & I & & \\ \lambda A_1 + A_0 & \lambda I & & & \end{bmatrix}_{5n \times 5n},$$

which is a T-palindromic companion form [59]. Many more palindromic companion forms are constructed in [14] and [20], all for odd grade polynomials. Indeed, all the known structured companion forms arising from Fiedler pencils are for odd grade matrix polynomials.

The lack of any Fiedler-based structured companion forms for even grade polynomials is curious; is this just an oddity[2] of Fiedler pencils, or is it a sign of some intrinsic limitation on all pencils?

12.5.3 *Existence: Leave It to Smith*

"The first thing to do," said Psmith, "is to ascertain that such a place as Clapham Common really exists. One has heard of it, of course, but has its existence ever been proved? I think not."

– *P.G. Wodehouse, Psmith in the City* [84]

Several phenomena now contribute to the suspicion that structured even grade polynomials may be intrinsically "harder" to linearize (at least by a structured companion form) than structured matrix polynomials of odd grade. Among these are the plenitude of Fiedler-based structured companion forms for odd grade as contrasted with the absence of any known for even grade; another is the presence of "problematic eigenvalues" that block the existence of any structured linearization in the ansatz spaces for certain even grade structured matrix polynomials. The resolution of this issue was finally achieved by the detailed investigation of the Smith forms of various types of structured matrix polynomials in the Smith form trilogy [58–60], described at the end of Sect. 12.4.

A structured companion form for even grade would be able to simultaneously provide a structured linearization for *every* structured polynomial of that even grade. But the Smith form results of [58] and [59] show that the admissible Jordan characteristics of even and odd grade polynomials in the palindromic (or the alternating) structure class are not the same. Consequently, for each structure class there are always structured polynomials of each even grade whose elementary divisor structure is incompatible with that of *every* pencil in that structure class. This elementary divisor incompatibility thus precludes the existence of any structured companion form for any even grade, for either palindromic or alternating matrix polynomials.

The existence or non-existence of Hermitian or symmetric companion forms for even grades cannot be settled by a similar argument; for these structures there are no comparable elementary divisor incompatibilities between even and odd grade. Nonetheless, the impossibility of such structured companion forms for even grades has recently been shown in [22]; the argument given there is based on minimal index incompatibilities between even and odd grade structured polynomials that are singular.

[2]Pun intended.[3]

[3]The previous footnote[2], and this footnote[3] to that footnote[2], are here especially for Volker.

The impossibility of any even grade structured companion form, for any of these three most important structure classes, suggests that a reduction to a spectrally equivalent *quadratic* matrix polynomial might be a more natural alternative to linearization for even grade structured polynomials. This is one motivation to investigate the possibility of *structure-preserving quadratifications*, as part of a wider investigation of the properties of quadratic matrix polynomials [23, 54], quadratifications more generally, and the development of algorithms that work directly on a quadratic polynomial, without any intervening linearization. Some initial work in this direction can be found in [44] for palindromic structure. From a characterization of the possible elementary divisor and singular structures of quadratic palindromic polynomials [24], it has been recently shown that *every* even grade palindromic polynomial has a palindromic (strong) quadratification. Similar results are also now known to hold for even grade alternating polynomials [24], and for even grade Hermitian matrix polynomials [63].

12.6 Impact on Numerical Practice

In order to analyze the numerical properties of algorithms for the polynomial eigenproblem, both left *and* right eigenvectors of a matrix polynomial P must be considered. In this context, then, the polynomial eigenproblem is more properly formulated as

$$P(\lambda)x = 0, \quad y^* P(\lambda) = 0\,, \tag{12.20}$$

where $x \neq 0$ is a right eigenvector, and $y \neq 0$ is a left eigenvector for $P(\lambda)$. For this analysis, it is also usually assumed that P is regular, which we do throughout this section. The associated generalized eigenvalue problem

$$L(\lambda)z = 0, \quad w^* L(\lambda) = 0 \tag{12.21}$$

for a linearization L of P can now be solved using standard techniques and readily available software. In particular, if the size of L is not very large, dense transformation-based methods can be used to solve (12.21), such as the QZ algorithm [72], or a structure-preserving algorithm when L is a structured linearization [28, 47, 57, 65, 75, 79]. Krylov methods can be used for large sparse problems [9, 64, 68, 70, 79]. Among the infinitely many linearizations L of P, we are interested in those which preserve the structure, if any, and whose right and left eigenvectors permit easy recovery of the corresponding eigenvectors of P. So all the linearizations described in Sects. 12.3 and 12.5 are obvious candidates.

The introduction of these new structured and unstructured linearizations in the last decade has led to not only the development of structure-preserving algorithms, but also the development of techniques to analyze the influence of the linearization process on the accuracy and stability of the computed solution, so as to guide us in our choice of linearization. To indicate the key idea we assume that P is expressed in the monomial basis as in (12.1). Let x and y denote right and left eigenvectors of P, and let z and w denote right and left eigenvectors of L, all corresponding to a simple, nonzero, finite eigenvalue λ. Eigenvalue condition numbers are given, in the 2-norm, by the following expressions [78, Thm. 5]:

$$\kappa_P(\lambda) = \frac{\left(\sum_{i=0}^{k} |\lambda|^i \, \|A_i\|_2\right) \|y\|_2 \|x\|_2}{|\lambda| \, |y^* P'(\lambda) x|}, \quad \kappa_L(\lambda) = \frac{\left(|\lambda| \|X\|_2 + \|Y\|_2\right) \|w\|_2 \|z\|_2}{|\lambda| \, |w^* L'(\lambda) z|}.$$

These condition numbers measure the sensitivity of the eigenvalue λ of P and L, respectively, to small perturbations of P and L measured in a normwise relative fashion. Different linearizations of the same matrix polynomial can have widely varying eigenvalue condition numbers. Unless the block structure of the linearization is respected (and it is not by standard algorithms), the conditioning of the larger linear problem can be worse than that of the original matrix polynomial, since the class of admissible perturbations is larger. For example, eigenvalues that are well-conditioned for $P(\lambda)$ may be ill-conditioned for $L(\lambda)$ [39, 41, 78]. Ideally, when solving (12.20) via (12.21) we would like to have $\kappa_P(\lambda) \approx \kappa_L(\lambda)$. Most linearizations in Sects. 12.3 and 12.5 satisfy one-sided factorizations of the form

$$L(\lambda)F(\lambda) = G(\lambda)P(\lambda), \quad E(\lambda)L(\lambda) = P(\lambda)H(\lambda), \tag{12.22}$$

where $G(\lambda), H^T(\lambda), F(\lambda)$ and $E(\lambda)^T$ are $kn \times n$ matrix functions. Assume that $F(\lambda)$ is of full rank in a neighborhood of a finite eigenvalue λ of P and L, and that $y := G(\lambda)^* w \neq 0$. Then it follows from (12.22) that $z = F(\lambda)x$ is a right eigenvector of L, y is a left eigenvector of P, and $w^* L'(\lambda) z = y^* P'(\lambda) x$ (see [34, Lemma 3.2]) so that

$$\frac{\kappa_L(\lambda)}{\kappa_P(\lambda)} = \frac{|\lambda| \|X\|_2 + \|Y\|_2}{\sum_{j=0}^{k} |\lambda|^j \, \|A_j\|_2} \cdot \frac{\|w\|_2 \|z\|_2}{\|y\|_2 \|x\|_2}. \tag{12.23}$$

This expression can now be used to investigate the size of the ratio $\kappa_L(\lambda)/\kappa_P(\lambda)$ as L varies, for fixed P, where the L-dependent terms are X, Y, w, and z. This is done for example in [39] for pencils $L \in \mathbb{DL}(P)$, where minimization of the ratio over L is considered.

Backward errors characterize the stability of a numerical method for solving a problem by measuring how far the problem has to be perturbed for an approximate solution to be an exact solution of the perturbed problem. Let (x, λ) be an approximate right eigenpair for $P(\lambda)$ obtained from an approximate right eigenpair (z, λ) for $L(\lambda) = \lambda X + Y$. The relative backward errors for (x, λ) and (z, λ) are given in the 2-norm by

$$\eta_P(x,\lambda) = \frac{\|P(\lambda)x\|_2}{\left(\sum_{i=0}^{k} |\lambda^i| \|A_i\|_2\right) \|x\|_2}, \quad \eta_L(z,\lambda) = \frac{\|L(\lambda)z\|_2}{(|\lambda| \|X\|_2 + \|Y\|_2)\, \|z\|_2}. \tag{12.24}$$

There are analogous formulae for approximate left eigenpairs.

We would like the linearization L that we use to lead, after recovering an approximate eigenpair of P from one of L, to a backward error for P of the same order of magnitude as that for L. To relate backward errors for L and P we need to assume that the pencil L satisfies a left-sided factorization as in the right hand-side of (12.22), with $E(\lambda)$ of full rank, and that x is recovered from z via $x = H(\lambda)z$. Then $E(\lambda)L(\lambda)z = P(\lambda)x$ so that

$$\frac{\eta_P(x,\lambda)}{\eta_L(z,\lambda)} \le \frac{|\lambda| \|X\|_2 + \|Y\|_2}{\sum_{i=0}^{k} |\lambda^i| \|A_i\|_2} \cdot \frac{\|E(\lambda)\|_2 \|z\|_2}{\|x\|_2}. \tag{12.25}$$

This bound, which largely separates the dependence on L, P, and λ (in the first term) from the dependence on E and z (in the second term), can then be analyzed for a given linearization. This was done for Frobenius companion linearizations and $\mathbb{DL}(P)$ linearizations in [37].

For Frobenius companion linearizations, a straightforward analysis of the ratio (12.23) and the upper bound (12.25) shows that if $\|A_i\|_2 \approx 1$ for $i = 0, \ldots, k$, then $\kappa_L(x) \approx \kappa_P(\lambda)$ and the upper bound in (12.25) will be of order 1; this suggests that scaling the polynomial eigenproblem to try to achieve this condition before computing the eigenpairs via a Frobenius companion linearization could be numerically advantageous. Fan, Lin, and Van Dooren [27] considered the following scaling strategy for quadratics, which converts $P(\lambda) = \lambda^2 A_2 + \lambda A_1 + A_0$ to $\widetilde{P}(\mu) = \mu^2 \widetilde{A}_2 + \mu \widetilde{A}_1 + \widetilde{A}_0$, where

$$\lambda = \gamma\mu, \quad P(\lambda)\delta = \mu^2(\gamma^2 \delta A_2) + \mu(\gamma\delta A_1) + \delta A_0 \equiv \widetilde{P}(\mu),$$

and is dependent on two nonzero scalar parameters γ and δ. They showed that when A_0 and A_2 are nonzero, then taking $\gamma = \sqrt{\|A_0\|_2 / \|A_2\|_2}$ and $\delta = 2/(\|A_0\|_2 + \|A_1\|_2 \gamma)$ solves the problem of minimizing the maximum distance of the coefficient matrix norms from 1:

$$\min_{\gamma,\delta} \max\{\|\widetilde{A}_0\|_2 - 1, \|\widetilde{A}_1\|_2 - 1, \|\widetilde{A}_2\|_2 - 1\}.$$

It is shown in [37] that with this choice of parameters and for not too heavily damped quadratics, that is, $\|A_1\|_2^2 \lesssim \|A_0\|_2 \|A_2\|_2$, then $\kappa_P \approx \kappa_L$ for all eigenvalues and $\eta_P \approx \eta_L$ for both left and right eigenpairs. Hence, with this scaling the linearization process does not affect the eigenvalue condition numbers, and if the generalized eigenvalue problem (12.21) is solved by a backward stable algorithm such as the QZ algorithm, then the computed eigenpairs for P will have small backward errors.

These ideas have been implemented in an algorithm for the complete solution of quadratic eigenvalue problems by Hammarling, Munro, and Tisseur [36]. The case of heavily damped quadratics has been addressed by Zeng and Su [86].

It is now well established that for structured polynomial eigenproblems, it is important to use algorithms that preserve the structure of the problem when computing its eigenvalues, so that the eigenvalue pairings are preserved. This has lead to the development of a number of structure-preserving algorithms for structured linearizations of structured eigenproblems [45, 47, 65, 68, 70, 75, 79], as well as the derivation of structured backward errors and structured condition numbers corresponding to structured perturbations [1–3, 11, 12].

12.7 Related Recent Developments

The linearization strategy for the polynomial eigenproblem continues to be actively developed for more types of matrix polynomials; this strategy is even beginning to be extended to other types of nonlinear eigenproblems.

In recent research on matrix polynomials, for example, a new theme has started to attract increasing attention – finding simple, template-like ways to construct linearizations when the polynomial

$$P(\lambda) = \sum_{i=0}^{k} A_i \phi_i(\lambda) \qquad (12.26)$$

is expressed in some non-standard basis $\{\phi_i(\lambda)\}$. Particularly important for numerical computation are the classical examples of such bases, e.g., those associated with the names Chebyshev, Newton, Hermite, Lagrange, and Bernstein. It is tempting to simply convert $P(\lambda)$ in (12.26) to the standard basis, and then leverage the existing body of knowledge about linearizations. However, it is important to avoid reformulating P into the standard basis, since a change of basis has the potential to introduce numerical errors not present in the original problem. Instead we should look for templates that construct linearizations for $P(\lambda)$ *directly* from the coefficients A_i in (12.26), without any matrix additions, multiplications, or inverses. This could be viewed as another kind of structure preservation, i.e., a preservation of the polynomial basis.

Although there are precedents for doing this for scalar polynomials in [10], and even earlier in [33], the first serious effort in this direction for matrix polynomials was [5] and the earlier [17], where concrete templates for producing strong linearizations were provided, one for each of several classical polynomial bases, including Chebyshev, Newton, Lagrange, and Bernstein bases. This work has been used in [26], as part of a Chebyshev interpolation method for solving non-polynomial nonlinear eigenproblems. Additional examples for the Hermite and Lagrange bases have been developed and used in [81, 82]. More systematic methods for constructing large families of template-like linearizations for matrix polynomials expressed in non-standard bases can be found in the very recent papers [62, 74, 80].

The linearization strategy has been so effective for polynomial eigenproblems that researchers have started to consider ways to extend this strategy to other nonlinear eigenproblems, especially to *rational* eigenproblems $P(\lambda)x = 0$, where the scalar $\phi_i(\lambda)$ functions in $P(\lambda)$ as in (12.26) are now rational functions of λ rather than just polynomials. Significant advances in this direction have been made in [77], and more recently in [35].

12.8 Concluding Remarks

Wer wirklich Neues erdenken will, muss hin und wieder ein wenig spinnen.
– Quote on Room MA 466, TU Berlin.

We hope this review has shown how the discovery of new families of linearizations in the last decade has propelled research on polynomial eigenproblems forward, with significant advances made in the development of theory and algorithms for structured problems. Volker has contributed much to this effort, as a researcher and, equally importantly, as a stimulating mentor. There is still more waiting to be discovered, and more fun to be had in uncovering it. As Volker taught us to say to one another, “Es gibt viel zu tun, fangt schon mal an!”

References

1. Adhikari, B.: Backward perturbation and sensitivity analysis of structured polynomial eigenvalue problems. PhD thesis, Indian Institute of Technology Guwahati (2008)
2. Adhikari, B., Alam, R., Kressner, D.: Structured eigenvalue condition numbers and linearizations for matrix polynomials. Linear Algebra Appl. **435**(9), 2193–2221 (2011)
3. Ahmad, Sk.S., Mehrmann, V.: Perturbation analysis for complex symmetric, skew symmetric even and odd matrix polynomials. Electron. Trans. Numer. Anal. **38**, 275–302 (2011)
4. Al-Ammari, M., Tisseur, F.: Hermitian matrix polynomials with real eigenvalues of definite type. Part I: classification. Linear Algebra Appl. **436**, 3954–3973 (2012)
5. Amiraslani, A., Corless, R.M., Lancaster, P.: Linearization of matrix polynomials expressed in polynomial bases. IMA J. Numer. Anal. **29**, 141–157 (2009)
6. Antoniou, E.N., Vologiannidis, S.: A new family of companion forms of polynomial matrices. Electron. J. Linear Algebra **11**, 78–87 (2004)
7. Apel, T., Mehrmann, V., Watkins, D.: Structured eigenvalue methods for the computation of corner singularities in 3D anisotropic elastic structures. Comput. Methods Appl. Mech. Eng. **191**, 4459–4473 (2002)
8. Apel, T., Mehrmann, V., Watkins, D.: Numerical solution of large scale structured polynomial or rational eigenvalue problems. In: Cucker, F., Olver, P. (eds.) Foundations of Computational Mathematics. London Mathematical Society Lecture Note Series, vol. 312. Cambridge University Press, Cambridge, pp. 137–157 (2004)

9. Bai, Z., Demmel, J.W., Dongarra, J., Ruhe, A., van der Vorst, H.A. (eds.): Templates for the Solution of Algebraic Eigenvalue Problems: A Practical Guide. Society for Industrial and Applied Mathematics, Philadelphia (2000)
10. Barnett, S.: Congenial matrices. Linear Algebra Appl. **41**, 277–298 (1981)
11. Bora, S.: Structured eigenvalue condition number and backward error of a class of polynomial eigenvalue problems. SIAM J. Matrix Anal. Appl. **31**(3), 900–917 (2009)
12. Bora, S., Karow, M., Mehl, C., Sharma, P.: Structured eigenvalue backward errors of matrix pencils and polynomials with Hermitian and related structures. SIAM J. Matrix Anal. Appl. **35**(2), 453–475 (2014)
13. Bueno, M.I., De Terán, F., Dopico, F.M.: Recovery of eigenvectors and minimal bases of matrix polynomials from generalized Fiedler linearizations. SIAM J. Matrix Anal. Appl. **32**, 463–483 (2011)
14. Bueno, M.I., Furtado, S.: Palindromic linearizations of a matrix polynomial of odd degree obtained from Fiedler pencils with repetition. Electron. J. Linear Algebra **23**, 562–577 (2012)
15. Byers, R., Mackey, D.S., Mehrmann, V., Xu, H.: Symplectic, BVD, and palindromic approaches to discrete-time control problems. In: Petkov, P., Christov, N. (eds.) A Collection of Papers Dedicated to the 60th Anniversary of Mihail Konstantinov, Sofia, pp. 81–102. Publishing House Rodina (2009). Also available as MIMS EPrint 2008.35, Manchester Institute for Mathematical Sciences, Manchester, Mar 2008
16. Byers, R., Mehrmann, V., Xu, H.: Trimmed linearizations for structured matrix polynomials. Linear Algebra Appl. **429**, 2373–2400 (2008)
17. Corless, R.M.: Generalized companion matrices in the Lagrange basis. In: Gonzalez-Vega, L., Recio, T., (eds.) Proceedings EACA, Santander, pp. 317–322 (2004)
18. De Terán, F., Dopico, F.M., Mackey, D.S.: Linearizations of singular matrix polynomials and the recovery of minimal indices. Electron. J. Linear Algebra **18**, 371–402 (2009)
19. De Terán, F., Dopico, F.M., Mackey, D.S.: Fiedler companion linearizations and the recovery of minimal indices. SIAM J. Matrix Anal. Appl. **31**, 2181–2204 (2010)
20. De Terán, F., Dopico, F.M., Mackey, D.S.: Palindromic companion forms for matrix polynomials of odd degree. J. Comput. Appl. Math. **236**, 1464–1480 (2011)
21. De Terán, F., Dopico, F.M., Mackey, D.S.: Fiedler companion linearizations for rectangular matrix polynomials. Linear Algebra Appl. **437**, 957–991 (2012)
22. De Terán, F., Dopico, F.M., Mackey, D.S.: Spectral equivalence of matrix polynomials and the index sum theorem. Linear Algebra Appl. **459**, 264–333 (2014)
23. De Terán, F., Dopico, F.M., Mackey, D.S.: A quasi-canonical form for quadratic matrix polynomials, Part 2: the singular case (2014, in preparation)
24. De Terán, F., Dopico, F.M., Mackey, D.S., Perović, V.: Quadratic realizability for palindromic matrix polynomials (2014, in preparation)
25. Duffin, R.J.: The Rayleigh-Ritz method for dissipative and gyroscopic systems. Q. Appl. Math. **18**, 215–221 (1960)
26. Effenberger, C., Kressner, D.: Chebyshev interpolation for nonlinear eigenvalue problems. BIT Numer. Math. **52**, 933–951 (2012)
27. Fan, H.-Y., Lin, W.-W., Van Dooren, P.: Normwise scaling of second order polynomial matrices. SIAM J. Matrix Anal. Appl. **26**(1), 252–256 (2004)
28. Fassbender, H., Mackey, D.S., Mackey, N., Schröder, C.: Structured polynomial eigenproblems related to time-delay systems. Electron. Trans. Numer. Anal. **31**, 306–330 (2008)
29. Fiedler, M.: A note on companion matrices. Linear Algebra Appl. **372**, 325–331 (2003)
30. Frobenius, G.: Theorie der linearen Formen mit ganzen Coefficienten. J. Reine Angew. Math. (Crelle) **86**, 146–208 (1878)
31. Gladwell, G.M.L.: Inverse Problems in Vibration, 2nd edn. Kluwer Academic, Dordrecht (2004)
32. Gohberg, I., Lancaster, P., Rodman, L.: Matrix Polynomials. Academic, New York (1982)
33. Good, I.J.: The colleague matrix, a Chebyshev analogue of the companion matrix. Q. J. Math. Oxf. Ser. **12**, 61–68 (1961)

34. Grammont, L., Higham, N.J., Tisseur, F.: A framework for analyzing nonlinear eigenproblems and parametrized linear systems. Linear Algebra Appl. **435**, 623–640 (2011)
35. Güttel, S., Van Beeumen, R., Meerbergen, K., Michiels, W.: NLEIGS: a class of robust fully rational Krylov methods for nonlinear eigenvalue problems. SIAM J. Sci. Comput. **36**(6), A2842–A2864 (2014). Also available as MIMS EPrint 2013.49. Manchester Institute for Mathematical Sciences, The University of Manchester (2013)
36. Hammarling, S., Munro, C.J., Tisseur, F.: An algorithm for the complete solution of quadratic eigenvalue problems. ACM Trans. Math. Softw. **39**(3), 18:1–18:19 (2013)
37. Higham, N.J., Li, R.-C., Tisseur, F.: Backward error of polynomial eigenproblems solved by linearization. SIAM J. Matrix Anal. Appl. **29**, 1218–1241 (2007)
38. Higham, N.J., Mackey, D.S., Mackey, N., Tisseur, F.: Symmetric linearizations for matrix polynomials. SIAM J. Matrix Anal. Appl. **29**, 143–159 (2006)
39. Higham, N.J., Mackey, D.S., Tisseur, F.: The conditioning of linearizations of matrix polynomials. SIAM J. Matrix Anal. Appl. **28**, 1005–1028 (2006)
40. Higham, N.J., Mackey, D.S., Tisseur, F.: Definite matrix polynomials and their linearization by definite pencils. SIAM J. Matrix Anal. Appl. **31**, 478–502 (2009)
41. Higham, N.J., Mackey, D.S., Tisseur, F., Garvey, S.D.: Scaling, sensitivity and stability in the numerical solution of quadratic eigenvalue problems. Int. J. Numer. Methods Eng. **73**, 344–360 (2008)
42. Hilliges, A., Mehl, C., Mehrmann, V.: On the solution of palindromic eigenvalue problems. In: Proceedings of the 4th European Congress on Computational Methods in Applied Sciences and Engineering (ECCOMAS), Jyväskylä (2004) CD-ROM
43. Hofer, M., Finger, N., Schöberl, J., Zaglmayr, S., Langer, U., Lerch, R.: Finite element simulation of wave propagation in periodic piezoelectric SAW structures. IEEE Trans. UFFC **53**(6), 1192–1201 (2006)
44. Huang, T.-M., Lin, W.-W., Su, W.-S.: Palindromic quadratization and structure-preserving algorithm for palindromic matrix polynomials of even degree. Numer. Math. **118**, 713–735 (2011)
45. Hwang, T.-M., Lin, W.-W., Mehrmann, V.: Numerical solution of quadratic eigenvalue problems with structure-preserving methods. SIAM J. Sci. Comput. **24**, 1283–1302 (2003)
46. Ipsen, I.C.F.: Accurate eigenvalues for fast trains. SIAM News **37**(9), 1–2 (2004)
47. Kressner, D., Schröder, C., Watkins, D.S.: Implicit QR algorithms for palindromic and even eigenvalue problems. Numer. Algorithms **51**(2), 209–238 (2009)
48. Lancaster, P.: Lambda-matrices and vibrating systems. Pergamon, Oxford (1966)
49. Lancaster, P.: Strongly stable gyroscopic systems. Electron. J. Linear Algebra, **5**, 53–66 (1999)
50. Lancaster, P., Rodman, L.: The Algebraic Riccati Equation. Oxford University Press, Oxford (1995)
51. Lazebnik, F.: On systems of linear diophantine equations. Math. Mag. **69**(4), 261–266 (1996)
52. Mackey, D.S.: Structured linearizations for matrix polynomials. PhD thesis, The University of Manchester, Manchester (2006). Available as MIMS EPrint 2006.68. Manchester Institute for Mathematical Sciences.
53. Mackey, D.S.: The continuing influence of Fiedler's work on companion matrices. Linear Algebra Appl. **439**(4), 810–817 (2013)
54. Mackey, D.S.: A quasi-canonical form for quadratic matrix polynomials, Part 1: the regular case (2014, in preparation)
55. Mackey, D.S., Mackey, N., Mehl, C., Mehrmann, V.: Vector spaces of linearizations for matrix polynomials. SIAM J. Matrix Anal. Appl. **28**, 971–1004 (2006)
56. Mackey, D.S., Mackey, N., Mehl, C., Mehrmann, V.: Structured polynomial eigenvalue problems: good vibrations from good linearizations. SIAM J. Matrix Anal. Appl. **28**, 1029–1051 (2006)
57. Mackey, D.S., Mackey, N., Mehl, C., Mehrmann, V.: Numerical methods for palindromic eigenvalue problems: computing the anti-triangular Schur form. Numer. Linear Algebra Appl. **16**, 63–86 (2009)

58. Mackey, D.S., Mackey, N., Mehl, C., Mehrmann, V.: Jordan structures of alternating matrix polynomials. Linear Algebra Appl. **432**(4), 867–891 (2010)
59. Mackey, D.S., Mackey, N., Mehl, C., Mehrmann, V.: Smith forms of palindromic matrix polynomials. Electron. J. Linear Algebra **22**, 53–91 (2011)
60. Mackey, D.S., Mackey, N., Mehl, C., Mehrmann, V.: Skew-symmetric matrix polynomials and their Smith forms. Linear Algebra Appl. **438**(12), 4625–4653 (2013)
61. Mackey, D.S., Mackey, N., Mehl, C., Mehrmann, V.: Möbius transformations of matrix polynomials. Article in Press in Linear Algebra Appl. **470**, 120–184 (2015). http://dx.doi.org/10.1016/j.laa.2014.05.013
62. Mackey, D.S., Perović, V.: Linearizations of matrix polynomials in Bernstein basis. Submitted for publication. Available as MIMS EPrint 2014.29, Manchester Institute for Mathematical Sciences, The University of Manchester, UK (2014)
63. Mackey, D.S., Tisseur, F.: The Hermitian quadratic realizability problem (2014, in preparation)
64. Meerbergen, K.: The quadratic Arnoldi method for the solution of the quadratic eigenvalue problem. SIAM J. Matrix Anal. Appl. **30**(4), 1463–1482 (2008)
65. Mehl, C.: Jacobi-like algorithms for the indefinite generalized Hermitian eigenvalue problem. SIAM J. Matrix Anal. Appl. **25**(4), 964–985 (2004)
66. Mehl, C., Mehrmann, V., Xu, H.: On doubly structured matrices and pencils that arise in linear response theory. Linear Algebra Appl. **380**, 3–51 (2004)
67. Mehrmann, V.: A step toward a unified treatment of continuous and discrete time control problems. Linear Algebra Appl. **241–243**, 749–779 (1996)
68. Mehrmann, V., Schröder, C., Simoncini, V.: An implicitly-restarted Krylov method for real symmetric/skew-symmetric eigenproblems. Linear Algebra Appl. **436**(10), 4070–4087 (2012)
69. Mehrmann, V., Voss, H.: Nonlinear eigenvalue problems: a challenge for modern eigenvalue methods. GAMM Mitt. Ges. Angew. Math. Mech. **27**(2), 121–152 (2004)
70. Mehrmann, V., Watkins, D.: Structure-preserving methods for computing eigenpairs of large sparse skew-Hamiltonian/Hamiltonian pencils. SIAM J. Sci. Comput. **22**, 1905–1925 (2001)
71. Mehrmann, V., Watkins, D.: Polynomial eigenvalue problems with Hamiltonian structure. Electron. Trans. Numer. Anal. **13**, 106–118 (2002)
72. Moler, C.B., Stewart, G.W.: An algorithm for generalized matrix eigenvalue problems. SIAM J. Numer. Anal. **10**(2), 241–256 (1973)
73. Noferini, V.: The behavior of the complete eigenstructure of a polynomial matrix under a generic rational transformation. Electron. J. Linear Algebra **23**, 607–624 (2012)
74. Perović, V., Mackey, D.S.: Linearizations of matrix polynomials in Newton basis. In preparation (2014)
75. Schröder, C.: Palindromic and Even Eigenvalue Problems – Analysis and Numerical Methods. PhD thesis, Technische Universität Berlin (2008)
76. Smith, H.J.S.: On systems of linear indeterminate equations and congruences. Philos. Trans. R. Soc. Lond. **151**, 293–326 (1861)
77. Su, Y., Bai, Z.: Solving rational eigenvalue problems via linearization. SIAM J. Matrix Anal. Appl. **32**, 201–216 (2011)
78. Tisseur, F.: Backward error and condition of polynomial eigenvalue problems. Linear Algebra Appl. **309**, 339–361 (2000)
79. Tisseur, F., Meerbergen, K.: The quadratic eigenvalue problem. SIAM Rev. **43**, 235–286 (2001)
80. Townsend, A., Noferini, V., Nakatsukasa, Y.: Vector spaces of linearizations of matrix polynomials: a bivariate polynomial approach. Submitted for publication. Available as MIMS EPrint 2012.118. Manchester Institute for Mathematical Sciences, The University of Manchester (2012)
81. Van Beeumen, R., Meerbergen, K., Michiels, W.: A rational Krylov method based on Hermite interpolation for nonlinear eigenvalue problems. SIAM. J. Sci. Comput. **35**, 327–350 (2013)
82. Van Beeumen, R., Meerbergen, K., Michiels, W.: Linearization of Lagrange and Hermite interpolating matrix polynomials. IMA J. Numer. Anal. (2014). doi: 10.1093/imanum/dru019 (First published online on 7 May 2014)

83. Vologiannidis, S., Antoniou, E.N.: A permuted factors approach for the linearization of polynomial matrices. Math. Control Signals Syst. **22**, 317–342 (2011)
84. Wodehouse, P.G.: Psmith in the City. Adam & Charles Black, London (1910)
85. Zaglmayr, S.: Eigenvalue problems in SAW-filter simulations. Diplomarbeit, Johannes Kepler Universität Linz (2002)
86. Zeng, L., Su, Y.: A backward stable algorithm for quadratic eigenvalue problems. SIAM J. Matrix Anal. Appl. **35**(2), 499–516 (2014)

Chapter 13
Stability in Matrix Analysis Problems

André C. M. Ran and Leiba Rodman*

Abstract Let there be given a class of matrices $\mathscr{A}$, and for each $X \in \mathscr{A}$, a set $\mathscr{G}(X)$. The following two broadly formulated problems are addressed: 1. An element $Y_0 \in \mathscr{G}(X_0)$ will be called stable with respect to $\mathscr{A}$ and the collection $\{\mathscr{G}(X)\}_{X \in \mathscr{A}}$, if for every $X \in \mathscr{A}$ which is sufficiently close to X_0 there exists an element $Y \in \mathscr{G}(X)$ that is as close to Y_0 as we wish. Give criteria for existence of a stable Y_0, and describe all of them. 2. Fix $\alpha \geq 1$. An element $Y_0 \in \mathscr{G}(X_0)$ will be called α-stable with respect to $\mathscr{A}$ and the collection $\{\mathscr{G}(X)\}_{X \in \mathscr{A}}$, if for every $X \in \mathscr{A}$ which is sufficiently close to X_0 there exists an element $Y \in \mathscr{G}(X)$ such that the distance between Y and Y_0 is bounded above by a constant times the distance between X and X_0 raised to the power $1/\alpha$ where the constant does not depend on X. Give criteria for existence of an α-stable Y_0, and describe all of them. We present an overview of several basic results and literature guide concerning various stability notions, including the concept of conditional stability, as well as prove several new results and state open problems of interest. Large part of the work leading to this chapter was done while the second author visited Vrije Universiteit, Amsterdam, whose hospitality is gratefully acknowledged.

13.1 Introduction

Many problems in applied mathematics, numerical analysis, engineering, and science require for their solutions that certain quantities associated with given matrices be computed. (All matrices in this chapter are assumed to be real, complex,

A.C.M. Ran (✉)
Department of Mathematics, Vrije Universiteit Amsterdam, De Boelelaan 1081a, 1081 HV Amsterdam, The Netherlands

Unit for BMI, North-West University, Potchefstroom, South Africa
e-mail: ran@few.vu.nl

L. Rodman
Department of Mathematics, College of William and Mary, P. O. Box 8795, Williamsburg, VA 23187-8795, USA
e-mail: lxrodm@gmail.com

P. Benner et al. (eds.), *Numerical Algebra, Matrix Theory, Differential-Algebraic Equations and Control Theory*, DOI 10.1007/978-3-319-15260-8_13

or quaternion.) For example, these quantities may be solutions of a matrix equation with specified additional properties, invariant subspaces, Cholesky factorizations of positive semidefinite matrices, or the set of singular values. Matrix equations and inequalities are of particular interest in this context. The literature on this subject is voluminous, and we mention only the books [1, 10, 26, 33, 54] on matrix equations and inequalities that arise in linear control systems and other applications. In particular, there are many important contributions by V. Mehrmann and his co-authors, see in particular [16, 28, 33].

From the point of view of computation or approximation of a solution of the problem at hand it is therefore important to know whether or not the required quantities can in principle be computed more or less accurately; for example, is backward error analysis applicable? This and related problems in matrix computations are addressed by matrix perturbation theory of which there is a voluminous literature; we mention only the books [21, 25, 56]; the book [25] by V. Mehrmann and co-authors has been a source of inspiration for some of the lines of investigation reported on in this chapter.

In 2006 the authors started a long term collaboration with V. Mehrmann on the topic of perturbation analysis of structured matrices [28–32]. In this chapter, we will not review the theory of rank one perturbations as this is expounded in Chap. 8 by Bora and Karow. Instead we will focus on the theme investigated in the paper [28], where several stability concepts for Lagrangian subspaces of symplectic matrices were considered. This research was in line with many other earlier research papers by the authors of this chapter, in particular, with the expository paper [41] and it is the aim of this chapter to give an overview over these stability concepts in various areas in matrix analysis.

We will not review here many well known and widely used results in matrix perturbation theory, and concentrate instead on the less known and, perhaps, new aspects of the theory. In particular, we focus on some developments that stem from the following general problem. Can one represent the computed solution of the problem as the exact solution of a nearby problem? If this is possible, this is known as backward stability, and as strong backward stability (if the computation involved is done in a structure preserving way). These are key concepts in Numerical Analysis. Among much important work by V. Mehrmann and his co-authors in that direction we mention design of structure-preserving algorithms for Hamiltonian matrices [6, 7, 12].

Thus, studies of backward stability and strong backward stability involve approximation of the required quantities of a matrix by the corresponding quantities of matrices that are perturbations (often restricted to a certain class of matrices) of the given matrix. To formalize this notion, the following metaproblem was formulated in [41]:

Metaproblem 1 *Let there be given a class of matrices $\mathscr{A}$, and for each $X \in \mathscr{A}$, a set of mathematical quantities $\mathscr{G}(X)$ is given. An element $Y_0 \in \mathscr{G}(X_0)$ will be called stable, or robust, with respect to $\mathscr{A}$ and the collection $\{\mathscr{G}(X)\}_{X\in\mathscr{A}}$, if for every $X \in \mathscr{A}$ which is sufficiently close to X_0 there exists an element $Y \in \mathscr{G}(X)$*

that is as close to Y_0 as we wish. Give criteria for existence of a stable Y_0, and describe all of them.

Note that we do not exclude the case when $\mathscr{G}(X)$ is empty for some $X \in \mathscr{A}$.

The statement of Metaproblem 1 presumes a topology on $\mathscr{A}$ and on $\cup_{X\in\mathscr{A}}\mathscr{G}(X)$. In all cases, these will be natural, or standard, topologies. For example, if $\mathscr{A}$ is a subset of $n \times n$ matrices with complex entries, the topology induced by the operator norm will be assumed on $\mathscr{A}$. If $\cup_{X\in\mathscr{A}}\mathscr{F}(X)$ is a set of subspaces in the real or complex finite dimensional vector space, then the topology induced by the gap metric will be given on $\cup_{X\in\mathscr{A}}\mathscr{F}(X)$ (see [18, 24] and references given there for more details, as well as Sect. 13.6).

If, under the notation of Metaproblem 1, $\mathscr{G}(X)$ is empty for some $X \in \mathscr{A}$ which are arbitrarily close to X_0, then Y_0 is not stable. This simple observation allows us to rule out stability in some situations.

The literature concerning various aspects of Metaproblem 1, and its applications and connections, is extensive. The research in this direction continues to generate interest among researchers. In particular, we mention the expository paper [41] where many results in this direction, often with full proofs, are exposed.

Assuming that Y_0 is stable, one might be interested in the degree of stability, in other words, comparison of magnitudes between approximation of Y_0 by Y and approximation of X_0 by X. This approach leads to a more refined scale of stability properties, as follows.

Metaproblem 2 *Let $\mathscr{A}$, $X \in \mathscr{A}$, and $\mathscr{G}(X)$ be as in Metaproblem 1. Fix $\alpha \geq 1$. An element $Y_0 \in \mathscr{G}(X_0)$ will be called α-stable with respect to $\mathscr{A}$ and the collection $\{\mathscr{G}(X)\}_{X\in\mathscr{A}}$, if for every $X \in \mathscr{A}$ which is sufficiently close to X_0 there exists an element $Y \in \mathscr{G}(X)$ such that the distance between Y and Y_0 is bounded above by*

$$M \cdot (\text{distance between } X \text{ and } X_0)^{1/\alpha}, \tag{13.1}$$

where the positive constant M does not depend on X. Give criteria for existence of an α-stable Y_0, and describe all of them.

Again, Metaproblem 2 presumes that some metric is introduced in the sets $\mathscr{A}$ and $\cup_{X\in\mathscr{A}}\mathscr{G}(X)$. These metrics will be the standard ones.

Clearly, if Y_0 is α-stable, then Y_0 is β-stable for every $\beta \geq \alpha$. Also, if Y_0 is α-stable for some $\alpha \geq 1$, then Y_0 is stable.

The particular case of 1-stability is known as *Lipschitz stability* or *well-posedness* (in the terminology of [25]).

A related, and somewhat more nuanced, notion of stability has to do with conditional stability. Under the hypotheses and notation of Metaproblem 1, we say that $Y_0 \in \mathscr{G}(X_0)$ is *conditionally stable*, with respect to $\mathscr{A}$ and the collection $\{\mathscr{G}(X)\}_{X\in\mathscr{A}}$, if for every $X \in \mathscr{A}$ which is sufficiently close to X_0 and such that $\mathscr{G}(X) \neq \emptyset$ there exists an element $Y \in \mathscr{G}(X)$ that is as close to Y_0 as we wish. Thus, a provision is made in the concept of conditional stability to the effect that only those X with $\mathscr{G}(X) \neq \emptyset$ are being considered. Analogously, we say $Y_0 \in \mathscr{G}(X_0)$

is *conditionally* α*-stable*, or *conditionally Lipschitz stable* if $\alpha = 1$, with respect to $\mathcal{A}$ and the collection $\{\mathcal{G}(X)\}_{X\in\mathcal{A}}$, if for every $X \in \mathcal{A}$ which is sufficiently close to X_0 and such that $\mathcal{G}(X) \neq \emptyset$ there exists an element $Y \in \mathcal{G}(X)$ such that the distance between Y and Y_0 is bounded above by (13.1).

In this chapter we focus mostly on Metaproblem 2 as well as conditional α-stability. We present an overview of several basic results and a literature guide concerning various stability notions, as well as prove some new results and state open problems.

Of particular interest are estimates of the constant M in Metaproblem 2, as well as in the definition of conditional α-stability. We call M a *constant of* α*-stability*, or a *constant of conditional* α*-stability* etc. as the case may be. For many cases of Lipschitz stability, estimates of the constant of Lipschitz stability, are well known (see, e.g., [8, 56]). For canonical bases of selfadjoint matrices with respect to indefinite inner products and for Hamiltonian matrices Lipschitz stability, including estimation of the constant of Lipschitz stability, was established in [47]. Note that in the context of [47], see also [5], the consideration is restricted to those selfadjoint (or Hamiltonian) matrices that have a fixed Jordan structure.

Other related notions of stability in matrix analysis problems have been studied in the literature, such as strong stability and strong α-stability. We do not mention these notions here, and refer the reader to the relevant papers, e.g., [45, 51].

Besides the introduction and conclusions, this chapter consists of seven sections. In Sect. 13.2 linear equations $A\mathbf{x} = \mathbf{b}$ are considered, and stability, Lipschitz stability and conditional stability of the solution $\mathbf{x}$ are discussed. In Sect. 13.3 roots of matrices are considered. A characterization of stable pth roots for complex matrices is given. Section 13.4 is concerned with generalized inverses. Among other things it is shown that if a matrix is invertible at least from one side, then every generalized inverse is Lipschitz stable, and otherwise none of its generalized inverses is stable. In Sect. 13.5 matrix decompositions are considered, in particular the Cholesky decomposition for positive semidefinite matrices, and the singular value decomposition. Sections 13.6 and 13.7 discuss stability of invariant subspaces of matrices. Section 13.6 treats the general case, while in Sect. 13.7 the matrices are assumed to have special structure with respect to an indefinite inner product, and the subspaces considered are either maximal nonnegative or maximal neutral in the indefinite inner product. Applications to algebraic Riccati equations are discussed in this section as well. Finally, in Sect. 13.8 certain classes of nonlinear matrix equations are studied.

In the sequel, $\mathbf{F}$ stands for the real field $\mathbf{R}$, the complex field $\mathbf{C}$, or the skew field of quaternions $\mathbf{H}$.

13.2 Linear Equations

We start with linear equations. Basic linear algebra yields the following results.

Theorem 1 *The following statements are equivalent for a solution* $\mathbf{x} \in \mathbf{F}^n$ *of a linear system*

$$A\mathbf{x} = \mathbf{b}, \qquad A \in \mathbf{F}^{m \times n}, \quad \mathbf{b} \in \mathbf{F}^m. \tag{13.2}$$

(1) $\mathbf{x}$ *is stable;*
(2) $\mathbf{x}$ *is Lipschitz stable;*
(3) *The matrix* A *is right invertible.*

To put Theorem 1 in the context of Metaproblem 2, let $\mathscr{A}$ be the set of $m \times (n+1)$ matrices partitioned as $[A \ \ \mathbf{b}]$, where A is $m \times n$ and $\mathbf{b}$ is $m \times 1$, and for every $X = [A \ \ \mathbf{b}] \in \mathscr{A}$, let $\mathscr{G}(X)$ be the solution set (possibly empty) of $A\mathbf{x} = \mathbf{b}$.

Proof The implication (2) $\Rightarrow$ (1) is trivial.

Proof of (3) $\Rightarrow$ (2): Using the rank decomposition for A, we may assume without loss of generality that $A = [I_m \ \ 0]$. Then the solutions of (13.2) have the form $\mathbf{x} = \begin{bmatrix} \mathbf{x}_1 \\ \mathbf{x}_2 \end{bmatrix}$ with $\mathbf{x}_1 = \mathbf{b}$. Consider nearby equations

$$\tilde{\mathbf{x}} = \tilde{\mathbf{b}}, \qquad \tilde{A} = [B \ \ C], \qquad \tilde{\mathbf{x}} = \begin{bmatrix} \tilde{\mathbf{x}}_1 \\ \tilde{\mathbf{x}}_2 \end{bmatrix}, \tag{13.3}$$

where B and C are close to I and 0, respectively. Clearly,

$$\tilde{\mathbf{x}}_1 = B^{-1}(\tilde{\mathbf{b}} - C\tilde{\mathbf{x}}_2), \tag{13.4}$$

and taking $\tilde{\mathbf{x}}_2 = \mathbf{x}_2$, we see that for any solution $\mathbf{x}$ of (13.2), a solution $\tilde{\mathbf{x}}$ of (13.3) exists such that

$$\|\tilde{\mathbf{x}} - \mathbf{x}\| \leq M(\|\tilde{A} - A\| + \|\tilde{\mathbf{b}} - \mathbf{b}\|),$$

where the constant M depends on $A, \mathbf{b}$ and $\mathbf{x}$ only. This proves (13.2).

It remains to prove (1) $\Rightarrow$ (3). Suppose A is not right invertible. We shall prove that (13.2) has no stable solutions. We may assume without loss of generality that $A = \begin{bmatrix} A_1 \\ 0 \end{bmatrix}$, where 0 stands for the zero $1 \times n$ row. If $\mathbf{x}$ is a solution of (13.2), then necessarily the bottom component of $\mathbf{b}$ is zero. Replacing this bottom component by $\epsilon \neq 0$, we obtain a nearby linear system which is inconsistent, hence $\mathbf{x}$ is not stable. □

Theorem 2 *The following statements are equivalent for a solution* $\mathbf{x} \in \mathbf{F}^n$ *of a linear system* (13.2)*:*

(1) $\mathbf{x}$ *is conditionally stable;*
(2) $\mathbf{x}$ *is conditionally Lipschitz stable;*
(3) *The matrix* A *is invertible at least from one side.*

Proof The part (2) ⇒ (1) is trivial. To prove (3) ⇒ (2), in view of Theorem 1 and the proof of (3) ⇒ (1) there, we may assume that $A = \begin{bmatrix} I_n \\ 0 \end{bmatrix}$. Then the solutions of every nearby *consistent* system

$$\tilde{A}\tilde{\mathbf{x}} = \tilde{\mathbf{b}}, \qquad \tilde{A} = \begin{bmatrix} B \\ C \end{bmatrix}, \tag{13.5}$$

have the form

$$\tilde{\mathbf{x}} = B^{-1} \cdot \{\text{the first } n \text{ components of } \tilde{\mathbf{b}}\},$$

and the conditional stability of $\mathbf{x}$ follows.

Finally, we prove that if A is not invertible from either side, then no solution of (13.2) is conditionally stable. We assume that

$$A = \begin{bmatrix} I_p & 0 \\ 0 & 0 \end{bmatrix}, \quad p < \min\{m, n\}.$$

Let $\mathbf{x}$ be a solution of (13.2). For every $\epsilon \neq$ close to zero, consider the following nearby equation:

$$\begin{bmatrix} I_p & 0 & 0 \\ 0 & \epsilon & 0 \\ 0 & 0 & 0_{(m-p-1)\times(n-p-1)} \end{bmatrix} \tilde{\mathbf{x}} = \tilde{\mathbf{b}}, \tag{13.6}$$

where $\tilde{\mathbf{b}}$ is obtained from $\mathbf{b}$ by adding $q\epsilon$ to its $(p+1)$th component; here $q \in \mathbf{F}$ is any fixed number different from the $(p+1)$th component of $\mathbf{x}$. Then (13.6) is consistent, but the $(p+1)$th component of any solution $\tilde{\mathbf{x}}$ is not close to the the $(p+1)$th component of $\mathbf{x}$. □

It is not difficult to estimate the constants of Lipschitz stability and conditional Lipschitz stability in Theorems 1 and 2. Indeed, assuming A is one-sided invertible, let U and V be invertible matrices such that $UAV = \tilde{A}$, where $\tilde{A} = [I \;\; 0]$ or $\tilde{A} = \begin{bmatrix} I \\ 0 \end{bmatrix}$. Clearly, $\mathbf{x}$ is a solution of (13.2) if and only if $\tilde{\mathbf{x}} := V^{-1}\mathbf{x}$ is a solution of

$$\tilde{A}\tilde{\mathbf{x}} = \tilde{\mathbf{b}}, \tag{13.7}$$

where $\tilde{\mathbf{b}} = U\mathbf{b}$. Let $\tilde{Q}$ be a Lipschitz constant of (13.7), in other words,

$$\|\tilde{\mathbf{x}}' - \tilde{\mathbf{x}}\| \leq \tilde{Q}(\|\tilde{A}' - \tilde{A}\| + \|\tilde{\mathbf{b}}' - \tilde{\mathbf{b}}\|)$$

for some solution $\tilde{\mathbf{x}}'$ of any system $\tilde{A}'\tilde{\mathbf{x}}' = \tilde{\mathbf{b}}'$ which is sufficiently close to (13.7). Letting

$$A' = U^{-1}\tilde{A}V^{-1}, \quad \mathbf{b}' = U^{-1}\tilde{\mathbf{b}}', \quad \mathbf{x}' = V\tilde{\mathbf{x}}',$$

we have

$$\begin{aligned}\|\mathbf{x} - \mathbf{x}'\| \leq \|V\| \cdot \|\tilde{\mathbf{x}} - \tilde{\mathbf{x}}'\| &\leq \|V\| \cdot \tilde{Q}(\|\tilde{A}' - \tilde{A}\| + \|\tilde{\mathbf{b}}' - \tilde{\mathbf{b}}\|) \\ &\leq \|V\| \cdot \tilde{Q}(\|U\| \, \|V\| \, \|A - A'\| + \|U\| \, \|\mathbf{b} - \mathbf{b}'\|),\end{aligned}$$

and so

$$Q := \|V\| \, \tilde{Q} \, \max\{\|U\|\|V\|, \|U\|\}$$

can be taken as a Lipschitz constant for (13.2). In turn, to estimate $\tilde{Q}$, we consider two cases: (1) $A = [I_m \;\; 0]$; (2) $A = \begin{bmatrix} I_n \\ 0 \end{bmatrix}$. In the second case, if $\mathbf{x}'$ is the solution of a system $A'\mathbf{x}' = \mathbf{b}'$ which is sufficiently close to (13.2), then

$$\|\mathbf{x} - \mathbf{x}'\| = \|\mathbf{b}_n - B^{-1}\mathbf{b}_n'\|,$$

where B is the top $n \times n$ submatrix of A' and $\mathbf{y}_n$ stands for the vector formed by the top n components of $\mathbf{y}$, for $\mathbf{y} = \mathbf{b}$ or $\mathbf{y} = \mathbf{b}'$. Now fix $\beta > 1$; for A' sufficiently close to A so that $\|I - B^{-1}\| \leq \beta\|A' - A\|$ and $\|B^{-1}\| \leq \beta$, we have

$$\begin{aligned}\|\mathbf{b}_n - B^{-1}\mathbf{b}_n'\| &\leq \|\mathbf{b}_n - B^{-1}\mathbf{b}_n\| + \|B^{-1}\| \cdot \|\mathbf{b}_n - \mathbf{b}_n'\| \\ &\leq \beta\|\mathbf{b}_n\| \cdot \|A' - A\| + \|B^{-1}\| \cdot \|\mathbf{b}_n - \mathbf{b}_n'\|,\end{aligned}$$

so we can take $\tilde{Q} = \beta \max\{\|\mathbf{b}\|, 1\}$. In the case (13.1), using notation (13.3) and formula (13.4), we have

$$\|\mathbf{x} - \tilde{\mathbf{x}}\| = \|\mathbf{b} - B^{-1}(\tilde{\mathbf{b}} - C\mathbf{x}_2)\|,$$

and arguing similarly to the case (13.2), we see that we can take

$$\tilde{Q} = \beta\,(\max\{\|\mathbf{b}\|, 1\} + \|\mathbf{x}\|).$$

13.3 Matrix Roots

For a fixed integer $p \geq 2$, consider the equation $Y^p = X_0$, where $X_0 \in \mathbf{F}^{n \times n}$ is a given $n \times n$ matrix over $\mathbf{F}$, and $Y \in \mathbf{F}^{n \times n}$ is a matrix to be found. We say that a solution Y_0 of $Y^p = X_0$ is *stable* if for every $\epsilon > 0$ there is $\delta > 0$ such that every equation $Y^p = X$ has a solution Y for which $\|Y - Y_0\| < \epsilon$ provided $\|X - X_0\| < \delta$. We say that a solution Y_0 of $Y^p = X_0$ is *α-stable* if there is $\delta > 0$ with the property that every equation $Y^p = X$ has a solution Y such that

$$\|Y - Y_0\| \leq M\|X - X_0\|^{1/\alpha},$$

provided $\|X - X_0\| < \delta$, where $M > 0$ is independent of X.

Recall that a pth root Y of $X \in \mathbf{F}^{n\times n}$ is called a *primary root* if Y is a polynomial of X with coefficients in $\mathbf{F}$ (if $\mathbf{F} = \mathbf{C}$ or $\mathbf{F} = \mathbf{R}$) or with real coefficients (if $\mathbf{F} = \mathbf{H}$). Note (see, for example, [22, 23], or [27]) that in the case $\mathbf{F} = \mathbf{C}$, a pth root Y of an invertible matrix X is a primary root if and only if

$$Y = \frac{1}{2\pi i}\int_{\Gamma} z^{1/p}(zI - X)^{-1}dz, \tag{13.8}$$

for a suitable simple contour Γ in the complex plane that surrounds the eigenvalues of X, where $z^{1/p}$ is any branch of the pth root function which is analytic on and inside Γ. (Note that Γ need not enclose a connected set.)

Conjecture 1 *Assume $X_0 \in \mathbf{F}^{n\times n}$ is invertible. Then the following statements are equivalent for a solution $Y_0 \in \mathbf{F}^{n\times n}$ of $Y^p = X_0 \in \mathbf{F}^{n\times n}$:*

(1) *Y_0 is stable;*
(2) *Y_0 is conditionally stable;*
(3) *Y_0 is Lipschitz stable, i.e., 1-stable;*
(4) *Y_0 is conditionally Lipschitz stable;*
(5) *Y_0 is a primary pth root of X_0.*

Note that the implications (3) $\Rightarrow$ (1); (4) $\Rightarrow$ (2); (3) $\Rightarrow$ (4); and (1) $\Leftrightarrow$ (2) are trivial (since every invertible complex matrix has a p-th root, the stability is the same as conditional stability in Conjecture 1). In the complex case, (5) $\Rightarrow$ (3) follows from formula (13.8).

Conjecture 1 is false if the invertibility hypothesis is removed: The pth root of $0 \in \mathbf{C}$ is p-stable but is not Lipschitz stable (if $p > 1$).

For the case where $\mathbf{F} = \mathbf{C}$ Conjecture 1 holds.

Theorem 3 *Assume $X_0 \in \mathbf{C}^{n\times n}$ is invertible. Then the five statements in Conjecture 1 are equivalent for a solution $Y_0 \in \mathbf{C}^{n\times n}$ of $Y^p = X_0 \in \mathbf{C}^{n\times n}$.*

Proof Because of the remarks above it remains to show that (1) $\Rightarrow$ (5). Assume a nonprimary root is stable. Without loss of generality we may assume that X_0 is in Jordan normal form

$$X_0 = \bigoplus_{j=1}^{m}\bigoplus_{i=1}^{k_j} J_{n_{j,i}}(\lambda_j),$$

where $J_p(\lambda)$ stands for the $p \times p$ Jordan block with eigenvalue λ. Since X_0 has a nonprimary root, there are at least two Jordan blocks corresponding to some eigenvalue, so there is at least one $k_j > 1$. Assume that the eigenvalues and Jordan blocks are ordered in such a way that $k_1, \cdots, k_l$ are bigger than one, and $k_{l+1}, \cdots, k_m$ are one (when $l < m$).

Now consider a sequence of perturbations X_n that "glue together" the Jordan blocks by putting small (going to zero) entries in the usual places that one needs to make the matrix nonderogatory:

$$X_n = X_0 + \bigoplus_{j=1}^{m} \frac{1}{n} Z_j$$

where for $j = 1, \cdots, l$, the matrix Z_j has ones in the positions

$$(n_{j,1}, n_{j,1} + 1), \cdots (n_{j,1} + \cdots + n_{j,k_j-1}, n_{j,1} + \cdots + n_{j,k_j-1} + 1),$$

and zeros elsewhere, and Z_j is the one by one zero matrix for $j = l + 1, \cdots, m$. Then X_n is nonderogatory and $\sigma(X_n) = \sigma(X)$ for all n; here and elsewhere in the chapter, we denote by $\sigma(A)$ the set of eigenvalues of a square size matrix A.

For those perturbations X_n there are only primary roots. So, assuming stability of a nonprimary root Y of X_0 for each n there is a contour (a-priori also depending on n) and one of the pth root branches, say $f_{j(n),n}$, where $f_{j,n}(z) = |z|^{1/p} \exp(ij\pi/p)$ for $j = 1, 2, \ldots, p$, so that the root $Y_n = \frac{1}{2\pi i} \int_{\Gamma_n} f_{j(n),n}(zI - X_n)^{-1} dz$ converges to Y. Now first observe that we may fix the contour independent of n (as the integral is contour-independent and the spectra of all X_n's are the same as the spectrum of X). Next, because there is only a finite number of choices for the branch of the pth root, there is at least one branch that occurs infinitely often for the Y_n. Taking a subsequence if necessary, we may assume then that the branch $f_{j(n),n}$ is actually independent of n. But then Y_n converges to a primary root of X, which contradicts the assumption that Y is a nonprimary root. □

Characterizations of stability and Lipschitz stability of selfadjoint square roots in the context of indefinite inner products on $\mathbf{R}^n$ and on $\mathbf{C}^n$ are given in [57]. The results there are motivated by a study of polar decompositions in indefinite inner product spaces, see e.g., [58] and the references quoted there.

Let us call a matrix $A \in \mathbf{F}^{n \times n}$ *almost invertible* if it is either invertible or singular with zero being a simple eigenvalue (algebraic multiplicity one). Clearly, every almost invertible complex or quaternion matrix has pth roots for every integer $p \geq 2$ (for quaternion matrices this follows for example from the Jordan form which may be chosen complex). Also, the set of almost invertible matrices is open. Thus, for the cases $\mathbf{F} = \mathbf{C}$ or $\mathbf{F} = \mathbf{H}$, a concept of stability of pth roots and the corresponding concept of conditional stability are the same for almost invertible matrices. On the other hand, for other matrices we have the following result:

Proposition 1 *Assume that $A \in \mathbf{F}^{n\times n}$ is not almost invertible. Then for every $\epsilon > 0$ there exists $B_\epsilon \in \mathbf{F}^{n\times n}$ such that B_ϵ has no pth roots for any integer $p \geq 2$ and $\|A - B_\epsilon\| < \epsilon$.*

Proof We use the fact that a single Jordan block with eigenvalue zero has no pth roots for $p \geq 2$. We may assume that A is in a Jordan form, and let

$$A = J_{k_1}(0) \oplus \cdots \oplus J_{k_s}(0) \oplus A_0,$$

where A_0 is invertible and $J_k(0)$ is the $k \times k$ nilpotent Jordan block. Replacing the zeros in the $(k_1, k_1 + 1), \cdots, (k_1 + \cdots + k_{s-1}, k_1 + \cdots + k_{s-1} + 1)$ positions by ϵ, we obtain B_ϵ with the desired properties. (If $s = 1$, then we take $B_\epsilon = A$.) □

Corollary 1 *If $A \in \mathbf{F}^{n\times n}$ is not almost invertible, then none of its pth roots (if they exist) are stable.*

The next example considers nonprimary roots.

Example 1 Let $X_0 = I_2$, and let $Y_0 = \begin{bmatrix} 1 & 0 \\ 0 & -1 \end{bmatrix}$. Then Y_0 is a nonprimary square root of X_0. Consider the following perturbation of X_0: $X(\epsilon) = \begin{bmatrix} 1+\epsilon & \epsilon \\ \epsilon & 1+\epsilon \end{bmatrix}$, where $\epsilon \neq 0$ is close to zero. Then there are no quaternion square roots of $X(\epsilon)$ which are close to Y_0. Indeed, the eigenvalues of $X(\epsilon)$ are 1 (with eigenvector $\begin{bmatrix} 1 \\ -1 \end{bmatrix}$) and $1 + 2\epsilon$ (with eigenvector $\begin{bmatrix} 1 \\ 1 \end{bmatrix}$). Hence all four quaternion square roots of $X(\epsilon)$ are primary and given by

$$\delta_1 \begin{bmatrix} 1 \\ -1 \end{bmatrix} \begin{bmatrix} 1 & -1 \end{bmatrix} + \delta_2 \sqrt{1+2\epsilon} \begin{bmatrix} 1 \\ 1 \end{bmatrix} \begin{bmatrix} 1 & 1 \end{bmatrix}, \qquad \delta_1, \delta_2 \in \{1, -1\}. \tag{13.9}$$

This follows from a general statement: If $A \in \mathbf{F}^{n\times n}$ is positive semidefinite with distinct eigenvalues $\lambda_1, \ldots, \lambda_n$ and corresponding normalized eigenvectors $\mathbf{x}_1, \ldots, \mathbf{x}_n \in \mathbf{F}^n$, then all square roots of A are hermitian and have the form

$$\sum_{j=1}^{n} \delta_j \sqrt{\lambda_j} x_j x_j^*, \qquad \delta_j \in \{1, -1\}.$$

None of (13.9) is close to Y_0.

For primary pth roots of invertible complex matrices, it is easy to estimate the constants of Lipschitz stability. Indeed, if $X \in \mathbf{C}^{n\times n}$ is invertible and Y is a primary pth root given by (13.8), then for any $X' \in \mathbf{C}^{n\times n}$ sufficiently close to X, the pth primary root Y' of X' given by (13.8) (with X replaced by X') satisfies

$$\|Y' - Y\| \leq \frac{1}{2\pi} \cdot (\max_{z \in \Gamma} |z|)^{1/p} \cdot (\text{length of } \Gamma) \cdot \beta M^2, \tag{13.10}$$

where $\beta > 1$ is any fixed number, and

$$M := \max_{z \in \Gamma} \|(zI - X)^{-1}\|.$$

To obtain (13.10), use the formulas

$$(zI - Y')^{-1} - (zI - Y)^{-1} = (zI - Y')^{-1}(Y' - Y)(zI - Y)^{-1},$$
$$\max_{z \in \Gamma} \|(zI - X')^{-1}\| \leq \beta \max_{z \in \Gamma} \|(zI - X)^{-1}\|,$$

and the standard upper bound for integrals.

13.4 Generalized Inverses

We consider in this section stability of generalized inverses. Recall that $Y \in \mathbf{F}^{n \times m}$ is said to be a *generalized inverse* of $X \in \mathbf{F}^{m \times n}$ if the equalities $YXY = Y$ and $XYX = X$ hold true. Letting $\mathscr{A} = \mathbf{F}^{m \times n}$ (for a fixed m and n) and $\mathscr{G}(X)$ the set of all generalized inverses of $X \in \mathscr{A}$, we obtain various notions of stability of generalized inverses as in Metaproblems 1 and 2. Note that since $\mathscr{G}(X)$ is not empty for every X, the notion of conditional stability here is the same as the corresponding notion of stability.

Theorem 4 *Assume $X \in \mathbf{F}^{m \times n}$ is invertible at least from one side. Then every generalized inverse (in this case, one-sided inverse) of X is Lipschitz stable. Conversely, if $X \in \mathbf{F}^{m \times n}$ is not one-sided invertible from neither side, then none of its generalized inverses is stable.*

Proof The direct statement follows from the well known results on perturbation analysis of one-sided inverses (see, e.g., [55]).

To prove the converse statement, we can assume without loss of generality that

$$X = \begin{bmatrix} I_p & 0 \\ 0 & 0 \end{bmatrix}, \quad p < \min\{m, n\}. \tag{13.11}$$

Say, $m - p \leq n - p$. Every generalized inverse Y of X has the form (partitioned conformably with (13.11))

$$Y = \begin{bmatrix} I & B \\ C & CB \end{bmatrix}, \tag{13.12}$$

where $B \in \mathbf{F}^{p\times(m-p)}$ and $C \in \mathbf{F}^{(n-p)\times p}$ are arbitrary matrices. Fix Y of the form (13.12). We shall prove that Y is not a stable generalized inverse of X. Replacing X and Y with

$$\begin{bmatrix} I & B \\ 0 & I \end{bmatrix} X \begin{bmatrix} I & 0 \\ C & I \end{bmatrix} = X$$

and

$$\begin{bmatrix} I & 0 \\ -C & I \end{bmatrix} Y \begin{bmatrix} I & -B \\ 0 & I \end{bmatrix} = \begin{bmatrix} I & 0 \\ 0 & 0 \end{bmatrix}$$

respectively, we may assume that $B = 0$ and $C = 0$, i.e., $Y = X$. Consider the following matrix:

$$X(\epsilon) = \begin{bmatrix} I_p & 0 & 0 \\ 0 & \epsilon I_{m-p} & 0 \end{bmatrix},$$

where $\epsilon \neq 0$ is close to zero. Then the right inverses of $X(\epsilon)$ tend to infinity as $\epsilon \to 0$, hence the generalized inverse Y of X is not stable. □

The Moore-Penrose inverse of $X \in \mathbf{F}^{m\times n}$ is the unique (for a given X) matrix $Y \in \mathbf{F}^{n\times m}$ that satisfies the following four conditions:

(a) $XYX = X$;
(b) $YXY = Y$;
(c) YX is an orthogonal projection;
(d) XY is an orthogonal projection.

Theorem 5 *Assume $X \in \mathbf{F}^{m\times n}$ is invertible at least from one side. Then the Moore-Penrose inverse of X is Lipschitz stable. Conversely, if $X \in \mathbf{F}^{m\times n}$ is not one-sided invertible from neither side, then none of its generalized inverses is stable.*

Proof Using the singular value decomposition (valid for quaternion matrices as well), we can assume that X is a diagonal matrix $X = \operatorname{diag}(d_1, \dots, d_p)$, where $p = \min\{m, n\}$ and $d_j > 0$ for $j = 1, 2, \dots, r$ and $d_j = 0$ for $j > r$; here $r = \operatorname{rank} X$. Then the converse statement follows by arguments analogous to those in the proof of the converse part of Theorem 4. The direct statement follows (in the real and complex cases) from a formula for the Moore-Penrose inverse which is valid as long as the rank of the matrix X is constant [55]. The quaternion case is easily reduced to the real or complex case by using the standard real or complex matrix representations of quaternion matrices (see [53, Sections 3.3 and 3.4], for example). □

We show that, for the case of Lipschitz stability of the Moore-Penrose inverse of X, a constant of Lipschitz stability can be chosen any number larger than

$$s_r^{-1} + s_1 s_r^{-4}(2s_1 + 1) \tag{13.13}$$

where $s_r > 0$ and s_1 are the smallest and the largest singular values of X, respectively. (Compare also Theorem 3.3 in [55].) To verify this statement, in view of Theorem 5, and using the singular value decomposition of X, without loss of generality we may assume that

$$X = [D \;\; 0] \quad \text{or} \quad X = \begin{bmatrix} D \\ 0 \end{bmatrix}, \qquad D = \operatorname{diag}(s_1, \ldots, s_r),$$

where $s_1 \geq s_2 \geq \cdots \geq s_r > 0$. Indeed, if X' is close to X, then the Moore-Penrose inverse Y' of X' is given by the formula

$$Y' = (X')^*(X'(X')^*)^{-1} \quad \text{or} \quad Y' = ((X')^* X')^{-1}(X')^*, \tag{13.14}$$

as the case may be (see [11], for example). Say, $X = \begin{bmatrix} D \\ 0 \end{bmatrix}$ (if $X = [D \;\; 0]$, the proof is analogous, or else apply the result for X^* rather than X.) Then partition X' accordingly: $X' = \begin{bmatrix} D' \\ E' \end{bmatrix}$, where D' is close to D and E' is close to zero. Now, (13.14) gives

$$Y' = ((D')^* D' + (E')^* E')^{-1}[(D')^* \;\; (E')^*].$$

So, for Y the Moore-Penrose inverse of X, we have

$$\begin{aligned} \|Y - Y'\| &= \|((D')^* D' + (E')^* E')^{-1}(X')^* - (D^* D)^{-1} X^*\| \\ &\leq \|((D')^* D' + (E')^* E')^{-1}\| \, \|(X')^* - X^*\| \\ &+ \|X\| \, \|((D')^* D' + (E')^* E')^{-1} - (D^* D)^{-1}\|. \end{aligned} \tag{13.15}$$

Using the formulas (for X' sufficiently close to X) $\|X\| = s_1$,

$$\|((D')^* D' + (E')^* E')^{-1}\| \leq \beta \|(D^* D)^{-1}\| \leq \beta s_r^{-2},$$

$$\begin{aligned} \|((D')^* D' + (E')^* E') - (D^* D)\| &\leq \|(D')^*(D' - D) + ((D')^* - D^*)D\| + \|E'\| \\ &\leq \beta \|X\| \, \|X' - X\| + \|(X')^* - X^*\| \, \|X\| + \|X' - X\| \\ &\leq (\beta \|X\| + \|X\| + 1)\|X' - X\| \\ &= ((\beta + 1)s_1 + 1) \, \|X' - X\|, \end{aligned}$$

and

$$\begin{aligned} \|((D')^* D' + (E')^* E')^{-1} - (D^* D)^{-1}\| &\leq \|((D')^* D' + (E')^* E')^{-1}\| \\ &\cdot \; \|(D')^* D' + (E')^* E' - D^* D\| \|(D^* D)^{-1}\| \\ &\leq (\beta s_r^{-2} \cdot ((\beta + 1)s_1 + 1) \cdot s_r^{-2}) \cdot \|X' - X\|, \end{aligned}$$

(13.15) yields

$$\|Y - Y'\| \le (\beta s_r^{-1} + s_1 \beta s_r^{-4}((\beta + 1)s_1 + 1)) \cdot \|X' - X\|,$$

and (13.13) follows.

Open problem 1 *Characterize stability for other classes of generalized inverses, for example those that satisfy some, but perhaps not all, conditions* (a)–(d) *above, and find constants of Lipschitz stability for one-sided inverses which are not Moore-Penrose.*

13.5 Matrix Decompositions

Many matrix decompositions have stability properties, under suitable hypotheses. We present in this section several results of this nature.

It is well-known that every positive semidefinite $n \times n$ matrix A with entries in $\mathbf{F}$ admits a factorization

$$A = R^* R, \tag{13.16}$$

where R is an upper triangular $n \times n$ matrix with entries in $\mathbf{F}$ and such that the diagonal entries of R are all real nonnegative. Such factorizations will be called *Cholesky decompositions*. Note that a Cholesky decomposition of a given positive semidefinite matrix is unique if A is positive definite, but in general it is not unique.

A criterion for uniqueness of Cholesky decompositions will be given. For a positive semidefinite $n \times n$ matrix A let $\alpha_j(A)$ be the rank of the $j \times j$ upper left block of A; $j = 1, \ldots, n$.

Theorems 6 and 7 below were proved in [41] for the real and complex cases. In the quaternion case the proof is essentially the same.

Theorem 6 *A Cholesky decomposition* (13.16) *of a positive semidefinite A is unique if and only if either A is invertible or*

$$\alpha_{j_0}(A) = \alpha_{j_0+1}(A) = \cdots = \alpha_n(A),$$

where j_0 is the smallest index such that $\alpha_{j_0}(A) < j_0$.

A Cholesky decomposition (13.16) is called *stable* if for every $\varepsilon > 0$ there is a $\delta > 0$ such that $\|A - B\| < \delta$ and B positive semidefinite with entries in $\mathbf{F}$ implies the existence of a Cholesky decomposition $B = S^*S$ such that $\|S - R\| < \varepsilon$; (13.16) is called *α-stable* if there is a positive constant K such that every positive semidefinite B (with entries in $\mathbf{F}$) sufficiently close to A admits a Cholesky decomposition $B = S^*S$ in which $\|S - R\| < K\|B - A\|^{\frac{1}{\alpha}}$. In the case of 1-stability, we simply say that the Cholesky decomposition is *Lipschitz stable*.

Theorem 7 (i) *A Cholesky decomposition of $A \in \mathbf{F}^{n\times n}$ is Lipschitz stable if and only if A is positive definite.*
(ii) *A Cholesky decomposition of A is 2-stable if and only if it is unique, i.e., the conditions of Theorem 6 are satisfied.*
(iii) *In all other cases, no Cholesky decomposition of A is stable.*

Open problem 2 *Provide estimates for Lipschitz constants in Theorem 7.*

Next, we consider singular value decompositions. In what follows, we denote by $|A| \in \mathbf{F}^{n\times n}$ the unique positive semidefinite square root of A^*A, where $A \in \mathbf{F}^{m\times n}$. Recall that a decomposition $A = UDV$ is called a *singular value decomposition* of a matrix $A \in \mathbf{F}^{m\times n}$ if $U \in \mathbf{F}^{m\times m}$ and $V \in \mathbf{F}^{n\times n}$ are unitary (real orthogonal in the real case) and D is a positive semidefinite diagonal matrix with the diagonal entries ordered in nonincreasing order. Obviously, such a decomposition is not unique: the columns of U and of V^* form orthonormal bases of eigenvectors of $|A^*|$ and $|A|$, respectively, which allows for considerable freedom. Note also that the singular values $s_1(A) \geq \cdots \geq s_{\min\{m,n\}}(A)$ of A, i.e., the leading $\min\{m,n\}$ eigenvalues of $|A|$, or what is the same, of $|A^*|$, are always well-behaved with respect to perturbations:

$$\max\left\{|s_j(A) - s_j(B)| : 1 \leq j \leq \min\{m,n\}\right\} \leq \|A - B\|, \quad \forall\ A, B \in \mathbf{F}^{m\times n}. \tag{13.17}$$

(This fact is well-known, see, e.g., p. 78 in [8], where a proof is given for the complex case. The proof is essentially the same in the quaternion case.)

Lemma 1 *Let $A \in \mathbf{F}^{m\times n}$, where $m \geq n$. If $|A|$ has n distinct eigenvalues, then every orthonormal eigenbasis of $|A|$ is Lipschitz stable with respect to A; in other words, if $f_1, \ldots, f_n$ is an orthonormal basis of $F^{n\times 1}$ consisting of eigenvectors of $|A|$, then there exists a constant $K > 0$ such that for every $B \in \mathbf{F}^{m\times n}$ sufficiently close to A there exists an orthonormal basis $g_1, \ldots, g_n$ of eigenvectors of $|B|$ such that $\|f_j - g_j\| \leq K\|A - B\|$.*

If $|A|$ has fewer than n distinct eigenvalues, then no orthonormal eigenbasis of $|A|$ is stable with respect to A.

In the case $m = n$ and $\mathbf{F}$ the real or complex field, Lemma 1 was proved in [41].

Proof The direct statement follows as in the proof of Lemma 4.7 of [41]: it is a direct consequence of well-known results on the perturbation of eigenspaces of Hermitian matrices combined with the inequality

$$\||A| - |B|\|_2 \leq \sqrt{2}\|A - B\|_2.$$

Conversely, assume that $|A|$ has fewer than n distinct eigenvalues. Then the singular value decomposition of A is given by $A = U\mathrm{diag}(s_j)_{j=1}^n V$ with $s_i = s_{i+1}$ for some i. Assume first that all other singular values are different.

First consider perturbations

$$A(\varepsilon) = U\mathrm{diag}(s_1, \cdots, s_{i-1}, s_i + \varepsilon, s_{i+1}, \cdots, s_n)V.$$

Letting $\varepsilon \to 0$ we see that the only orthonormal basis of eigenvectors of $|A|$ that is possibly stable is given by the columns of V. Next, consider perturbations

$$\begin{aligned} B(\varepsilon) &= U\mathrm{diag}(s_i, \cdots, s_{i-1}, \begin{bmatrix} s_i & \varepsilon \\ \varepsilon & s_i \end{bmatrix}, s_{i+2}, \cdots, s_n)V \\ &= U(I_{i-1} \oplus \begin{bmatrix} \sqrt{2}/2 & \sqrt{2}/2 \\ \sqrt{2}/2 & -\sqrt{2}/2 \end{bmatrix} \oplus I_{n-i-1}) \cdot \\ &\quad \cdot \ \mathrm{diag}(s_1, \cdots, s_{i-1},\ s_i + \varepsilon,\ s_i - \varepsilon,\ s_{i+2}, \cdots, s_n) \\ &\quad \cdot \ (I_{i-1} \oplus \begin{bmatrix} \sqrt{2}/2 & \sqrt{2}/2 \\ \sqrt{2}/2 & -\sqrt{2}/2 \end{bmatrix} \oplus I_{n-i-1})V. \end{aligned}$$

Denote $\tilde{V} = (I_{i-1} \oplus \begin{bmatrix} \sqrt{2}/2 & \sqrt{2}/2 \\ \sqrt{2}/2 & -\sqrt{2}/2 \end{bmatrix} \oplus I_{n-i-1})V$. Letting $\varepsilon \to 0$ we see that the only orthonormal basis of eigenvectors of $|A|$ that is possibly stable is given by the columns of $\tilde{V}$. Since these are not the columns of V, we conclude that there is no stable orthonormal eigenbasis of $|A|$.

When there are more than two eigenvalues of $|A|$ which are equal it is obvious that doing similar perturbations will lead to the same conclusion. □

As it was done in [41], the lemma above may be applied immediately to give a result on Lipschitz stability of the singular value decomposition. As the proof is verbatim the same, we only state the result.

Theorem 8 *Let $A \in \mathbf{F}^{m\times n}$, with $m \geq n$, and let $A = UDV$ be its singular value decomposition. If A has n distinct singular values then there exists a constant $K > 0$ such that every matrix B has a singular value decomposition $B = U'D'V'$ for which*

$$\|U - U'\| + \|D - D'\| + \|V - V'\| \leq K \cdot \|A - B\|.$$

If A does not have n distinct singular values, then there is an $\varepsilon > 0$ and a sequence of matrices B_m such that for every singular value decomposition $B_m = U_m D_m V_m$ of B_m and for every m

$$\|U - U_m\| + \|D - D_m\| + \|V - V_m\| > \varepsilon.$$

Stability and Lipschitz stability of polar decompositions, including polar decompositions in the context of indefinite inner products, have been studied in [57, 58] for real and complex matrices; see also [41]. We will not reproduce these results here.

13.6 Invariant Subspaces

Let $A \in \mathbf{F}^{n\times n}$, where $\mathbf{F}$ is $\mathbf{R}$, $\mathbf{C}$, or $\mathbf{H}$. We consider A naturally as a linear transformation on $\mathbf{F}^{n\times 1}$ (understood as a right quaternion vector space if $\mathbf{F} = \mathbf{H}$).

For two subspaces $\mathscr{N}$ and $\mathscr{M}$ in $\mathbf{F}^{n\times 1}$, the *gap* $\theta(\mathscr{M}, \mathscr{N})$ is defined as follows: $\theta(\mathscr{M}, \mathscr{N}) = \|P_{\mathscr{M}} - P_{\mathscr{N}}\|$, where $P_{\mathscr{M}}$, respectively $P_{\mathscr{N}}$, denote the orthogonal projection onto $\mathscr{M}$, respectively, $\mathscr{N}$. For convenience, here and elsewhere in the chapter we use the operator norm $\|X\|$, i.e., the maximal singular value of the matrix X, although the results presented here turn out to be independent of the choice of the matrix norm. It is well known that $\theta(\mathscr{M}, \mathscr{N})$ is a metric on the set of subspaces in $\mathbf{F}^{n\times 1}$ which makes this set a complete compact metric space (see, e.g., [18] for more details in the real and complex cases).

Let $A \in \mathbf{F}^{n\times n}$, and let $\mathscr{M} \subseteq \mathbf{F}^{n\times 1}$ be an invariant subspace of A. Then $\mathscr{M}$ is called *stable* if for every $\varepsilon > 0$ there is a $\delta > 0$ such that $\|A - B\| < \delta$ implies the existence of a B-invariant subspace $\mathscr{N}$ such that $\theta(\mathscr{M}, \mathscr{N}) < \varepsilon$.

A characterization of stable invariant subspaces is well known (in the real and complex cases). It involves root subspaces of A. To define the root subspaces in a unified framework for $\mathbf{R}$, $\mathbf{C}$, or $\mathbf{H}$, let $p(t)$ be the minimal polynomial of A, with real coefficients (if $\mathbf{F} = \mathbf{R}$ or $\mathbf{F} = \mathbf{H}$) or complex coefficients (if $\mathbf{F} = \mathbf{C}$). Factor

$$p(t) = p_1(t) \cdot \ldots \cdot p_s(t), \tag{13.18}$$

where each $p_j(t)$ is a power of an irreducible (over $\mathbf{R}$ if $\mathbf{F} = \mathbf{R}$ or $\mathbf{F} = \mathbf{H}$, or over $\mathbf{C}$ if $\mathbf{F} = \mathbf{C}$) polynomial $q_j(t)$, and we assume that $q_1(t), \ldots, q_s(t)$ are all distinct. Then

$$\mathscr{R}_j := \{x \in \mathbf{F}^{n\times 1} \ : \ q_j(A)x = 0\}, \quad j = 1, 2, \ldots, s,$$

are the *root subspaces* of A. It is well known that the root subspaces are A-invariant, they form a direct sum decomposition of $\mathbf{F}^{n\times 1}$, and every A-invariant subspace is a direct sum of its intersection with the root subspaces.

Before stating the results, let us introduce the following notation. For two positive integers k and n, with $k < n$, we introduce a number $\gamma(k, n)$, as follows: $\gamma(k, n) = n$, whenever there is no set of k distinct nth roots of unity that sum to zero, while $\gamma(k, n) = n - 1$ if such a set of k distinct nth roots of unity does exist.

Theorem 9 (The complex and quaternion cases) *Let $A \in \mathbf{F}^{n\times n}$, where $\mathbf{F} = \mathbf{C}$ or $\mathbf{F} = \mathbf{H}$, and let $\mathscr{M} \subseteq \mathbf{F}^{n\times 1}$ be a nontrivial A-invariant subspace. In the case $\mathbf{F} = \mathbf{H}$, assume in addition that the following hypothesis is satisfied:*

(ℵ) *for every root subspace $\mathscr{R}$ of A corresponding to $p_j(t)$ in (13.18) with a real root, either*

$$\dim(\mathscr{M} \cap \mathscr{R}) \leq 1 \quad \text{or} \quad \dim(\mathscr{M} \cap \mathscr{R}) \geq \dim \mathscr{R} - 1.$$

(The dimensions here are understood in the quaternion sense.)

Then the following statements are equivalent:

(1) $\mathscr{M}$ *is stable, equivalently conditionally stable;*
(2) $\mathscr{M}$ *is* α*-stable, equivalently conditionally* α*-stable for some positive* α*;*
(3) *If* $\mathscr{R}$ *is a root subspace of* A *such that*

$$\{0\} \neq \mathscr{M} \cap \mathscr{R} \neq \mathscr{R}, \tag{13.19}$$

then the geometric multiplicity of $\mathscr{R}$ *is equal to one, i.e., there is only one (up to scaling) eigenvector of* A *in* $\mathscr{R}$*.*

In the case (1)–(3) *hold true, then* $\mathscr{M}$ *is* α*-stable if and only if*

$$\gamma(\dim(\mathscr{M} \cap \mathscr{R}), \dim \mathscr{R}) \leq \alpha,$$

for every root subspace $\mathscr{R}$ *of* A *such that* (13.19) *holds true. If there are no such root subspaces, then* $\mathscr{M}$ *is* 1*-stable.*

The equivalence of stability and of the corresponding notion of conditional stability in Theorem 9 is trivial, because every matrix has invariant subspaces.

We do not know whether or not hypothesis ($\aleph$) is essential in Theorem 9. Theorem 9 can be found in [43], see also [41]. The quaternion case was proved in [50].

The real analogue of Theorem 9 is somewhat more involved. We will not state or prove the result here, and refer the reader to [41] and [40].

We remark that estimates for constants of α-stability in Theorem 9 (except for Lipschitz stability) do not seem to be known. For Lipschitz stability, such estimates are developed in e.g., [56].

Applications of stability of invariant subspaces, both in the complex and real case, to stability and Lipschitz stability of factorizations of rational matrix functions is treated in [3], Chaps. 13–15. An application to a nonsymmetric algebraic Riccati equation is discussed there as well.

There is rather extensive literature on various notions of stability of generalized invariant subspaces. Much of this literature is reviewed in [41]. Later works in this area include [19, 20, 34]. We mention here also related research on Lipschitz stability of canonical bases [5, 47, 49] and Lipschitz properties of matrix group actions ([48] and references there).

13.7 Invariant Subspaces with Symmetries

Let $A \in \mathbf{C}^{n\times n}$ be H-selfadjoint, where $H = H^* \in \mathbf{C}^{n\times n}$ is invertible. To avoid trivialities we assume that H is indefinite. A subspace $\mathscr{M} \subseteq \mathbf{C}^{n\times n}$ is said to be H-*nonnegative* if $x^*Hx \geq 0$ for all $x \in \mathscr{M}$. A maximal A-invariant H-nonnegative subspace $\mathscr{M}$ is called α-stable if there exists a positive constant K (depending only on A and H) such that for every H-selfadjoint B there exists a maximal B-invariant H-nonnegative subspace $\mathscr{N}$ such that

$$\operatorname{gap}(\mathscr{N}, \mathscr{M}) \leq K\|B - A\|^{1/\alpha}.$$

The index of stability of $\mathscr{M}$ is the minimal α for which $\mathscr{M}$ is α-stable (if the set of all such $\alpha \geq 1$ is not empty).

Theorem 10 *Let $\mathscr{M}$ be a maximal A-invariant H-nonnegative subspace.*

(a) *$\mathscr{M}$ is α-stable for some $\alpha \geq 1$ if and only if $\mathscr{M} \cap \mathscr{L} = \{0\}$ or $\mathscr{M} \supseteq \mathscr{L}$ for every root subspace $\mathscr{L}$ of A corresponding to an eigenvalue of geometric multiplicity ≥ 2.*
(b) *If the condition in* (a) *holds true, then the index of stability of $\mathscr{M}$ is equal to the maximal dimension γ of the root subspace $\mathscr{L}$ of A such that $\{0\} \neq \mathscr{M} \cap \mathscr{L} \neq \mathscr{L}$, or to one if no such $\mathscr{L}$ exists.*

Proof By the general theory of the index of strong stability (without symmetries) [45] we know that the index of stability of $\mathscr{M}$ does not exceed γ.

To prove that the index is equal to γ we exhibit a perturbation of A in which no α-stability is possible with $\alpha < \gamma$. It suffices to consider the case when $A = J_n(0)$, the nilpotent Jordan block of size n, and $H = \pm[\delta_{i+j,n+1}]_{i,j=1}^n$ (the sip matrix in the terminology of [17]). We assume $n \geq 2$. Also, $\dim(\mathscr{M}) = n/2$ if n is even, and $\dim(\mathscr{M}) = (n \pm 1)/2$ if n is odd.

Let B be obtained from A by adding $\epsilon > 0$ in the lower left corner entry. The eigenvalues of B are the distinct nth roots $\lambda_1|\epsilon|^{1/n}, \ldots, \lambda_n|\epsilon|^{1/n}$ of ϵ, where $\lambda_1, \ldots, \lambda_n$ are the distinct nth roots of 1. Let a maximal B-invariant H-nonnegative subspace $\mathscr{M}'$ (necessarily of the same dimension as $\mathscr{M}$) be spanned by the eigenvectors corresponding to $\lambda_{i_1}, \ldots, \lambda_{i_p}$, $p = \dim \mathscr{M}$. A calculation ([18], Section 15.5) shows that there exists $k > 0$ such that

$$\operatorname{gap}(\mathscr{M}, \mathscr{M}') \geq k\epsilon^{1/n} \tag{13.20}$$

for all $\epsilon \in (0, 1]$, unless

$$\lambda_{i_1} + \cdots + \lambda_{i_p} = 0. \tag{13.21}$$

We show that equality (13.21) cannot happen, thereby proving (13.20) which in turn implies that no α-stability is possible with $\alpha < n$.

Consider first the case when n is even; then $p = n/2$. Since the signature of H is zero, the canonical form of (B, H) implies that exactly one of the two root subspaces of B corresponding to the eigenvalues $\pm|\epsilon|^{1/n}$ is H-nonnegative. Say, it is the root subspace corresponding to $|\epsilon|^{1/n}$ which is H-nonnegative (actually H positive). Arguing by contradiction, assume (13.21) holds true. Among $\lambda_{i_1}, \ldots, \lambda_{i_p}$ there cannot be nonreal complex conjugate pairs, because the sum of root subspaces corresponding to such a pair is H-indefinite, which would contradict $\mathcal{M}'$ being H-nonnegative. Also, $1 \in \{\lambda_{i_1}, \ldots, \lambda_{i_p}\}$ and $-1 \notin \{\lambda_{i_1}, \ldots, \lambda_{i_p}\}$, so let $1 = \lambda_{i_1}$. We have

$$1 + \sum_{j=2}^{p} \lambda_{i_j} = 0, \quad 1 + \sum_{j=2}^{p} \overline{\lambda_{i_j}} = 0, \tag{13.22}$$

so

$$2 + \sum_{j=2}^{p} \lambda_{i_j} + \sum_{j=2}^{p} \overline{\lambda_{i_j}} = 0. \tag{13.23}$$

But the list

$$\{\lambda_{i_2}, \ldots, \lambda_{i_p}, \overline{\lambda_{i_2}}, \ldots, \overline{\lambda_{i_p}}\} \tag{13.24}$$

comprises all distinct nth roots of unity except 1 and -1. Since all distinct roots of unity sum up to zero, we must have

$$\sum_{j=2}^{p} \lambda_{i_j} + \sum_{j=2}^{p} \overline{\lambda_{i_j}} = 0,$$

a contradiction with (13.23).

Assume now n is odd. If $p = (n+1)/2$, then 1 must be among $\{\lambda_{i_1}, \ldots, \lambda_{i_p}\}$ (otherwise, the set $\{\lambda_{i_1}, \ldots, \lambda_{i_p}\}$ would contain a pair of nonreal complex conjugate numbers, a contradiction with the H-nonnegativeness of $\mathcal{M}'$). Letting $1 = \lambda_{i_1}$ and arguing by contradiction, we obtain equalities (13.22) and (13.23) as before. Now the list (13.24) comprises all distinct roots of unity except 1, and we must have

$$\sum_{j=2}^{p} \lambda_{i_j} + \sum_{j=2}^{p} \overline{\lambda_{i_j}} = -1,$$

a contradiction with (13.23). If $p = (n-1)/2$, then either 1 is among $\{\lambda_{i_1}, \ldots, \lambda_{i_p}\}$, or $1 \notin \{\lambda_{i_1}, \ldots, \lambda_{i_p}\}$. In the latter case arguing by contradiction we would have

$$\sum_{j=1}^{p} \lambda_{i_j} + \sum_{j=1}^{p} \overline{\lambda_{i_j}} = 0, \tag{13.25}$$

but on the other hand $\{\lambda_{i_j}, \overline{\lambda_{i_j}} \ : \ j = 1, \ldots p\}$ comprises all nth roots of unity except 1, so

$$\sum_{j=2}^{p} \lambda_{i_j} + \sum_{j=2}^{p} \overline{\lambda_{i_j}} = -1,$$

and (13.25) is impossible.

Finally, suppose (still assuming n is odd and $p = (n-1)/2$) that $1 = \lambda_{i_1}$. Arguing by contradiction, we assume (13.21) holds. Then (13.23) is obtained. On the other hand, the list (13.24) comprises $n-3$ distinct roots of unity, and 1 is not on this list. Let $1, \mu, \overline{\mu}$ be the three distinct nth roots of unity not on the list (13.24). Then

$$1 + \mu + \overline{\mu} + \sum_{j=2}^{p} \lambda_{i_j} + \sum_{j=2}^{p} \overline{\lambda_{i_j}} = 0,$$

and comparing with (13.23) we see that $\mu + \overline{\mu} = 1$. Thus,

$$\mu = \cos(\pi/3) \pm i \sin(\pi/3).$$

The condition $\mu^n = 1$ gives that $n\pi/3$ is an integer multiple of 2π, a contradiction with the assumption that n is odd. □

A subspace $\mathscr{M} \subseteq \mathbf{C}^n$ is said to be *H-neutral* if $x^* H x = 0$ for every $x \in \mathscr{M}$. An analogous result holds (with essentially the same proof) for maximal invariant neutral subspaces, i.e. those subspaces that are both invariant and neutral, and are maximal with respect to this property. We need to consider conditional stability here, and assume that n (as in the proof of Theorem 10) is odd:

Theorem 11 *Let A be H-selfadjoint, and let $\mathscr{M}$ be a maximal A-invariant H-neutral subspace.*

(a) *$\mathscr{M}$ is conditionally α-stable for some $\alpha \geq 1$ if and only if $\mathscr{M} \cap \mathscr{L} = \{0\}$ or $\mathscr{M} \supseteq \mathscr{L}$ for every root subspace $\mathscr{L}$ of A corresponding to an eigenvalue of geometric multiplicity ≥ 2.*
(b) *Assume the condition in (a) holds true, and suppose that the maximal dimension γ of the root subspace $\mathscr{L}$ of A such that $\{0\} \neq \mathscr{M} \cap \mathscr{L} \neq \mathscr{L}$ is either attained at a root subspace corresponding to a nonreal eigenvalue of A, or it is not attained at any root subspace corresponding to a nonreal eigenvalue of A, in which case we assume that γ is odd or equal to 2. Then the index of conditional stability of $\mathscr{M}$ is equal γ, or to one if no such $\mathscr{L}$ exists.*

We note that Theorems 10 and 11 are valid also in the case when H is subject to change.

Open problem 3 (a) *Characterize α-stability of maximal A-invariant H-nonnegative subspaces, and conditional α-stability of maximal A-invariant*

H-neutral subspaces, for real H-selfadjoint matrices, where H is assumed to be real as well, with the perturbations of A are restricted to real H-selfadjoint matrices. Analogous problems in the context of quaternion matrices.

(b) *State and prove results analogous to Theorems* 10 *and* 11 *for other classes of interest of invariant subspaces with symmetries. For example:*

   (1) *Lagrangian invariant subspaces of real, complex, or quaternion Hamiltonian and symplectic matrices. In the real and complex cases, the structure of such subspaces was studied in* [16, 52], *and their perturbation analysis is given in* [37, 38, 42]*; see in particular* [28] *for perturbation analysis of Lagrangian invariant subspaces in the context of symplectic matrices*

   (2) *Invariant nonnegative subspaces for real, complex, and quaternion dissipative and expansive matrices. See* [13, 44, 46] *for information about the structure of such subspaces (in the complex case), their applications and perturbation theory. For the real case see* [14, 15].

Invariant Lagrangian subspaces play an important role in several problems. The algebraic Riccati equations in continuous and discrete time, coming from optimal control theory, can be solved by finding specific invariant Lagrangian subspaces. For example, consider the matrix equation

$$XBR^{-1}B^*X - XA - A^*X - Q = 0 \tag{13.26}$$

where R and Q are Hermitian, R is positive definite. Under the assumptions that (A, B) is controllable and (Q, A) is detectable there is a unique Hermitian solution X such that $A - BR^{-1}B^*X$ has all its eigenvalues in the open left half plane (see, e.g., [26, 33] for the real and complex cases, [53] for the quaternion case).

Introduce the matrices

$$H = \begin{bmatrix} A & -BR^{-1}B^* \\ -Q & -A^* \end{bmatrix}, \qquad J = \begin{bmatrix} 0 & I \\ -I & 0 \end{bmatrix}.$$

The Hermitian solutions of (13.26) are in one-one correspondence with J-Lagrangian H-invariant subspaces as follows: if X is a Hermitian solution, then $\mathcal{M} = \operatorname{im}\begin{bmatrix} I \\ X \end{bmatrix}$ is H-invariant and J-Lagrangian. Conversely, under the assumptions stated above, any such subspace is of this form [26].

Equation (13.26) plays an important role in LQ-optimal control, in H^∞ control and in factorization problems for rational matrix functions with symmetries, see e.g., [1, 4, 26, 33]. It is therefore of interest to study its behavior under perturbations of the matrices A, B, R and Q. It turns out that under the conditions of controllability of (A, B) and detectability of (Q, A) the Hermitian solution X for which $\sigma(A - BR^{-1}B^*X)$ is in the open left half plane is Lipschitz stable (see [39] for the real case and [36] for the complex case).

Dropping the condition of detectability on the pair (Q, A), there may be many Hermitian solutions that are conditionally stable. See, e.g., [36].

Open problem 4 *Characterize α-stability for general Hermitian solutions of* (13.26).

13.8 Some Nonlinear Matrix Equations and Stability of Their Solutions

Consider matrix equations of the form $X = Q \pm A^* F(X) A$, on the set of complex Hermitian matrices. Here $Q = Q^*$ is an $n \times n$ matrix, and A is an $m \times n$ matrix, and $F : \mathbf{C}^{n\times n} \to \mathbf{C}^{m\times m}$. One is interested in situations where there is a unique solution. The solution is viewed as a fixed point of the map $G : \mathbf{C}^{n\times n} \to \mathbf{C}^{n\times n}$ given by $G(X) = Q \pm A^* F(X) A$.

We start with a general theorem on contractions. Let (X, d) be a complete metric space, and let Ω be a closed subset of X. For $\alpha \in (0, 1)$ fixed, consider the set $M(\Omega, \alpha)$ of all maps $\Phi : \Omega \to \Omega$ such that for all $x, y \in \Omega$ we have $d(\Phi(x), \Phi(y)) \leq \alpha d(x, y)$. It is well-known (Banach's fixed point theorem) that each such function Φ has a unique fixed point $x^*(\Phi)$ in Ω, and that for every starting point $x_0 \in \Omega$ the sequence of iterates $\Phi^m(x)$ converges to $x^*(\Phi)$, with the additional property that the rate of convergence is given by $d(\Phi^m(x_0), x^*(\Phi)) \leq \frac{\alpha^m}{1-\alpha} d(\Phi(x_0), x_0)$.

The following theorem complements Theorem 2.2 in [35]. The proof is based on the ideas in the proof of that theorem.

Theorem 12 *Let $\Phi \in M(\Omega, \alpha)$. Then for all $\Psi \in M(\Omega, \alpha)$ we have*

$$d(x^*(\Phi), x^*(\Psi)) \leq \frac{1}{1-\alpha} \sup_{x\in\Omega} d(\Phi(x), \Psi(x)).$$

Proof Denote $\Upsilon = \sup_{x\in\Omega} d(\Phi(x), \Psi(x))$, and let $x_0 \in \Omega$ be arbitrary. Then for every $k = 1, 2, 3, \cdots$ we have

$$d(\Phi^k(x_0), x^*(\Phi)) \leq \tfrac{\alpha^k}{1-\alpha} d(\Phi(x_0), x_0),$$
$$d(\Psi^k(x_0), x^*(\Psi)) \leq \tfrac{\alpha^k}{1-\alpha} d(\Psi(x_0), x_0).$$

Denote $c_m = \sup_{x\in\Omega} d(\Phi^m(x), \Psi^m(x))$. Note that for all $m = 2, 3, \cdots$ and for all $x \in \Omega$

$$\begin{aligned} d(\Phi^m(x), \Psi^m(x)) &\leq d(\Phi^{m-1}(\Phi(x)), \Phi^{m-1}(\Psi(x)) + d(\Phi^{m-1}(\Psi(x)), \Psi^{m-1}(\Psi(x)) \\ &\leq \alpha^{m-1} d(\Phi(x), \Psi(x)) + c_{m-1} \leq \alpha^{m-1}\Upsilon + c_{m-1}. \end{aligned}$$

Taking the supremum over all $x \in \Omega$ gives

$$c_m \leq \alpha^{m-1}\Upsilon + c_{m-1}, \qquad m = 2, 3, \cdots ,$$

and so

$$c_m \le (\alpha^{m-1} + \cdots + \alpha)\Upsilon + c_1 = (\alpha^{m-1} + \cdots + 1)\Upsilon = \frac{1-\alpha^m}{1-\alpha}\Upsilon.$$

Next, consider

$$\begin{aligned} d(x^*(\Phi), x^*(\Psi)) &\le d(\Phi^k(x_0), x^*(\Phi)) + d(\Psi^k(x_0), x^*(\Psi)) + d(\Phi^k(x_0), \Psi^k(x_0)) \\ &\le \frac{\alpha^k}{1-\alpha} d(\Phi(x_0), x_0) + \frac{\alpha^k}{1-\alpha} d(\Psi(x_0), x_0) + c_k \\ &\le \frac{\alpha^k}{1-\alpha} (d(\Phi(x_0), x_0) + d(\Psi(x_0), x_0)) + \frac{1-\alpha^k}{1-\alpha}\Upsilon. \end{aligned}$$

Taking the limit $k \to \infty$, we obtain $d(x^*(\Phi), x^*(\Psi)) \le \frac{1}{1-\alpha}\Upsilon$ as desired. □

As a sample result, let us consider the equation $X = Q + A^* X^m A$, where Q is positive definite, on the set of positive definite $n \times n$ Hermitian matrices $P(n)$ with the usual metric $d(X, Y) = \|X - Y\|$. We first assume $m \le 1$. Denote $\Phi_{Q,A}(X) = Q + A^* X^m A$. Define $\Omega = \{X \in P(n) \mid X \ge I_n\}$. Then combining results in [2] and [9] (see also [35], Theorem 4.1 and the text following it) we have

$$d(\Phi_{Q,A}(X), \Phi_{Q,A}(Y)) \le \|A\|^2 \sup_{Z \in L_{X,Y}} |m| \|Z^{m-1}\| d(X, Y)$$

where $L_{X,Y}$ is the line segment joining X and Y. Now, since we are considering X and Y in Ω the whole line segment joining them is in Ω, and for matrices in Ω we have $\|Z^{m-1}\| \le 1$ because $m \le 1$. So, for any A with $\|A\| \le r$ we have

$$d(\Phi_{Q,A}(X), \Phi_{Q,A}(Y)) \le r^2 |m|\, d(X, Y).$$

Put $\alpha = r^2|m|$. Assume r is chosen smaller than $1/\sqrt{|m|}$, so that $\alpha < 1$. Then the equation $X = Q + A^* X^m A$ has a unique positive definite solution. Denote the solution by $X_{Q,A}$. Applying Theorem 12 to this case we see that for every $\tilde{A}$ with $\|\tilde{A}\| < r$ and every positive definite $\tilde{Q}$ one has

$$\|X_{Q,A} - X_{\tilde{Q},\tilde{A}}\| \le \frac{1}{1-\alpha} \sup_{X \ge I} \|\tilde{Q} - Q + \tilde{A}^* X^m \tilde{A} - A^* X^m A\|.$$

Now restrict m to $m < 0$, then for $X \ge I$ we have $X^m \le I$, and then

$$\begin{aligned} \|X_{Q,A} - X_{\tilde{Q},\tilde{A}}\| &\le \tfrac{1}{1-\alpha} \left(\|\tilde{Q} - Q\| + \sup_{X \ge I} \|\tilde{A}^* X^m \tilde{A} - A^* X^m A\|\right) \\ &\le \tfrac{1}{1-\alpha} \left(\|\tilde{Q} - Q\| + \sup_{X \ge I} (\|\tilde{A}^* X^m \tilde{A} - \tilde{A}^* X^m A\| + \|\tilde{A}^* X^m A - A^* X^m A\|)\right) \\ &\le \tfrac{1}{1-\alpha} \left(\|\tilde{Q} - Q\| + \sup_{X \ge I} (\|\tilde{A}\| + \|A\|)\|X^m\| \cdot \|\tilde{A} - A\|\right) \\ &\le \tfrac{1}{1-\alpha} \left(\|\tilde{Q} - Q\| + 2r\|\tilde{A} - A\|\right). \end{aligned}$$

This result complements those in [35] and the references cited there. Thus, $X_{Q,A}$ is Lipschitz stable with respect to perturbations of both Q and A (assuming $r < 1/\sqrt{|m|}$).

Clearly the result of Theorem 12 can be applied to many other nonlinear matrix equations in a similar way, including analogous equations with real and quaternion matrices.

13.9 Conclusions

In this chapter, we review several notions of stability in a variety of problems in matrix stability. These include linear equations, matrix roots, generalized inverses, and two types of matrix decompositions: Cholesky decompositions of positive semidefinite matrices and singular value decompositions. In different directions, we also discuss stability results for invariant subspaces of matrices, with or without additional special structures, and indicate applications to algebraic Riccati equations. New results are presented with full proofs, and several known results are reviewed. Finally, we address the problem of stability for certain classes of nonlinear matrix equations. Open problems, hopefully of interest to researchers, are formulated.

Acknowledgements We thank C. Mehl for very careful reading of the chapter and many useful comments and suggestions.

References

1. Abou-Kandil, H., Freiling, G., Ionescu, V., Jank, G.: Matrix Riccati Equations in Control and Systems Theory. Birkhäuser Verlag, Basel (2003)
2. Ambrostetti, A., Prodi, G.: A Primer of Nonlinear Analysis. Cambridge Studies in Advanced Mathematics, vol. 34. Cambridge University Press, Cambridge/New York (1995)
3. Bart, H., Gohberg, I., Kaashoek, M.A., Ran, A.C.M.: Factorization of Matrix and Operator Functions: The State Space Method. Operator Theory: Advances and Applications, vol. 178. Birkhäuser, Basel/Boston (2008)
4. Bart, H., Gohberg, I., Kaashoek, M.A., Ran, A.C.M.: A State Space Approach to Canonical Factorization with Applications. Operator Theory: Advances and Applications, vol. 200. Birkhäuser, Basel (2010)
5. Bella, T., Olshevsky, V., Prasad, U.: Lipschitz stability of canonical Jordan bases of H-selfadjoint matrices under structure-preserving perturbations. Linear Algebra Appl. **428**, 2130–2176 (2008)
6. Benner, P., Mehrmann, V., Xu, H.: A new method for computing the stable invariant subspace of a real Hamiltonian matrix. Special issue dedicated to William B. Gragg (Monterey, 1996). J. Comput. Appl. Math. **86**, 17–43 (1997)
7. Benner, P., Mehrmann, V., Xu, H.: A numerically stable, structure preserving method for computing the eigenvalues of real Hamiltonian or symplectic pencils. Numer. Math. **78**, 329–358 (1998)

8. Bhatia, R.: Matrix Analysis. Springer, New York (1997)
9. Bhatia, R., Sinha, K.B.: Variation of real powers of positive operators. Indiana Univ. Math. J. **43**, 913–925 (1994)
10. Boyd, S., El Ghaoui, L., Feron, E., Balakrishnan, V.: Linear Matrix Inequalities in System and Control Theory. SIAM Studies in Applied Mathematics, vol. 15. SIAM, Philadelphia (1994)
11. Campbell, S.L., Meyer, C.D.: Generalized Inverses of Linear Transformations. Pitman, London (1979)
12. Chu, D., Liu, X., Mehrmann, V.: A numerical method for computing the Hamiltonian Schur form. Numer. Math. **105**, 375–412 (2007)
13. Fourie, J.H., Groenewald, G.J., Ran, A.C.M.: Positive real matrices in indefinite inner product spaces and invariant maximal semi definite subspaces. Linear Algebra Appl. **424**, 346–370 (2007)
14. Fourie, J.H., Groenewald, G.J., Janse van Rensburg, D.B., Ran, A.C.M.: Real and complex invariant subspaces for matrices which are positive real in an indefinite inner product space. Electron. Linear Algebra **27**, 124–145 (2014)
15. Fourie, J.H., Groenewald, G.J., Janse van Rensburg, D.B., Ran, A.C.M.: Simple forms and invariant subspaces of H-expansive matrices. Linear Algebra Appl. (to appear 2015), doi:10.1016/j.laa.2014.11.022
16. Freiling, G., Mehrmann, V., Xu, H.: Existence, uniqueness, and parametrization of Lagrangian invariant subspaces. SIAM J. Matrix Anal. Appl. **23**, 1045–1069 (2002)
17. Gohberg, I., Lancaster, P., Rodman, L.: Indefinite Linear Algebra. Birkhäuser, Basel (2005)
18. Gohberg, I., Lancaster, P., Rodman, L.: Invariant Subspaces of Matrices. Wiley, New York (1986); republication SIAM, Philadelphia (2009)
19. Gracia, J.-M., Velasco, F.E.: Stability of controlled invariant subspaces. Linear Algebra Appl. **418**, 416–434 (2006)
20. Gracia, J.-M., Velasco, F.E.: Lipschitz stability of controlled invariant subspaces. Linear Algebra Appl. **434**, 1137–1162 (2011)
21. Higham, N.J.: Accuracy and Stability of Numerical Algorithms. 2nd edn. SIAM, Philadelphia (2002)
22. Higham, N.J.: Functions of Matrices. Theory and Applications. SIAM, Philadelphia (2008)
23. Horn, R., Johnson, C.R.: Topics in Matrix Analysis. Cambridge University Press, Cambridge (1991)
24. Kato, T.: Perturbation Theory for Linear Operators. Springer, Berlin/Heidelberg/New York (1966)
25. Konstantinov, M., Gu, D.-W., Mehrmann, V., Petkov, P.: Perturbation Theory for Matrix Equations. North-Holland, Amsterdam/London (2003)
26. Lancaster, P., Rodman, L.: Algebraic Riccati Equations. Oxford University Press, New York (1995)
27. Lancaster, P., Tismenetsky, M.: The Theory of Matrices. Academic, Orlando (1988)
28. Mehl, C., Mehrmann, V., Ran, A.C.M., Rodman, L.: Perturbation analysis of Lagrangian invariant subspaces of symplectic matrices. Linear Multilinear Algebra **57**, 141–184 (2009)
29. Mehl, C., Mehrmann, V., Ran, A.C.M., Rodman, L.: Eigenvalue perturbation theory of classes of structured matrices under generic structured rank one perturbations. Linear Algebra Appl. **435**, 687–716 (2011)
30. Mehl, C., Mehrmann, V., Ran, A.C.M., Rodman, L.: Perturbation theory of selfadjoint matrices and sign characteristics under generic structured rank one perturbations. Linear Algebra Appl. **436**, 4027–4042 (2012)
31. Mehl, C., Mehrmann, V., Ran, A.C.M., Rodman, L.: Jordan forms of real and complex matrices under rank one perturbations. Oper. Matrices **7**, 381–398 (2013)
32. Mehl, C., Mehrmann, V., Ran, A.C.M., Rodman, L.: Eigenvalue perturbation theory of symplectic, orthogonal, and unitary matrices under generic structured rank one perturbations. BIT **54**, 219–255 (2014)
33. Mehrmann, V.: The Autonomous Linear Quadratic Control Problem. Lecture Notes in Control and Information Systems, vol. 163. Springer, Berlin (1991)

34. Puerta, F., Puerta, X.: On the Lipschitz stability of (A,B)-invariant subspaces. Linear Algebra Appl. **438**, 182–190 (2013)
35. Ran, A.C.M., Reurings, M.C.B., Rodman, L.: A perturbation analysis for nonlinear selfadjoint operator equations. SIAM J. Matrix Anal. Appl. **28**, 89–104 (2006)
36. Ran, A.C.M., Rodman, L.: Stability of invariant maximal semidefinite subspaces II. Applications: self-adjoint rational matrix functions, algebraic Riccati equations. Linear Algebra Appl. **63**, 133–173 (1984)
37. Ran, A.C.M., Rodman, L.: Stability of invariant Lagrangian subspaces I. In: Topics in Operator Theory, Constantin Apostol Memorial Issue. Operator Theory Advances and Applications, vol. 32, pp. 181–218. Birkhäuser, Basel (1988)
38. Ran, A.C.M., Rodman, L.: Stability of invariant Lagrangian subspaces II. In: Dym, H., Goldberg, S., Kaashoek, M.A., Lancaster, P. (eds.) The Gohberg Anniversary Collection. Operator Theory Advances and Applications, vol. 40, pp. 391–425. Birkhäuser, Basel (1989)
39. Ran, A.C.M., Rodman, L.: Stable solutions of real algebraic Riccati equations. SIAM J. Control Optim. **30**, 63–81 (1992)
40. Ran, A.C.M., Rodman, L.: The rate of convergence of real invariant subspaces. Linear Algebra Appl. **207**, 194–224 (1994)
41. Ran, A.C.M., Rodman, L.: A class of robustness problems in matrix analysis. In: The Harry Dym Anniversary Volume. Operator Theory Advances and Applications, vol. 134, pp. 337–383. Birkhäuser, Basel (2002)
42. Ran, A.C.M., Rodman, L.: On the index of conditional stability of stable invariant Lagrangian subspaces. SIAM J. Matrix Anal. **29**, 1181–1190 (2007)
43. Ran, A.C.M., Rodman, L., Rubin, A.L.: Stability index of invariant subspaces of matrices. Linear Multilinear Algebra **36**, 27–39 (1993)
44. Ran, A.C.M., Rodman, L., Temme, D.: Stability of pseudospectral factorizations. In: Operator Theory and Analysis, The M.A. Kaashoek Anniversary Volume. Operator Theory Advances and Applications, vol. 122, pp. 359–383. Birkhäuser Verlag, Basel (2001)
45. Ran, A.C.M., Roozemond, L.: On strong α-stability of invariant subspaces of matrices. In: The Gohberg Anniversary Volume. Operator Theory Advances and Applications, vol. 40, pp. 427–435. Birkhäuser, Basel (1989)
46. Ran, A.C.M., Temme, D.: Invariant semidefinite subspaces of dissipative matrices in an indefinite inner product space, existence, construction and uniqueness. Linear Algebra Appl. **212/213**, 169–214 (1994)
47. Rodman, L.: Similarity vs unitary similarity and perturbation analysis of sign characteristics: complex and real indefinite inner products. Linear Algebra Appl. **416**, 945–1009 (2006)
48. Rodman, L.: Remarks on Lipschitz properties of matrix groups actions. Linear Algebra Appl. **434**, 1513–1524 (2011)
49. Rodman, L.: Lipschitz properties of structure preserving matrix perturbations. Linear Algebra Appl. **437**, 1503–1537 (2012)
50. Rodman, L.: Stability of invariant subspaces of quaternion matrices. Complex Anal. Oper. Theory **6**, 1069–1119 (2012)
51. Rodman, L.: Strong stability of invariant subspaces of quaternion matrices. In: Advances in Structure Operator Theory and Related Areas. Operator Theory, Advances and Applications, vol. 237, pp. 221–239. Birkhäuser, Basel (2013)
52. Rodman, L.: Invariant neutral subspaces for Hamiltonian matrices. Electron. J. Linear Algebra **27**, 55–99 (2014)
53. Rodman, L.: Topics in Quaternion Linear Algebra. Princeton University Press, Princeton/Oxford (2014)
54. Saberi, A., Stoorvogel, A.A., Sannuti, P.: Filtering Theory. With Applications to Fault Detection, Isolation, and Estimation. Birkhäuser, Boston (2007)
55. Stewart, G.W.: On the perturbation of pseudo-inverses, projections and linear least squares problems. SIAM Rev. **19**, 634–662 (1977)
56. Stewart, G.W., Sun, J.-g.: Matrix Perturbation Theory. Academic, Boston (1990)

57. van der Mee, C.V.M., Ran, A.C.M., Rodman, L.: Stability of self-adjoint square roots and polar decompositions in indefinite scalar product spaces. Linear Algebra Appl. **302/303**, 77–104 (1999)
58. van der Mee, C.V.M., Ran, A.C.M., Rodman, L.: Stability of polar decompositions. Dedicated to the memory of Branko Najman. Glas. Mat. Ser. III. **35**(55), 137–148 (2000)

Chapter 14
Low-Rank Approximation of Tensors

Shmuel Friedland* **and Venu Tammali**

Abstract In many applications such as data compression, imaging or genomic data analysis, it is important to approximate a given tensor by a tensor that is sparsely representable. For matrices, i.e. 2-tensors, such a representation can be obtained via the singular value decomposition, which allows to compute best rank k-approximations. For very big matrices a low rank approximation using SVD is not computationally feasible. In this case different approximations are available. It seems that variants of the CUR-decomposition are most suitable. For d-mode tensors $\mathscr{T} \in \otimes_{i=1}^{d}\mathbb{R}^{n_i}$, with $d > 2$, many generalizations of the singular value decomposition have been proposed to obtain low tensor rank decompositions. The most appropriate approximation seems to be best $(r_1, \ldots, r_d)$-approximation, which maximizes the ℓ_2 norm of the projection of $\mathscr{T}$ on $\otimes_{i=1}^{d}\mathbf{U}_i$, where $\mathbf{U}_i$ is an r_i-dimensional subspace $\mathbb{R}^{n_i}$. One of the most common methods is the *alternating maximization method* (AMM). It is obtained by maximizing on one subspace $\mathbf{U}_i$, while keeping all other fixed, and alternating the procedure repeatedly for $i = 1, \ldots, d$. Usually, AMM will converge to a local best approximation. This approximation is a fixed point of a corresponding map on Grassmannians. We suggest a Newton method for finding the corresponding fixed point. We also discuss variants of CUR-approximation method for tensors. The first part of the paper is a survey on low rank approximation of tensors. The second new part of this paper is a new Newton method for best $(r_1, \ldots, r_d)$-approximation. We compare numerically different approximation methods.

*This work was supported by NSF grant DMS-1216393.

S. Friedland (✉) • V. Tammali
Department of Mathematics, Statistics and Computer Science, University of Illinois at Chicago, 322 Science and Engineering Offices (M/C 249), 851 S. Morgan Street, Chicago, IL 60607-7045, USA
e-mail: friedlan@uic.edu; vtamma2@uic.edu

P. Benner et al. (eds.), *Numerical Algebra, Matrix Theory, Differential-Algebraic Equations and Control Theory*, DOI 10.1007/978-3-319-15260-8_14

14.1 Introduction

Let $\mathbb{R}$ be the field of real numbers. Denote by $\mathbb{R}^{\mathbf{n}} = \mathbb{R}^{n_1 \times \ldots \times n_d} := \otimes_{i=1}^{d} \mathbb{R}^{n_j}$, where $\mathbf{n} = (n_1, \ldots, n_d)$, the tensor products of $\mathbb{R}^{n_1}, \ldots, \mathbb{R}^{n_d}$. $\mathscr{T} = [t_{i_1,\ldots,i_d}] \in \mathbb{R}^{\mathbf{n}}$ is called a *d-mode tensor*. Note that the number of coordinates of $\mathscr{T}$ is $N = n_1 \ldots n_d$. A tensor $\mathscr{T}$ is called a *sparsely representable tensor* if it can represented with a number of coordinates that is much smaller than N.

Apart from sparse matrices, the best known example of a sparsely representable 2-tensor is a low rank approximation of a matrix $A \in \mathbb{R}^{n_1 \times n_2}$. A rank k-approximation of A is given by $A_{\text{appr}} := \sum_{i=1}^{k} \mathbf{u}_i \mathbf{v}_i^\top$, which can be identified with $\sum_{i=1}^{k} \mathbf{u}_i \otimes \mathbf{v}_i$. To store A_{appr} we need only the $2k$ vectors $\mathbf{u}_1, \ldots, \mathbf{u}_k \in \mathbb{R}^{n_1}$, $\mathbf{v}_1, \ldots, \mathbf{v}_k \in \mathbb{R}^{n_2}$. A best rank k-approximation of $A \in \mathbb{R}^{n_1 \times n_2}$ can be computed via the *singular value decomposition*, abbreviated here as SVD, [19]. Recall that if A is a real symmetric matrix, then the best rank k-approximation must be symmetric, and is determined by the spectral decomposition of A.

The computation of the SVD requires $\mathscr{O}(n_1 n_2^2)$ operations and at least $\mathscr{O}(n_1 n_2)$ storage, assuming that $n_2 \leq n_1$. Thus, if the dimensions n_1 and n_2 are very large, then the computation of the SVD is often infeasible. In this case other type of low rank approximations are considered, see e.g. [1, 5, 7, 11, 13, 18, 21].

For d-tensors with $d > 2$ the situation is rather unsatisfactory. It is a major theoretical and computational problem to formulate good generalizations of low rank approximation for tensors and to give efficient algorithms to compute these approximations, see e.g. [3, 4, 8, 13, 15, 29, 30, 33, 35, 36, 43].

We now discuss briefly the main ideas of the approximation methods for tensors discussed in this paper. We need to introduce (mostly) standard notation for tensors. Let $[n] := \{1, \ldots, n\}$ for $n \in \mathbb{N}$. For $\mathbf{x}_i := (x_{1,i}, \ldots, x_{n_i,i})^\top \in \mathbb{R}^{n_i}, i \in [d]$, the tensor $\otimes_{i\in[d]} \mathbf{x}_i = \mathbf{x}_1 \otimes \cdots \otimes \mathbf{x}_d = \mathscr{X} = [x_{j_1,\ldots,j_d}] \in \mathbb{R}^{\mathbf{n}}$ is called a decomposable tensor, or rank one tensor if $\mathbf{x}_i \neq \mathbf{0}$ for $i \in [d]$. That is, $x_{j_1,\ldots,j_d} = x_{j_1,1} \cdots x_{j_d,d}$ for $j_i \in [n_i], i \in [d]$. Let $\langle \mathbf{x}_i, \mathbf{y}_i \rangle_i := \mathbf{y}_i^\top \mathbf{x}_i$ be the standard inner product on $\mathbb{R}^{n_i}$ for $i \in [d]$. Assume that $\mathscr{S} = [s_{j_1,\ldots,j_d}]$ and $\mathscr{T} = [t_{j_1,\ldots,j_d}]$ are two given tensors in $\mathbb{R}^{\mathbf{n}}$. Then $\langle \mathscr{S}, \mathscr{T} \rangle := \sum_{j_i \in [n_i], i \in [d]} s_{j_1,\ldots,j_d} t_{j_1,\ldots,j_d}$ is the standard inner product on $\mathbb{R}^{\mathbf{n}}$. Note that

$$\langle \otimes_{i\in[d]} \mathbf{x}_i, \otimes_{i\in[d]} \mathbf{y}_i \rangle = \prod_{i\in[d]} \langle \mathbf{x}_i, \mathbf{y}_i \rangle_i,$$

$$\langle \mathscr{T}, \otimes_{i\in[d]} \mathbf{x}_i \rangle = \sum_{j_i \in [n_i], i\in[d]} t_{j_1,\ldots,j_d} x_{j_1,1} \cdots x_{j_d,d}.$$

The norm $\|\mathscr{T}\| := \sqrt{\langle \mathscr{T}, \mathscr{T} \rangle}$ is called the Hilbert-Schmidt norm. (For matrices, i.e. $d = 2$, it is called the Frobenius norm.)

Let $I = \{1 \le i_1 < \cdots < i_l \le d\} \subset [d]$. Assume that $\mathscr{X} = [x_{j_{i_1},\cdots,j_{i_l}}] \in \otimes_{k\in[l]}\mathbb{R}^{n_{i_k}}$. Then the contraction $\mathscr{T} \times \mathscr{X}$ on the set of indices I is given by:

$$\mathscr{T} \times \mathscr{X} = \sum_{j_{i_k}\in[n_{i_k}],k\in[l]} t_{j_1,\ldots,j_d} x_{j_{i_1},\ldots,j_{i_l}} \in \otimes_{p\in[d]\setminus I}\mathbb{R}^{n_p}.$$

Assume that $\mathbf{U}_i \subset \mathbb{R}^{n_i}$ is a subspace of dimension r_i with an orthonormal basis $\mathbf{u}_{1,i},\ldots,\mathbf{u}_{r_i,i}$ for $i \in [d]$. Let $\mathbf{U} := \otimes_{i=1}^d \mathbf{U}_i \subset \mathbb{R}^{\mathbf{n}}$. Then $\otimes_{i=1}^d \mathbf{u}_{j_i,i}$, where $j_i \in [n_i], i \in [d]$, is an orthonormal basis in $\mathbf{U}$. We are approximating $\mathscr{T} \in \mathbb{R}^{n_1\times\cdots\times n_d}$ by a tensor

$$\mathscr{S} = \sum_{j_i\in[r_i],i\in[d]} s_{j_1,\ldots,j_d} \mathbf{u}_{j_1,1} \otimes \cdots \otimes \mathbf{u}_{j_d,d} \in \mathbb{R}^{\mathbf{n}} \tag{14.1}$$

The tensor $\mathscr{S}' = [s_{j_1,\ldots,j_d}] \in \mathbb{R}^{r_1\times\cdots\times r_d}$ is the *core tensor* corresponding to $\mathscr{S}$ in the terminology of [42].

There are two major problems: The first one is how to choose the subspaces $\mathbf{U}_1,\ldots,\mathbf{U}_d$. The second one is the choice of the core tensor $\mathscr{S}'$. Suppose we already made the choice of $\mathbf{U}_1,\ldots,\mathbf{U}_d$. Then $\mathscr{S} = P_{\mathbf{U}}(\mathscr{T})$ is the orthogonal projection of $\mathscr{T}$ on $\mathbf{U}$:

$$P_{\otimes_{i\in[d]}\mathbf{U}_i}(\mathscr{T}) = \sum_{j_i\in[r_i],i\in[d]} \langle \mathscr{T}, \otimes_{i\in[d]}\mathbf{u}_{j_i,i}\rangle \otimes_{i\in[d]} \mathbf{u}_{j_i,i}. \tag{14.2}$$

If the dimensions of $n_1,\ldots,n_d$ are not too big, then this projection can be explicitly carried out. If the dimension $n_1,\ldots,n_d$ are too big to compute the above projection, then one needs to introduce other approximations. That is, one needs to compute the core tensor $\mathscr{S}'$ appearing in (14.1) accordingly. The papers [1, 5, 7, 11, 13, 18, 21, 29, 33, 35, 36] essentially choose $\mathscr{S}'$ in a particular way.

We now assume that the computation of $P_{\mathbf{U}}(\mathscr{T})$ is feasible. Recall that

$$\| P_{\otimes_{i\in[d]}\mathbf{U}_i}(\mathscr{T})\|^2 = \sum_{j_i\in[r_i],i\in[d]} |\langle \mathscr{T}, \otimes_{i=1}^d \mathbf{u}_{j_i,i}\rangle|^2. \tag{14.3}$$

The best $\mathbf{r}$-approximation of $\mathscr{T}$, where $\mathbf{r} = (r_1,\ldots,r_d)$, in Hilbert-Schmidt norm is the solution of the minimal problem:

$$\min_{\mathbf{U}_i,\dim \mathbf{U}_i=r_i,i\in[d]} \ \min_{\mathscr{X}\in\otimes_{i\in[d]}^d\mathbf{U}_i} \|\mathscr{T} - \mathscr{X}\|. \tag{14.4}$$

This problem is equivalent to the following maximum

$$\max_{\mathbf{U}_i,\dim \mathbf{U}_i=r_i,i\in[d]} \| P_{\otimes_{i\in[d]}\mathbf{U}_i}(\mathscr{T})\|^2. \tag{14.5}$$

The standard *alternating maximization method*, denoted by AMM, for solving (14.5) is to solve the maximum problem, where all but the subspace $\mathbf{U}_i$ is fixed. Then this maximum problem is equivalent to finding an r_i-dimensional subspace of $\mathbf{U}_i$ containing the r_i biggest eigenvalues of a corresponding nonnegative definite matrix $A_i(\mathbf{U}_1, \ldots, \mathbf{U}_{i-1}, \mathbf{U}_{i+1}, \ldots, \mathbf{U}_d) \in \mathrm{S}_{n_i}$. Alternating between $\mathbf{U}_1, \mathbf{U}_2, \ldots, \mathbf{U}_d$ we obtain a nondecreasing sequence of norms of projections which converges to v. Usually, v is a critical value of $\|P_{\otimes_{i\in[d]}\mathbf{U}_i}(\mathscr{T})\|$. See [4] for details.

Assume that $r_i = 1$ for $i \in [d]$. Then $\dim \mathbf{U}_i = 1$ for $i \in [d]$. In this case the minimal problem (14.4) is called a best rank one approximation of $\mathscr{T}$. For $d = 2$ a best rank one approximation of a matrix $\mathscr{T} = T \in \mathbb{R}^{n_1 \times n_2}$ is accomplished by the first left and right singular vectors and the corresponding maximal singular value $\sigma_1(T)$. The complexity of this computation is $\mathscr{O}(n_1 n_2)$ [19]. Recall that the maximum (14.5) is equal to $\sigma_1(T)$, which is also called the spectral norm $\|T\|_2$. For $d > 2$ the maximum (14.5) is called the spectral norm of $\mathscr{T}$, and denoted by $\|\mathscr{T}\|_\sigma$. The fundamental result of Hillar-Lim [27] states that the computation of $\|\mathscr{T}\|_\sigma$ is NP-hard in general. Hence the computation of best $\mathbf{r}$-approximation is NP-hard in general.

Denote by $\mathrm{Gr}(r, \mathbb{R}^n)$ the variety of all r-dimensional subspaces in $\mathbb{R}^n$, which is called Grassmannian or Grassmann manifold. Let

$$\mathbf{1}_d := \underbrace{(1, \ldots, 1)}_{d}, \quad \mathrm{Gr}(\mathbf{r}, \mathbf{n}) := \mathrm{Gr}(r_1, n_1) \times \cdots \times \mathrm{Gr}(r_d, n_d).$$

Usually, the AMM for best $\mathbf{r}$-approximation of $\mathscr{T}$ will converge to a fixed point of a corresponding map $\mathbf{F}_{\mathscr{T}} : \mathrm{Gr}(\mathbf{r}, \mathbf{n}) \to \mathrm{Gr}(\mathbf{r}, \mathbf{n})$. This observation enables us to give a new Newton method for finding a best $\mathbf{r}$-approximation to $\mathscr{T}$. For best rank one approximation the map $\mathbf{F}_{\mathscr{T}}$ and the corresponding Newton method was stated in [14].

This paper consists of two parts. The first part surveys a number of common methods for low rank approximation methods of matrices and tensors. We did not cover all existing methods here. We were concerned mainly with the methods that the first author and his collaborators were studying, and closely related methods. The second part of this paper is a new contribution to Newton algorithms related to best $\mathbf{r}$-approximations. These algorithms are different from the ones given in [8, 39, 43]. Our Newton algorithms are based on finding the fixed points corresponding to the map induced by the AMM. In general its known that for big size problem, where each n_i is big for $i \in [d]$ and $d \geq 3$, Newton methods are not efficient. The computation associate the matrix of derivatives (Jacobian) is too expensive in computation and time. In this case AMM or MAMM (modified AMM) are much more cost effective. This well known fact is demonstrated in our simulations.

We now briefly summarize the contents of this paper. In Sect. 14.2 we review the well known facts of singular value decomposition (SVD) and its use for best rank k-approximation of matrices. For large matrices approximation methods using SVD are computationally unfeasible. Section 14.3 discusses a number of approximation methods of matrices which do not use SVD. The common feature of these methods is sampling of rows, or columns, or both to find a low rank approximation. The basic observation in Sect. 14.3.1 is that, with high probability, a best k-rank approximation of a given matrix based on a subspace spanned by the sampled row is with in a relative ϵ error to the best k-rank approximation given by SVD. We list a few methods that use this observation. However, the complexity of finding a particular k-rank approximation to an $m \times n$ matrix is still $\mathcal{O}(kmn)$, as the complexity truncated SVD algorithms using Arnoldi or Lanczos methods [19, 31]. In Sect. 14.3.2 we recall the CUR-approximation introduced in [21]. The main idea of CUR-approximation is to choose k columns and rows of A, viewed as matrices C and R, and then to choose a square matrix U of order k in such a way that CUR is an optimal approximation of A. The matrix U is chosen to be the inverse of the corresponding $k \times k$ submatrix A' of A. The quality of CUR-approximation can be determined by the ratio of $|\det A'|$ to the maximum possible value of the absolute value of all $k \times k$ minors of A. In practice one searches for this maximum using a number of random choices of such minors. A modification of this search algorithm is given in [13]. The complexity of storage of C, R, U is $\mathcal{O}(k \max(m,n))$. The complexity of finding the value of each entry of CUR is $\mathcal{O}(k^2)$. The complexity of computation of CUR is $\mathcal{O}(k^2mn)$. In Sect. 14.4 we survey CUR-approximation of tensors given in [13]. In Sect. 14.5 we discuss preliminary results on best **r**-approximation of tensors. In Sect. 14.5.1 we show that the minimum problem (14.4) is equivalent to the maximum problem (14.5). In Sect. 14.5.2 we discuss the notion of singular tuples and singular values of a tensor introduced in [32]. In Sect. 14.5.3 we recall the well known solution of maximizing $\|P_{\otimes_{i\in[d]}\mathbf{U}_i}(\mathscr{T})\|^2$ with respect to one subspace, while keeping other subspaces fixed. In Sect. 14.6 we discuss AMM for best **r**-approximation and its variations. (In [4, 15] AMM is called *alternating least squares*, abbreviated as ALS.) In Sect. 14.6.1 we discuss the AMM on a product space. We mention a *modified alternating maximization method* and and *2-alternating maximization method*, abbreviated as MAMM and 2AMM respectively, introduced in [15]. The MAMM method consists of choosing the one variable which gives the steepest ascend of AMM. 2AMM consists of maximization with respect to a pair of variables, while keeping all other variables fixed. In Sect. 14.6.2 we discuss briefly AMM and MAMM for best **r**-approximations for tensors. In Sect. 14.6.3 we give the complexity analysis of AMM for $d = 3, r_1 \approx r_2 \approx r_3$ and $n_1 \approx n_2 \approx n_3$. In Sect. 14.7 we state a working assumption of this paper that AMM converges to a fixed point of the induced map, which satisfies certain smoothness assumptions. Under these assumptions we can apply the Newton method, which can be stated in the standard form in $\mathbb{R}^L$. Thus, we first do a number of AMM iterations and then switch to the Newton method. In Sect. 14.7.1 we give a simple application of these ideas to state a Newton method for best rank one approximation. This Newton method was suggested in [14]. It is different from the Newton method in [43]. The new contribution of this paper is the Newton method which is discussed in

Sects. 14.8 and 14.9. The advantage of our Newton method is its simplicity, which avoids the notions and tools of Riemannian geometry as for example in [8, 39]. In simulations that we ran, the Newton method in [8] was 20 % faster than our Newton method for best **r**-approximation of 3-mode tensors. However, the number of iterations of our Newton method was 40 % less than in [8]. In the last section we give numerical results of our methods for best **r**-approximation of tensors. In Sect. 14.11 we give numerical simulations of our different methods applied to a real computer tomography (CT) data set (the so-called MELANIX data set of OsiriX). The summary of these results are given in Sect. 14.12.

14.2 Singular Value Decomposition

Let $A \in \mathbb{R}^{m\times n} \setminus \{0\}$. We now recall well known facts on the SVD of A [19]. See [40] for the early history of the SVD. Assume that $r = \operatorname{rank} A$. Then there exist r-orthonormal sets of vectors $\mathbf{u}_1, \ldots, \mathbf{u}_r \in \mathbb{R}^m, \mathbf{v}_1, \ldots, \mathbf{v}_r \in \mathbb{R}^n$ such that we have:

$$A\mathbf{v}_i = \sigma_i(A)\mathbf{u}_i, \quad \mathbf{u}_i^\top A = \sigma_i(A)\mathbf{v}_i^\top, \quad i \in [r], \quad \sigma_1(A) \geq \cdots \geq \sigma_r(A) > 0,$$
$$A_k = \sum_{i\in[k]} \sigma_i(A)\mathbf{u}_i\mathbf{v}_i^\top, \quad k \in [r], \; A = A_r. \tag{14.6}$$

The quantities $\mathbf{u}_i$, $\mathbf{v}_i$ and $\sigma_i(A)$ are called the left, right i-th singular vectors and i-th singular value of A respectively, for $i \in [r]$. Note that $\mathbf{u}_k$ and $\mathbf{v}_k$ are uniquely defined up to ± 1 if and only if $\sigma_{k-1}(A) > \sigma_k(A) > \sigma_{k+1}(A)$. Furthermore for $k \in [r-1]$ the matrix A_k is uniquely defined if and only if $\sigma_k(A) > \sigma_{k+1}(A)$. Denote by $\mathscr{R}(m,n,k) \subset \mathbb{R}^{m\times n}$ the variety of all matrices of rank at most k. Then A_k is a best rank-k approximation of A:

$$\min_{B\in\mathscr{R}(m,n,k)} \|A - B\| = \|A - A_k\|.$$

Let $\mathbf{U} \in \operatorname{Gr}(p, \mathbb{R}^m), \mathbf{V} \in \operatorname{Gr}(q, \mathbb{R}^n)$. We identify $\mathbf{U} \otimes \mathbf{V}$ with

$$\mathbf{U}\mathbf{V}^\top := \operatorname{span}\{\mathbf{u}\mathbf{v}^\top, \quad \mathbf{u} \in \mathbf{U}, \mathbf{v} \in \mathbf{V}\} \subset \mathbb{R}^{m\times n}. \tag{14.7}$$

Then $P_{\mathbf{U}\otimes\mathbf{V}}(A)$ is identified with the projection of A on $\mathbf{U}\mathbf{V}^\top$ with respect to the standard inner product on $\mathbb{R}^{m\times n}$ given by $\langle X, Y\rangle = \operatorname{tr} XY^\top$. Observe that

$$\operatorname{Range} A = \mathbf{U}_r^\star, \; \mathbb{R}^m = \mathbf{U}_r^\star \oplus (\mathbf{U}_r^\star)^\perp, \; \operatorname{Range} A^\top = \mathbf{V}_r^\star, \; \mathbb{R}^n = \mathbf{V}_r^\star \oplus (\mathbf{V}_r^\star)^\perp.$$

Hence

$$P_{\mathbf{U}\otimes\mathbf{V}}(A) = P_{(\mathbf{U}\cap\mathbf{U}_r^\star)\otimes(\mathbf{V}\cap\mathbf{V}_r^\star)}(A) \Rightarrow \operatorname{rank} P_{\mathbf{U}\otimes\mathbf{V}}(A) \leq \min(\dim \mathbf{U}, \dim \mathbf{V}, r).$$

Thus

$$\max_{\mathbf{U}\in \mathrm{Gr}(p,\mathbb{R}^m),\mathbf{V}\in \mathrm{Gr}(q,\mathbb{R}^n)} \|P_{\mathbf{U}\otimes\mathbf{V}}(A)\|^2 = \|P_{\mathbf{U}_l^\star\otimes\mathbf{V}_l^\star}(A)\|^2 = \sum_{j\in[l]} \sigma_j(A)^2,$$

$$\min_{\mathbf{U}\in \mathrm{Gr}(p,\mathbb{R}^m),\mathbf{V}\in \mathrm{Gr}(q,\mathbb{R}^n)} \|A - P_{\mathbf{U}\otimes\mathbf{V}}(A)\|^2 = \|A - P_{\mathbf{U}_l^\star\otimes\mathbf{V}_l^\star}(A)\|^2 = \sum_{j\in[r]\setminus[l]} \sigma_j(A)^2,$$

$$l = \min(p,q,r). \tag{14.8}$$

To compute $\mathbf{U}_l^\star, \mathbf{V}_l^\star$ and $\sigma_1(A),\ldots,\sigma_l(A)$ of a large scale matrix A one can use Arnoldi or Lanczos methods [19, 31], which are implemented in the partial singular value decomposition. This requires a substantial number of matrix-vector multiplications with the matrix A and thus a complexity of at least $\mathscr{O}(lmn)$.

14.3 Sampling in Low Rank Approximation of Matrices

Let $A = [a_{i,j}]_{i=j=1}^{m,n} \in \mathbb{R}^{m\times n}$ be given. Assume that $\mathbf{b}_1,\ldots,\mathbf{b}_m \in \mathbb{R}^n, \mathbf{c}_1,\ldots,\mathbf{c}_n \in \mathbb{R}^m$ are the columns of $A^\top$ and A respectively. ($\mathbf{b}_1^\top,\ldots \mathbf{b}_m^\top$ are the rows of A.) Most of the known fast rank k-approximation are using sampling of rows or columns of A, or both.

14.3.1 Low Rank Approximations Using Sampling of Rows

Suppose that we sample a set

$$I := \{1 \le i_1 < \ldots < i_s \le m\} \subset [m], \quad |I| = s, \tag{14.9}$$

of rows $\mathbf{b}_{i_1}^\top,\ldots,\mathbf{b}_{i_s}^\top$, where $s \ge k$. Let $\mathbf{W}(I) := \mathrm{span}(\mathbf{b}_{i_1},\ldots,\mathbf{b}_{i_s})$. Then with high probability the projection of the first i-th right singular vectors $\mathbf{v}_i$ on $\mathbf{W}(I)$ is very close to $\mathbf{v}_i$ for $i \in [k]$, provided that $s \gg k$. In particular, [5, Theorem 2] claims:

Theorem 1 (Deshpande-Vempala) *Any $A \in \mathbb{R}^{m\times n}$ contains a subset I of $s = \frac{4k}{\epsilon} + 2k\log(k+1)$ rows such that there is a matrix $\tilde{A}_k$ of rank at most k whose rows lie in $\mathbf{W}(I)$ and*

$$\|A - \tilde{A}_k\|^2 \le (1+\epsilon)\|A - A_k\|^2.$$

To find a rank-k approximation of A, one projects each row of A on $\mathbf{W}(I)$ to obtain the matrix $P_{\mathbb{R}^m\otimes\mathbf{W}(I)}(A)$. Note that we can view $P_{\mathbb{R}^m\otimes\mathbf{W}(I)}(A)$ as an $m \times$

s' matrix, where $s' = \dim \mathbf{W}(I) \le s$. Then find a best rank k-approximation to $P_{\mathbb{R}^m \otimes \mathbf{W}(I)}(A)$, denoted as $P_{\mathbb{R}^m \otimes \mathbf{W}(I)}(A)_k$. Theorem 1 and the results of [18] yield that

$$\|A - P_{\mathbb{R}^m \otimes \mathbf{W}(I)}(A)_k\|^2 \le (1+\epsilon)\|A - A_k\|^2 + \eta\|A - P_{\mathbb{R}^m \otimes \mathbf{W}(I)}(A)\|^2.$$

Here η is proportional to $\frac{k}{s}$, and can be decreased with more rounds of sampling. Note that the complexity of computing $P_{\mathbb{R}^m \otimes \mathbf{W}(I)}(A)_k$ is $\mathscr{O}(ks'm)$. The key weakness of this method is that to compute $P_{\mathbb{R}^m \otimes \mathbf{W}(I)}(A)$ one needs $\mathscr{O}(s'mn)$ operations. Indeed, after having computed an orthonormal basis of $\mathbf{W}(I)$, to compute the projection of each row of A on $\mathbf{W}(I)$ one needs $s'n$ multiplications.

An approach for finding low rank approximations of A using random sampling of rows or columns is given in Friedland-Kave-Niknejad-Zare [11]. Start with a random choice of I rows of A, where $|I| \ge k$ and $\dim \mathbf{W}(I) \ge k$. Find $P_{\mathbb{R}^m \otimes \mathbf{W}(I)}(A)$ and $B_1 := P_{\mathbb{R}^m \otimes \mathbf{W}(I)}(A)_k$ as above. Let $\mathbf{U}_1 \in \mathrm{Gr}(k, \mathbb{R}^m)$ be the subspace spanned by the first k left singular vectors of B_1. Find $B_2 = P_{\mathbf{U}_1 \otimes \mathbb{R}^n}(A)$. Let $\mathbf{V}_2 \in \mathrm{Gr}(k, \mathbb{R}^n)$ correspond to the first k right singular vectors of B_2. Continuing in this manner we obtain a sequence of rank k-approximations $B_1, B_2, \ldots$. It is shown in [11] that $\|A - B_1\| \ge \|A - B_2\| \ge \ldots$ and $\|B_1\| \le \|B_2\| \le \ldots$. One stops the iterations when the relative improvement of the approximation falls below the specified threshold. Assume that $\sigma_k(A) > \sigma_{k+1}(A)$. Since best rank-$k$ approximation is a unique local minimum for the function $\|A - B\|$, $B \in \mathscr{R}(m,n,k)$ [19], it follows that in general the sequence $B_j, j \in \mathbb{N}$ converges to A_k. It is straightforward to show that this algorithm is the AMM for low rank approximations given in Sect. 14.6.2. Again, the complexity of this method is $\mathscr{O}(kmn)$.

Other suggested methods as [1, 7, 37, 38] seem to have the same complexity $\mathscr{O}(kmn)$, since they project each row of A on some k-dimensional subspace of $\mathbb{R}^n$.

14.3.2 CUR-Approximations

Let

$$J := \{1 \le j_1 < \ldots < j_t \le n\} \subset [n], \quad |J| = t, \tag{14.10}$$

and $I \subset [m]$ as in (14.9) be given. Denote by $A[I, J] := [a_{i_p, j_q}]_{p=q=1}^{s,t} \in \mathbb{R}^{s \times t}$. CUR-approximation is based on sampling simultaneously the set of I rows and J columns of A and the approximation matrix to $A(I, J, U)$ given by

$$A(I, J, U) := CUR, \quad C := A[[m], J], \ R := A[I, [n]], \ U \in \mathbb{R}^{t \times s}. \tag{14.11}$$

Once the sets I and J are chosen the approximation $A(I, J, U)$ depends on the choice of U. Clearly the row and the column spaces of $A(I, J, U)$ are contained in the row and column spaces of $A[I,[n]]$ and $A[[m],J]$ respectively. Note that to store the approximation $A(I,J,U)$ we need to store the matrices C, R and U. The number of these entries is $tm+sn+st$. So if n,m are of order 10^5 and s,t are of order 10^2 the storages of C,R,U can be done in Random Access Memory (RAM), while the entries of A are stored in external memory. To compute an entry of $A(I,J,U)$, which is an approximation of the corresponding entry of A, we need st flops.

Let $\mathbf{U}$ and $\mathbf{V}$ be subspaces spanned by the columns of $A[[m],J]$ and $A[I,[n]]^\top$ respectively. Then $A(I,J,U)\in \mathbf{U}\mathbf{V}^\top$, see (14.7).

Clearly, a best CUR approximation is chosen by the least squares principle:

$$
\begin{aligned}
&A(I,J,U^\star) = A([m],J)U^\star A(I,[n]),\\
&U^\star = \arg\min\{\|A - A([m],J)UA(I,[n])\|,\ U\in\mathbb{R}^{|J|\times|I|}\}. \qquad (14.12)
\end{aligned}
$$

The results in [17] show that the least squares solution of (14.12) is given by:

$$
U^\star = A([m],J)^\dagger A A(I,[n])^\dagger. \qquad (14.13)
$$

Here $F^\dagger$ denotes the Moore-Penrose pseudoinverse of a matrix F. Note that $U^\star$ is unique if and only if

$$
\operatorname{rank} A[[m],J] = |J|, \quad \operatorname{rank} A[I,[n]] = |I|. \qquad (14.14)
$$

The complexity of computation of $A([m],J)^\dagger$ and $A(I,[n])^\dagger$ are $\mathcal{O}(t^2m)$ and $\mathcal{O}(s^2n)$ respectively. Because of the multiplication formula for $U^\star$, the complexity of computation of $U^\star$ is $\mathcal{O}(stmn)$.

One can significantly improve the computation of U, if one tries to best fit the entries of the submatrix $A[I',J']$ for given subsets $I'\subset[m]$, $J'\subset[n]$. That is, let

$$
\begin{aligned}
&U^\star(I',J') := \arg\min\{\|A[I',J'] - A(I',J)UA(I,J')\|,\ U\in\mathbb{R}^{|J|\times|I|}\} =\\
&A[I',J]^\dagger A[I',J']A^\dagger[I,J']. \qquad (14.15)
\end{aligned}
$$

(The last equality follows from (14.13).) The complexity of computation of $U^\star(I',J')$ is $\mathcal{O}(st|I'||J'|)$.

Suppose finally, that $I' = I$ and J'. Then (14.15) and the properties of the Moore-Penrose inverse yield that

$$
U^\star(I,J) = A[I,J]^\dagger, \quad A(I,J,U^\star(I,J)) = B(I,J) := A[[m],J]A[I,J]^\dagger A[I,[n]]. \qquad (14.16)
$$

In particular $B(I,J)[I,J] = A[I,J]$. Hence

$$A[[m],J] = B(I,J)[[m],J], \quad A[I,[n]] = B(I,J)[I,[n]],$$
$$B(I,J) = A[[m],J]A[I,J]^{-1}A[I,[n]] \quad \text{if } |I| = |J| = k \text{ and } \det A[I,J] \neq 0. \tag{14.17}$$

The original *CUR* approximation of rank k has the form $B(I,J)$ given by (14.17) [21].

Assume that rank $A \geq k$. We want to choose an approximation $B(I,J)$ of the form (14.17) which gives a good approximation to A. It is possible to give an upper estimate for the maximum of the absolute values of the entries of $A - B(I,J)$ in terms of $\sigma_{k+1}(A)$, provided that $\det A[I,J]$ is relatively close to

$$\mu_k := \max_{I\subset[m],J\subset[n],|I|=|J|=} |\det A[I,J]| > 0. \tag{14.18}$$

Let

$$\|F\|_{\infty,e} := \max_{i\in[m],j\in[n]} |f_{i,j}|, \quad F = [f_{i,j}] \in \mathbb{R}^{m\times n}. \tag{14.19}$$

The results of [21, 22] yield:

$$\|A - B(I,J)\|_{\infty,e} \leq \frac{(k+1)\mu_k}{\det A[I,J]}\sigma_{p+1}(A). \tag{14.20}$$

(See also [10, Chapter 4, §13].)

To find μ_k is probably an NP-hard problem in general [9]. A standard way to find μ_k is either a random search or greedy search [9, 20]. In the special case when A is a symmetric positive definite matrix one can give the exact conditions when the greedy search gives a relatively good result [9].

In the paper by Friedland-Mehrmann-Miedlar-Nkengla [13] a good approximation $B(I,J)$ of the form (14.17) is obtained by a random search on the maximum value of the product of the significant singular values of $A[I,J]$. The approximations found in this way are experimentally better than the approximations found by searching for μ_k.

14.4 Fast Approximation of Tensors

The fast approximation of tensors can be based on several decompositions of tensors such as: Tucker decomposition [42]; matricizations of tensors, as unfolding and applying SVD one time or several time recursively, (see below); higher order singular value decomposition (HOSVD) [3], Tensor-Train decompositions [34, 35];

hierarchical Tucker decomposition [23, 25]. A very recent survey [24] gives an overview on this dynamic field. In this paper we will discuss only the CUR-approximation.

14.4.1 CUR-Approximations of Tensors

Let $\mathscr{T} \in \mathbb{R}^{n_1 \times \ldots n_d}$. In this subsection we denote the entries of $\mathscr{T}$ as $\mathscr{T}(i_1, \ldots, i_d)$ for $i_j \in [n_j]$ and $j \in [d]$. CUR-approximation of tensors is based on matricizations of tensors. The unfolding of $\mathscr{T}$ in the mode $l \in [d]$ consists of rearranging the entries of $\mathscr{T}$ as a matrix $T_l(\mathscr{T}) \in \mathbb{R}^{n_l \times N_l}$, where $N_l = \frac{\prod_{i \in [d]} n_i}{n_l}$. More general, let $K \cup L = [d]$ be a partition of $[d]$ into two disjoint nonempty sets. Denote $N(K) = \prod_{i \in K} n_i$, $N(L) = \prod_{j \in L} n_j$. Then unfolding $\mathscr{T}$ into the two modes K and L consists of rearranging the entires of $\mathscr{T}$ as a matrix $T(K, L, \mathscr{T}) \in \mathbb{R}^{N(K) \times N(L)}$.

We now describe briefly the *CUR*-approximation of 3 and 4-tensors as described by Friedland-Mehrmann-Miedlar-Nkengla [13]. (See [33] for another approach to CUR-approximations for tensors.) We start with the case $d = 3$. Let I_i be a nonempty subset of $[n_i]$ for $i \in [3]$. Assume that the following conditions hold:

$$|I_1| = k^2, \quad |I_2| = |I_3| = k, \quad J := I_2 \times I_3 \subset [n_2] \times [n_3].$$

We identify $[n_2] \times [n_3]$ with $[n_2 n_3]$ using a lexicographical order. We now take the CUR-approximation of $T_1(\mathscr{T})$ as given in (14.17):

$$B(I_1, J) = T_1(\mathscr{T})[[n_1], J] T_1(\mathscr{T})[I_1, J]^{-1} T_1(\mathscr{T})[I_1, [n_2 n_3]].$$

We view $T_1(\mathscr{T})[[n_1], J]$ as an $n_1 \times k^2$ matrix. For each $\alpha_1 \in I_1$ we view $T_1(\mathscr{T})[\{\alpha_1\}, [n_2 n_3]]$ as an $n_2 \times n_3$ matrix $Q(\alpha_1) := [\mathscr{T}(\alpha_1, i_2, i_3)]_{i_2 \in [n_2], i_3 \in [n_3]}$. Let $R(\alpha_1)$ be the *CUR*-approximation of $Q(\alpha_1)$ based on the sets I_2, I_3:

$$R(\alpha_1) := Q(\alpha_1)[[n_2], I_3] Q(\alpha_1)[I_2, I_3]^{-1} Q(\alpha_1)[I_2, [n_3]].$$

Let $F := T_1(\mathscr{T})[I_1, J]^{-1} \in \mathbb{R}^{k^2 \times k^2}$. We view the entries of this matrix indexed by the row $(\alpha_2, \alpha_3) \in I_2 \times I_3$ and column $\alpha_1 \in I_1$. We write these entries as $\mathscr{F}(\alpha_1, \alpha_2, \alpha_3), \alpha_j \in I_j, j \in [3]$, which represent a tensor $\mathscr{F} \in \mathbb{R}^{I_1 \times I_2 \times I_3}$. The entries of $Q(\alpha_1)[I_2, I_3]^{-1}$ are indexed by the row $\alpha_3 \in I_3$ and column $\alpha_2 \in I_2$. We write these entries as $\mathscr{G}(\alpha_1, \alpha_2, \alpha_3), \alpha_2 \in I_2, \alpha_3 \in I_3$, which represent a tensor $\mathscr{G} \in \mathbb{R}^{I_1 \times I_2 \times I_3}$. Then the approximation tensor $\mathscr{B} = [\mathscr{B}(j_1, j_2, j_3)] \in \mathbb{R}^{n_1 \times n_2 \times n_3}$ is given by:

$$\mathscr{B}(i_1, i_2, i_3) = \sum_{\alpha_1 \in I_1, \alpha_j, \beta_j \in I_j, j=2,3} \mathscr{T}(i_1, \alpha_2, \alpha_3) \mathscr{F}(\alpha_1, \alpha_2, \alpha_3) \mathscr{T}(\alpha_1, j_2, \beta_3) \mathscr{G}(\alpha_1, \beta_2, \beta_3) \mathscr{T}(\alpha_1, \beta_2, j_3).$$

We now discuss a CUR-approximation for 4-tensors, i.e. $d = 4$. Let $\mathcal{T} \in \mathbb{R}^{n_1 \times n_2 \times n_3 \times n_4}$ and $K = \{1, 2\}, L = \{3, 4\}$. The rows and columns of $X := T(K, L, \mathcal{T}) \in \mathbb{R}^{(n_1 n_2) \times (n_3 n_4)}$ are indexed by pairs (i_1, i_2) and (i_3, i_4) respectively. Let

$$I_j \subset [n_j],\ |I_j| = k,\ j \in [4], \quad J_1 := I_1 \times I_2,\ J_2 := I_3 \times I_4.$$

First consider the CUR-approximation $X[[n_1 n_2], J_2] X[J_1, J_2]^{-1} X[J_1, [n_3 n_4]]$ viewed as tensor $\mathcal{C} \in \mathbb{R}^{n_1 \times n_2 \times n_3 \times n_4}$. Denote by $\mathcal{H}(\alpha_1, \alpha_2, \alpha_3, \alpha_4)$ the $((\alpha_3, \alpha_4), (\alpha_1, \alpha_2))$ entry of the matrix $X[J_1, J_2]^{-1}$. So $\mathcal{H} \in \mathbb{R}^{I_1 \times I_2 \times I_3 \times I_4}$. Then

$$\mathcal{C}(i_1, i_2, i_3, i_4) = \sum_{\alpha_j \in I_j, j \in [4]} \mathcal{T}(i_1, i_2, \alpha_3, \alpha_4) \mathcal{H}(\alpha_1, \alpha_2, \alpha_3, \alpha_4) \mathcal{T}(\alpha_1, \alpha_2, i_3, i_4).$$

For $\alpha_j \in I_j, j \in [4]$ view vectors $X[[n_1 n_2], (\alpha_3, \alpha_4)]$ and $X[(\alpha_1, \alpha_2), [n_3 n_4]]$ as matrices $Y(\alpha_3, \alpha_4) \in \mathbb{R}^{n_1 \times n_2}$ and $Z(\alpha_1, \alpha_2) \in \mathbb{R}^{n_3 \times n_4}$ respectively. Next we find the CUR-approximations to these two matrices using the subsets (I_1, I_2) and (I_3, I_4) respectively:

$$Y(\alpha_3, \alpha_4)[[n_1], I_2] Y(\alpha_3, \alpha_4)[I_1, I_2]^{-1} Y(\alpha_3, \alpha_4)[I_1, [n_2]],$$
$$Z(\alpha_1, \alpha_2)[[n_3], I_4] Z(\alpha_1, \alpha_2)[I_3, I_4]^{-1} Z(\alpha_1, \alpha_2)[I_3, [n_4]].$$

We denote the entries of $Y(\alpha_3, \alpha_4)[I_1, I_2]^{-1}$ and $Z(\alpha_1, \alpha_2)[I_3, I_4]^{-1}$ by

$$\mathcal{F}(\alpha_1, \alpha_2, \alpha_3, \alpha_4), \quad \alpha_1 \in I_1, \alpha_2 \in I_2,$$
$$\text{and } \mathcal{G}(\alpha_1, \alpha_2, \alpha_3, \alpha_4), \quad \alpha_3 \in I_3, \alpha_4 \in I_4,$$

respectively. Then the CUR-approximation tensor $\mathcal{B}$ of $\mathcal{T}$ is given by:

$$\mathcal{B}(i_1, i_2, i_3, i_4) = \sum_{\alpha_j, \beta_j \in I_j, j \in [4]} \mathcal{T}(i_1, \beta_2, \alpha_3, \alpha_4) \mathcal{F}(\beta_1, \beta_2, \alpha_3, \alpha_4) \mathcal{T}(\beta_1, i_2, \alpha_3, \alpha_4)$$
$$\mathcal{H}(\alpha_1, \alpha_2, \alpha_3, \alpha_4) \mathcal{T}(\alpha_1, \alpha_2, i_3, \beta_4) \mathcal{G}(\alpha_1, \alpha_2, \beta_3, \beta_4) \mathcal{T}(\alpha_1, \alpha_2, \beta_3, i_4).$$

We now discuss briefly the complexity of the storage and computing an entry of the CUR-approximation $\mathcal{B}$. Assume first that $d = 3$. Then we need to store k^2 columns of the matrices $T_1(\mathcal{T})$, k^3 columns of $T_2(\mathcal{T})$ and $T_3(\mathcal{T})$, and k^4 entries of the tensors $\mathcal{F}$ and $\mathcal{G}$. The total storage space is $k^2 n_1 + k^3(n_2 + n_3) + 2k^4$. To compute each entry of $\mathcal{B}$ we need to perform $4k^6$ multiplications and k^6 additions.

Assume now that $d = 4$. Then we need to store k^3 columns of $T_l(\mathcal{T}), l \in [4]$ and k^4 entries of $\mathcal{F}, \mathcal{G}, \mathcal{H}$. Total storage needed is $k^3(n_1 + n_2 + n_3 + n_4 + 3k)$. To compute each entry of $\mathcal{B}$ we need to perform $6k^8$ multiplications and k^8 additions.

14.5 Preliminary Results on Best r-Approximation

14.5.1 The Maximization Problem

We first show that the best approximation problem (14.4) is equivalent to the maximum problem (14.5), see [4] and [26, §10.3]. The Pythagoras theorem yields that

$$\|\mathcal{T}\|^2 = \|P_{\otimes_{i=1}^d \mathbf{U}_i}(\mathcal{T})\|^2 + \|P_{(\otimes_{i=1}^d \mathbf{U}_i)^\perp}(\mathcal{T})\|^2,$$
$$\|\mathcal{T} - P_{\otimes_{i=1}^d \mathbf{U}_i}(\mathcal{T})\|^2 = \|P_{(\otimes_{i=1}^d \mathbf{U}_i)^\perp}(\mathcal{T})\|^2.$$

(Here $(\otimes_{i=1}^d \mathbf{U}_i)^\perp$ is the orthogonal complement of $\otimes_{i=1}^d \mathbf{U}_i$ in $\otimes_{i=1}^d \mathbb{R}^{n_i}$.) Hence

$$\min_{\mathbf{U}_i \in \mathrm{Gr}(r_i,\mathbb{R}^{n_i}), i\in[d]} \|\mathcal{T} - P_{\otimes_{i=1}^d \mathbf{U}_i}(\mathcal{T})\|^2 = \|\mathcal{T}\|^2 - \max_{\mathbf{U}_i \in \mathrm{Gr}(r_i,\mathbb{R}^{n_i}), i\in[d]} \|P_{\otimes_{i=1} \mathbf{U}_i}(\mathcal{T})\|^2. \tag{14.21}$$

This shows the equivalence of (14.4) and (14.5).

14.5.2 Singular Values and Singular Tuples of Tensors

Let $\mathrm{S}(n) = \{\mathbf{x} \in \mathbb{R}^n,\ \|\mathbf{x}\| = 1\}$. Note that one dimensional subspace $\mathbf{U} \in \mathrm{Gr}(1,\mathbb{R}^n)$ is $\mathrm{span}(\mathbf{u})$, where $\mathbf{u} \in \mathrm{S}(n)$. Let $\mathrm{S}(\mathbf{n}) := \mathrm{S}(n_1) \times \cdots \times \mathrm{S}(n_d)$. Then best rank one approximation problem for $\mathcal{T} \in \mathbb{R}^{\mathbf{n}}$ is equivalent to finding

$$\|\mathcal{T}\|_\sigma := \max_{(\mathbf{x}_1,\ldots,\mathbf{x}_d)\in \mathrm{S}(\mathbf{n})} \mathcal{T} \times (\otimes_{i\in[d]}\mathbf{x}_i). \tag{14.22}$$

Let $f_{\mathcal{T}} : \mathbb{R}^{\mathbf{n}} \to \mathbb{R}$ is given by $f_{\mathcal{T}}(\mathcal{X}) = \langle \mathcal{X}, \mathcal{T}\rangle$. Denote by $\mathrm{S}'(\mathbf{n}) \subset \mathbb{R}^{\mathbf{n}}$ all rank one tensors of the form $\otimes_{i\in[d]}\mathbf{x}_i$, where $(\mathbf{x}_1,\ldots,\mathbf{x}_n) \in \mathrm{S}(\mathbf{n})$. Let $f_{\mathcal{T}}(\mathbf{x}_1,\ldots,\mathbf{x}_d) := f_{\mathcal{T}}(\otimes_{i\in[d]}\mathbf{x}_i)$. Then the critical points of $f_{\mathcal{T}}|\mathrm{S}'(\mathbf{n})$ are given by the Lagrange multipliers formulas [32]:

$$\mathcal{T} \times (\otimes_{j\in[d]\setminus\{i\}}\mathbf{u}_j) = \lambda \mathbf{u}_i, \quad i \in [d], \quad (\mathbf{u}_1,\ldots,\mathbf{u}_d) \in \mathrm{S}(\mathbf{n}). \tag{14.23}$$

One calls λ and $(\mathbf{u}_1,\ldots,\mathbf{u}_d)$ a singular value and singular tuple of $\mathcal{T}$. For $d = 2$ these are the singular values and singular vectors of $\mathcal{T}$. The number of *complex* singular values of a generic $\mathcal{T}$ is given in [16]. This number increases exponentially with d. For example for $n_1 = \cdots = n_d = 2$ the number of distinct singular values is $d!$. (The number of real singular values as given by (14.23) is bounded by the numbers given in [16].)

Consider first the maximization problem of $f_{\mathcal{T}}(\mathbf{x}_1,\ldots,\mathbf{x}_d)$ over $\mathrm{S}(\mathbf{n})$ where we vary $\mathbf{x}_i \in \mathrm{S}(n_i)$ and keep the other variables fixed. This problem is equivalent to the maximization of the linear form $\mathbf{x}_i^\top(\mathscr{T} \times (\otimes_{j\in[d]\setminus\{i\}}\mathbf{x}_j))$. Note that if $\mathscr{T} \times (\otimes_{j\in[d]\setminus\{i\}}\mathbf{x}_j) \neq \mathbf{0}$ then this maximum is achieved for $\mathbf{x}_i = \frac{1}{\|\mathscr{T}\times(\otimes_{j\in[d]\setminus\{i\}}\mathbf{x}_j)\|}\mathscr{T} \times (\otimes_{j\in[d]\setminus\{i\}}\mathbf{x}_j)$.

Consider second the maximization problem of $f_{\mathscr{T}}(\mathbf{x}_1,\ldots,\mathbf{x}_d)$ over $\mathrm{S}(\mathbf{n})$ where we vary $(\mathbf{x}_i,\mathbf{x}_j) \in \mathrm{S}(n_i)\times\mathrm{S}(n_j), 1 \le i < j \le d$ and keep the other variables fixed. This problem is equivalent to finding the first singular value and the corresponding right and left singular vectors of the matrix $\mathscr{T} \times (\otimes_{k\in[d]\setminus\{i,j\}}\mathbf{x}_k)$. This can be done by using use Arnoldi or Lanczos methods [19, 31]. The complexity of this method is $\mathscr{O}(n_i n_j)$, given the matrix $\mathscr{T} \times (\otimes_{k\in[d]\setminus\{i,j\}}\mathbf{x}_k)$.

14.5.3 A Basic Maximization Problem for Best r-Approximation

Denote by $\mathrm{S}_n \subset \mathbb{R}^{n\times n}$ the space of real symmetric matrices. For $A \in \mathrm{S}_n$ denote by $\lambda_1(A) \ge \ldots \ge \lambda_n(A)$ the eigenvalues of A arranged in a decreasing order and repeated according to their multiplicities. Let $\mathrm{O}(n,k) \subset \mathbb{R}^{n\times k}$ be the set of all $n\times k$ matrices X with k orthonormal columns, i.e. $X^\top X = I_k$, where I_k is $k\times k$ identity matrix. We view $X \in \mathbb{R}^{n\times k}$ as composed of k-columns $[\mathbf{x}_1\ldots\mathbf{x}_k]$. The column space of $X \in \mathrm{O}(n,k)$ corresponds to a k-dimensional subspace $\mathbf{U} \subset \mathbb{R}^n$. Note that $\mathbf{U} \in \mathrm{Gr}(k,\mathbb{R}^n)$ is spanned by the orthonormal columns of a matrix $Y \in O(n,k)$ if and only if $Y = XO$, for some $O \in \mathrm{O}(k,k)$.

For $A \in \mathrm{S}_n$ one has the Ky-Fan maximal characterization [28, Cor. 4.3.18]

$$\max_{[\mathbf{x}_1\ldots\mathbf{x}_k]\in\mathrm{O}(n,k)} \sum_{i=1}^{k} \mathbf{x}_i^\top A\mathbf{x}_i = \sum_{i=1}^{k} \lambda_i(A). \tag{14.24}$$

Equality holds if and only if the column space of $X = [\mathbf{x}_1\ldots\mathbf{x}_k]$ is a subspace spanned by k eigenvectors corresponding to k-largest eigenvalues of A.

We now reformulate the maximum problem (14.5) in terms of orthonormal bases of $\mathbf{U}_i, i \in [d]$. Let $\mathbf{u}_{1,i},\ldots,\mathbf{u}_{n_i,i}$ be an orthonormal basis of $\mathbf{U}_i$ for $i \in [d]$. Then $\otimes_{i=1}^d \mathbf{u}_{j_i,i}, j_i \in [n_i], i \in [d]$ is an orthonormal basis of $\otimes_{i=1}^d \mathbf{U}_i$. Hence

$$\|P_{\otimes_{i=1}^d \mathbf{U}_i}(\mathscr{T})\|^2 = \sum_{j_i\in[n_i], i\in[d]} \langle \mathscr{T}, \otimes_{i=1}^d \mathbf{u}_{j_i,i}\rangle^2.$$

Hence (14.5) is equivalent to

$$\max_{[\mathbf{u}_{1,i}\ldots\mathbf{u}_{r_i,i}]\in\mathrm{O}(n_i,r_i), i\in[d]} \sum_{j_i\in[n_i], i\in[d]} \langle \mathscr{T}, \otimes_{i=1}^d \mathbf{u}_{j_i,i}\rangle^2 = \tag{14.25}$$

$$\max_{\mathbf{U}_i\in\mathrm{Gr}(r_i,\mathbb{R}^{n_i}), i\in[d]} \|P_{\otimes_{i=1}^d \mathbf{U}_i}(\mathscr{T})\|^2.$$

A simpler problem is to find

$$\max_{[\mathbf{u}_{1,i}\ldots\mathbf{u}_{r_i,i}]\in \mathrm{O}(n_i,r_i)} \sum_{j_i\in[n_i],i\in[d]} \langle \mathscr{T}, \otimes_{i=1}^d \mathbf{u}_{j_i,i}\rangle^2 = \tag{14.26}$$

$$\max_{\mathbf{U}_i\in \mathrm{Gr}(r_i,\mathbb{R}^{n_i})} \|P_{\otimes_{i=1}^d \mathbf{U}_i}(\mathscr{T})\|^2,$$

for a fixed $i \in [d]$. Let

$$\underline{U} := (\mathbf{U}_1,\ldots,\mathbf{U}_d) \in \mathrm{Gr}(\mathbf{r},\mathbf{n}),$$

$$\mathrm{Gr}_i(\mathbf{r},\mathbf{n}) := \mathrm{Gr}(r_1,n_1)\times\ldots\times \mathrm{Gr}(r_{i-1},n_{i-1})\times \mathrm{Gr}(r_{i+1},n_{i+1})\times\ldots\times \mathrm{Gr}(r_d,n_d),$$

$$\underline{U}_i := (\mathbf{U}_1,\ldots,\mathbf{U}_{i-1},\mathbf{U}_{i+1},\ldots,\mathbf{U}_d) \in \mathrm{Gr}_i(\mathbf{r},\mathbf{n}),$$

$$A_i(\underline{U}_i) := \sum_{j_l\in[r_l],l\in[d]\setminus\{i\}} (\mathscr{T}\times\otimes_{k\in[d]\setminus\{i\}}\mathbf{u}_{j_k,k})(\mathscr{T}\times\otimes_{k\in[d]\setminus\{i\}}\mathbf{u}_{j_k,k})^\top. \tag{14.27}$$

The maximization problem (14.25) reduces to the maximum problem (14.24) with $A = A_i(\underline{U}_i)$. Note that each $A_i(\underline{U}_i)$ is a positive semi-definite matrix. Hence $\sigma_j(A_i(\underline{U}_i)) = \lambda_j(A_i(\underline{U}_i))$ for $j \in [n_i]$. Thus the complexity to find the first r_i eigenvectors of $A_i(\underline{U}_i)$ is $\mathscr{O}(r_i n_i^2)$. Denote by $\mathbf{U}_i(\underline{U}_i) \in \mathrm{Gr}(r_i,\mathbb{R}^{n_i})$ a subspace spanned by the first r_i eigenvectors of $A_i(\underline{U}_i)$. Note that this subspace is unique if and only if

$$\lambda_{r_i}(A_i(\underline{U}_i)) > \lambda_{r_i+1}(A_i(\underline{U}_i)). \tag{14.28}$$

Finally, if $\mathbf{r} = \mathbf{1}_d$ then each $A_i(\underline{U}_i)$ is a rank one matrix. Hence $\mathbf{U}_i(\underline{U}_i) = \mathrm{span}(\mathscr{T}\times\otimes_{k\in[d]\setminus\{i\}}\mathbf{u}_{1,k})$. For more details see [12].

14.6 Alternating Maximization Methods for Best r-Approximation

14.6.1 General Definition and Properties

Let Ψ_i be a compact smooth manifold for $i \in [d]$. Define

$$\Psi := \Psi_1\times\cdots\times\Psi_d,\ \hat{\Psi}_i = (\Psi_1\times\cdots\times\Psi_{i-1}\times\Psi_{i+1}\times\cdots\times\Psi_d) \text{ for } i\in[d].$$

We denote by ψ_i, $\psi = (\psi_1,\ldots,\psi_d)$ and $\hat{\psi}_i = (\psi_1,\ldots,\psi_{i-1},\psi_{i+1},\ldots,\psi_d)$ the points in Ψ_i, Ψ and $\hat{\Psi}_i$ respectively. Identify ψ with $(\psi_i,\hat{\psi}_i)$ for each $i \in [d]$. Assume that $f : \Psi \to \mathbb{R}$ is a continuous function with continuous first and second partial derivatives. (In our applications it may happen that f has discontinuities in first and second partial derivatives.) We want to find the maximum value of f and a corresponding maximum point $\psi^\star$:

$$\max_{\psi \in \Psi} f(\psi) = f(\psi^\star). \tag{14.29}$$

Usually, this is a hard problem, where f has many critical points and a number of these critical points are local maximum points. In some cases, as best $\mathbf{r}$ approximation to a given tensor $\mathscr{T} \in \mathbb{R}^{\mathbf{n}}$, we can solve the maximization problem with respect to one variable ψ_i for any fixed $\hat{\psi}_i$:

$$\max_{\psi_i \in \Psi_i} f((\psi_i, \hat{\psi}_i)) = f((\psi_i^\star(\hat{\psi}_i), \hat{\psi}_i)), \tag{14.30}$$

for each $i \in [d]$.

Then the *alternating maximization method*, abbreviated as AMM, is as follows. Assume that we start with an initial point $\psi^{(0)} = (\psi_1^{(0)}, \ldots, \psi_d^{(0)}) = (\psi_1^{(0)}, \hat{\psi}_1^{(0,1)})$. Then we consider the maximal problem (14.30) for $i = 1$ and $\hat{\psi}_1 := \hat{\psi}_1^{(0,1)}$. This maximum is achieved for $\psi_1^{(1)} := \psi_1^\star(\hat{\psi}_1^{(0,1)})$. Assume that the coordinates $\psi_1^{(1)}, \ldots, \psi_j^{(1)}$ are already defined for $j \in [d-1]$. Let $\hat{\psi}_{j+1}^{(0,j+1)} := (\psi_1^{(1)}, \ldots, \psi_j^{(1)}, \psi_{j+2}^{(0)}, \ldots, \psi_d^{(0)})$. Then we consider the maximum problem (14.30) for $i = j+1$ and $\hat{\psi}_{j+1} := \hat{\psi}_{j+1} j^{(0,j+1)}$. This maximum is achieved for $\psi_{j+1}^{(1)} := \psi_{j+1}^\star(\hat{\psi}_{j+1}^{(0,j+1)})$. Executing these d iterations we obtain $\psi^{(1)} := (\psi_1^{(1)}, \ldots, \psi_d^{(1)})$. Note that we have a sequence of inequalities:

$$f(\psi^{(0)}) \le f(\psi_1^{(1)}, \hat{\psi}_1^{(0,1)}) \le f(\psi_2^{(1)}, \hat{\psi}_2^{(0,2)}) \le \cdots \le f(\psi_d^{(1)}, \hat{\psi}_d^{(0,d)}) = f(\psi^{(1)}).$$

Replace $\psi^{(0)}$ with $\psi^{(1)}$ and continue these iterations to obtain a sequence $\psi^{(l)} = (\psi_1^{(l)}, \ldots, \psi_d^{(l)})$ for $l = 0, \ldots, N$. Clearly,

$$f(\psi^{(l-1)}) \le f(\psi^{(l)}) \text{ for } l \in \mathbb{N} \Rightarrow \lim_{l \to \infty} f(\psi^{(l)}) = M. \tag{14.31}$$

Usually, the sequence $\psi^{(l)}, l = 0, \ldots,$ will converge to 1-semi maximum point $\phi = (\phi_1, \ldots, \phi_d) \in \Psi$. That is, $f(\phi) = \max_{\psi_i \in \Psi} f((\psi_i, \hat{\phi}_i))$ for $i \in [d]$. Note that if f is differentiable at ϕ then ϕ is a critical point of f. Assume that f is twice differentiable at ϕ. Then ϕ does not have to be a local maximum point [15, Appendix].

The *modified alternating maximization method*, abbreviated as MAMM, is as follows. Assume that we start with an initial point $\psi^{(0)} = (\psi_1^{(0)}, \ldots, \psi_d^{(0)})$. Let $\psi^{(0)} = (\psi_i^{(0)}, \hat{\psi}_i^{(0)})$ for $i \in [d]$. Compute $f_{i,0} = \max_{\psi_i \in \Psi_i} f((\psi_i, \hat{\psi}_i^{(0)}))$ for $i \in [d]$. Let $j_1 \in \arg\max_{i \in [d]} f_{i,0}$. Then $\psi^{(1)} = (\psi_j^\star(\hat{\psi}_{j_1}^{(0)}), \hat{\psi}_{j_1}^{(0)})$ and $f_1 = f_{1,j_1} = f(\psi^{(1)})$. Note that it takes d iterations to compute $\psi^{(1)}$. Now replace $\psi^{(0)}$ with $\psi^{(1)}$ and compute $f_{i,1} = \max_{\psi_i \in \Psi_i} f((\psi_i, \hat{\psi}_i^{(1)}))$ for $i \in [d] \setminus \{j_1\}$. Continue as above to find

$\psi^{(l)}$ for $l = 2, \ldots, N$. Note that for $l \geq 2$ it takes $d - 1$ iterations to determine $\psi^{(l)}$. Clearly, (14.31) holds. It is shown in [15] that the limit ϕ of each convergent subsequence of the points $\psi^{(j)}$ is 1-semi maximum point of f.

In certain very special cases, as for best rank one approximation, we can solve the maximization problem with respect to any pair of variables ψ_i, ψ_j for $1 \leq i < j \leq d$, where $d \geq 3$ and all other variables are fixed. Let

$$\hat{\Psi}_{i,j} := \Psi_1 \times \cdots \times \Psi_{i-1} \times \Psi_{i+1} \times \cdots \times \Psi_{j-1} \times \Psi_{j+1} \times \cdots \times \Psi_d,$$
$$\hat{\psi}_{i,j} = (\psi_1, \ldots, \psi_{i-1}, \psi_{i+1}, \ldots, \psi_{j-1}, \psi_{j+1}, \ldots, \psi_d) \in \hat{\Psi}_{i,j},$$
$$\psi_{i,j} = (\psi_i, \psi_j) \in \Psi_i \times \Psi_j.$$

View $\psi = (\psi_1, \ldots, \psi_d)$ as $(\psi_{i,j}, \hat{\psi}_{i,j})$ for each pair $1 \leq i < j \leq d$. Then

$$\max_{\psi_{i,j} \in \Psi_i \times \Psi_j} f((\psi_{i,j}, \hat{\psi}_{i,j})) = f((\psi^{\star}_{i,j}(\hat{\psi}_{i,j}), \hat{\psi}_{i,j})). \tag{14.32}$$

A point ψ is called *2-semi maximum point* if the above maximum equals to $f(\psi)$ for each pair $1 \leq i < j \leq d$.

The *2-alternating maximization method*, abbreviated here as 2AMM, is as follows. Assume that we start with an initial point $\psi^{(0)} = (\psi_1^{(0)}, \ldots, \psi_d^{(0)})$. Suppose first that $d = 3$. Then we consider the maximization problem (14.32) for $i = 2, j = 3$ and $\hat{\psi}_{2,3} = \psi_1^{(0)}$. Let $(\psi_2^{(0,1)}, \psi_3^{(0,1)}) = \psi^{\star}_{2,3}(\psi_1^{(0)})$. Next let $i = 1, j = 3$ and $\psi_{1,3} = \psi_2^{(0,1)}$. Then $(\psi_1^{(0,2)}, \psi_3^{(0,2)}) = \psi^{\star}_{1,3}(\psi_2^{(0,1)})$. Next let $i = 1, 2$ and $\hat{\psi}_{1,2} = \psi_3^{(0,2)}$. Then $\psi^{(1)} = (\hat{\psi}^{\star}_{1,2}(\psi_3^{(0,2)}), \psi_3^{(0,2)})$. Continue these iterations to obtain $\psi^{(l)}$ for $l = 2, \ldots$. Again, (14.31) holds. Usually the sequence $\psi^{(l)}, l \in \mathbb{N}$ will converge to a 2-semi maximum point ϕ. For $d \geq 4$ the 2AMM can be defined appropriately see [15].

A *modified 2-alternating maximization method*, abbreviated here as M2AMM, is as follows. Start with an initial point $\psi^{(0)} = (\psi_1^{(0)}, \ldots, \psi_d^{(0)})$ viewed as $(\psi_{i,j}^{(0)}, \hat{\psi}_{i,j}^{(0)})$, for each pair $1 \leq i < j \leq d$. Let $f_{i,j,0} := \max_{\psi_{i,j} \in \Psi_i \times \Psi_j} f((\psi_{i,j}, \hat{\psi}_{i,j}^{(0)}))$. Assume that $(i_1, j_1) \in \arg\max_{1 \leq i < j \leq d} f_{i,j,0}$. Then $\psi^{(1)} = (\psi^{\star}_{i_1,j_1}(\hat{\psi}^{(0)}_{i_1,j_1}), \hat{\psi}^{(0)}_{i_1,j_1})$. Let $f_{i_1,j_1,1} := f(\psi^{(1)})$. Note that it takes $\binom{d}{2}$ iterations to compute $\psi^{(1)}$. Now replace $\psi^{(0)}$ with $\psi^{(1)}$ and compute $f_{i,j,1} = \max_{\psi_{i,j} \in \Psi_i \times \Psi_j} f((\psi_{i,j}, \hat{\psi}^{(1)}_{i,j}))$ for all pairs $1 \leq i < j \leq d$ except the pair (i_1, j_1). Continue as above to find $\psi^{(l)}$ for $l = 2, \ldots, N$. Note that for $l \geq 2$ it takes $\binom{d}{2} - 1$ iterations to determine $\psi^{(l)}$. Clearly, (14.31) holds. It is shown in [15] that the limit ϕ of each convergent subsequence of the points $\psi^{(j)}$ is 2-semi maximum point of f.

14.6.2 AMM for Best r-Approximations of Tensors

Let $\mathscr{T} \in \mathbb{R}^{\mathbf{n}}$. For best rank one approximation one searches for the maximum of the function $f_{\mathscr{T}} = \mathscr{T} \times (\otimes_{i\in[d]}\mathbf{x}_i)$ on $\mathrm{S}(\mathbf{n})$, as in (14.22). For best $\mathbf{r}$-approximation one searches for the maximum of the function $f_{\mathscr{T}} = \|P_{\otimes_{i\in[d]}\mathbf{U}_i}\|^2$ on $\mathrm{Gr}(\mathbf{r},\mathbf{n})$, as in (14.26). A solution to the basic maximization problem with respect to one subspace $\mathbf{U}_i$ is given in Sect. 14.5.3.

The AMM for best $\mathbf{r}$-approximation were studied first by de Lathauwer-Moor-Vandewalle [4]. The AMM is called in [4] *alternating least squares*, abbreviated as ALS. A crucial problem is the starting point of AMM. A high order SVD, abbreviated as HOSVD, for $\mathscr{T}$, see [3], gives a good starting point for AMM. That is, let $T_l(\mathscr{T}) \in \mathbb{R}^{n_l\times N_l}$ be the unfolded matrix of $\mathscr{T}$ in the mode l, as in Sect. 14.4.1. Then $\mathbf{U}_l$ is the subspace spanned by the first l-left singular vectors of $T_l(\mathscr{T})$. The complexity of computing $\mathbf{U}_l$ is $\mathscr{O}(r_l N)$, where $N = \prod_{i\in[d]} n_i$. Hence for large N the complexity of computing partial HOSVD is high. Another approach is to choose the starting subspaces at random, and repeat the AMM for several choices of random starting points.

MAMM for best rank one approximation of tensors was introduced by Friedland-Mehrmann-Pajarola-Suter in [15] by the name *modified alternating least squares*, abbreviated as MALS. 2AMM for best rank one approximation was introduced in [15] by the name *alternating SVD*, abbreviated as ASVD. It follows from the observation that $A := \mathscr{T}\times(\otimes_{l\in[d]\setminus\{i,j\}}\mathbf{x}_l)$ is an $n_i \times n_j$ matrix. Hence the maximum of the bilinear form $\mathbf{x}^\top A\mathbf{y}$ on $\mathrm{S}((n_i, n_j))$ is $\sigma_1(A)$. See Sect. 14.2. M2AMM was introduced in [15] by the names MASVD.

We now introduce the following variant of 2AMM for best $\mathbf{r}$-rank approximation, called 2-*alternating maximization method variant* and abbreviated as 2AMMV. Consider the maximization problem for a pair of variables as in (14.32). Since for $\mathbf{r} \neq \mathbf{1}_d$ we do not have a closed solution to this problem, we apply the AMM for two variables ψ_i and ψ_j, while keeping $\hat{\psi}_{i,j}$ fixed. We then continue as in 2AMM method.

14.6.3 Complexity Analysis of AMM for Best r-Approximation

Let $\underline{U} = (\mathbf{U}_1, \ldots, \mathbf{U}_d)$. Assume that

$$\mathbf{U}_i = \mathrm{span}(\mathbf{u}_{1,i}, \ldots, \mathbf{u}_{r_i,n_i}), \quad \mathbf{U}_i^\perp = \mathrm{span}(\mathbf{u}_{r_i+1}, \ldots, \mathbf{u}_{n_i,i}),$$
$$\mathbf{u}_{j,i}^\top \mathbf{u}_{k,i} = \delta_{j,k}, \; j,k \in [n_i], \quad i \in [d]. \tag{14.33}$$

For each $i \in [d]$ compute the symmetric positive semi-definite matrix $A_i(\underline{U}_i)$ given by (14.27). For simplicity of exposition we give the complexity analysis for $d = 3$. To compute $A_1(\underline{U}_1)$ we need first to compute the vectors $\mathscr{T}\times(\mathbf{u}_{j_2,2}\otimes\mathbf{u}_{j_3,3})$ for $j_2 \in [r_2]$ and $j_3 \in [r_3]$. Each computation of such a vector has complexity $\mathscr{O}(N)$, where

$N = n_1n_2n_3$. The number of such vectors is r_2r_3. To form the matrix $A_i(\underline{U}_i)$ we need $\mathcal{O}(r_2r_3n_1^2)$ flops. To find the first r_1 eigenvectors of $A_1(\underline{U}_1)$ we need $\mathcal{O}(r_1n_1^2)$ flops. Assuming that $n_1, n_2, n_3 \approx n$ and $r_1, r_2, r_3 \approx r$ we deduce that we need $O(r^2n^3)$ flops to find the first r_1 orthonormal eigenvectors of $A_1(\underline{U}_1)$ which span $\mathbf{U}_1(\underline{U}_1)$. Hence the complexity of finding orthonormal bases of $\mathbf{U}_1(\underline{U}_1)$ is $\mathcal{O}(r^2n^3)$, which is the complexity of computing $A_1(\underline{U}_1)$. Hence the complexity of each step of AMM for best **r**-approximation, i.e. computing $\psi^{(l)}$, is $\mathcal{O}(r^2n^3)$.

It is possible to reduce the complexity of AMM for best **r**-approximation is to $\mathcal{O}(rn^3)$ if we compute and store the matrices $\mathcal{T} \times \mathbf{u}_{j_1,1}, \mathcal{T} \times \mathbf{u}_{j_2,2}, \mathcal{T} \times \mathbf{u}_{j_3,3}$. See Sect. 14.10.

We now analyze the complexity of AMM for rank one approximation. In this case we need only to compute the vector of the form $\mathbf{v}_i := \mathcal{T} \times (\otimes_{j\in[d]\setminus\{i\}}\mathbf{u}_j)$ for each $i \in [d]$, where $\mathbf{U}_j = \mathrm{span}(\mathbf{u}_j)$ for $j \in [i]$. The computation of each $\mathbf{v}_i$ needs $\mathcal{O}((d-2)N)$ flops, where $N = \prod_{j\in[d]}$. Hence each step of AMM for best rank one approximation is $\mathcal{O}(d(d-2)N)$. So for $d = 3$ and $n_1 \approx n_2 \approx n_3$ the complexity is $\mathcal{O}(n^3)$, which is the same complexity as above with $r = 1$.

14.7 Fixed Points of AMM and Newton Method

Consider the AMM as described in Sect. 14.6.1. Assume that the sequence $\psi^{(l)}, l \in \mathbb{N}$ converges to a point $\phi \in \Psi$. Then ϕ is a fixed point of the map:

$$\tilde{\mathbf{F}} : \Psi \to \Psi, \quad \tilde{\mathbf{F}} = (\tilde{F}_1, \ldots, \tilde{F}_d), \; \tilde{F}_i : \Psi \to \Psi_i, \; \tilde{F}_i(\psi) = \psi_i^\star(\hat{\psi}_i), \; \psi = (\psi_i, \hat{\psi}_i), \; i \in [d]. \tag{14.34}$$

In general, the map $\tilde{\mathbf{F}}$ is a multivalued map, since the maximum given in (14.30) may be achieved at a number of points denoted by $\psi_i^\star(\hat{\psi}_i)$. In what follows we assume:

Assumption 1 *The AMM converges to a fixed point ϕ of $\tilde{\mathbf{F}}$ i.e. $\tilde{\mathbf{F}}(\phi) = \phi$, such that the following conditions hold:*

1. *There is a connected open neighborhood $O \subset \Psi$ such that $\tilde{\mathbf{F}} : O \to O$ is one valued map.*
2. *$\tilde{\mathbf{F}}$ is a contraction on O with respect to some norm on O.*
3. *$\tilde{\mathbf{F}} \in \mathrm{C}^2(O)$, i.e. $\tilde{\mathbf{F}}$ has two continuous partial derivatives in O.*
4. *O is diffeomorphic to an open subset in $\mathbb{R}^L$. That is, there exists a smooth one-to-one map $H : O \to \mathbb{R}^L$ such that the Jacobian $D(H)$ is invertible at each point $\psi \in O$.*

Assume that the conditions of Assumption 1 hold. Then the map $\tilde{\mathbf{F}} : O \to O$ can be represented as

$$\mathbf{F} : O_1 \to O_1, \quad \mathbf{F} = H \circ \tilde{\mathbf{F}} \circ H^{-1}, \quad O_1 = H(O).$$

Hence to find a fixed point of $\tilde{\mathbf{F}}$ in O it is enough to find a fixed point of $\mathbf{F}$ in O_1. A fixed point of $\mathbf{F}$ is a zero point of the system

$$\mathbf{G}(\mathbf{x}) = \mathbf{0}, \quad G(\mathbf{x}) := \mathbf{x} - \mathbf{F}(\mathbf{x}). \tag{14.35}$$

To find a zero of $\mathbf{G}$ we use the standard Newton method.

In this paper we propose new Newton methods. We make a few iterations of AMM and switch to a Newton method assuming that the conditions of Assumption 1 hold as explained above. A fixed point of the map $\tilde{\mathbf{F}}$ for best rank one approximation induces a fixed point of map $\mathbf{F} : \mathbb{R}^{\mathbf{n}} \to \mathbb{R}^{\mathbf{n}}$ [15, Lemma 2]. Then the corresponding Newton method to find a zero of $\mathbf{G}$ is straightforward to state and implement, as explained in the next subsection. This Newton method was given in [14, §5]. See also Zhang-Golub [43] for a different Newton method for best $(1, 1, 1)$ approximation.

Let $\tilde{\mathbf{F}} : \mathrm{Gr}(\mathbf{r}, \mathbf{n}) \to \mathrm{Gr}(\mathbf{r}, \mathbf{n})$ be the induced map AMM. Each $\mathrm{Gr}(r, \mathbb{R}^n)$ can be decomposed as a compact manifold to a finite number of charts as explained in Sect. 14.8. These charts induce standard charts of $\mathrm{Gr}(\mathbf{r}, \mathbf{n})$. After a few AMM iterations we assume that the neighborhood O of the fixed point of $\tilde{\mathbf{F}}$ lies in one the charts of $\mathrm{Gr}(\mathbf{r}, \mathbf{n})$. We then construct the corresponding map $\mathbf{F}$ in this chart. Next we apply the standard Newton method to $\mathbf{G}$. The papers by Eldén-Savas [8] and Savas-Lim [39] discuss Newton and quasi-Newton methods for (r_1, r_2, r_3) approximation of 3-tensors using the concepts of differential geometry.

14.7.1 Newton Method for Best Rank One Approximation

Let $\mathcal{T} \in \mathbb{R}^{\mathbf{n}} \setminus \{0\}$. Define:

$$\Psi_i = \mathbb{R}^{n_i},\ i \in [d], \quad \Psi = \mathbb{R}^{n_1} \times \cdots \times \mathbb{R}^{n_d}, \quad \psi = (\mathbf{x}_1, \ldots, \mathbf{x}_d) \in \Psi,$$

$$f_{\mathcal{T}} : \Psi \to \mathbb{R}, \quad f_{\mathcal{T}}(\psi) = \mathcal{T} \times (\otimes_{j \in [d]} \mathbf{x}_j), \tag{14.36}$$

$$\mathbf{F} = (F_1, \ldots, F_d) : \Psi \to \Psi, \quad F_i(\psi) = \mathcal{T} \times (\otimes_{j \in [d] \setminus \{i\}} \mathbf{x}_j), \quad i \in [d]. \tag{14.37}$$

Recall the results of Sect. 14.5.2: Any critical point of $f_{\mathcal{T}}|\mathrm{S}(\mathbf{n})$ satisfies (14.23). Suppose we start the AMM with $\psi^{(0)} = (\mathbf{x}_1^{(0)}, \ldots, \mathbf{x}_d^{(0)}) \in \mathrm{S}(\mathbf{n})$ such that $f_{\mathcal{T}}(\psi^{(0)}) \neq 0$. Then it is straightforward to see that $f_{\mathcal{T}}(\psi^{(l)}) > 0$ for $l \in \mathbb{N}$. Assume that $\lim_{l \to \infty} \psi^{(l)} = \omega = (\mathbf{u}_1, \ldots, \mathbf{u}_d) \in \mathrm{S}(\mathbf{n})$. Then ω is the singular tuple of $\mathcal{T}$ satisfying (14.23). Clearly, $\lambda = f_{\mathcal{T}}(\omega) > 0$. Let

$$\phi = (\mathbf{y}_1, \ldots, \mathbf{y}_d) := \lambda^{-\frac{1}{d-2}} \omega = \lambda^{-\frac{1}{d-2}} (\mathbf{u}_1, \ldots, \mathbf{u}_d). \tag{14.38}$$

Then ϕ is a fixed point of $\mathbf{F}$.

Our Newton algorithm for finding the fixed point ϕ of $\mathbf{F}$ corresponding to a fixed point ω of AMM is as follows. We do a number of iterations of AMM to obtain $\psi^{(m)}$. Then we renormalize $\psi^{(m)}$ according to (14.38):

$$\phi_0 := (f_{\mathscr{T}}(\psi^{(m)}))^{-\frac{1}{d-2}}\psi^{(m)}. \tag{14.39}$$

Let $D\mathbf{F}(\psi)$ denote the Jacobian of $\mathbf{F}$ at ψ, i.e. the matrix of partial derivatives of $\mathbf{F}$ at ψ. Then we perform Newton iterations of the form:

$$\phi^{(l)} = \phi^{(l-1)} - (I - D\mathbf{F}(\phi^{(l-1)}))^{-1}(\phi^{(l-1)} - \mathbf{F}(\phi^{(l-1)})), \quad l \in \mathbb{N}. \tag{14.40}$$

After performing a number of Newton iterations we obtain $\phi^{(m')} = (\mathbf{z}_1, \ldots, \mathbf{z}_d)$ which is an approximation of ϕ. We then renormalize each $\mathbf{z}_i$ to obtain $\omega^{(m')} := (\frac{1}{\|\mathbf{z}_1\|}\mathbf{z}_1, \ldots, \frac{1}{\|\mathbf{z}_d\|}\mathbf{z}_d)$ which is an approximation to the fixed point ω. We call this Newton method *Newton-1*.

We now give the explicit formulas for 3-tensors, where $n_1 = m, n_2 = n, n_3 = l$. First

$$\mathbf{F}(\mathbf{u}, \mathbf{v}, \mathbf{w}) := (\mathscr{T}\times(\mathbf{v}\otimes\mathbf{w}), \mathscr{T}\times(\mathbf{u}\otimes\mathbf{w}), \mathscr{T}\times(\mathbf{u}\otimes\mathbf{v})), \quad \mathbf{G} := (\mathbf{u}, \mathbf{v}, \mathbf{w}) - \mathbf{F}(\mathbf{u}, \mathbf{v}, \mathbf{w}). \tag{14.41}$$

Then

$$D\mathbf{G}(\mathbf{u}, \mathbf{v}, \mathbf{w}) = \begin{bmatrix} I_m & -\mathscr{T}\times\mathbf{w} & -\mathscr{T}\times\mathbf{v} \\ -(\mathscr{T}\times\mathbf{w})^\top & I_n & -\mathscr{T}\times\mathbf{u} \\ -(\mathscr{T}\times\mathbf{v})^\top & -(\mathscr{T}\times\mathbf{u})^\top & I_l \end{bmatrix}. \tag{14.42}$$

Hence Newton-1 iteration is given by the formula

$$(\mathbf{u}_{i+1}, \mathbf{v}_{i+1}, \mathbf{w}_{i+1}) = (\mathbf{u}_i, \mathbf{v}_i, \mathbf{w}_i) - (D\mathbf{G}(\mathbf{u}_i, \mathbf{v}_i, \mathbf{w}_i))^{-1}\mathbf{G}(\mathbf{u}_i, \mathbf{v}_i, \mathbf{w}_i),$$

for $i = 0, 1, \ldots$,. Here we abuse notation by viewing $(\mathbf{u}, \mathbf{v}, \mathbf{w})$ as a column vector $(\mathbf{u}^\top, \mathbf{v}^\top, \mathbf{w}^\top)^\top \in \mathbb{C}^{m+n+l}$.

Numerically, to find $(D\mathbf{G}(\mathbf{u}_i, \mathbf{v}_i, \mathbf{w}_i))^{-1}\mathbf{G}(\mathbf{u}_i, \mathbf{v}_i, \mathbf{w}_i)$ one solves the linear system

$$(D\mathbf{G}(\mathbf{u}_i, \mathbf{v}_i, \mathbf{w}_i))(\mathbf{x}, \mathbf{y}, \mathbf{z}) = \mathbf{G}(\mathbf{u}_i, \mathbf{v}_i, \mathbf{w}_i).$$

The final vector $(\mathbf{u}_j, \mathbf{v}_j, \mathbf{w}_j)$ of Newton-1 iterations is followed by a scaling to vectors of unit length $\mathbf{x}_j = \frac{1}{\|\mathbf{u}_j\|}\mathbf{u}_j, \mathbf{y}_j = \frac{1}{\|\mathbf{v}_j\|}\mathbf{v}_j, \mathbf{z}_j = \frac{1}{\|\mathbf{w}_j\|}\mathbf{w}_j$.

We now discuss the complexity of Newton-1 method for $d = 3$. Assuming that $m \approx n \approx l$ we deduce that the computation of the matrix $D\mathbf{G}$ is $\mathscr{O}(n^3)$. As the dimension of $D\mathbf{G}$ is $m + n + l$ it follows that the complexity of each iteration of Newton-1 method is $\mathscr{O}(n^3)$.

14.8 Newton Method for Best r-Approximation

Recall that an r-dimensional subspace $\mathbf{U} \in \mathrm{Gr}(r, \mathbb{R}^n)$ is given by a matrix $U = [u_{ij}]_{i,j=1}^{n,r} \in \mathbb{R}^{n\times r}$ of rank r. In particular there is a subset $\alpha \subset [n]$ of cardinality r so that $\det U[\alpha, [r]] \neq 0$. Here $\alpha = (\alpha_1, \ldots, \alpha_r), 1 \le \alpha_1 < \ldots < \alpha_r \le n$. So $U[\alpha, [r]] := [u_{\alpha_i j}]_{i,j=1}^{r} \in \mathbf{GL}(r, \mathbb{R})$, (the group of invertible matrices). Clearly, $V := UU[\alpha, [r]]^{-1}$ represents another basis in $\mathbf{U}$. Note that $V[\alpha, [r]] = I_r$. Hence the set of all $V \in \mathbb{R}^{n\times r}$ with the condition: $V[\alpha, [r]] = I_r$ represent an open cell in $\mathrm{Gr}(r, n)$ of dimension $r(n - r)$ denoted by $\mathrm{Gr}(r, \mathbb{R}^n)(\alpha)$. (The number of free parameters in all such V's is $(n - r)r$.) Assume for simplicity of exposition that $\alpha = [r]$. Note that $V_0 = \begin{bmatrix} I_r \\ 0 \end{bmatrix} \in \mathrm{Gr}(r, \mathbb{R}^n)([r])$. Let $\mathbf{e}_i = (\delta_{1i}, \ldots, \delta_{ni})^\top \in \mathbb{R}^n, i = 1, \ldots, n$ be the standard basis in $\mathbb{R}^n$. So $\mathbf{U}_0 = \mathrm{span}(\mathbf{e}_1, \ldots, \mathbf{e}_r) \in \mathrm{Gr}(r, \mathbb{R}^n)([r])$, and V_0 is the unique representative of $\mathbf{U}_0$. Note that $\mathbf{U}_0^\perp$, the orthogonal complement of $\mathbf{U}_0$, is $\mathrm{span}(\mathbf{e}_{r+1}, \ldots, \mathbf{e}_n)$. It is straightforward to see that $\mathbf{V} \in \mathrm{Gr}(r, \mathbb{R}^n)([r])$ if and only if $\mathbf{V} \cap \mathbf{U}_0^\perp = \{\mathbf{0}\}$.

The following definition is a geometric generalization of $\mathrm{Gr}(r, \mathbb{R}^n)(\alpha)$:

$$\mathrm{Gr}(r, \mathbb{R}^n)(\mathbf{U}) := \{\mathbf{V} \in \mathrm{Gr}(r, \mathbb{R}^n),\ \mathbf{V} \cap \mathbf{U}^\perp = \{\mathbf{0}\}\} \text{ for } \mathbf{U} \in \mathrm{Gr}(r, \mathbb{R}^n). \tag{14.43}$$

A basis for $\mathrm{Gr}(r, \mathbb{R}^n)(\mathbf{U})$, which can be identified the tangent hyperplane $T_{\mathbf{U}}\,\mathrm{Gr}(r, \mathbb{R}^n)$, can be represented as $\oplus^r \mathbf{U}^\perp$: Let $\mathbf{u}_1, \ldots, \mathbf{u}_r$ and $\mathbf{u}_{r+1}, \ldots, \mathbf{u}_n$ be orthonormal bases of $\mathbf{U}$ and $\mathbf{U}^\perp$ respectively Then each subspace $\mathbf{V} \in \mathrm{Gr}(r, \mathbb{R}^n)(\mathbf{U})$ has a unique basis of the form $\mathbf{u}_1 + \mathbf{x}_1, \ldots, \mathbf{u}_r + \mathbf{x}_r$ for unique $\mathbf{x}_1, \ldots, \mathbf{x}_r \in \mathbf{U}^\perp$. Equivalently, every matrix $X \in \mathbb{R}^{(n-r)\times r}$ induces a unique subspace $\mathbf{V}$ using the equality

$$[\mathbf{x}_1\ \ldots\ \mathbf{x}_r] = [\mathbf{u}_1\ \ldots\ \mathbf{u}_{n-r}]X \text{ for each } X \in \mathbb{R}^{(n-r)\times r}. \tag{14.44}$$

Recall the results of Sect. 14.5.3. Let $\underline{U} = (U_1, \ldots, U_d) \in \mathrm{Gr}(\mathbf{r}, \mathbf{n})$. Then

$$\tilde{\mathbf{F}} = (\tilde{F}_1, \ldots, \tilde{F}_d) : \mathrm{Gr}(\mathbf{r}, \mathbf{n}) \to \mathrm{Gr}(\mathbf{r}, \mathbf{n}), \quad \tilde{F}_i(\underline{U}) = \mathbf{U}_i(\underline{U}_{\hat{i}}),\ i \in [d], \tag{14.45}$$

where $\mathbf{U}_i(\underline{U}_{\hat{i}})$ a subspace spanned by the first r_i eigenvectors of $A_i(\underline{U}_{\hat{i}})$. Assume that $\tilde{\mathbf{F}}$ is one valued at $\underline{U}$, i.e. (14.28) holds. Then it is straightforward to show that $\tilde{\mathbf{F}}$ is smooth (real analytic) in neighborhood of $\underline{U}$. Assume next that there exists a neighborhood O of $\underline{U}$ such that

$$O \subset \mathrm{Gr}(\mathbf{r}, \mathbf{n})(\underline{U}) := \mathrm{Gr}(r_1, \mathbb{R}^{n_1})(\mathbf{U}_1) \times \cdots \times \mathrm{Gr}(r_d, \mathbb{R}^{n_d})(\mathbf{U}_d), \quad \underline{U} = (\mathbf{U}_1, \ldots, \mathbf{U}_d), \tag{14.46}$$

such that the conditions *1–3* of Assumption 1 hold. Observe next that $\mathrm{Gr}(\mathbf{r},\mathbf{n})(\underline{U})$ is diffeomorphic to

$$\mathbb{R}^L := \mathbb{R}^{(n_1-r_1)\times r_1} \ldots \times \mathbb{R}^{(n_d-r_d)\times r_d}, \quad L = \sum_{i\in[d]} (n_i - r_i) r_i$$

We say that $\tilde{\mathbf{F}}$ is *regular* at $\underline{U}$ if in addition to the above condition the matrix $I - D\tilde{\mathbf{F}}(\underline{U})$ is invertible. We can view $X = [X_1 \ \ldots \ X_d] \in \mathbb{R}^{(n_1-r_1)\times r_1} \ldots \times \mathbb{R}^{(n_d-r_d)\times r_d}$. Then $\tilde{\mathbf{F}}$ on O can be viewed as

$$\mathbf{F}: O_1 \to O_1, \quad O_1 \subset \mathbb{R}^L, \quad \mathbf{F}(X) = [F_1(X), \ldots, F_d(X)], \ X = [X_1 \ \ldots \ X_d] \in \mathbb{R}^L. \tag{14.47}$$

Note that $F_i(X)$ does not depend on X_i for each $i \in [d]$. In our numerical simulations we first do a small number of AMM and then switch to Newton method given by (14.40). Observe that $\underline{U}$ corresponds to $X(\underline{U}) = [X_1(\underline{U}), \ldots, X_d(\underline{U})]$. When referring to (14.40) we identify $X = [X_1, \ldots, X_d]$ with $\phi = (\phi_1, \ldots, \phi_d)$ and no ambiguity will arise.

Note that the case $\mathbf{r} = \mathbf{1}_d$ corresponds to best rank one approximation. The above Newton method in this case is different from Newton method given in Sect. 14.7.1.

14.9 A Closed Formula for $D\mathbf{F}(X(\underline{U}))$

Recall the definitions and results of Sect. 14.5.3. Given $\underline{U}$ we compute $\tilde{F}_i(\underline{U}) = \mathbf{U}_i(\underline{U}_i)$, which is the subspace spanned by the first r_i eigenvectors of $A_i(\underline{U}_i)$, which is given by (14.27), for $i \in [d]$. Assume that (14.33) holds. Let

$$\begin{aligned} &\mathbf{U}_i(\underline{U}_i) = \mathrm{span}(\mathbf{v}_{1,i}, \ldots, \mathbf{v}_{r_i,n_i}), \quad \mathbf{U}_i(\underline{U}_i)^\perp = \mathrm{span}(\mathbf{v}_{r_i+1,i}, \ldots, \mathbf{v}_{n_i,i}), \\ &\mathbf{v}_{j,i}^\top \mathbf{v}_{k,i} = \delta_{jk}, \ j,k \in [n_i], \quad i \in [d]. \end{aligned} \tag{14.48}$$

With each $X = [X_1, \ldots, X_d] \in \mathbb{R}^L$ we associate the following point $(\mathbf{W}_1, \ldots, \mathbf{W}_d) \in \mathrm{Gr}(\mathbf{r},\mathbf{n})(\underline{U})$. Suppose that $X_i = [x_{pq,i}] \in \mathbb{R}^{(n_i-r_i)\times r_i}$. Then $\mathbf{W}_i$ has a basis of the form

$$\mathbf{u}_{j_i,i} + \sum_{k_i\in[n_i-r_i]} x_{k_i j_i,i} \mathbf{u}_{r_i+k_i,i}, \quad j_i \in [r_i].$$

One can use the following notation for a basis $\mathbf{w}_{1,i}, \ldots, \mathbf{w}_{r_i,i}$, written as a vector with vector coordinates $[\mathbf{w}_{1,i} \cdots \mathbf{w}_{r_i,i}]$:

$$[\mathbf{w}_{1,i} \cdots \mathbf{w}_{r_i,i}] = [\mathbf{u}_{1,i} \cdots \mathbf{u}_{r_i,i}] + [\mathbf{u}_{r_i+1,i} \cdots \mathbf{u}_{n_i,i}] X_i, \quad i \in [d]. \tag{14.49}$$

Note that to the point $\underline{U} \in \mathrm{Gr}(\mathbf{r},\mathbf{n})(\underline{U})$ corresponds the point $X = 0$. Since $\mathbf{u}_{1,i},\ldots,\mathbf{u}_{n_i,i}$ is a basis in $\mathbb{R}^{n_i}$ it follows that

$$[\mathbf{v}_{1,i}\cdots\mathbf{v}_{r_i,i}] = [\mathbf{u}_{1,i}\cdots\mathbf{u}_{r_i,i}]Y_{i,0} + [\mathbf{u}_{r_i+1,i},\ldots,\mathbf{u}_{n_i,i}]X_{i,0} = [\mathbf{u}_{1,i}\cdots\mathbf{u}_{n_i,i}]Z_{i,0},$$

$$Y_{i,0} \in \mathbb{R}^{r_i\times r_i},\ X_{i,0} \in \mathbb{R}^{(n_i-r_i)\times r_i},\ Z_{i,0} = \begin{bmatrix} Y_{i,0} \\ X_{i,0} \end{bmatrix} \in \mathbb{R}^{n_i\times r_i},\ \text{for } i \in [d]. \quad (14.50)$$

View $[\mathbf{v}_{1,i}\cdots\mathbf{v}_{r_i,i}]$, $[\mathbf{u}_{1,i}\cdots\mathbf{u}_{n_i,i}]$ as $n_i \times r_i$ and $n_i \times n_i$ matrices with orthonormal columns. Then

$$Z_{i,0} = [\mathbf{u}_{1,i}\cdots\mathbf{u}_{n_i,i}]^\top[\mathbf{v}_{1,i}\cdots\mathbf{v}_{r_i,i}], \quad i \in [d]. \quad (14.51)$$

The assumption that $\tilde{\mathbf{F}} : O \to O$ implies that $Y_{i,0}$ is an invertible matrix. Hence $[\mathbf{v}_{1,i}\cdots\mathbf{v}_{r_i,i}]Y_{i,0}^{-1}$ is also a basis in $\mathbf{U}_i(\underline{U}_i)$. Clearly,

$$[\mathbf{v}_{1,i}\cdots\mathbf{v}_{r_i,i}]Y_{i,0}^{-1} = [\mathbf{u}_{1,i}\cdots\mathbf{u}_{r_i,n_i}] + [\mathbf{u}_{r_i+1,i},\ldots,\mathbf{u}_{n_i,i}]X_{i,0}Y_{i,0}^{-1}, \quad i \in [d].$$

Hence $\tilde{\mathbf{F}}(\underline{U})$ corresponds to $\mathbf{F}(0)$ where

$$F_i(0) = X_{i,0}Y_{i,0}^{-1},\ i \in [d], \quad \mathbf{F}(0) = (F_1(0),\ldots,F_d(0)). \quad (14.52)$$

We now find the matrix of derivatives. So $D_iF_j \in \mathbb{R}^{((n_i-r_i)r_i\times((n_j-r_j)r_j))}$ is the partial derivative matrix of $(n_j - r_j)r_j$ coordinates of F_j with respect to $(n_i - r_i)r_i$ the coordinates of $\mathbf{U}_i$ viewed as the matrix $\begin{bmatrix} I_{r_i} \\ G_i \end{bmatrix}$. So $G_i \in \mathbb{R}^{(n_i-r_i)\times r_i}$ are the variables representing the subspace $\mathbf{U}_i$. Observe first that $D_iF_i = 0$ just as in Newton method for best rank one approximation in Sect. 14.7.1.

Let us now find $D_iF_j(0)$. Recall that $D_iF_j(0)$ is a matrix of size $(n_i - r_i)r_i \times (n_j - r_j)r_j$. The entries of $D_iF_j(0)$ are indexed by $((p,q),(s,t))$ as follows: The entries of $G_i = [g_{pq,i}] \in \mathbb{R}^{(n_i-r_i)\times r_i}$ are viewed as $(n_i - r_i)r_i$ variables, and are indexed by (p,q), where $p \in [n_i - r_i], q \in [r_i]$. F_j is viewed as a matrix $G_j \in \mathbb{R}^{(n_j-r_j)\times r_j}$.The entries of F_j are indexed by (s,t), where $s \in [n_j - r_j]$ and $t \in [r_j]$. Since $\underline{U} \in \mathrm{Gr}(\mathbf{r},\mathbf{n})(\underline{U})$ corresponds to $0 \in \mathbb{R}^L$ we denote by $A_j(0)$ the matrix $A_j(\underline{U}_j)$ for $j \in [d]$. We now give the formula for $\frac{\partial A_j(0)}{\partial g_{pq,i}}$. This is done by noting that we vary $\mathbf{U}_i$ by changing the orthonormal basis $\mathbf{u}_{1,i},\ldots,\mathbf{u}_{r_i,i}$ up to the first perturbation with respect to the real variable ε to

$$\hat{\mathbf{u}}_{1,i} = \mathbf{u}_{1,i},\ldots,\hat{\mathbf{u}}_{q-1,i} = \mathbf{u}_{q-1,i},\hat{\mathbf{u}}_{q,i} = \mathbf{u}_{q,i} + \varepsilon\mathbf{u}_{r_i+p,i},\hat{\mathbf{u}}_{q+1,i} = \mathbf{u}_{q+1,i},\ldots,\hat{\mathbf{u}}_{r_i,i} = \mathbf{u}_{r_i,i}$$

We denote the subspace spanned by these vectors as $\mathbf{U}_i(\varepsilon,p,q)$. That is, we change only the q orthonormal vector of the standard basis in $\mathbf{U}_i$, for $q = 1,\ldots,r_i$. The new basis is an orthogonal basis, and up order ε, the vector $\mathbf{u}_{q,i} + \varepsilon\mathbf{u}_{r_i+p,i}$ is also of length 1. Let $\underline{U}(\varepsilon,i,p,q) = (\mathbf{U}_1,\ldots,\mathbf{U}_{i-1},\mathbf{U}_i(\varepsilon,p,q),\mathbf{U}_{i+1},\ldots,\mathbf{U}_d)$. Then

$\underline{U}(\varepsilon, i, p, q)_j$ is obtained by dropping the subspace $\mathbf{U}_j$ from $\underline{U}(\varepsilon, i, p, q)$. We will show that

$$A_j(\underline{U}(\varepsilon, i, p, q)_j) = A_j(\underline{U}_j) + \varepsilon B_{j,i,p,q} + O(\varepsilon^2). \tag{14.53}$$

We now give a formula to compute $B_{j,i,p,q}$. Assume that $i, j \in [d]$ is a pair of different integers. Let J be a set of $d-2$ pairs $\cup_{l\in[d]\setminus\{i,j\}}\{(k_l, l)\}$, where $k_l \in [r_l]$. Denote by $\mathscr{J}_{ij}$ the set of all such J's. Note that $\mathscr{J}_{ij} = \mathscr{J}_{ji}$. Furthermore, the number of elements in $\mathscr{J}_{ij}$ is $R_{ij} = \prod_{l\in[d]\setminus\{i,j\}} r_l$. We now introduce the following matrices

$$C_{ij}(J) := \mathscr{T} \times (\otimes_{(k,l)\in J}\mathbf{u}_{k,l}) \in \mathbb{R}^{n_i\times n_j}, \quad J \in \mathscr{J}_{ij}. \tag{14.54}$$

Note that $C_{ij}(J) = C_{ji}(J)^\top$.

Lemma 1 *Let $i, j \in [d], i \neq j$. Assume that $p \in [n_i - r_i], q \in [r_i]$. Then* (14.53) *holds. Furthermore*

$$A_j(\underline{U}_j) = \sum_{k\in[r_i],J\in\mathscr{J}_{ji}} (C_{ji}(J)\mathbf{u}_{k,i})(C_{ji}(J)\mathbf{u}_{k,i})^\top, \tag{14.55}$$

$$B_{j,i,p,q} = \sum_{J\in\mathscr{J}_{ji}} (C_{ji}(J)\mathbf{u}_{k_i+p,i})(C_{ji}(J)\mathbf{u}_{q,i})^\top + (C_{ji}(J)\mathbf{u}_{q,i})(C_{ji}(J)\mathbf{u}_{k_i+p,i})^\top \tag{14.56}$$

Proof The identity of (14.55) is just a restatement of (14.27). To compute $A_j(\underline{U}(\varepsilon, i, p,$
$q)_j)$ use (14.54) by replacing $u_{k,i}$ with $\hat{u}_{k,i}$ for $k \in [n_i]$. Deduce first (14.53) and then (14.56). □

Recall that $\mathbf{v}_{1,j}, \ldots, \mathbf{v}_{r_j,j}$ is an orthonormal basis of $\mathbf{U}_j(\underline{U}_j)$, and these vectors are the eigenvectors $A_j(\underline{U}_j)$ corresponding its first r_j eigenvalues. Let $\mathbf{v}_{r_j+1,j}, \ldots, \mathbf{v}_{n_j,i}$ be the last $n_j - r_j$ orthonormal eigenvectors of $A_j(\underline{U}_j)$. We now find the first perturbation of the first r_ι eigenvectors for the matrix $A_j(\underline{U}_j) + \varepsilon B_{j,i,p,q}$. Assume first, for simplicity of exposition, that each $\lambda_k(A_j(\underline{U}_j))$ is simple for $k \in [r_j]$: Then it is known, e.g. [10, Chapter 4, §19, (4.19.2)]:

$$\mathbf{v}_{k,j}(\varepsilon, i, p, q) = \mathbf{v}_{k,j} + \varepsilon(\lambda_k(A_j(\underline{U}_j))I_{n_j} - A_j(\underline{U}_j))^\dagger B_{j,i,p,q}\mathbf{v}_{k,j} + O(\varepsilon^2), k \in [r_j]. \tag{14.57}$$

The assumption that $\lambda_k(A_j(\underline{U}_j))$ is a simple eigenvalue for $k \in [r_j]$ yields

$$(\lambda_k(A_j(\underline{U}_i))I_{n_j} - A_j(\underline{U}_j))^\dagger \mathbf{y} = \sum_{l\in[n_j]\setminus\{k\}} \frac{1}{\lambda_k(A_j(\underline{U}_j)) - \lambda_l(A_j(\underline{U}_j))}(\mathbf{v}_{l,j}^\top \mathbf{y})\mathbf{v}_{l,j},$$

for $\mathbf{y} \in \mathbb{R}^{n_j}$.

Since we are interested in a basis of $\mathbf{U}_j(\underline{U}(\varepsilon,i,p,q)_j)$ up to the order of ε we can assume that this basis is of the form

$$\tilde{\mathbf{v}}_{k,j}(\varepsilon,i,p,q)=\mathbf{v}_{k,j}+\varepsilon\mathbf{w}_{k,j}(i,p,q),\quad \mathbf{w}_{k,j}(i,p,q)\in\mathrm{span}(\mathbf{v}_{r_j+1,j}\ldots,\mathbf{v}_{n_j,j}).$$

Hence

$$\mathbf{w}_{k,j}(i,p,q)=\sum_{l\in[n_j]\setminus[r_j]}\frac{1}{\lambda_k(A_j(\underline{U}_j))-\lambda_l(A_j(\underline{U}_j))}(\mathbf{v}_{l,j}^\top\mathbf{c}_{k,j,i,p,q})\mathbf{v}_{l,j},$$

$$\mathbf{c}_{k,j,i,p,q}:=B_{j,i,p,q}\mathbf{v}_{k,j}. \tag{14.58}$$

Note that the assumption (14.28) yields that $\mathbf{w}_{k,j}$ is well defined for $k\in[r_j]$. Let

$$W_j(i,p,q)=[\mathbf{w}_{1,j}(i,p,q)\cdots\mathbf{w}_{r_j,j}(i,p,q)]=\begin{bmatrix}V_j(i,p,q)\\U_j(i,p,q)\end{bmatrix},$$

$$V_j(i,p,q)\in\mathbb{R}^{r_j\times r_j},\quad U_j(i,p,q)\in\mathbb{R}^{(n_j-r_j)\times r_j}.$$

Up to the order of ε we have that a basis of $\mathbf{U}_j(\underline{U}(\varepsilon,i,p,q)_j)$ is given by columns of matrix $Z_{j,0}+\varepsilon W_j(i,p,q)=\begin{bmatrix}Y_{j,0}+\varepsilon V_j(i,p,q)\\X_{j,0}+\varepsilon U_j(i,p,q)\end{bmatrix}$. Note that

$$\begin{aligned}&(Z_{j,0}+\varepsilon W_j(i,p,q))(Y_{j,0}+\varepsilon V_j(i,p,q))^{-1}\\&=\begin{bmatrix}I_{r_j}\\(X_{j,0}+\varepsilon U_j(i,p,q))(Y_{j,0}+\varepsilon V_j(i,p,q))^{-1}\end{bmatrix}.\end{aligned}$$

Observe next

$$\begin{aligned}&Y_{j,0}+\varepsilon V_j(i,p,q)=Y_{j,0}(I_{r_j}+\varepsilon Y_{j,0}^{-1}V_j(i,p,q)),\\&(Y_{j,0}+\varepsilon V_j(i,p,q))^{-1}=(I_{r_j}+\varepsilon Y_{j,0}^{-1}V_j(i,p,q))^{-1}Y_{j,0}^{-1}=\\&Y_{j,0}^{-1}-\varepsilon Y_{j,0}^{-1}V_j(i,p,q)Y_{j,0}^{-1}+O(\varepsilon^2),\\&(X_{j,0}+\varepsilon U_j(i,p,q))(Y_{j,0}+\varepsilon V_j(i,p,q))^{-1}=\\&X_{j,0}Y_{j,0}^{-1}+\varepsilon(U_j(i,p,q)Y_{j,0}^{-1}-X_{j,0}Y_{j,0}^{-1}V_j(i,p,q)Y_{j,0}^{-1})+O(\varepsilon^2).\end{aligned}$$

Hence

$$\frac{\partial F_j}{\partial g_{pq,i}}(0)=U_j(i,p,q)Y_{j,0}^{-1}-X_{j,0}Y_{j,0}^{-1}V_j(i,p,q)Y_{j,0}^{-1}. \tag{14.59}$$

Thus $D\mathbf{F}(0) = [D_i F_J]_{i,j\in[d]} \in \mathbb{R}^{L\times L}$. We now make one iteration of Newton method given by (14.40) for $l = 1$, where $\phi^{(0)} = 0$:

$$\phi^{(1)} = -(I - D\mathbf{F}(0))^{-1}F(0), \quad \phi^{(1)} = [X_{1,1}, \ldots, X_{d,1}] \in \mathbb{R}^L. \tag{14.60}$$

Let $\mathbf{U}_{i,1} \in \mathrm{Gr}(r_i, \mathbb{R}^{n_i})$ be the subspace represented by the matrix $X_{i,1}$:

$$\mathbf{U}_{i,1} = \mathrm{span}(\tilde{\mathbf{u}}_{1,i,1}, \ldots, \tilde{\mathbf{u}}_{r_i,i,1}), \ [\tilde{\mathbf{u}}_{1,i,1}, \ldots, \tilde{\mathbf{u}}_{n_i,i,1}] = [\mathbf{u}_{1,i}, \ldots, \mathbf{u}_{n_i,i}] \begin{bmatrix} I_{r_i} \\ X_{j,1} \end{bmatrix} \tag{14.61}$$

for $i \in [d]$. Perform the Gram-Schmidt process on $\tilde{\mathbf{u}}_{1,i,1}, \ldots, \tilde{\mathbf{u}}_{r_i,i,1}$ to obtain an orthonormal basis $\mathbf{u}_{1,i,1}, \ldots, \mathbf{u}_{r_i,i,1}$ of $\mathbf{U}_{i,1}$. Let $\underline{U} := (\mathbf{U}_{1,1}, \ldots, \mathbf{U}_{d,1})$ and repeat the algorithm which is described above. We call this Newton method *Newton-2*.

14.10 Complexity of Newton-2

In this section we assume for simplicity that $d = 3$, $r_1 = r_2 = r_3 = r$, $n_i \approx n$ for $i \in [3]$. We assume that executed a number of times the AMM for a given $\mathcal{T} \in \mathbb{R}^{\mathbf{n}}$. So we are given $\underline{U} = (\mathbf{U}_1, \ldots, \mathbf{U}_d)$, and an orthonormal basis $\mathbf{u}_{1,i}, \ldots, \mathbf{u}_{r,i}$ of $\mathbf{U}_i$ for $i \in [d]$. First we complete each $\mathbf{u}_{1,i}, \ldots, \mathbf{u}_{r,i}$ to an orthonormal basis $\mathbf{u}_{1,i}, \ldots, \mathbf{u}_{n_i,i}$ of $\mathbb{R}^{n_i}$, which needs $\mathcal{O}(n^3)$ flops. Since $d = 3$ we still need only $\mathcal{O}(n^3)$ to carry out this completion for each $i \in [3]$.

Next we compute the matrices $C_{ij}(J)$. Since $d = 3$, we need n flops to compute each entry of $C_{ij}(J)$. Since we have roughly n^2 entries, the complexity of computing $C_{ij}(J)$ is $\mathcal{O}(n^3)$. As the cardinality of $\mathcal{J}_{ij}$ is r we need $\mathcal{O}(rn^3)$ flops to compute all $C_{ij}(J)$ for $J \in \mathcal{J}_{ij}$. As the number of pairs in [3] is 3 it follows that the complexity of computing all $C_{ij}(J)$ is $\mathcal{O}(rn^3)$.

The identity (14.55) yields that the complexity of computing $A_j(\underline{U}_j)$ is $\mathcal{O}(r^2n^2)$. Recall next that $A_j(\underline{U}_j)$ is $n_j \times n_j$ symmetric positive semi-definite matrix. The complexity of computations of the eigenvalues and the orthonormal eigenvectors of $A_j(\underline{U}_j)$ is $\mathcal{O}(n^3)$. Hence the complexity of computing $\underline{U}$ is $\mathcal{O}(rn^3)$, as we pointed out at the end of Sect. 14.6.3.

The complexity of computing $B_{j,i,p.q}$ using (14.56) is $\mathcal{O}(rn^2)$. The complexity of computing $\mathbf{w}_{k,j}(i, p, q)$, given by (14.58) is $\mathcal{O}(n^2)$. Hence the complexity of computing $W_j(i, p, q)$ is $\mathcal{O}(rn^2)$. Therefore the complexity of computing $D_i F_j$ is $\mathcal{O}(r^2n^3)$. Since $d = 3$, the complexity of computing the matrix $D\mathbf{F}(0)$ is also $\mathcal{O}(r^2n^3)$.

As $D\mathbf{F}(0) \in \mathbb{R}^{L\times L}$, where $L \approx 3rn$, the complexity of computing $(I - D\mathbf{F}(0))^{-1}$ is $\mathcal{O}(r^3n^3)$. In summary, the complexity of one step in Newton-2 is $\mathcal{O}(r^3n^3)$.

14.11 Numerical Results

We have implemented a Matlab library tensor decomposition using Tensor Toolbox given by [30]. The performance was measured via the actual CPU-time (seconds) needed to compute. All performance tests have been carried out on a 2.8 GHz Quad-Core Intel Xeon Macintosh computer with 16 GB RAM. The performance results are discussed for real data sets of third-order tensors. We worked with a real computer tomography (CT) data set (the so-called MELANIX data set of OsiriX) [15].

Our simulation results are averaged over 10 different runs of the each algorithm. In each run, we changed the initial guess, that is, we generated new random start vectors. We always initialized the algorithms by random start vectors, because this is cheaper than the initialization via HOSVD. We note here that for Newton methods our initial guess is the subspaces returned by one iteration of AMM method.

All the alternating algorithms have the same stopping criterion where convergence is achieved if one of the two following conditions are met: $iterations > 10; fitchange < 0.0001$ is met. All the Newton algorithms have the same stopping criterion where convergence is achieved if one of the two following conditions are met: $iterations > 10; change < \exp(-10)$.

Our numerical simulations demonstrate the well known fact that for large size tensors Newton methods are not efficient. Though the Newton methods converge in fewer iterations than alternating methods, the computation associated with the matrix of derivatives (Jacobian) in each iteration is too expensive making alternating maximization methods much more cost effective. Our simulations also demonstrate that our Newton-1 for best rank one approximation is as fast as AMM methods. However our Newton-2 is much slower than alternating methods. We also give a comparison between our Newton-2 and the Newton method based on Grassmannian manifold by [8], abbreviated as Newton-ES.

We also observe that for large tensors and large rank approximation two alternating maximization methods, namely MAMM and 2AMMV, seem to outperform the other alternating maximization methods. We would recommend Newton-1 for rank one approximation in case of rank one approximation both for large and small sized tensors. For higher rank approximation we recommend 2AMMV in case of large size tensors and AMM or MAMM in case of small size tensors.

Our Newton-2 performs a bit slower than Newton-ES, however we would like to point couple of advantages. Our method can be easily extendable to higher dimensions (for $d > 3$ case) both analytically and numerically compared to Newton-ES. Our method is also highly parallelizable which can bring down the computation time drastically. Computation of $D_i F_j$ matrices in each iteration contributes to about 50 % of the total time, which however can be parallelizable. Finally the number of iterations in Newton-2 is at least 30 % less than in Newton-ES (Figs. 14.1–14.3).

It is not only important to check how fast the different algorithms perform but also what quality they achieve. This was measured by checking the Hilbert-Schmidt

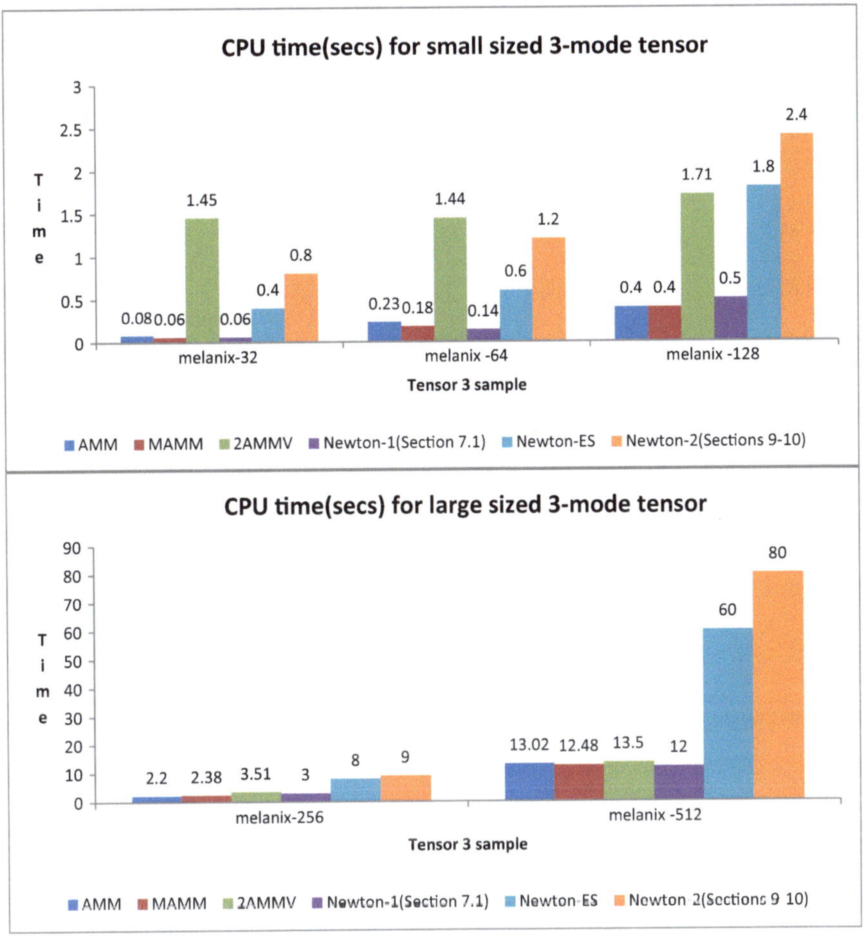

Fig. 14.1 Rank (1, 1, 1) average CPU times taken over 10 initial random guesses

norm, abbreviated as HS norm, of the resulting decompositions, which serves as a measure for the quality of the approximation. In general, we can say that the higher the HS norm, the more likely it is that we find a global maximum. Accordingly, we compared the HS norms to say whether the different algorithms converged to the same stationary point. In Figs. 14.4 and 14.5, we show the average HS norms achieved by different algorithms and compared them with the input norm. We observe all the algorithms seem to attain the same local maximum.

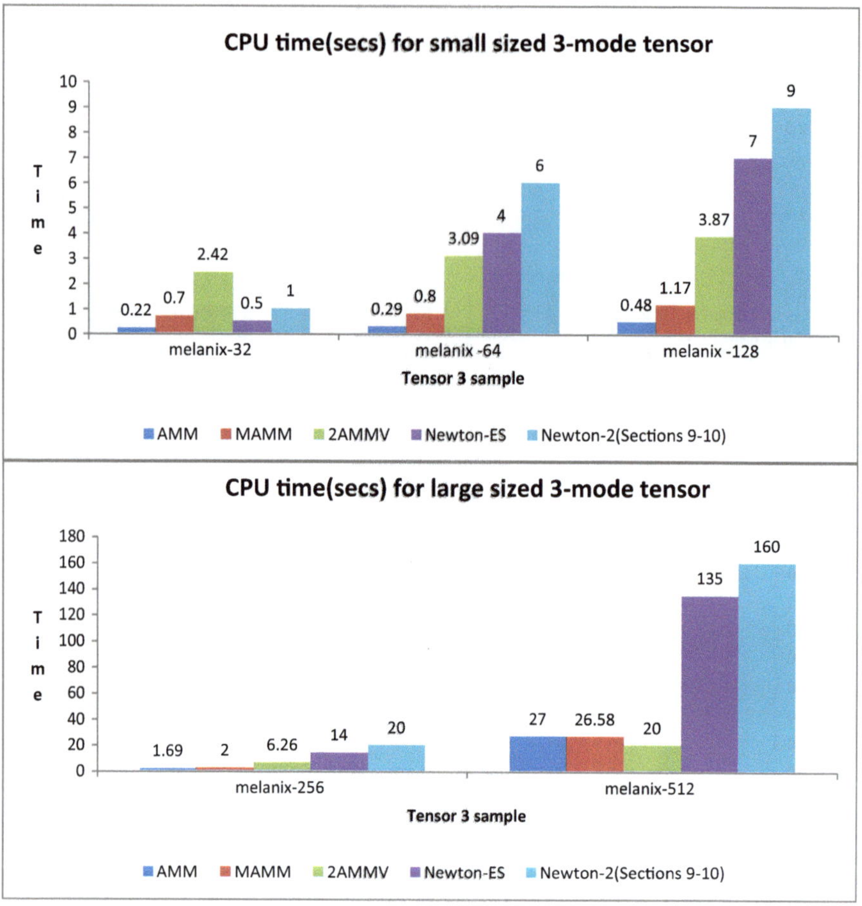

Fig. 14.2 Rank (2, 2, 2) average CPU times taken over 10 initial random guesses

14.11.1 Best (2,2,2) and Rank Two Approximations

Assume that $\mathscr{T}$ is a 3-tensor of rank three at least and let $\mathscr{S}$ be a best $(2,2,2)$-approximation to $\mathscr{T}$ given by (14.1). It is easy to show that $\mathscr{S}$ has at least rank 2. Let $\mathscr{S}' = [s_{j_1,j_2,j_3}] \in \mathbb{R}^{2\times 2\times 2}$ be the core tensor corresponding to $\mathscr{S}$. Clearly rank $\mathscr{S}$ = rank $\mathscr{S}' \geq 2$. Recall that a real nonzero $2\times 2\times 2$ tensor has rank one, two or three [41]. So rank $\mathscr{S} \in \{2,3\}$. Observe next that if rank $\mathscr{S}$ = rank $\mathscr{S}' = 2$ then $\mathscr{S}$ is also a best rank two approximation of $\mathscr{T}$. Recall that a best rank two approximation of $\mathscr{T}$ may not always exist. In particular where rank $\mathscr{T} > 2$ and the border rank of $\mathscr{T}$ is 2 [6]. In all our numerical simulations for best $(2,2,2)$-approximation we performed on random large tensors, the tensor $\mathscr{S}'$ had rank two.

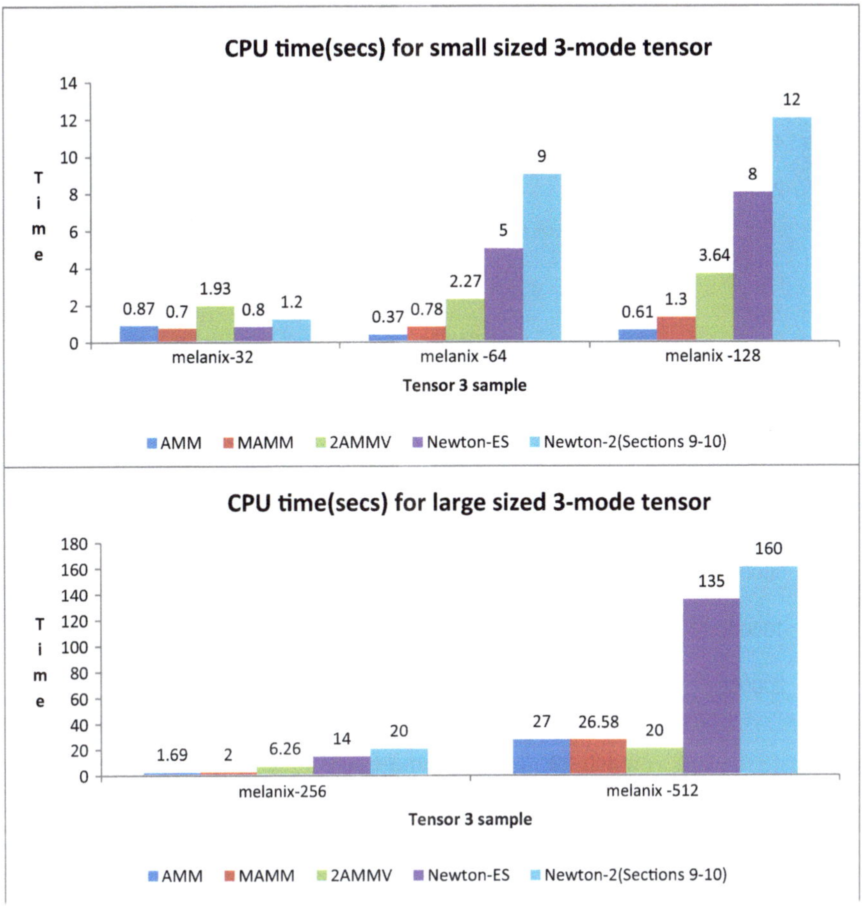

Fig. 14.3 Rank (3, 3, 3) average CPU times taken over 10 initial random guesses

Note that the probability of $2 \times 2 \times 2$ tensors, with entries normally distributed with mean 0 and variance 1, to have rank 2 is $\frac{\pi}{4}$ [2].

14.12 Conclusions

We have extended the alternating maximization method (AMM) and modified alternating maximization method (MAMM) given in [15] for the computation of best rank one approximation to best **r**-approximations. We have also presented new algorithms such as 2-alternating maximization method variant (2AMMV) and Newton method for best **r**-approximation (Newton-2). We have provided closed form solutions for computing the DF matrix in Newton-2. We implemented Newton-1

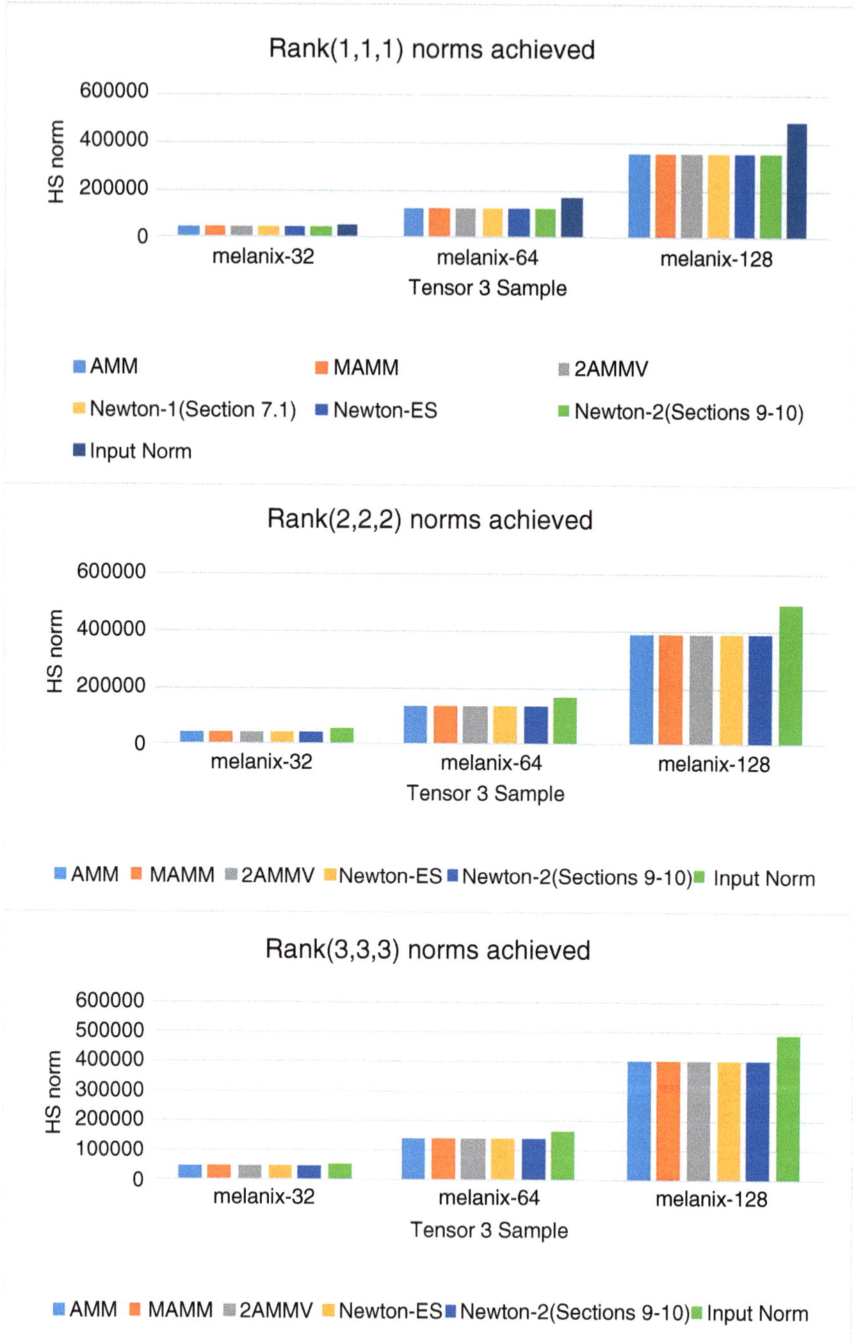

Fig. 14.4 Average Hilbert-Schmidt (HS) norms per algorithm and per data set taken over 10 different initial random guesses for small and medium size tensors

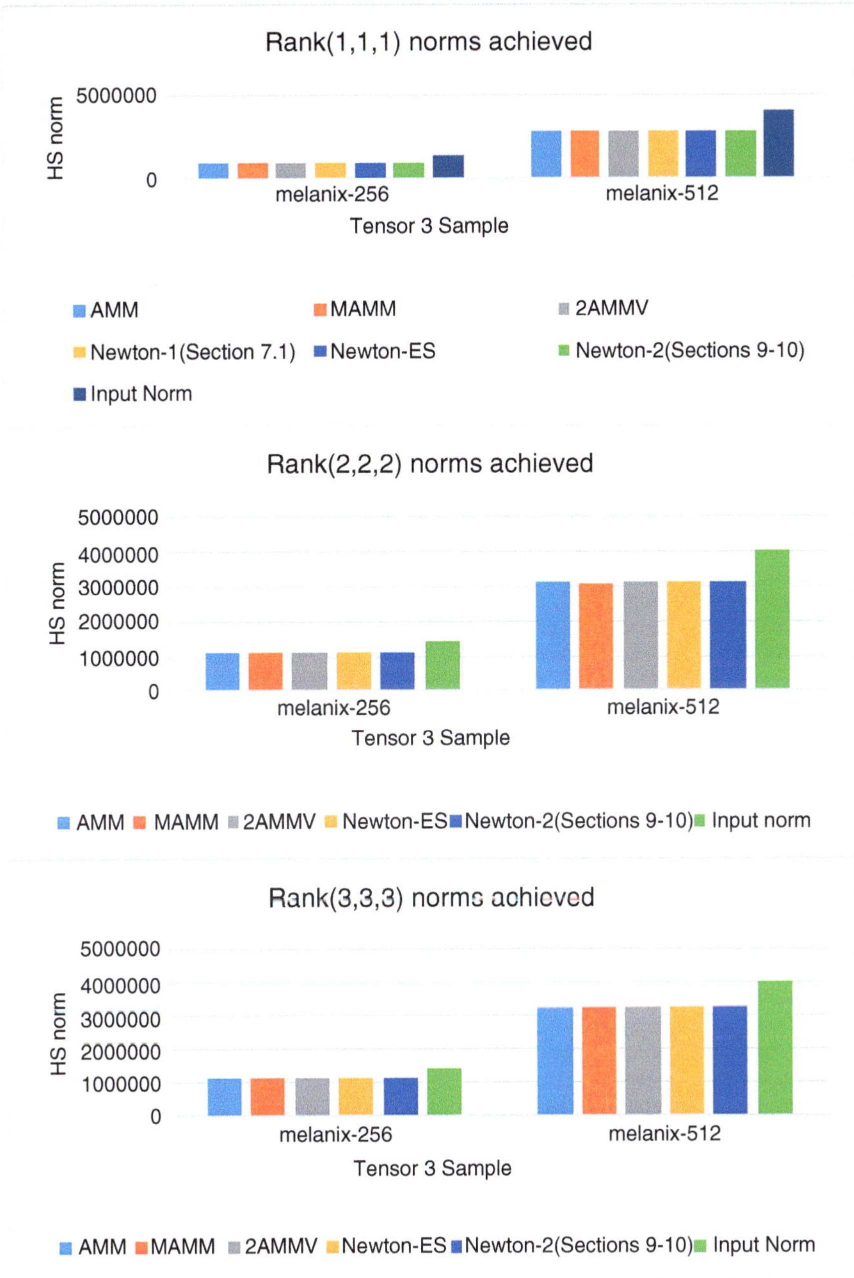

Fig. 14.5 Average Hilbert-Schmidt (HS) norms per algorithm and per data set taken over 10 different initial random guesses for large size tensors

for best rank one approximation [14] and Newton-2. From the simulations, we have found out that for rank one approximation of both large and small sized tensors, Newton-1 performed the best. For higher rank approximation, the best performers were 2AMMV in case of large size tensors and AMM or MAMM in case of small size tensors.

Acknowledgements We thank Daniel Kressner for his remarks.

References

1. Achlioptas, D., McSherry, F.: Fast computation of low rank approximations. In: Proceedings of the 33rd Annual Symposium on Theory of Computing, Heraklion, pp. 1–18 (2001)
2. Bergqvist, G.: Exact probabilities for typical ranks of $2 \times 2 \times 2$ and $3 \times 3 \times 2$ tensors. Linear Algebra Appl. **438**, 663–667 (2013)
3. de Lathauwer, L., de Moor, B., Vandewalle, J.: A multilinear singular value decomposition. SIAM J. Matrix Anal. Appl. **21**, 1253–1278 (2000)
4. de Lathauwer, L., de Moor, B., Vandewalle, J.: On the best rank-1 and rank-$(R_1, R_2, \ldots, R_N)$ approximation of higher-order tensors. SIAM J. Matrix Anal. Appl. **21**, 1324–1342 (2000)
5. Deshpande, A., Vempala, S.: Adaptive sampling and fast low-rank matrix approximation, Electronic Colloquium on Computational Complexity, Report No. 42 pp. 1–11 (2006)
6. de Silva, V., Lim, L.-H.: Tensor rank and the ill-posedness of the best low-rank approximation problem. SIAM J. Matrix Anal. Appl. **30**, 1084–1127 (2008)
7. Drineas, P., Kannan, R., Mahoney, M.W.: Fast Monte Carlo algorithms for matrices I–III: computing a compressed approximate matrix decomposition. SIAM J. Comput. **36**, 132–206 (2006)
8. Eldén, L., Savas, B.: A Newton-Grassmann method for computing the best multilinear rank-(r_1, r_2, r_3) approximation of a tensor. SIAM J. Matrix Anal. Appl. **31**, 248–271 (2009)
9. Friedland, S.: Nonnegative definite hermitian matrices with increasing principal minors. Spec. Matrices **1**, 1–2 (2013)
10. Friedland, S.: MATRICES, a book draft in preparation. http://homepages.math.uic.edu/~friedlan/bookm.pdf, to be published by World Scientific
11. Friedland, S., Kaveh, M., Niknejad, A., Zare, H.: Fast Monte-Carlo low rank approximations for matrices. In: Proceedings of IEEE Conference SoSE, Los Angeles, pp. 218–223 (2006)
12. Friedland, S., Mehrmann, V.: Best subspace tensor approximations, arXiv:0805.4220v1
13. Friedland, S., Mehrmann, V., Miedlar, A., Nkengla, M.: Fast low rank approximations of matrices and tensors. J. Electron. Linear Algebra **22**, 1031–1048 (2011)
14. Friedland, S., Mehrmann, V., Pajarola, R., Suter, S.K.: On best rank one approximation of tensors. http://arxiv.org/pdf/1112.5914v1.pdf
15. Friedland, S., Mehrmann, V., Pajarola, R., Suter, S.K.: On best rank one approximation of tensors. Numer. Linear Algebra Appl. **20**, 942–955 (2013)
16. Friedland, S., Ottaviani, G.: The number of singular vector tuples and uniqueness of best rank one approximation of tensors. Found. Comput. Math. **14**, 1209–1242 (2014)
17. Friedland, S., Torokhti, A.: Generalized rank-constrained matrix approximations. SIAM J. Matrix Anal. Appl. **29**, 656–659 (2007)
18. Frieze, A., Kannan, R., Vempala, S.: Fast Monte-Carlo algorithms for finding low-rank approximations. J. ACM **51**, 1025–1041 (2004)
19. Golub, G.H., Van Loan, C.F.: Matrix Computation, 3rd edn. John Hopkins University Press, Baltimore (1996)

20. Goreinov, S.A., Oseledets, I.V., Savostyanov, D.V., Tyrtyshnikov, E.E., Zamarashkin, N.L.: How to find a good submatrix. In: Matrix Methods: Theory, Algorithms and Applications, pp. 247–256. World Scientific Publishing, Hackensack (2010)
21. Goreinov, S.A., Tyrtyshnikov, E.E., Zamarashkin, N.L.: A theory of pseudo-skeleton approximations of matrices. Linear Algebra Appl. **261**, 1–21 (1997)
22. Goreinov, S.A., Tyrtyshnikov, E.E.: The maximum-volume concept in approximation by low-rank matrices. Contemp. Math. **280**, 47–51 (2001)
23. Grasedyck, L.: Hierarchical singular value decomposition of tensors. SIAM J. Matrix Anal. Appl. **31**, 2029–2054 (2010)
24. Grasedyck, L., Kressner, D., Tobler, C.: A literature survey of low-rank tensor approximation techniques. GAMM-Mitteilungen **36**, 53–78 (2013)
25. Hackbusch, W.: Tensorisation of vectors and their efficient convolution. Numer. Math. **119**, 465–488 (2011)
26. Hackbusch, W.: Tensor Spaces and Numerical Tensor Calculus. Springer, Heilderberg (2012)
27. Hillar, C.J., Lim, L.-H.: Most tensor problems are NP-hard. J. ACM **60**, Art. 45, 39 pp. (2013)
28. Horn, R.A., Johnson, C.R.: Matrix Analysis. Cambridge University Press, Cambridge/New York (1988)
29. Khoromskij, B.N.: Methods of Tensor Approximation for Multidimensional Operators and Functions with Applications, Lecture at the workschop, Schnelle Löser für partielle Differentialgleichungen, Oberwolfach, 18.-23.05 (2008)
30. Kolda, T.G., Bader, B.W.: Tensor decompositions and applications. SIAM Rev. **51**, 455–500 (2009)
31. Lehoucq, R.B., Sorensen, D.C., Yang, C.: ARPACK User's Guide: Solution of Large-Scale Eigenvalue Problems With Implicitly Restarted Arnoldi Methods (Software, Environments, Tools). SIAM, Philadelphia (1998)
32. Lim, L.-H.: Singular values and eigenvalues of tensors: a variational approach. In: Proceedings of IEEE International Workshop on Computational Advances in Multi-Sensor Adaptive Processing (CAMSAP '05), Puerto Vallarta, vol. 1, pp. 129–132 (2005)
33. Mahoney, M.W., Maggioni, M., Drineas, P.: Tensor-CUR decompositions for tensor-based data. In: Proceedings of the 12th Annual ACM SIGKDD Conference, Philadelphia, pp. 327–336 (2006)
34. Oseledets, I.V.: On a new tensor decomposition. Dokl. Math. **80**, 495–496 (2009)
35. Oseledets, I.V.: Tensor-Train decompositions. SIAM J. Sci. Comput. **33**, 2295–2317 (2011)
36. Oseledets, I.V., Tyrtyshnikov, E.E.: Breaking the curse of dimensionality, or how to use SVD in many dimensions. SIAM J. Sci. Comput. **31**, 3744–3759 (2009)
37. Rudelson, M., Vershynin, R.: Sampling from large matrices: an approach through geometric functional analysis. J. ACM **54**, Art. 21, 19 pp. (2007)
38. Sarlos, T.: Improved approximation algorithms for large matrices via random projections. In: Proceedings of the 47th Annual IEEE Symposium on Foundations of Computer Science (FOCS), Berkeley, pp. 143–152 (2006)
39. Savas, B., Lim, L.-H.: Quasi-Newton methods on Grassmannians and multilinear approximations of tensors. SIAM J. Sci. Comput. **32**, 3352–3393 (2010)
40. Stewart, G.W.: On the early history of the singular value decomposition. SIAM Rev. **35**, 551–566 (1993)
41. ten Berge, J.M.F., Kiers, H.A.L.: Simplicity of core arrays in three-way principal component analysis and the typical rank of $p \times q \times 2$ arrays. Linear Algebra Appl. **294**, 169–179 (1999)
42. Tucker, L.R.: Some mathematical notes on three-mode factor analysis. Psychometrika **31**, 279–311 (1966)
43. Zhang, T., Golub, G.H.: Rank-one approximation to high order tensors. SIAM J. Matrix Anal. Appl. **23**, 534–550 (2001)

Part III
Differential-Algebraic Equations and Control Theory

Chapter 15
Regularization of Descriptor Systems

Nancy K. Nichols and Delin Chu

Abstract Implicit dynamic-algebraic equations, known in control theory as descriptor systems, arise naturally in many applications. Such systems may not be regular (often referred to as singular). In that case the equations may not have unique solutions for consistent initial conditions and arbitrary inputs and the system may not be controllable or observable. Many control systems can be "regularized" by proportional and/or derivative feedback. We present an overview of mathematical theory and numerical techniques for regularizing descriptor systems using feedback controls. The aim is to provide stable numerical techniques for analyzing and constructing regular control and state estimation systems and for ensuring that these systems are robust. State and output feedback designs for regularizing linear time-invariant systems are described, including methods for disturbance decoupling and mixed output problems. Extensions of these techniques to time-varying linear and nonlinear systems are discussed in the final section.

15.1 Introduction

Singular systems of differential equations, known in control theory as *descriptor systems* or *generalized state-space systems*, have fascinated Volker Mehrmann throughout his career. His early research, starting with his habilitation [33, 35], concerned autonomous linear-quadratic control problems constrained by descriptor systems. Descriptor systems arise naturally in many applications, including aircraft guidance, chemical processing, mechanical body motion, power generation, network fluid flow and many others, and can be considered as continuous or discrete implicit dynamic-algebraic systems [32, 41]. Such systems may not be regular (often

N.K. Nichols (✉)
Department of Mathematics, University of Reading, Box 220 Whiteknights, Reading, RG6 6AX, UK
e-mail: n.nichols@rdg.ac.uk

D. Chu
Department of Mathematics, National University of Singapore, 2 Science Drive 2, Singapore 117543, Singapore
e-mail: matchudl@nus.edu.sg

P. Benner et al. (eds.), *Numerical Algebra, Matrix Theory, Differential-Algebraic Equations and Control Theory*, DOI 10.1007/978-3-319-15260-8_15

referred to as singular). In that case unique solutions to initial value problems consistent with the system may not exist and the system may not be controllable or observable. An important aspect of control system design is therefore to ensure regularity of the system.

In this chapter we review the work of Volker and his colleagues on mathematical theory and numerical techniques for regularizing descriptor systems using feedback controls. Two key elements contributed initially to the research: the establishment of conditions for the regularizability of descriptor systems by feedback [25, 30] and the development of stable numerical techniques for the reduction of descriptor systems to condensed matrix forms [33, 34, 36]. Following a stimulating meeting at the International Conference on Linear Algebra and Applications in Valencia in 1987, these two research threads were brought together in a report on feedback design for descriptor systems [5] and later published in [6] and [7].

Since that time, Volker has contributed to a whole sequence of exciting results on the regularization of descriptor systems [3, 8–12, 15, 20–22, 24, 31, 37]. The development of sound numerical methods for system design, as well as techniques for guaranteeing the *robustness* of the systems to model uncertainties and disturbances, has formed the main emphasis throughout this research. We describe some of this work in the next sections.

We start with preliminary definitions and properties of descriptor systems and then discuss regularization by state feedback for linear time-invariant systems. Disturbance decoupling by state feedback is also discussed. The problem of regularization by output feedback is then considered. Further developments involving mixed output feedback regularization are given next, and finally work on time-varying and nonlinear systems is briefly described.

15.2 System Design for Descriptor Systems

We consider linear dynamical control systems of the form

$$\begin{aligned} E\dot{x}(t) &= Ax(t) + Bu(t), \quad x(t_0) = x_0, \\ y(t) &= Cx(t), \end{aligned} \tag{15.1}$$

or, in the discrete-time case,

$$\begin{aligned} Ex(k+1) &= Ax(k) + Bu(k), \quad x(0) = x_0, \\ y(k) &= Cx(k), \end{aligned} \tag{15.2}$$

where $E, A \in \mathbb{R}^{n \times n}$, $B \in \mathbb{R}^{n \times m}$, $C \in \mathbb{R}^{p \times n}$. Here $x(\cdot)$ is the state, $y(\cdot)$ is the output, and $u(\cdot)$ is the input or control of the system. It is assumed that $m, p \leq n$ and that the matrices B, C are of full rank. The matrix E may be *singular*. Such systems are known as *descriptor* or *generalized state-space* systems. In the case $E = I$, the identity matrix, we refer to (15.1) or (15.2) as a *standard* system.

We assume initially that the system is time-invariant; that is, the system matrices E, A, B, C are constant, independent of time. In this context, we are interested in proportional and derivative feedback control of the form $u(t) = Fy(t) - G\dot{y}(t) + v(t)$ or $u(k) = Fy(k) - Gy(k+1) + v(k)$, where $F, G \in \mathbb{R}^{m \times p}$ are selected to give the closed-loop system

$$(E + BGC)\dot{x}(t) = (A + BFC)x(t) + Bv(t) \tag{15.3}$$

or

$$(E + BGC)x(k+1) = (A + BFC)x(k) + Bv(k) \tag{15.4}$$

desired properties. *Proportional output* feedback control is achieved in the special case $G = 0$. *Derivative output* feedback control corresponds to the special case $F = 0$ and derivative and proportional *state* feedback control corresponds to the special case $C = I$. The dual of the control system, an *observer* (or state-estimator), is attained with an appropriate choice for v in the special case $B = I$. The aim of the feedback designs is to alter the behaviour of the system response. Proportional feedback acts to modify the system matrix A, whilst derivative feedback alters the system matrix E. Different properties of the system can, therefore, be achieved using different feedback combinations.

15.2.1 Structure of the System Response

The response of the descriptor system (15.1) or (15.2) can be described in terms of the eigenstructure of the matrix pencil $\alpha E - \beta A$, which we denote by (E, A). The system is *regular* if the pencil (E, A) is regular, that is,

$$\det(\alpha E - \beta A) \neq 0 \quad \text{for some} \quad (\alpha, \beta) \in \mathbb{C}^2. \tag{15.5}$$

The generalized eigenvalues of a regular pencil are defined by the pairs $(\alpha_j, \beta_j) \in \mathbb{C}^2 \backslash \{0, 0\}$ such that

$$\det(\alpha_j E - \beta_j A) = 0, \quad j = 1, 2, \ldots, n. \tag{15.6}$$

If $\beta_j \neq 0$, the eigenvalue pair is said to be *finite* with value given by $\lambda_j = \alpha_j / \beta_j$ and otherwise, if $\beta_j = 0$, then the pair is said to be an *infinite* eigenvalue. The maximum number of finite eigenvalues that a pencil can have is less than or equal to the rank of E.

If the system (15.1) or (15.2) is regular, then the existence and uniqueness of classical smooth solutions to the dynamical equations is guaranteed for sufficiently smooth inputs and consistent initial conditions [14, 43]. The solutions are characterized in terms of the Kronecker Canonical Form (KCF) [26]. Nonsingular matrices

X and Y (representing right and left generalized eigenvectors and principal vectors of the system pencil, respectively) then exist such that

$$XEY = \begin{bmatrix} I & 0 \\ 0 & N \end{bmatrix}, \quad XAY = \begin{bmatrix} J & 0 \\ 0 & I \end{bmatrix}, \tag{15.7}$$

where the eigenvalues of the Jordan matrix J coincide with the finite eigenvalues of the pencil and N is a nilpotent Jordan matrix such that $N^i = 0$, $N^{i-1} \neq 0$, $i > 0$, corresponding to the infinite eigenvalues. The *index* of a descriptor system, denoted by $\text{ind}(E, A)$, is defined to be the degree i of nilpotency of the matrix N, that is, the index of the system is the dimension of the largest Jordan block associated with an infinite eigenvalue of the KCF (15.7). The index is a fundamental characteristic of a descriptor system, determining the existence and smoothness of solutions.

By convention, a descriptor system is regular and of index 0 if and only if E is nonsingular. In this case the system can be reformulated as a standard system. However, the reduction to standard form can be numerically unreliable if E is ill-conditioned with respect to inversion. Therefore it is desirable to work directly with the generalized state-space form even where E is nonsingular.

A descriptor system is regular and has index at most one if and only if it has exactly $q = \text{rank}(E)$ finite eigenvalues and $n-q$ *non-defective* infinite eigenvalues. Conditions for the system to be regular and of index ≤ 1 are given by the following important result.

Theorem 1 ([25, 30]) *Let $E, A \in \mathbb{R}^{n \times n}$ and let $S_\infty(E)$ and $T_\infty(E)$ be full rank matrices whose columns span the null spaces $\mathcal{N}(E)$ and $\mathcal{N}(E^T)$ respectively. Then the following are equivalent:*

(i) $\alpha E - \beta A$ *is regular and of index* ≤ 1;
(ii) $rank([E, AS_\infty(E)]) = n$;
(iii) $rank(\begin{bmatrix} E \\ T_\infty^T(E)A \end{bmatrix}) = n$;
(iv) $rank(T_\infty^T(E)AS_\infty(E)) = n - rank(E)$.

Systems that are regular and of index at most one can be separated into purely dynamical and algebraic parts (fast and slow modes) [14, 23] and in theory the algebraic part can be eliminated to give a reduced-order standard system. The reduction process, however, may be ill-conditioned for numerical computation and lead to large errors in the reduced order system [28]. If the system is not regular or if $\text{ind}(E, A) > 1$, then impulses can arise in the response of the system if the control is not sufficiently smooth [27, 42]. Since the linear constant coefficient system is usually only a model that approximates a nonlinear model, disturbances in the real application will in general lead to impulsive solutions if the system is of index higher than one.

15.2.2 Controllability and Observability

If the descriptor system (15.1) or (15.2) is *regular*, then the following controllability and observability conditions are sufficient for most classical design aims. To simplify the notation, we hereafter denote a matrix with orthonormal columns spanning the right nullspace of the matrix M by $S_\infty(M)$ and a matrix with orthonormal columns spanning the left nullspace of M by $T_\infty(M)$. The controllability conditions are defined to be:

C0: $\operatorname{rank}([\alpha E - \beta A, B]) = n$ for all $(\alpha, \beta) \in \mathbb{C}^2 \backslash \{(0,0)\}$.
C1: $\operatorname{rank}([\lambda E - A, B]) = n$ for all $\lambda \in \mathbb{C}$.
C2: $\operatorname{rank}([E, AS_\infty(E), B]) = n$, where the columns of $S_\infty(E)$ span the null space of E.

(15.8)

The observability conditions are defined as the dual of the controllability conditions:

O0: $\operatorname{rank}(\begin{bmatrix} \alpha E - \beta A \\ C \end{bmatrix}) = n$ for all $(\alpha, \beta) \in \mathbb{C}^2 \backslash \{(0,0)\}$.
O1: $\operatorname{rank}(\begin{bmatrix} \lambda E - A \\ C \end{bmatrix}) = n$ for all $\lambda \in \mathbb{C}$.
O2: $\operatorname{rank}(\begin{bmatrix} E \\ T_\infty^T(E)A \\ C \end{bmatrix}) = n$, where the columns of $T_\infty(E)$ span the right null space of E.

(15.9)

For systems that are regular, these conditions characterize the controllability of the system. The condition **C0** ensures that for any given initial and final states of the system, x_0, x_f, there exists an admissible control that transfers the system from x_0 to x_f in finite time [43]. Condition **C1** ensures the same for any given initial and final states x_0, x_f belonging to the solution space of the descriptor system [5, 7]. A regular system that satisfies the conditions **C0** and **O0** is said to be *completely controllable* (C–controllable) and *completely observable* (C–observable) and has properties similar to those of standard control systems. A regular system is *strongly controllable* (S–controllable) if **C1** and **C2** hold and *strongly observable* (S–observable) if **O1** and **O2** hold. Regular systems that satisfy condition **C2** are *controllable at infinity* or *impulse controllable* [27, 42]. For these systems, impulsive modes can be excluded. Condition **C2** is closely related to the second condition in Theorem 1, which characterizes regular systems of index at most one. By the definition, a regular descriptor system of index at most one is controllable at infinity.

The controllability and observability conditions **C0**, **C1**, **C2**, and **O0**, **O1**, **O2** are all preserved under non-singular "equivalence" transformations of the pencil and under proportional state and output feedback, but **C2** is not necessarily preserved under derivative feedback. Therefore, if derivative feedback is used to modify the system dynamics, it is necessary to avoid losing controllability at infinity [5, 7].

Whilst regularity is required for controllability and observability, it is *not* needed in order to regularize the system by feedback. Many descriptor systems that are not regular can be regularized by proportional and/or derivative feedback. Conversely, systems that are regular can easily be transformed by feedback into closed-loop systems that are not regular. It is important, therefore, to establish conditions that ensure the regularity of systems under feedback and to develop numerically reliable techniques for constructing regular feedback systems of index at most one.

Theorem 1 defines conditions that must be satisfied by a closed-loop system pencil (15.3) or (15.4) for it to be regular and of index ≤ 1. These conditions are closely related to the properties **C1, C2, O1, O2**, but regularity is needed for controllability and observability, whereas it is not required for regularization. In [25, 30] it was first shown that these conditions can be used to determine a closed-loop descriptor feedback system that is both regular and of index at most one, using proportional feedback. The system itself does not need to be regular to achieve this result.

In a standard system, derivative feedback does not alter the system behaviour in any way that could not be achieved by proportional feedback alone. However, for descriptor systems, it is possible that derivative feedback can decrease the susceptibility to noise and change the dynamic order of the descriptor system. One of the applications of derivative feedback is to shift infinite frequencies to finite frequencies in order to regularize and control the system. These possibilities together with the implications of Theorem 1, provided a challenge to Volker and his colleagues and motivated their initial work on feedback design for descriptor systems [5–7]. The work is based on numerically stable methods for reducing descriptor systems to condensed forms using unitary transformations. In the next section we summarize this research.

15.3 Regularization by Feedback for Time-Invariant Systems

The problem of regularizing a descriptor system of form (15.1) or (15.2) by feedback is defined as:

Problem 1 Given real system matrices E, A, B, C, find real matrices F and G such that the closed-loop pencil

$$(E + BGC, A + BFC) \tag{15.10}$$

is regular and $\mathrm{ind}(E + BGC, A + BFC) \leq 1$.

If $C = I$ this is the *state* feedback regularization problem and otherwise it is the *output* regularization feedback problem.

In the report [5], both the output and the state feedback regularization problems are investigated initially, but the published version [7] treats only the state feedback problem. A complete solution to the state feedback problem was achieved, but the output case proved to be more elusive, and a number of papers tackling this problem followed later. The state feedback problem has its own importance in real applications, so here we consider first the state feedback problem and then the output feedback problem separately.

15.3.1 Regularization by State Feedback

In the papers [5–7], two major contributions are made. The first provides conditions for the existence of solutions to the state feedback regularization problem. This is achieved by numerically stable transformations to condensed forms that enable the required feedback matrices to be constructed accurately in practice. The second establishes 'robust' system design techniques for ensuring that the properties of the closed-loop system pencil are insensitive to perturbations in the system matrices $E + BG, A + BF, B$.

The following theorem gives the complete solution to the state feedback regularization problem.

Theorem 2 ([7]) *Given a system of the form (15.1) or (15.2), if rank*$([E, AS_\infty(E), B])=n$, *that is, if* **C2** *holds, then there exist real feedback matrices* $F, G \in \mathbb{R}^{m\times n}$ *such that the pencil* $(E + BG, A + BF)$ *is regular,* $\operatorname{ind}(E + BG, A + BF) \le 1$, *and rank*$(E + BG) = r$, *where* $0 \le \mathit{rank}([E, B]) - \mathit{rank}(B) \le r \le \mathit{rank}([E, B])$.

To establish the theorem, we compute the QR factorization of B and the URV factorization [28] of $T_\infty^T(B)E$ to obtain orthogonal matrices P and Q such that

$$PEQ = \begin{bmatrix} E_{11} & 0 & 0 \\ E_{21} & E_{22} & 0 \\ 0 & 0 & 0 \end{bmatrix}, \quad PB = \begin{bmatrix} 0 \\ B_2 \\ 0 \end{bmatrix}, \quad PAQ = \begin{bmatrix} A_{11} & A_{12} & A_{13} \\ A_{21} & A_{22} & A_{23} \\ A_{31} & A_{32} & A_{33} \end{bmatrix}. \tag{15.11}$$

Here E_{11} and B_2 are nonsingular and E_{22} is full column rank. Both E_{11} and B_2 can be further reduced by orthogonal transformations to full-rank positive diagonal matrices. The theorem then follows by selecting feedback matrices to ensure that the closed-loop pencil

$$(E + BG, A + BF) \tag{15.12}$$

satisfies condition (ii) of Theorem 1. If **C1** holds as well as **C2**, the resulting closed-loop system is then strongly controllable [7]. This system could be reduced further to a standard system, but in this case the feedback matrices would have to be selected with care to ensure that the reduction is numerically stable.

Additional results on state feedback regularization using only proportional or derivative feedback are also given in [5–7]. The existence of regularizing proportional state feedback designs is easily shown in the case where **C2** holds using the condensed form (15.11). For the derivative feedback case, the results are the same as in Theorem 2, with the exception that the potential rank of the matrix $(E + BG)$ is now restricted from below. The maximum rank that can be obtained remains equal to rank$([E, B])$.

In general the feedback designs that regularize the system (15.1) or (15.2) are not uniquely determined by Theorem 2 and additional degrees of freedom in the design can be exploited to obtain robustness and stability of the system as well as regularity. For robustness we want the system to remain regular and of index at most one under perturbations to the closed-loop system matrices. From Theorem 1 the closed-loop pencil (15.12) is regular and of index ≤ 1 if and only if

$$\text{rank}\left(\begin{bmatrix} E + BG \\ T_\infty^T(E + BG)(A + BF) \end{bmatrix}\right) = n. \tag{15.13}$$

It is well-known that for a matrix with full rank, the distance to the nearest matrix of lower rank is equal to its minimum singular value [28]. Hence for robustness of the closed-loop pencil (15.12) we aim to select F and G such that the pencil is unitarily equivalent to a pencil of the form $\alpha S_1 - \beta S_2$ where

$$S_1 = \begin{bmatrix} \Sigma_R & 0 \\ 0 & 0 \end{bmatrix}, \quad S_2 = \begin{bmatrix} A_{11} & A_{12} \\ A_{21} & \Sigma_L \end{bmatrix}, \tag{15.14}$$

and the assigned singular values of Σ_R, Σ_L are such that the condition numbers of Σ_R and Σ_L are minimal. This choice ensures regularity of the system and maximizes a lower bound on the minimum singular value of (15.13), whilst retaining an upper bound on the magnitude of the gains F and G. Details of the algorithm to achieve these results are given in [5, 7, 39]. This choice also ensures that the reduction of the closed-loop descriptor system to a standard form is as well-conditioned as possible. In practice such robust systems also have improved performance characteristics (see [40]).

In addition to regularity, it is desirable to ensure that a system design has stability and even that it has specified finite eigenvalues. The following result, shown in [5, 7], holds for descriptor systems.

Theorem 3 ([5, 7]) *Given a system of the form (15.1) or (15.2), if the conditions* **C1** *and* **C2** *hold and r is an integer such that $0 \leq rank([E, B]) - rank(B) \leq r \leq rank([E, B])$, then for any arbitrary set $\mathscr{S}$ of r self-conjugate finite poles there exist feedback matrices $F, G \in \mathbb{R}^{m\times n}$ such that the pencil $(E + BG, A + BF)$ is regular,* $\text{ind}(E + BG, A + BF) \leq 1$, *$rank(E + BG) = r$ and all pairs in $\mathscr{S}$ are the finite generalized eigenvalues of the pencil $(E + BG, A + BF)$.*

For robustness of the closed-loop system, we require the maximum number of finite eigenvalues to be assigned and both the finite and infinite eigenvalues to be insensitive to perturbations in the closed-loop system matrices. One strategy for obtaining a robust solution to the eigenvalue assignment problem for a descriptor system is to apply derivative feedback alone to obtain a robust, regular index-one system with $\operatorname{rank}(E + BG) = r = \operatorname{rank}([E, B])$ using singular value assignment, and then to use *robust* proportional state feedback to assign r finite eigenvalues to the system. The problem of eigenvalue assignment by proportional state feedback in descriptor systems is treated in [17, 25, 30]. Techniques for robust eigenstructure assignment ensuring that the assigned eigenvalues of the closed-loop system are insensitive to perturbations in the system matrices are established in [29, 30, 38].

The problem of designing an observer, or state-estimator, is the dual of the state feedback control problem. An observer is an auxiliary dynamical system designed to provide estimates $\hat{x}$ of all the states x of the system (15.1) or (15.2) using measured output data y and $\dot{y}$. The estimator is a closed-loop system that is driven by the differences between the measured outputs and derivatives of the system and their estimated values. The system pencil is given by

$$(E + GC, A + FC), \tag{15.15}$$

where the matrices F and G must be selected to ensure that the response $\hat{x}$ of the observer converges to the system state x for any arbitrary starting condition; that is, the system must be asymptotically stable. By duality with the state feedback problem, it follows that if the condition **O2** holds, then the matrices F and G can be chosen such that the corresponding closed-loop pencil (15.15) is regular and of index at most one. If condition **O1** also holds, then the closed-loop system is S-observable. Furthermore, the remaining freedom in the system can be selected to ensure the stability and robustness of the system and the finite eigenvalues of the system pencil can be assigned explicitly by the techniques described for the state feedback control problem.

15.3.2 Disturbance Decoupling by State Feedback

In practice control systems are subject to disturbances that may include modelling or measurement errors, higher order terms from linearization, or unknown inputs to the system. For such systems it is important to design feedback controllers and observers that suppress the disturbance so that it does not affect the input-output of the system. In research strongly inspired by the earlier work of Volker and his colleagues on state feedback regularization, the problem of disturbance decoupling is treated in [20, 21].

In the case that disturbances are present, the linear time-invariant system takes the form

$$\begin{aligned} E\dot{x}(t) &= Ax(t) + Bu(t) + Hq(t), \quad x(t_0) = x_0, \\ y(t) &= Cx(t), \end{aligned} \tag{15.16}$$

or

$$\begin{aligned} Ex(k+1) &= Ax(k) + Bu(k) + Hq(k), \quad x(0) = x_0, \\ y(k) &= Cx(k), \end{aligned} \tag{15.17}$$

where $E, A \in \mathbb{R}^{n \times n}$, $B \in \mathbb{R}^{n \times m}$, $C \in \mathbb{R}^{q \times n}$, $H \in \mathbb{R}^{n \times p}$, and $q(\cdot)$ represents a vector of disturbances.

To suppress the disturbances, a state feedback controller is used to modify the input-output map, or transfer function, of the system. The disturbance decoupling problem for the descriptor system (15.16) or (15.17) is then to find proportional and derivative feedback matrices F, G such that the closed-loop pencil $(E+BG, A+BF)$ is regular and of index at most one and

$$T(s) \equiv C(s(E+BG) - (A+BF))^{-1}H \equiv 0, \tag{15.18}$$

where $T(s)$ defines the transfer function of the closed-loop system from the input disturbance $q(\cdot)$ to the output $y(\cdot)$. This condition ensures that the disturbance does not affect the input-output response of the closed-loop system for any choice of the input control $u(\cdot)$. Necessary and sufficient conditions for the existence of a solution to this problem are established in [21]. In addition, conditions are derived under which the feedback matrices can be chosen such that the closed-loop system is also stable. The derivations are constructive and a numerically stable algorithm is given for implementing the procedure.

In [20] the problem of designing a disturbance-decoupled observer system for estimating (a subset of) the states of the system (15.16) or (15.17) is developed. The aim is to select feedback matrices such that the closed-loop observer is regular and of index at most one and such that the disturbances have no influence on the error in the estimated states of the system. Necessary and sufficient conditions are derived for the existence of disturbance-decoupled observers of this form and also for the observer to be stable, ensuring that the estimated states converge over time to the corresponding states of the original system. The main results are established constructively and are again based on a condensed form that can be computed in a numerically stable way using unitary matrix transformations.

15.3.3 Regularization by Output Feedback

The output feedback regularization problem is to find derivative and state output feedback matrices F, G such that the closed-loop system pencil (15.10) is regular and has index at most one.

Meeting at the Institute for Mathematics and Its Applications in Minnesota in 1992 and following up the earlier research on regularization, Volker and his colleagues tackled the difficult output feedback problem in earnest. The results of the research are published in an extensive report [8] and in later papers [9, 10]. In these papers, a condensed form of the descriptor system pencil is derived that displays the conditions under which the system can be transformed into a regular system of index at most one by output feedback using numerically stable orthogonal transformations. For proportional output feedback the solution to the design problem follows immediately from this condensed form. Necessary and sufficient conditions for a feedback matrix $F \in \mathbb{R}^{m \times p}$ to exist such that the pencil $(E, A + BFC)$ is regular and has index at most one are given by **C2** and **O2**. The closed-loop system is then S-controllable and S-observable if **C1** and **O1** also hold [8, 10].

For combined derivative and proportional output feedback, it is also established in [8, 10], using the condensed form, that if **C2** and **O2** hold, then there exist matrices $F, G \in \mathbb{R}^{m \times p}$ such that the closed-loop pencil $(E + BGC, A + BFC)$ is regular, has index at most one, and $\mathrm{rank}(E + BGC)$ lies in a given range. Techniques such as those used for the state feedback problem to ensure optimal conditioning, or robustness of the closed-loop system to perturbations, are also described in [8, 39].

With proportional output feedback alone, if the system has index ≤ 1, then the number of finite eigenvalues of the closed-loop pencil $(E, A + BFC)$ is fixed at $r = \mathrm{rank}(E)$. With derivative and proportional feedback, the system pencil becomes $(E + BGC, A + BFC)$ and the system properties that depend on the left and right null spaces of E, such as **C2** and **O2**, may be altered and the rank of $E + BGC$ may be increased or decreased from that of E. If the closed-loop system is regular with index $= 1$, then the system may be separated into $r = \mathrm{rank}(E + BGC)$ differential or difference equations and $n - r$ purely algebraic equations. In applications, it may be useful to have more or fewer differential or difference equations. A complete characterization of the achievable ranks r for systems that are regular and of index at most one is, therefore, desirable.

Variations of the condensed form of [8, 10] that can be obtained by stable orthogonal transformations have subsequently been derived in [11, 18, 19, 22] and different approaches to the output feedback problem have been developed. A comprehensive summary of the extended results, based on these condensed forms, is given in [3]. The main result can be expressed as follows.

Theorem 4 ([3, 11, 18, 19, 22]) *Let* $T_a = T_\infty(ES_\infty(C))$, $S_a = S_\infty(T_\infty^T(B)E)$, *and*

$$T_b = T_\infty([E, AS_\infty(\begin{bmatrix} E \\ C \end{bmatrix}), B]), \quad S_b = S_\infty(\begin{bmatrix} E \\ T_\infty^T([E, B])A) \\ C \end{bmatrix}).$$

Then the following statements are equivalent:

(i) There exist feedback matrices $F, G \in \mathbb{R}^{m \times p}$ *such that the closed-loop pencil* $(E + BGC, A + BFC)$ *is regular and of index at most one.*

(ii) $T_a^T AS_b$ *has full column rank,* $T_b^T AS_a$ *has full row rank and*

$$rank(T_\infty^T([E, B])AS_\infty(\begin{bmatrix} E \\ C \end{bmatrix})) \geq n - rank(\begin{bmatrix} E & B \\ C & 0 \end{bmatrix}).$$

Moreover, if the closed-loop pencil $(E + BGC, A + BFC)$ *is regular and of index at most one with* $r = rank(E + BGC)$ *then*

$$rank\ ([E, B]) + rank(\begin{bmatrix} E \\ C \end{bmatrix}) - rank(\begin{bmatrix} E & B \\ C & 0 \end{bmatrix}) \leq r \leq$$

$$\leq rank([E, B]) - rank(T_a^T AS_b) \equiv rank(\begin{bmatrix} E \\ C \end{bmatrix}) - rank(T_b^T AS_a).$$

The matrices in the theorem and their ranks are easily obtained from the following condensed form [3, 18, 22], where $U, V, \in \mathbb{R}^{n \times n}$, $P \in \mathbb{R}^{m \times m}$ and $W \in \mathbb{R}^{p \times p}$ are orthogonal matrices:

$$UEV = \begin{matrix} & t_1 & t_2 & t_3 & s_4 & s_5 \\ t_1 \\ t_2 \\ t_3 \\ t_4 \\ t_5 \end{matrix} \!\!\!\! \begin{matrix} \\ \begin{bmatrix} E_{11} & 0 & 0 & 0 & 0 \\ E_{21} & E_{22} & 0 & 0 & 0 \\ E_{31} & E_{32} & E_{33} & E_{34} & 0 \\ E_{41} & E_{42} & 0 & E_{44} & 0 \\ 0 & 0 & 0 & 0 & 0 \end{bmatrix} \end{matrix},$$

$$UBP = \begin{matrix} & t_3 & t_4 \\ t_1 \\ t_2 \\ t_3 \\ t_4 \\ t_5 \end{matrix} \!\!\!\! \begin{matrix} \\ \begin{bmatrix} 0 & 0 \\ 0 & 0 \\ B_{31} & B_{32} \\ 0 & B_{42} \\ 0 & 0 \end{bmatrix} \end{matrix}, \qquad (15.19)$$

$$WCV = \begin{matrix} & t_1 & t_2 & t_3 & s_4 & s_5 \\ s_4 \\ t_1 \end{matrix} \!\!\!\! \begin{matrix} \\ \begin{bmatrix} C_{11} & C_{12} & 0 & C_{14} & 0 \\ C_{21} & 0 & 0 & 0 & 0 \end{bmatrix} \end{matrix},$$

where the blocks E_{11}, C_{21}, E_{22}, E_{33}, B_{31}, B_{42}, and C_{14} are nonsingular.

Theorem 4 follows directly from the condensed form (15.19). The theorem gives a complete characterization of the possible ranks of $E + BGC$ for systems that are regular and of index at most one. Additional results on output feedback regularization using only proportional or derivative feedback are also presented in the references. Corresponding results for observer designs can be determined directly by duality.

In practice, it is desirable not only that the closed-loop descriptor system is regular and has index at most one, but also that it is robust in the sense that it is insensitive to perturbations in the system matrices. As in the state feedback case, the aim is to choose F and G such that the closed-loop pencil is unitarily equivalent to a pencil of the form (15.14) where the matrices Σ_R and Σ_L are well-conditioned for inversion. This choice ensures that the reduction of the closed-loop system to a standard system is computationally reliable. Partial solutions to this problem are provided in [8, 9], based on the results of [24], and an algorithm is given for minimizing upper bounds on the conditioning of Σ_R and Σ_L using unitary transforms to condensed forms. This procedure generally improves the conditioning of the closed-loop system.

15.3.4 Regularization by Mixed Output Feedback

Systems where different states and derivatives can be output arise commonly in mechanical multi-body motion. In such systems, velocities and accelerations can often be measured more easily than states (e.g. by tachometers or accelerometers). Time-invariant systems of this type can be written in the form:

$$\begin{aligned} E\dot{x}(t) &= Ax(t) + Bu(t), \quad x(t_0) = x_0, \\ y_1(t) &= Cx(t), \\ y_2(t) &= \Gamma\dot{x}, \end{aligned} \tag{15.20}$$

or, in the discrete time case

$$\begin{aligned} Ex(k+1) &= Ax(k) + Bu(k), \quad x(0) = x_0, \\ y_1(k) &= Cx(k), \\ y_2(k+1) &= \Gamma x(k+1), \end{aligned} \tag{15.21}$$

where $E, A \in \mathbb{R}^{n\times n}, B \in \mathbb{R}^{n\times m}, C \in \mathbb{R}^{p\times n}, \Gamma \in \mathbb{R}^{q\times n}$. In this case we are interested in proportional and derivative control of the form $u(t) = Fy_1(t) - G\dot{y}_2(t)$ or $u(k) = Fy_1(k) - Gy_2(k+1)$, where F and G are chosen to give the closed-loop system pencil

$$(E + BG\Gamma, A + BFC) \tag{15.22}$$

desired properties. In particular the aim is to ensure that the closed-loop system is regular and of index at most one. The mixed output feedback regularization problem for this system is stated explicitly as follows.

Problem 2 For a system of the form (15.20) or (15.21), give necessary and sufficient conditions to ensure the existence of feedback matrices $F \in \mathbb{R}^{m\times p}$ and $G \in$

$\mathbb{R}^{m\times q}$ such that the closed-loop system pencil $(E + BG\Gamma, A + BFC)$ is regular and $\mathrm{ind}(E + BG\Gamma, A + BFC) \le 1$.

The mixed feedback regularization problem and its variants, which are significantly more difficult than the state and output feedback regularization problems, have been studied systematically by Volker and his colleagues in [22, 37]. These have not been investigated elsewhere, although systems where different states and derivatives are output arise commonly in practice.

Examples frequently take the second order form

$$M\,\ddot{z} + K\dot{z} + Pz = B_1\dot{u} + B_2 u \tag{15.23}$$

and can be written in the generalized state space form

$$\begin{bmatrix} M & 0 \\ K & I \end{bmatrix}\begin{bmatrix} \dot{z} \\ \dot{v} \end{bmatrix} = \begin{bmatrix} 0 & I \\ -P & 0 \end{bmatrix}\begin{bmatrix} z \\ v \end{bmatrix} + \begin{bmatrix} B_1 \\ B_2 \end{bmatrix} u. \tag{15.24}$$

If the velocities $\dot{z}$ of the states of the system can be measured, then the states $v = M\dot{z} - B_1 u$ are also available and the outputs

$$y_1 = Cx = \begin{bmatrix} 0 & I \end{bmatrix}\begin{bmatrix} z \\ v \end{bmatrix}, \; y_2 = \Gamma\dot{x} = \begin{bmatrix} I & 0 \end{bmatrix}\begin{bmatrix} \dot{z} \\ \dot{v} \end{bmatrix} \tag{15.25}$$

can be used separately to modify the system by either proportional or derivative feedback, respectively. The corresponding closed-loop state-space system matrices then take the form

$$E + BG\Gamma = \begin{bmatrix} M + B_1G & 0 \\ K + B_2G & I \end{bmatrix}, \quad A + BFC = \begin{bmatrix} 0 & I + B_1F \\ -P & B_2F \end{bmatrix}. \tag{15.26}$$

Different effects can, therefore, be achieved by feeding back either the derivatives $\dot{z}$ or the states v. In particular, in the case where M is singular, but $\mathrm{rank}[M, B_1] = n$, the feedback G can be chosen such that $M + B_1G$ is invertible and well-conditioned [7], giving a *robust* closed-loop system that is regular and of index zero. The feedback matrix F can be chosen separately to assign the eigenvalues of the system [30], for example, or to achieve other objectives.

The complete solution to the mixed output feedback regularization problem is given in [22]. The theorem and its proof are very technical. Solvability is established using condensed forms derived in the paper. The solution to the output feedback problem given in Theorem 4 is a special case of the complete result for the mixed output case given in [22]. The required feedback matrices are constructed directly from the condensed forms using numerically stable transformations.

Usually the design of the feedback matrices still contains freedom, however, which can be resolved in many different ways. One choice is to select the feedbacks such that the closed-loop system is robust, or insensitive to perturbations, and, in particular, such that it remains regular and of index at most one under perturbations (due, for example, to disturbances or parameter variations). This choice can also be shown to maximize a lower bound on the stability radius of the closed-loop system [13]. Another natural choice would be to use minimum norm feedbacks, which would be a least squares approach based on the theory in [24]. This approach is also investigated in [22, 37]. The conclusion is that although minimum norm feedbacks are important in other control problems, such as eigenvalue assignment or stabilization because they remove ambiguity in the solution in a least squares sense, for the problem of regularization they do not lead to a useful solution, unless the rank of E is decreased. Heuristic procedures for obtaining a system by output feedback that is robustly regular and of index at most one are discussed in [8, 9, 39].

15.4 Regularization of Time-Varying and Nonlinear Descriptor Systems

Feedback regularization for time-varying and nonlinear descriptor systems provided the next target for Volker's research. Extending the previous work to the time-varying case was enabled primarily by the seminal paper on the analytic singular value decomposition (ASVD) published by Volker and colleagues in 1991 [4]. The ASVD allows condensed forms to be derived for the time-varying problem, just as the SVD does for the time-invariant case, and it provides numerically stable techniques for determining feedback designs.

The continuous form of the time-varying descriptor system is given by the implicit system

$$\begin{aligned} E(t)\dot{x}(t) &= A(t)x(t) + B(t)u(t), \quad x(t_0) = x_0, \\ y(t) &= C(t)x(t), \end{aligned} \tag{15.27}$$

where $E(t), A(t) \in \mathbb{R}^{n\times n}$, $B(t) \in \mathbb{R}^{n\times m}$, $C(t) \in \mathbb{R}^{p\times n}$ are all *continuous* functions of time and $x(t)$ is the state, $y(t)$ is the output, and $u(t)$ is the input or control of the system. (Corresponding discrete-time systems with time-varying coefficients can also be defined, but these are not considered here.)

In this general form, complex dynamical systems including constraints can be modelled. Such systems arise, in particular, as linearizations of a general nonlinear control system of the form

$$\begin{aligned} \mathscr{F}(t, x, \dot{x}, u) &= 0, \qquad x(t_0) = x_0, \\ y &= \mathscr{G}(t, x), \end{aligned} \tag{15.28}$$

where the linearized system is such that $E(t), A(t), B(t)$ are given by the Jacobians of $\mathscr{F}$ with respect to $\dot{x}, x, u$, respectively, and $C(t)$ is given by the Jacobian of $\mathscr{G}$ with respect to x (see [31]).

For the time-varying system (15.27) and the nonlinear system (15.28), the system properties can be modified by time-varying state and output feedback as in the time-invariant case, but the characterization of the system, in particular the solvability and regularity of the system, is considerably more complicated to define than in the time-invariant case and it is correspondingly more difficult to analyse the feedback problem. The ultimate goal remains, however, to obtain stable numerical approaches to the problem using time-varying orthogonal transformations to condensed forms.

If time-varying orthogonal transformations $U(t), V(t), W(t), Y(t)$ are applied to the system (15.27), and all variables are assumed to be time-dependent, then the system becomes

$$\begin{aligned} U^T EV\dot{z} &= (U^T AV - U^T EVS)z + U^T BWw, \\ \tilde{y} &= YCVz, \end{aligned} \tag{15.29}$$

where $x(t) = V(t)z(t)$, $u(t) = W(t)w(t)$, $\tilde{y}(t) = Y(t)y(t)$ and $S(t) = V(t)^T \dot{V}(t)$ is a skew-symmetric matrix. We see that applying time-varying transformations alters the system matrix A, and this must be taken into account where reducing the system to equivalent condensed forms.

In [1, 2] it is shown that the ASVD can be used to produce a condensed form for system (15.27), similar to the form derived in [10]. A time-varying system is defined here to be regular and of index at most one if the conditions of Theorem 1 hold for all t and the system can be decoupled into purely dynamic and algebraic parts. In order to establish regularizability of system (15.27), the strong assumption is made that rank$(E(t))$ is constant and that ranks in the condensed form are also constant. Time-varying output feedback matrices are then constructed to produce a closed-loop pointwise regular pencil of the form (15.10) with index at most one. The rank assumptions ensure the solvability of the closed-loop system. The system matrices E, A, B, C, are assumed to be analytic functions of t, but these conditions can be relaxed provided the ASVD decompositions remain sufficiently smooth.

In the papers [12, 31], a much deeper analysis of the regularization problem is developed. Detailed solvability conditions for the time-varying system (15.27) are established and different condensed forms are derived, again using the ASVD. Constant rank assumptions do not need to be applied, although the existence of smooth ASVDs are required. The analysis covers a plethora of different possible behaviours of the system. One of the tasks of the analysis is to determine redundancies and inconsistencies in the system in order that these may be excluded from the design process. The reduction to the condensed forms displays all the invariants that determine the existence and uniqueness of the solution. The descriptor system is then defined to be regularizable if there exist proportional or derivative feedback matrices such that the closed-loop system is uniquely solvable for every consistent initial state vector and any given (sufficiently smooth) control. Conditions for the system to be regularizable then follow directly from the condensed forms.

In [31] a behaviour approach is taken to the linear time-varying problem where state, input and output variables are all combined into one system vector and the combined system is studied. This approach allows inhomogeneous control problems also to be analysed. Instead of forming a derivative array from which the system invariants and the solutions of the original system can be determined, as in [14, 16], the behaviour approach allows the invariants to be found without differentiating the inputs and thus avoids restrictions on the set of admissible controls. Reduction of the behaviour system to condensed form enables an underlying descriptor system to be extracted and the conditions under which this system can be regularized by proportional and derivative feedback are determined. The construction of the feedback matrices is also described. The reduction and construction methods rely on numerically stable equivalence transformations.

More recent work of Volker and his colleagues [15] extends the behaviour approach to a general implicit nonlinear model of the form

$$\mathscr{F}(t, x, \dot{x}, u, y) = 0, \quad x(t_0) = x_0. \tag{15.30}$$

The property of 'strangeness-index' is defined and used in the analysis. This property corresponds to 'index', as defined for a linear time-invariant descriptor system, and 'strangeness-free' corresponds to the condition that a time-invariant system is of index at most one. Conditions are established under which a behaviour system can be reduced to a differential-algebraic system, and after reinterpretation of the variables, to a typical implicit nonlinear system consisting of differential and algebraic parts. Locally linear state feedback can then be applied to ensure that the system is regular and strangeness-free. Standard simulation, control, and optimization techniques can be applied to the reformulated feedback system. Further details of Volker's work on nonlinear differential–algebraic systems can be found in other chapters in this text.

15.5 Conclusions

We have given here a broad-brush survey of the work of Volker Mehrmann on the problems of regularizing descriptor systems. The extent of this work alone is formidable and forms only part of his research during his career. We have concentrated specifically on results from Volker's own approaches to the regularity problem. The primary aim of his work has been to provide stable numerical techniques for analyzing and constructing control and state estimation systems and for ensuring that these systems are robust. The reduction of systems to condensed forms using orthogonal equivalence transformations forms the major theme in this work. Whilst some of the conclusions described here can also be obtained via other canonical or condensed forms published in the literature, these cannot be derived by sound numerical methods and the required feedbacks cannot be generated from these by backward stable algorithms. Volker's work has therefore had a real practical impact on control system design in engineering as well as producing some beautiful theory. It has been a pleasure for us to be involved in this work.

References

1. Bell, S.J.G.: Numerical techniques for smooth transformation and regularisation of time-varying linear descriptor systems. Ph.D. thesis, Department of Mathematics, University of Reading (1995). http://www.reading.ac.uk/maths-and-stats/research/theses/maths-phdtheses.aspx
2. Bell, S.J.G, Nichols, N.K.: Regularization of time-varying descriptor systems by the ASVD. In: Lewis, J. (ed.) Applied Linear Algebra, pp. 172–176. SIAM, Philadelphia (1994)
3. Bunse-Gerstner, A., Byers, R., Mehrmann, V., Nichols, N.K.: Feedback design for regularizing descriptor systems. Linear Algebra Appl. **299**, 119–151 (1999)
4. Bunse-Gerstner, A., Byers, R., Mehrmann, V., Nichols, N.K.: Numerical computation of an analytic singular value decomposition of a matrix valued function. Numer. Math. **60**, 1–40 (1991)
5. Bunse-Gerstner, A., Mehrmann, V., Nichols, N.K.: Derivative feedback for descriptor systems. Technical report, Materialien LVIII, FSP Mathematisierung, Universität Bielefeld, Bielefeld (1989). Presented at the International Workshop on Singular Systems, Prague, Sept 1989
6. Bunse-Gerstner, A., Mehrmann, V., Nichols, N.K.: On derivative and proportional feedback design for descriptor systems. In: Kaashoek, M.A., et al. (eds.) Proceedings of the International Symposium on the Mathematical Theory of Networks and Systems, Juni 1989, Amsterdam, pp. 437–447. Birkhäuser (1990)
7. Bunse-Gerstner, A., Mehrmann, V., Nichols, N.K.: Regularization of descriptor systems by derivative and proportional state feedback. SIAM J. Matrix Anal. Appl. **13**, 46–67 (1992)
8. Bunse-Gerstner, A., Mehrmann, V., Nichols, N.K.: Numerical methods for the regularization of descriptor system by output feedback. Technical report 987, Institute for Mathematics and Its Applications, University of Minnesota (1992)
9. Bunse-Gerstner, A., Mehrmann, V., Nichols, N.K.: Output feedback in descriptor systems. In: Van Dooren, P., Wyman, B. (eds.) Linear Algebra for Control Theory, pp. 43–54, Springer-Verlag (1993)
10. Bunse-Gerstner, A., Mehrmann, V., Nichols, N.K.: Regularization of descriptor systems by output feedback. IEEE Trans. Automat. Control **AC–39**, 742–1747 (1994)
11. Byers, R., Geerts, T., Mehrmann, V.: Descriptor systems without controllability at infinity. SIAM J. Control Optim. **35**, 462–479 (1997)
12. Byers, R., Kunkel, P., Mehrmann, V.: Regularization of linear descriptor systems with variable coefficients. SIAM J. Control Optim. **35**, 117–133 (1997)
13. Byers, R., Nichols, N.K.: On the stability radius of generalized state-space systems. Linear Algebra Appl. **188/189**, 113–134 (1993)
14. Campbell, S.L.: Singular Systems of Differential Equations. Pitman, San Francisco (1980)
15. Campbell, S.L., Kunkel, P., Mehrmann, V.: Regularization of linear and nonlinear descriptor systems. In: Biegler, L.T., Campbell, S.L., Mehrmann, V. (eds.) Control and Optimization with Differential-Algebraic Constraints, pp. 17–34. SIAM, Philadelphia (2012)
16. Campbell, S.L., Nichols, N.K., Terrell, W.J.: Duality, observability and controllability for linear time-varying systems. Circuits Syst. Signal Process. **10**, 455–470 (1991)
17. Chu, D., Chan, H.C., Ho, D.W.C.: A general framework for state feedback pole assignment of singular systems. Int. J. Control **67**, 135–152 (1997)
18. Chu, D., Chan, H.C., Ho, D.W.C.: Regularization of singular systems by derivative and proportional output feedback. SIAM J. Matrix Anal. Appl. **19**, 21–38 (1998)
19. Chu, D., Ho, D.W.C.: Necessary and sufficient conditions for the output feedback regularization of descriptor systems. IEEE Trans. Autom. Control **AC-44**, 405–412 (1999)
20. Chu, D., Mehrmann, V.: Disturbance decoupled observer design for descriptor systems. Syst. Control Lett. **38**, 37–48 (1999)
21. Chu, D., Mehrmann, V.: Disturbance decoupling for descriptor systems by state feedback. SIAM J. Control Optim. **38**, 1830–1858 (2000)

22. Chu, D., Mehrmann, V., Nichols, N.K.: Minimum norm regularization of descriptor systems by mixed output feedback. Linear Algebra Appl. **296**, 39–77 (1999)
23. Dai, L.: Singular Control Systems. Lecture Notes in Control and Information Sciences, vol. 118. Springer, Berlin (1989)
24. Elsner, L., He, H., Mehrmann, V.: Completion of a matrix so that the inverse has minimum norm. Application to the regularization of descriptor control problems. In: Van Dooren, P., Wyman, B. (eds.) Linear Algebra for Control Theory, pp. 75–86. Springer, New York (1983)
25. Fletcher, L.R., Kautsky, J., Nichols, N.K.: Eigenstructure assignment in descriptor systems. IEEE Trans. Autom. Control **AC-31**, 1138–1141 (1986)
26. Gantmacher, F.R.: Theory of Matrices, vol. II. Chelsea, New York (1959)
27. Geerts, T.: Solvability conditions, consistency, and weak consistency for linear differential-algebraic equations and time-invariant linear systems: the general case. Linear Algebra Appl. **181**, 111–130 (1993)
28. Golub, G.H., Van Loan, C.F.: Matrix Computations, 2nd edn. The Johns Hopkins University Press, Baltimore (1989)
29. Kautsky, J., Nichols, N.K.: Algorithms for robust pole assignment in singular systems. In: American Control Conference 1986, Seattle, vol. 1, pp. 433–436 (1986)
30. Kautsky, J., Nichols, N.K., Chu, E.K.-W.: Robust pole assignment in singular control systems. Linear Algebra Appl. **121**, 9–37 (1989)
31. Kunkel, P., Mehrmann, V., Rath, W.: Analysis and numerical solution of control problems in descriptor form. Math. Control Signals Syst. **14**, 29–61 (2001)
32. Luenberger, D.G.: Dynamic equations in descriptor form. IEEE Trans. Autom. Control **AC-22**, 312–321 (1977)
33. Mehrmann, V.: The Linear Quadratic Control Problem: Theory and Numerical Algorithms. Habilitation Thesis, Universität Bielefeld (1987)
34. Mehrmann, V.: Existence, uniqueness and stability of solutions to singular linear quadratic control problems. Linear Algebra Appl. **121**, 291–331 (1989)
35. Mehrmann, V.: The Autonomous Linear Quadratic Control Problem: Theory and Numerical Algorithms. Lecture Notes in Control and Information Sciences, vol. 163. Springer, Heidelberg (1991)
36. Mehrmann, V., Krause, G.: Linear transformations which leave controllable multi-input descriptor systems controllable. Linear Algebra Appl. **120**, 47–69 (1989)
37. Mehrmann, V., Nichols, N.K.: Mixed output feedback for descriptor systems. Technical report SPC 95-37, DFG Forschergruppe SPC, Fakultät für Mathematik, TU Chemnitz-Zwickau, D-09107 Chemnitz (1995)
38. Nichols, N.K.: Robust control system design for generalised state-space systems. In: 25th IEEE Conference on Decision and Control, Athens, vol. 1, pp. 538–540. Institute of Electrical and Electronic Engineers (1986)
39. Nichols, N.K.: Dynamic-algebraic equations and control system design. In: Griffith, D.F., Watson, G.A. (eds.) Numerical Analysis, 1993, pp. 208–224. Longman Scientific and Technical, Harlow (1994)
40. Pearson, D.W., Chapman, M.J., Shields, D.N.: Partial singular value assignment in the design of robust observers for discrete time descriptor systems. IMA. J. Math. Control Inf. **5**, 203–213 (1988)
41. Rosenbrock, H.H.: Structural properties of linear dynamic systems. Int. J. Control **20**, 191–202 (1974)
42. Verghese, G.C., Van Dooren, P., Kailath, T.: Properties of the system matrix of a generalized state space system. Int. J. Control **30**, 235–243 (1979)
43. Yip, E.L., Sincovec, R.F.: Solvability, controllability and observability of continuous descriptor systems. IEEE Trans. Autom. Control **AC–26**, 702–707 (1981)

Chapter 16
Differential-Algebraic Equations: Theory and Simulation

Peter Kunkel

Abstract We give an overview of the theory of unstructured nonlinear DAEs of arbitrary index. The approach is extended to overdetermined consistent DAEs in order to be able to include known first integrals. We then discuss various computational issues for the numerical solution of corresponding DAE problems. These include the design of special Gauß-Newton techniques as well as the treatment of parametrized nonlinear systems in the context of DAEs. Examples demonstrate their applicability and performance.

16.1 Preface

It was in the year 1988. My contract at the *Sonderforschungsbereich 123* of the University of Heidelberg as research assistant was about to expire and could not be prolonged. My supervisor at that time was W. Jäger. While I was searching a new position, aiming at the possibility to earn a habilitation, he organized a fellowship at the research center of IBM in Heidelberg. I seized this opportunity and signed a contract for 9 months. On the first day of the new job, I met two other colleagues starting at the same day. One of them had a permanent contract. The other one was Volker Mehrmann who was on leave from the University of Bielefeld to spend the same 9 months at the research center of IBM. Our common head R. Janßen put us three into the same office. This was the beginning of Volker's and my joint venture. I therefore want to express my sincere thanks to W. Jäger and R. Janßen for their support which brought me in contact with Volker.

P. Kunkel (✉)
Fakultät für Mathematik und Informatik, Mathematisches Institut, Universität Leipzig, Augustusplatz 10, D-04109 Leipzig, Germany
e-mail: kunkel@math.uni-leipzig.de

P. Benner et al. (eds.), *Numerical Algebra, Matrix Theory, Differential-Algebraic Equations and Control Theory*, DOI 10.1007/978-3-319-15260-8_16

16.2 Introduction

Differential-algebraic equations (DAEs) arise if physical systems are modeled that contain constraints restricting the possible states of the systems. Moreover, in modern hierarchical modeling tools like [5], even if the submodels are ordinary differential equations (ODEs), the equations describing how the submodels are linked yield DAEs as overall models.

The general form of a DAE is given by

$$F(t, x, \dot{x}) = 0, \tag{16.1}$$

with $F \in C(\mathbb{I} \times \mathbb{D}_x \times \mathbb{D}_{\dot{x}}, \mathbb{R}^m)$ sufficiently smooth, $\mathbb{I} \subseteq \mathbb{R}$ (compact) interval, and $\mathbb{D}_x, \mathbb{D}_{\dot{x}} \subseteq \mathbb{R}^n$ open. In this paper, we will not assume any further structure of the equations. It should, however, be emphasized that additional structure should, if possible, be utilized in the numerical treatment when efficiency is an issue. On the other hand, a general approach is of advantage when it is desirable to have no restrictions in the applicability of the numerical procedure.

It is the aim of the present paper to give an overview of the relevant theory of general unstructured nonlinear DAEs with arbitrary index and its impact on the design of numerical techniques for their approximate solution. We will concentrate mainly on the quadratic case, i.e., on the case $m = n$, but also address the overdetermined case $m \geq n$ assuming consistency of the equations. The attractivity of the latter case lies in the fact that we may add known properties of the solution like first integrals to the system, thus enforcing that the generated numerical solution will respect these properties as well. In the discussion of numerical techniques, we focus on two families of Runge-Kutta type one-step methods and the development of appropriate techniques for the solution of the arising nonlinear systems. Besides the mentioned issues on DAE techniques for treating first integrals, we include a discussion on numerical path following and turning point determination in the area of parametrized nonlinear equations, which can also be treated in the context of DAEs combined with root finding. Several examples demonstrate the performance of the presented numerical approaches.

The paper is organized as follows. In Sect. 16.3, we give an overview of the analysis of unstructured regular nonlinear DAEs of arbitrary index. In particular, we present existence and uniqueness results. We discuss how these results can be extended to overdetermined consistent DAEs, thus allowing for the treatment of known first integrals. Section 16.4 is then dedicated to various computational issues. We first present possible one-step methods, develop Gauß-Newton like processes for the treatment of the arising nonlinear systems, which includes a modification to stabilize the numerical solution. After some remarks on the use of automatic differentiation, we show how problems with first integrals and parametrized nonlinear equations can be treated in the context of DAEs. We close with some conclusions in Sect. 16.5.

16.3 Theory of Nonlinear DAEs

Dealing with nonlinear problems, the first step is to require a suitable kind of regularity. In the special case of an ODE $\dot{x} = f(t, x)$, obviously no additional properties besides smoothness must be required to obtain (local) existence and uniqueness of solutions for the corresponding initial value problem. In the special case of a pure algebraic (parametrized) system $F(t, x) = 0$, the typical requirement is given by assuming that $F_x(t, x)$, denoting the Jacobian of F with respect to x, is nonsingular for all relevant arguments. The regularity then corresponds to the applicability of the implicit function theorem allowing to (locally) solve for x in terms of t. In the general case of DAEs, we of course want to include these extreme cases into the definition of a regular problem. Moreover, we want to keep the conditions as weak as possible. The following example gives an idea, how the conditions for regularity should look like.

Example 1 The system

$$\begin{aligned}
\dot{x}_1 &= x_4, & \dot{x}_4 &= 2x_1x_7, \\
\dot{x}_2 &= x_5, & \dot{x}_5 &= 2x_2x_7, \\
\dot{x}_3 &= x_6, & \dot{x}_6 &= -1 - x_7, \\
0 &= x_3 - x_1^2 - x_2^2,
\end{aligned}$$

see [16], describes the movement of a mass point on a paraboloid under the influence of gravity.

Differentiating the constraint twice and eliminating the arising derivatives of the unknowns yields

$$\begin{aligned}
0 &= x_6 - 2x_1x_4 - 2x_2x_5, \\
0 &= -1 - x_7 - 2x_4^2 - 4x_1^2x_7 - 2x_5^2 - 4x_2^2x_7.
\end{aligned}$$

In particular, the so collected three constraints can be solved for x_3, x_6, and x_7 in terms of the other unknowns, leaving, if eliminated, ODEs for these other unknowns. Hence, we may replace the original problem by

$$\begin{aligned}
\dot{x}_1 &= x_4, & \dot{x}_4 &= 2x_1x_7, \\
\dot{x}_2 &= x_5, & \dot{x}_5 &= 2x_2x_7, \\
0 &= x_3 - x_1^2 - x_2^2, \\
0 &= x_6 - 2x_1x_4 - 2x_2x_5, \\
0 &= -1 - x_7 - 2x_4^2 - 4x_1^2x_7 - 2x_5^2 - 4x_2^2x_7.
\end{aligned}$$

◇

From this example, we deduce the following. The solution process may require to differentiate part of the equations such that the solution may depend on the derivatives of the data. Without assuming structure, it is not known in advance which equations should be differentiated. By the differentiation process, we obtain additional constraints that must be satisfied by a solution.

16.3.1 A Hypothesis

In order to include differentiated data, we follow an idea of Campbell, see [1], and define so-called derivative array equations

$$F_\ell(t, x, \dot{x}, \ddot{x}, \ldots, x^{(\ell+1)}) = 0, \tag{16.2}$$

where the functions $F_\ell \in C(\mathbb{I} \times \mathbb{D}_x \times \mathbb{D}_{\dot{x}} \times \mathbb{R}^n \times \cdots \times \mathbb{R}^n, \mathbb{R}^{(l+1)m})$ are defined by stacking the original function F together with its formal time derivatives up to order ℓ, i.e.,

$$F_\ell(t, x, \dot{x}, \ddot{x}, \ldots, x^{(\ell+1)}) = \begin{bmatrix} F(t, x, \dot{x}) \\ \frac{d}{dt} F(t, x, \dot{x}) \\ \vdots \\ (\frac{d}{dt})^\ell F(t, x, \dot{x}) \end{bmatrix}. \tag{16.3}$$

Jacobians of F_l with respect to the selected variables x, y will be denoted by $F_{l;x,y}$ in the following. A similar notation will be used for other functions.

The desired regularity condition should include that the original DAE implies a certain number of constraints, that these constraints should be independent, and that given an initial value satisfying these constraints can always be extended to a local solution. In the case $m = n$, this leads to the following hypothesis.

Hypothesis 1 *There exist (nonnegative) integers μ, a, and d such that the set*

$$\mathbb{L}_\mu = \{(t, x, y) \in \mathbb{R}^{(\mu+2)n+1} \mid F_\mu(t, x, y) = 0\} \tag{16.4}$$

associated with F is nonempty and such that for every point $(t_0, x_0, y_0) \in \mathbb{L}_\mu$, there exists a (sufficiently small) neighborhood $\mathbb{V}$ in which the following properties hold:

1. *We have* $\operatorname{rank} F_{\mu;y} = (\mu + 1)n - a$ *on $\mathbb{L}_\mu \cap \mathbb{V}$ such that there exists a smooth matrix function Z_2 of size $((\mu + 1)n, a)$ and pointwise maximal rank, satisfying $Z_2^T F_{\mu;y} = 0$ on $\mathbb{L}_\mu \cap \mathbb{V}$.*
2. *We have* $\operatorname{rank} Z_2^T F_{\mu;x} = a$ *on $\mathbb{V}$ such that there exists a smooth matrix function T_2 of size (n, d), $d = n - a$, and pointwise maximal rank, satisfying $Z_2^T F_{\mu;x} T_2 = 0$.*
3. *We have* $\operatorname{rank} F_{\dot{x}} T_2 = d$ *on $\mathbb{V}$ such that there exists a smooth matrix function Z_1 of size (n, d) and pointwise maximal rank, satisfying* $\operatorname{rank} Z_1^T F_{\dot{x}} T_2 = d$.

Note that the local existence of functions Z_2, T_2, Z_1 is guaranteed by the following theorem, see, e.g., [13, Theorem 4.3]. Moreover, it shows that we may assume that they possess (pointwise) orthonormal columns.

Theorem 1 *Let $E \in C^\ell(\mathbb{D}, \mathbb{R}^{m,n})$, $\ell \in \mathbb{N}_0 \cup \{\infty\}$, and assume that* rank $E(x) = r$ *for all $x \in \mathbb{M} \subseteq \mathbb{D}$, $\mathbb{D} \subseteq \mathbb{R}^k$ open. For every $\hat{x} \in \mathbb{M}$ there exists a sufficiently small neighborhood $\mathbb{V} \subseteq \mathbb{D}$ of $\hat{x}$ and matrix functions $T \in C^\ell(\mathbb{V}, \mathbb{R}^{n,n-r})$, $Z \in C^\ell(\mathbb{V}, \mathbb{R}^{m,m-r})$, with pointwise orthonormal columns such that*

$$ET = 0, \quad Z^T E = 0 \tag{16.5}$$

on $\mathbb{M}$.

The quantity μ denotes how often we must differentiate the original DAE in order to be able to make conclusions about existence and uniqueness of solutions. Typically, such a quantity is called index. To distinguish it from other indices, the quantity μ, if chosen minimally, is called strangeness index of the given DAE.

For linear DAEs, the above hypothesis is equivalent (for sufficiently smooth data) to the assumption of a well-defined differentiation index and thus to regularity of the given linear DAE, see [13]. In the nonlinear case, the hypothesis, of course, should imply some kind of regularity of the given problems.

In the following, we say that F satisfies Hypothesis 1 with (μ, a, d), if Hypothesis 1 holds with the choice μ, a, and d for the required integers.

16.3.2 Implications

In order to show that Hypothesis 1 implies a certain kind of regularity for the given DAE, we revise the approach first given in [12], see also [13].

Let $(t_0, x_0, y_0) \in \mathbb{L}_\mu$ and

$$T_{2,0} = T_2(t_0, x_0, y_0), \quad Z_{1,0} = Z_1(t_0, x_0, y_0), \quad Z_{2,0} = Z_2(t_0, x_0, y_0).$$

Furthermore, let $Z'_{2,0}$ be chosen such that $[Z'_{2,0}\, Z_{2,0}]$ is orthogonal. By Hypothesis 1, the matrices $Z_{2,0}^T F_{\mu;x}(t_0, x_0, y_0)$ and $Z_{2,0}'^T F_{\mu;y}(t_0, x_0, y_0)$ have full row rank. Thus, we can split the variables x and y, without loss of generalization according to $x = (x_1, x_2)$ and $y = (y_1, y_2)$, such that $Z_{2,0}^T F_{\mu;x_2}(t_0, x_0, y_0)$ and $Z_{2,0}'^T F_{\mu;y_2}(t_0, x_0, y_0)$ are nonsingular. Because of

$$\operatorname{rank} F_{\mu;x_2,y_2} = \operatorname{rank} \begin{bmatrix} Z_{2,0}'^T F_{\mu;x_2} & Z_{2,0}'^T F_{\mu;y_2} \\ Z_{2,0}^T F_{\mu;x_2} & Z_{2,0}^T F_{\mu;y_2} \end{bmatrix}$$

and $Z_{2,0}^T F_{\mu;y_2}(t_0, x_0, y_0) = 0$, this implies that $F_{\mu;x_2,y_2}(t_0, x_0, y_0)$ is nonsingular. The implicit function theorem then yields that the equation $F_\mu(t, x_1, x_2, y_1, y_2) = 0$

is locally solvable for x_2 and y_2. Hence, there are locally defined functions $\mathscr{G}$ and $\mathscr{H}$ with

$$F_\mu(t, x_1, \mathscr{G}(t, x_1, y_1), y_1, \mathscr{H}(t, x_1, y_1)) \equiv 0, \tag{16.6}$$

implying the following structure of $\mathbb{L}_\mu$.

Theorem 2 *The set $\mathbb{L}_\mu$ forms a manifold of dimension $n+1$ that can be locally parametrized by variables (t, x_1, y_1), where x_1 consists of d variables from x and y_1 consists of a variables from y.*

In order to examine the implicitly defined functions in more detail, we consider the system of nonlinear equations $H(t, x, y, \alpha) = 0$ with $\alpha \in \mathbb{R}^a$ given by

$$H(t, x, y, \alpha) = \begin{bmatrix} F_\mu(t, x, y) - Z_{2,0}\alpha \\ T_{1,0}^T(y - y_0) \end{bmatrix}, \tag{16.7}$$

where the columns of $T_{1,0}$ form an orthonormal basis of kernel $F_{\mu;y}(t_0, x_0, y_0)$. Obviously, we have that $H(t_0, x_0, y_0, 0) = 0$. Choosing $T_{1,0}'$ such that $[\, T_{1,0}' \;\; T_{1,0} \,]$ is orthogonal, we get

$$\operatorname{rank} H_{y,\alpha} = \operatorname{rank} \begin{bmatrix} F_{\mu;y} & -Z_{2,0} \\ T_{1,0}^T & 0 \end{bmatrix} = \operatorname{rank} \begin{bmatrix} Z_{2,0}'^T F_{\mu;y} T_{1,0}' & Z_{2,0}'^T F_{\mu;y} T_{1,0} & * \\ Z_{2,0}^T F_{\mu;y} T_{1,0}' & Z_{2,0}^T F_{\mu;y} T_{1,0} & -I_a \\ * & I_d & 0 \end{bmatrix},$$

where here and in the following I_k denotes the identity matrix in $\mathbb{R}^{k,k}$ and its counterpart as constant matrix function. It follows that

$$\operatorname{rank} H_{y,\alpha}(t_0, x_0, y_0, 0) = \operatorname{rank} \begin{bmatrix} Z_{2,0}'^T F_{\mu;y}(t_0, x_0, y_0) T_{1,0}' & 0 & 0 \\ 0 & 0 & -I_a \\ 0 & I_d & 0 \end{bmatrix}$$

and $H_{y,\alpha}(t_0, x_0, y_0, 0)$ is nonsingular because $Z_{2,0}'^T F_{\mu;y}(t_0, x_0, y_0) T_{1,0}'$, representing the linear map obtained by the restriction of $F_{\mu;y}(t_0, x_0, y_0)$ to the linear map from its cokernel onto its range, is nonsingular. Thus, the nonlinear equation (16.7) is locally solvable with respect to (y, α), i.e., there are locally defined functions $\hat{F}_2$ and $\mathscr{Y}$ such that

$$F_\mu(t, x, \mathscr{Y}(t, x)) - Z_{2,0}\hat{F}_2(t, x) \equiv 0, \quad T_{1,0}^T(\mathscr{Y}(t, x) - y_0) \equiv 0. \tag{16.8}$$

If we then define $\hat{F}_1$ by

$$\hat{F}_1(t, x, \dot{x}) = Z_{1,0}^T F(t, x, \dot{x}), \tag{16.9}$$

we obtain a DAE

$$\begin{aligned} \hat{F}_1(t,x,\dot{x}) &= 0, \ (d \text{ differential equations}) \\ \hat{F}_2(t,x) &= 0, \quad (a \text{ algebraic equations}) \end{aligned} \tag{16.10}$$

whose properties shall be investigated.

Differentiating (16.8) with respect to x gives

$$F_{\mu;x} + F_{\mu;y}\mathscr{Y}_x - Z_{2,0}\hat{F}_{2;x} = 0.$$

Multiplying with $Z_{2,0}^T$ form the left and evaluating at (t_0, x_0) then yields

$$\hat{F}_{2;x}(t_0, x_0) = Z_{2,0}^T F_{\mu;x}(t_0, x_0, y_0).$$

With the above splitting for x, we have that $\hat{F}_2(t_0, x_0) = 0$ due to the construction of $\hat{F}_2$ and $\hat{F}_{2;x_2}(t_0, x_0)$ being nonsingular due to the choice of the splitting. Hence, we can apply the implicit function theorem once more to obtain a locally defined function $\mathscr{R}$ satisfying

$$\hat{F}_2(t, x_1, \mathscr{R}(t, x_1)) \equiv 0. \tag{16.11}$$

In particular, the set $\mathbb{M} = \hat{F}_2^{-1}(\{0\})$ forms a manifold of dimension $d + 1$.

Lemma 1 *Let* $(t_0, x_0, y_0) \in \mathbb{L}_\mu$. *Then there is a neighborhood of* (t_0, x_0, y_0) *such that*

$$\mathscr{R}(t, x_1) = \mathscr{G}(t, x_1, y_1) \tag{16.12}$$

for all (t, x, y) *in this neighborhood.*

Proof We choose the neighborhood of (t_0, x_0, y_0) to be a ball with center (t_0, x_0, y_0) and sufficiently small radius. In particular, we assume that all implicitly defined functions can be evaluated for the stated arguments.

Differentiating (16.6) with respect to y_1 gives

$$F_{\mu;x_2}\mathscr{G}_{y_1} + F_{\mu;y_1} + F_{\mu;y_2}\mathscr{H}_{y_1} = 0,$$

where we omitted the argument $(t_1, x_1, \mathscr{G}(t, x_1, y_1), y_1, \mathscr{H}(t, x_1, y_1))$. If we multiply this with $Z_2(t_1, x_1, \mathscr{G}(t, x_1, y_1), y_1, \mathscr{H}(t, x_1, y_1))^T$, defined according to Hypothesis 1, we get $Z_2^T F_{\mu;x_2}\mathscr{G}_{y_1} = 0$. Since $Z_2^T F_{\mu;x_2}$ is nonsingular for a sufficiently small radius of the neighborhood, it follows that $\mathscr{G}_{y_1}(t, x_1, y_1) = 0$.

Inserting $x_2 = \mathscr{R}(t, x_1)$ into the first relation of (16.8) and splitting $\mathscr{Y}$ according to y, we obtain

$$F_\mu(t, x_1, \mathscr{R}(t, x_1), \mathscr{Y}_1(t, x_1, \mathscr{R}(t, x_1)), \mathscr{Y}_2(t, x_1, \mathscr{R}(t, x_1))) = 0.$$

Comparing with (16.6), this yields

$$\mathscr{R}(t,x_1) = \mathscr{G}(t,x_1,\mathscr{Y}_1(t,x_1,\mathscr{R}(t,x_1))).$$

With this, we further obtain, setting $\tilde{y}_1 = \mathscr{Y}_1(t,x_1,\mathscr{R}(t,x_1))$ for short, that

$$\begin{aligned}\mathscr{G}(t,x_1,y_1) - \mathscr{R}(t,x_1) &= \mathscr{G}(t,x_1,y_1) - \mathscr{G}(t,x_1,\tilde{y}_1)\\ &= \mathscr{G}(t,x_1,\tilde{y}_1 + s(y_1-\tilde{y}_1))|_0^1\\ &= \textstyle\int_0^1 \mathscr{G}_{y_1}(t,x_1,\tilde{y}_1 + s(y_1-\tilde{y}_1))(y_1-\tilde{y}_1)\,ds = 0.\end{aligned}$$

□

With the help of Lemma 1, we can simplify the relation (16.6) to

$$F_\mu(t,x_1,\mathscr{R}(t,x_1),y_1,\mathscr{H}(t,x_1,y_1)) \equiv 0. \tag{16.13}$$

Theorem 3 *Consider a sufficiently small neighborhood of $(t_0,x_0,y_0) \in \mathbb{L}_\mu$. Let $\hat{F}_2$ and $\mathscr{R}$ be well-defined according to the above construction and let (t,x) with $x = (x_1,x_2)$ be given such that (t,x) is in the domain of $\hat{F}_2$ and (t,x_1) is in the domain of $\mathscr{R}$. Then the following statements are equivalent:*

(a) There exists y such that $F_\mu(t,x,y) = 0$.
(b) $\hat{F}_2(t,x) = 0$.
(c) $x_2 = \mathscr{R}(t,x_1)$.

Proof The statements (b) and (c) are equivalent due to the implicit function theorem defining $\mathscr{R}$. Assuming (a), let there be y such that $F_\mu(t,x,y) = 0$. Then, $x_2 = \mathscr{G}(t,x_1,y_1) = \mathscr{R}(t,x_1)$ due to the implicit function theorem defining $\mathscr{G}$ and Lemma 1. Assuming (c), we set $y = \mathscr{Y}(t,x)$. With $\hat{F}_2(t,x) = 0$, the relation (16.8) yields $F_\mu(t,x,y) = 0$. □

Theorem 4 *Let F from (16.1) satisfy Hypothesis 1 with (μ,a,d). Then, $\hat{F} = (\hat{F}_1,\hat{F}_2)$ satisfies Hypothesis 1 with $(0,a,d)$.*

Proof Let $\hat{\mathbb{L}}_0 = \hat{F}^{-1}(\{0\})$ and let $\hat{Z}_2, \hat{T}_2, \hat{Z}_1$ denote the matrix functions belonging to $\hat{F}$ as addressed by Hypothesis 1.

For $(t_0,x_0,y_0) \in F_\mu^{-1}(\{0\})$, the above construction yields $\hat{F}_2(t_0,x_0) = 0$. If $\dot{x}_0$ denotes the first n components of y_0, then $F(t_0,x_0,\dot{x}_0) = 0$ holds as first block of $F_\mu(t_0,x_0,y_0) = 0$ implying $\hat{F}_1(t_0,x_0,\dot{x}_0) = 0$. Hence, $(t_0,x_0,\dot{x}_0) \in \hat{\mathbb{L}}_0$ and $\hat{\mathbb{L}}_0$ is not empty.

Since $Z_{1,0}^T F_{\dot{x}}(t_0,x_0,\dot{x}_0)$ possesses full row rank due to Hypothesis 1, we may choose $\hat{Z}_2^T = [\,0 \;\; I_a\,]$. Differentiating (16.8) with respect to x yields

$$F_{\mu;x} + F_{\mu;y}\mathscr{Y}_x - Z_{2,0}\hat{F}_{2;x} = 0.$$

Multiplying with Z_2^T from the left, we get $Z_2^T Z_{2,0} \hat{F}_{2;x} = Z_2^T F_{\mu;x}$, where $Z_2^T Z_{2,0}$ is nonsingular in a neighborhhood of (t_0, x_0, y_0). Hence, we have

$$\operatorname{kernel} \hat{F}_{2;x} = \operatorname{kernel} Z_2^T F_{\mu;x}$$

such that we can choose $\hat{T}_2 = T_2$. The claim then follows since $\hat{F}_{1;\dot{x}} T_2 = Z_{1,0}^T F_{\dot{x}} T_2$ possesses full column rank due to Hypothesis 1. □

Since (16.10) has vanishing strangeness index, it is called a reduced DAE belonging to the original possibly higher index DAE (16.1). Note that a reduced DAE is defined in a neighborhood of every $(t_0, x_0, y_0) \in \mathbb{L}_\mu$, but also that it is not uniquely determined by the original DAE even for a fixed $(t_0, x_0, y_0) \in \mathbb{L}_\mu$. What is uniquely determined for a fixed $(t_0, x_0, y_0) \in \mathbb{L}_\mu$ is (at least when treating it as a function germ) the function $\mathscr{R}$.

Every continuously differentiable solution of (16.10) will satisfy $x_2 = \mathscr{R}(t, x_1)$ pointwise. Thus, it will also satisfy $\dot{x}_2 = \mathscr{R}_t(t, x_1) + \mathscr{R}_{x_1}(t, x_1)\dot{x}_1$ pointwise. Using these two relations, we can reduce the relation $\hat{F}_1(t, x_1, x_2, \dot{x}_1, \dot{x}_2) = 0$ of (16.10) to

$$\hat{F}_1(t, x_1, \mathscr{R}(t, x_1), \dot{x}_1, \mathscr{R}_t(t, x_1) + \mathscr{R}_{x_1}(t, x_1)\dot{x}_1) = 0. \tag{16.14}$$

If we now insert $x_2 = \mathscr{R}(t, x_1)$ into (16.8), we obtain

$$F_\mu(t, x_1, \mathscr{R}(t, x_1), \mathscr{Y}(t, x_1, \mathscr{R}(t, x_1))) = 0. \tag{16.15}$$

Differentiating this with respect to x_1 yields

$$F_{\mu;x_1} + F_{\mu;x_2}\mathscr{R}_{x_1} + F_{\mu;y}(\mathscr{Y}_{x_1} + \mathscr{Y}_{x_2}\mathscr{R}_{x_1}) = 0.$$

Multiplying with Z_2^T from the left, we get

$$Z_2^T [\, F_{\mu;x_1} \;\; F_{\mu;x_2} \,] \begin{bmatrix} I_d \\ \mathscr{R}_{x_1} \end{bmatrix} = 0.$$

Comparing with Hypothesis 1, we see that we may choose

$$T_2 = \begin{bmatrix} I_d \\ \mathscr{R}_{x_1} \end{bmatrix}. \tag{16.16}$$

Differentiating now (16.14) with respect to $\dot{x}_1$ and using the definition of $\hat{F}_1$, we find

$$Z_{1,0}^T F_{\dot{x}_1} + Z_{1,0}^T F_{\dot{x}_2}\mathscr{R}_{x_1} = Z_{1,0}^T F_{\dot{x}} T_2,$$

which is nonsingular due to Hypothesis 1. In order to apply the implicit function theorem, we need to require that $(t_0, x_{10}, \dot{x}_{10})$ solves (16.14). Note that this is not a consequence of $(t_0, x_0, y_0) \in \mathbb{L}_\mu$. Under this additional requirement, the implicit function theorem implies the local existence of a function $\mathscr{L}$ satisfying

$$\hat{F}_1(t, x_1, \mathscr{R}(t, x_1), \mathscr{L}(t, x_1), \mathscr{R}_t(t, x_1) + \mathscr{R}_{x_1}(t, x_1)\mathscr{L}(t, x_1)) \equiv 0. \tag{16.17}$$

With the help of the functions $\mathscr{L}$ and $\mathscr{R}$, we can formulate a further DAE of the form

$$\begin{aligned} \dot{x}_1 &= \mathscr{L}(t, x_1), \ (d \text{ differential equations}) \\ x_2 &= \mathscr{R}(t, x_1). \ (a \text{ algebraic equations}) \end{aligned} \tag{16.18}$$

Note that this DAE consists of a decoupled ODE for x_1, where we can freely impose an initial condition as long as we remain in the domain of $\mathscr{L}$. Having so fixed x_1, the part x_2 follows directly from the second relation. In this sense, (16.18) can be seen as a prototype for a regular DAE.

The further discussion is now dedicated to the relation between (16.18) and the original DAE.

We start with the assumption that the original DAE (16.1) possesses a smooth local solution x^* in the sense that there is a continuous path $(t, x^*(t), \mathscr{P}(t)) \in \mathbb{L}_\mu$ defined on a neighborhood of t_0, where the first block of $\mathscr{P}$ coincides with $\dot{x}^*$. Note that if x^* is $(\mu+1)$-times continuously differentiable we can just take the path given by $\mathscr{P} = (\dot{x}^*, \ddot{x}^*, \ldots, (d/dt)^{\mu+1}x^*)$. Setting $(t_0, x_0, y_0) = (t_0, x^*(t_0), \mathscr{P}(t_0))$, Theorem 3 yields that $x_2^*(t) = \mathscr{R}(t, x_1^*(t))$. Hence, $\dot{x}_2^*(t) = \mathscr{R}_t(t, x_1^*(t)) + \mathscr{R}_{x_1}(t, x_1^*(t))\dot{x}_1^*(t)$. In particular, Eq. (16.14) is solved by $(t, x_1, \dot{x}_1) = (t, x_1^*, \dot{x}_1^*)$. Thus, it follows also that $\dot{x}_1^*(t) = \mathscr{L}(t, x_1^*(t))$. In this way, we have proven the following theorem.

Theorem 5 *Let F from (16.1) satisfy Hypothesis 1 with (μ, a, d). Then every local solution x^* of (16.1) in the sense that it extends to a continuous local path $(t, x^*(t), \mathscr{P}(t)) \in \mathbb{L}_\mu$, where the first block of $\mathscr{P}$ coincides with $\dot{x}^*$, also solves the reduced problems (16.10) and (16.18).*

16.3.3 The Way Back

To show a converse result to Theorem 5, we need to require the solvability of (16.14) for the local existence of the function $\mathscr{L}$. For this, we assume that F not only satisfies Hypothesis 1 with (μ, a, d), but also with $(\mu + 1, a, d)$. Let now $(t_0, x_0, y_0, z_0) \in \mathbb{L}_{\mu+1}$. Due to the construction of F_ℓ, we have

$$F_{\mu+1} = \begin{bmatrix} F_\mu \\ (\frac{d}{dt})^{\mu+1} F \end{bmatrix}, \quad F_{\mu+1;y,z} = \begin{bmatrix} F_{\mu;y} & 0 \\ ((\frac{d}{dt})^{\mu+1} F)_y & ((\frac{d}{dt})^{\mu+1} F)_z \end{bmatrix}, \tag{16.19}$$

where the independent variable z is a short-hand notation for $x^{(\mu+2)}$. Since $F_{\mu;y}$ and $F_{\mu+1;y,z}$ are assumed to have the same rank drop, we find that Z_2 belonging to F_μ satisfies

$$[\,Z_2^T\ \ 0\,]F_{\mu+1;y,z} = [\,Z_2^T\ \ 0\,]\begin{bmatrix} F_{\mu;y} & 0 \\ ((\frac{d}{dt})^{\mu+1}F)_y & ((\frac{d}{dt})^{\mu+1}F)_z \end{bmatrix} = [\,0\ \ 0\,].$$

Consequently, in Hypothesis 1 considered for $F_{\mu+1}$, we may choose $[\,Z_2^T\ \ 0\,]$ describing the left nullspace of $F_{\mu+1;y,z}$ such that the same choices are possible for T_2 and Z_1.

Observing that we may write the independent variables (t,x,y,z) also as $(t,x,\dot{x},\dot{y})$ by simply changing the partitioning of the blocks, and that the equation $F_{\mu+1}=0$ contains $F_\mu=0$ as well as $\frac{d}{dt}F_\mu=0$, which has the form

$$\tfrac{d}{dt}F_\mu = F_{\mu;t} + F_{\mu;x}\dot{x} + F_{\mu;y}\dot{y} = 0,$$

we get

$$Z_2^T F_{\mu;t} + Z_2^T F_{\mu;x}\dot{x} = 0.$$

Using the same splitting $x=(x_1,x_2)$ as above and $\dot{x}=(\dot{x}_1,\dot{x}_2)$ accordingly, we obtain

$$Z_2^T F_{\mu;t} + Z_2^T F_{\mu;x_1}\dot{x}_1 + Z_2^T F_{\mu;x_2}\dot{x}_2 = 0,$$

which yields

$$\dot{x}_2 = -(Z_2^T F_{\mu;x_2})^{-1}(Z_2^T F_{\mu;t} + Z_2^T F_{\mu;x_1}\dot{x}_1). \tag{16.20}$$

On the other hand, differentiation of (16.13) with respect to t yields

$$F_{\mu;t} + F_{\mu;x_1}\dot{x}_1 + F_{\mu;x_2}(\mathscr{R}_t + \mathscr{R}_{x_1}\dot{x}_1) + F_{\mu;y_1}\dot{y}_1 + F_{\mu;y_2}(\mathscr{H}_t + \mathscr{H}_{x_1}\dot{x}_1 + \mathscr{H}_{y_1}\dot{y}_1) = 0$$

and thus

$$Z_2^T F_{\mu;t} + Z_2^T F_{\mu;x_1}\dot{x}_1 = -Z_2^T F_{\mu;x_2}(\mathscr{R}_t + \mathscr{R}_{x_1}\dot{x}_1).$$

Inserting this into (16.20) yields

$$\dot{x}_2 = \mathscr{R}_t + \mathscr{R}_{x_1}\dot{x}_1. \tag{16.21}$$

Hence, the given point $(t_0, x_0, \dot{x}_0, \dot{y}_0)$ satisfies

$$\dot{x}_{20} = \mathscr{R}_t(t_0, x_{10}) + \mathscr{R}_{x_1}(t_0, x_{10})\dot{x}_{10}.$$

It then follows that $(t_0, x_{10}, \dot{x}_{10})$ solves (16.14). In particular, this guarantees that the implicit function theorem is applicable to (16.14) leading to a locally defined $\mathscr{L}$. Thus, the reduced system (16.18) is locally well-defined. Moreover, for every initial value for x_1 near x_{10}, the initial value problem for x_1 in (16.18) possesses a solution x_1^*. The second equation in (16.18) then yields a locally defined x_2^* such that $x^* = (x_1^*, x_2^*)$ forms a solution of (16.18).

For the same reasons as for $\mathbb{L}_\mu$, the set $\mathbb{L}_{\mu+1}$ can be locally parametrized by $n+1$ variables. Among these variables are again t and x_1. But since x_2, $\dot{x}_1$, and $\dot{x}_2$ are all functions of (t, x_1), the remaining variables, say p, are now from $\dot{y}$. In particular, there is a locally defined function $\mathscr{Z}$ satisfying

$$F_{\mu+1}(t, x_1, \mathscr{R}(t, x_1), \mathscr{L}(t, x_1), \mathscr{R}_t(t, x_1) + \mathscr{R}_{x_1}(t, x_1)\mathscr{L}(t, x_1), \mathscr{Z}(t, x_1, p)) \equiv 0.$$

Choosing now $x_1^*(t)$ for x_1 and $p^*(t)$ arbitrarily within the domain of $\mathscr{Z}$, for example $p^*(t) = p_0$, where p_0 is the matching part of $\dot{y}_0$, yields

$$F_{\mu+1}(t, x_1^*(t), x_2^*(t), \dot{x}_1^*(t), \dot{x}_2^*(t), \mathscr{Z}(t, x_1^*(t), p^*(t))) \equiv 0,$$

which contains

$$F(t, x_1^*(t), x_2^*(t), \dot{x}_1^*(t), \dot{x}_2^*(t)) \equiv 0 \tag{16.22}$$

in the first block. But this means nothing else than that $x^* = (x_1^*, x_2^*)$ locally solves the original problem. Moreover, locally there is a continuous function $\mathscr{P}$ such that its first block coincides with $\dot{x}^*$ and $(t, x^*(t), \mathscr{P}(t)) \in \mathbb{L}_\mu$. Summarizing, we have proven the following statement.

Theorem 6 *If F satisfies Hypothesis 1 with (μ, a, d) and $(\mu+1, a, d)$ then every local solution x^* of the reduced DAE (16.18) is also a local solution of the original DAE. Moreover, it extends to a continuous local path $(t, x^*(t), \mathscr{P}(t)) \in \mathbb{L}_\mu$, where the first block of $\mathscr{P}$ coincides with $\dot{x}^*$.*

The numerical treatment of DAEs is usually based on the assumption that there is a solution to be computed. In view of Theorem 5 it is therefore sufficient to work with the derivative array F_μ. However, we must assume in addition that the given point $(t_0, x_0, y_0) \in \mathbb{L}_\mu$ provides suitable starting values for the nonlinear system solvers being part of the numerical procedure. Note that this corresponds to the assumption that we may apply the implicit function theorem for the definition of $\mathscr{L}$.

16.3.4 Overdetermined Consistent DAEs

Hypothesis 1 can be generalized in various ways. For example, we may include underdetermined problems which would cover control problems by treating states and controls as indistinguishable parts of the unknown. We may also allow overdetermined problems or problems with redundant equations. The main problem in the formulation of corresponding hypotheses is for which points to require properties of the Jacobians of the derivative array equation. Note that the restriction in Hypothesis 1 to points in the solution set of the derivative array equation leads to better covariance properties of the hypothesis, see [13], but it excludes problems where this set is empty, e.g., linear least-squares problems. In the following, we want to present a generalization to overdetermined, but consistent (i.e., solvable) DAEs. Such DAEs may arise by extending a given DAE by some or all hidden constraints, i.e., relations contained in $\hat{F}_2(t,x) = 0$ that require the differentiation of the original DAE, or by extending a given DAE or even an ODE by known first integrals.

Hypothesis 2 *There exist (nonnegative) integers μ, a, d, and v such that the set*

$$\mathbb{L}_\mu = \{(t,x,y) \in \mathbb{R}^{(\mu+2)n+1} \mid F_\mu(t,x,y) = 0\} \tag{16.23}$$

associated with F is nonempty and such that for every point $(t_0, x_0, y_0) \in \mathbb{L}_\mu$, there exists a (sufficiently small) neighborhood $\mathbb{V}$ in which the following properties hold:

1. *We have* rank $F_{\mu;y} = (\mu+1)m - v$ *on $\mathbb{L}_\mu \cap \mathbb{V}$ such that there exists a smooth matrix function Z_2 of size $((\mu+1)m, v)$ and pointwise maximal rank, satisfying $Z_2^T F_{\mu;y} = 0$ on $\mathbb{L}_\mu \cap \mathbb{V}$.*
2. *We have* rank $Z_2^T F_{\mu;x} = a$ *on $\mathbb{V}$ such that there exists a smooth matrix function T_2 of size (n,d), $d = n - a$, and pointwise maximal rank, satisfying $Z_2^T F_{\mu;x} T_2 = 0$.*
3. *We have* rank $F_{\dot{x}} T_2 = d$ *on $\mathbb{V}$ such that there exists a smooth matrix function Z_1 of size (m,d) and pointwise maximal rank, satisfying* rank $Z_1^T F_{\dot{x}} T_2 = d$.

A corresponding construction as for Hypothesis 1 shows that Hypothesis 2 implies a reduced DAE of the form (16.10) with the same properties as stated there. In particular, a result similar to Theorem 5 holds. Due to the assumed consistency, the omitted relations (the reduced DAEs are $m - n$ scalar relations short) do not contradict these equations. Thus, the solutions fixed by the reduced DAE will be solutions of the original overdetermined DAE under assumptions similar to those of Theorem 6. Since the arguments are along the same lines as presented above, we omit details here.

An example for a problem covered by Hypothesis 2 is given by Example 1 when we just add the two equations obtained by differentiation and elimination to the original DAE leading to a problem consisting of 9 equations in 7 unknowns. A second example, which we will also address in the numerical experiments, consists of an ODE with known first integral.

Example 2 A simple predator/prey model is described by the so-called Lotka/Volterra system

$$\dot{x}_1 = x_1(1-x_2), \quad \dot{x}_2 = -c\,x_2(1-x_1),$$

where $c > 0$ is some given constant, see, e.g., [14]. It is well-known that

$$H(x_1, x_2) = c(x_1 - \log x_1) + (x_2 - \log x_2)$$

is a first integral of this system implying that the positive solutions are periodic. The combined overdetermined system

$$\begin{aligned}\dot{x}_1 &= x_1(1 - x_2),\\ \dot{x}_2 &= -c\, x_2(1 - x_1),\\ c(x_1 - \log x_1) + (x_2 - \log x_2) &= H_0,\end{aligned}$$

where $H_0 = H(x_{10}, x_{20})$ for given initial values $x_1(t_0) = x_{10}$, $x_2(t_0) = x_{20}$, is therefore consistent. Moreover, it can be shown to satisfy Hypothesis 2 with $\mu = 0$, $a = 1$, $d = 1$, and $v = 1$. In contrast to Example 1, we cannot decide in advance which of the two differential equations should be used together with the algebraic constraint. For stability reasons, we should rather use an appropriate linear combination of the two differential equations. But this just describes the role of Z_1 in Hypothesis 2. ◇

16.4 Integration of Nonlinear DAEs

In this section, we discuss several issues that play a role when one wants to integrate DAE systems numerically in an efficient way.

16.4.1 Discretizations

The idea for developing methods for the numerical solution of unstructured DAEs is to discretize not the original DAE (16.1) but the reduced DAE (16.10) because of its property that it does not contain hidden constraints, i.e., that we do not need to differentiate the functions in the reduced DAE. Of course, the functions in the reduced DAE are themselves defined by relations that contain differentiations. But these are differentiations of the original function F which may be obtained by hand or by means of automatic differentiation.

A well-known discretization of DAEs are the BDF methods, see, e.g., [6]. We want to concentrate here on two families of one-step methods that are suitable for the integration of DAEs of the form (16.10). In the following, we denote the initial value at t_0 by x_0 and the stepsize by h. The discretization should then fix an approximate solution x_1 at the point $t_1 = t_0 + h$.

The first family of methods are the Radau IIa methods, which are collocation methods based on the Radau nodes

$$0 < \gamma_1 < \cdots < \gamma_s = 1, \tag{16.24}$$

where $s \in \mathbb{N}$ denotes the number of stages, see, e.g., [10]. The discretization of (16.10) then reads

$$\begin{aligned} &\hat{F}_1(t_0 + \gamma_j h, X_j, \tfrac{1}{h}(v_{j0}x_0 + \textstyle\sum_{l=1}^{s} v_{jl}X_l)) = 0, \\ &\hat{F}_2(t_0 + \gamma_j h, X_j) = 0, \qquad\qquad j = 1, \ldots, s, \end{aligned} \tag{16.25}$$

together with $x_1 = X_s$, where X_j, $j = 1, \ldots, s$, denote the stage values of the Runge-Kutta scheme. The coefficients v_{jl} are determined by the nodes (16.24). For details and the proof of the following convergence result, see, e.g., [13].

Theorem 7 *The Radau IIa methods (16.25) applied to a reduced DAE (16.10) are convergent of order* $p = 2s - 1$.

Note that the Radau IIa methods exhibit the same convergence order as in the special case of an ODE. The produced new value x_1 satisfies all the constraints due to the included relation $\hat{F}_2(t_1, x_1) = 0$.

The second family of methods consists of partitioned collocation methods, which use Gauß nodes for the differential equations and Lobatto nodes for the algebraic equations given by

$$0 < \rho_1 < \cdots < \rho_k < 1, \quad 0 = \sigma_0 < \cdots < \sigma_k = 1, \tag{16.26}$$

with $k \in \mathbb{N}$. Observe that we use one more Lobatto node equating thus the order of the corresponding collocation methods for ODEs. The discretization of (16.10) then reads

$$\begin{aligned} &\hat{F}_1(t_0 + \rho_j h, u_{j0}x_0 + \textstyle\sum_{l=1}^{k} u_{jl}X_l, \tfrac{1}{h}(v_{j0}x_0 + \textstyle\sum_{l=1}^{k} v_{jl}X_l)) = 0, \\ &\hat{F}_2(t_0 + \sigma_j h, X_j) = 0, \qquad\qquad j = 1, \ldots, k, \end{aligned} \tag{16.27}$$

together with $x_1 = X_k$. The coefficients u_{jl} and v_{jl} are determined by the nodes (16.26). For details and the proof of the following convergence result, see again [13].

Theorem 8 *The Gauß-Lobatto methods (16.27) applied to a reduced DAE (16.10) are convergent of order* $p = 2k$.

Note that in contrast to the Radau IIa methods, the Gauß-Lobatto methods are symmetric. Thus, they may be prefered when symmetry of the method is an issue, e.g., in the solution of boundary value problems. In the case of an ODE, the Gauß-Lobatto methods reduce to the corresponding Gauß collocation methods. As for the Radau IIa methods, the produced new value x_1 satisfies all the constraints due to the included relation $\hat{F}_2(t_1, x_1) = 0$.

For the actual computation, we lift the discretization from the reduced DAE to the original DAE by using Theorem 3. In particular, we replace every relation of the form $\hat{F}_2(t,x) = 0$ by $F_\mu(t,x,y) = 0$ with the help of an additional unknown y. Note that by this process the system describing the discretization becomes underdetermined. Nevertheless, the desired value x_1 will still (at least locally) be uniquely fixed. The Radau IIa methods then read

$$\begin{aligned} &Z_{1,0}^T F(t_0 + \gamma_j h, X_j, \tfrac{1}{h}(v_{j0}x_0 + \textstyle\sum_{l=1}^s v_{jl}X_l)) = 0, \\ &F_\mu(t_0 + \gamma_j h, X_j, Y_j) = 0, \qquad j = 1,\ldots,s, \end{aligned} \tag{16.28}$$

and the Gauß-Lobatto methods then read

$$\begin{aligned} &Z_{1,0}^T F(t_0 + \rho_j h, u_{j0}x_0 + \textstyle\sum_{l=1}^k u_{jl}X_l, \tfrac{1}{h}(v_{j0}x_0 + \textstyle\sum_{l=1}^k v_{jl}X_l)) = 0, \\ &F_\mu(t_0 + \sigma_j h, X_j, Y_j) = 0, \qquad j = 1,\ldots,k. \end{aligned} \tag{16.29}$$

In the case of overdetermined DAEs governed by Hypothesis 2, the discretizations look the same.

In order to perform a step with the above one-step methods given an initial value $(t_0, x_0, y_0) \in \mathbb{L}_\mu$, we can determine $Z_{1,0}$ along the lines of the above hypotheses. We then must provide starting values for a suitable nonlinear system solver for the solution of the nonlinear systems describing the discretization, typically the Gauß-Newton method or a variant of it. Upon convergence, we obtain a final value (t_1, x_1, y_1) as part of the overall solution (which includes the internal stages), which will then be the initial value for the next step. Note that for performing a Gauß-Newton-like method for these problems, which we will write as $\mathscr{F}(z) = 0$ for short in the following, we must be able to evaluate the function $\mathscr{F}$ and its Jacobian $\mathscr{F}_z$ at given points. Thus, we must be able to evaluate F and F_μ and their Jacobians, which can be done by using automatic differentiation, see below.

16.4.2 Gauß-Newton-Like Processes

The design of the Gauß-Newton-like method is crucial for the efficiency of the approach. Note that we had to replace $\hat{F}_2$ by F_μ thus increasing the number of equations and unknowns significantly. However, there is some structure in the equations that can be utilized in order to improve the efficiency. We will sketch this approach in the following for the case of the Radau IIa discretization. Similar techniques can be applied to the case of the Gauß-Lobatto discretization.

Linearizing the equation $\mathscr{F}(z) = 0$ around some given z yields the linear problem $\mathscr{F}(z) + \mathscr{F}_z(z)\Delta z = 0$ for the correction Δz. The ordinary Gauß-Newton method is then characterized by solving for Δz by means of the Moore-Penrose pseudoinverse $\mathscr{F}_z(z)^+$ of $\mathscr{F}_z(z)$, i.e.,

$$\Delta z = -\mathscr{F}_z(z)^+ \mathscr{F}(z). \tag{16.30}$$

Instead of the Moore-Penrose pseudoinverse, we are allowed to use any other equation-solving generalized inverse of $\mathscr{F}_z(z)$. Due to the consistency of the nonlinear problem to be solved, we are also allowed to perturb the Jacobian as long as the perturbation is sufficiently small or even tends to zero during the iteration.

In the case (16.28), linearization leads to

$$\begin{aligned} Z_{1,0}^T F_x^j \Delta X_j + Z_{1,0}^T F_{\dot{x}}^j \tfrac{1}{h} \textstyle\sum_{l=1}^s v_{jl} \Delta X_l &= -Z_{1,0}^T F^j, \\ F_{\mu;x}^j \Delta X_j + F_{\mu;y}^j \Delta Y_j &= -F_\mu^j, \qquad j = 1, \ldots, s. \end{aligned} \tag{16.31}$$

which is to be solved for $(\Delta X_j, \Delta Y_j)$, $j = 1, \ldots, s$. The superscript j indicates, that the corresponding function is evaluated at the argument occurring in the j-th equation, i.e., at $(t_0 + \gamma_j h, X_j, \frac{1}{h}(v_{j0} x_0 + \sum_{l=1}^s v_{jl} X_l))$ in the case of F and $(t_0 + \gamma_j h, X_j, Y_j)$ in the case of F_μ. Since (16.28) contains $F_\mu^j = 0$, we will have $\operatorname{rank} F_{\mu;y}^j = (\mu + 1)n - a$ at a solution of (16.28) due to Hypothesis 1. Near the solution, the matrix $F_{\mu;y}^j$ is thus a perturbation of a matrix with rank drop a. The idea therefore is to perturb $F_{\mu;y}^j$ to a matrix M_j with $\operatorname{rank} M_j = (\mu + 1)n - a$. Such a perturbation can be obtained by rank revealing QR decomposition or by singular value decomposition, see, e.g., [7]. The second part of (16.31) then consists of equations of the form

$$F_{\mu;x}^j \Delta X_j + M_j \Delta Y_j = -F_\mu^j. \tag{16.32}$$

With the help of an orthogonal matrix $[\, Z_{2,j}' \;\; Z_{2,j} \,]$, where the columns of $Z_{2,j}$ form an orthonormal basis of the left nullspace of M_j, we can split (16.32) into

$$Z_{2,j}'^T F_{\mu;x}^j \Delta X_j + Z_{2,j}'^T M_j \Delta Y_j = -Z_{2,j}'^T F_\mu^j, \quad Z_{2,j}^T F_{\mu;x}^j \Delta X_j = -Z_{2,j}^T F_\mu^j. \tag{16.33}$$

The first part can be solved for ΔY_j via the Moore-Penrose pseudoinverse

$$\Delta Y_j = -(Z_{2,j}'^T M_j)^+ Z_{2,j}'^T (F_\mu^j + F_{\mu;x}^j \Delta X_j) \tag{16.34}$$

in terms of ΔX_j, thus fixing a special equation-solving pseudoinverse of the Jacobian under consideration. In order to determine the corrections ΔX_j, we take an orthogonal matrix $[\, T_{2,j}' \;\; T_{2,j} \,]$, where the columns of $T_{2,j}$ form an orthonormal basis of the right nullspace of $Z_{2,j}'^T F_{\mu;x}^j$, which is of full row rank near the solution due to Hypothesis 1. Defining the transformed corrections

$$\Delta V_j' = T_{2,j}'^T \Delta X_j, \quad \Delta V_j = T_{2,j}^T \Delta X_j, \tag{16.35}$$

we have $\Delta X_j = T_{2,j}' \Delta V_j' + T_{2,j} \Delta V_j$ and the second part of (16.33) becomes

$$Z_{2,j}^T F_{\mu;x}^j T_{2,j}' \Delta V_j' = -Z_{2,j}^T F_\mu^j. \tag{16.36}$$

Due to Hypothesis 1, the square matrix $Z_{2,j}^T F_{\mu;x}^j T_{2,j}'$ is nonsingular near a solution such that we can solve for $\Delta V_j'$ to get

$$\Delta V_j' = -(Z_{2,j}^T F_{\mu;x}^j T_{2,j}')^{-1} Z_{2,j}^T F_\mu^j. \tag{16.37}$$

Finally, transforming the equation in the first part of (16.31) to the variables $(\Delta V_j', \Delta V_j)$ and eliminating the terms $\Delta V_j'$ leaves a system in the unknowns ΔV_j, which is of the same size and form as if we would discretize an ODE of d equations by means of the Radau IIa method. This means that we actually have reduced the complexity to that of solving an ODE of the size of the differential part. Solving this system for the quantities ΔV_j and combining these with the already obtained values $\Delta V_j'$ then yields the corrections ΔX_j.

The overall Gauß-Newton-like process, which can be written as

$$\Delta z = -\mathscr{J}(z)^+ \mathscr{F}(z) \tag{16.38}$$

with $\mathscr{J}(z) \to \mathscr{F}_z(z)$ when z converges to a solution, can be shown to be locally and quadratically convergent, see again [13]. Using such a process is indispensable for the efficient numerical solution of unstructured DAEs.

16.4.3 Minimal-Norm-Corrected Gauß-Newton Method

We have implemented the approach of the previous section both for the Radau IIa methods and for the Gauß-Lobatto methods. Experiments show that one can successfully solve nonlinear DAEs even for larger values of μ without having to assume a special structure. Applying it to the problem of Example 1, however, reveals a drawback of the approach described so far. In particular, we observe the following. Trying to solve the problem of Example 1 on a larger time interval starting at $t = 0$, one realizes that the integration terminates at about $t = 14.5$ because the nonlinear system solver fails, cp. Fig. 16.1. A closer look shows that the reason for this is that the undetermined components y, which are not relevant for the solution one is interested in, run out of scale. Scaling techniques cannot avoid the effect. They can only help to make use of the whole range provided by the floating point arithmetic. Using diagonal scaling, the iteration terminates then at about $t = 71.4$, cp. again Fig. 16.1.

Actually, proceeding from numerical approximations (x_i, y_i) at t_i to numerical approximations (x_{i+1}, y_{i+1}) at t_{i+1} consists of two mechanisms. First, we must provide a starting value z for the nonlinear system solver. We call this predictor and write

$$z = \mathfrak{P}(x_i, y_i). \tag{16.39}$$

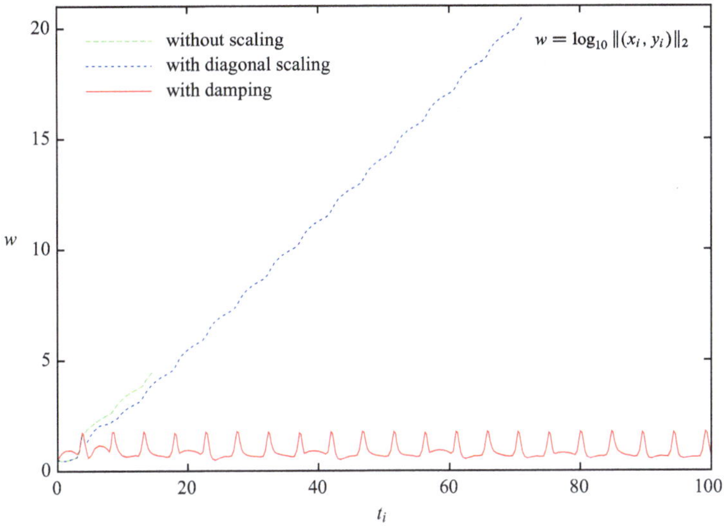

Fig. 16.1 Decadic logarithm of the Euclidean norm of the generated numerical solution (x_i, y_i)

Then, the nonlinear system solver, called corrector in this context, yields the new approximation according to

$$(x_{i+1}, y_{i+1}) = \mathfrak{C}(z). \tag{16.40}$$

Thus the numerical flow Φ of our method effectively has the form

$$(x_{i+1}, y_{i+1}) = \Phi(x_i, y_i), \quad \Phi = \mathfrak{C} \circ \mathfrak{P}. \tag{16.41}$$

The problem can then be described as follows. Even if the actual solution and the numerical approximations x_i are bounded, there is no guaranty that the overall numerical solutions (x_i, y_i) stay bounded.

In [3], it was examined how different predictors $\mathfrak{P}$, in particular extrapolation of some order, influence the overall behavior of the process. The result was that linear extrapolation should be prefered to higher order extrapolation. However, even linear extrapolation cannot avoid the blow-up.

The idea here is to modify the corrector $\mathfrak{C}$, in particular to introduce damping into the nonlinear system solver. Recall that the nonlinear system to be solved does in general not have a unique solution but that the part one is interested in, namely x_{i+1}, is unique. Consider the iteration given by

$$\Delta z = -\alpha z - \mathscr{F}_z(z)^+ (\mathscr{F}(z) - \alpha \mathscr{F}_z(z) z) \tag{16.42}$$

with $\alpha \in [0, 1]$ replacing (16.30). For $\alpha = 0$, we rediscover (16.30). For $\alpha = 1$, we have

$$z + \Delta z = \mathscr{F}_z(z)^+(\mathscr{F}_z(z)z - \mathscr{F}(z)),$$

which in the linear case $\mathscr{F}(z) = \mathbf{A}z - \mathbf{b}$ leads to $z + \Delta z = \mathbf{A}^+\mathbf{b}$ and thus to the shortest solution with respect to the Euclidean norm. In this sense, the process defined by (16.42) contains some damping. Moreover, if $\alpha \to 0$ quadratically during the iteration, we maintain the quadratic convergence of the Gauß-Newton process. The following result is due to [2].

Theorem 9 *Consider the problem $\mathscr{F}(z) = 0$ and assume that the Jacobians $\mathscr{F}_z(z)$ have full row rank. Furthermore, consider the iteration defined by (16.42) and assume that $\alpha \to 0$ quadratically during the iteration. Then the so defined process yields iterates that converge locally and quadratically to a solution of the given problem.*

Observe that replacing (16.30) by (16.42) only consists of a slight modification of the original process. The main computational effort, namely the representation of $\mathscr{F}_z(z)^+$, stays the same. Moreover, using a perturbed Jacobian $\mathscr{J}(z)$ instead of $\mathscr{F}_z(z)$ is still possible and does not influence the convergence behavior. Figure 16.1 shows that with this modified nonlinear system solver we are now able to produce bounded overall solutions in the case of Example 1.

16.4.4 Automatic Differentiation

In order to integrate (unstructured) DAEs, we must provide procedures for the evaluation of F and F_μ together with their Jacobians. As already mentioned this can be done by exploiting techniques from automatic differentiation, see, e.g., [9].

The simplest approach is to evaluate the functions on the fly, i.e., by using special classes and overloaded operators, a call of a template function which implements F can produce the needed evaluations just by changing the class of the variables. The drawback in this approach is that there may be a lot of trivial computations when the derivatives are actually zero. Moreover, no code optimization is possible.

An alternative approach consists of two phases. First, one uses automatic differentiation to produce code for the evaluation of the needed functions. This code can then be easily compiled using optimization. The drawback here is that one has to adapt the automatic differentiation process or the produced code to the form one needs for the following integration of the DAE. Nevertheless, one can expect this approach to be more efficient for the actual integration of the DAE, especially for larger values of μ. Actually, one would prefer the first approach while a model is developed. If the model is finalized, one would then prefer the second approach.

As an example, we have run the problem from Example 1 with both approaches on the interval $[0, 100]$ using the Gauß-Lobatto method for $k = 3$ and the minimal-norm-corrected Gauß-Newton-like method starting with $\alpha = 0.1$ and using successive squaring. The computing time in the first case exploiting automatic differentiation on the fly was 2.8 s. The computing time in the second case exploiting optimized code produced by automatic differentiation was 0.6 s.

16.4.5 Exploiting First Integrals

If for a given ODE or DAE model first integrals are known, they should be included into the model thus enforcing the produced numerical approximations to obey these first integrals. The enlarged system is of course overdetermined but consistent. In general, it is not clear how to deduce a square system from the overdetermined one in order to apply standard integration procedures, cp. Example 2.

In Example 1, there are two hidden constraints which were found by differentiation. As already mentioned there, it is in this case possible to reduce the problem consisting of the original equations and the two additional constraints to a square system by just omitting two equations of the original system. Sticking to automatic differentiation and using the same setting as above, we can solve the overdetermined system in 0.9 s and the reduced square system in 0.7 s.

For Example 2, such a beforehand reduction is not so obvious, but still possible due to the simple structure of this specific problem. We solved the overdetermined problem by means of the implicit Euler method (which is the Radau IIa method for $s = 1$) as well as the original ODE by means of the explicit and implicit Euler method performing 1,000 steps with stepsize $h = 0.02$. The results are shown in Fig. 16.2. As one would expect, the numerical solution for the ODE

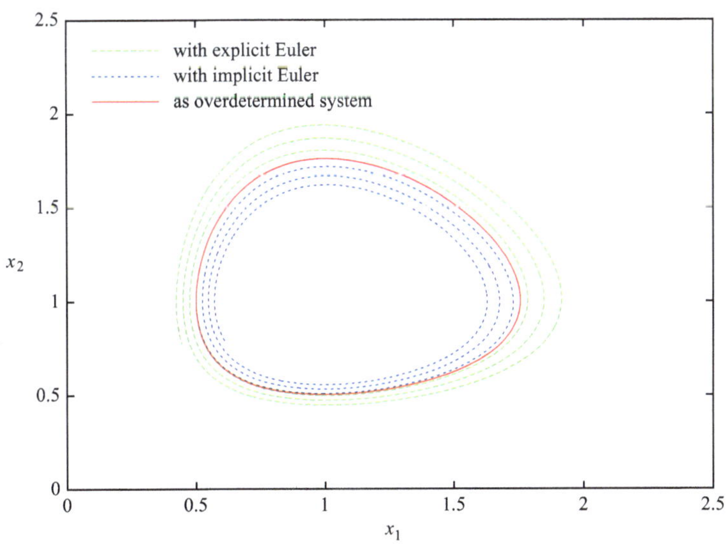

Fig. 16.2 Numerical solutions for the Lotka/Volterra model

produced by the explicit Euler method spirals outwards thus increasing the energy while the numerical solution for the ODE produced by the implicit Euler method spirals inwards thus decreasing energy. The numerical solution obtained from the overdetermined system, of course, conserves the energy by construction.

16.4.6 Path Following by Arclength Parametrization

There are two extreme cases of DAEs, the case of ODEs $\dot{x} = f(t,x)$ on the one hand and the case of nonlinear equations $f(x) = 0$ on the other hand. For $F(t,x,\dot{x}) = \dot{x} - f(t,x)$, Hypothesis 1 is trivially satisfied with $\mu = 0$, $a = 0$, and $d = n$. For $F(t,x,\dot{x}) = f(x)$, Hypothesis 1 is satisfied with $\mu = 0$, $a = n$, and $d = 0$, provided $f_x(x)$ is nonsingular for all $x \in \mathbb{L}_0$. Since t does neither occur as an argument nor via differentiated variables, the solutions are constant in time and thus, as solutions of a DAE, not so interesting. This changes if one considers parameter dependent nonlinear equations $f(x,\tau) = 0$, where τ shall be a scalar parameter. The problem is now underdetermined. Thus, it cannot satisfy one of the above hypotheses. Under the assumption that $[\ f_x \quad f_\tau\ ]$ has full row rank for all $(x,\tau) \in \mathbb{M} = f^{-1}(\{0\}) \neq \emptyset$, the solution set forms a one-dimensional manifold. If one is interested in tracing this manifold, one can use path following techniques, see, e.g., [4, 17]. However, it is also possible to treat such problems with solution techniques for DAEs. A first choice would be to interpret the parameter τ as time t of the DAE. This would, however, imply that the parameter τ is strictly monotone along the one-dimensional manifold. But there are applications, where this is not the case. It may even happen that the points where the parameter τ is extremal are of special interest. In order to treat such problems, we are in need of defining a special type of time which is monotone in any case. Such a quantity is given as the arclength of the one-dimensional manifold, measured say from the initial point we start off. Since the arclength parametrization of a path is characterized by the property that the derivative with respect to the parametrization has Euclidean length one, we consider the DAE

$$f(x,\tau) = 0, \quad \|\dot{x}\|_2^2 + |\dot{\tau}|^2 = 1 \tag{16.43}$$

for the unknown (x,τ). If $(x_0,\tau_0) \in \mathbb{M}$ and $[\ f_x \quad f_\tau\ ]$ is of full row rank on $\mathbb{M}$, the implicit function theorem yields that there is a local solution path $(\hat{x}(t),\hat{\tau}(t))$ passing through (x_0,τ_0). Moreover, $\|\dot{\hat{x}}(t)\|_2^2 + |\dot{\hat{\tau}}(t)|^2 = 1$, when we parametrize by arclength. Hence, the DAE (16.43) possesses a solution. Moreover, writing (16.43) as $F(z,\dot{z}) = 0$ with $z = (x,\tau)$, we have

$$\mathbb{L}_0 = \{(z,\dot{z}) \mid z = (\hat{x}(t),\hat{\tau}(t)),\ \dot{z} = (\dot{\hat{x}}(t),\dot{\hat{\tau}}(t))\}$$

in Hypothesis 1. Because of

$$F_{0;\dot z} = \begin{bmatrix} 0 & 0 \\ 2\dot x^T & 2\dot\tau \end{bmatrix}, \quad F_{0;z} = \begin{bmatrix} f_x & f_\tau \\ 0 & 0 \end{bmatrix},$$

we may choose

$$Z_2 = \begin{bmatrix} I_n \\ 0 \end{bmatrix}.$$

By assumption, $Z_2^T F_{0;z} = [\ f_x \ \ f_\tau\ ]$ has full row rank and we may choose T_2 as a normalized vector in kernel$[\ f_x \ \ f_\tau\ ]$, which is one-dimensional. In particular, we may choose

$$T_2 = \begin{bmatrix} \dot{\hat x} \\ \dot{\hat\tau} \end{bmatrix}$$

on $\mathbb{L}_0$. Finally, we observe that

$$F_{\dot z} T_2 = \begin{bmatrix} 0 & 0 \\ 2\dot{\hat x}^T & 2\dot{\hat\tau} \end{bmatrix} \begin{bmatrix} \dot{\hat x} \\ \dot{\hat\tau} \end{bmatrix} = \begin{bmatrix} 0 \\ 2 \end{bmatrix}$$

has full column rank at the solution and thus in a neighborhood of it. Hence, the DAE (16.43) satisfies Hypothesis 1 with $\mu = 0$, $a = n$, and $d = 1$, where n denotes the size of x. We can then use DAE solution techniques to solve (16.43) thus tracing the solution path of the original parametrized system of nonlinear equations.

In order to determine points along the path, where the parameter τ is extremal, we may combine the DAE (16.43) with a root finding procedure, e.g., along the lines of [18] or the references therein. The points of interests are characterized by the condition $\dot\tau = 0$. We therefore augment the DAE (16.43) according to

$$f(x,\tau) = 0, \quad \|\dot x\|_2^2 + |\dot\tau|^2 = 1, \quad w - \dot\tau = 0, \tag{16.44}$$

and try to locate points along the solution satisfying $w = 0$. Writing the DAE (16.44) again as $F(z) = 0$, where now $z = (x, \tau, w)$, we have

$$\mathbb{L}_0 = \{(z,\dot z) \mid z = (\hat x(t), \hat\tau(t), \dot{\hat\tau}(t)),\ \dot z = (\dot{\hat x}(t), \dot{\hat\tau}(t), \ddot{\hat\tau}(t))\}$$

in Hypothesis 1. Because of

$$F_{0;\dot z} = \begin{bmatrix} 0 & 0 & 0 \\ 2\dot x^T & 2\dot\tau & 0 \\ 0 & -1 & 0 \end{bmatrix}, \quad F_{0;z} = \begin{bmatrix} f_x & f_\tau & 0 \\ 0 & 0 & 0 \\ 0 & 0 & 1 \end{bmatrix},$$

we may choose

$$Z_2 = \begin{bmatrix} I_n \\ 0 \\ 0 \end{bmatrix}.$$

Along the same lines as above, we may now choose

$$T_2 = \begin{bmatrix} \dot{\hat{x}} & 0 \\ \dot{\hat{\tau}} & 0 \\ 0 & 1 \end{bmatrix}$$

on $\mathbb{L}_0$. We then observe that

$$F_{\dot{z}} T_2 = \begin{bmatrix} 0 & 0 & 0 \\ 2\dot{\hat{x}}^T & 2\dot{\hat{\tau}} & 0 \\ 0 & -1 & 0 \end{bmatrix} \begin{bmatrix} \dot{\hat{x}} & 0 \\ \dot{\hat{\tau}} & 0 \\ 0 & 1 \end{bmatrix} = \begin{bmatrix} 0 & 0 \\ 2 & 0 \\ -\dot{\hat{\tau}} & 0 \end{bmatrix}$$

fails to have full column rank at the solution. Thus, Hypothesis 1 cannot hold with $\mu = 0$. We therefore consider Hypothesis 1 for $\mu = 1$. Starting from

$$F_{1;\dot{z},\ddot{z}} = \left[\begin{array}{ccc|ccc} 0 & 0 & 0 & & & \\ 2\dot{x}^T & 2\dot{\tau} & 0 & & & \\ 0 & -1 & 0 & & & \\ \hline f_x & f_\tau & 0 & 0 & 0 & 0 \\ * & * & 0 & 2\dot{x}^T & 2\dot{\tau} & 0 \\ 0 & 0 & 1 & 0 & -1 & 0 \end{array}\right], \quad F_{1;z} = \left[\begin{array}{ccc} f_x & f_\tau & 0 \\ 0 & 0 & 0 \\ 0 & 0 & 1 \\ \hline * & * & 0 \\ 0 & 0 & 0 \\ 0 & 0 & 0 \end{array}\right],$$

we use the fact that $0 \neq (\dot{x}^T, \tau)^T \in \operatorname{kernel}[\, f_x \;\; f_\tau \,]$ at a solution and therefore

$$\begin{bmatrix} f_x & f_\tau \\ \dot{x}^T & \dot{\tau} \end{bmatrix} \quad \text{nonsingular}$$

near the solution to deduce that $\operatorname{rank} F_{1;\dot{z},\ddot{z}} = n + 3$. Choosing

$$Z_2 = \left[\begin{array}{cc} I_n & 0 \\ 0 & * \\ 0 & 1 \\ \hline 0 & * \\ 0 & 0 \\ 0 & 0 \end{array}\right]$$

gives

$$Z_2^T F_{1;z} = \begin{bmatrix} f_x & f_\tau & 0 \\ * & * & 1 \end{bmatrix},$$

which has full row rank by assumption. Choosing

$$T_2 = \begin{bmatrix} \dot{\hat{x}} \\ \dot{\hat{\tau}} \\ * \end{bmatrix}$$

at the solution then yields

$$F_{\dot{z}} T_2 = \begin{bmatrix} 0 & 0 & 0 \\ 2\dot{\hat{x}}^T & 2\dot{\hat{\tau}} & 0 \\ 0 & -1 & 0 \end{bmatrix} \begin{bmatrix} \dot{\hat{x}} \\ \dot{\hat{\tau}} \\ * \end{bmatrix} = \begin{bmatrix} 0 \\ 2 \\ -\dot{\hat{\tau}} \end{bmatrix}.$$

Hence, the DAE (16.44) satisfies Hypothesis 1 with $\mu = 1$, $a = n + 1$, and $d = 1$, and we can treat (16.44) by the usual techniques. The location of points $\hat{t}$ with $\dot{\tau}(\hat{t}) = 0$ can now be seen as a root finding problem along solutions of (16.44) for the function g defined by

$$g(x, \tau, w) = w \tag{16.45}$$

In particualar, it can be treated by standard means of root finding techniques.

In order to be able to determine a root $\hat{t}$ of g, we need that this root is simple, i.e., that

$$\tfrac{d}{dt} g(\hat{x}(t), \hat{\tau}(t), \hat{w}(t))|_{t=\hat{t}} \neq 0, \quad \hat{w}(t) = \dot{\hat{\tau}}(t). \tag{16.46}$$

In the case of (16.45), this condition simply reads

$$\ddot{\hat{\tau}}(\hat{t}) \neq 0. \tag{16.47}$$

In order to determine $\ddot{\hat{\tau}}(\hat{t})$, we start with $f(\hat{x}(t), \hat{\tau}(t)) = 0$ along the solution. Differentiating twice yields (omitting arguments)

$$f_x \dot{\hat{x}} + f_\tau \dot{\hat{\tau}} = 0 \tag{16.48}$$

and

$$f_{xx}(\dot{\hat{x}}, \dot{\hat{x}}) + f_{x\tau}(\dot{\hat{x}})(\dot{\hat{\tau}}) + f_x \ddot{\hat{x}} + f_{x\tau}(\dot{\hat{x}})(\dot{\hat{\tau}}) + f_{\tau\tau}(\dot{\hat{\tau}}, \dot{\hat{\tau}}) + f_\tau \ddot{\hat{\tau}} = 0. \tag{16.49}$$

Since $\dot{\hat{\tau}}(\hat{t}) = 0$, the relation (16.48) gives

$$f_x(x^*, \tau^*)v = 0, \quad v = \dot{\hat{x}}(\hat{t}) \neq 0, \tag{16.50}$$

with $x^* = \hat{x}(\hat{t})$ and $\tau^* = \hat{\tau}(\hat{t})$ for short. Thus, the square matrix $f_x(x^*, \tau^*)$ is rank-deficient such that there is a vector $u \neq 0$ with

$$u^T f_x(x^*, \tau^*) = 0. \tag{16.51}$$

Multiplying (16.49) with u^T from the left and evaluating at $\hat{t}$ yields

$$u^T f_{xx}(x^*, \tau^*)(v, v) + u^T f_\tau(x^*, \tau^*)\ddot{\hat{\tau}}(\hat{t}) = 0. \tag{16.52}$$

Assuming now that

$$u^T f_{xx}(x^*, \tau^*)(v, v) \neq 0, \quad u^T f_\tau(x^*, \tau^*) \neq 0 \tag{16.53}$$

guarantees

$$\ddot{\hat{\tau}}(\hat{t}) = -(u^T f_\tau(x^*, \tau^*))^{-1}(u^T f_{xx}(x^*, \tau^*)(v, v)) \neq 0. \tag{16.54}$$

Note that the assumptions for (x^*, τ^*) we have required here are just those that characterize a so-called simple turning point, see, e.g., [8, 15].

Example 3 Consider the example

$$\begin{aligned}
&\tau(1 - x_3)\exp(10x_1)/(1 + 0.01x_1) - x_3 = 0,\\
&22\tau(1 - x_3)\exp(10x_1)/(1 + 0.01x_1) - 30x_1 = 0,\\
&x_3 - x_4 + \tau(1 - x_3)\exp(10x_2)/(1 + 0.01x_2) = 0,\\
&10x_1 - 30x_2 + 22\tau(1 - x_4)\exp(10x_2)/(1 + 0.01x_2) = 0,
\end{aligned}$$

from [11]. Starting from the trivial solution into the positive cone, the solution path exhibits six turning points before the solution becomes nearly independent of τ, see Fig. 16.3, which has been produced by solving the corresponding DAE (16.44) by the implicit Euler method combined with standard root finding techniques. ◇

16.5 Conclusions

We revised the theory of regular nonlinear DAEs of arbitrary index and gave some extensions to overdetermined but consistent DAEs. We also discussed several computational issues in the numerical treatment of such DAEs, namely suitable

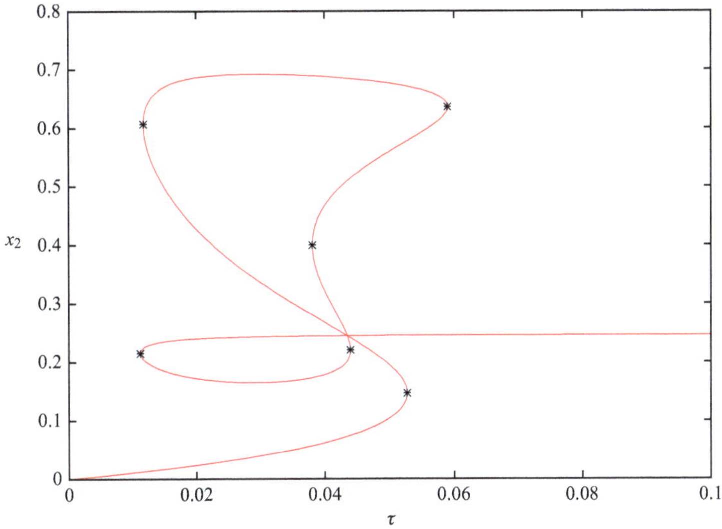

Fig. 16.3 Solution path for Example 3 projected into the (τ, x_2)-plane

discretizations, efficient nonlinear system solvers and their stabilization, as well as automatic differentiation. We finally presented a DAE approach for numerical path following for parametrized systems of nonlinear equations including the detection and determination of (simple) turning points.

References

1. Campbell, S.L.: A general form for solvable linear time varying singular systems of differential equations. SIAM J. Math. Anal. **18**, 1101–1115 (1987)
2. Campbell, S.L., Kunkel, P., Bobinyec, K.: A minimal norm corrected underdetermined Gauß-Newton procedure. Appl. Numer. Math. **62**, 592–605 (2012)
3. Campbell, S.L., Yeomans, K.D.: Behavior of the nonunique terms in general DAE integrators. Appl. Numer. Math. **28**, 209–226 (1998)
4. Deuflhard, P., Fiedler, B., Kunkel, P.: Efficient numerical path following beyond critical points. SIAM J. Numer. Anal. **24**, 912–927 (1987)
5. Fritzson, P.: Principles of Object-Oriented Modeling and Simulation with Modelica. Wiley/IEEE, Hoboken/Piscataway (2003)
6. Gear, C.W.: The simultaneous numerical solution of differential-algebraic equations. IEEE Trans. Circuit Theory **CT-18**, 89–95 (1971)
7. Golub, G.H., Van Loan, C.F.: Matrix Computations, 2nd edn. The Johns Hopkins University Press, Baltimore (1989)
8. Golubitsky, M., Schaeffer, D.: Singularities and Groups in Bifurcation Theory, vol. I. Springer, New York (1984)
9. Griewank, A., Walther, A.: Evaluating Derivatives: Principles and Techniques of Algorithmic Differentiation, 2nd edn. SIAM, Philadelphia (2008)

10. Hairer, E., Wanner, G.: Solving Ordinary Differential Equations II. Springer, Berlin (1991)
11. Kubiček, M.: Algorithm 502. Dependence of solutions of nonlinear systems on a parameter. ACM Trans. Math. Softw. **2**, 98–107 (1976)
12. Kunkel, P., Mehrmann, V.: Regular solutions of nonlinear differential-algebraic equations and their numerical determination. Numer. Math. **79**, 581–600 (1998)
13. Kunkel, P., Mehrmann, V.: Differential-Algebraic Equations – Analysis and Numerical Solution. EMS Publishing House, Zürich (2006)
14. Lotka, A.J.: Analytical note on certain rhythmic relations in organic systems. Proc. Natl. Acad. Sci. U.S.A. **6**, 410–415 (1920)
15. Pönisch, G., Schwetlick, H.: Computing turning points of curves implicitly defined by nonlinear equations depending on a parameter. Computing **26**, 107–121 (1981)
16. Rheinboldt, W.C.: Differential-algebraic systems as differential equations on manifolds. Math. Comput. **43**, 473–482 (1984)
17. Rheinboldt, W.C., Burkardt, J.V.: A locally parametrized continuation process. ACM Trans. Math. Softw. **9**, 236–246 (1983)
18. Shampine, L.F., Gladwell, I., Brankin, R.W.: Reliable solution of special event location problems for ODEs. ACM Trans. Math. Softw. **17**, 11–25 (1991)

Chapter 17
DAEs in Applications

Lena Scholz and Andreas Steinbrecher

Abstract Differential-algebraic equations (DAEs) arise naturally in many technical and industrial applications. By incorporating the special structure of the DAE systems arising in certain physical domains, the general approach for the regularization of DAEs can be efficiently adapted to the system structure. We will present the analysis and regularization approaches for DAEs arising in mechanical multibody systems, electrical circuit equations, and flow problems. In each of these cases the DAEs exhibit a certain structure that can be used for an efficient analysis and regularization. Moreover, we discuss the numerical treatment of hybrid DAE systems, that also occur frequently in industrial applications. For such systems, the framework of DAEs provides essential information for a robust numerical treatment.

17.1 Introduction

In the simulation and control of constrained dynamical systems *differential-algebraic equations* (DAEs) are widely used, since they naturally arise in the modeling process. In particular, the automatic modeling using coupling of modularized subcomponents is frequently used in industrial applications yielding large-scale (but often sparse) DAE systems. An important aspect in the simulation of these systems is that conservation laws (e.g. conservation of mass or momentum, mass or population balances, etc.) included in the model equations should be preserved during the numerical integration. These algebraic relations pose constraints on the solution and may lead to so-called *hidden constraints* (for higher index DAEs). Also path-following constraints can be considered as additional algebraic constraints. The occurrence of hidden constraints leads to difficulties in the numerical solution as

---

The authors have been supported by the *European Research Council* through ERC Advanced Grant MODSIMCONMP.

L. Scholz • A. Steinbrecher (✉)
Institut für Mathematik, Technische Universität Berlin, Sekretariat MA 4-5, Straße des 17. Juni 136, D-10623 Berlin, Germany
e-mail: lscholz@math.tu-berlin.de; anst@math.tu-berlin.de

P. Benner et al. (eds.), *Numerical Algebra, Matrix Theory, Differential-Algebraic Equations and Control Theory*, DOI 10.1007/978-3-319-15260-8_17

instabilities or order reduction can occur. Therefore, it is necessary to regularize or remodel the model equations to ensure a robust numerical integration.

The theory of general nonlinear DAEs developed by P. Kunkel and V. Mehrmann [28] (see also Chap. 16) provides a unified concept for modeling, simulation, control and optimization. However, usually such a general approach for the regularization of DAEs is not efficient enough for the usage in industrial applications. By incorporating the special structure of the DAE systems arising in certain physical domains, this general approach can be adapted efficiently to the system structure to construct robust regularizations in an efficient manner. Also, the structure of a system often comprises certain physical properties, that should be preserved as far as possible during the regularization and simulation process.

Another aspect that arises in numerous industrial applications is switching or hybrid behavior of the system model, e.g., mechanical systems with dry friction [10] or impact phenomena, electrical circuits with switching elements like diodes or switches [46], or control systems where the value of a control switches [59]. For such systems, the framework of DAEs provides essential information for a robust numerical treatment.

In the following, we present regularization approaches for some of the most important physical structures that occur in industrial applications, that is, for mechanical multibody systems, for electrical circuit equations and for flow problems. Moreover, we explain how hybrid DAE systems can be handled numerically in a robust and efficient manner using the concept of DAEs.

17.2 Preliminaries

Notation 1 *For a differentiable time depending function x, the i-th (total) derivative of x with respect to t is denoted by $x^{(i)}(t) = \mathrm{d}^i x(t)/\mathrm{d}t^i$ for $i \in \mathbb{N}$, using the convention $x^{(1)}(t) = \dot{x}(t)$, and $x^{(2)}(t) = \ddot{x}(t)$. For a differentiable function f depending on x, the (partial) derivative of f with respect to x is denoted by $f_{,x}(x) = \frac{\partial}{\partial x} f(x)$. The same notation is used for differentiable vector- and matrix-valued functions. Partial derivatives of a function $\mathscr{F}(t, x, \ldots, x^{(i+1)})$ with respect to selected variables p from $(t, x, \ldots, x^{(i+1)})$ are denoted by $\mathscr{F}_{,p}$, e.g.,*

$$\mathscr{F}_{,[\dot{x},\ldots,x^{(i+1)}]} = \left[\tfrac{\partial}{\partial \dot{x}} \mathscr{F} \;\cdots\; \tfrac{\partial}{\partial x^{(i+1)}} \mathscr{F} \right].$$

Let $F = [F_i] \in \mathbb{R}^n$ and let $I = \{i_1, i_2, \ldots, i_p\} \subseteq \{1, \ldots, n\}$ be an index set. We use the notation F_I for the vector

$$F_I = \begin{bmatrix} F_{i_1} \\ F_{i_2} \\ \vdots \\ F_{i_p} \end{bmatrix}.$$

For a matrix $A \in \mathbb{R}^{m,n}$, $\operatorname{im}(A)$ *denotes the image of A,* $\ker(A)$ *denotes the kernel of A, and* $\operatorname{coker}(A)$ *denotes the cokernel of A. Furthermore,* $\operatorname{rank}(A)$ *denotes the rank of the matrix A and* $\operatorname{corank}(A)$ *denotes the corank of A that is defined as the codimension of the image.*

In this section, we shortly recapitulate the basic facts of the general theory for nonlinear DAEs following the presentation in [28]. We consider a nonlinear DAE of the form

$$F(t, x, \dot{x}) = 0, \tag{17.1}$$

with sufficiently smooth function $F : \mathbb{I} \times \mathbb{D}_x \times \mathbb{D}_{\dot{x}} \to \mathbb{R}^m$, $\mathbb{I} \subset \mathbb{R}$ (compact) interval, and $\mathbb{D}_x, \mathbb{D}_{\dot{x}} \subset \mathbb{R}^n$ open. Together with an initial condition

$$x(t_0) = x_0, \quad t_0 \in \mathbb{I},\ x_0 \in \mathbb{D}_x, \tag{17.2}$$

we obtain an initial value problem consisting of (17.1) and (17.2). The *derivative array* $\mathscr{F}_i$ of level i stacks the original equations of the DAE and all its derivatives up to level i into one large system

$$\mathscr{F}_i(t, x, \dot{x}, \ldots, x^{(i+1)}) = \begin{bmatrix} F(t, x, \dot{x}) \\ \frac{d}{dt} F(t, x, \dot{x}) \\ \vdots \\ \frac{d^i}{dt^i} F(t, x, \dot{x}) \end{bmatrix} = 0. \tag{17.3}$$

Hypothesis 1 ([28]) *Consider a nonlinear DAE (17.1). There exist integers μ, r, a, d and v such that the solution set*

$$\mathbb{L}_\mu = \left\{(t, x, \dot{x}, \ldots, x^{(\mu+1)}) \in \mathbb{I} \times \mathbb{R}^{(\mu+2)n} \mid \mathscr{F}_\mu(t, x, \dot{x}, \ldots, x^{(\mu+1)}) = 0\right\} \tag{17.4}$$

of the derivative array $\mathscr{F}_\mu$ is nonempty, and the following properties hold:

1. *The set $\mathbb{L}_\mu \subset \mathbb{R}^{(\mu+2)n+1}$ forms a manifold of dimension $(\mu + 2)n + 1 - r$;*
2. $\operatorname{rank}(\mathscr{F}_{\mu,[x,\dot{x},\ldots,x^{(\mu+1)}]}) = r$ *on $\mathbb{L}_\mu$;*
3. $\operatorname{corank}(\mathscr{F}_{\mu,[x,\dot{x},\ldots,x^{(\mu+1)}]}) - \operatorname{corank}(\mathscr{F}_{\mu-1,[x,\dot{x},\ldots,x^{(\mu)}]}) = v$ *on $\mathbb{L}_\mu$ (We use the convention that* $\operatorname{corank}(\mathscr{F}_{-1,x}) = 0$.*);*

4. $\text{rank}(\mathscr{F}_{\mu,[\dot{x},\ldots,x^{(\mu+1)}]}) = r - a$ *on* $\mathbb{L}_\mu$ *such that there are smooth full rank matrix-valued functions* Z_2 *and* T_2 *defined on* $\mathbb{L}_\mu$ *of size* $((\mu+1)m, a)$ *and* $(n, n-a)$, *respectively, satisfying*

$$Z_2^T \mathscr{F}_{\mu,[\dot{x},\ldots,\dot{x}^{(\mu+1)}]} = 0, \quad \text{rank}(Z_2^T \mathscr{F}_{\mu,x}) = a, \quad Z_2^T \mathscr{F}_{\mu,x} T_2 = 0$$

on $\mathbb{L}_\mu$;

5. $\text{rank}(F_{,\dot{x}} T_2) = d = m - a - v$ *on* $\mathbb{L}_\mu$ *such that there exists a smooth matrix-valued function* Z_1 *defined on* $\mathbb{L}_\mu$ *of size* (m, d) *with* $Z_1^T F_{,\dot{x}} T_2$ *having full rank.*

The smallest possible μ in Hypothesis 1 is called the *strangeness index* (or *s-index*) ν_s of the DAE (17.1) and a system with vanishing strangeness index is called *strangeness-free*. Further, a DAE (17.1) that satisfies Hypothesis 1 with $n = m = d + a$ is called *regular*. The corresponding numbers d and a are the numbers of differential and algebraic equations of the DAE. It has been shown in [26], that Hypothesis 1 implies (locally) the existence of a reduced system (in the original variables) of the form

$$\hat{F}_1(t, x, \dot{x}) = 0, \tag{17.5a}$$

$$\hat{F}_2(t, x) = 0, \tag{17.5b}$$

with $\hat{F}_1 = Z_1^T F$ and $\hat{F}_2 = Z_2^T \mathscr{F}_\mu$. An initial value $x_0 \in \mathbb{R}^n$ is *consistent* with the DAE, if it satisfies the algebraic equation $\hat{F}_2(t_0, x_0) = 0$. From part 4 of Hypothesis 1 we know that there exists a partitioning of x into $\begin{bmatrix} x_1^T & x_2^T & x_3^T \end{bmatrix}^T$, with x_1 of dimension d, x_3 of dimension a, and x_2 of dimension $u = n - d - a$, such that (17.5b) is locally equivalent to a formulation of the form

$$x_3 = \mathscr{R}(t, x_1, x_2). \tag{17.6}$$

Eliminating x_3 and $\dot{x}_3$ in (17.5a) with the help of (17.6) and its derivative leads to a system

$$\hat{F}_1(t, x_1, x_2, \mathscr{R}, \dot{x}_1, \dot{x}_2, \mathscr{R}_{,t} + \mathscr{R}_{,x_1}\dot{x}_1 + \mathscr{R}_{,x_2}\dot{x}_2) = 0,$$

omitting the arguments of $\mathscr{R}$. By part 5 of Hypothesis 1 we may assume w.l.o.g. that this system can (locally) be solved for $\dot{x}_1$ leading to a system of the form

$$\dot{x}_1 = \mathscr{L}(t, x_1, x_2, \dot{x}_2), \tag{17.7a}$$

$$x_3 = \mathscr{R}(t, x_1, x_2). \tag{17.7b}$$

Remark 1 The matrix-valued function T_2 obtained from Hypothesis 1 can be used to extract the differential components of the DAE. Since $Z_1^T F_{,\dot{x}} T_2$ has full rank, the matrix-valued function T_2 of size $(n, n-a)$ can be partitioned into $\begin{bmatrix} T_{21} & T_{22} \end{bmatrix}$ with T_{21} of size (n, d) and T_{22} of size (n, u), respectively, in such a way that $Z_1^T F_{,\dot{x}} T_{21}$

is nonsingular. Thus, using the Implicit Function Theorem [3] equation(17.5a) can locally be solved for the differential components $\dot{x}_1$ leading to (17.7a).

Obviously, the component x_2 in (17.7) can be chosen arbitrarily (at least when staying in the domain of definition of $\mathscr{R}$ and $\mathscr{L}$), i.e., it plays the role of a control. When x_2 has been chosen and a consistent initial condition $x(t_0) = x_0$ is given, then the resulting system has a unique solution for x_1 and x_3, locally in a neighborhood of (t_0, x_0), provided that $\mathscr{L}$ is sufficiently smooth such that the Theorem of Picard-Lindelöf can be applied to (17.7a). In particular, the equations(17.5b) and (17.7b) contain all constraints that exist in the systems, i.e., all explicit constraints as well as all (formerly) hidden constraints. We define the set

$$\mathbb{S} = \{(t, x) \in \mathbb{I} \times \mathbb{D}_x \mid \hat{F}_2(t, x) = 0\} \tag{17.8}$$

as the *set of consistency* of the DAE (17.1). It has been shown in [26], that every solution of (17.1) also solves the reduced problems (17.5) and (17.7).

Definition 1 We call the regularized equivalent system formulation (17.5) of the DAE (17.1) the *Kunkel-Mehrmann formulation* or short *KM-formulation* of (17.1). The regularization approach consisting of the application of Hypothesis 1 to the DAE (17.1) is called *Kunkel-Mehrmann regularization* or short *KM-regularization*.

For the numerical solution, the regularized formulations (17.5) or (17.7) are to be preferred, since it can be guaranteed that all constraints are satisfied using suitable integration methods.

Remark 2 In the reduced systems (17.5) and (17.7) we have not used the quantity v. This quantity measures the number of equations in the original system that give rise to trivial equations $0 = 0$, i.e., it counts the number of redundancies in the system. Together with a and d it gives a complete classification of the m equations into d differential equations, a algebraic equations and v trivial equations. Of course, trivial equations can be simply removed without altering the solution set. ◁

For specially structured systems, it may be known in advance which equations are responsible for hidden constraints. In this case, it is sufficient to consider only a *reduced derivative array*, where only a subset of equations has been differentiated and added to the system equations. A reduced derivative array of level i is given by

$$\hat{\mathscr{F}}_i(t, x, \dot{x}, \ldots, x^{(i+1)}) = \begin{bmatrix} F(t, x, \dot{x}) \\ \frac{d}{dt} F_{I_1}(t, x, \dot{x}) \\ \vdots \\ \frac{d^i}{dt^i} F_{I_i}(t, x, \dot{x}) \end{bmatrix} = 0 \tag{17.9}$$

with certain index sets $I_1 \subseteq \{1, \ldots, m\}$, $I_k \subseteq I_{k-1}$ for $k = 2, \ldots, i$.

Example 1 We consider the DAE given by

$$\begin{aligned} \dot{p} &= v, \\ \dot{v} &= f(p, v) - H^T \zeta, \\ 0 &= Hv, \end{aligned}$$

with $p(t), v(t) \in \mathbb{R}^{n_p}$, $\zeta(t) \in \mathbb{R}^{n_\zeta}$, $f : \mathbb{R}^{n_p} \times \mathbb{R}^{n_p} \to \mathbb{R}^{n_p}$ with $n_\zeta < n_p$, and $H \in \mathbb{R}^{n_\zeta, n_p}$ of full row rank. Note that this is a special case of the system considered in Sect. 17.3. There are hidden constraints in the system. Moreover, the special structure allows us to work with a reduced derivative array instead of the whole inflated system (17.3). Adding the first derivative of the third equation we obtain the reduced derivative array (of level 1) as

$$\hat{\mathscr{F}}_1 = \begin{bmatrix} -\dot{p} + v \\ -\dot{v} + f(p, v) - H^T \zeta \\ Hv \\ H\dot{v} \end{bmatrix} = 0. \tag{17.10}$$

Now, it can be shown that Hypothesis 1 is satisfied with the vector $x^T = \begin{bmatrix} p^T & v^T & \zeta^T \end{bmatrix}$, $m = n = 2n_p + n_\zeta$ and integers $\mu = 1$, $a = 2n_\zeta$, $d = 2n_p - n_\zeta$ and $v = 0$ for the derivative array $\mathscr{F}_1$, but also for the reduced derivative array $\hat{\mathscr{F}}_1$ given in (17.10). ◁

The possibility to work with a reduced derivative array for specially structured systems enables us to use more efficient approaches for index reduction and regularization. This idea is used in the *index reduction by minimal extension* [27] that can be applied e.g. for mechanical systems (see Sect. 17.3), electrical circuit equations (see Sect. 17.4), or flow problems (see Sect. 17.5).

For a general nonlinear DAE (17.1), the idea of regularization by minimal extension can be described by the following iterative process (starting with $i = 0$ and $F^0(t, x^0, \dot{x}^0) = 0$ given by the original system (17.1)).

1. For $F^i(t, x^i, \dot{x}^i) = 0$ determine a full rank matrix-valued function $Z^i(t, x^i, \dot{x}^i)$ such that the system transforms to

$$\begin{aligned} F_1^i(t, x^i, \dot{x}^i) &= 0, \\ F_2^i(t, x^i) &= 0, \\ F_3^i(t, x^i) &= 0, \end{aligned}$$

with $\operatorname{rank}\left(\begin{bmatrix} F^i_{1,\dot{x}} \\ F^i_{3,x} \end{bmatrix}\right) = \operatorname{rank}(F^i_{1,\dot{x}})$ and $\operatorname{rank}\left(\begin{bmatrix} F^i_{1,\dot{x}} \\ F^i_{2,x} \end{bmatrix}\right) = \operatorname{rank}(F^i_{1,\dot{x}}) + \operatorname{rank}(F^i_{2,x})$.

This means that the equations $F_3^i(t, x^i) = 0$ are the "strange" equations of level i, since their derivatives are redundant to parts of the first equations

$F_1^i(t, x^i, \dot{x}^i) = 0$, while the algebraic equations given by $F_2^i(t, x^i) = 0$ are not "strange".

2. Find a nonsingular constant matrix $\Pi = \begin{bmatrix} \Pi_1 & \Pi_2 \end{bmatrix} \in \mathbb{R}^{n,n}$ such that $F_{3,x}^i \Pi_2$ is regular. Let $\bar{\Pi}$ denote the inverse of Π.
3. Add the first time derivative of $F_3^i(t, x^i) = 0$ to the system equations leading to the augmented system

$$\begin{aligned} F_1^i(t, x^i, \dot{x}^i) &= 0, \\ F_2^i(t, x^i) &= 0, \\ F_3^i(t, x^i) &= 0, \\ F_{3,t}^i(t, x^i) + F_{3,x}^i(t, x^i)\dot{x}^i &= 0. \end{aligned}$$

4. With the coordinate transformation $x^i = \Pi \tilde{x}^i$ we obtain the splitting of $\tilde{x}^i = \bar{\Pi} x^i$ into

$$\begin{bmatrix} \tilde{x}_1^i \\ \tilde{x}_2^i \end{bmatrix} = \begin{bmatrix} \bar{\Pi}_1 x^i \\ \bar{\Pi}_2 x^i \end{bmatrix}.$$

Introducing the new algebraic variable w^i for $\dot{\tilde{x}}_2^i$, we get the new extended system

$$\begin{aligned} \tilde{F}_1^i(t, \tilde{x}_1^i, \tilde{x}_2^i, \dot{\tilde{x}}_1^i, w^i) &= 0, \\ \tilde{F}_2^i(t, \tilde{x}_1^i, \tilde{x}_2^i) &= 0, \\ \tilde{F}_3^i(t, \tilde{x}_1^i, \tilde{x}_2^i) &= 0, \\ \tilde{F}_{3,t}^i(t, \tilde{x}_1^i, \tilde{x}_2^i) + \begin{bmatrix} \tilde{F}_{3,x}^i(t, \tilde{x}_1^i, \tilde{x}_2^i)\Pi_1 & \tilde{F}_{3,x}^i(t, \tilde{x}_1^i, \tilde{x}_2^i)\Pi_2 \end{bmatrix} \begin{bmatrix} \dot{\tilde{x}}_1 \\ w^i \end{bmatrix} &= 0, \end{aligned} \tag{17.11}$$

where

$$\begin{aligned} \tilde{F}_1^i(t, \tilde{x}_1^i, \tilde{x}_2^i, \dot{\tilde{x}}_1^i, \dot{\tilde{x}}_2^i) &= F_1^i(t, \Pi \tilde{x}^i, \Pi \dot{\tilde{x}}^i), \\ \tilde{F}_2^i(t, \tilde{x}_1^i, \tilde{x}_2^i) &= F_2^i(t, \Pi \tilde{x}^i), \\ \tilde{F}_3^i(t, \tilde{x}_1^i, \tilde{x}_2^i) &= F_3^i(t, \Pi \tilde{x}^i). \end{aligned}$$

5. System (17.11) is defined to be the new extended DAE system $F^{i+1}(t, x^{i+1}, \dot{x}^{i+1}) = 0$ with extended vector of unknown variables $(x^{i+1})^T = \begin{bmatrix} (\tilde{x}_1^i)^T & (\tilde{x}_2^i)^T & (w^i)^T \end{bmatrix}$.

The variables $\tilde{x}_2^i$ determined in the third step of the iterative process can be defined as the "strange" variables corresponding to the "strange" equations F_3^i.

From a numerical perspective the described approach can be applied very efficiently for the computation of a regularization if the first step, i.e., the determination of the transformation matrix Z, can be easily achieved from the structure of the problem. In the ideal case, the matrix Π can be determined as a simple permutation matrix, see Sect. 17.4.2. Note, however, that the described iterative process may hold only locally, since only coordinate transformations with constant matrix Π are used.

In the following, we will explain how this general theory can be applied and efficiently adapted to DAE systems that exhibit a special structure of the system equations. In particular, we will consider mechanical multibody systems, electrical circuit equations, and flow problems.

17.3 Mechanical Systems

The dynamical behavior of *mechanical systems* or of *multibody systems* (MBS) is of great importance in many fields of mechanical engineering, like robotics, road and rail vehicle construction, air and space craft design, see [10, 24, 42–44]. Often, mechanical systems are part of more complex dynamical systems, so-called multiphysics systems. Nevertheless, in this section we are interested in purely mechanical systems in form of multibody systems.

We mainly consider multibody systems from the dynamical point of view in which a multibody system is regarded as a number of mass points and rigid or elastic bodies, subject to possibly existing interconnections and constraints of various kinds, e.g., joints, springs, dampers, and actuators.

In a large part of the literature, equations of motion in standard form including the dynamical equations of motion subject to some holonomic constraints are discussed in detail. However, in industrial applications, more complex equations arise which include friction effects, contact force laws, dynamical force elements, nonholonomic constraints, and in some cases the existing constraints are even redundant. Therefore, we will focus our investigation on slightly more complex model equations that includes holonomic as well as nonholonomic constraints, see Sect. 17.3.1. Examples of multibody systems with nonholonomic constraints are the sliding of blades, knives, or skates, or the rolling of balls, cylinders, or wheels without sliding. For the investigation of even more general model equations used in industrial applications, we refer to [49].

It is well known that the direct numerical integration of equations of motion is a nontrivial problem. Thus, a numerical method should combine a discretization method with a suitable regularization technique. First, we analyze the equations of motion in Sect. 17.3.2 with respect to important quantities like existence and uniqueness of a solution followed by the discussion of some important regularization techniques in Sect. 17.3.3.

17.3.1 Equations of Motion

The interest in the dynamical behavior, i.e., the movement, of mechanical systems and, in particular, in model equations goes back far in history. Let us mention only Newton [36] (*Newtonian mechanics*), Euler (*Euler equations* for rigid bodies or free multibody systems), d'Alembert (*d'Alembert's principle of virtual displacements*), Lagrange [32] (*Lagrangian mechanics* and *Euler-Lagrange equations* for constrained multibody systems), and Hamilton (*Hamilton principle of least action*).

Following these historical approaches, e.g., Euler-Lagrange equations, and even more modern approaches in modeling multibody system, one can formulate the model equations in form of the *equations of motion*. The equations of motion usually form a nonlinear system of DAEs with a very special structure that can and should be exploited in the numerical treatment [10, 11, 21, 41, 42]. As mentioned above, we will focus on slightly more general equations of motion given in the form

$$\dot{p} = v, \tag{17.12a}$$

$$M(p)\dot{v} = f(p,v,t) - G^T(p,t)\lambda - H^T(p,t)\zeta, \tag{17.12b}$$

$$0 = g(p,t), \tag{17.12c}$$

$$0 = H(p,t)v + \breve{h}(p,t). \tag{17.12d}$$

Here, the *configuration* of the multibody system is described by the *position variables* $p(t)$ of dimension n_p, while the *velocities* or the change of the configuration of the multibody system is described by the *velocity variables* $v(t)$ of dimension n_p. The dynamical behavior is affected by n_λ *holonomic constraints* (17.12c) and n_ζ *nonholonomic constraints* (17.12d). Nonholonomic constraints (17.12d) are linear in v, as discussed in [7, 23, 40]. Furthermore, the n_p equations (17.12a) are called *kinematical equations of motion*. The n_p equations (17.12b) are called *dynamical equations of motion*. They follow from the equilibrium of forces and momenta and include the *mass matrix* $M(p)$, the vector $f(p,v,t)$ of the *applied and gyroscopic forces*, the *holonomic constraint matrix* $G(p,t) = g_{,p}(p,t)$ of the holonomic constraints, the associated *holonomic constraint forces* $G^T(p,t)\lambda$, and the *holonomic Lagrange multipliers* λ, as well as the *nonholonomic constraint matrix* $H(p,t)$ of the nonholonomic constraints, the associated *nonholonomic constraint forces* $H^T(p,t)\zeta$, and the *nonholonomic Lagrange multipliers* ζ. The mass matrix $M(p)$ is positive semi-definite, since the kinetic energy is a positive semi-definite quadratic form, and it includes the inertia properties of the multibody system. The columns of the matrices $G^T(p,t)$ and $H^T(p,t)$ describe the inaccessible directions of the motion.

In connection with initial conditions

$$p(t_0) = p_0,\ v(t_0) = v_0,\ \lambda(t_0) = \lambda_0,\ \zeta(t_0) = \zeta_0 \tag{17.13}$$

we have the initial value problem (17.12), (17.13) for the equations of motion on the domain $\mathbb{I} = [t_0, t_f]$.

For reasons of readability and simplicity, we will often omit the dependency on p, v, λ, ζ, and t in the notation unless we want to focus on some of those dependencies. Furthermore, we will anticipate the following assumption. In the investigations below it will become clear why these assumptions are justified and necessary.

Assumption 1 *Consider the equations of motion (17.12). We assume that*

$$M \text{ is nonsingular and} \tag{17.14a}$$

$$\begin{bmatrix} M & G^T & H^T \\ G & 0 & 0 \\ H & 0 & 0 \end{bmatrix} \text{ is nonsingular} \tag{17.14b}$$

and have bounded inverses for all consistent p and t. Furthermore, it is assumed that all functions in (17.12) are smooth enough.

Remark 3 (a) From Assumption 1 it follows that

$$\begin{bmatrix} GM^{-1}G^T & GM^{-1}H^T \\ HM^{-1}G^T & HM^{-1}H^T \end{bmatrix} \text{ is nonsingular with a bounded inverse and} \tag{17.15a}$$

$$\begin{bmatrix} G \\ H \end{bmatrix} \text{ has full (row) rank} \tag{17.15b}$$

for all consistent p and t. In particular, (17.15b) guarantees that the constraints are not redundant. In case of redundant constraints the investigations below can be adapted and yield similar results as for non-redundant constraints apart from the uniqueness of the solution.

(b) The non-singularity of the mass matrix M is not necessarily to assume. But for reasons of simplicity, we will restrict our investigations to that case. Nevertheless, the following results remain valid even for singular M as long as (17.14b) holds.

◁

17.3.2 Analysis

In this section, we will analyze the equations of motion (17.12) using the strangeness-index concept [28] revisited in Sect. 17.2. In particular, we will investigate consistency conditions and the existence and uniqueness of a solution of the equations of motion (17.12).

Theorem 1 *The equations of motion (17.12) satisfying Assumption 1 satisfy Hypothesis 1 with integers* $\mu = 2$, $r = 3(n_p + n_p + n_\lambda + n_\zeta)$, $a = 3n_\lambda + 2n_\zeta$, $d = 2n_p - 2n_\lambda - n_\zeta$, *and* $v = 0$.

Proof For the equations of motion (17.12) we have $m = n = n_p + n_p + n_\lambda + n_\zeta$. The derivative array of level 2 for the equations of motion (17.12) with $x^T = \begin{bmatrix} p^T & v^T & \lambda^T & \zeta^T \end{bmatrix}$ of size n is given by

$$\mathscr{F}_2(t, x, \dot{x}, \ddot{x}, x^{(3)})$$
$$= \left[\begin{array}{c} -\dot{p} + v \\ -M(p)\dot{v} + f(p, v, t) - G^T(p, t)\lambda - H^T(p, t)\zeta \\ g(p, t) \\ H(p, t)v + \breve{h}(p, t) \\ \hline -\ddot{p} + \dot{v} \\ \tilde{d}^I(p, v, \lambda, \zeta, \dot{p}, \dot{v}, t) - G^T(p, t)\dot{\lambda} - H^T(p, t)\dot{\zeta} - M(p)\ddot{v} \\ G(p, t)\dot{p} + g_{,t}(p, t) \\ \tilde{h}^I(p, v, \dot{p}, t) + H(p, t)\dot{v} \\ \hline -p^{(3)} + \ddot{v} \\ \tilde{d}^{II}(p, v, \lambda, \zeta, \dot{p}, \dot{v}, \dot{\lambda}, \dot{\zeta}, \ddot{p}, \ddot{v}, t) - G^T(p, t)\ddot{\lambda} - H^T(p, t)\ddot{\zeta} - M(p)v^{(3)} \\ \tilde{g}^{II}(p, \dot{p}, t) + G(p, t)\ddot{p} \\ \tilde{h}^{II}(p, v, \dot{p}, \dot{v}, \ddot{p}, t) + H(p, t)\ddot{v} \end{array}\right] \tag{17.16}$$

with

$$d(p, v, \lambda, \zeta, \dot{v}, t) = -M(p)\dot{v} + f(p, v, t) - G^T(p, t)\lambda - H^T(p, t)\zeta,$$
$$h(p, v, t) = H(p, t)v + \breve{h}(p, t),$$

and

$$\begin{aligned}
\tilde{d}^{I}(p,v,\lambda,\zeta,\dot{p},\dot{v},t) &= d_{,p}(p,v,\lambda,\zeta,\dot{v},t)\dot{p} + f_{,v}(p,v,t)\dot{v} + d_{,t}(p,v,\lambda,\zeta,\dot{v},t),\\
\tilde{h}^{I}(p,v,\dot{p},t) &= h_{,p}(p,v,t)\dot{p} + h_{,t}(p,v,t),\\
\tilde{d}^{II}(p,v,\lambda,\zeta,\dot{p},\dot{v},\dot{\lambda},\dot{\zeta},\ddot{p},\ddot{v},t) &= [\tilde{d}^{I}(p,v,\lambda,\zeta,\dot{p},\dot{v},t) - G^{T}(p,t)\dot{\lambda} - H^{T}(p,t)\dot{\zeta} - M(p)\ddot{v}]_{,p}\dot{p}\\
&\quad + \tilde{d}^{I}_{,v}(p,v,\lambda,\zeta,\dot{p},\dot{v},t)\dot{v} + \tilde{d}^{I}_{,\lambda}(p,v,\lambda,\zeta,\dot{p},\dot{v},t)\dot{\lambda}\\
&\quad + \tilde{d}^{I}_{,\zeta}(p,v,\lambda,\zeta,\dot{p},\dot{v},t)\dot{\zeta} + \tilde{d}^{I}_{,\dot{p}}(p,v,\lambda,\zeta,\dot{p},\dot{v},t)\ddot{p}\\
&\quad + \tilde{d}^{I}_{,\dot{v}}(p,v,\lambda,\zeta,\dot{p},\dot{v},t)\ddot{v}\\
&\quad + [\tilde{d}^{I}(p,v,\lambda,\zeta,\dot{p},\dot{v},t) - G^{T}(p,t)\dot{\lambda} - H^{T}(p,t)\dot{\zeta} - M(p)\ddot{v}]_{,t},\\
\tilde{h}^{II}(p,v,\dot{p},\dot{v},\ddot{p},t) &= [\tilde{h}^{I}(p,v,\dot{p},t) + H(p,t)\dot{v}]_{,p}\dot{p} + \tilde{h}^{I}_{,v}(p,v,\dot{p},t)\dot{v}\\
&\quad + \tilde{h}^{I}_{,\dot{p}}(p,v,\dot{p},t)\ddot{p} + [\tilde{h}^{I}(p,v,\dot{p},t) + H(p,t)\dot{v}]_{,t},\\
\tilde{g}^{II}(p,\dot{p},t) &= [G(p,t)\dot{p} + g_{,t}(p,t)]_{,p}\dot{p} + [G(p,t)\dot{p} + g_{,t}(p,t)]_{,t}.
\end{aligned}$$

It follows that the solution set $\mathbb{L}_2$, see (17.4), is not empty and due to Assumption 1 $\mathbb{L}_2$ forms a manifold of dimension $n+1$ such that we get $r = 3n$. We get

$$\mathscr{F}_{2,[x,\dot{x},\ddot{x},x^{(3)}]}$$
$$= \left[\begin{array}{cccc|cccc|cccc|cccc}
0 & I & 0 & 0 & -I & 0 & 0 & 0 & 0 & 0 & 0 & 0 & 0 & 0 & 0 & 0\\
d_{,p} & d_{,v} & -G^T & -H^T & 0 & -M & 0 & 0 & 0 & 0 & 0 & 0 & 0 & 0 & 0 & 0\\
G & 0 & 0 & 0 & 0 & 0 & 0 & 0 & 0 & 0 & 0 & 0 & 0 & 0 & 0 & 0\\
h_{,p} & H & 0 & 0 & 0 & 0 & 0 & 0 & 0 & 0 & 0 & 0 & 0 & 0 & 0 & 0\\
\hline
0 & 0 & 0 & 0 & 0 & I & 0 & 0 & -I & 0 & 0 & 0 & 0 & 0 & 0 & 0\\
\dot{d}_{,p} & \dot{d}_{,v} & \dot{d}_{,\lambda} & \dot{d}_{,\zeta} & \dot{d}_{,\dot{p}} & \dot{d}_{,\dot{v}} & -G^T & -H^T & 0 & -M & 0 & 0 & 0 & 0 & 0 & 0\\
\dot{g}_{,p} & 0 & 0 & 0 & G & 0 & 0 & 0 & 0 & 0 & 0 & 0 & 0 & 0 & 0 & 0\\
\dot{h}_{,p} & \dot{h}_{,v} & 0 & 0 & \dot{h}_{,\dot{p}} & H & 0 & 0 & 0 & 0 & 0 & 0 & 0 & 0 & 0 & 0\\
\hline
0 & 0 & 0 & 0 & 0 & 0 & 0 & 0 & 0 & I & 0 & 0 & -I & 0 & 0 & 0\\
\ddot{d}_{,p} & \ddot{d}_{,v} & \ddot{d}_{,\lambda} & \ddot{d}_{,\zeta} & \ddot{d}_{,\dot{p}} & \ddot{d}_{,\dot{v}} & \ddot{d}_{,\dot{\lambda}} & \ddot{d}_{,\dot{\zeta}} & \ddot{d}_{,\ddot{p}} & \ddot{d}_{,\ddot{v}} & -G^T & -H^T & 0 & -M & 0 & 0\\
\ddot{g}_{,p} & 0 & 0 & 0 & \ddot{g}_{,\dot{p}} & 0 & 0 & 0 & G & 0 & 0 & 0 & 0 & 0 & 0 & 0\\
\ddot{h}_{,p} & \ddot{h}_{,v} & 0 & 0 & \ddot{h}_{,\dot{p}} & \ddot{h}_{,\dot{v}} & 0 & 0 & \ddot{h}_{,\ddot{p}} & H & 0 & 0 & 0 & 0 & 0 & 0
\end{array}\right].$$

From Assumption 1 we have M and $\begin{bmatrix} M & G^T & H^T \\ H & 0 & 0 \\ G & 0 & 0 \end{bmatrix}$ are nonsingular and, therefore, $\begin{bmatrix} M & G^T & H^T \\ H & 0 & 0 \end{bmatrix}$ and $\begin{bmatrix} G & 0 & 0 \\ H & 0 & 0 \end{bmatrix}$ have full (row) rank. Therefore, we get $\operatorname{rank}(\mathscr{F}_{2,[x,\dot{x},\ddot{x},x^{(3)}]}) = 3n$ and, in particular, $\operatorname{corank}(\mathscr{F}_{2,[x,\dot{x},\ddot{x},x^{(3)}]}) = 0$. Furthermore, it can be shown that $\operatorname{rank}(\mathscr{F}_{1,[x,\dot{x},\ddot{x}]}) = 2n$ and, thus, $\operatorname{corank}(\mathscr{F}_{1,[x,\dot{x},\ddot{x}]}) = 0$.

Consequently, $\operatorname{corank}(\mathscr{F}_{2,[x,\dot{x},\ddot{x},x^{(3)}]}) - \operatorname{corank}(\mathscr{F}_{1,[x,\dot{x},\ddot{x}]}) = 0 = v$. For proceeding with the Hypothesis 1 we have

$$\mathscr{F}_{2,[\dot{x},\ddot{x},x^{(3)}]} = \left[\begin{array}{cccc|cccc|cccc}
-I & 0 & 0 & 0 & 0 & 0 & 0 & 0 & 0 & 0 & 0 & 0 \\
0 & -M & 0 & 0 & 0 & 0 & 0 & 0 & 0 & 0 & 0 & 0 \\
0 & 0 & 0 & 0 & 0 & 0 & 0 & 0 & 0 & 0 & 0 & 0 \\
0 & 0 & 0 & 0 & 0 & 0 & 0 & 0 & 0 & 0 & 0 & 0 \\
\hline
0 & I & 0 & 0 & -I & 0 & 0 & 0 & 0 & 0 & 0 & 0 \\
\dot{d}_{,\dot{p}} & \dot{d}_{,\dot{v}} & -G^T & -H^T & 0 & -M & 0 & 0 & 0 & 0 & 0 & 0 \\
G & 0 & 0 & 0 & 0 & 0 & 0 & 0 & 0 & 0 & 0 & 0 \\
\dot{h}_{,\dot{p}} & H & 0 & 0 & 0 & 0 & 0 & 0 & 0 & 0 & 0 & 0 \\
\hline
0 & 0 & 0 & 0 & 0 & I & 0 & 0 & -I & 0 & 0 & 0 \\
\ddot{d}_{,\dot{p}} & \ddot{d}_{,\dot{v}} & \ddot{d}_{,\dot{\lambda}} & \ddot{d}_{,\dot{\zeta}} & \ddot{d}_{,\ddot{p}} & \ddot{d}_{,\ddot{v}} & -G^T & -H^T & 0 & -M & 0 & 0 \\
\ddot{g}_{,\dot{p}} & 0 & 0 & 0 & G & 0 & 0 & 0 & 0 & 0 & 0 & 0 \\
\ddot{h}_{,\dot{p}} & \ddot{h}_{,\dot{v}} & 0 & 0 & \ddot{h}_{,\ddot{p}} & H & 0 & 0 & 0 & 0 & 0 & 0
\end{array}\right]$$

as a submatrix of $\mathscr{F}_{2,[x,\dot{x},\ddot{x},x^{(3)}]}$. Again, with M nonsingular and $\begin{bmatrix} M & G^T & H^T \\ H & 0 & 0 \end{bmatrix}$ of full (row) rank we get $\operatorname{rank}(\mathscr{F}_{2,[\dot{x},\ddot{x},x^{(3)}]}) = 6n_p + n_\zeta$. Thus, there exists a matrix-valued function Z_2 with full rank and size $(3n, a)$ with $a = 3n - (6n_p + n_\zeta) = 3n_\lambda + 2n_\zeta$ such that $Z_2^T \mathscr{F}_{2,[\dot{x},\ddot{x},x^{(3)}]} = 0$. We can choose

$$Z_2^T = \left[\begin{array}{cccc|cccc|cccc}
0 & 0 & I_{n_\lambda} & 0 & 0 & 0 & 0 & 0 & 0 & 0 & 0 & 0 \\
0 & 0 & 0 & I_{n_\zeta} & 0 & 0 & 0 & 0 & 0 & 0 & 0 & 0 \\
\hline
G & 0 & 0 & 0 & 0 & 0 & I_{n_\lambda} & 0 & 0 & 0 & 0 & 0 \\
\dot{h}_{,\dot{p}} & HM^{-1} & 0 & 0 & 0 & 0 & 0 & I_{n_\zeta} & 0 & 0 & 0 & 0 \\
\hline
\ddot{g}_{,\dot{p}} & GM^{-1} & 0 & 0 & G & 0 & 0 & 0 & 0 & 0 & I_{n_\lambda} & 0
\end{array}\right]. \tag{17.17}$$

With this choice of Z_2 we get

$$Z_2^T \mathscr{F}_{2,x} = \left[\begin{array}{cccc}
G & 0 & 0 & 0 \\
h_{,p} & H & 0 & 0 \\
\hline
\dot{g}_{,p} & G & 0 & 0 \\
HM^{-1}d_{,p} + \dot{h}_{,p} & \dot{h}_{,\dot{p}} + HM^{-1}d_{,v} + \dot{h}_{,v} & -HM^{-1}G^T & -HM^{-1}H^T \\
\hline
GM^{-1}d_{,p} + \ddot{g}_{,p} & \ddot{g}_{,\dot{p}} + GM^{-1}d_{,v} & -GM^{-1}G^T & -GM^{-1}H^T
\end{array}\right],$$

which has rank $3n_\lambda + 2n_\zeta$ due to the full (row) rank of H and G and even of $\begin{bmatrix} G \\ H \end{bmatrix}$, as well as the nonsingularity of $\begin{bmatrix} GM^{-1}G^T & GM^{-1}H^T \\ HM^{-1}G^T & HM^{-1}H^T \end{bmatrix}$, compare with Remark 3.

Thus, there exists a matrix-valued function T_2 of size (n,d) with $d = 2n_p - 2n_\lambda - n_\zeta$ having full rank such that $Z_2^T \mathscr{F}_{2,x} T_2 = 0$. We can choose

$$T_2 = \begin{bmatrix} K_G & 0 \\ -B_{GH} Y^{-1} S K_G & K_{GH} \\ X^{-1}(D B_{GH} Y^{-1} S - C) K_G & -X^{-1} D K_{GH} \end{bmatrix}$$

with matrix-valued functions

$$\begin{aligned} &K_G \text{ of size } (n_p, n_p - n_\lambda) && \text{with } \operatorname{im}(K_G) = \ker(G), \\ &K_{GH} \text{ of size } (n_p, n_p - n_\lambda - n_\zeta) && \text{with } \operatorname{im}(K_{GH}) = \ker(\begin{bmatrix} G \\ H \end{bmatrix}), \\ &B_{GH} \text{ of size } (n_p, n_\lambda - n_\zeta) && \text{with } \operatorname{im}(B_{GH}) = \operatorname{coker}(\begin{bmatrix} G \\ H \end{bmatrix}), \end{aligned}$$

and

$$C = \begin{bmatrix} HM^{-1} d_{,p} + \dot{h}_{,p} \\ GM^{-1} d_{,p} + \ddot{g}_{,p} \end{bmatrix}, \qquad D = \begin{bmatrix} \dot{h}_{,\dot{p}} + HM^{-1} d_{,v} + \dot{h}_{,v} \\ \ddot{g}_{,\dot{p}} + GM^{-1} d_{,v} \end{bmatrix}, \quad S = \begin{bmatrix} h_{,p} \\ \dot{g}_{,p} \end{bmatrix},$$
$$X = \begin{bmatrix} -HM^{-1}G^T - HM^{-1}H^T \\ -GM^{-1}G^T - GM^{-1}H^T \end{bmatrix}, \quad Y = \begin{bmatrix} H \\ G \end{bmatrix} B_{GH}.$$

In particular, it holds that $GK_G = 0$, $\begin{bmatrix} G \\ H \end{bmatrix} K_{GH} = 0$, and $\begin{bmatrix} G \\ H \end{bmatrix} B_{GH}$ is nonsingular. With this T_2 we get

$$F_{,\dot{x}} T_2 = \begin{bmatrix} -K_G & 0 \\ M B_{GH} Y^{-1} S K_G & -M K_{GH} \\ 0 & 0 \\ 0 & 0 \end{bmatrix}$$

with $\operatorname{rank}(F_{,\dot{x}} T_2) = \operatorname{rank}(K_G) + \operatorname{rank}(K_{GH}) = (n_p - n_\lambda) + (n_p - n_\lambda - n_\zeta) = 2n_p - 2n_\lambda - n_\zeta$. Therefore, there exists a full rank matrix-valued function Z_1 of size (n,d) such that $Z_1^T F_{,\dot{x}} T_2$ has full rank d, i.e., $Z_1^T F_{,\dot{x}} T_2$ is nonsingular. We can choose

$$Z_1^T = \begin{bmatrix} K_G^T & 0 & 0 & 0 \\ 0 & K_{GH}^T M^T & 0 & 0 \end{bmatrix} \tag{17.18}$$

such that

$$Z_1^T F_{,\dot{x}} T_2 = \begin{bmatrix} -K_G^T K_G & 0 \\ K_{GH}^T M^T M B_{GH} Y^{-1} S K_G & -K_{GH}^T M^T M K_{GH} \end{bmatrix}$$

is nonsingular due to the full rank of K_G and MK_{GH}. □

Remark 4 (a) The matrix-valued function Z_1^T is not uniquely determined. For instance, instead of $K_{GH}^T M^T$ in Z_1^T in the proof it is also possible to choose, e.g., $K_{GH}^T M^{-1}$ or since M is assumed to be nonsingular and, therefore, positive definite, K_{GH}^T is also possible.

(b) The equations of motion (17.12) satisfying Assumption 1 form a DAE with strangeness-index

$$\nu_s = \begin{cases} 2 & \text{for } n_\lambda > 0, \\ 1 & \text{for } n_\lambda = 0 \text{ and } n_\zeta > 0, \\ 0 & \text{for } n_\lambda = 0 \text{ and } n_\zeta = 0. \end{cases}$$

This means that, if there appear holonomic constraints (17.12d), i.e., $n_\lambda > 0$, then the equations of motion (17.12) form a system of DAEs of s-index $\nu_s = 2$. This follows from the fact that the Hypothesis 1 is satisfied for $\mu = 2$ but not for $\mu = 1$. In the case of purely nonholonomic systems, i.e., $n_\lambda = 0$ and $n_\zeta > 0$, the equations of motion (17.12) form a system of DAEs (17.12a), (17.12b), (17.12d) (without $G^T(p,t)\lambda$) which is of s-index $\nu_s = 1$, since in that case it can be shown that the Hypothesis 1 is satisfied for $\mu = 1$ but not for $\mu = 0$. For an example, see Example 1. Furthermore, if neither holonomic nor nonholonomic constraints exist, i.e., $n_\lambda = 0$ and $n_\zeta = 0$, the equations of motion (17.12) appear as set of ordinary differential equations (17.12a), (17.12b) (without $G^T(p,t)\lambda$ and $H^T(p,t)\zeta$) which has s-index $\nu_s = 0$.

(c) Due to the zero columns in the matrix-valued function Z_1^T (17.18) it would be sufficient to use the reduced derivative array (17.9) in form

$$\hat{\mathscr{F}}_2(t,x,\dot{x},\ddot{x},x^{(3)}) = \left[\begin{array}{c} -\dot{p}+v \\ -M(p)\dot{v}+f(p,v,t)-G^T(p,t)\lambda-H^T(p,t)\zeta \\ g(p,t) \\ H(p,t)v+\breve{h}(p,t) \\ \hline -\ddot{p}+\dot{v} \\ G(p,t)\dot{p}+g_{,t}(p,t) \\ \tilde{h}^I(p,v,\dot{p},t)+H(p,t)\dot{v} \\ \hline \tilde{g}^{II}(p,\dot{p},t)+G(p,t)\ddot{p} \end{array}\right]$$

as basis to analyze the equations of motion (17.12) with use of the Hypothesis 1. ◁

Lemma 1 *Let the equations of motion (17.12) satisfy the Assumption 1. Then the hidden constraints are given by*

$$0 = g^I(p,v,t) \qquad = Gv+g_{,t}, \tag{17.19a}$$

$$0 = h^I(p,v,\lambda,\zeta,t) = (Hv+\breve{h})_{,p}v+HM^{-1}(f-G^T\lambda-H^T\zeta)+(Hv+\breve{h})_{,t}, \tag{17.19b}$$

$$0 = g^{II}(p,v,\lambda,\zeta,t) = (Gv+g_{,t})_{,p}v+GM^{-1}(f-G^T\lambda-H^T\zeta)+(Gv+g_{,t})_{,t}. \tag{17.19c}$$

Proof Due to Theorem 1 we have the existence of a matrix-valued function Z_2 (see (17.17)) which allows to extract the (hidden) constraints as in (17.5b) from the derivative array (17.16) according to [28]. We get

$$Z_2^T\mathscr{F}_2=\left[\begin{array}{c} g \\ Hv+\breve{h} \\ \hline G(-\dot{p}+v)+(G\dot{p}+g_{,t}) \\ \dot{h}_{,\dot{p}}(-\dot{p}+v)+HM^{-1}(-M\dot{v}+f-G^T\lambda-H^T\zeta)+(\tilde{h}^I+H\dot{v}) \\ \hline \ddot{g}_{,\dot{p}}(-\dot{p}+v)+GM^{-1}(-M\dot{v}+f-G^T\lambda-H^T\zeta)+G(-\ddot{p}+\dot{v})+(\tilde{g}^{II}+G\ddot{p}) \end{array}\right].$$

This can be reformulated as

$$0=Z_2^T\mathscr{F}_2=\left[\begin{array}{c} g \\ Hv+\breve{h} \\ \hline Gv+g_{,t} \\ (Hv+\breve{h})_{,p}v+HM^{-1}(f-G^T\lambda-H^T\zeta)+(Hv+\breve{h})_{,t} \\ \hline (Gv+g_{,t})_{,p}v+GM^{-1}(f-G^T\lambda-H^T\zeta)+(Gv+g_{,t})_{,t} \end{array}\right] \tag{17.20}$$

using algebraic manipulations. While the first two block equations are already stated explicitly as equations in (17.12) the last three block equations form the hidden constraints. □

Let the equations of motion (17.12) satisfy Assumption 1. Let the hidden constraints be defined as in (17.19). Then, the initial values p_0, v_0, λ_0, and ζ_0 are consistent with (17.12), if and only if they fulfill the constraints (17.12c), (17.12d), and (17.19). Furthermore, from Lemma 1 we get the set of consistency $\mathbb{S}$ as

$$\begin{aligned}\mathbb{S}=\{(p,v,\lambda,\zeta,t)\in\mathbb{R}^{n_p+n_p+n_\lambda+n_\zeta}\times\mathbb{I}\ :\ & 0=g^I(p,v,t),\\ & 0=h^I(p,v,\lambda,\zeta,t),\\ & 0=g^{II}(p,v,\lambda,\zeta,t)\}.\end{aligned} \tag{17.21}$$

The first two block constraints in (17.20) are extracted from the original equations of motion, i.e., from the derivative array $\mathscr{F}_0 = F$ of level 0. Therefore, these constraints are called *constraints on level 0*. The third and fourth block constraints in (17.20) are known as the *holonomic constraints on velocity level* and *nonholonomic constraints on acceleration level*. They are extracted from the original equations of motion (17.12) and the first time derivative, i.e., from the derivative array $\mathscr{F}_1 = \left[\, F^T \,\middle|\, \frac{\mathrm{d}}{\mathrm{d}t}F^T \,\right]^T$ of level 1. Therefore, these constraints are also called *hidden constraints on level 1*. The last block constraint in (17.20) is known as the *holonomic constraints on acceleration level*. They are extracted from the original equations of motion (17.12) and the first and second time derivatives, i.e., from the derivative array $\mathscr{F}_2$ of level 2, see (17.16). Therefore, these constraints are called *hidden constraints on level 2* in accordance to the investigations in [49]. These hidden constraints (17.19) lead to the difficulties in the direct numerical treatment of the equations of motion in form (17.12). Therefore, they have to be treated very carefully. In the following theorem, we discuss the existence and uniqueness of the solution of the equations of motion (17.12).

Theorem 2 *Let the equations of motion (17.12) satisfy Assumption 1 and let the initial values in (17.13) be consistent. Assume further that* $\tilde{f}$ *given by*

$$\tilde{f}(p, v, t) = f - \begin{bmatrix} G^T & H^T \end{bmatrix} \begin{bmatrix} GM^{-1}G^T & GM^{-1}H^T \\ HM^{-1}G^T & HM^{-1}H^T \end{bmatrix}$$

$$\cdot \begin{bmatrix} GM^{-1}f + (Gv + g_{,t})_{,p}v + (Gv + g_{,t})_{,t} \\ HM^{-1}f + (Hv + \breve{h})_{,p}v + (Hv + \breve{h})_{,t} \end{bmatrix}$$

as well as M^{-1} *are continuous and bounded on* $\mathbb{S}$ *and Lipschitz continuous with respect to* p *and* v *on* $\mathbb{S}$. *Then in a neighborhood of* $(p_0, v_0, \lambda_0, \zeta_0, t_0)$ *there exists a unique solution for* p, v, λ, *and* ζ *of the initial value problem (17.12), (17.13).*

Proof In [49] it is shown that the solution of the initial value problem (17.12) with (17.13) and the solution of the initial value problem consisting of (17.12a), (17.12b), (17.19b), (17.19c) with (17.13) (for the same initial values) are identical. Therefore, the proof reduces to show that the solution of the latter one exists and is unique.

Because of the assumption (17.15a) we get the unique solvability of (17.19b) and (17.19c) with respect to λ and ζ as functions of p, v, and t from the Implicit Function Theorem [3]. Therefore, with assumption (17.15a) the equations (17.12a) and (17.12b) correspond to the ODE

$$\begin{bmatrix} \dot{p} \\ \dot{v} \end{bmatrix} = \begin{bmatrix} v \\ M^{-1}(p)\tilde{f}(p, v, t) \end{bmatrix}.$$

Because of the assumed boundedness, continuity, and Lipschitz continuity of $\tilde{f}$ and M^{-1} the conditions of the Theorem of Picard-Lindelöf [25] are satisfied and the existence of a unique solution for p and v then follows. From that also the Lagrange multiplier λ and ζ are uniquely defined by (17.19b) and (17.19c) with known p and v. □

Remark 5 In [33, 49] the influence of redundant constraints on the existence and uniqueness of solutions for the equations of motion is discussed. It was shown that redundant constraints only influence the uniqueness of the Lagrange multipliers. The constraint forces as well as the solutions for p and v are unique. ◁

17.3.3 Regularization

In general, the numerical integration of the equations of motion (17.12) is substantially more difficult and prone to intensive numerical computation than that of ODEs, see [8, 19, 21, 28, 37]. As mentioned above, the index of a DAE provides a measure of the difficulty to solve the DAE. A lower index is to be preferred for the numerical simulation. However, simple differentiation of the constraints decreases the index, but simultaneously the drift-off effects are increased as shown in [14, 47–49].

Several approaches have been introduced in order to stabilize the integration process. For an overview, see [8, 9, 21, 28, 47, 49]. Frequently used index reduction techniques or regularizations for equations of motion are the *Baumgarte stabilization* [6], lowering the index by differentiation of the constraints [16], the *Gear-Gupta-Leimkuhler formulation* [17], or overdetermined formulation [15, 49].

In the following, we will focus on regularization techniques based on the strangeness concept. In particular, these are the *Kunkel-Mehrmann regularization*, the *regularization by minimal extension*, and the *regularization using overdetermined formulations*.

Kunkel-Mehrmann Regularization In Sect. 17.2, we have seen that Hypothesis 1 implies (locally) the existence of a reduced system (in the original variables) of the form (17.5). Applying Hypothesis 1 to the equations of motion (17.12), see Sect. 17.3.2, we get the matrix-valued functions Z_2^T as, e.g., (17.17) and Z_1^T as, e.g., (17.18). While Z_2^T extracts the algebraic part, i.e., the (hidden) constraints (17.20) of the dynamical system, the matrix-valued function Z_1^T allows the extraction of differential equations of minimal number describing the dynamics of the dynamical system. With Z_1^T as in (17.18) we get the differential part as

$$Z_1^T F = \begin{bmatrix} K_G^T(-\dot{p} + v) \\ K_{GH}^T M^T(M\dot{v} + f - G^T\lambda - H^T\zeta) \end{bmatrix}.$$

Then, the KM-regularization is defined as in (17.5), and we get it in the specific form

$$K_G^T\dot{p} = K_G^T v, \tag{17.22a}$$

$$-K_{GH}^T M^T M\dot{v} = K_{GH}^T M^T(f - G^T\lambda - H^T\zeta), \tag{17.22b}$$

$$0 = g, \tag{17.22c}$$

$$0 = Hv + \breve{h}, \tag{17.22d}$$

$$0 = Gv + g_{,t}, \tag{17.22e}$$

$$0 = (Hv + \breve{h})_{,p}v + HM^{-1}(f - G^T\lambda - H^T\zeta) + (Hv + \breve{h})_{,t}, \tag{17.22f}$$

$$0 = (Gv + g_{,t})_{,p}v + GM^{-1}(f - G^T\lambda - H^T\zeta) + (Gv + g_{,t})_{,t}. \tag{17.22g}$$

This formulation is strangeness-free with the same solution set as the original equations of motion (17.12) regardless of any initial values. Therefore, this KM-formulation (17.22) corresponds to a regularization of the equations of motion and, in particular, is suitable for the numerical treatment using implicit numerical methods for stiff ordinary differential equations. For more details, we refer to [52].

Regularization by Minimal Extension In [27], a further regularization approach is discussed, the so-called *Regularization by Minimal Extension*. The idea of the minimal extension is illustrated in Sect. 17.2. Its application to the equations of motion (17.12) yields the following.

Find a (locally constant and) nonsingular matrix

$$\Pi = \begin{bmatrix} \Pi_1 \; \Pi_2 \; \Pi_3 \end{bmatrix} \text{ with } \Pi^{-1} = \begin{bmatrix} \bar{\Pi}_1 \\ \bar{\Pi}_2 \\ \bar{\Pi}_3 \end{bmatrix},$$

such that

$$\begin{bmatrix} G(p,t) \\ H(p,t) \end{bmatrix} \Pi = \begin{bmatrix} G(p,t)\Pi_1 & G(p,t)\Pi_2 & G(p,t)\Pi_3 \\ H(p,t)\Pi_1 & H(p,t)\Pi_2 & H(p,t)\Pi_3 \end{bmatrix} \tag{17.23a}$$

$$\text{where } G(p,t)\Pi_3 \text{ and } H(p,t)\Pi_2 \text{ are pointwise nonsingular} \tag{17.23b}$$

in a neighborhood of a set of consistent values $\bar{p}$, $\bar{v}$, $\bar{\lambda}$, $\bar{\zeta}$, and $\bar{t}$. Then from the approach of regularization by minimal extension we get by adding the hidden constraints (17.20) and introducing new algebraic variables w_3^p, w_2^v, and w_3^v for $\bar{\Pi}_3\dot{p}$, $\bar{\Pi}_2\dot{v}$, and $\bar{\Pi}_3\dot{v}$, respectively, the minimally extended formulation

$$\Pi \begin{bmatrix} \bar{\Pi}_1\dot{p} \\ \bar{\Pi}_2\dot{p} \\ w_3^p \end{bmatrix} = v, \tag{17.24a}$$

$$M\Pi \begin{bmatrix} \bar{\Pi}_1\dot{v} \\ w_2^v \\ w_3^v \end{bmatrix} = f - G^T\lambda - H^T\zeta, \tag{17.24b}$$

$$0 = g, \tag{17.24c}$$

$$0 = Hv + \breve{h}, \tag{17.24d}$$

$$0 = Gv + g_{,t}, \tag{17.24e}$$

$$0 = (Hv + \breve{h})_{,p}v + HM^{-1}(f - G^T\lambda - H^T\zeta) + (Hv + \breve{h})_{,t}, \tag{17.24f}$$

$$0 = (Gv + g_{,t})_{,p}v + GM^{-1}(f - G^T\lambda - H^T\zeta) + (Gv + g_{,t})_{,t}. \tag{17.24g}$$

This formulation is again strangeness-free. In comparison to the KM-Regularization above, it is not necessary to compute the matrix-valued functions K_G and K_{GH}, but the number of unknown variables is increased by $2n_\lambda + n_\zeta$. We have the vector of unknowns $\left[\, p^T\ v^T\ \lambda^T\ \zeta^T\ (w_3^p)^T\ (w_2^v)^T\ (w_3^v)^T \,\right]^T$ with p, v, λ, and ζ as in the original equations of motion (17.12). Therefore, every p, v, λ, and ζ that solve the formulation (17.24) together with w_3^p, w_2^v, and w_3^v solves also the original equations of motion (17.12). Conversely, every solution p, v, λ, and ζ of the equations of motion (17.12) solves also the formulation (17.24) together with $w_3^p = \bar{\Pi}_3\dot{p}$, $w_2^v = \bar{\Pi}_2\dot{v}$, and $w_3^v = \bar{\Pi}_3\dot{v}$, independently of any initial values. Therefore, this formulation (17.24) corresponds to a regularization of the equations of motion and, in particular, is suitable for the numerical treatment using implicit numerical methods for stiff ordinary differential equations. For more details, we refer to [45].

Regularization by Overdetermined Formulations While the regularization techniques that are discussed in the previous sections mostly lead to an equivalent regularized form of the equations of motion of square size, i.e., with the same number of unknowns and equations, in [10, 15, 49] an approach is proposed which adds all hidden constraints to the equations of motion. This approach leads to an overdetermined system consisting of the equations of motion (17.12) and all hidden constraints (17.19), i.e., we get the overdetermined DAE

$$
\begin{aligned}
\dot{p} &= v, && (17.25a)\\
M\dot{v} &= f - G^T\lambda - H^T\zeta, && (17.25b)\\
0 &= g, && (17.25c)\\
0 &= Hv + \breve{h}, && (17.25d)\\
0 &= Gv + g_{,t}, && (17.25e)\\
0 &= (Hv + \breve{h})_{,p}v + HM^{-1}(f - G^T\lambda - H^T\zeta) + (Hv + \breve{h})_{,t}, && (17.25f)\\
0 &= (Gv + g_{,t})_{,p}v + GM^{-1}(f - G^T\lambda - H^T\zeta) + (Gv + g_{,t})_{,t}. && (17.25g)
\end{aligned}
$$

From Hypothesis 1 it follows that the DAE (17.25) has s-index $\nu_s = 1$. In particular, this means that the overdetermined formulation (17.25) is not strangeness-free. But, all necessary information is contained in the system in an explicit way. Consequently, the overdetermined formulation (17.25) has the maximal constraint level $\nu_c = 0$, see [49]. This means no hidden constraints are obtained in formulation (17.25).

In comparison to the KM-regularization and the regularization by minimal extension, no effort for selecting a reduced differential part nor new variables have to be introduced. Therefore, this regularization approach is very simple, the unknown variables are unchanged, but this approach results in an overdetermined system. Nevertheless, this overdetermined formulation (17.25) is suitable for the numerical treatment by use of adapted implicit numerical methods for stiff ordinary differential equations. A further advantage of the regularization by overdetermined formulations is that solution invariants, like the invariance of the total energy, momentum, or impulse, simply can be added to the overdetermined formulation (17.25). For more details, see [49–51].

Furthermore, note that the KM-formulation can be determined from the overdetermined formulation (17.25) by selecting some differential equations with use of the matrix-valued function Z_1, see (17.18), determined in Sect. 17.3.2. Another approach for selecting the differential equations for multibody systems, is developed in [49]. Furthermore, the minimally extended formulation (17.24) can be determined from the overdetermined formulation (17.25) by detecting strange variables and replacing the derivative of these strange variables by newly introduced algebraic variables.

17.4 Electrical Systems

Another important problem class are electrical circuits or electrical subcomponents embedded in multi-physical systems. In this section, we consider connected electrical circuits containing (possibly nonlinear) resistors, capacitors, and inductors as well as (independent) voltage sources and (independent) current sources.

17.4.1 Model Equations and Analysis

A common way for the modeling of electrical circuits is the *Modified Nodal Analysis (MNA)* [55]. Using Kirchhoff's laws and the constitutive element relations for inductors, capacitors, and resistors we get a DAE system of the form

$$A_{\mathcal{C}}\frac{d}{dt}q(A_{\mathcal{C}}^T\eta) + A_{\mathcal{L}}\imath_{\mathcal{L}} + A_{\mathcal{R}}g(A_{\mathcal{R}}^T\eta) + A_{\mathcal{V}}\imath_{\mathcal{V}} + A_{\mathcal{I}}\mathcal{I}_s(t) = 0, \quad (17.26a)$$

$$\frac{d}{dt}\phi(\imath_{\mathcal{L}}) - A_{\mathcal{L}}^T\eta = 0, \quad (17.26b)$$

$$A_{\mathcal{V}}^T\eta - \mathcal{V}_s(t) = 0, \quad (17.26c)$$

see also [20]. Equations (17.26) are also known as the *MNA equations*. Here,

$$A = \left[A_{\mathcal{C}}\ A_{\mathcal{L}}\ A_{\mathcal{R}}\ A_{\mathcal{V}}\ A_{\mathcal{I}}\right] \quad (17.27)$$

denotes the reduced incidence matrix of the directed graph describing the circuit topology, assuming that the branches are ordered by the type of component, such that $A_{\mathcal{C}} \in \mathbb{R}^{n_\eta,n_{\mathcal{C}}}$, $A_{\mathcal{L}} \in \mathbb{R}^{n_\eta,n_{\mathcal{L}}}$, $A_{\mathcal{R}} \in \mathbb{R}^{n_\eta,n_{\mathcal{R}}}$, $A_{\mathcal{V}} \in \mathbb{R}^{n_\eta,n_{\mathcal{V}}}$, and $A_{\mathcal{I}} \in \mathbb{R}^{n_\eta,n_{\mathcal{I}}}$. Furthermore, $n_{\mathcal{V}}$ denotes the number of voltage sources, $n_{\mathcal{I}}$ the number of current sources, $n_{\mathcal{C}}$ the number of capacitors, $n_{\mathcal{L}}$ the number of inductors, and $n_{\mathcal{R}}$ the number of resistors in the circuit, respectively. Moreover, $g : \mathbb{R}^{n_{\mathcal{R}}} \to \mathbb{R}^{n_{\mathcal{R}}}$ is a vector-valued function composed of the functions $g_i : \mathbb{R} \to \mathbb{R}$, $i = 1, \dots, n_{\mathcal{R}}$, describing the conductance for each resistor, $q : \mathbb{R}^{n_{\mathcal{C}}} \to \mathbb{R}^{n_{\mathcal{C}}}$ composed of the functions $q_i : \mathbb{R} \to$

$\mathbb{R}$, $i = 1, \ldots, n_{\mathcal{C}}$, describes the charges for each capacitor, and $\phi : \mathbb{R}^{n_{\mathcal{L}}} \to \mathbb{R}^{n_{\mathcal{L}}}$ composed of functions $\phi_i : \mathbb{R} \to \mathbb{R}$, $i = 1, \ldots, n_{\mathcal{L}}$, describes the magnetic flux for each inductor. The vectors $\imath_{\mathcal{L}}(t) \in \mathbb{R}^{n_{\mathcal{L}}}$ and $\imath_{\mathcal{V}}(t) \in \mathbb{R}^{n_{\mathcal{V}}}$ denote the currents of all inductive branches, and branches corresponding to voltage sources, respectively, and $\eta(t) \in \mathbb{R}^{n_\eta}$ denotes the vector of all node potentials, where n_η is the number of nodes in the directed graph (excluding the ground node). In (17.26) we have restricted to the case of independent current and voltage sources described by the source functions $\mathcal{I}_s(t)$ and $\mathcal{V}_s(t)$, respectively. In general, also controlled sources are possible, see [4, 12]. For details on the constitutive element relations and on the derivation of the MNA equations, see also Chap. 18 or [12, 20, 38].

Due to the special structure of the MNA equations (17.26), it is possible to determine the index and those parts of the system that lead to hidden constraints by graph theoretical considerations, see also [12]. We say that the DAE system (17.26) is *well-posed* if it satisfies the following assumptions:

(A1) The circuit contains no $\mathcal{V}$-loops, i.e., $A_{\mathcal{V}}$ has full column rank.
(A2) The circuit contains no $\mathcal{I}$-cutsets, i.e., $\left[\, A_{\mathcal{C}}\; A_{\mathcal{L}}\; A_{\mathcal{R}}\; A_{\mathcal{V}} \right]$ has full row rank.
(A3) The charge functions $q_i : \mathbb{R} \to \mathbb{R}$, $i = 1, \ldots n_{\mathcal{C}}$, are strictly monotonically increasing and continuously differentiable.
(A4) The flux functions $\phi_i : \mathbb{R} \to \mathbb{R}$, $i = 1, \ldots n_{\mathcal{L}}$, are strictly monotonically increasing and continuously differentiable.
(A5) The conductance functions $g_i : \mathbb{R} \to \mathbb{R}$ for $i = 1, \ldots n_{\mathcal{R}}$ are strictly monotonically increasing and continuously differentiable.

Here, a $\mathcal{V}$-loop is defined as a loop in the circuit graph consisting only of branches corresponding to voltage sources. In the same way, a $\mathcal{CV}$-loop means a loop in the circuit graph consisting only of branches corresponding to capacitances and/or voltage sources. Likewise, an $\mathcal{I}$-cutset is a cutset in the circuit graph consisting only of branches corresponding to current sources, and an $\mathcal{LI}$-cutset is a cutset in the circuit graph consisting only of branches corresponding to inductances and/or current sources. Assumption **(A1)** implies that there are no short-circuits. In a similar manner, the occurrence of $\mathcal{I}$-cutsets may lead to contradictions in the Kirchhoff laws (source functions may not sum up to zero), which is excluded by assumption **(A2)**. The assumptions **(A3)**, **(A4)** and **(A5)** imply that all circuit elements are passive, i.e., they do not generate energy. Furthermore, we introduce matrices $Z_{\mathcal{C}} \in \mathbb{R}^{n_\eta, p_{\mathcal{C}}}$, $Z_{\mathcal{V}-\mathcal{C}} \in \mathbb{R}^{p_{\mathcal{C}}, p_b}$, $Z_{\mathcal{R}-\mathcal{CV}} \in \mathbb{R}^{p_b, p_a}$, $\bar{Z}_{\mathcal{V}-\mathcal{C}} \in \mathbb{R}^{n_{\mathcal{V}}, p_{\mathcal{V}-\mathcal{C}}}$ each of full column rank, such that

$$\operatorname{im}(Z_{\mathcal{C}}) = \ker(A_{\mathcal{C}}^T), \qquad \operatorname{im}(Z_{\mathcal{V}-\mathcal{C}}) = \ker(A_{\mathcal{V}}^T Z_{\mathcal{C}}),$$
$$\operatorname{im}(Z_{\mathcal{R}-\mathcal{CV}}) = \ker(A_{\mathcal{R}}^T Z_{\mathcal{C}} Z_{\mathcal{V}-\mathcal{C}}), \quad \operatorname{im}(\bar{Z}_{\mathcal{V}-\mathcal{C}}) = \ker(Z_{\mathcal{C}}^T A_{\mathcal{V}}),$$

and we define $Z_{\mathcal{CRV}} := Z_{\mathcal{C}} Z_{\mathcal{V}-\mathcal{C}} Z_{\mathcal{R}-\mathcal{CV}} \in \mathbb{R}^{n_\eta, p_a}$. Note that in [12] projectors were used, while here we use matrices whose columns span the corresponding subspaces.

Theorem 3 ([12, 38]) *Consider an electrical circuit with circuit equations as in* (17.26)*. Assume that the assumptions* ***(A1)–(A5)*** *hold.*

1. *If the circuit contains no voltage sources, i.e.,* $n_{\mathcal{V}} = 0$*, and* $\operatorname{rank}(A_{\mathcal{C}}) = n_\eta$*, i.e., there are no* $\mathcal{RLJ}$*-cutsets, then* (17.26) *is an ODE (in implicit form) with s-index* $\nu_s = 0$.
2. *If the circuit contains neither* $\mathcal{LJ}$*-cutsets nor* $\mathcal{CV}$*-loops (except for* $\mathcal{C}$*-loops), then the DAE* (17.26) *has s-index* $\nu_s = 0$ *and the* algebraic constraints *are given by*

$$0 = Z_{\mathcal{C}}^T \left(A_{\mathcal{R}} g(A_{\mathcal{R}}^T \eta) + A_{\mathcal{L}} \iota_{\mathcal{L}} + A_{\mathcal{V}} \iota_{\mathcal{V}} + A_{\mathcal{J}} \mathcal{J}_s(t) \right), \tag{17.28a}$$

$$0 = A_{\mathcal{V}}^T \eta - \mathcal{V}_s(t). \tag{17.28b}$$

3. *If the circuit contains* $\mathcal{LJ}$*-cutsets or* $\mathcal{CV}$*-loops which are no pure* $\mathcal{C}$*-loops, then the DAE* (17.26) *has s-index* $\nu_s = 1$*. We distinguish the following three cases:*

   a. *If the circuit contains* $\mathcal{CV}$*-loops, but no* $\mathcal{LJ}$*-cutsets, then, in addition to* (17.28)*, there exist* hidden constraints

$$0 = \bar{Z}_{\mathcal{V}-\mathcal{C}}^T \left(A_{\mathcal{V}}^T \frac{d}{dt}\eta - \frac{d}{dt}\mathcal{V}_s(t) \right). \tag{17.29}$$

   b. *If the circuit contains* $\mathcal{LJ}$*-cutsets, but no* $\mathcal{CV}$*-loops, then, in addition to* (17.28)*, there exist* hidden constraints

$$0 = Z_{\mathcal{CRV}}^T \left(A_{\mathcal{L}} \frac{d}{dt}\iota_{\mathcal{L}} + A_{\mathcal{J}} \frac{d}{dt}\mathcal{J}_s(t) \right). \tag{17.30}$$

   c. *If the circuit contains both* $\mathcal{CV}$*-loops and* $\mathcal{LJ}$*-cutsets, then all of the constraints* (17.28)–(17.30) *have to be fulfilled.*

Note that the source functions that belong to $\mathcal{CV}$-loops or $\mathcal{LJ}$-cutsets have to be differentiable if the DAE has s-index $\nu_s = 1$. In [4] it has been shown that the multiplication with the matrices $\bar{Z}_{\mathcal{V}-\mathcal{C}}$ and $Z_{\mathcal{CRV}}$ for the determination of the hidden constraints can be interpreted in graph theoretical ways.

17.4.2 *Regularization by Minimal Extension*

If the electrical circuit contains $\mathcal{CV}$-loops or $\mathcal{LJ}$-cutsets, a regularization of the MNA equations (17.26) is required. The equations that are responsible for a higher index are given by the projected equations (17.29) and (17.30). Thus, the reduced derivative array (17.9) consists of the MNA equations (17.26) together with the hidden constraints (17.29) and (17.30) resulting in

$$A_{\mathcal{C}} \frac{d}{dt} q(A_{\mathcal{C}}^T \eta) + A_{\mathcal{L}} \iota_{\mathcal{L}} + A_{\mathcal{R}} g(A_{\mathcal{R}}^T \eta) + A_{\mathcal{V}} \iota_{\mathcal{V}} + A_{\mathcal{J}} \mathcal{J}_s(t) = 0, \tag{17.31a}$$

$$\frac{d}{dt}\phi(\iota_{\mathcal{L}}) - A_{\mathcal{L}}^T \eta = 0, \tag{17.31b}$$

$$A_{\mathcal{V}}^T \eta - \mathcal{V}_s(t) = 0, \quad (17.31c)$$

$$\bar{Z}_{\mathcal{V}-C}^T A_{\mathcal{V}}^T \frac{d}{dt} \eta - \bar{Z}_{\mathcal{V}-C}^T \frac{d}{dt} \mathcal{V}_s(t) = 0, \quad (17.31d)$$

$$Z_{C\mathcal{R}\mathcal{V}}^T A_{\mathcal{L}} \frac{d}{dt} \iota_{\mathcal{L}} + Z_{C\mathcal{R}\mathcal{V}}^T A_{\mathcal{J}} \frac{d}{dt} \mathcal{J}_s(t) = 0. \quad (17.31e)$$

The concept of index reduction by minimal extension for electrical circuit equations (17.26) has been introduced in [4, 27]. In the MNA equations (17.26), the vector of unknowns consists of the node potentials η, the currents through inductors $\iota_{\mathcal{L}}$ and the currents through voltage sources $\iota_{\mathcal{V}}$. Thus, the state vector has dimension $n = n_\eta + n_{\mathcal{L}} + n_{\mathcal{V}}$. On the other hand, the reduced derivative array consists of $n_\eta + n_{\mathcal{L}} + n_{\mathcal{V}} + p_{\mathcal{V}-C} + p_a$ equations. In order to obtain a regular square system, we have to introduce new variables. To determine this minimal extension of the original system (17.26), we have to identify those differential variables that are "strange" and, therefore, have to be replaced by new algebraic variables.

For hidden constraints due to $C\mathcal{V}$-loops or $\mathcal{L}\mathcal{J}$-cutsets given by (17.31d) and (17.31e), we have to find permutations Π_η and $\Pi_{\iota_{\mathcal{L}}}$ such that

$$\bar{Z}_{\mathcal{V}-C}^T A_{\mathcal{V}}^T \Pi_\eta^T = \begin{bmatrix} B_1 & B_2 \end{bmatrix}, \quad Z_{C\mathcal{R}\mathcal{V}}^T A_{\mathcal{L}} \Pi_{\iota_{\mathcal{L}}}^T = [F_1 \ F_2],$$

with $B_1 \in \mathbb{R}^{p_{\mathcal{V}-C}, p_{\mathcal{V}-C}}$ and $F_1 \in \mathbb{R}^{p_a, p_a}$ nonsingular. Then, we can partition the vector η and $\iota_{\mathcal{L}}$ as

$$\Pi_\eta \eta = \begin{bmatrix} \tilde{\eta}_1 \\ \tilde{\eta}_2 \end{bmatrix} = \tilde{\eta}, \quad \Pi_{\iota_{\mathcal{L}}} \iota_{\mathcal{L}} = \begin{bmatrix} \tilde{\iota}_{\mathcal{L}1} \\ \tilde{\iota}_{\mathcal{L}2} \end{bmatrix} = \tilde{\iota}_{\mathcal{L}}, \quad (17.32)$$

accordingly, and introduce the new variables

$$\hat{\eta}_1(t) = \frac{d}{dt} \tilde{\eta}_1(t), \quad \hat{\iota}_{\mathcal{L}1}(t) = \frac{d}{dt} \tilde{\iota}_{\mathcal{L}1}(t). \quad (17.33)$$

Now, the extended system (17.31) with the newly introduced variables $\hat{\eta}_1$ and $\hat{\iota}_1$ can be written as

$$\tilde{A}_C \frac{d}{dv_C} q(\tilde{A}_C^T \tilde{\eta}) \tilde{A}_C^T \begin{bmatrix} \hat{\eta}_1 \\ \frac{d}{dt} \tilde{\eta}_2 \end{bmatrix} + \tilde{A}_{\mathcal{L}} \tilde{\iota}_{\mathcal{L}} + \tilde{A}_{\mathcal{R}} g(\tilde{A}_{\mathcal{R}}^T \tilde{\eta}) + \tilde{A}_{\mathcal{V}} \iota_{\mathcal{V}} + \tilde{A}_{\mathcal{J}} \mathcal{J}_s(t) = 0, \quad (17.34a)$$

$$\Pi_{\iota_{\mathcal{L}}} \frac{d}{d\iota_{\mathcal{L}}} \phi(\Pi_{\iota_{\mathcal{L}}}^T \tilde{\iota}_{\mathcal{L}}) \Pi_{\iota_{\mathcal{L}}}^T \begin{bmatrix} \hat{\iota}_{\mathcal{L}1} \\ \frac{d}{dt} \tilde{\iota}_{\mathcal{L}2} \end{bmatrix} - \tilde{A}_{\mathcal{L}}^T \tilde{\eta} = 0, \quad (17.34b)$$

$$\tilde{A}_{\mathcal{V}}^T \tilde{\eta} - \mathcal{V}_s(t) = 0, \quad (17.34c)$$

$$B_1\hat{\eta}_1 + B_2\frac{d}{dt}\tilde{\eta}_2 - \bar{Z}^T_{\mathcal{V}-C}\frac{d}{dt}\mathcal{V}_s(t) = 0, \tag{17.34d}$$

$$F_1\hat{\imath}_{\mathcal{L}1} + F_2\frac{d}{dt}\tilde{\imath}_{\mathcal{L}2} + Z^T_{C\mathcal{R}\mathcal{V}}A_{\mathcal{I}}\frac{d}{dt}\mathcal{I}_s(t) \;= 0 \tag{17.34e}$$

with $\nu_C = \tilde{A}^T_C\tilde{\eta}$ and $\tilde{A}_* := \Pi_\eta A_*$ for $* \in \{C, \mathcal{R}, \mathcal{I}, \mathcal{V}\}$, $\tilde{A}_\mathcal{L} := \Pi_\eta A_\mathcal{L}\Pi^T_{\imath_\mathcal{L}}$. In (17.34) we have added exactly as many equations as needed, thus, the extension is minimal. The minimally extended system (17.34) now consists of $n_\eta + n_\mathcal{L} + n_\mathcal{V} + p_{\mathcal{V}-C} + p_a$ equations in the unknowns

$$\tilde{\eta} = \begin{bmatrix}\tilde{\eta}_1\\ \tilde{\eta}_2\end{bmatrix} \in \mathbb{R}^{n_\eta}, \quad \tilde{\imath}_\mathcal{L} = \begin{bmatrix}\tilde{\imath}_{\mathcal{L}1}\\ \tilde{\imath}_{\mathcal{L}2}\end{bmatrix} \in \mathbb{R}^{n_\mathcal{L}}, \quad \imath_\mathcal{V} \in \mathbb{R}^{n_\mathcal{V}}, \quad \hat{\eta}_1 \in \mathbb{R}^{p_{\mathcal{V}-C}}, \quad \hat{\imath}_{\mathcal{L}1} \in \mathbb{R}^{p_a}.$$

Since typically $p_{\mathcal{V}-C}$ and p_a are small, the minimally extended system is only slightly larger than the original system. Note, that the computation of Π_η and $\Pi_{\imath_\mathcal{L}}$ is possible with very small computational effort and very accurately, since $Z^T_{C\mathcal{R}\mathcal{V}}A_\mathcal{L}$ and $A_\mathcal{V}\bar{Z}_{\mathcal{V}-C}$ are incidence-like matrices, see [4, 12].

Theorem 4 *Consider an electrical circuit described by the MNA equations* (17.26) *that satisfies the Assumption (A1)–(A5). Assume that the DAE* (17.26) *is of s-index* $\nu_s = 1$, *i.e., there exists* $C\mathcal{V}$*-loops and/or* $\mathcal{L}\mathcal{I}$*-cutsets. Then, the minimally extended MNA system* (17.34) *is strangeness-free.*

Proof See [27, 28]. □

Remark 6 Another approach to obtain a regularization of (17.26) in case of $C\mathcal{V}$-loops or $\mathcal{L}\mathcal{I}$-cutsets has been proposed in [4, 5]. Here, the basic idea is that once the critical configurations ($C\mathcal{V}$-loops or $\mathcal{L}\mathcal{I}$-cutsets) are identified some electrical components can be replaced by controlled sources that provide the same characteristic voltage-current behavior. At the same time, critical configurations in the circuit graphs are eliminated by the replacements. The advantage of this approach is, that one can work directly on the netlist defining the circuit topology. ◁

17.5 Flow Problems

Navier-Stokes equations are commonly used to describe the motion of fluid substances [53]. For an incompressible flow, the non-stationary Navier-Stokes equations are given by the partial differential equations

$$\frac{\partial u}{\partial t} + (u \cdot \nabla)u - \frac{1}{Re}\Delta u + \nabla p = f \quad \text{in } \Omega \times \mathbb{I}, \tag{17.35a}$$

$$\nabla \cdot u = 0 \quad \text{in } \Omega \times \mathbb{I} \tag{17.35b}$$

on a domain $\Omega \subset \mathbb{R}^n, n \in \{2,3\}$ and a time interval $\mathbb{I}$ together with initial and boundary conditions

$$u(x,t_0) = a \quad \text{for all } x \in \Omega, \quad u = 0 \quad \text{on } \partial\Omega \times \mathbb{I}.$$

Equations (17.35) describe the evolution of the velocity field $u : \Omega \times \mathbb{I} \to \mathbb{R}^n$ and the pressure $p : \Omega \times \mathbb{I} \to \mathbb{R}$, i.e., $u(x,t)$ is the flow velocity at position $x \in \Omega$ and time $t \in \mathbb{I}$ and $p(x,t)$ is the corresponding pressure of the fluid. Furthermore, the Reynolds number $Re > 0$ is a given parameter, $a \in \mathbb{R}^n$ is the initial value at time t_0 and the function $f : \Omega \times \mathbb{I} \to \mathbb{R}^n$ denotes a given volume force.

Using the method of lines, a spatial semi-discretization of the Navier-Stokes equations (17.35) leads to a nonlinear DAE. It has been shown in [56] that if the non-uniqueness of a free constant in the pressure is fixed by the discretization method, then the s-index of the semi-discretized system is well-defined and for most discretization methods, the s-index of the semi-discretized system is $\nu_s = 1$. However, it has been shown in [1] that the semi-explicit structure of the equations (17.35) allows for a variant of minimal extension, and that a specific discretization of the spatial domain (in particular, a specific splitting of the ansatz spaces in the spatial discretization by finite elements) can lead directly to a semi-discretized strangeness-free system.

First, one considers an operator form of (17.35), i.e., we want to find $u : \mathbb{I} \to \mathscr{V}$ and $p : \mathbb{I} \to \mathscr{Q}$ such that

$$\dot{u}(t) + \mathscr{K}u(t) - \mathscr{B}'p(t) = \mathscr{F}(t) \quad \text{in } \mathscr{V}', \tag{17.36a}$$

$$\mathscr{B}u(t) = \mathscr{G}(t) \quad \text{in } \mathscr{Q}', \tag{17.36b}$$

$$u(t_0) = a \quad \text{in } \mathscr{H}, \tag{17.36c}$$

holds a.e. in $\mathbb{I}$. Here, $\mathscr{V}$ and $\mathscr{Q}$ are suitable ansatz spaces, where $\mathscr{V}$ is densely and continuously embedded in a Hilbert space $\mathscr{H}$ (for details, see [1]). Moreover, $\mathscr{V}'$ and $\mathscr{Q}'$ denote the dual spaces of $\mathscr{V}$ and $\mathscr{Q}$, respectively. The operators $\mathscr{K} : \mathscr{V} \to \mathscr{V}'$ and $\mathscr{B} : \mathscr{V} \to \mathscr{Q}'$ are defined via

$$\langle \mathscr{K}u, v\rangle = \int_\Omega (u\cdot\nabla)u\cdot v\,\mathrm{d}x + \frac{1}{Re}\int_\Omega \nabla u\cdot\nabla v\,\mathrm{d}x, \quad u, v \in \mathscr{V},$$

$$\langle \mathscr{B}u, q\rangle = \int_\Omega (\nabla\cdot u)q\,\mathrm{d}x, \quad u \in \mathscr{V},\ q \in \mathscr{Q},$$

and $\mathscr{B}' : \mathscr{Q} \to \mathscr{V}'$ denotes the dual operator of $\mathscr{B}$. Note that the equalities in (17.36) should be understood pointwise in L^1_{loc} in the corresponding dual products. Also the time derivative in (17.36) is to be considered in the generalized sense as weak time derivative, i.e., an equality of the form $\dot{v}(t) = \mathscr{F}(t)$ in $\mathscr{V}'$ holds if

$$-\int_0^T \langle v(t), w\rangle \dot{\phi}(t)\,\mathrm{d}t = \int_0^T \langle \mathscr{F}(t), w\rangle \phi(t)\,\mathrm{d}t$$

for all $w \in \mathscr{V}$ and $\phi \in \mathscr{C}_0^\infty(0, T)$, see [60]. The inhomogeneity $\mathscr{G}(t)$ in (17.36b) is considered in order to cover also more general cases. $\mathscr{G}(t) \not\equiv 0$ may appear in discretization schemes or for more general boundary conditions.

Now, the second equation(17.36b) poses a constraint on the velocity field $u(t)$. Since the operator $\mathscr{B}$ is independent of time t, we get

$$\mathscr{B}\dot{u}(t) = \dot{\mathscr{G}}(t), \tag{17.37}$$

assuming sufficient regularity of $\mathscr{G}(t)$. A splitting of the ansatz space $\mathscr{V}$ into the divergence-free space $\mathscr{V}_{df}$ and its orthogonal complement $\mathscr{V}_{df}^{\perp}$, i.e.,

$$\mathscr{V}_{df} := \ker(\mathscr{B}) = \{u \in \mathscr{V} \mid \nabla \cdot u = 0\}, \quad \mathscr{V} = \mathscr{V}_{df} \oplus \mathscr{V}_{df}^{\perp},$$

allows for a unique decomposition of $u \in \mathscr{V}$ into $u = u_1 + u_2$ with $u_1 \in \mathscr{V}_{df}$ and $u_2 \in \mathscr{V}_{df}^{\perp}$. Following the idea of minimal extension, we can now add (17.37) to the operator formulation (17.36). Using the splitting into $u = u_1 + u_2$ and introducing a new variable $\tilde{u}_2 := \dot{u}_2$ yields the extended problem

$$\begin{aligned} \dot{u}_1(t) + \tilde{u}_2(t) + \mathscr{K}(u_1(t) + u_2(t)) - \mathscr{B}' p(t) &= \mathscr{F}(t) \quad \text{in } \mathscr{V}', && (17.38a)\\ \mathscr{B}u_2(t) &= \mathscr{G}(t) \quad \text{in } \mathscr{Q}', && (17.38b)\\ \mathscr{B}\tilde{u}_2(t) &= \dot{\mathscr{G}}(t) \quad \text{in } \mathscr{Q}', && (17.38c)\\ u_1(t_0) &= a_1 \quad \text{in } \mathscr{H}. && (17.38d) \end{aligned}$$

Note that here we have used that $\mathscr{B}u_1(t) = 0$ and $\mathscr{B}\dot{u}_1(t) = 0$.

For the spatial discretization of (17.38) by finite elements, finite dimensional subspaces V_h and Q_h of $\mathscr{V}$ and $\mathscr{Q}$ have to be constructed based on a triangulation of the domain Ω. Similar as before, the finite dimensional space V_h is decomposed corresponding to the decomposition of $\mathscr{V}$ into $\mathscr{V}_{df} \oplus \mathscr{V}_{df}^{\perp}$ and we denote by V_h^1 the approximation space of $\mathscr{V}_{df}$ and by V_h^2 the approximation space of $\mathscr{V}_{df}^{\perp}$ assuming that $V_h = V_h^1 \oplus V_h^2$. Note that the considered spatial discretization need not to be conform, i.e., we allow for $V_h^1 \not\subset \mathscr{V}_{df}$ and $V_h^2 \not\subset \mathscr{V}_{df}^{\perp}$, although $V_h \subset \mathscr{V}$. Let $\{\Psi_j\}$ be a basis of V_h and $\{\varphi_i\}$ be a basis of Q_h. We assume that the basis of V_h is ordered according to the decomposition into V_h^1 and V_h^2. Now, the semi-discretization of (17.38) leads to a DAE of the form

$$M \begin{bmatrix} \dot{q}_1(t) \\ \tilde{q}_2(t) \end{bmatrix} + K\left(\begin{bmatrix} q_1(t) \\ q_2(t) \end{bmatrix}\right) - B^T p_h(t) = f(t), \quad q_1(t_0) = a, \tag{17.39a}$$

$$B \begin{bmatrix} q_1(t) \\ q_2(t) \end{bmatrix} = g(t), \tag{17.39b}$$

$$B \begin{bmatrix} \dot{q}_1(t) \\ \tilde{q}_2(t) \end{bmatrix} = \dot{g}(t). \tag{17.39c}$$

Here, $q_1(t)$, $q_2(t)$ and $\tilde{q}_2(t)$ denote the semi-discretizations of u_1, u_2 and $\tilde{u}_2$, respectively, and $p_h(t) \in \mathbb{R}^{n_p}$ denotes the semi-discrete representation of the pressure p. Furthermore, $M = [m_{jk}] \in \mathbb{R}^{n_q,n_q}$ denotes the positive definite mass matrix obtained from the FEM discretization

$$m_{jk} = \int_\Omega \Psi_j \cdot \Psi_k \,\mathrm{d}x,$$

and K is a nonlinear function describing the discrete version of operator $\mathcal{K}$ defined by

$$K_j(q(t)) = \int_\Omega (q(t)\cdot\nabla)q(t)\cdot\Psi_j \,\mathrm{d}x + \frac{1}{Re}\int_\Omega \nabla q(t)\cdot\nabla\Psi_j \,\mathrm{d}x.$$

The functions f and g denote the finite dimensional approximations of $\mathcal{F}(t)$ and $\mathcal{G}(t)$, and $B = [b_{ij}] \in \mathbb{R}^{n_p,n_q}$ is defined by

$$b_{ij} = \int_\Omega \varphi_i \operatorname{div}(\Psi_j)\,\mathrm{d}x. \tag{17.40}$$

Note that for the discrete version of u_1 the terms $\mathcal{B}u_1(t) = 0$ and $\mathcal{B}\dot{u}_1(t) = 0$ from the operator formulation may not vanish for non-conforming finite elements.

It has been shown in [1] that a finite element discretization of (17.38) can be constructed that directly leads to a DAE of s-index 0. Hereby, finite element spaces V_h^1, V_h^2 and Q_h are chosen in such a way that the matrix representation B as defined in (17.40) has the block structure $B = \begin{bmatrix} B_1 & B_2 \end{bmatrix}$ with nonsingular square matrix B_2 containing the columns corresponding to V_h^2. Such a reordering of the basis of V_h always exits under the standard stability conditions for the spatial discretization (see [1]) and it basically consists of a permutation of the velocity variables. Examples for finite element discretizations that directly satisfy these conditions are given in [1].

One of the great advantage of this approach is that since $\mathcal{B}\dot{u}(t) = \mathcal{G}(t)$ is added to the system, instabilities are reduced. During the splitting into the divergence-free and non divergence-free part the variables are transformed via a simple permutation and, thus, all variables keep their physical meaning. Thus, in particular, the pressure p remains a physically valid part of the system, rather than being eliminated or functioning as a velocity correction. Moreover, the sole application of permutations preserves the sparsity structure of the system and ensures that the transformation to the strangeness-free system (17.39) is well-conditioned. The increase of the system size is compensated by the direct applicability of efficient time stepping schemes. Unlike in penalization or in projection methods [18], this approach does not require time step restrictions or artificial boundary conditions for the pressure. Furthermore, consistency and stability of half-explicit methods [2] can be ensured.

17.6 Hybrid Systems

A particular feature of many complex dynamical systems is that they are so-called *switched systems* or *hybrid systems*, i.e., the mathematical model changes with time depending on certain indicators which leads to different *operation modes*. The continuous dynamics in the different operation modes can be described by DAEs and the change between different operation modes is usually modeled by discrete transitions. As the discrete and continuous dynamics interact, they must be analyzed simultaneously. One of the main difficulties in hybrid systems is that, after a mode switch takes place, the structure of the system as well as its properties may change, such as the model dimension, the index, or the number of algebraic or differential equations or redundancies. Furthermore, special phenomena that can occur during the simulation of hybrid systems, as, e.g., the problem of event detection and reinitialization, or numerical chattering, have to be treated in an appropriate way to ensure an efficient numerical integration.

17.6.1 Hybrid System Formulation

The general theory for DAEs as presented in [28] and revisited in Sect. 17.2 can be applied for hybrid DAEs as shown in [22, 34, 57]. We assume that the discrete and continuous dynamics only interact via instantaneous discrete transitions at distinct points in time called *events*. Let $\mathbb{I} = [t_0, t_f] \subset \mathbb{R}$ be decomposed into subintervals $\mathbb{I}_i = [\tau_i, \tau_i')$ with $\tau_i < \tau_i'$, $\tau_i' = \tau_{i+1}$ for $i = 1, \dots, N_{\mathbb{I}} - 1$, $N_{\mathbb{I}} \in \mathbb{N}$, as well as $\mathbb{I}_{N_{\mathbb{I}}} = [\tau_{N_{\mathbb{I}}}, \tau_{N_{\mathbb{I}}}']$, $\tau_{N_{\mathbb{I}}} < \tau_{N_{\mathbb{I}}}'$, such that $\mathbb{I} = \bigcup_{i=1}^{N_{\mathbb{I}}} \mathbb{I}_i$, (i.e., $\tau_1 = t_0$, $\tau_{N_{\mathbb{I}}}' = t_f$). Note that the number of subintervals $N_{\mathbb{I}}$ and the event times τ_i are, in general, not known a priori, but determined during the simulation. The event times can depend on the state and, in particular, on the initial conditions, such that the partitioning into the intervals $\mathbb{I}_i$ is not known in advance. However, we assume that the number of subintervals is finite. Furthermore, let $\mathbb{M} := \{1, \dots, N_F\}$, $N_F \in \mathbb{N}$ be the *set of modes* and for each $\ell \in \mathbb{M}$, let D_ℓ be the union of certain intervals $\mathbb{I}_i$ such that $\bigcup_{\ell \in \mathbb{M}} D_\ell = \mathbb{I}$ and $D_\ell \cap D_k = \emptyset$ for $\ell, k \in \mathbb{M}$, $\ell \neq k$.

Definition 2 A *hybrid system of DAEs* is defined as a collection of

- a set of nonlinear DAEs

$$F^\ell(t, x^\ell, \dot{x}^\ell) = 0, \quad \ell \in \mathbb{M}, \tag{17.41}$$

  with sufficiently smooth functions $F^\ell : D_\ell \times \mathbb{R}^{n_\ell} \times \mathbb{R}^{n_\ell} \to \mathbb{R}^{m_\ell}$;
- an index set of autonomous transitions $J^\ell = \{1, 2, \dots, n_T^\ell\}$ for each mode $\ell \in \mathbb{M}$, where $n_T^\ell \in \mathbb{N}$ is the number of possible transitions of mode ℓ;
- *switching functions* $g_j^\ell : D_\ell \times \mathbb{R}^{n_\ell} \times \mathbb{R}^{n_\ell} \to \mathbb{R}$ for all $j \in J^\ell$, with

$$g_j^\ell(t, x^\ell, \dot{x}^\ell) > 0 \quad \text{in mode } \ell;$$

- *mode allocation functions* $S^\ell : J^\ell \to \mathbb{M}$ with $S^\ell(j) = k$ for all $\ell \in \mathbb{M}$; and
- *transition functions* $T_\ell^k : \mathbb{R}^{n_\ell} \times \mathbb{R}^{n_\ell} \to \mathbb{R}^{n_k} \times \mathbb{R}^{n_k}$ of the form

$$T_\ell^k(x^\ell(\tau_i'), \dot{x}^\ell(\tau_i')) = (x^k(\tau_{i+1}), \dot{x}^k(\tau_{i+1})), \quad \tau_i' = \tau_{i+1} \in D_k, \tag{17.42}$$

for all $\ell \in \mathbb{M}$ with successor mode $k \in \mathbb{M}$.

The hybrid system changes between different modes at the event times that are defined as roots of the switching functions $g_j^\ell(t, x^\ell, \dot{x}^\ell)$. If $g_j^\ell(t, x^\ell, \dot{x}^\ell) > 0$ for all $j \in J^\ell$, then the system stays in the current mode ℓ, but if $g_j^\ell(t, x^\ell, \dot{x}^\ell) \leq 0$ for some $j \in J^\ell$, then the system switches to a new mode. For each switching function, the associated *switching surface* is given by

$$\Gamma_j^\ell = \left\{(t, x^\ell, \dot{x}^\ell) \in D_\ell \times \mathbb{R}^{n_\ell} \times \mathbb{R}^{n_\ell} \mid g_j^\ell(t, x^\ell, \dot{x}^\ell) = 0\right\}, \quad j \in J^\ell, \ \ell \in \mathbb{M}. \tag{17.43}$$

The mode allocation functions S^ℓ are used to determine the successor mode k after a mode change in mode $\ell \in \mathbb{M}$. Finally, the transition functions T_ℓ^k map the final values and derivatives of the variables in mode $\ell \in \mathbb{M}$ to the corresponding initial values in the successor mode $k \in \mathbb{M}$ at the event time $\tau_i' = \tau_{i+1} \in D_k$.

Remark 7 Here, we restrict to hybrid systems where mode switches occur whenever a switching surfaces Γ_j^ℓ is crossed by the solution trajectory. In general, one can also define the mode switching strategy on the basis of *transition conditions*

$$L_j^\ell : D_\ell \times \mathbb{R}^{n_\ell} \times \mathbb{R}^{n_\ell} \to \{\mathit{TRUE}, \mathit{FALSE}\}$$

for all transitions $j \in J^\ell$ and all modes $\ell \in \mathbb{M}$. Then, the system switches to another mode whenever there exists a $j \in J^\ell$ such that $L_j^\ell(\hat{t}, x^\ell(\hat{t}), \dot{x}^\ell(\hat{t})) = \mathit{TRUE}$ at $\hat{t}$. The transition conditions L_j^ℓ can be composed of logical combinations of several separate switching functions $g_{j,i}^\ell$, $i = 1, \ldots, n_j^\ell$. Thus, switching functions $g_{j,i}^\ell$ can be chosen as simple as possible, allowing an efficient and reliable event detection by root finding methods. See also [34, 57]. ◁

For hybrid DAE systems certain problems arise concerning the existence and uniqueness of solutions. The piecewise continuous functions $x^\ell : D_\ell \to \mathbb{R}^{n_\ell}$ describe the continuous state of the hybrid system in mode ℓ. Let $x^\ell(\tau_i')$ be the smooth extension of x^ℓ to the interval boundary $\tau_i' = \tau_{i+1}$ of $\mathbb{I}_i \in D_\ell$. The state transfer to the successor mode k defined by (17.42) may result in jumps in the state vector of the hybrid system. In order to obtain a solution in the new mode, the initial values obtained by the transition function have to be consistent with the DAE in mode k. In general, it is a modeling task to design the transition functions correctly, i.e., in such a way that continuity (if required) and consistency conditions of the initial values after mode switching are fulfilled. However, if the number of equations or the number of free variables changes at a mode change, these conditions may be difficult to realize. In particular, we may face the situation that the solution is not unique after a mode change. For higher index DAEs correct initial values that satisfy all hidden constraints may not be known in advance. Thus, detecting discontinuities in the state variables of a hybrid system may also be used to verify the mathematical model itself. Moreover, the solution of a hybrid system depends on the initial mode and on the initial conditions that can also influence the mode switching behavior. For details on solution concepts for hybrid DAEs, see [34, 57].

The theory of DAEs turns out to be very useful for the treatment of hybrid systems. In the following, we consider a hybrid system as in Definition 2 with $n_\ell = n$ for the DAEs (17.41) in each mode $\ell \in \mathbb{M}$. We assume that F^ℓ is sufficiently smooth in $[\tau_i, \tau_i' + \epsilon]$ for small $\epsilon > 0$ for each interval $\mathbb{I}_i = [\tau_i, \tau_i') \in D_\ell$ (such that the Implicit Function Theorem can locally be applied), and that the strangeness index ν_s^ℓ of (17.41) is well-defined for all modes $\ell \in \mathbb{M}$. Thus, in each mode $\ell \in \mathbb{M}$ and each domain D_ℓ, the DAE (17.41) in mode ℓ satisfies Hypothesis 1 with constant characteristic values $\mu^\ell, r^\ell, a^\ell, d^\ell$, and v^ℓ. If the characteristic values are not constant for some mode $\ell \in \mathbb{M}$, we can introduce new modes and further switch points to satisfy Hypothesis 1 locally for each mode. Then, the *maximal strangeness index* ν_s^{max} of a hybrid system is defined by

$$\nu_s^{max} := \max_{\ell \in \mathbb{M}} \{\nu_s^\ell\},$$

and a hybrid system is called *strangeness-free* if $\nu_s^{max} = 0$. If the hybrid system is not strangeness-free, then reduced systems as in (17.5) or (17.7) can be extracted independently in each mode. Collecting all these reduced systems, we obtain an equivalent *reduced hybrid system* which is strangeness-free.

Theorem 5 ([34]) *Consider a hybrid system as in Definition 2 with sufficiently smooth functions F^ℓ in each mode $\ell \subset \mathbb{M}$ that satisfy Hypothesis 1 locally for each mode $\ell \in \mathbb{M}$ and each domain D_ℓ. Then, every sufficiently smooth solution of the hybrid system is also a solution of the reduced hybrid system that consists of reduced DAE systems of the form* (17.5) *corresponding to* (17.41) *in each mode* $\ell \in \mathbb{M}$.

17.6.2 Sliding Mode Regularization

Conceptually, the numerical integration of hybrid DAE systems can be realized similar to the numerical integration of general DAEs, by generating, locally in each integration step, an equivalent strangeness-free formulation for the DAE (17.41) in the mode ℓ. This integration is continued until a switching function crosses zero and a mode switch occurs. Once the event time is determined as a root of a switching function (within a certain error tolerance), the system is transferred to the next mode via the transition function (17.42), and the numerical integration is continued in the new mode. Thus, for the simulation of a hybrid DAE system, besides the pure integration process (that includes index reduction), we need a root finding procedure to determine the event times, and a process to compute consistent initial values after mode change. The time stepping procedure for the numerical integration in each mode can be achieved via any method suitable for strangeness-free DAEs (e.g., BDF methods or Radau IIA methods).

Another special phenomena that can occur during the simulation of hybrid systems is a fast changing between different modes, called *chattering*. Such oscillations around different modes may be real properties of the physical model due to hysteresis, delays and other dynamic non-idealities. On the other hand, numerical errors may lead to numerical chattering, since switching conditions may be satisfied due to local errors. The numerical treatment of a hybrid system exhibiting chattering behavior requires high computational costs as small stepsizes are required to restart the integration after each mode change. In the worst case, the numerical integration breaks down, as it does not proceed in time, but chatters between modes. To prevent chattering and to reduce the computational costs, a regularization of the system dynamics along the switching surfaces has been derived in [34, 57]. Here, the basic idea is to introduce an additional mode that describes the dynamic behavior in cases where chattering occur.

In the following, for simplicity, we consider regular hybrid systems with the same state dimension $n = n_\ell = m_\ell$ for all $\ell \in \mathbb{M}$. Furthermore, we restrict to switching functions independent of the state derivative $\dot{x}$. Note that if we have transformed the DAE in every mode to a reduced system of the form (17.7) we can always insert the differential components into the corresponding switching function. Then, we assume that the hybrid DAE system switches from mode ℓ to mode k along the switching surface

$$\Gamma_j^\ell = \{(t,x) \in D_\ell \times \mathbb{R}^n \mid g_j^\ell(t,x) = 0\},$$

for some $j \in J^\ell$. Furthermore, we assume that there exists a mode transition $\tilde{j} \in J^k$, such that $\Gamma_j^\ell = \Gamma_{\tilde{j}}^k = \{(t,x) \in D_k \times \mathbb{R}^n \mid g_{\tilde{j}}^k(t,x) = 0\}$, i.e., $g_j^\ell(t,x) = -g_{\tilde{j}}^k(t,x)$. Now, chattering around the switching surface Γ_j^ℓ will occur, if all solutions near the surface Γ_j^ℓ approach it from both sides, see Fig. 17.1. This means that, if we consider the system at a fixed time point t and project the velocity

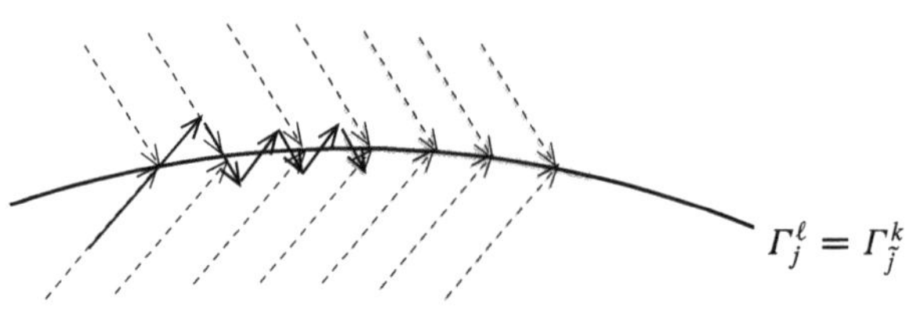

Fig. 17.1 Chattering behavior along a switching surface

vectors of both systems onto the surface gradient, then these projections are of opposite signs and directed towards the surface from both sides in a neighborhood of the switching surface. Under the assumption of regularity and well-definedness of the strangeness index in each mode, the DAEs in the adjacent modes ℓ and k can be transformed to the corresponding reduced systems of the form (17.7). Assuming that $g^{\ell}_{j,x}(t,x) \neq 0$ in a neighborhood of the switching surface Γ^{ℓ}_j, the conditions for chattering can be formulated as

$$g^{\ell}_{j,x^{\ell}_1}(t,x^{\ell}_1,x^{\ell}_2)\mathscr{L}^{\ell}(t,x^{\ell}_1) < 0 \quad \text{and} \quad g^{k}_{\tilde{j},x^{k}_1}(t,x^{k}_1,x^{k}_2)\mathscr{L}^{k}(t,x^{k}_1) < 0 \tag{17.44}$$

for fixed t, where $\mathscr{L}^{\ell}$ and $\mathscr{L}^{k}$ describe the dynamical parts of the reduced systems. The conditions (17.44) are called the *sliding conditions*. In particular, the sliding conditions describe the directional derivatives of $g^{\ell}_j = -g^{k}_{\tilde{j}}$ in direction of $\mathscr{L}^{\ell}$, $\mathscr{L}^{k}$, respectively, which correspond to the projections of the vectors $\mathscr{L}^{\ell}(t,x^{\ell}_1)$ and $\mathscr{L}^{k}(t,x^{k}_1)$ onto the gradient of the switching surface Γ^{ℓ}_j.

In the numerical simulation of a hybrid system, an immediate switch back to mode ℓ after one or a few integration steps in mode k would result if the sliding condition (17.44) is satisfied. To avoid this, we add an additional mode defining the dynamics of the system during sliding, i.e., we define equivalent dynamics such that the solution trajectory move along the *sliding surface* $\Gamma^{\ell}_S \subseteq \Gamma^{\ell}_j$ defined by

$$\Gamma^{\ell}_S := \{(t,x) \in \Gamma^{\ell}_j \mid g^{\ell}_{j,x^{\ell}_1}(t,x^{\ell}_1,x^{\ell}_2)\mathscr{L}^{\ell}(t,x^{\ell}_1) < 0,\ g^{k}_{\tilde{j},x^{k}_1}(t,x^{k}_1,x^{k}_2)\mathscr{L}^{k}(t,x^{k}_1) < 0\}.$$

Here, Γ^{ℓ}_S corresponds to the part of the switching surface, where chattering will occur. Thus, whenever the sliding condition (17.44) is satisfied at a point (t,x) we switch to the so-called *sliding mode*. The system should stay in this sliding mode as long as the solution trajectory stays on the sliding surface, and resume in mode ℓ or k, depending on the sign of the directional derivatives, if the solution leaves Γ^{ℓ}_S. To define the system behavior during sliding, the dynamics are approximated in such a way that the state trajectory moves along the switching surface.

Let d^{ℓ}, d^{k} and a^{ℓ}, a^{k} denote the number of differential and algebraic equations in mode ℓ and mode k, i.e., the dimension of x^{ℓ}_1, x^{k}_1 and x^{ℓ}_2, x^{k}_2 in the corresponding reduced systems (17.7). Further, let $d^{\ell} + a^{\ell} = d^{k} + a^{k} = n$ and without loss of generality assume that $d^{\ell} \geq d^{k}$ and $a^{\ell} \leq a^{k}$. Then, x^{ℓ}_1 and x^{k}_2 can be partitioned into

$$x^{\ell}_1 = \begin{bmatrix} x^{\ell}_{1,1} \\ x^{\ell}_{1,2} \end{bmatrix}, \quad x^{k}_2 = \begin{bmatrix} x^{k}_{2,1} \\ x^{k}_{2,2} \end{bmatrix},$$

with $x_{1,1}^{\ell} \in \mathbb{R}^{d^k}$, $x_{1,2}^{\ell} \in \mathbb{R}^{d^{\ell}-d^k}$ and $x_{2,1}^{k} \in \mathbb{R}^{a^{\ell}}$, $x_{2,2}^{k} \in \mathbb{R}^{a^k-a^{\ell}}$. The corresponding reduced systems as in (17.7) can be partitioned accordingly into

$$\begin{bmatrix} \dot{x}_{1,1}^{\ell} \\ \dot{x}_{1,2}^{\ell} \end{bmatrix} = \begin{bmatrix} \mathscr{L}_1^{\ell}(t, x_{1,1}^{\ell}, x_{1,2}^{\ell}) \\ \mathscr{L}_2^{\ell}(t, x_{1,1}^{\ell}, x_{1,2}^{\ell}) \end{bmatrix}, \qquad \dot{x}_1^k = \mathscr{L}^k(t, x_1^k),$$
$$x_2^{\ell} = \mathscr{R}^{\ell}(t, x_{1,1}^{\ell}, x_{1,2}^{\ell}), \qquad \begin{bmatrix} x_{2,1}^{k} \\ x_{2,2}^{k} \end{bmatrix} = \begin{bmatrix} \mathscr{R}_1^{k}(t, x_1^k) \\ \mathscr{R}_2^{k}(t, x_1^k) \end{bmatrix}. \tag{17.45}$$

Now, we can define the *differential-algebraic system in sliding motion* as

$$\dot{x}_1 = \alpha \begin{bmatrix} \mathscr{L}_1^{\ell}(t, x_1) \\ \mathscr{L}_2^{\ell}(t, x_1) \end{bmatrix} + (1-\alpha) \begin{bmatrix} \mathscr{L}^{k}(t, x_1) \\ 0 \end{bmatrix}, \tag{17.46a}$$

$$x_2 = \alpha \begin{bmatrix} \mathscr{R}^{\ell}(t, x_1) \\ 0 \end{bmatrix} + (1-\alpha) \begin{bmatrix} \mathscr{R}_1^{k}(t, x_1) \\ \mathscr{R}_2^{k}(t, x_1) \end{bmatrix}, \tag{17.46b}$$

$$0 = g_j^{\ell}(t, x_1, x_2). \tag{17.46c}$$

The velocity vector of the sliding motion is approximated as convex combination of the velocity vectors on both side of the switching surface in such a way that it lies on a plane tangential to the switching surface. Here, the additional algebraic variable α ensures that the solution stays on the manifold Γ_j^{ℓ} described by the algebraic constraint $g_j^{\ell}(t, x_1, x_2) = 0$. The differential equation (17.46a) describes the equivalent dynamics of the system during sliding motion, and the algebraic equations (17.46b) is defined as a transformation of the constraint manifolds in the two modes such that $\mathscr{R}^{\ell}(t, x_1)$ is turned into $\mathscr{R}^{k}(t, x_1)$ across the discontinuity or vice versa depending on the direction of the discontinuity crossing. See also Fig. 17.2. The DAE in sliding mode (17.46) consists of $d^{\ell} + a^k + 1$ equations with unknowns $x_1 \in \mathbb{R}^{d^{\ell}}$, $x_2 \in \mathbb{R}^{a^k}$, and $\alpha \in \mathbb{R}$. The construction of the DAE in

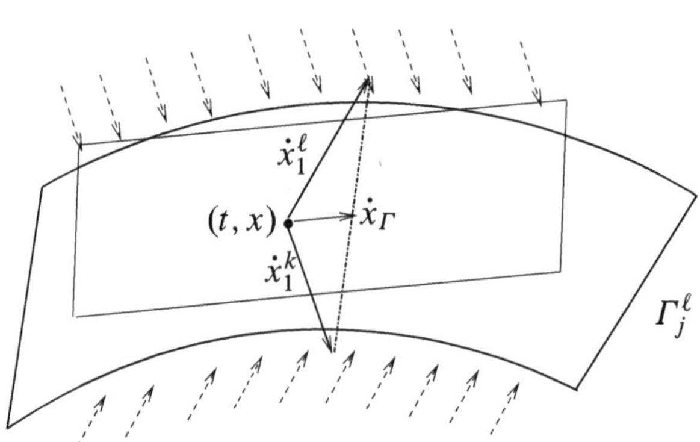

Fig. 17.2 Sliding mode regularization

sliding mode can be seen as a regularization, since a potentially unsolvable problem is replaced by a solvable one. *Sliding mode regularization* is a generalization of the *Filippov regularization* for ODEs, see e.g. [13, 54].

Theorem 6 *Consider a reduced hybrid system with regular DAEs in mode ℓ and k partitioned as in* (17.45)*, that switches from mode ℓ to mode k along the smooth switching surface $\Gamma_j^\ell = \{(t,x) \in \mathbb{D}_\ell \times \mathbb{R}^n \mid g_j^\ell(t,x) = 0\}$, $j \in J^\ell$. If*

$$g^{\ell}_{j,x_2}(t,x_1,x_2)\left(\begin{bmatrix}\mathscr{R}^\ell(t,x_1) - \mathscr{R}^k_1(t,x_1)\\ -\mathscr{R}^k_2(t,x_1)\end{bmatrix}\right)$$

is nonsingular for all $(t,x_1,x_2) \in \mathbb{D}_\ell \times \mathbb{R}^n$, then the DAE in sliding mode (17.46) *is regular and strangeness-free.*

Proof Since the DAEs in mode ℓ and k are regular, also (17.46) is regular due to construction. Inserting relation (17.46b) into (17.46c) yields

$$0 = g_j^\ell(t, x_1, \alpha(\mathscr{R}_1(t,x_1) - \mathscr{R}_2(t,x_1)) + \mathscr{R}_2(t,x_1)), \tag{17.47}$$

where $\mathscr{R}_1(t,x_1) := \begin{bmatrix}\mathscr{R}^\ell(t,x_1)\\ 0\end{bmatrix}$ and $\mathscr{R}_2(t,x_1) := \begin{bmatrix}\mathscr{R}^k_1(t,x_1)\\ \mathscr{R}^k_2(t,x_1)\end{bmatrix}$. Thus, if

$$g^{\ell}_{j,x_2}(t,x_1,x_2)\,(\mathscr{R}_1(t,x_1) - \mathscr{R}_2(t,x_1))$$

is nonsingular for all relevant points (t,x_1,x_2), equation(17.47) can be solved for α using the Implicit Function Theorem. □

17.7 Numerical Integration

As discussed in the previous sections, numerical methods have to combine appropriate discretization methods with suitable regularization methods. Furthermore, the numerical integration methods should exploit the properties and the structure of the model equations to avoid arising problematic effects. In this context, a large number of numerical methods has been developed for the numerical integration of general DAEs and DAEs arising in several applications. For an overview of suitable numerical methods, we refer to [21, 28, 35, 39, 47, 49].

While most of the implemented codes have limitations to the index that they can handle, which is typically a strangeness index of at most two, several codes that are based on the strangeness-concept have been implemented that allow to handle DAEs of arbitrary high index. Here, we will only focus on codes that are based on the theory presented in the previous sections.

For general nonlinear DAEs (17.1) with $m = n$ of arbitrary s-index ν_s the code GENDA[1] has been developed by the authors of [30]. As model equations the derivative array (17.3) at least of level ν_s has to be provided. GENDA uses an adaption of the code DASSL, i.e., BDF methods, for the discretization. Based on the provided derivative array GENDA determines (locally) the s-index and the KM-regularization that is used for the numerical integration.

The code GELDA[2] implemented by the authors of [29], is suited for linear DAEs of the form $E(t)\dot{x} = A(t)x + f(t)$ of arbitrary s-index ν_s and (possibly) of non-square size, i.e., m and n arbitrary. As model equations the matrix-valued functions E and A, their time-derivatives up to order (at least) ν_s as well as the inhomogeneity f and its time-derivatives up to order (at least) ν_s have to be provided. Based on the provided model equations GELDA determines the s-index as well as the KM-regularization for the numerical integration. For the discretization an adaption of both, RADAU5, i.e., Runge-Kutta method Radau IIa of order 5, and DASSL, i.e., BDF methods, can be used as integrator.

For both GELDA and GENDA the Matlab interface SolveDAE[3] has been implemented by the authors of [31], including symbolic generation of the derivative array or the derivatives of E, A, and f.

For the numerical integration of the equations of motion (17.12) for multibody systems, the code GEOMS[4] has been developed by the author of [49–51]. This code is based on the Runge-Kutta method Radau IIa of order 5 for the direct discretization of the overdetermined formulation (17.25). This direct discretization of (17.25) leads to an overdetermined set of nonlinear equations to be solved. Because of truncation errors, the discretized equations become contradictory and can only be solved in a generalized sense which is done in GEOMS using adapted solvers for the overdetermined set of nonlinear equations.

For the numerical simulation of electrical circuit equations the code qPsim [4, 5] has been implemented that is based on the regularization approach presented in Sect. 17.4.2. The regularized system equations are solved by the use of RADAU5 or DASPK3.1.

For the numerical simulation of switched systems, not only the numerical integration but also the modeling has to be taken into account. Based on the theory presented in Sect. 17.6 the hybrid mode controller GESDA[5] [58] has been implemented that enables the solution of general hybrid systems consisting of linear or nonlinear DAEs of arbitrary high index in each mode. The numerical integration routines of GESDA are based on the solvers GELDA and GENDA. A particular feature

[1] GENDA - http://www3.math.tu-berlin.de/multiphysics/Software/GENDA/

[2] GELDA - http://www3.math.tu-berlin.de/multiphysics/Software/GELDA/

[3] SolveDAE - http://www3.math.tu-berlin.de/multiphysics/Software/SolveDAE/

[4] GEOMS - http://page.math.tu-berlin.de/~anst/Software/GEOMS/

[5] GESDA - http://www3.math.tu-berlin.de/multiphysics/Software/GESDA/

of the solver GESDA is the use of sliding mode simulation that allows an efficient treatment of chattering behavior during the numerical simulation.

Acknowledgements This work is dedicated to Prof. Dr. Volker Mehrmann. We would like to express our gratitude to him for initiating our work on DAEs and their applications many years ago. We would like to thank him for his support, his valuable criticism and discussions, but also for the freedom he gave us in our research.

References

1. Altmann, R., Heiland, J.: Finite element decomposition and minimal extension for flow equations. Preprint 2013-11, Institute of Mathematics, TU Berlin (2013)
2. Arnold, M.: Half-explicit Runge-Kutta methods with explicit stages for differential-algebraic systems of index 2. BIT **38**(3), 415–438 (1998)
3. Auslander, L., MacKenzie, R.E.: Introduction to Differentiable Manifolds. Dover, New York (1977). Corrected reprinting
4. Bächle, S.: Numerical solution of differential-algebraic systems arising in circuit simulation. Ph.D. thesis, TU Berlin (2007)
5. Bächle, S., Ebert, F.: Index reduction by element-replacement for electrical circuits. In: G. Ciprina and D. Ioan (eds): Scientific Computing in Electrical Engineering. Mathematics in Industry, vol. 11, pp. 191–198. Springer, Berlin/Heidelberg (2007)
6. Baumgarte, J.: Asymptotische Stabilisierung von Integralen bei gewöhnlichen Differentialgleichungen 1. Ordnung. ZAMM **53**, 701–704 (1973)
7. Bremer, H.: Dynamik und Regelung mechanischer Systeme. Teubner Studienbücherei, Stuttgart (1988)
8. Brenan, K.E., Campbell, S.L., Petzold, L.R.: Numerical Solution of Initial-Value Problems in Differential Algebraic Equations. Classics in Applied Mathematics, vol. 14. SIAM, Philadelphia (1996)
9. Eich, E., Führer, C., Leimkuhler, B., Reich, S.: Stabilization and projection methods for multibody dynamics. Technical Report A281, Helsinki University of Technology (1990)
10. Eich-Soellner, E., Führer, C.: Numerical Methods in Multibody Dynamics. Teubner, B.G., Stuttgart (1998)
11. Eichberger, A.: Simulation von Mehrkörpersystemen auf parallelen Rechnerarchitekturen. Number 332 in Fortschritt-Berichte VDI, Reihe 8: Meß-, Steuerungs- und Regelungstechnik. VDI-Verlag Düsseldorf (1993)
12. Estévez Schwarz, D., Tischendorf, C.: Structural analysis of electric circuits and consequences for MNA. Int. J. Circuit Theory Appl. **28**(2), 131–162 (2000)
13. Filippov, A.F.: Differential Equations with Discontinuous Righthand Sides. Kluwer Academic, Dordrecht (1988)
14. Führer, C.: Differential-algebraische Gleichungssysteme in mechanischen Mehrkörpersystemen - Theorie, numerische Ansätze und Anwendungen. Ph.D. thesis, TU München (1988)
15. Führer, C., Leimkuhler, B.J.: Numerical solution of differential-algebraic equations for constrained mechanical motion. Numer. Math. **59**, 55–69 (1991)
16. Gear, C.W.: Differential-algebraic equation index transformations. SIAM J. Sci. Stat. Comput. **9**, 39–47 (1988)
17. Gear, C.W., Leimkuhler, B., Gupta, G.K.: Automatic integration of Euler-Lagrange equations with constraints. J. Comput. Appl. Math. **12/13**, 77–90 (1985)
18. Gresho, P.M., Sani, R.L.: Incompressible Flow and the Finite Element Method. Isothermal Laminar Flow, vol. 2. Wiley, Chichester (2000)

19. Griepentrog, E., März, R.: Differential-Algebraic Equations and Their Numerical Treatment. Teubner-Texte zur Mathematik, vol. 88. BSB B.G.Teubner Verlagsgesellschaft, Leipzig (1986)
20. Günther, M., Feldmann, U.: CAD-based electric-circuit modeling in industry, I. Mathematical structure and index of network equations. Surv. Math. Ind. **8**, 97–129 (1999)
21. Hairer, E., Wanner, G.: Solving Ordinary Differential Equations II: Stiff and Differential-Algebraic Problems, 2nd edn. Springer, Berlin (1996)
22. Hamann, P., Mehrmann, V.: Numerical solution of hybrid systems of differential-algebraic equations. Comput. Methods Appl. Mech. Eng. **197**(6–8), 693–705 (2008)
23. Hamel, G.: Theoretische Mechanik. Springer, Berlin (1967)
24. Haug, E.J.: Computer Aided Kinematics and Dynamics of Mechanical Systems. Basic Methods, vol. 1. Allyn & Bacon, Boston (1989)
25. Heuser, H.: Gewöhnliche Differentialgleichungen. B.G. Teubner, Stuttgart (1991)
26. Kunkel, P., Mehrmann, V.: Analysis of over- and underdetermined nonlinear differential-algebraic systems with application to nonlinear control problems. Math. Control Signals Syst. **14**, 233–256 (2001)
27. Kunkel, P., Mehrmann, V.: Index reduction for differential-algebraic equations by minimal extension. Zeitschrift für Angewandte Mathematik und Mechanik **84**, 579–597 (2004)
28. Kunkel, P., Mehrmann, V.: Differential-Algebraic Equations—Analysis and Numerical Solution. EMS Publishing House, Zürich (2006)
29. Kunkel, P., Mehrmann, V., Rath, W., Weickert, J.: GELDA: A software package for the solution of general linear differential algebraic equations. SIAM J. Sci. Comput. **18**, 115–138 (1997)
30. Kunkel, P., Mehrmann, V., Seufer, I.: GENDA: A software package for the numerical solution of general nonlinear differential-algebraic equations. Report 730-02, Institute of Mathematics, TU Berlin (2002)
31. Kunkel, P., Mehrmann, V., Seidel, S.: A MATLAB package for the numerical solution of general nonlinear differential-algebraic equations. Preprint, Institute of Mathematics, TU Berlin(2005)
32. Lagrange, J.L.: Méchanique analytique. Desaint, Paris (1788)
33. Lötstedt, P.: Mechanical systems of rigid bodies subject to unilateral constraints. SIAM J. Appl. Math. **42**, 281–296 (1982)
34. Mehrmann, V., Wunderlich, L.: Hybrid systems of differential-algebraic equations – Analysis and numerical solution. J. Process Control **19**, 1218–1228 (2009)
35. Mosterman, P.J.: An overview of hybrid simulation phenomena and their support by simulation packages. In: F.W. Vaandrager and J.H. van Schuppen (eds): Hybrid Systems: Computation and Control. Lecture Notes in Computer Science, vol. 1569, pp. 165–177. Springer, Berlin(1999)
36. Newton, S.I.: Philosophiae naturalis principia mathematica. (1687)
37. Petzold, L.R.: Differential/algebraic equations are not ODEs. SIAM J. Sci. Stat. Comput. **3**, 367–384 (1982)
38. Reis, T.: Mathematical modeling and analysis of nonlinear time-invariant RLC circuits. In: P. Benner, R. Findeisen, D. Flockerzi, U. Reichl and K. Sundmacher (eds): Large-Scale Networks in Engineering and Life Sciences, pp. 125–198. Modeling and Simulation in Science, Engineering and Technology, Birkhäuser, Basel, (2014)
39. Rentrop, P., Strehmel, K., Weiner, R.: Ein Überblick über Einschrittverfahren zur numerischen Integration in der technischen Simulation. GAMM-Mitteilungen **1**, 9–43 (1996)
40. Roberson, R.E., Schwertassek, R.: Dynamics of Multibody Systems. Springer, Berlin (1988)
41. Rulka, W.: Effiziente Simulation der Dynamik mechatronischer Systeme für industrielle Anwendungen. Technical Report IB 532–01–06, DLR German Aerospace Center, Institute of Aeroelasticity, Vehicle System Dynamics Group (2001)
42. Schiehlen, W.O.: Multibody System Handbook. Springer, Berlin (1990)
43. Schiehlen, W.O.: Advanced Multibody System Dynamics. Kluwer Academic, Dordrecht (1993)
44. Schiehlen, W.O.: Multibody system dynamics: Roots and perspectives. Multibody Syst. Dyn. **1**, 149–188 (1997)
45. Scholz, L., Steinbrecher, A.: Regularization by minimal extension for multibody systems. Preprint, Institute of Mathematics, TU Berlin (in preparation)

46. De Schutter, B., Heemels, W.P.M.H., Lunze, J., Prieur, C.: Survey of modeling, analysis, and control of hybrid systems. In: J. Lunze, and F. Lamnabhi-Lagarrigue, (eds): Handbook of Hybrid Systems Control – Theory, Tools, Applications. Cambridge University Press, Cambridge, UK, 31–55 (2009). doi:10.1017/CBO9780511807930.003
47. Simeon, B.: MBSPACK – Numerical integration software for constrained mechanical motion. Surv. Math. Ind. **5**(3), 169–202 (1995)
48. Steinbrecher, A.: Regularization of nonlinear equations of motion of multibody systems by index reduction with preserving the solution manifold. Preprint 742-02, Institute of Mathematics, TU Berlin (2002)
49. Steinbrecher, A.: Numerical solution of quasi-linear differential-algebraic equations and industrial simulation of multibody systems. Ph.D. thesis, TU Berlin (2006)
50. Steinbrecher, A.: GEOMS: A software package for the numerical integration of general model equations of multibody systems. Preprint 400, MATHEON – DFG Research Center "Mathematics for key techonolgies", Weierstraß-Institut für Angewandte Analysis und Stochastik, Berlin (2007)
51. Steinbrecher, A.: GEOMS: A new software package for the numerical simulation of multibody systems. In: 10th International Conference on Computer Modeling and Simulation (EUROSIM/UKSIM 2008), Cambridge, pp. 643–648. IEEE (2008)
52. Steinbrecher, A.: Strangeness-index concept for multibody systems. Preprint, Institute of Mathematics, TU Berlin (in preparation)
53. Temam, R.: Navier-Stokes Equations. Theory and Numerical Analysis. North-Holland, Amsterdam (1977)
54. Utkin, V.I., Zhao, F.: Adaptive simulation and control of variable-structure control systems in sliding regimes. Automatica **32**(7), 1037–1042 (1996)
55. Vlach, J., Singhal, K.: Computer Methods for Circuit Analysis and Design. Van Nostrand Reinhold, New York (1994)
56. Weickert, J.: Applications of the theory of differential-algebraic equations to partial differential equations of fluid dynamics. Ph.D. thesis, TU Chemnitz, Fakultät Mathematik, Chemnitz (1997)
57. Wunderlich, L.: Analysis and numerical solution of structured and switched differential-algebraic systems. Ph.D. thesis, TU Berlin (2008)
58. Wunderlich, L.: GESDA: A software package for the numerical solution of general switched differential-algebraic equations. Preprint 576-2009, MATHEON – DFG Research Center "Mathematics for key techonolgies", TU Berlin (2009)
59. Yu, M., Wang, L., Chu, T., Xie, G.: Stabilization of networked control systems with data packet dropout and network delays via switching system approach. In: IEEE Conference on Decision and Control, Nassau, pp. 3539–3544 (2004)
60. Zeidler, E.: Nonlinear Functional Analysis and its Applications II/A: Linear Monotone Operators. Springer, New York (1990)

Chapter 18
A Condensed Form for Nonlinear Differential-Algebraic Equations in Circuit Theory

Timo Reis and Tatjana Stykel

Abstract We consider nonlinear differential-algebraic equations arising in modelling of electrical circuits using modified nodal analysis and modified loop analysis. A condensed form for such equations under the action of a constant block diagonal transformation will be derived. This form gives rise to an extraction of over- and underdetermined parts and an index analysis by means of the circuit topology. Furthermore, for linear circuits, we construct index-reduced models which preserve the structure of the circuit equations.

18.1 Introduction

One of the most important structural quantities in the theory of differential-algebraic equations (DAEs) is the *index*. Roughly speaking, the index measures the order of derivatives of the inhomogeneity entering to the solution. Since (numerical) differentiation is an ill-posed problem, the index can – inter alia – be regarded as a quantity that expresses the difficulty in numerical solution of DAEs. In the last three decades various index concepts have been developed in order to characterize different properties of DAEs. These are the *differentiation index* [7], the *geometric index* [26], the *perturbation index* [13], the *strangeness index* [22], and the *tractability index* [24], to mention only a few. We refer to [25] for a recent survey on all these index concepts and their role in the analysis and numerical treatment of DAEs.

T. Reis
Department of Mathematics, University of Hamburg, Bundesstraße 55, D-20146 Hamburg, Germany
e-mail: timo.reis@math.uni-hamburg.de

T. Stykel (✉)
Institut für Mathematik, Universität Augsburg, Universitätsstraße 14, D-86159 Augsburg, Germany
e-mail: stykel@math.uni-augsburg.de

P. Benner et al. (eds.), *Numerical Algebra, Matrix Theory, Differential-Algebraic Equations and Control Theory*, DOI 10.1007/978-3-319-15260-8_18

In this paper, we present a structure-preserving condensed form for DAEs modelling electrical circuits with possibly nonlinear components. This form is inspired by the canonical forms for linear DAEs developed by KUNKEL and MEHRMANN [17, 22]. The latter forms give rise to the so-called strangeness index concept which has been successfully applied to the analysis and simulation of structural DAEs from different application areas, see the doctoral theses [2, 6, 14, 27, 31–33, 36, 38] supervised by VOLKER MEHRMANN. The great advantage of the strangeness index is that it can be defined for over- and underdetermined DAEs. Our focus is on circuit DAEs arising from *modified nodal analysis* [9, 15, 35] and *modified loop analysis* [9, 29]. We show that such DAEs have a very special structure which is preserved in the developed condensed form. In the linear case, we can, furthermore, construct index-reduced models which also preserve the special structure of circuit equations.

Nomenclature

Throughout this paper, the identity matrix of size $n \times n$ is denoted by I_n, or simply by I if it is clear from context. We write $M > N$ $(M \geq N)$ if the square real matrix $M - N$ is symmetric and positive (semi-)definite. The symbol $\|x\|$ stands for the Euclidean norm of $x \in \mathbb{R}^n$. For a subspace $\mathcal{V} \subset \mathbb{R}^n$, $\mathcal{V}^\perp$ denotes the orthogonal complement of $\mathcal{V}$ with respect to the Euclidean inner product. The image and the kernel of a matrix A are denoted by $\operatorname{im} A$ and $\ker A$, respectively, and $\operatorname{rank} A$ stands for the rank of A.

18.2 Differential-Algebraic Equations

Consider a nonlinear DAE in general form

$$\mathcal{F}(\dot{x}(t), x(t), t) = 0, \tag{18.1}$$

where $\mathcal{F} : \mathbb{D}_{\dot{x}} \times \mathbb{D}_x \times \mathbb{I} \to \mathbb{R}^k$ is a continuous function, $\mathbb{D}_{\dot{x}}, \mathbb{D}_x \subseteq \mathbb{R}^n$ are open, $\mathbb{I} = [t_0, t_f] \subset \mathbb{R}$, $x : \mathbb{I} \to \mathbb{D}_x$ is a continuously differentiable unknown function, and $\dot{x}$ denotes the derivative of x with respect to t.

Definition 1 A function $x : \mathbb{I} \to \mathbb{D}_x$ is said to be a *solution* of the DAE (18.1) if it is continuously differentiable for all $t \in \mathbb{I}$ and (18.1) is fulfilled pointwise for all $t \in \mathbb{I}$. This function is called a *solution of the initial value problem* (18.1) and $x(t_0) = x_0$ with $x_0 \in \mathbb{D}_x$ if x is the solution of (18.1) and satisfies additionally $x(t_0) = x_0$. An initial value $x_0 \in \mathbb{D}_x$ is called *consistent*, if the initial value problem (18.1) and $x(t_0) = x_0$ has a solution.

If the function $\mathcal{F}$ has the form $\mathcal{F}(\dot{x}, x, t) = \dot{x} - f(x, t)$ with $f : \mathbb{D}_x \times \mathbb{I} \to \mathbb{R}^n$, then (18.1) is an ordinary differential equation (ODE). In this case, the assumption of continuity of f gives rise to the consistency of any initial value. If, moreover, f is locally Lipschitz continuous with respect to x then any initial condition determines the local solution uniquely [1, Section 7.3].

Let $\mathcal{F}(\hat{\dot{x}}, \hat{x}, \hat{t}) = 0$ for some $(\hat{\dot{x}}, \hat{x}, \hat{t}) \in \mathbb{D}_{\dot{x}} \times \mathbb{D}_x \times \mathbb{I}$. If $\mathcal{F}$ is partially differentiable with respect to $\dot{x}$ and the derivative $\frac{\partial}{\partial \dot{x}} \mathcal{F}(\hat{\dot{x}}, \hat{x}, \hat{t})$ is an invertible matrix, then by the implicit function theorem [34, Section 17.8] equation (18.1) can locally be solved for $\dot{x}$ resulting in an ODE $\dot{x}(t) = f(x(t), t)$. For general DAEs, however, the solvability theory is much more complex and still not as well understood as for ODEs.

A powerful framework for analysis of DAEs is provided by the derivative array approach introduced in [8]. For the DAE (18.1) with a sufficiently smooth function $\mathcal{F}$, the *derivative array of order* $l \in \mathbb{N}_0$ is defined by stacking equation (18.1) and all its formal derivatives up to order l, that is,

$$\mathcal{F}_l(x^{(l+1)}(t), x^{(l)}(t), \ldots, \dot{x}(t), x(t), t) = \begin{bmatrix} \mathcal{F}(\dot{x}(t), x(t), t) \\ \frac{d}{dt}\mathcal{F}(\dot{x}(t), x(t), t) \\ \vdots \\ \frac{d^l}{dt^l}\mathcal{F}(\dot{x}(t), x(t), t) \end{bmatrix} = 0. \qquad (18.2)$$

Loosely speaking, the DAE (18.1) is said to have the *differentiation index* $\mu_d \in \mathbb{N}_0$ if $l = \mu_d$ is the smallest number of differentiations required to determine $\dot{x}$ from (18.2) as a function of x and t. If the differentiation index is well-defined, one can extract from the derivative array (18.2) a so-called *underlying ODE* $\dot{x}(t) = \phi(x(t), t)$ with the property that every solution of the DAE (18.1) also solves the underlying ODE.

Another index concept, called *strangeness index*, was first introduced by KUNKEL and MEHRMANN for linear DAEs [17, 19, 23] and then extended to the nonlinear case [20, 22]. The strangeness index is closely related to the differentiation index and, unlike the latter, can also be defined for over- and underdetermined DAEs [21]. For our later proposes, we restrict ourselves to a linear time-varying DAE

$$\mathcal{E}(t)\dot{x}(t) = \mathcal{A}(t)x(t) + f(t), \qquad (18.3)$$

where $\mathcal{E}, \mathcal{A} : \mathbb{I} \to \mathbb{R}^{k,n}$ and $f : \mathbb{I} \to \mathbb{R}^k$ are sufficiently smooth functions. Such a system can be viewed as a linearization of the nonlinear DAE (18.1) along a trajectory. Two pairs $(\mathcal{E}_1(t), \mathcal{A}_1(t))$ and $(\mathcal{E}_2(t), \mathcal{A}_2(t))$ of matrix-valued functions are called *globally equivalent* if there exist a pointwise nonsingular continuous matrix-valued function $U : \mathbb{I} \to \mathbb{R}^{k,k}$ and a pointwise nonsingular continuously differentiable matrix-valued function $V : \mathbb{I} \to \mathbb{R}^{n,n}$ such that

$$\mathcal{E}_2(t) = U(t)\mathcal{E}_1(t)V(t), \qquad \mathcal{A}_2(t) = U(t)\mathcal{A}_1(t)V(t) - U(t)\mathcal{E}_1(t)\dot{V}(t).$$

For $(\mathcal{E}(t), \mathcal{A}(t))$ at a fixed point $t \in \mathbb{I}$, the local characteristic values r, a and s are defined as

$$r = \operatorname{rank}(\mathcal{E}), \qquad a = \operatorname{rank}(Z^\top \mathcal{A} T), \qquad s = \operatorname{rank}(S^\top Z^\top \mathcal{A} T'),$$

where the columns of Z, T, T', and S span $\ker \mathcal{E}^\top$, $\ker \mathcal{E}$, $\operatorname{im} \mathcal{E}^\top$, and $\ker T^\top \mathcal{A}^\top Z$, respectively. Considering these values pointwise, we obtain functions $r, a, s : \mathbb{I} \to \mathbb{N}_0$. It was shown in [17] that under the constant rank conditions $r(t) \equiv r$, $a(t) \equiv a$ and $s(t) \equiv s$, the DAE (18.3) can be transformed into the globally equivalent system

$$\begin{bmatrix} I_s & 0 & 0 & 0 & 0 \\ 0 & I_d & 0 & 0 & 0 \\ 0 & 0 & 0 & 0 & 0 \\ 0 & 0 & 0 & 0 & 0 \\ 0 & 0 & 0 & 0 & 0 \end{bmatrix} \begin{bmatrix} \dot{x}_1(t) \\ \dot{x}_2(t) \\ \dot{x}_3(t) \\ \dot{x}_4(t) \\ \dot{x}_5(t) \end{bmatrix} = \begin{bmatrix} 0 & A_{12}(t) & 0 & A_{14}(t) & A_{15}(t) \\ 0 & 0 & 0 & A_{24}(t) & A_{25}(t) \\ 0 & 0 & I_a & 0 & 0 \\ I_s & 0 & 0 & 0 & 0 \\ 0 & 0 & 0 & 0 & 0 \end{bmatrix} \begin{bmatrix} x_1(t) \\ x_2(t) \\ x_3(t) \\ x_4(t) \\ x_5(t) \end{bmatrix} + \begin{bmatrix} f_1(t) \\ f_2(t) \\ f_3(t) \\ f_4(t) \\ f_5(t) \end{bmatrix}. \tag{18.4}$$

Note that the component x_1 satisfies the pure algebraic equation (the fourth equation in (18.4)) and its derivative is also involved in (18.4). Adding the differentiated fourth equation to the first one, we eliminate the derivative $\dot{x}_1$ from the first equation. The resulting system can again be transformed into the form (18.4) with new global characteristic values r, a and s. This procedure is repeated until s becomes zero. The minimal number μ_s of steps required to extract a DAE with $s = 0$ is called the *strangeness index* of the DAE (18.3). By construction, the strangeness index reduces by one for each elimination step described above. A DAE with vanishing strangeness index is called *strangeness-free*. Since the characteristic values are invariant under global equivalence transformations, μ_s is also invariant under global equivalence transformations. One can also show that the strangeness index μ_s is one below the differentiation index μ_d provided that both indices exist (except for the case, where the differentiation index is zero, then the strangeness index vanishes as well), see [17, 22].

This index reduction procedure has a rather theoretical character since the global equivalence transformations are difficult to determine numerically. It was shown in [19] that the solvability properties of the DAE (18.3) can also be established from the associated derivative array given by

$$\mathcal{M}_l(t)\dot{z}_l(t) = \mathcal{N}_l(t) z_l(t) + g_l(t),$$

where

$$\begin{aligned} \left[\mathcal{M}_l\right]_{ij} &= \binom{i}{j} \mathcal{E}^{(i-j)} - \binom{i}{j+1} \mathcal{A}^{(i-j-1)}, \quad i, j = 0, \ldots, l, \\ \left[\mathcal{N}_l\right]_{ij} &= \begin{cases} \mathcal{A}^{(i)} & \text{for } i = 0, \ldots, l, \; j = 0, \\ 0 & \text{else}, \end{cases} \\ \left[z_l\right]_i &= x^{(i)}, \quad \left[g_l\right]_i = f^{(i)}, \quad i = 0, \ldots, l, \end{aligned}$$

with the convention that $\binom{i}{j} = 0$ for $i < j$. If the strangeness index μ_s is well-defined, then the DAE (18.3) satisfies the following hypothesis.

Hypothesis 1 *There exist integers μ, a, d and w such that the pair $(\mathcal{M}_\mu, \mathcal{N}_\mu)$ associated with $(\mathcal{E}, \mathcal{A})$ has the following properties:*

1. *For all $t \in \mathbb{I}$, we have $\operatorname{rank} \mathcal{M}_\mu(t) = (\mu+1)k - a - w$. This implies the existence of a smooth full rank matrix-valued function Z of size $((\mu+1)k, a+w)$ satisfying $Z^\top \mathcal{M}_\mu = 0$.*
2. *For all $t \in \mathbb{I}$, we have $\operatorname{rank}\big(Z(t)^\top \mathcal{N}_\mu(t) \begin{bmatrix} I_n & 0 & \dots & 0 \end{bmatrix}^\top\big) = a$ and without loss of generality Z can be partitioned as $\begin{bmatrix} Z_2 & Z_3 \end{bmatrix}$ with Z_2 of size $((\mu+1)k, a)$ and Z_3 of size $((\mu+1)k, w)$ such that $\mathcal{A}_2 = Z_2^\top \mathcal{N}_\mu \begin{bmatrix} I_n & 0 & \dots & 0 \end{bmatrix}^\top$ has full row rank and $Z_3^\top \mathcal{N}_\mu \begin{bmatrix} I_n & 0 & \dots & 0 \end{bmatrix}^\top = 0$. Furthermore, there exists a smooth full rank matrix-valued function T_2 of size $(n, n-a)$ satisfying $\mathcal{A}_2 T_2 = 0$.*
3. *For all $t \in \mathbb{I}$, we have $\operatorname{rank}\big(\mathcal{E}(t) T_2(t)\big) = d$, where $d = k - a - w_\mu$ and*

$$w_\mu = k - \operatorname{rank}\begin{bmatrix} \mathcal{M}_\mu & \mathcal{N}_\mu \end{bmatrix} + \operatorname{rank}\begin{bmatrix} \mathcal{M}_{\mu-1} & \mathcal{N}_{\mu-1} \end{bmatrix}$$

with the convention that $\operatorname{rank}\begin{bmatrix} \mathcal{M}_{-1} & \mathcal{N}_{-1} \end{bmatrix} = 0$. This implies the existence of a smooth full rank matrix function Z_1 of size (k, d) such that $\mathcal{E}_1 = Z_1^\top \mathcal{E}$ has full row rank.

The smallest possible μ in Hypothesis 1 is the strangeness index of the DAE (18.3) and $u = n - d - a$ defines the number of underdetermined components. Introducing $\mathcal{A}_1 = Z_1^\top \mathcal{A}$, $f_1(t) = Z_1^\top f(t)$, $f_2(t) = Z_2^\top g_\mu(t)$ and $f_3(t) = Z_3^\top g_\mu(t)$, we obtain a strangeness-free DAE system

$$\begin{bmatrix} \mathcal{E}_1(t) \\ 0 \\ 0 \end{bmatrix} \dot{x}(t) = \begin{bmatrix} \mathcal{A}_1(t) \\ \mathcal{A}_2(t) \\ 0 \end{bmatrix} x(t) + \begin{bmatrix} f_1(t) \\ f_2(t) \\ f_3(t) \end{bmatrix} \tag{18.5}$$

which has the same solutions as (18.3). The DAE (18.3) is solvable if $f_3(t) \equiv 0$ in (18.5). Moreover, an initial condition $x(t_0) = x_0$ is consistent if $\mathcal{A}_2(t_0)x_0 + f_2(t_0) = 0$. The initial value problem with consistent initial condition has a unique solution if $u = 0$.

18.3 Modified Nodal and Modified Loop Analysis

In this section, we consider the modelling of electrical circuits by DAEs based on the Kirchhoff laws and the constitutive relations for the electrical components. Derivations of these relations from Maxwell's equations can be found in [28].

A general electrical circuit with voltage and current sources, resistors, capacitors and inductors can be modelled as a directed graph whose nodes correspond to the nodes of the circuit and whose branches correspond to the circuit elements [9–11, 28]. We refer to the aforementioned works for the graph theoretic preliminaries related to circuit theory. Let n_{n}, n_{b} and n_{l} be, respectively, the number of nodes, branches and loops in this graph. Moreover, let $i(t) \in \mathbb{R}^{n_{\mathrm{b}}}$ be the vector of currents and let $v(t) \in \mathbb{R}^{n_{\mathrm{b}}}$ be the vector of corresponding voltages. Then Kirchhoff's current law [11, 28] states that at any node, the sum of flowing-in currents is equal to the sum of flowing-out currents, see Fig. 18.1. Equivalently, this law can be written as $A_0 i(t) = 0$, where $A_0 = [a_{kl}] \in \mathbb{R}^{n_{\mathrm{n}} \times n_{\mathrm{b}}}$ is an *all-node incidence matrix* with

$$a_{kl} = \begin{cases} 1, & \text{if branch } l \text{ leaves node } k, \\ -1, & \text{if branch } l \text{ enters node } k, \\ 0, & \text{otherwise.} \end{cases}$$

Furthermore, Kirchhoff's voltage law [11, 28] states that the sum of voltages along the branches of any loop vanishes, see Fig. 18.2. This law can equivalently be written as $B_0 v(t) = 0$, where $B_0 = [b_{kl}] \in \mathbb{R}^{n_{\mathrm{l}} \times n_{\mathrm{b}}}$ is an *all-loop matrix* with

$$b_{kl} = \begin{cases} 1, & \text{if branch } l \text{ belongs to loop } k \text{ and has the same orientation}, \\ -1, & \text{if branch } l \text{ belongs to loop } k \text{ and has the contrary orientation}, \\ 0, & \text{otherwise.} \end{cases}$$

The following proposition establishes a relation between the incidence and loop matrices A_0 and B_0.

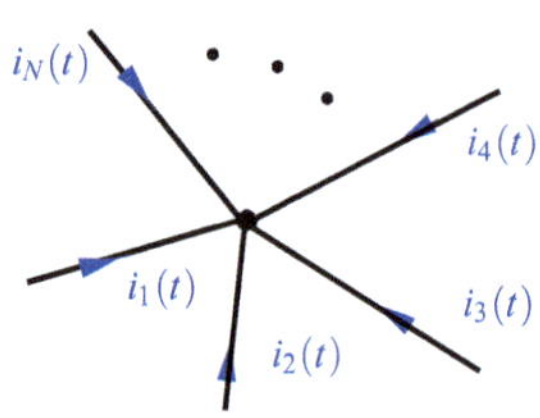

$\Rightarrow i_1(t) + \ldots + i_N(t) = 0$

Fig. 18.1 Kirchhoff's current law

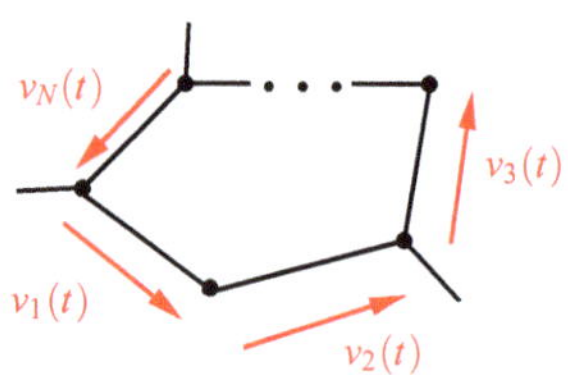

$\Rightarrow v_1(t) + \ldots + v_N(t) = 0$

Fig. 18.2 Kirchhoff's voltage law

Proposition 1 ([10, p. 213]) *Let $A_0 \in \mathbb{R}^{n_\mathrm{n} \times n_\mathrm{b}}$ be an all-node incidence matrix and let $B_0 \in \mathbb{R}^{n_\mathrm{l} \times n_\mathrm{b}}$ be an all-loop matrix of a connected graph. Then*

$$\ker B_0 = \operatorname{im} A_0^\top, \qquad \operatorname{rank} A_0 = n_\mathrm{n} - 1, \qquad \operatorname{rank} B_0 = n_\mathrm{b} - n_\mathrm{n} + 1.$$

We now consider the full rank matrices $A \in \mathbb{R}^{n_\mathrm{n}-1 \times n_\mathrm{b}}$ and $B \in \mathbb{R}^{n_\mathrm{b}-n_\mathrm{n}+1 \times n_\mathrm{b}}$ obtained from A_0 and B_0, respectively, by removing linear dependent rows. The matrices A and B are called the *reduced incidence* and *reduced loop matrices*, respectively. Then the Kirchhoff laws are equivalent to

$$A\, i(t) = 0, \qquad B\, v(t) = 0. \tag{18.6}$$

Due to the relation $\ker B = \operatorname{im} A^\top$, we can reformulate Kirchhoff's laws as follows: there exist vectors $\eta(t) \in \mathbb{R}^{n_\mathrm{n}-1}$ and $\iota(t) \in \mathbb{R}^{n_\mathrm{b}-n_\mathrm{n}+1}$ such that

$$i(t) = B^\top \iota(t), \qquad v(t) = A^\top \eta(t). \tag{18.7}$$

The vectors $\eta(t)$ and $\iota(t)$ are called the vectors of *node potentials* and *loop currents*, respectively. We partition the voltage and current vectors

$$v(t) = \left[v_\mathcal{C}^\top(t)\ v_\mathcal{L}^\top(t)\ v_\mathcal{R}^\top(t)\ v_\mathcal{V}^\top(t)\ v_\mathcal{I}^\top(t)\right]^\top,$$
$$i(t) = \left[\, i_\mathcal{C}^\top(t)\ \ i_\mathcal{L}^\top(t)\ \ i_\mathcal{R}^\top(t)\ \ i_\mathcal{V}^\top(t)\ \ i_\mathcal{I}^\top(t)\right]^\top$$

into voltage and current vectors of capacitors, inductors, resistors, voltage and current sources of dimensions $n_\mathcal{C}$, $n_\mathcal{L}$, $n_\mathcal{R}$, $n_\mathcal{V}$ and $n_\mathcal{I}$, respectively. Furthermore, partitioning the incidence and loop matrices

$$A = \left[A_\mathcal{C}\ A_\mathcal{L}\ A_\mathcal{R}\ A_\mathcal{V}\ A_\mathcal{I}\right], \qquad B = \left[B_\mathcal{C}\ B_\mathcal{L}\ B_\mathcal{R}\ B_\mathcal{V}\ B_\mathcal{I}\right], \tag{18.8}$$

the Kirchhoff laws (18.6) and (18.7) can now be represented in two alternative ways, namely, in the incidence-based formulation

$$A_\mathcal{C} i_\mathcal{C}(t) + A_\mathcal{L} i_\mathcal{L}(t) + A_\mathcal{R} i_\mathcal{R}(t) + A_\mathcal{V} i_\mathcal{V}(t) + A_\mathcal{I} i_\mathcal{I}(t) = 0, \tag{18.9}$$

$$v_\mathcal{C}(t) = A_\mathcal{C}^\top \eta(t), \qquad v_\mathcal{L}(t) = A_\mathcal{L}^\top \eta(t), \qquad v_\mathcal{R}(t) = A_\mathcal{R}^\top \eta(t), \tag{18.10}$$

$$v_\mathcal{V}(t) = A_\mathcal{V}^\top \eta(t), \qquad v_\mathcal{I}(t) = A_\mathcal{I}^\top \eta(t), \tag{18.11}$$

or in the loop-based formulation

$$B_\mathcal{R} v_\mathcal{R}(t) + B_\mathcal{C} v_\mathcal{C}(t) + B_\mathcal{L} v_\mathcal{L}(t) + B_\mathcal{V} v_\mathcal{V}(t) + B_\mathcal{I} v_\mathcal{I}(t) = 0, \tag{18.12}$$

$$i_\mathcal{C}(t) = B_\mathcal{C}^\top \iota(t), \qquad i_\mathcal{L}(t) = B_\mathcal{L}^\top \iota(t), \qquad i_\mathcal{R}(t) = B_\mathcal{R}^\top \iota(t), \tag{18.13}$$

$$i_\mathcal{V}(t) = B_\mathcal{V}^\top \iota(t), \qquad i_\mathcal{I}(t) = B_\mathcal{I}^\top \iota(t). \tag{18.14}$$

The dynamics of electrical circuits are not only relying on the Kirchhoff laws, but their behaviour is also determined by the components being located at the branches. The branch constitutive relations for capacitors, inductors and resistors are given by

$$i_{\mathcal{C}}(t) = \frac{d}{dt}q(v_{\mathcal{C}}(t)), \tag{18.15}$$

$$v_{\mathcal{L}}(t) = \frac{d}{dt}\psi(i_{\mathcal{L}}(t)), \tag{18.16}$$

$$i_{\mathcal{R}}(t) = g(v_{\mathcal{R}}(t)), \tag{18.17}$$

respectively, where $q : \mathbb{R}^{n_{\mathcal{C}}} \to \mathbb{R}^{n_{\mathcal{C}}}$ is the *charge function*, $\psi : \mathbb{R}^{n_{\mathcal{L}}} \to \mathbb{R}^{n_{\mathcal{L}}}$ is the *flux function*, and $g : \mathbb{R}^{n_{\mathcal{R}}} \to \mathbb{R}^{n_{\mathcal{R}}}$ is the *conductance function*. We now give our general assumptions on the considered circuit elements. For an interpretation of these assumptions in terms of total energy of the circuit, we refer to [28].

(A1) The charge, flux and conductance functions are continuously differentiable.

(A2) The Jacobian of the charge function

$$\mathcal{C}(v_{\mathcal{C}}) := \frac{d}{dv_{\mathcal{C}}}q(v_{\mathcal{C}})$$

is symmetric and pointwise positive definite.

(A3) The Jacobian of the flux function

$$\mathcal{L}(i_{\mathcal{L}}) := \frac{d}{di_{\mathcal{L}}}\psi(i_{\mathcal{L}})$$

is symmetric and pointwise positive definite.

(A4) The conductance function satisfies $g(0) = 0$ and there exists a constant $c > 0$ such that

$$(v_{\mathcal{R},1} - v_{\mathcal{R},2})^\top \big(g(v_{\mathcal{R},1}) - g(v_{\mathcal{R},2})\big) \geq c\|v_{\mathcal{R},1} - v_{\mathcal{R},2}\|^2 \tag{18.18}$$

for all $v_{\mathcal{R},1}, v_{\mathcal{R},2} \in \mathbb{R}^{n_{\mathcal{R}}}$.

Using the chain rule, the relations (18.15) and (18.16) can equivalently be written as

$$i_{\mathcal{C}}(t) = \mathcal{C}(v_{\mathcal{C}}(t))\frac{d}{dt}v_{\mathcal{C}}(t), \tag{18.19}$$

$$v_{\mathcal{L}}(t) = \mathcal{L}(i_{\mathcal{L}}(t))\frac{d}{dt}i_{\mathcal{L}}(t). \tag{18.20}$$

Furthermore, the property (18.18) implies that the Jacobian of the conductance function

$$\mathcal{G}(v_{\mathcal{R}}) := \frac{d}{dv_{\mathcal{R}}}\mathcal{g}(v_{\mathcal{R}})$$

fulfils

$$\mathcal{G}(v_{\mathcal{R}}) + \mathcal{G}^{\top}(v_{\mathcal{R}}) \geq 2\,c\,I > 0 \qquad \text{for all } v_{\mathcal{R}} \in \mathbb{R}^{n_{\mathcal{R}}}. \tag{18.21}$$

Thus, the matrix $\mathcal{G}(v_{\mathcal{R}})$ is invertible for all $v_{\mathcal{R}} \in \mathbb{R}^{n_{\mathcal{R}}}$. Applying the Cauchy-Schwarz inequality to (18.18) and taking into account that $\mathcal{g}(0) = 0$, we have

$$\|\mathcal{g}(v_{\mathcal{R}})\|\,\|v_{\mathcal{R}}\| \geq v_{\mathcal{R}}^{\top}\mathcal{g}(v_{\mathcal{R}}) \geq c\,\|v_{\mathcal{R}}\|^2 \quad \text{for all } v_{\mathcal{R}} \in \mathbb{R}^{n_{\mathcal{R}}}$$

and, hence, $\|\mathcal{g}(v_{\mathcal{R}})\| \geq c\,\|v_{\mathcal{R}}\|$. Then it follows from [37, Corollary, p. 201] that $\mathcal{g}$ has a global inverse function. This inverse is denoted by $r = \mathcal{g}^{-1}$ and referred to as the *resistance function*. Consequently, the relation (18.17) is equivalent to

$$v_{\mathcal{R}}(t) = r(i_{\mathcal{R}}(t)). \tag{18.22}$$

Moreover, we obtain from (18.18) that

$$\begin{aligned}
\big(i_{\mathcal{R},1} - i_{\mathcal{R},2}\big)^{\top}\big(r(i_{\mathcal{R},1}) - r(i_{\mathcal{R},2})\big) &= \big(\mathcal{g}(r(i_{\mathcal{R},1})) - \mathcal{g}(r(i_{\mathcal{R},2}))\big)^{\top}\big(r(i_{\mathcal{R},1}) - r(i_{\mathcal{R},2})\big)\\
&= \big(r(i_{\mathcal{R},1}) - r(i_{\mathcal{R},2})\big)^{\top}\big(\mathcal{g}(r(i_{\mathcal{R},1})) - \mathcal{g}(r(i_{\mathcal{R},2}))\big) \geq c\,\|r(i_{\mathcal{R},1}) - r(i_{\mathcal{R},2})\|^2
\end{aligned}$$

holds for all $i_{\mathcal{R},1}, i_{\mathcal{R},2} \in \mathbb{R}^{n_{\mathcal{R}}}$. Then the inverse function theorem implies that the Jacobian

$$\mathcal{R}(i_{\mathcal{R}}) := \frac{d}{di_{\mathcal{R}}}r(i_{\mathcal{R}})$$

fulfils $\mathcal{R}(i_{\mathcal{R}}) = (\mathcal{G}(r(i_{\mathcal{R}})))^{-1}$. In particular, $\mathcal{R}(i_{\mathcal{R}})$ is invertible for all $i_{\mathcal{R}} \in \mathbb{R}^{n_{\mathcal{R}}}$, and the relation (18.21) yields

$$\mathcal{R}(i_{\mathcal{R}}) + \mathcal{R}^{\top}(i_{\mathcal{R}}) > 0 \qquad \text{for all } i_{\mathcal{R}} \in \mathbb{R}^{n_{\mathcal{R}}}.$$

Having collected all physical laws for an electrical circuit, we are now able to set up a circuit model. This can be done in two different ways. The first approach is based on the formulation of Kirchhoff's laws via the incidence matrices given in (18.9)–(18.11), whereas the second approach relies on the equivalent representation of Kirchhoff's laws with the loop matrices given in (18.12)–(18.14).

(a) **Modified nodal analysis (MNA)**

Starting with Kirchhoff's current law (18.9), we eliminate the resistive and capacitive currents and voltages by using (18.17) and (18.19) as well as Kirchhoff's voltage law in (18.10) for resistors and capacitors. This results in

$$A_{\mathcal{C}}\mathcal{C}(A_{\mathcal{C}}^{\top}\eta(t))A_{\mathcal{C}}^{\top}\tfrac{d}{dt}\eta(t) + A_{\mathcal{R}}g(A_{\mathcal{R}}^{\top}\eta(t)) + A_{\mathcal{L}}i_{\mathcal{L}}(t) + A_{\mathcal{V}}i_{\mathcal{V}}(t) + A_{\mathcal{I}}i_{\mathcal{I}}(t) = 0.$$

Kirchhoff's voltage law in (18.10) for the inductive voltages and the component relation (18.20) for the inductors give

$$-A_{\mathcal{L}}^{\top}\eta(t) + \mathcal{L}(i_{\mathcal{L}}(t))\tfrac{d}{dt}i_{\mathcal{L}}(t) = 0.$$

Using Kirchhoff's voltage law in (18.11) for voltage sources, we obtain finally the MNA system

$$\begin{aligned} A_{\mathcal{C}}\mathcal{C}(A_{\mathcal{C}}^{\top}\eta(t))A_{\mathcal{C}}^{\top}\tfrac{d}{dt}\eta(t) + A_{\mathcal{R}}g(A_{\mathcal{R}}^{\top}\eta(t)) + A_{\mathcal{L}}i_{\mathcal{L}}(t) + A_{\mathcal{V}}i_{\mathcal{V}}(t) + A_{\mathcal{I}}i_{\mathcal{I}}(t) &= 0,\\ -A_{\mathcal{L}}^{\top}\eta(t) + \mathcal{L}(i_{\mathcal{L}}(t))\tfrac{d}{dt}i_{\mathcal{L}}(t) &= 0,\\ -A_{\mathcal{V}}^{\top}\eta(t) + v_{\mathcal{V}}(t) &= 0. \end{aligned} \tag{18.23}$$

In this formulation, voltages of voltage sources $v_{\mathcal{V}}$ and currents of current sources $i_{\mathcal{I}}$ are assumed to be given, whereas node potentials η, inductive currents $i_{\mathcal{L}}$ and currents of voltage sources $i_{\mathcal{V}}$ are unknown. The remaining physical variables such as voltages of the resistive, capacitive and inductive elements as well as resistive and capacitive currents can be algebraically reconstructed from the solution of system (18.23).

(b) **Modified loop analysis (MLA)**

Using the loop matrix based formulation of Kirchhoff's voltage law (18.12), the constitutive relations (18.20) and (18.22) for inductors and resistors, and the loop matrix based formulation of Kirchhoff's current law in (18.13) for the inductive and resistive currents, we obtain

$$B_{\mathcal{L}}\mathcal{L}(B_{\mathcal{L}}^{\top}\iota(t))B_{\mathcal{L}}^{\top}\tfrac{d}{dt}\iota(t) + B_{\mathcal{R}}r(B_{\mathcal{R}}^{\top}\iota(t)) + B_{\mathcal{C}}v_{\mathcal{C}}(t) + B_{\mathcal{I}}v_{\mathcal{I}}(t) + B_{\mathcal{V}}v_{\mathcal{V}}(t) = 0.$$

Moreover, Kirchhoff's voltage law in (18.13) for capacitors together with the component relation (18.19) for capacitors gives

$$-B_{\mathcal{C}}^{\top}\iota(t) + \mathcal{C}(v_{\mathcal{C}}(t))\tfrac{d}{dt}v_{\mathcal{C}}(t) = 0.$$

Combining these two relations together with Kirchhoff's voltage law in (18.14) for voltage sources, we obtain the MLA system

$$\begin{aligned} B_{\mathcal{L}}\mathcal{L}(B_{\mathcal{L}}^{\top}\iota(t))B_{\mathcal{L}}^{\top}\tfrac{d}{dt}\iota(t) + B_{\mathcal{R}}r(B_{\mathcal{R}}^{\top}\iota(t)) + B_{\mathcal{C}}v_{\mathcal{C}}(t) + B_{\mathcal{I}}v_{\mathcal{I}}(t) + B_{\mathcal{V}}v_{\mathcal{V}}(t) &= 0, \\ -B_{\mathcal{C}}^{\top}\iota(t) + \mathcal{C}(v_{\mathcal{C}}(t))\tfrac{d}{dt}v_{\mathcal{C}}(t) &= 0, \\ -B_{\mathcal{I}}^{\top}\iota(t) + i_{\mathcal{I}}(t) &= 0. \end{aligned}$$

Here, the unknown variables are loop currents ι, capacitive voltages $v_{\mathcal{C}}$ and voltages of current sources $v_{\mathcal{I}}$, and, as before, $v_{\mathcal{V}}$ and $i_{\mathcal{I}}$ are assumed to be known.

Thus, the overall circuit is described by the resistance law $i_{\mathcal{R}}(t) = g(v_{\mathcal{R}}(t))$ or $v_{\mathcal{R}}(t) = r(i_{\mathcal{R}}(t))$, the differential equations (18.19) and (18.20) for capacitors and inductors, and the Kirchhoff laws either in the form (18.9)–(18.11) or (18.12)–(18.14). By setting

$$x(t) = \begin{bmatrix} \eta(t) \\ i_{\mathcal{L}}(t) \\ i_{\mathcal{V}}(t) \end{bmatrix} \qquad \left(\text{resp. } x(t) = \begin{bmatrix} \iota(t) \\ v_{\mathcal{C}}(t) \\ v_{\mathcal{I}}(t) \end{bmatrix} \right)$$

in the MNA (resp. MLA) case, we obtain a nonlinear DAE of the form (18.1).

In the linear case, the capacitance matrix $\mathcal{C}(v_{\mathcal{C}}(t)) \equiv \mathcal{C}$ and the inductance matrix $\mathcal{L}(i_{\mathcal{L}}(t)) \equiv \mathcal{L}$ are both constant, and the component relations (18.17) and (18.22) for resistors read

$$i_{\mathcal{R}}(t) = \mathcal{G}v_{\mathcal{R}}(t), \qquad v_{\mathcal{R}}(t) = \mathcal{R}i_{\mathcal{R}}(t),$$

respectively, with $\mathcal{R} = \mathcal{G}^{-1} \in \mathbb{R}^{n_{\mathcal{R}} \times n_{\mathcal{R}}}$, $\mathcal{G} + \mathcal{G}^{\top} > 0$ and $\mathcal{R} + \mathcal{R}^{\top} > 0$. Then the circuit equations can be written as a linear DAE system

$$\mathcal{E}\dot{x}(t) = \mathcal{A}x(t) + \mathcal{B}u(t), \tag{18.24}$$

where $u(t) = \left[i_{\mathcal{I}}^{\top}(t),\ v_{\mathcal{V}}^{\top}(t) \right]^{\top}$, and the system matrices have the form

$$\mathcal{E} = \begin{bmatrix} A_{\mathcal{C}}\mathcal{C}A_{\mathcal{C}}^{\top} & 0 & 0 \\ 0 & \mathcal{L} & 0 \\ 0 & 0 & 0 \end{bmatrix}, \quad \mathcal{A} = \begin{bmatrix} -A_{\mathcal{R}}\mathcal{G}A_{\mathcal{R}}^{\top} & -A_{\mathcal{L}} & -A_{\mathcal{V}} \\ A_{\mathcal{L}}^{\top} & 0 & 0 \\ A_{\mathcal{V}}^{\top} & 0 & 0 \end{bmatrix}, \quad \mathcal{B} = \begin{bmatrix} -A_{\mathcal{I}} & 0 \\ 0 & 0 \\ 0 & -I_{n_{\mathcal{V}}} \end{bmatrix} \tag{18.25}$$

in the MNA case and

$$\mathcal{E} = \begin{bmatrix} B_{\mathcal{L}}\mathcal{L}B_{\mathcal{L}}^{\top} & 0 & 0 \\ 0 & \mathcal{C} & 0 \\ 0 & 0 & 0 \end{bmatrix}, \quad \mathcal{A} = \begin{bmatrix} -B_{\mathcal{R}}\mathcal{R}B_{\mathcal{R}}^{\top} & -B_{\mathcal{C}} & -B_{\mathcal{I}} \\ B_{\mathcal{C}}^{\top} & 0 & 0 \\ B_{\mathcal{I}}^{\top} & 0 & 0 \end{bmatrix}, \quad \mathcal{B} = \begin{bmatrix} 0 & -B_{\mathcal{V}} \\ 0 & 0 \\ -I_{n_{\mathcal{I}}} & 0 \end{bmatrix} \tag{18.26}$$

in the MLA case.

18.4 Differential-Algebraic Equations of Circuit Type

In this section, we study a special class of DAEs. First of all note that both the MNA and MLA systems can be written in a general form as

$$\begin{aligned} 0 &= E\Phi\big(E^\top x_1(t)\big)E^\top \dot{x}_1(t) + F\rho\big(F^\top x_1(t)\big) + G_2x_2(t) + G_3x_3(t) + f_1(t), \\ 0 &= \qquad \Psi\big(x_2(t)\big)\,\dot{x}_2(t) - G_2^\top x_1(t) \qquad\qquad\qquad\qquad + f_2(t), \\ 0 &= \qquad\qquad\qquad\qquad - G_3^\top x_1(t) \qquad\qquad\qquad\qquad + f_3(t), \end{aligned} \tag{18.27}$$

with the matrices $E \in \mathbb{R}^{n_1 \times m_1}$, $F \in \mathbb{R}^{n_1 \times m_2}$, $G_2 \in \mathbb{R}^{n_1 \times n_2}$, $G_3 \in \mathbb{R}^{n_1 \times n_3}$ and the continuously differentiable functions $\Phi : \mathbb{R}^{m_1} \to \mathbb{R}^{m_1 \times m_1}$, $\Psi : \mathbb{R}^{n_2} \to \mathbb{R}^{n_2 \times n_2}$ and $\rho : \mathbb{R}^{m_2} \to \mathbb{R}^{m_2}$ satisfying

$$\Phi(z_1) > 0 \qquad \text{for all } z_1 \in \mathbb{R}^{m_1}, \tag{18.28}$$

$$\Psi(z_2) > 0 \qquad \text{for all } z_2 \in \mathbb{R}^{n_2}, \tag{18.29}$$

$$\frac{d}{dz}\rho(z) + \left(\frac{d}{dz}\rho(z)\right)^\top > 0 \qquad \text{for all } z \in \mathbb{R}^{m_2}. \tag{18.30}$$

We now investigate the differentiation index of the DAE (18.27). The following result has been proven in [28] with the additional assumption $f_2(t) = 0$. However, this assumption has not been required in the proof.

Theorem 1 ([28, Theorem 6.6]) *Let a DAE* (18.27) *be given and assume that the functions* $\Phi : \mathbb{R}^{m_1} \to \mathbb{R}^{m_1 \times m_1}$, $\Psi : \mathbb{R}^{n_2} \to \mathbb{R}^{n_2 \times n_2}$ *and* $\rho : \mathbb{R}^{m_2} \to \mathbb{R}^{m_2}$ *satisfy* (18.28)–(18.30). *Further, assume that the matrices* $E \in \mathbb{R}^{n_1 \times m_1}$, $F \in \mathbb{R}^{n_1 \times m_2}$, $G_2 \in \mathbb{R}^{n_1 \times n_2}$ *and* $G_3 \in \mathbb{R}^{n_1 \times n_3}$ *fulfil*

$$\operatorname{rank}\big[E\ \ F\ \ G_2\ \ G_3\big] = n_1, \qquad \operatorname{rank} G_3 = n_3. \tag{18.31}$$

Then the differentiation index μ_d *of* (18.27) *is well-defined and it holds*

(a) $\mu_d = 0$, *if and only if* $n_3 = 0$ *and* $\operatorname{rank} E = n_1$.
(b) $\mu_d = 1$, *if and only if it is not zero and*

$$\operatorname{rank}\big[E\ \ F\ \ G_3\big] = n_1, \qquad \ker\big[E\ \ G_3\big] = \ker E \times \{0\}. \tag{18.32}$$

(c) $\mu_d = 2$, *if and only if* $\mu_d \notin \{0, 1\}$.

The additional assumptions (18.31) ensure that the DAE (18.27) is neither over- nor underdetermined, i.e., a solution of (18.27) exists for sufficiently smooth f_1, f_2 and f_3, and it is unique for any consistent initial value. Note that the assumptions (18.31) will not be made in the following. We will show that from any DAE of the form (18.27) one can extract a DAE of differentiation index one which has the same structure as (18.27). This extraction will be done by a special linear coordinate transformation.

To this end, we first introduce the matrices W_1, W_1', W_{11}, W_{11}', W_{12}, W_{12}', W_2, W_2', W_3, W_3', W_{31}, W_{31}', W_{32} and W_{32}' which have full column rank and satisfy the following conditions:

$$
\begin{array}{lll}
\text{(C1)} & \operatorname{im} W_1 = \ker E^\top, & \operatorname{im} W_1' = \operatorname{im} E, \\
\text{(C2)} & \operatorname{im} W_{11} = \ker \left[F\ G_3\right]^\top W_1, & \operatorname{im} W_{11}' = \operatorname{im} W_1^\top \left[F\ G_3\right], \\
\text{(C3)} & \operatorname{im} W_{12} = \ker G_2^\top W_1 W_{11}, & \operatorname{im} W_{12}' = \operatorname{im} W_{11}^\top W_1^\top G_2, \\
\text{(C4)} & \operatorname{im} W_2 = \ker W_{11}^\top W_1^\top G_2, & \operatorname{im} W_2' = \operatorname{im} G_2^\top W_1 W_{11}, \\
\text{(C5)} & \operatorname{im} W_3 = \ker W_1^\top G_3, & \operatorname{im} W_3' = \operatorname{im} G_3^\top W_1, \\
\text{(C6)} & \operatorname{im} W_{31} = \ker G_3 W_3, & \operatorname{im} W_{31}' = \operatorname{im} W_3^\top G_3^\top, \\
\text{(C7)} & \operatorname{im} W_{32} = \ker W_3^\top G_3^\top W_1', & \operatorname{im} W_{32}' = \operatorname{im} W_1'^\top G_3 W_3.
\end{array}
$$

The following lemma provides some useful properties for these matrices.

Lemma 1 *Let $E \in \mathbb{R}^{n_1 \times m_1}$, $F \in \mathbb{R}^{n_1 \times m_2}$, $G_2 \in \mathbb{R}^{n_1 \times n_2}$ and $G_3 \in \mathbb{R}^{n_1 \times n_3}$ be given, and let W_j and W_j' for $j \in J := \{1, 11, 12, 2, 3, 31, 32\}$ be matrices of full column rank satisfying the conditions* (C1)–(C7). *Then the following holds true:*

(a) The relations $(\operatorname{im} W_j)^\perp = \operatorname{im} W_j'$ are fulfilled for $j \in J$.
(b) The matrix $W_1 W_{11}$ has full column rank with

$$\operatorname{im} W_1 W_{11} = \ker \left[E\ F\ G_3\right]^\top. \tag{18.33}$$

(c) The matrix $W_1 W_{11} W_{12}$ has full column rank with

$$\operatorname{im} W_1 W_{11} W_{12} = \ker \left[E\ F\ G_2\ G_3\right]^\top. \tag{18.34}$$

(d) The matrix $W_3 W_{31}$ has full column rank with

$$\operatorname{im} W_3 W_{31} = \ker G_3. \tag{18.35}$$

(e) The matrix $W_{31}'^\top W_3^\top G_3^\top W_1' W_{32}'$ is square and invertible.
(f) The matrix $W_{12}'^\top W_{11}^\top W_1^\top G_2 W_2'$ is square and invertible.

Proof The proof mainly relies on the simple fact that $\ker M^\top = (\operatorname{im} M)^\perp$ holds for any matrix $M \in \mathbb{R}^{m \times n}$.

(a) The case $j = 1$ simply follows from

$$(\operatorname{im} W_1)^\perp = (\ker E^\top)^\perp = \operatorname{im} E = \operatorname{im} W_1'.$$

The remaining relations can be proved analogously.

(b) The matrix $W_1 W_{11}$ has full column rank as a product of matrices with full column rank. Furthermore, the subset relation "$\subseteq$" in (18.33) is a consequence of $\left[E\ F\ G_3\right]^\top W_1 W_{11} = 0$ which follows from (C1) and (C2). To prove the reverse inclusion, assume that $x \in \ker\left[E\ F\ G_3\right]^\top$. Then

$$x \in \ker E^\top = \operatorname{im} W_1 \qquad \text{and} \qquad x \in \ker\left[F\ G_3\right]^\top.$$

Hence, there exists a vector y such that $x = W_1 y$. We have

$$\left[F\ G_3\right]^\top W_1 y = \left[F\ G_3\right]^\top x = 0.$$

The definition of W_{11} gives rise to the existence of a vector z satisfying $y = W_{11} z$. Thus, $x = W_1 y = W_1 W_{11} z \in \operatorname{im} W_1 W_{11}$.

(c) The matrix $W_1 W_{11} W_{12}$ has full column rank as a product of matrices with full column rank. The inclusion "$\subseteq$" in (18.34) follows from

$$\left[E\ F\ G_2\ G_3\right]^\top W_1 W_{11} W_{12} = 0$$

which can be proved using (C1)–(C3). For the proof of the reverse inclusion, assume that $x \in \ker\left[E\ F\ G_2\ G_3\right]^\top$. Then $x \in \ker\left[E\ F\ G_3\right]^\top$. Hence, due (b) there exists a vector y such that $x = W_1 W_{11} y$. Consequently,

$$G_2^\top W_1 W_{11} y = G_2^\top x = 0.$$

The definition of W_{12} gives rise to the existence of a vector z such hat $y = W_{12} z$, and, thus, $x = W_1 W_{11} y = W_1 W_{11} W_{12} z \in \operatorname{im} W_1 W_{11} W_{12}$.

(d) The matrix $W_3 W_{31}$ has full column rank as a product of matrices with full column rank. The inclusion "$\subseteq$" in (18.35) follows from $G_3 W_3 W_{31} = 0$. For the proof of the reverse inclusion, assume that $x \in \ker G_3$. Then $x \in \ker W_1^\top G_3$, whence, by definition of W_3, there exists a vector y with $x = W_3 y$. Then $0 = G_3 x = G_3 W_3 y$ and, by definition of W_{31}, there exists a vector z such that $y = W_{31} z$. This gives $x = W_3 y = W_3 W_{31} z \in \operatorname{im} W_3 W_{31}$.

(e) First, we show that

$$\ker W_{31}'^\top W_3^\top G_3^\top W_1' W_{32}' = \{0\}. \tag{18.36}$$

Assume that $x \in \ker W_{31}'^\top W_3^\top G_3^\top W_1' W_{32}'$. Then

$$W_3^\top G_3^\top W_1' W_{32}' x \in \ker W_{31}'^\top = (\operatorname{im} W_{31}')^\perp = (\operatorname{im} W_3^\top G_3^\top)^\perp,$$

and, hence, $W_3^\top G_3^\top W_1' W_{32}' x \in \operatorname{im} W_3^\top G_3^\top \cap (\operatorname{im} W_3^\top G_3^\top)^\perp = \{0\}$. Thus, we have

$$W_{32}' x \in \ker W_3^\top G_3^\top W_1' = (\operatorname{im} W_1'^\top G_3 W_3)^\perp = (\operatorname{im} W_{32}')^\perp,$$

and, therefore, $W_{32}' x = 0$. Since W_{32}' has full column rank, we obtain that $x = 0$.

Next, we show that

$$\ker W_{32}'^\top W_1'^\top G_3 W_3 W_{31}' = \{0\}. \tag{18.37}$$

Assume that $x \in \ker W_{32}'^\top W_1'^\top G_3 W_3 W_{31}'$. Then

$$W_1'^\top G_3 W_3 W_{31}' x \in \ker W_{32}'^\top = (\operatorname{im} W_{32}')^\perp = (\operatorname{im} W_1'^\top G_3 W_3)^\perp$$

and, therefore, $W_1'^\top G_3 W_3 W_{31}' x = 0$. This gives

$$G_3 W_3 W_{31}' x \in \ker W_1'^\top = (\operatorname{im} W_1')^\perp = \operatorname{im} W_1 = \ker W_3^\top G_3^\top = (\operatorname{im} G_3 W_3)^\perp,$$

whence $G_3 W_3 W_{31}' x = 0$. From this we obtain

$$W_{31}' x \in \ker G_3 W_3 = (\operatorname{im} W_3^\top G_3^\top)^\perp = (\operatorname{im} W_{31}')^\perp.$$

Thus, $W_{31}' x = 0$. The property that W_{31}' has full column rank leads to $x = 0$. Finally, (18.36) and (18.37) together imply that $W_{31}'^\top W_3^\top G_3^\top W_1' W_{32}'$ is nonsingular.

(f) First, we show that

$$\ker W_{12}'^\top W_{11}^\top W_1^\top G_2 W_2' = \{0\}. \tag{18.38}$$

Assuming that $x \in \ker W_{12}'^\top W_{11}^\top W_1^\top G_2 W_2'$, we have

$$W_{11}^\top W_1^\top G_2 W_2' x \in \ker W_{12}'^\top = (\operatorname{im} W_{12}')^\perp = (\operatorname{im} W_{11}^\top W_1^\top G_2)^\perp,$$

whence $W_{11}^\top W_1^\top G_2 W_2' x = 0$. This gives rise to

$$W_2' x \in \ker W_{11}^\top W_1^\top G_2 = (\operatorname{im} G_2^\top W_1 W_{11})^\perp = (\operatorname{im} W_2')^\perp,$$

and, therefore, $W_2' x = 0$. The fact that W_2' has full column rank leads to $x = 0$.

We now show that

$$\ker W_2'^\top G_2^\top W_1 W_{11} W_{12}' = \{0\}. \tag{18.39}$$

Let $x \in \ker W_2'^{\top} G_2^{\top} W_1 W_{11} W_{12}'$. Then

$$G_2^{\top} W_1 W_{11} W_{12}' x \in \ker W_2'^{\top} = (\operatorname{im} W_2')^{\perp} = (\operatorname{im} G_2^{\top} W_1 W_{11})^{\perp},$$

and, thus, $G_2^{\top} W_1 W_{11} W_{12}' x = 0$. Then we have

$$W_{12}' x \in \ker G_2^{\top} W_1 W_{11} = (\operatorname{im} W_{11}^{\top} W_1^{\top} G_2)^{\perp} = (\operatorname{im} W_{12}')^{\perp},$$

whence $W_{12}' x = 0$. Since W_{12}' has full column rank, we obtain that $x = 0$. Finally, it follows from (18.38) and (18.39) that $W_{12}'^{\top} W_{11}^{\top} W_1^{\top} G_2 W_2'$ is nonsingular. □

We use the previously introduced matrices and their properties to decompose the vectors $x_1(t)$, $x_2(t)$ and $x_3(t)$ in the DAE (18.27) as

$$\begin{aligned} x_1(t) &= W_1' W_{32}' x_{11}(t) + W_1' W_{32} x_{21}(t) + W_1 W_{11}' x_{31}(t) \\ &\quad + W_1 W_{11} W_{12}' (W_2'^{\top} G_2^{\top} W_1 W_{11} W_{12}')^{-1} x_{41}(t) + W_1 W_{11} W_{12} x_{51}(t), \\ x_2(t) &= W_2' x_{12}(t) + W_2 x_{22}(t), \\ x_3(t) &= W_3' x_{13}(t) + W_3 W_{31}' (W_{32}'^{\top} W_1'^{\top} G_3 W_3 W_{31}')^{-1} x_{23}(t) + W_3 W_{31} x_{33}(t). \end{aligned} \tag{18.40}$$

Introducing the vector-valued functions and matrices

$$\tilde{x}_1(t) = \begin{bmatrix} x_{11}(t) \\ x_{21}(t) \\ x_{31}(t) \\ x_{41}(t) \\ x_{51}(t) \end{bmatrix}, \quad T_1 = \begin{bmatrix} W_{32}'^{\top} W_1'^{\top} \\ W_{32}^{\top} W_1'^{\top} \\ W_{11}'^{\top} W_1^{\top} \\ (W_{12}'^{\top} W_{11}^{\top} W_1^{\top} G_2 W_2')^{-1} W_{12}'^{\top} W_{11}^{\top} W_1^{\top} \\ W_{12}^{\top} W_{11}^{\top} W_1^{\top} \end{bmatrix}, \tag{18.41}$$

$$\tilde{x}_2(t) = \begin{bmatrix} x_{12}(t) \\ x_{22}(t) \end{bmatrix}, \quad T_2 = \begin{bmatrix} W_2'^{\top} \\ W_2^{\top} \end{bmatrix},$$

$$\tilde{x}_3(t) = \begin{bmatrix} x_{13}(t) \\ x_{23}(t) \\ x_{33}(t) \end{bmatrix}, \quad T_3 = \begin{bmatrix} W_3'^{\top} \\ (W_{31}'^{\top} W_3^{\top} G_3^{\top} W_1' W_{32}')^{-1} W_{31}'^{\top} W_3^{\top} \\ W_{31}^{\top} W_3^{\top} \end{bmatrix}, \tag{18.42}$$

equations (18.40) can be written as

$$x_1(t) = T_1^{\top} \tilde{x}_1(t), \qquad x_2(t) = T_2^{\top} \tilde{x}_2(t), \qquad x_3(t) = T_3^{\top} \tilde{x}_3(t). \tag{18.43}$$

Note that, by construction of the matrices W_j and W'_j, $j \in J$, the matrices T_1, T_2 and T_3 are nonsingular, and, hence, the vectors $\tilde{x}_1(t)$, $\tilde{x}_2(t)$ and $\tilde{x}_3(t)$ are uniquely determined by $x_1(t)$, $x_2(t)$ and $x_3(t)$, respectively. Further, we define

$$\tilde{f}_1(t)=\begin{bmatrix} f_{11}(t)\\ f_{21}(t)\\ f_{31}(t)\\ f_{41}(t)\\ f_{51}(t)\end{bmatrix}=T_1 f_1(t),\quad \tilde{f}_2(t)=\begin{bmatrix} f_{12}(t)\\ f_{22}(t)\end{bmatrix}=T_2 f_2(t),\quad \tilde{f}_3(t)=\begin{bmatrix} f_{13}(t)\\ f_{23}(t)\\ f_{33}(t)\end{bmatrix}=T_3 f_3(t).$$

Multiplying the DAE (18.27) from the left by $\operatorname{diag}(T_1, T_2, T_3)$ and substituting the vectors $x_1(t)$, $x_2(t)$ and $x_3(t)$ as in (18.43), we obtain an equivalent DAE

$$\begin{aligned}
0 &= \tilde{E}\Phi\big(\tilde{E}^\top\tilde{x}_1(t)\big)\tilde{E}^\top\dot{\tilde{x}}_1(t) + \tilde{F}\rho\big(\tilde{F}^\top\tilde{x}_1(t)\big) + \tilde{G}_2\tilde{x}_2(t) + \tilde{G}_3\tilde{x}_3(t) + \tilde{f}_1(t),\\
0 &= \Psi\big(\tilde{x}_2(t)\big)\dot{\tilde{x}}_2(t) - \tilde{G}_2^\top\tilde{x}_1(t) + \tilde{f}_2(t),\\
0 &= -\tilde{G}_3^\top\tilde{x}_1(t) + \tilde{f}_3(t),
\end{aligned} \tag{18.44}$$

with the matrices

$$\tilde{E} = \begin{bmatrix} E_1\\ E_2\\ 0\\ 0\\ 0\end{bmatrix},\quad \tilde{F} = \begin{bmatrix} F_1\\ F_2\\ F_3\\ 0\\ 0\end{bmatrix},\quad \tilde{G}_2 = \begin{bmatrix} G_{2,11} & G_{2,12}\\ G_{2,21} & G_{2,22}\\ G_{2,31} & G_{2,32}\\ I & 0\\ 0 & 0\end{bmatrix},\quad \tilde{G}_3 = \begin{bmatrix} G_{3,11} & I & 0\\ G_{3,21} & 0 & 0\\ G_{3,31} & 0 & 0\\ 0 & 0 & 0\\ 0 & 0 & 0\end{bmatrix}, \tag{18.45}$$

which are partitioned according to the partition of $\tilde{x}_i(t)$ in (18.41) and (18.42). The matrix blocks in (18.45) have the form

$$\begin{aligned}
E_1 &= W_{32}'^\top W_1'^\top E, & E_2 &= W_{32}^\top W_1'^\top E, & &\\
F_1 &= W_{32}'^\top W_1'^\top F, & F_2 &= W_{32}^\top W_1'^\top F, & F_3 &= W_{11}'^\top W_1^\top F,\\
G_{2,11} &= W_{32}'^\top W_1'^\top G_2 W_2', & G_{2,21} &= W_{32}^\top W_1'^\top G_2 W_2', & G_{2,31} &= W_{11}'^\top W_1^\top G_2 W_2',\\
G_{2,12} &= W_{32}'^\top W_1'^\top G_2 W_2, & G_{2,22} &= W_{32}^\top W_1'^\top G_2 W_2, & G_{2,32} &= W_{11}'^\top W_1^\top G_2 W_2,\\
G_{3,11} &= W_{32}'^\top W_1'^\top G_3 W_3', & G_{3,21} &= W_{32}^\top W_1'^\top G_3 W_3', & G_{3,31} &= W_{11}'^\top W_1^\top G_3 W_3'.
\end{aligned} \tag{18.46}$$

This leads to the following condensed form of the DAE (18.27):

$$
\begin{bmatrix} 0\\0\\0\\0\\0\\0\\0\\0\\0\\0 \end{bmatrix}
=
\begin{bmatrix}
E_1\Phi\big(E_1^\top x_{11}(t)+E_2^\top x_{21}(t)\big)E_1^\top \dot{x}_{11}(t)+E_1\Phi\big(E_1^\top x_{11}(t)+E_2^\top x_{21}(t)\big)E_2^\top \dot{x}_{21}(t)\\
E_2\Phi\big(E_1^\top x_{11}(t)+E_2^\top x_{21}(t)\big)E_1^\top \dot{x}_{11}(t)+E_2\Phi\big(E_1^\top x_{11}(t)+E_2^\top x_{21}(t)\big)E_2^\top \dot{x}_{21}(t)\\
0\\
0\\
0\\
W_2'^\top\Psi\big(W_2'x_{12}(t)+W_2x_{22}(t)\big)W_2'\dot{x}_{12}(t)+W_2'^\top\Psi\big(W_2'x_{12}(t)+W_2x_{22}(t)\big)W_2\dot{x}_{22}(t)\\
W_2^\top\Psi\big(W_2'x_{12}(t)+W_2x_{22}(t)\big)W_2'\dot{x}_{12}(t)+W_2^\top\Psi\big(W_2'x_{12}(t)+W_2x_{22}(t)\big)W_2\dot{x}_{22}(t)\\
0\\
0\\
0
\end{bmatrix}
$$

$$
+\begin{bmatrix}
F_1\rho\big(F_1^\top x_{11}(t)+F_2^\top x_{21}(t)+F_3^\top x_{31}(t)\big)\\
F_2\rho\big(F_1^\top x_{11}(t)+F_2^\top x_{21}(t)+F_3^\top x_{31}(t)\big)\\
F_3\rho\big(F_1^\top x_{11}(t)+F_2^\top x_{21}(t)+F_3^\top x_{31}(t)\big)\\
0\\0\\0\\0\\0\\0\\0
\end{bmatrix}
+\begin{bmatrix}
0 & 0 & 0 & 0 & 0 & G_{2,11} & G_{2,12} & G_{3,11} & I & 0\\
0 & 0 & 0 & 0 & 0 & G_{2,21} & G_{2,22} & G_{3,21} & 0 & 0\\
0 & 0 & 0 & 0 & 0 & G_{2,31} & G_{2,32} & G_{3,31} & 0 & 0\\
0 & 0 & 0 & 0 & 0 & I & 0 & 0 & 0 & 0\\
0 & 0 & 0 & 0 & 0 & 0 & 0 & 0 & 0 & 0\\
-G_{2,11}^\top & -G_{2,21}^\top & -G_{2,31}^\top & -I & 0 & 0 & 0 & 0 & 0 & 0\\
-G_{2,12}^\top & -G_{2,22}^\top & -G_{2,32}^\top & 0 & 0 & 0 & 0 & 0 & 0 & 0\\
-G_{3,11}^\top & -G_{3,21}^\top & -G_{3,31}^\top & 0 & 0 & 0 & 0 & 0 & 0 & 0\\
-I & 0 & 0 & 0 & 0 & 0 & 0 & 0 & 0 & 0\\
0 & 0 & 0 & 0 & 0 & 0 & 0 & 0 & 0 & 0
\end{bmatrix}
\begin{bmatrix} x_{11}(t)\\x_{21}(t)\\x_{31}(t)\\x_{41}(t)\\x_{51}(t)\\x_{12}(t)\\x_{22}(t)\\x_{13}(t)\\x_{23}(t)\\x_{33}(t)\end{bmatrix}
+\begin{bmatrix} f_{11}(t)\\f_{21}(t)\\f_{31}(t)\\f_{41}(t)\\f_{51}(t)\\f_{12}(t)\\f_{22}(t)\\f_{13}(t)\\f_{23}(t)\\f_{33}(t)\end{bmatrix}
\qquad (18.47)
$$

The following facts can be seen from this structure:

(a) The components $x_{51}(t)$ and $x_{33}(t)$ are actually not involved. As a consequence, they can be chosen freely. It follows from (18.40) and Lemma 1(c) that the vector $x_{51}(t)$ is trivial, i.e., it evolves in the zero-dimensional space, if and only if $\ker\begin{bmatrix} E & F & G_2 & G_3 \end{bmatrix}^\top = \{0\}$. Furthermore, by Lemma 1(d) the vector $x_{33}(t)$ is trivial if and only if $\ker G_3 = \{0\}$.

(b) The components $f_{51}(t)$ and $f_{33}(t)$ have to vanish in order to guarantee solvability. Due to Lemma 1(c), the equation $f_{51}(t) = 0$ does not appear if and only if $\ker\begin{bmatrix} E & F & G_2 & G_3 \end{bmatrix}^\top = \{0\}$. Moreover, Lemma 1(d) implies that the equation $f_{33}(t) = 0$ does not appear if and only if $\ker G_3 = \{0\}$.

(c) We see from (a) and (b) that over- and underdetermined parts occur in pairs. This is a consequence of the symmetric structure of the DAE (18.27).

(d) The remaining components fulfil the reduced DAE

$$
\begin{aligned}
0 &= \tilde{E}_r \Phi(\tilde{E}_r^\top \tilde{x}_{1r}(t))\tilde{E}_r^\top \dot{\tilde{x}}_{1r}(t) + \tilde{F}_r \rho(\tilde{F}_r^\top \tilde{x}_{1r}(t)) + \tilde{G}_{2r}\tilde{x}_{2r}(t) + \tilde{G}_{3r}\tilde{x}_{3r}(t) + \tilde{f}_{1r}(t),\\
0 &= \Psi(\tilde{x}_{2r}(t))\dot{\tilde{x}}_{2r}(t) - \tilde{G}_{2r}^\top \tilde{x}_{1r}(t) + \tilde{f}_{2r}(t),\\
0 &= - \tilde{G}_{3r}^\top \tilde{x}_{1r}(t) + \tilde{f}_{3r}(t),
\end{aligned}
\tag{18.48}
$$

with the matrices, functions and vectors

$$
\tilde{E}_r = \begin{bmatrix} E_1 \\ E_2 \\ 0 \\ 0 \end{bmatrix}, \quad \tilde{F}_r = \begin{bmatrix} F_1 \\ F_2 \\ F_3 \\ 0 \end{bmatrix}, \quad \tilde{G}_{2r} = \begin{bmatrix} G_{2,11} & G_{2,12} \\ G_{2,21} & G_{2,22} \\ G_{2,31} & G_{2,32} \\ I & 0 \end{bmatrix}, \quad \tilde{G}_{3r} = \begin{bmatrix} G_{3,11} & I \\ G_{3,21} & 0 \\ G_{3,31} & 0 \\ 0 & 0 \end{bmatrix},
\tag{18.49}
$$

$$
\begin{aligned}
\tilde{x}_{1r}(t) &= \begin{bmatrix} x_{11}(t) \\ x_{21}(t) \\ x_{31}(t) \\ x_{41}(t) \end{bmatrix}, & \tilde{f}_{1r}(t) &= \begin{bmatrix} f_{11}(t) \\ f_{21}(t) \\ f_{31}(t) \\ f_{41}(t) \end{bmatrix},\\
\tilde{x}_{2r}(t) &= \begin{bmatrix} x_{12}(t) \\ x_{22}(t) \end{bmatrix} = \tilde{x}_2(t), & \tilde{f}_{2r}(t) &= \begin{bmatrix} f_{12}(t) \\ f_{22}(t) \end{bmatrix} = \tilde{f}_2(t),\\
\tilde{x}_{3r}(t) &= \begin{bmatrix} x_{13}(t) \\ x_{23}(t) \end{bmatrix}, & \tilde{f}_{3r}(t) &= \begin{bmatrix} f_{13}(t) \\ f_{23}(t) \end{bmatrix}.
\end{aligned}
\tag{18.50}
$$

Note that this DAE has the same structure as (18.27) and (18.44). It is obtained from (18.44) by cancelling the components $x_{51}(t)$ and $x_{33}(t)$ and the equations $f_{51}(t) = 0$ and $f_{33}(t) = 0$.

We now analyze the reduced DAE (18.48). In particular, we show that it satisfies the preliminaries of Theorem 1. For this purpose, we prove the following auxiliary result.

Lemma 2 *Let $E \in \mathbb{R}^{n_1 \times m_1}$, $F \in \mathbb{R}^{n_1 \times m_2}$, $G_2 \in \mathbb{R}^{n_1 \times n_2}$ and $G_3 \in \mathbb{R}^{n_1 \times n_3}$ be given. Assume that the matrices W_j and W_j', $j \in J$, are of full column rank and satisfy the conditions* (C1)–(C7). *Then for the matrices in* (18.46), *the following holds true:*

(a) $\ker\begin{bmatrix} E_1^\top & E_2^\top \end{bmatrix} = \{0\}$;
(b) $\ker\begin{bmatrix} F_3 & G_{3,31} \end{bmatrix}^\top = \{0\}$;
(c) $\ker G_{3,31} = \{0\}$.

Proof (a) First, we show that the matrix $E^\top W_1'$ has full column rank. Assume that there exists a vector x such that $E^\top W_1' x = 0$. Then

$$W_1' x \in \ker E^\top = \operatorname{im} W_1 = (\operatorname{im} W_1')^\perp,$$

and, hence, $W_1' x = 0$. Since W_1' has full column rank, we obtain that $x = 0$. Consider now an accordingly partitioned vector

$$\begin{bmatrix} x_1 \\ x_2 \end{bmatrix} \in \ker\begin{bmatrix} E_1^\top & E_2^\top \end{bmatrix}.$$

From the first two relations in (18.46) we have

$$\begin{bmatrix} W_{32}' & W_{32} \end{bmatrix}\begin{bmatrix} x_1 \\ x_2 \end{bmatrix} \in \ker E^\top W_1' = \{0\}.$$

Then Lemma 1(a) yields $x_1 = 0$ and $x_2 = 0$.

(b) Let $x \in \ker\begin{bmatrix} F_3 & G_{3,31} \end{bmatrix}^\top$. Then, $0 = G_{3,31}^\top x = W_3'^\top G_3^\top W_1 W_{11}' x$, which gives

$$G_3^\top W_1 W_{11}' x \in \ker W_3'^\top = (\operatorname{im} W_3')^\perp = (\operatorname{im} G_3^\top W_1)^\perp.$$

Hence, $G_3^\top W_1 W_{11}' x = 0$. It follows from $F_3^\top x = 0$, that $\begin{bmatrix} F & G_3 \end{bmatrix}^\top W_1 W_{11}' x = 0$, and, therefore,

$$W_{11}' x \in \ker\begin{bmatrix} F & G_3 \end{bmatrix}^\top W_1 = (\operatorname{im} W_1^\top \begin{bmatrix} F & G_3 \end{bmatrix})^\perp = (\operatorname{im} W_{11}')^\perp.$$

This yields $W_{11}' x = 0$. Since W_{11}' has full column rank, we obtain $x = 0$.

(c) Assume that $x \in \ker G_{3,31}$. Then $0 = G_{3,31} x = W_{11}'^\top W_1^\top G_3 W_3' x$, which gives

$$W_1^\top G_3 W_3' x \in \ker W_{11}'^\top = (\operatorname{im} W_{11}')^\perp = (\operatorname{im} W_1^\top \begin{bmatrix} F & G_3 \end{bmatrix})^\perp \subset (\operatorname{im} W_1^\top G_3)^\perp.$$

Thus, we obtain $W_1^\top G_3 W_3' x = 0$, which is equivalent to

$$W_3' x \in \ker W_1^\top G_3 = (\operatorname{im} G_3^\top W_1)^\perp = (\operatorname{im} W_3')^\perp.$$

As a consequence, we have $W_3' x = 0$, and the property of W_3' to be of full column rank gives $x = 0$. □

It follows from Lemma 2(a) and (b) that $\ker\begin{bmatrix} \tilde{E}_r & \tilde{F}_r & \tilde{G}_{2r} & \tilde{G}_{3r} \end{bmatrix}^\top = \{0\}$, whereas Lemma 2(c) implies that $\ker \tilde{G}_{3r} = \{0\}$. In this case, the index of the DAE (18.48) can be established using Theorem 1.

Theorem 2 *Let a reduced DAE* (18.48) *be given with matrices and functions as in* (18.49) *and* (18.50)*, respectively. Then the differentiation index* $\tilde{\mu}_d$ *of* (18.48) *fulfils*

(a) $\tilde{\mu}_d = 0$ *if and only if* $\tilde{E}_r = E_2$, $\tilde{F} = F_2$, $\tilde{G}_{2r} = G_{2,22}$ *and the matrix* $\tilde{G}_{3r}$ *is empty.*

(b) $\tilde{\mu}_d = 1$ *if and only if it is not zero and*

$$\tilde{E}_r = \begin{bmatrix} E_2 \\ 0 \end{bmatrix}, \quad \tilde{F}_r = \begin{bmatrix} F_2 \\ F_3 \end{bmatrix}, \quad \tilde{G}_{2r} = \begin{bmatrix} G_{2,22} \\ G_{2,32} \end{bmatrix}, \quad \tilde{G}_{3r} = \begin{bmatrix} G_{3,21} \\ G_{3,31} \end{bmatrix}. \tag{18.51}$$

(c) $\tilde{\mu}_d = 2$ *if and only if* $\tilde{\mu}_d \notin \{0, 1\}$.

Proof (a) If $\tilde{E}_r = E_2$ and the matrix $\tilde{G}_{3r}$ is empty, then Lemma 2(a) implies that $\tilde{E}_r$ has full row rank. Then Theorem 1(a) implies $\tilde{\mu}_d = 0$. On the other hand, if $\tilde{\mu}_d = 0$, then Theorem 1(a) yields that the lower two blocks of $\tilde{E}_r$ in (18.49) vanish. Hence, the identity matrix in $\tilde{G}_{2r}$ has zero columns and rows meaning that the first block column in $\tilde{G}_{2r}$ vanishes. Furthermore, the absence of $\tilde{G}_{3r}$ implies that the first row in $\tilde{E}_r$, $\tilde{F}_r$ and $\tilde{G}_{2r}$ vanishes, which gives $\tilde{E}_r = E_2$, $\tilde{F} = F_2$, and $\tilde{G}_{2r} = G_{2,22}$.

(b) First assume that $\tilde{\mu}_d > 0$ and (18.51) holds true. Then it follows from Lemma 2(a) and (b) that

$$\left[\tilde{E}_r \ \tilde{F}_r \ \tilde{G}_{3r}\right] = \begin{bmatrix} E_2 & F_2 & G_{3,21} \\ 0 & F_3 & G_{3,31} \end{bmatrix}$$

has full row rank. We can further conclude from Lemma 2(c) that

$$\ker\left[\tilde{E}_r \ \tilde{G}_{3r}\right] = \ker \begin{bmatrix} E_2 & G_{3,21} \\ 0 & G_{3,31} \end{bmatrix} = \ker E_2 \times \{0\} = \ker \tilde{E}_r \times \{0\}.$$

Theorem 1(b) implies $\tilde{\mu}_d = 1$.

To prove the converse implication, assume that $\tilde{\mu}_d = 1$. Seeking for a contradiction, assume that the second block column of $\tilde{G}_{3r}$ in (18.49) has r columns for $r > 0$. Then there exists a vector $x_3 \in \mathbb{R}^r \setminus \{0\}$. Lemma 2(a) implies that there exists a vector x_1 such that

$$\begin{bmatrix} E_1 \\ E_2 \end{bmatrix} x_1 = \begin{bmatrix} x_3 \\ 0 \end{bmatrix}.$$

Then using Theorem 1(b) we have

$$\begin{bmatrix} -x_1 \\ 0 \\ x_3 \end{bmatrix} \in \ker\left[\tilde{E}_r \ \tilde{G}_{3r}\right] = \ker \tilde{E}_r \times \{0\}.$$

This is a contradiction.

It remains to prove that the forth block row of $\tilde{E}_r$, $\tilde{F}_r$, $\tilde{G}_{2r}$ and $\tilde{G}_{3r}$ vanishes. Seeking for a contradiction, assume that the forth block row has $r > 0$ rows. Then there exists some $x_3 \in \mathbb{R}^r \setminus \{0\}$, and

$$\begin{bmatrix} 0 \\ 0 \\ x_3 \end{bmatrix} \in \ker \begin{bmatrix} \tilde{E}_r & \tilde{F}_r & \tilde{G}_{3r} \end{bmatrix}^\top = \ker \begin{bmatrix} E_2^\top & 0 & 0 \\ F_2^\top & F_3^\top & 0 \\ G_{3,21}^\top & G_{3,31}^\top & 0 \end{bmatrix}.$$

Hence, $\begin{bmatrix} \tilde{E}_r & \tilde{F}_r & \tilde{G}_{3r} \end{bmatrix}$ does not have full row rank. Then Theorem 1(b) implies that $\tilde{\mu}_d > 1$, which is a contradiction. □

It follows from Lemma 2, Theorem 2 and the construction of the matrices $\tilde{E}_r$, $\tilde{F}_r$, $\tilde{G}_{2r}$ and $\tilde{G}_{3r}$ that $\tilde{\mu}_d = 0$ if and only if

$$\operatorname{rank} \begin{bmatrix} E & F & G_3 \end{bmatrix} = \operatorname{rank} \begin{bmatrix} E & F & G_2 & G_3 \end{bmatrix} \quad \text{and} \quad G_3 = 0.$$

Furthermore, we have $\tilde{\mu}_d = 1$ if and only if

$$\operatorname{rank} \begin{bmatrix} E & F & G_3 \end{bmatrix} = \operatorname{rank} \begin{bmatrix} E & F & G_2 & G_3 \end{bmatrix} \quad \text{and} \quad \ker \begin{bmatrix} E & G_3 \end{bmatrix} = \ker E \times \ker G_3.$$

Remark 1 Theorem 2 essentially states that the blocks in (18.47) corresponding to identity matrices are responsible for the index rising to $\tilde{\mu}_d = 2$. The equations in (18.47) corresponding to these blocks are algebraic constraints on variables whose derivatives are also involved in the overall DAE. KUNKEL and MEHRMANN call this phenomenon *strangeness* [17, 18, 22].

18.5 Index Reduction for Linear DAEs of Circuit Type

In this section, we consider index reduction of the DAE (18.27) based on the representation (18.48) in which the over- and underdetermined parts are already eliminated. We restrict ourselves to linear time-invariant systems. Roughly speaking, index reduction is a manipulation of the DAE such that another DAE with lower index is obtained whose solution set does not differ from the original one. Our approach is strongly inspired by the index reduction approach by KUNKEL and MEHRMANN for linear DAEs with time-varying coefficients [17, 22] briefly described in Sect. 18.2.

Consider the DAE (18.27), where we assume that the functions $\Phi : \mathbb{R}^{m_1} \to \mathbb{R}^{m_1 \times m_1}$ and $\Psi : \mathbb{R}^{n_2} \to \mathbb{R}^{n_2 \times n_2}$ are constant, that is,

$$\Phi(z_1) = \Phi \quad \text{for all } z_1 \in \mathbb{R}^{m_1} \quad \text{and} \quad \Psi(z_2) = \Psi \quad \text{for all } z_2 \in \mathbb{R}^{n_2}$$

with symmetric, positive definite matrices $\Phi \in \mathbb{R}^{m_1 \times m_1}$ and $\Psi \in \mathbb{R}^{n_2 \times n_2}$. Furthermore, we assume that the function $\rho : \mathbb{R}^{m_2} \to \mathbb{R}^{m_2}$ is linear, that is, $\rho(z) = Pz$ for some $P \in \mathbb{R}^{m_2 \times m_2}$ with $P + P^\top > 0$. Then by Remark 1 we can apply the index reduction technique proposed in [17]. To this end, we perform the following steps:

(i) Multiply the ninth equation in (18.47) from the left by $E_1 \Phi E_1^\top$, differentiate it and add to the first equation;
(ii) Multiply the ninth equation in (18.47) from the left by $E_2 \Phi E_1^\top$, differentiate it and add to the second equation;
(iii) Replace $x_{23}(t)$ by a new variable

$$\tilde{x}_{23}(t) = E_1 \Phi E_2^\top \dot{x}_{21}(t) + E_1 \Phi E_1^\top \dot{f}_{23}(t) + x_{23}(t).$$

(iv) Multiply the fourth equation in (18.47) from the left by $W_2'^\top \Psi W_2'$, differentiate it and subtract from the sixth equation;
(v) Multiply the fourth equation in (18.47) from the left by $W_2^\top \Psi W_2'$, differentiate it and subtract from the seventh equation;
(vi) Replace $x_{41}(t)$ by a new variable

$$\tilde{x}_{41}(t) = -W_2'^\top \Psi W_2 \dot{x}_{22}(t) + W_2'^\top \Psi W_2' \dot{f}_{41}(t) + x_{41}(t).$$

Thereby, we obtain the DAE

$$\begin{bmatrix} 0\\0\\0\\0\\0\\0\\0\\0\\0\\0 \end{bmatrix} = \begin{bmatrix} 0\\ E_2 \Phi E_2^\top \dot{x}_{21}(t)\\ 0\\0\\0\\0\\ W_2^\top \Psi W_2 \dot{x}_{22}(t)\\ 0\\0\\0 \end{bmatrix} + \begin{bmatrix} F_1 P F_1^\top x_{11}(t) + F_1 P F_2^\top x_{21}(t) + F_1 P F_3^\top x_{31}(t)\\ F_2 P F_1^\top x_{11}(t) + F_2 P F_2^\top x_{21}(t) + F_2 P F_3^\top x_{31}(t)\\ F_3 P F_1^\top x_{11}(t) + F_3 P F_2^\top x_{21}(t) + F_3 P F_3^\top x_{31}(t)\\ 0\\0\\0\\0\\0\\0\\0 \end{bmatrix}$$

$$+ \begin{bmatrix} 0 & 0 & 0 & 0 & 0 & G_{2,11} & G_{2,12} & G_{3,11} & I & 0\\ 0 & 0 & 0 & 0 & 0 & G_{2,21} & G_{2,22} & G_{3,21} & 0 & 0\\ 0 & 0 & 0 & 0 & 0 & G_{2,31} & G_{2,32} & G_{3,31} & 0 & 0\\ 0 & 0 & 0 & 0 & 0 & I & 0 & 0 & 0 & 0\\ 0 & 0 & 0 & 0 & 0 & 0 & 0 & 0 & 0 & 0\\ -G_{2,11}^\top & -G_{2,21}^\top & -G_{2,31}^\top & -I & 0 & 0 & 0 & 0 & 0 & 0\\ -G_{2,12}^\top & -G_{2,22}^\top & -G_{2,32}^\top & 0 & 0 & 0 & 0 & 0 & 0 & 0\\ -G_{3,11}^\top & -G_{3,21}^\top & -G_{3,31}^\top & 0 & 0 & 0 & 0 & 0 & 0 & 0\\ -I & 0 & 0 & 0 & 0 & 0 & 0 & 0 & 0 & 0\\ 0 & 0 & 0 & 0 & 0 & 0 & 0 & 0 & 0 & 0 \end{bmatrix} \begin{bmatrix} x_{11}(t)\\ x_{21}(t)\\ x_{31}(t)\\ \tilde{x}_{41}(t)\\ x_{51}(t)\\ x_{12}(t)\\ x_{22}(t)\\ x_{13}(t)\\ \tilde{x}_{23}(t)\\ x_{33}(t) \end{bmatrix} + \begin{bmatrix} f_{11}(t)\\ \tilde{f}_{21}(t)\\ f_{31}(t)\\ f_{41}(t)\\ f_{51}(t)\\ f_{12}(t)\\ \tilde{f}_{22}(t)\\ f_{13}(t)\\ f_{23}(t)\\ f_{33}(t) \end{bmatrix} \tag{18.52}$$

with $\tilde{f}_{21}(t) = f_{21}(t) + E_2\Phi E_1^{\top} \dot{f}_{23}(t)$ and $\tilde{f}_{22}(t) = f_{22}(t) - W_2^{\top}\Psi W_2' \dot{f}_{41}(t)$ which is again of type (18.27). Furthermore, it follows from Theorem 1 and Lemma 2 that the differentiation index of the resulting DAE obtained from (18.52) by removing the redundant variables $x_{51}(t)$ and $x_{33}(t)$ as well as the constrained equations for the inhomogeneity components $f_{51}(t) = 0$ and $f_{33}(t) = 0$ is at most one.

Remark 2 (a) We note that the previously introduced index reduction heavily uses linearity. In the case where, for instance, Φ depends on $x_{11}(t)$ and $x_{21}(t)$, the transformation (ii) would be clearly dependent on these variables as well. This causes that the unknown variables $x_{11}(t)$ and $x_{21}(t)$ enter the inhomogeneity $f_{21}(t)$.

(b) Structure-preserving index reduction for circuit equations has been considered previously in [3–5]. An index reduction procedure presented there provides a reduced model which can be interpreted as an electrical circuit containing controlled sources. As a consequence, the index-reduced system is not a DAE of type (18.27) anymore.

18.6 Consequences for Circuit Equations

In this section, we present a graph-theoretical interpretation of the previous results for circuit equations. First, we collect some basic concepts from the graph theory, which will be used in the subsequent discussion. For more details, we refer to [10].

Let $\mathcal{G} = (\mathcal{V}, \mathcal{B})$ be a directed graph with a finite set $\mathcal{V}$ of vertices and a finite set $\mathcal{B}$ of branches. For v_{k_1}, v_{k_2}, an ordered pair $b_{k_1} = \langle v_{k_1}, v_{k_2}\rangle$ denotes a branch leaving v_{k_1} and entering v_{k_2}. A tuple $(b_{k_1}, \dots, b_{k_{s-1}})$ of branches $b_{k_j} = \langle v_{k_j}, v_{k_{j+1}}\rangle$ in $\mathcal{G}$ is called a *path* connecting v_{k_1} and v_{k_s} if all vertices $v_{k_1}, \dots, v_{k_s}$ are different except possibly v_{k_1} and v_{k_s}. A path is *closed* if $v_{k_1} = v_{k_s}$, and *open*, otherwise. A closed path is called a *loop*. A graph $\mathcal{G}$ is called *connected* if for every two different vertices there exists an open path connecting them.

A *subgraph* $\mathcal{K} = (\mathcal{V}', \mathcal{B}')$ of $\mathcal{G} = (\mathcal{V}, \mathcal{B})$ is a graph with $\mathcal{V}' \subseteq \mathcal{V}$ and

$$\mathcal{B}' \subseteq \mathcal{B}\,|_{\mathcal{V}'} = \left\{ b_{k_1} = \langle v_{k_1}, v_{k_2}\rangle \in \mathcal{B} \;:\; v_{k_1}, v_{k_2} \in \mathcal{V}' \right\}.$$

A subgraph $\mathcal{K} = (\mathcal{V}', \mathcal{B}')$ is called *spanning* if $\mathcal{V}' = \mathcal{V}$. A spanning subgraph $\mathcal{K} = (\mathcal{V}, \mathcal{B}')$ is called a *cutset* of a connected graph $\mathcal{G} = (\mathcal{V}, \mathcal{B})$ if a complementary subgraph $\mathcal{G} - \mathcal{K} = (\mathcal{V}, \mathcal{B} \setminus \mathcal{B}')$ is disconnected and $\mathcal{K}$ is minimal with this property. For a spanning subgraph $\mathcal{K}$ of $\mathcal{G}$, a subgraph $\mathcal{L}$ of $\mathcal{G}$ is called a *$\mathcal{K}$-cutset*, if $\mathcal{L}$ is a cutset of $\mathcal{K}$. Furthermore, a path ℓ of $\mathcal{G}$ is called a *$\mathcal{K}$-loop*, if ℓ is a loop of $\mathcal{K}$.

For an electrical circuit, we consider an associated graph $\mathcal{G}$ whose vertices correspond to the nodes of the circuit and whose branches correspond to the circuit elements. Let $A \in \mathbb{R}^{n_{\mathrm{n}}-1\times n_{\mathrm{b}}}$ and $B \in \mathbb{R}^{n_{\mathrm{b}}-n_{\mathrm{n}}+1\times n_{\mathrm{b}}}$ be the reduced incidence and loop matrices as defined in Sect. 18.3. For a spanning graph $\mathcal{K}$ of $\mathcal{G}$, we denote by $A_{\mathcal{K}}$ (resp. $A_{\mathcal{G}-\mathcal{K}}$) a submatrix of A formed by the columns corresponding to the branches in $\mathcal{K}$ (resp. the complementary graph $\mathcal{G}-\mathcal{K}$). Analogously, we construct the loop matrices $B_{\mathcal{K}}$ and $B_{\mathcal{G}-\mathcal{K}}$. By a suitable reordering of the branches, the reduced incidence and loop matrices can be partitioned as

$$A = \begin{bmatrix} A_{\mathcal{K}} & A_{\mathcal{G}-\mathcal{K}} \end{bmatrix}, \qquad B = \begin{bmatrix} B_{\mathcal{K}} & B_{\mathcal{G}-\mathcal{K}} \end{bmatrix}. \tag{18.53}$$

The following lemma from [28] characterizes the absence of $\mathcal{K}$-loops and $\mathcal{K}$-cutsets in terms of submatrices of the incidence and loop matrices. It is crucial for our considerations. Note that this result has previously been proven for incidence matrices in [30].

Lemma 3 (Subgraphs, incidence and loop matrices [28, Lemma 4.10]) *Let $\mathcal{G}$ be a connected graph with the reduced incidence and loop matrices $A \in \mathbb{R}^{n_{\mathrm{n}}-1\times n_{\mathrm{b}}}$ and $B \in \mathbb{R}^{n_{\mathrm{b}}-n_{\mathrm{n}}+1\times n_{\mathrm{b}}}$. Further, let $\mathcal{K}$ be a spanning subgraph of $\mathcal{G}$. Assume that the branches of $\mathcal{G}$ are sorted in a way that* (18.53) *is satisfied.*

(a) The following three assertions are equivalent:

- *(i) $\mathcal{G}$ does not contain $\mathcal{K}$-cutsets;*
- *(ii)* $\ker A_{\mathcal{G}-\mathcal{K}}^{\top} = \{0\}$*;*
- *(iii)* $\ker B_{\mathcal{K}} = \{0\}$.

(b) The following three assertions are equivalent:

- *(i) $\mathcal{G}$ does not contain $\mathcal{K}$-loops;*
- *(ii)* $\ker A_{\mathcal{K}} = \{0\}$*;*
- *(iii)* $\ker B_{\mathcal{G}-\mathcal{K}}^{\top} = \{0\}$.

The next two auxiliary results are concerned with properties of subgraphs of subgraphs and give some equivalent characterizations in terms of their incidence and loop matrices. These statements have first been proven for incidence matrices in [30, Propositions 4.4 and 4.5]

Lemma 4 (Loops in subgraphs [28, Lemma 4.11]) *Let $\mathcal{G}$ be a connected graph with the reduced incidence and loop matrices $A \in \mathbb{R}^{n_{\mathrm{n}}-1\times n_{\mathrm{b}}}$ and $B \in \mathbb{R}^{n_{\mathrm{b}}-n_{\mathrm{n}}+1\times n_{\mathrm{b}}}$. Further, let $\mathcal{K}$ be a spanning subgraph of $\mathcal{G}$, and let $\mathcal{L}$ be a spanning subgraph of $\mathcal{K}$. Assume that the branches of $\mathcal{G}$ are sorted in a way that*

$$A = \begin{bmatrix} A_{\mathcal{L}} & A_{\mathcal{K}-\mathcal{L}} & A_{\mathcal{G}-\mathcal{K}} \end{bmatrix}, \qquad B = \begin{bmatrix} B_{\mathcal{L}} & B_{\mathcal{K}-\mathcal{L}} & B_{\mathcal{G}-\mathcal{K}} \end{bmatrix}. \tag{18.54}$$

Then the following three assertions are equivalent:

- *(i) $\mathcal{G}$ does not contain $\mathcal{K}$-loops except for $\mathcal{L}$-loops;*
- *(ii) For some (and hence any) matrix $Z_{\mathcal{L}}$ with* $\operatorname{im} Z_{\mathcal{L}} = \ker A_{\mathcal{L}}^{\top}$ *holds*

$$\ker Z_{\mathcal{L}}^{\top} A_{\mathcal{K}-\mathcal{L}} - \{0\};$$

- *(iii) For some (and hence any) matrix $Y_{\mathcal{K}-\mathcal{L}}$ with* $\operatorname{im} Y_{\mathcal{K}-\mathcal{L}} = \ker B_{\mathcal{K}-\mathcal{L}}^{\top}$ *holds*

$$Y_{\mathcal{K}-\mathcal{L}}^{\top} B_{\mathcal{G}-\mathcal{K}} = 0.$$

Lemma 5 (Cutsets in subgraphs [28, Lemma 4.12]) *Let $\mathcal{G}$ be a connected graph with the reduced incidence and loop matrices $A \in \mathbb{R}^{n_n-1\times n_b}$ and $B \in \mathbb{R}^{n_b-n_n+1\times n_b}$. Further, let $\mathcal{K}$ be a spanning subgraph of $\mathcal{G}$, and let $\mathcal{L}$ be a spanning subgraph of $\mathcal{K}$. Assume that the branches of $\mathcal{G}$ are sorted in a way that* (18.54) *is satisfied. Then the following three assertions are equivalent:*

(i) $\mathcal{G}$ does not contain $\mathcal{K}$-cutsets except for $\mathcal{L}$-cutsets;
(ii) For some (and hence any) matrix $Y_{\mathcal{L}}$ with $\operatorname{im} Y_{\mathcal{L}} = \ker B_{\mathcal{L}}^{\top}$ *holds*

$$\ker Y_{\mathcal{L}}^{\top} B_{\mathcal{K}-\mathcal{L}} = \{0\};$$

(iii) For some (and hence any) matrix $Z_{\mathcal{K}-\mathcal{L}}$ with $\operatorname{im} Z_{\mathcal{K}-\mathcal{L}} = \ker A_{\mathcal{K}-\mathcal{L}}^{\top}$ *holds*

$$Z_{\mathcal{K}-\mathcal{L}}^{\top} A_{\mathcal{G}-\mathcal{K}} = 0.$$

We use these results to analyze the condensed form (18.47) for the MNA equations (18.23). The MLA equations can be treated analogously. For a given electrical circuit whose corresponding graph is connected and has no self-loops (see [28]), we introduce the following matrices which take the role of the matrices W_i and W_i' defined in Sect. 18.4. Consider matrices of full column rank satisfying the following conditions:

$$
\begin{array}{lll}
(\mathrm{C1}') & \operatorname{im} Z_C = \ker A_C^{\top}, & \operatorname{im} Z_C' = \operatorname{im} A_C,\\
(\mathrm{C2}') & \operatorname{im} Z_{\mathcal{RV}-C} = \ker \left[A_{\mathcal{R}}\ A_{\mathcal{V}}\right]^{\top} Z_C, & \operatorname{im} Z_{\mathcal{RV}-C}' = \operatorname{im} Z_C^{\top}\left[A_{\mathcal{R}}\ A_{\mathcal{V}}\right],\\
(\mathrm{C3}') & \operatorname{im} Z_{\mathcal{L}-C\mathcal{RV}} = \ker A_{\mathcal{L}}^{\top} Z_C Z_{\mathcal{RV}-C}, & \operatorname{im} Z_{\mathcal{L}-C\mathcal{RV}}' = \operatorname{im} Z_{\mathcal{RV}-C}^{\top} Z_C^{\top} A_{\mathcal{L}},\\
(\mathrm{C4}') & \operatorname{im} \bar{Z}_{\mathcal{L}-C\mathcal{RV}} = \ker Z_{\mathcal{RV}-C}^{\top} Z_C^{\top} A_{\mathcal{L}}, & \operatorname{im} \bar{Z}_{\mathcal{L}-C\mathcal{RV}}' = \operatorname{im} A_{\mathcal{L}}^{\top} Z_C Z_{\mathcal{RV}-C},\\
(\mathrm{C5}') & \operatorname{im} \bar{Z}_{\mathcal{V}-C} = \ker Z_C^{\top} A_{\mathcal{V}}, & \operatorname{im} \bar{Z}_{\mathcal{V}-C}' = \operatorname{im} A_{\mathcal{V}}^{\top} Z_C,\\
(\mathrm{C6}') & \operatorname{im} \tilde{Z}_{\mathcal{V}-C} = \ker A_{\mathcal{V}} \bar{Z}_{\mathcal{V}-C}, & \operatorname{im} \tilde{Z}_{\mathcal{V}-C}' = \operatorname{im} \bar{Z}_{\mathcal{V}-C}^{\top} A_{\mathcal{V}}^{\top},\\
(\mathrm{C7}') & \operatorname{im} \tilde{Z}_{C\mathcal{V}C} = \ker \bar{Z}_{\mathcal{V}-C}^{\top} A_{\mathcal{V}}^{\top} Z_C', & \operatorname{im} \tilde{Z}_{C\mathcal{V}C}' = \operatorname{im} Z_C'^{\top} A_{\mathcal{V}} \bar{Z}_{\mathcal{V}-C}.
\end{array}
$$

Note that the introduced matrices can be determined by computationally cheap graph search algorithms [12, 16]. We have the following correspondences to the matrices W_i and W_i':

$$
\begin{array}{llll}
Z_C \hat{=} W_1, & Z_C' \hat{=} W_1', & Z_{\mathcal{RV}-C} \hat{=} W_{11}, & Z_{\mathcal{RV}-C}' \hat{=} W_{11}',\\
Z_{\mathcal{L}-C\mathcal{RV}} \hat{=} W_{12}, & Z_{\mathcal{L}-C\mathcal{RV}}' \hat{=} W_{12}', & \bar{Z}_{\mathcal{L}-C\mathcal{RV}} \hat{=} W_2, & \bar{Z}_{\mathcal{L}-C\mathcal{RV}}' \hat{=} W_2',\\
\bar{Z}_{\mathcal{V}-C} \hat{=} W_3, & \bar{Z}_{\mathcal{V}-C}' \hat{=} W_3', & \tilde{Z}_{\mathcal{V}-C} \hat{=} W_{31}, & \tilde{Z}_{\mathcal{V}-C}' \hat{=} W_{31}',\\
\tilde{Z}_{C\mathcal{V}C} \hat{=} W_{32}, & \tilde{Z}_{C\mathcal{V}C}' \hat{=} W_{32}'. & &
\end{array}
$$

Using Lemmas 1 and 3–5, we can characterize the absence of certain blocks in the condensed form (18.47) in terms of the graph structure of the circuit. Based on the definition of $\mathcal{K}$-loop and $\mathcal{K}$-cutset, we arrange the following way of speaking. An expression like "$\mathcal{CV}$-loop" indicates a loop in the circuit graph whose branch set consists only of branches corresponding to capacitors and/or voltage sources. Likewise, an "$\mathcal{LI}$-cutset" is a cutset in the circuit graph whose branch set consists only of branches corresponding to inductors and/or current sources.

(a) The matrix Z_C has zero columns if and only if the circuit does not contain any $\mathcal{RLVI}$-cutsets (Lemma 3(a)).
(b) The matrix Z'_C has zero columns if and only if the circuit does not contain any capacitors.
(c) The matrix $Z_{\mathcal{RV}-C}$ has zero columns if and only if the circuit does not contain any $\mathcal{LI}$-cutsets (Lemmas 1(b) and 3(a)).
(d) The matrix $Z'_{\mathcal{RV}-C}$ has zero columns if and only if the circuit does not contain any $\mathcal{CLI}$-cutsets except for $\mathcal{LI}$-cutsets (Lemma 5).
(e) The matrix $Z_{\mathcal{L}-\mathcal{CRV}}$ has zero columns if and only if the circuit does not contain any $\mathcal{I}$-cutsets (Lemmas 1(c) and 3(a)).
(f) The matrix $Z'_{\mathcal{L}-\mathcal{CRV}}$ (and by Lemma 1(f) also the matrix $\bar{Z}'_{\mathcal{L}-\mathcal{CRV}}$) has zero columns if and only if the circuit does not contain any $\mathcal{CRVI}$-cutsets except for $\mathcal{I}$-cutsets (Lemmas 1(b) and 5).
(g) The matrix $\bar{Z}_{\mathcal{L}-\mathcal{CRV}}$ has zero columns if and only if the circuit does not contain any $\mathcal{RCVL}$-loops except for $\mathcal{RCV}$-loops (Lemmas 1(b) and 4).
(h) The matrix $\bar{Z}_{\mathcal{V}-C}$ has zero columns if and only if the circuit does not contain any $\mathcal{CV}$-loops except for $\mathcal{C}$-loops (Lemma 4).
(i) The matrix $\bar{Z}'_{\mathcal{V}-C}$ has zero columns if and only if the circuit does not contain any $\mathcal{RCLI}$-cutsets except for $\mathcal{RLI}$-cutsets (Lemma 5).
(j) The matrix $\tilde{Z}_{\mathcal{V}-C}$ has zero columns if and only if the circuit does not contain any $\mathcal{V}$-loops (Lemmas 1(d) and 3(b)).
(k) The matrix $\tilde{Z}'_{CVC}$ (and by Lemma 1(e) also the matrix $\tilde{Z}'_{\mathcal{V}-C}$) has zero columns if and only if the circuit does not contain any $\mathcal{CV}$-loops except for $\mathcal{C}$-loops and $\mathcal{V}$-loops (this can be proven analogous to Lemma 4).

Exemplarily, we will show (a) only. Other assertions can be proved analogously. For the MNA system (18.23), we have $E = A_C$. Then by definition, the matrix Z_C has zero columns if and only if $\ker A_C^\top = \{0\}$. By Lemma 3(a), this condition is equivalent to the absence of $\mathcal{RLVI}$-cutsets.

In particular, we obtain from the previous findings that the condensed form (18.47) does not have any redundant variables and equations if and only if the circuit neither contains $\mathcal{I}$-cutsets nor $\mathcal{V}$-loops. We can also infer some assertions on the differentiation index of the reduced DAE (18.48) obtained from (18.47) by removing the redundant variables and equations. The DAE (18.48) has the differentiation index $\tilde{\mu}_d = 0$ if and only if the circuit does not contain voltage sources and $\mathcal{RLI}$-cutsets except for $\mathcal{I}$-cutsets. Furthermore, we have $\tilde{\mu}_d = 1$ if and only if and the circuit neither contains $\mathcal{CV}$-loops except for $\mathcal{C}$-loops and $\mathcal{V}$-loops nor $\mathcal{LI}$-cutsets except for $\mathcal{I}$-cutsets.

18.7 Conclusion

In this paper, we have presented a structural analysis for the MNA and MLA equations which are DAEs modelling electrical circuits with uncontrolled voltage and current sources, resistors, capacitors and inductors. These DAEs are shown to be of the same structure. A special condensed form under linear transformations has been introduced which allows to determine the differentiation index. In the linear case, we have presented an index reduction procedure which provides a DAE system of the differentiation index one and preserves the structure of the circuit DAE. Graph-theoretical characterizations of the condensed form have also been given.

References

1. Arnol'd, V.: Ordinary Differential Equations. Undergraduate Texts in Mathematics. Springer, Berlin/Heidelberg/New York (1992). Translated from the Russian by R. Cooke
2. Bächle, S.: Numerical solution of differential-algebraic systems arising in circuit simulation. Ph.D. thesis, Technische Universität Berlin, Berlin (2007)
3. Bächle, S., Ebert, F.: Element-based index reduction in electrical circuit simulation. In: PAMM – Proceedings of Applied Mathematics and Mechanics, vol. 6, pp. 731–732. Wiley, Weinheim (2006). doi: 10.1002/pamm.200610346
4. Bächle, S., Ebert, F.: A structure preserving index reduction method for MNA. In: PAMM – Proceedings of Applied Mathematics and Mechanics, vol. 6, pp. 727–728. Wiley, Weinheim (2006). doi: 10.1002/pamm.200610344
5. Bächle, S., Ebert, F.: Index reduction by element-replacement for electrical circuits. In: Ciuprina, G., Ioan, D. (eds.) Scientific Computing in Electrical Engineering. Mathematics in Industry, vol. 11, pp. 191–197. Springer, Berlin/Heidelberg (2007)
6. Baum, A.K.: A flow-on-manifold formulation of differential-algebraic equations. Application to positive systems. Ph.D. thesis, Technische Universität Berlin, Berlin (2014)
7. Brenan, K., Campbell, S., Petzold, L.: The Numerical Solution of Initial-Value Problems in Differential-Algebraic Equations. Classics in Applied Mathematics, vol. 14. SIAM, Philadelphia (1996)
8. Campbell, S.: A general form for solvable linear time varying singular systems of differential equations. SIAM J. Math. Anal. **18**(4), 1101–1115 (1987)
9. Chua, L., Desoer, C., Kuh, E.: Linear and Nonlinear Circuits. McGraw-Hill, New York (1987)
10. Deo, N.: Graph Theory with Application to Engineering and Computer Science. Prentice-Hall, Englewood Cliffs (1974)
11. Desoer, C., Kuh, E.: Basic Circuit Theory. McGraw-Hill, New York (1969)
12. Estévez Schwarz, D.: A step-by-step approach to compute a consistent initialization for the MNA. Int. J. Circuit Theory Appl. **30**(1), 1–16 (2002)
13. Hairer, E., Lubich, C., Roche, M.: The Numerical Solution of Differential-Algebraic Equations by Runge-Kutta Methods. Lecture Notes in Mathematics, vol. 1409. Springer, Berlin/Heidelberg (1989)
14. Heiland, J.: Decoupling and optimization of differential-algebraic equations with application in flow control. Ph.D. thesis, Technische Universität Berlin, Berlin (2014)
15. Ho, C.W., Ruehli, A., Brennan, P.: The modified nodal approach to network analysis. IEEE Trans. Circuits Syst. **22**(6), 504–509 (1975)

16. Ipach, H.: Graphentheoretische Anwendungen in der Analyse elektrischer Schaltkreise. Bachelor thesis, Universität Hamburg, Hamburg (2013)
17. Kunkel, P., Mehrmann, V.: Canonical forms for linear differential-algebraic equations with variable coefficients. J. Comput. Appl. Math. **56**, 225–259 (1994)
18. Kunkel, P., Mehrmann, V.: A new look at pencils of matrix valued functions. Linear Algebra Appl. **212/213**, 215–248 (1994)
19. Kunkel, P., Mehrmann, V.: Local and global invariants of linear differential-algebraic equations and their relation. Electron. Trans. Numer. Anal. **4**, 138–157 (1996)
20. Kunkel, P., Mehrmann, V.: Regular solutions of nonlinear differential-algebraic equations and their numerical determination. Numer. Math. **79**(4), 581–600 (1998)
21. Kunkel, P., Mehrmann, V.: Analysis of over- and underdetermined nonlinear differential-algebraic systems with application to nonlinear control problems. Math. Control Signals Syst. **14**(3), 233–256 (2001)
22. Kunkel, P., Mehrmann, V.: Differential-Algebraic Equations. Analysis and Numerical Solution. EMS Publishing House, Zürich (2006)
23. Kunkel, P., Mehrmann, V., Rath, W.: Analysis and numerical solution of control problems in descriptor form. Math. Control Signals Syst. **14**(1), 29–61 (2001)
24. Lamour, R., März, R., Tischendorf, C.: Differential Algebraic Equations: A Projector Based Analysis. Differential-Algebraic Equations Forum, vol. 1. Springer, Berlin/Heidelberg (2013)
25. Mehrmann, V.: Index concepts for differential-algebraic equations. Preprint 3-2012, Technische Universität Berlin, Berlin (2012). To appear in Encyclopedia of Applied and Computational Mathematics. Springer, Berlin (2016)
26. Rabier, P., Rheinboldt, W.: A geometric treatment of implicit differential-algebraic equations. J. Differ. Equ. **109**(1), 110–146 (1994)
27. Rath, W.: Feedback design and regularization for linear descriptor systems with variable coefficients. Ph.D. thesis, Technische Universität Chemnitz, Chemnitz (1996)
28. Reis, T.: Mathematical modeling and analysis of nonlinear time-invariant RLC circuits. In: Benner, P., Findeisen, R., Flockerzi, D., Reichl, U., Sundmacher, K. (eds.) Large Scale Networks in Engineering and Life Sciences, pp. 126–198. Birkhäuser, Basel (2014)
29. Reis, T., Stykel, T.: Lyapunov balancing for passivity-preserving model reduction of RC circuits. SIAM J. Appl. Dyn. Syst. **10**(1), 1–34 (2011)
30. Riaza, R., Tischendorf, C.: Qualitative features of matrix pencils and DAEs arising in circuit dynamics. Dyn. Syst. **22**(2), 107–131 (2007)
31. Seufer, I.: Generalized inverses of differential-algebraic equations and their discretization. Ph.D. thesis, Technische Universität Berlin, Berlin (2005)
32. Shi, C.: Linear differential-algebraic equations of higher-order and the regularity or singularity of matrix polynomials. Ph.D. thesis, Technische Universität Berlin, Berlin (2004)
33. Steinbrecher, A.: Numerical solution of quasi-linear differential-algebraic equations and industrial simulation of multibody systems. Ph.D. thesis, Technische Universität Berlin, Berlin (2006)
34. Tao, T.: Analysis II. Texts and Readings in Mathematics, vol. 38. Hindustan Book Agency, New Delhi (2009)
35. Wedepohl, L., Jackson, L.: Modified nodal analysis: an essential addition to electrical circuit theory and analysis. Eng. Sci. Educ. J. **11**(3), 84–92 (2002)
36. Weickert, J.: Applications of the theory of differential-algebraic equations to partial differential equations of fluid dynamics. Ph.D. thesis, Technische Universität Chemnitz, Chemnitz (1997)
37. Wu, F., Desoer, C.: Global inverse function theorem. IEEE Trans. Circuits Syst. **19**, 199–201 (1972)
38. Wunderlich, L.: Analysis and numerical solution of structured and switched differential-algebraic systems. Ph.D. thesis, Technische Universität Berlin, Berlin (2008)

Chapter 19
Spectrum-Based Robust Stability Analysis of Linear Delay Differential-Algebraic Equations

Vu Hoang Linh and Do Duc Thuan

Dedicated to Volker Mehrmann on the occasion of his 60th birthday

Abstract This paper presents a survey of results on the spectrum-based robust stability analysis of linear delay ordinary differential equations (DODEs) and linear delay differential-algebraic equations (DDAEs). We focus on the formulation of stability radii for continuous-time delay systems with coefficients subject to structured perturbations. First, we briefly overview important results on the stability radii for linear time-invariant DODEs and an extended result for linear time-varying DODEs. Then, we survey some recent results on the spectrum-based stability and robust stability analysis for general linear time-invariant DDAEs.

19.1 Introduction

This paper is concerned with the robust stability of homogeneous linear time-invariant delay systems of the form

$$E\dot{x}(t) = Ax(t) + Dx(t - \tau), \tag{19.1}$$

where $E, A, D \in \mathbb{K}^{n,n}$, $\mathbb{K} = \mathbb{R}$ or $\mathbb{K} = \mathbb{C}$, and $\tau > 0$ represents a time-delay. We study initial value problems with an initial function ϕ, so that

$$x(t) = \phi(t), \quad \text{for } -\tau \le t \le 0. \tag{19.2}$$

V.H. Linh (✉)
Faculty of Mathematics, Mechanics and Informatics, Vietnam National University, 334, Nguyen Trai Str., Thanh Xuan, Hanoi, Vietnam
e-mail: linhvh@vnu.edu.vn

D.D. Thuan
School of Applied Mathematics and Informatics, Hanoi University of Science and Technology, 1 Dai Co Viet Street, Hanoi, Vietnam
e-mail: ducthuank7@gmail.com

P. Benner et al. (eds.), *Numerical Algebra, Matrix Theory, Differential-Algebraic Equations and Control Theory*, DOI 10.1007/978-3-319-15260-8_19

When E is invertible, by multiplying both sides of the equation by E^{-1}, (19.1) becomes a delay ordinary differential equation (DODE). The more difficult case happens when E is singular. Then we have a delay differential-algebraic equations (DDAE), which is a generalization of both DODEs and non-delay DAEs. Non-delay differential-algebraic equations (DAEs) play an important role in many application areas, such as multibody mechanics, electrical circuit simulation, control theory, fluid dynamics, chemical engineering, see, e.g., [2, 5, 27, 37, 40, 51], the delay version is typically needed to model effects that do not arise instantaneously, see, e.g., [4, 24, 66].

One may also consider more general neutral delay DAEs

$$E\dot{x}(t) + F\dot{x}(t - \tau) = Ax(t) + Dx(t - \tau). \tag{19.3}$$

However, by introducing a new variable, (19.3) can be rewritten into the form (19.1) with double dimension, see [13]. For this reason here we only consider (19.1).

In this paper, we focus on the robust stability of DDAEs of the form (19.1), i.e. we investigate whether the asymptotic/exponential stability of a given system is preserved when the system coefficients are subject to structured perturbations. In particular, we are interested in computing the distance (measured in an appropriate metric) between the nominal stable system and the closest perturbed systems that loses the stability. This quantity is called the *stability radius* of the system, see [30–32].

The stability and robust stability analysis for DAEs is quite different from that of ordinary differential equations (ODEs), see, e.g., [33], and has recently received a lot of attention, see, e.g., [7, 8, 19, 38, 43, 46, 61, 62] and [17] for a recent survey. At the same time, extension of the stability and robust stability analysis from ODEs to DODEs is also well established, see, e.g., [28, 36, 54–58].

Up to our knowledge, the first work addressing the analysis of DDAEs was due to S.L. Campbell [10]. DDAEs arise, for instance, in the context of feedback control of DAE systems (where the feedback does not act instantaneously) or as limiting case for singularly perturbed ordinary delay systems, see e.g. [2, 3, 9–11, 13, 45, 53, 67]. Unlike the well-established theory of DODEs and DAEs, even the existence and uniqueness theory of DDAEs is much less well understood, see [25, 26] for a recent analysis and the discussion of many of the difficulties. Most of the existing results are obtained only for linear time-invariant regular DDAEs [20, 65] or DDAEs of special form [2, 44, 68]. Many of the results that are known for DODEs cannot be extended to the DDAE case. Even the well-known spectral analysis for the exponential stability or the asymptotic stability of linear time-invariant DDAEs (19.1) is much more complex than that for DAEs and DDEs, see [13, 63, 67] for some special cases.

For the illustration of difficulties that arise with DDAEs due to the simultaneous effects of the time-delay and the singular nature of the systems, let us consider the following simple examples.

Example 1 ([13]) Consider the system

$$\begin{bmatrix} 1 & 0 \\ 0 & 0 \end{bmatrix} \dot{x}(t) = \begin{bmatrix} -1 & 0 \\ 0 & -2 \end{bmatrix} x(t) + \begin{bmatrix} 1 & 1 \\ -1 & -1 \end{bmatrix} x(t-1), \quad (t \geq 0),$$

where $x = [x_1\ x_2]^T$, x_1 and x_2 are given by continuous functions on the initial interval $(-1, 0]$. The dynamics of x_1 is governed by a differential operator and continuity of x_1 is expected. The dynamics of x_2 is determined by a difference operator and unlike x_1, this component is expected to be only piecewise continuous.

Example 2 ([12]) Consider the following inhomogenous system

$$\begin{bmatrix} 1 & 0 \\ 0 & 0 \end{bmatrix} \dot{x}(t) = \begin{bmatrix} 0 & 0 \\ -1 & 0 \end{bmatrix} x(t) + \begin{bmatrix} 0 & 0 \\ 0 & 1 \end{bmatrix} x(t-1) + \begin{bmatrix} f(t) \\ g(t) \end{bmatrix}, \quad (t \geq 0).$$

The solution is given by

$$x_1(t) = \int_0^t f(s)ds + x_1(0), \quad x_2(t) = -g(t+1) + \int_0^{t+1} f(s)ds + x_1(0), \quad (t \geq 0).$$

The system dynamics is not causal. The solution component x_2 depends on future values of the input functions f and g. This interesting phenomenon should be noted in addition to the well-known fact in the DAE theory that the solution may depend on derivatives of the input.

Example 3 ([11]) Consider the DDAE

$$\begin{bmatrix} 0 & 1 \\ 0 & 0 \end{bmatrix} \dot{x}(t) = \begin{bmatrix} 1 & 0 \\ 0 & -1 \end{bmatrix} x(t) + \begin{bmatrix} 0 & 0 \\ 1 & 0 \end{bmatrix} x(t-1).$$

Obtaining x_2 from the second equation and substituting the result into the first equation, we have the delay ODE $\dot{x}_1(t-1) = x_1(t)$, which is of advanced type. Thus, $x_1(t) = x_1^{(m)}(t-m)$ for $m-1 \leq t < m$, $m \in \mathbb{N}$. Therefore, the solution is discontinuous in general and cannot be extended on $[0, \infty)$ unless the initial function is infinitely often differentiable.

For the stability and robust stability analysis of linear systems in general, there are two standard approaches: the spectrum-based approach, which relies upon the roots of associated characteristic polynomials, and the Lyapunov function approach which is also known as Lyapunov's second method. For the purpose of deriving a formula for the stability radius, the spectrum-based approach is more preferable.

The outline of the paper is as follows. In Sect. 19.2 we briefly summarize important results on the robust stability and stability radii for linear delay ODEs. In Sect. 19.3 we survey some very recent results on the stability and robust stability of linear time-invariant delay DAEs. A discussion of some further related results and topics for future research close the paper.

Notation Throughout this paper, we denote by $\mathbb{C}^-$ the set of complex numbers with negative real part, by $i\mathbb{R}$ the set of purely imaginary numbers, by $I_n \in \mathbb{C}^{n,n}$ the identity matrix, by 0 the zero matrix of appropriate size. We also denote by $C(\mathbb{I},\mathbb{K}^n)$ the space of continuous functions, by $AC(\mathbb{I},\mathbb{K}^n)$ the space of absolutely continuous functions, and by $C^k_{pw}(\mathbb{I},\mathbb{K}^n)$ the space of k-times piecewise continuously differentiable functions from $\mathbb{I} \subset [0,\infty)$ to $\mathbb{K}^n$. Given a matrix $W \in \mathbb{C}^{n,n}$ with elements w_{ij}, $|W|$ denotes the nonnegative matrix in $\mathbb{R}^{n,n}$ with element $|w_{ij}|$. For two real matrices $P = (p_{ij})$ and $Q = (q_{ij})$ in $\mathbb{R}^{m,n}$, we write $A \leq B$ if $p_{ij} \leq q_{ij}$ for all $i = 1,2,\ldots,m$, and $j = 1,2,\ldots,n$.

19.2 Robust Stability for Delay ODEs

We consider the linear time-invariant differential equation with delay

$$\dot{x}(t) = Ax(t) + Dx(t-\tau), \quad t \geq 0, \tag{19.4}$$

where $A,\ D \in \mathbb{K}^{n,n}$ and the initial condition $x(t) = \phi(t)$ for all $t \in [-\tau, 0]$, where $\phi(t)$ is a given continuous function. It is established, e.g. see [29], that this initial value problem has the unique solution $x(t,\phi)$ defined on $[0,\infty)$.

Definition 1 The trivial solution of Eq. (19.4) is called

1. *Stable* if for every $\epsilon > 0$ there exist $\delta > 0$ such that for all ϕ with $\|\phi\|_\infty < \delta$, the solution $x(t,\phi)$ satisfies $\|x(t,\phi)\| < \epsilon$ for all $t \geq 0$;
2. *Asymptotically stable* if it is stable and there exist $\delta > 0$ such that $\|\phi\|_\infty < \delta$ implies $\lim_{t\to\infty} \|x(t,\phi)\| = 0$;
3. *Exponentially stable* if there exist $\delta > 0, L > 0$ and $\gamma > 0$ such that with $\|\phi\|_\infty < \delta$, the solution $x(t,\phi)$ satisfies the estimate $\|x(t,\phi)\| < Le^{-\gamma t}$ for all $t \geq 0$.

If the trivial solution of (19.4) is (asymptotically, exponentially) stable, then we also say that the equation is (asymptotically, exponentially) stable.

It is also known that for linear time-invariant systems, the asymptotic stability and the exponential stability are equivalent. Furthermore, the local asymptotic stability implies the global asymptotic stability. In general, these statements are not true for time-varying or nonlinear systems.

19.2.1 *Stability Radii*

Define

$$H(s) = sI - A - De^{-\tau s}, \qquad \sigma(H) = \{s \in \mathbb{C} : \det(H(s)) = 0\}.$$

It is well known that equation (19.4) is exponentially stable if and only if $\sigma(H) \subset \mathbb{C}^-$, see [29]. Assume that (19.4) is subject to structured perturbations of the form

$$\dot{x}(t) = \tilde{A}x(t) + \tilde{D}x(t-\tau), \tag{19.5}$$

with

$$A \rightsquigarrow \tilde{A} = A + B_1\Delta_1 C_1, \qquad D \rightsquigarrow \tilde{D} = D + B_2\Delta_2 C_2, \tag{19.6}$$

where $B_i \in \mathbb{K}^{n,l_i}$, and $C_i \in \mathbb{K}^{q_i,n}$, $i = 1, 2$, are given structure matrices and $\Delta_i \in \mathbb{K}^{l_i,q_i}$, $i = 0, 1$, are unknown disturbance matrices.

Using the abbreviation $\underline{A} = [A, D]$, these perturbations can be described as a block-diagonal perturbation

$$\underline{A} \rightsquigarrow \underline{\tilde{A}} = \underline{A} + B\Delta_b C,$$

where $B = [B_1, B_2]$, $C = \begin{bmatrix} C_1 & 0 \\ 0 & C_2 \end{bmatrix}$ and $\Delta_b = \begin{bmatrix} \Delta_1 & 0 \\ 0 & \Delta_2 \end{bmatrix}$. We endow the linear space of block-diagonal perturbations with the norm

$$\|\Delta_b\| = \|\Delta_1\| + \|\Delta_2\|.$$

If Eq. (19.4) is exponentially stable, then we define

$$\Xi_{\mathbb{K}} = \{\Delta_b, : \Delta_i \in \mathbb{K}^{l_i,q_i}, (19.5) \text{ is not exponentially stable}\},$$

where $\mathbb{K} = \mathbb{R}$ or $\mathbb{K} = \mathbb{C}$, and the structured stability radius of (19.4) with respect to perturbations of the form (19.6) is defined as

$$r_{\mathbb{K}}^b(\underline{A}) = \inf\{\|\Delta_b\| \mid \Delta_b \in \Xi_{\mathbb{K}}\}.$$

Depending on $\mathbb{K} = \mathbb{R}$ or $\mathbb{K} = \mathbb{C}$, we have the real/complex stability radius. Define the transfer functions

$$G_{ij}(s) = C_i H(s)^{-1} B_j, \qquad 1 \le i, j \le 2.$$

The following result is established in [54].

Theorem 1 *Suppose that Eq.* (19.4) *is exponentially stable and subject to structured perturbations of the form* (19.6). *Then*

$$\frac{1}{\max\{\sup_{s\in i\mathbb{R}} \|G_{ij}(s)\| \mid 1 \le i, j \le 2\}} \le r_{\mathbb{C}}^b(\underline{A}) \le \frac{1}{\max\{\sup_{s\in i\mathbb{R}} \|G_{ii}(s)\| \mid 1 \le i \le 2\}}.$$

In particular, if $B_1 = B_2$ *(or* $C_1 = C_2$*) then*

$$r^b_{\mathbb{C}}(\underline{A}) = \frac{1}{\max\{\sup_{s\in i\mathbb{R}} \|G_{ii}(s)\| \mid 1 \le i \le 2\}}.$$

This result has also been extended for linear systems with multiple delays, and in a more general case, for linear functional differential equations with distribution delay of the form

$$\dot{x}(t) = Ax(t) + \int_{-\tau}^{0} d[\eta(\theta)]x(t+\theta), \tag{19.7}$$

where $\eta(\cdot)$ is a matrix-valued function of locally bounded variation defined on $[-\tau, 0]$, see [54, 57, 58].

Now, we consider the perturbed equation (19.5) subject to structured perturbations of the full-block form

$$\underline{\tilde{A}} = \underline{A} + B\Delta C, \tag{19.8}$$

where $B \in \mathbb{K}^{n,l}$, $C \in \mathbb{K}^{q,2n}$ are given structure matrices and $\Delta \in \mathbb{K}^{l,q}$ is an uncertain perturbation matrix. Introducing the set

$$\Delta_{\mathbb{K}} = \left\{\Delta \in \mathbb{K}^{l,q} \ : \ (19.5) \text{ is not exponentially stable}\right\}.$$

Then, the structured stability radius of (19.4) with respect to structured perturbations of the form (19.8) is defined via

$$r^{B,C}_{\mathbb{K}}(\underline{A}) = \{\|\Delta\| \mid \Delta \in \Delta_{\mathbb{K}}\}, \tag{19.9}$$

where $\|\cdot\|$ is an operator norm induced by a vector norm. Define

$$L(s) := \begin{bmatrix} I_n \\ e^{-\tau s} I_n \end{bmatrix}, \ C(s) := CL(s),$$

and the transfer function $G(s) = C(s)H(s)^{-1}B$. The following result is analogous to that for linear time-invariant ODEs of [31].

Theorem 2 *Suppose that Eq.* (19.4) *is exponentially stable and subject to structured perturbations of the form* (19.8). *Then, the complex stability radius of* (19.4) *is given by*

$$r^{B,C}_{\mathbb{C}}(\underline{A}) = \frac{1}{\sup_{s\in i\mathbb{R}} \|G(s)\|}.$$

We note that the proof can also be done by using the structured distance to non-surjectivity in [59, 60] and the same techniques as in [19, 31].

Unlike for the complex stability radius, a general formula for the real stability radius measured by an arbitrary matrix norm is not available. However, if we consider the Euclidean norm, then a computable formula for the real stability radius can be established. For a matrix $F \in \mathbb{K}^{q,l}$, the real structured singular value of F is defined by

$$\mu_{\mathbb{R}}(F) := (\inf\{\|\Delta\|_2 : \Delta \in \mathbb{R}^{l,q}, \text{ and } \det(I_l + \Delta F) = 0\})^{-1},$$

and it has been shown in [50] that the real structured singular value of M is given by

$$\mu_{\mathbb{R}}(F) = \inf_{\gamma \in (0,1]} \sigma_2 \begin{bmatrix} \mathrm{Re} F & -\gamma \,\mathrm{Im}\, F \\ \frac{1}{\gamma} \mathrm{Im}\, F & \mathrm{Re} F \end{bmatrix},$$

where $\sigma_2(P)$ denotes the second largest singular value of the matrix P.

Using this result, we obtain a formula for the real stability radius, see also [36].

Theorem 3 *Suppose that Eq.* (19.4) *is exponentially stable and subject to structured perturbations of the form* (19.8). *Then, the real stability radius of* (19.4) *(with respect to the Euclidean norm) is given by the formula*

$$r_{\mathbb{R}}^{B,C}(\underline{A}) = \left(\sup_{s \in i\mathbb{R}} \inf_{\gamma \in (0,1]} \sigma_2 \begin{bmatrix} \mathrm{Re} G(s) & -\gamma \,\mathrm{Im}\, G(s) \\ \frac{1}{\gamma} \mathrm{Im}\, G(s) & \mathrm{Re} G(s) \end{bmatrix} \right)^{-1}. \tag{19.10}$$

Remark 1 The formula of real stability radius has been extended in [36] to linear systems of neutral type under structured perturbations of the form (19.6), as well. However, one must be careful since the spectral condition $\sigma(H) \in \mathbb{C}^-$ does not necessarily imply the asymptotic stability of a neutral delay equation. A similar formula of the real stability radius has also been proven for linear time-invariant DAEs, see [17]. However, in the latter case, we must take the supremum on $\mathbb{C} \setminus \mathbb{C}^-$ instead of $i\mathbb{R}$ due to the singular nature of the system.

19.2.2 Positive Delay Systems

In the previous section, we see that the formula of a real stability radius is more sophisticated. Moreover, by the definition, it is easy to see that

$$r_{\mathbb{C}}^b(\underline{A}) \le r_{\mathbb{R}}^b(\underline{A}) \text{ and } r_{\mathbb{C}}^{B,C}(\underline{A}) \le r_{\mathbb{R}}^{B,C}(\underline{A}).$$

Therefore, it is a natural question to study when the real and complex stability radius are equal. The answer has been given in the case of positive systems, see [34, 35, 56, 58]. Now, we consider only systems with real coefficients and real initial functions.

Definition 2 Equation (19.4) is called positive if for any nonnegative continuous initial condition ϕ, i.e., $\phi(t) \geq 0$, $-\tau \leq t \leq 0$, then the corresponding solution $x(t, \phi)$ satisfies $x(t, \phi) \geq 0$ for all $t \geq 0$.

It is well known that Eq. (19.4) is positive if and only if A is a Metzler matrix and D is a positive matrix, see e.g. [56, 58]. We recall that a Metzler matrix is a matrix in which all the off-diagonal components are nonnegative. The following result is obtained in [56].

Theorem 4 *Assume that Eq.* (19.4) *is positive, exponentially stable and subject to structured perturbations of the form* (19.6) *with* $B_i \geq 0,\ C_i \geq 0, i = 1, 2$. *Then*

$$\frac{1}{\max\{\|G_{ij}(0)\| \mid 1 \leq i, j \leq 2\}} \leq r^b_{\mathbb{C}}(\underline{A}) = r^b_{\mathbb{R}}(\underline{A}) \leq \frac{1}{\max\{\|G_{ii}(0)\| \mid 1 \leq i \leq 2\}}.$$

In particular, if $D_1 = D_2$ *(or* $E_1 = E_2$*) then*

$$r^b_{\mathbb{C}}(\underline{A}) = r^b_{\mathbb{R}}(\underline{A}) = \frac{1}{\max\{\|G_{ii}(0)\| \mid 1 \leq i \leq 2\}}.$$

Remark 2 As a consequence, if Eq. (19.4) is positive and the structure matrices B_i, C_i are positive, then the real stability radius can be easily computed.

By extending this result, the authors of [58] have also showed that for the positive equation (19.7), the real and complex stability radius are equal and easy to calculate. By a similar proof as in [56], the following result is obtained.

Theorem 5 *Assume that Eq.* (19.4) *is positive, exponentially stable and subject to structured perturbations of the form* (19.8) *with* $B \geq 0,\ C \geq 0$. *Then, the complex stability radius of* (19.4) *is given by*

$$r^{B,C}_{\mathbb{C}}(\underline{A}) = r^{B,C}_{\mathbb{R}}(\underline{A}) = \frac{1}{\|G(0)\|}. \tag{19.11}$$

Example 4 Consider a positive linear delay equation in $\mathbb{R}^2$ described by $\dot{x}(t) = Ax(t) + Dx(t-1),\ t \geq 0$, where $A = \begin{bmatrix} -1 & 0 \\ 0 & -1 \end{bmatrix}$, $D = \begin{bmatrix} 0 & 1 \\ 0 & 0 \end{bmatrix}$. Then, the characteristic equation of the system is

$$\det(H(s)) = \det \begin{bmatrix} -1-s & e^{-s} \\ 0 & -1-s \end{bmatrix} = (1+s)^2 = 0,$$

and hence the equation is exponentially stable. Assume that the system is subject to structured perturbations of the form

$$[\tilde{A}, \tilde{D}] = \begin{bmatrix} -1+\delta_1 & \delta_1 & 0 & 1+\delta_2 \\ \delta_1 & \delta_1 & 0 & \delta_2 \end{bmatrix} = [A, D] + B\Delta C,$$

with

$$B = \begin{bmatrix} 1 \\ 1 \end{bmatrix}, \; C = \begin{bmatrix} 1 & 1 & 0 & 0 \\ 0 & 0 & 0 & 1 \end{bmatrix}, \; \Delta = [\delta_1 \; \delta_2],$$

and disturbance parameters $\delta_1, \delta_2 \in \mathbb{R}$ or $\mathbb{C}$. By simple algebraic manipulations we imply $G(0) = C(0)H(0)^{-1}B = \begin{bmatrix} -3 \\ -1 \end{bmatrix}$. Thus, if $\mathbb{C}^2$ is endowed with the 2-norm, then by (19.11), we obtain

$$r_{\mathbb{C}}^{B,C}(\underline{A}) = r_{\mathbb{R}}^{B,C}(\underline{A}) = \frac{1}{\sqrt{10}}.$$

19.2.3 Extension to Linear Time-Varying Systems with Delay

Consider the linear time-varying systems with delay

$$\dot{x}(t) = \bar{A}(t)x(t) + \bar{D}(t)x(t-\tau), \tag{19.12}$$

where $\bar{A}(\cdot), \bar{D}(\cdot) \in C([0,\infty), \mathbb{K}^{n,n})$, and the initial condition $x(t) = \phi(t)$ for all $t \in [-\tau, 0]$. Suppose that

$$M(\bar{A}(t)) \le A, \quad |\bar{D}(t)| \le D, \quad \text{for all } t \ge 0. \tag{19.13}$$

Here $M(\bar{A}(t)) = \big(m_{ij}(t)\big)$ denotes the Metzler matrix defined by

$$m_{ij}(t) = \begin{cases} |\bar{a}_{ij}(t)|, \; i \ne j, \\ \bar{a}_{ij}(t), \; i = j, \end{cases} \quad (1 \le i, \; j \le n).$$

If Eq. (19.4) is exponentially stable then Eq. (19.12) is exponentially stable, see [47]. Assume that (19.12) is subject to structured perturbations of the form

$$\dot{x}(t) = \tilde{A}(t)x(t) + \tilde{D}(t)x(t-\tau), \tag{19.14}$$

with

$$\tilde{A}(t) = A(t) + B_1(t)\Delta_1(t)C_1(t), \qquad \tilde{D}(t) = D(t) + B_2(t)\Delta_2(t)C_1(t), \tag{19.15}$$

where $B_i(t) \in \mathbb{K}^{n,l_i}, C_i(t) \in \mathbb{K}^{q_i,n}$, $i = 1, 2$, are given structure matrices and $\Delta_i(t) \in \mathbb{K}^{l_i,q_i}$, $i = 0, 1$ are unknown disturbance matrices such that

$$|B_i(t)| \le B_i, \quad |C_i(t)| \le C_i, \quad |\Delta_i(t)| \le \Delta_i \quad \text{for all } t \ge 0, i = 1, 2. \tag{19.16}$$

The following result is established recently in [48].

Theorem 6 *Assume that Eq.* (19.4) *is exponentially stable and Eq.* (19.12) *is subject to structured perturbations of the form* (19.15). *Then, if*

$$\|\Delta_1\| + \|\Delta_2\| < \frac{1}{\max\{\|G_{ij}(0)\| \mid 1 \le i, j \le 2\}}$$

then the perturbed equation (19.14) *is exponentially stable.*

Remark 3 The conditions (19.13) and (19.16) are rather strict. This conditions restrict us to consider the robust stability of only a special class of linear time-varying delay differential equations under structured perturbations.

Recently, there have been also many other results for the robust stability of delay differential equations which are based on the Lyapunov-Krasovskii functional approach and the use of linear matrix inequalities (LMIs), see [1, 39, 52]. Unfortunately, the Lyapunov-based approach does not allow us to calculate the stability radii explicitly as the spectrum-based approach does.

19.3 Robust Stability of Linear Time-Invariant Delay DAEs

In this section we survey some new results on the stability and robust stability of linear time-invariant DDAEs of the form (19.1).

19.3.1 Preliminary Notions and Transformations

First, we define solutions of the initial value problem (19.1)–(19.2).

Definition 3 A function $x(\cdot, \phi) : [0, \infty) \to \mathbb{C}^n$ is called *solution* of the initial value problem (19.1)–(19.2), if $x \in AC([0, \infty), \mathbb{C}^n)$ and $x(\cdot, \phi)$ satisfies (19.1) almost everywhere. An initial function ϕ is called *consistent* with (19.1) if the associated initial value problem (19.1) has at least one solution.

System (19.1) is called solvable if for every consistent initial function ϕ, the associated initial value problem (19.1)–(19.2) has a solution. It is called regular if it is solvable and the solution is unique.

Note that instead of seeking solutions in $AC([0, \infty), \mathbb{C}^n)$, alternatively we often consider the space $C^1_{pw}([0, \infty), \mathbb{C}^n)$. In fact, Eq. (19.1) may not be satisfied at (countably many) points, which usually arise at multiples of the delay time τ.

Definition 4 System (19.1)–(19.2) is called exponentially stable if there exist constants $K > 0, \omega > 0$ such that

$$\|x(t,\phi)\| \leq Ke^{-\omega t}\|\phi\|_\infty$$

for all $t \geq 0$ and all consistent initial functions ϕ, where $\|\phi\|_\infty = \sup_{-\tau \geq t \geq 0} \|\phi(t)\|$.

The following notions are well known in the theory of DAEs, see [5, 21, 23]. A matrix pair (E, A), $E, A \in \mathbb{C}^{n,n}$ is called regular if there exists $s \in \mathbb{C}$ such that $\det(sE - A)$ is different from zero. Otherwise, if $\det(sE - A) = 0$ for all $s \in \mathbb{C}$, then we say that (E, A) is singular. If (E, A) is regular, then a complex number s is called a (generalized finite) eigenvalue of (E, A) if $\det(sE - A) = 0$. The set of all (finite) eigenvalues of (E, A) is called the (finite) spectrum of the pencil (E, A) and denoted by $\sigma(E, A)$. If E is singular and the pair is regular, then we say that (E, A) has the eigenvalue ∞.

Regular pairs (E, A) can be transformed to Weierstraß-Kronecker canonical form, i.e., there exist nonsingular matrices $W, T \in \mathbb{C}^{n,n}$ such that

$$E = W\begin{bmatrix} I_r & 0 \\ 0 & N \end{bmatrix}T^{-1}, \quad A = W\begin{bmatrix} J & 0 \\ 0 & I_{n-r} \end{bmatrix}T^{-1}, \tag{19.17}$$

where I_r, I_{n-r} are identity matrices of indicated size, $J \in \mathbb{C}^{r,r}$, and $N \in \mathbb{C}^{(n-r),(n-r)}$ are matrices in Jordan canonical form and N is nilpotent. If E is invertible, then $r = n$, i.e., the second diagonal block does not occur. If $r < n$ and N has nilpotency index $\nu \in \{1, 2, \ldots\}$, i.e., $N^\nu = 0$, $N^i \neq 0$ for $i = 1, 2, \ldots, \nu-1$, then ν is called the index of the pair (E, A) and we write $\operatorname{ind}(E, A) = \nu$. If $r = n$ then the pair has index $\nu = 0$.

For system (19.1) with a regular pair (E, A), the existence and uniqueness of solutions has been studied in [10–12] and for the general case in [25]. It follows from Corollary 4.12 in [25] that (19.1)–(19.2) has a unique solution if and only if the initial condition ϕ is consistent and $\det(sE - A - e^{-s\tau}D) \not\equiv 0$.

For a matrix triple $(E, A, D) \in \mathbb{C}^{n,n} \times \mathbb{C}^{n,n} \times \mathbb{C}^{n,n}$, there always exists a nonsingular matrix $W \in \mathbb{C}^{n,n}$ such that

$$W^{-1}E = \begin{bmatrix} E_1 \\ 0 \\ 0 \end{bmatrix}, \quad W^{-1}A = \begin{bmatrix} A_1 \\ A_2 \\ 0 \end{bmatrix}, \quad W^{-1}D = \begin{bmatrix} D_1 \\ D_2 \\ D_3 \end{bmatrix}, \tag{19.18}$$

where $E_1, A_1, D_1 \in \mathbb{C}^{d,n}$, $A_2, D_2 \in \mathbb{C}^{a,n}$, $D_3 \in \mathbb{C}^{h,n}$ with $d+a+h = n$, $\operatorname{rank} E_1 = \operatorname{rank} E = d$, and $\operatorname{rank} A_2 = a$. Then, system (19.1) can be scaled by W^{-1} to obtain

$$\begin{aligned} E_1\dot{x}(t) &= A_1x(t) + D_1x(t-\tau), \\ 0 &= A_2x(t) + D_2x(t-\tau), \\ 0 &= D_3x(t-\tau). \end{aligned} \tag{19.19}$$

In practice, such a scaling matrix W and the corresponding transformed coefficient matrices can be easily constructed by using the singular value decomposition (SVD) of matrices, see the detailed procedure in [18].

Following the concept of strangeness-index in [37] we make the following definition, see also [25].

Definition 5 Equation (19.1) is called strangeness-free if there exists a nonsingular matrix $W \in \mathbb{C}^{n,n}$ that transforms the triple (E, A, D) to the form (19.18) and

$$\operatorname{rank}\begin{bmatrix} E_1 \\ A_2 \\ D_3 \end{bmatrix} = n.$$

It is easy to show that, although the transformed form (19.18) is not unique (any nonsingular matrix that operates blocks-wise in the three block-rows can be applied), the strangeness-free property is invariant with respect to the choice of W. Furthermore, the class of strangeness-free DDAEs includes all equations with regular pair (E, A) of index at most 1, but excludes all equations all equations with regular pair (E, A) of higher index. So, strangeness-free DDAEs generalizes index-1 regular DDAEs which have been discussed so far in the literature [13, 20, 45]. We refer to [18] for more details.

19.3.2 Spectrum-Based Stability Analysis

The stability analysis is usually based on the eigenvalues of the nonlinear function

$$H(s) = sE - A - e^{-s\tau} D, \tag{19.20}$$

associated with the Laplace transform of Eq. (19.1), i.e., the roots of the characteristic function

$$p_H(s) := \det H(s). \tag{19.21}$$

Let us define the spectral set $\sigma(H) = \{s : p_H(s) = 0\}$ and the spectral abscissa $\alpha(H) = \sup\{\mathrm{Re}(s) : p_H(s) = 0\}$. For linear time-invariant DODEs, i.e., if $E = I_n$, the exponential stability is equivalent to $\alpha(H) < 0$, see [28] and the spectral set $\sigma(H)$ is bounded from the right. However, for linear time-invariant DDAEs, the spectral set $\sigma(H)$ may not be bounded on the right as Example 1.1 in [18] shows. In some special cases, [45, 64], it has been shown that the exponential stability of DDAEs is equivalent to the spectral condition that $\alpha(H) < 0$. In general, however this spectral condition is necessary, but not sufficient for the exponential stability, see Examples 1.2 and 3.6 in [18]. In other words, linear time-invariant DDAEs may not be exponentially stable although all roots of the characteristic function are in the open left half complex plane. To characterize when the roots of the characteristic function allow the classification of stability, certain extra structural restrictions on (19.1) are to be taken into consideration. In the following we present some necessary and sufficient conditions for exponential stability of (19.1), which extend the results of [45, 64].

First, we obtain that for strangeness-free systems the spectral condition characterizes exponential stability.

Theorem 7 ([18, Theorem 3.1]) *Suppose that Eq. (19.1) is strangeness-free. Then Eq. (19.1) is exponentially stable if and only if* $\alpha(H) < 0$.

As an immediate consequence, the stability criterion for index-1 DDAEs proven in [45] follows.

Corollary 1 *Consider the DDAE (19.1)–(19.2) with a regular pair* (E, A), $\operatorname{ind}(E, A) \leq 1$, *and its associated spectral function* H. *Then Eq. (19.1) is exponentially stable if and only if* $\alpha(H) < 0$.

Now we consider the case when the pair (E, A) in Eq. (19.1) is regular and it is transformed into the Weierstraß-Kronecker canonical form (19.17). Setting

$$W^{-1}DT = \begin{bmatrix} D_{11} & D_{12} \\ D_{21} & D_{22} \end{bmatrix}, \quad T^{-1}x(t) = \begin{bmatrix} x_1(t) \\ x_2(t) \end{bmatrix}, \quad T^{-1}\phi(t) = \begin{bmatrix} \phi_1(t) \\ \phi_2(t) \end{bmatrix}, \tag{19.22}$$

with $D_{11} \in \mathbb{C}^{r,r}$, $D_{12} \in \mathbb{C}^{r,n-r}$, $D_{21} \in \mathbb{C}^{n-r,r}$, $D_{22} \in \mathbb{C}^{n-r,n-r}$, and x_1, x_2, ϕ_1, ϕ_2 partitioned correspondingly. Then Eq. (19.1) is equivalent to the system

$$\begin{aligned} \dot{x}_1(t) &= A_{11}x_1(t) + D_{11}x_1(t-\tau) + D_{12}x_2(t-\tau), \\ N\dot{x}_2(t) &= x_2(t) + D_{21}x_1(t-\tau) + D_{22}x_2(t-\tau), \end{aligned} \tag{19.23}$$

with initial conditions

$$x_i(t) = \phi_i(t), \text{ for } t \in [-\tau, 0], \; i = 1, 2.$$

From the explicit solution formula for linear time-invariant DAEs, see [9, 37], the second equation of (19.23) implies that

$$x_2(t) = -D_{21}x_1(t-\tau) - D_{22}x_2(t-\tau) - \sum_{i=1}^{\nu-1} \left(N^i D_{21} x_1^{(i)}(t-\tau) + N^i D_{22} x_2^{(i)}(t-\tau) \right), \tag{19.24}$$

and for $t \in [0, \tau)$, we get

$$x_2(t) = -D_{21}\phi_1(t) - D_{22}\phi_2(t) - \sum_{i=1}^{\nu-1} \left(N^i D_{21} \phi_1^{(i)}(t-\tau) + N^i D_{22} \phi_2^{(i)}(t-\tau) \right). \tag{19.25}$$

It follows that ϕ needs to be differentiable at least ν times if the coefficients D_{21} and D_{22} do not satisfy further conditions. Extending this argument to $t \in [\tau, 2\tau)$, $[2\tau, 3\tau)$, etc., the solution cannot be extended to the full real half-line unless the initial function ϕ is infinitely often differentiable or the coefficient associated with the delay is highly structured.

Let us now consider exponential stability for the case that $\operatorname{ind}(E, A) > 1$. In order to avoid an infinite number of differentiations of ϕ induced by (19.25), it is reasonable to assume that for a system in Weierstraß–Kronecker form (19.17) with transformed matrices as in (19.22) the allowable delay conditions $ND_{2i} = 0$, $i = 1, 2$ hold. Note that this condition is obviously holds for the index-1 case since then $N = 0$.

Choose any fixed $\hat{s} \in \mathbb{C}$ such that $\det(\hat{s}E - A) \neq 0$ and set

$$\hat{E} = (\hat{s}E - A)^{-1}E, \quad \hat{D} = (\hat{s}E - A)^{-1}D.$$

Proposition 3.4 in [18] shows that the allowable delay conditions $ND_{21} = 0$ and $ND_{22} = 0$ are simultaneously satisfied if and only if

$$(I - \hat{E}^D\hat{E})\hat{E}\hat{D} = 0, \tag{19.26}$$

where $\hat{E}^D$ denotes the Drazin inverse of $\hat{E}$.

Using this characterization of the allowable delay condition, we have the following characterization of exponential stability for DDAEs with regular pair (E, A) of arbitrary index.

Theorem 8 ([18, Theorem 3.5]) *Consider the DDAE (19.1)–(19.2) with a regular pair (E, A) satisfying (19.26). Then Eq.* (19.1) *is exponentially stable if and only if $\alpha(H) < 0$.*

We note that although the spectral criterion $\alpha(H) < 0$ was used for characterizing the exponential stability of DDAEs with regular pair (E, A) of arbitrary index in [13, 63, 67], a rigorous proof had not been available prior to the appearance [18].

We have seen that the spectral criterion $\alpha(H) < 0$ is necessary for the exponential stability of (19.1), but in general it is not sufficient. Introducing further restrictions on the delay term, we get that exponential stability is equivalent to the spectral criterion.

In general, the exponential stability of (19.1) depends on the delay parameter τ. If the exponential stability of (19.1) holds for all $\tau \geq 0$, we say that the DDAE (19.1) is delay-independently exponentially stable. Otherwise, the exponential stability of (19.1) is said to be delay-dependent. The following sufficient criterion was established in [13], which improved the one earlier given in [67], and later it was further developed in [63]. The following theorem for the delay-independent exponential stability of (19.1) is a special case of Theorem 1 in [13].

Theorem 9 *Suppose that (E, A) is regular and that*

$$\sigma(E, A) \in \mathbb{C}^- \text{ and } \sup_{\operatorname{Re}(s)\geq 0} \rho\left(|(sE - A)^{-1}D|\right) < 1. \tag{19.27}$$

Then the DDAE (19.1) is exponential stable for all values of the delay τ, i.e., the exponential stability of (19.1) is delay-independent.

By using the theory of nonnegativ matrices, practically checkable algebraic criteria were also derived from (19.27) for the delay-independent exponential stability of (19.1) in [13]. We note that Theorem 9 cannot apply directly to strangeness-free DDAEs, since the pair (E, A) of a strangeness-free DDAE may be singular. However, by shifting the argument of the third equation of (19.19), the result of Theorem 9 can be easily adapted to strangeness-free DDAEs. Finally, we emphasize a numerical consequence of delay-independent exponential stability. In [13, 63, 67], it is shown that for delay-independently exponentially stable DDAEs, numerical solutions by certain classes of A-stable discretization methods preserve the exponential stability.

19.3.3 Stability Radii

With the characterization of exponential stability at hand we also study the question of robust stability for linear time-invariant DDAEs, i.e., we discuss the structured stability radius of maximal perturbations that are allowed to the coefficients so that the system keeps its exponential stability. These results, which are presented in details in [18, Section 4], extend previous results on DODEs and DAEs in [7, 8, 17, 19, 36, 55].

Suppose that system (19.1) is exponentially stable and consider a perturbed system

$$(E + B_1\Delta_1 C)\dot{x}(t) = (A + B_2\Delta_2 C)x(t) + (D + B_3\Delta_3 C)x(t-\tau), \qquad (19.28)$$

where $\Delta_i \in \mathbb{C}^{p_i,q}$, $i = 1,2,3$ are perturbations and $B_i \in \mathbb{C}^{n,p_i}$, $i = 1,2,3$, $C \in \mathbb{C}^{q,n}$, are matrices that restrict the structure of the perturbations. We could also consider different matrices C_i in each of the coefficients but for simplicity, see Remark 6 below, we assume that the column structure in the perturbations is the same for all coefficients. Set

$$\Delta = \begin{bmatrix} \Delta_1 \\ \Delta_2 \\ \Delta_3 \end{bmatrix}, \quad B = \begin{bmatrix} B_1 & B_2 & B_3 \end{bmatrix}, \qquad (19.29)$$

and $p = p_1 + p_2 + p_3$ and consider the set of destabilizing perturbations

$$\mathscr{V}_{\mathbb{C}}(E, A, D; B, C) = \{\Delta \in \mathbb{C}^{p \times q} : \text{(19.28) is not exponentially stable}\}.$$

Then we define the structured complex stability radius of (19.1) subject to structured perturbations as in (19.28) as

$$r_{\mathbb{C}}(E, A, D; B, C) = \inf\{\|\Delta\| : \Delta \in \mathscr{V}_{\mathbb{C}}(E, A, D; B, C)\},$$

where $\|\cdot\|$ is a matrix norm induced by a vector norm. If only real perturbations Δ are considered, then we use the term structured real stability radius but here we focus on the complex stability radius.

With H as in (19.20), we introduce the transfer functions

$$G_1(s) = -sCH(s)^{-1}B_1,\ G_2(s) = CH(s)^{-1}B_2,\ G_3(s) = e^{-s\tau}CH(s)^{-1}B_3,$$

and with

$$G(s) = \left[G_1(s)\ G_2(s)\ G_3(s)\right], \tag{19.30}$$

we first obtain an upper bound for the structured complex stability radius.

Theorem 10 ([18, Theorem 4.1]) *Suppose that system* (19.1) *is exponentially stable. Then the structured stability radius of* (19.1) *subject to structured perturbations as in* (19.28) *satisfies the inequality*

$$r_{\mathbb{C}}(E, A, D; B, C) \leq \left(\sup_{\mathrm{Re}s\geq 0} \|G(s)\|\right)^{-1}.$$

For every perturbation Δ as in (19.29) we define

$$H_\Delta(s) = s(E + B_1\Delta_1C) - (A + B_2\Delta_2C) - e^{-s\tau}(D + B_3\Delta_3C).$$

and have the following proposition.

Proposition 1 ([18, Proposition 4.2]) *Consider system* (19.1) *and the perturbed system* (19.28). *If the associated spectral abscissa satisfy* $\alpha(H) < 0$ *and* $\alpha(H_\Delta) \geq 0$, *then we have*

$$\|\Delta\| \geq \left(\sup_{\mathrm{Re}s\geq 0} \|G(s)\|\right)^{-1}.$$

It is already known for the case of perturbed non-delay DAEs [8], see also [17], that it is necessary to restrict the perturbations in order to get a meaningful concept of the structured stability radius, since a DAE system may lose its regularity and/or stability under infinitesimal perturbations. We therefore introduce the following set of allowable perturbations.

Definition 6 Consider a strangeness-free system (19.1) and let $W \in \mathbb{C}^{n,n}$ be such that (19.18) holds. A structured perturbation as in (19.28) is called allowable if (19.28) is still strangeness-free with the same triple (d, a, h), i.e., there exists a nonsingular $\tilde{W} \in \mathbb{C}^{n,n}$ such that

$$\tilde{W}^{-1}(E + B_1\Delta_1 C) = \begin{bmatrix} \tilde{E}_1 \\ 0 \\ 0 \end{bmatrix}, \ \tilde{W}^{-1}(A + B_2\Delta_2 C) = \begin{bmatrix} \tilde{A}_1 \\ \tilde{A}_2 \\ 0 \end{bmatrix},$$

$$\tilde{W}^{-1}(D + B_3\Delta_3 C) = \begin{bmatrix} \tilde{D}_1 \\ \tilde{D}_2 \\ \tilde{D}_3 \end{bmatrix}, \qquad (19.31)$$

where $\tilde{E}_1, \tilde{A}_1, \tilde{D}_1 \in \mathbb{C}^{d,n}$, $\tilde{A}_2, \tilde{D}_2 \in \mathbb{C}^{a,n}$, $\tilde{D}_3 \in \mathbb{C}^{h,n}$, such that

$$\begin{bmatrix} \tilde{E}_1 \\ \tilde{A}_2 \\ \tilde{D}_3 \end{bmatrix}$$

is invertible.

Assume that the matrices B_i, $i = 1, 2, 3$, that are restricting the structure have the form

$$W^{-1}B_1 = \begin{bmatrix} B_{11} \\ B_{12} \\ B_{13} \end{bmatrix}, \ W^{-1}B_2 = \begin{bmatrix} B_{21} \\ B_{22} \\ B_{23} \end{bmatrix}, \ W^{-1}B_3 = \begin{bmatrix} B_{31} \\ B_{32} \\ B_{33} \end{bmatrix},$$

where $B_{j1} \in \mathbb{C}^{d,p_j}$, $B_{2j} \in \mathbb{C}^{a,p_j}$, and $B_{3,j} \in \mathbb{C}^{h,p_j}$, $j = 1, 2, 3$. According to [8, Lemma 3.3], if the structured perturbation is allowable then $B_{12}\Delta_1 C = 0$, $B_{13}\Delta_1 C = 0$, and $B_{23}\Delta_2 C = 0$. Thus, without loss of generality, we assume that

$$B_{12} = 0, \ B_{13} = 0, \text{ and } B_{23} = 0. \qquad (19.32)$$

Note that it can be shown that the condition (19.32) is invariant with respect to the choice of the transformation matrix W. Furthermore, it is easy to see that with all structured perturbations with B_i, $i = 1, 2, 3$, satisfying (19.32), if the perturbation Δ is sufficiently small, then the strangeness-free property is preserved with the same sizes of the blocks.

We denote the infimum of the norm of all perturbations Δ such that (19.28) is no longer strangeness-free or the sizes of the blocks d, a, h change, by $d^s_{\mathbb{C}}(E, A, D; B, C)$, and immediately have the following proposition.

Proposition 2 ([18, Proposition 4.4]) *Suppose that Eq.* (19.1) *is strangeness-free and subject to structured perturbations with* B_i, $i = 1, 2, 3$ *satisfying* (19.32). *Then*

$$d_{\mathbb{C}}^{s}(E, A, D; B, C) = \left\| C \begin{bmatrix} E_1 \\ A_2 \\ D_3 \end{bmatrix}^{-1} \begin{bmatrix} B_{11} & 0 & 0 \\ 0 & B_{22} & 0 \\ 0 & 0 & B_{33} \end{bmatrix} \right\|^{-1}.$$

Remark 4 It is not difficult to show that in fact the formula in Proposition 2 is independent of the choice of the transformation matrix W.

Proposition 3 ([18, Proposition 4.6]) *Consider system* (19.1) *with* $\alpha(H) < 0$. *If the system is strangeness-free and subject to structured perturbations as in* (19.28) *with structure matrices* B_1, B_2, B_3 *satisfying* (19.32) *and if the perturbation* Δ *satisfies*

$$\|\Delta\| < \left(\sup_{\mathrm{Re} s \geq 0} \|G(s)\| \right)^{-1},$$

then the structured perturbation is allowable, i.e., the perturbed equation (19.28) *is strangeness-free with the same block-sizes* d, a, *and* h.

We combine Theorem 7 and Propositions 1–3 to formulate the complex stability radius for strangeness-free DDAEs under suitable structured perturbations.

Theorem 11 ([18, Theorem 4.7]) *Suppose that Eq.* (19.1) *is exponentially stable and strangeness-free and subject to structured perturbations as in* (19.28) *with structure matrices* B_1, B_2, B_3 *satisfying* (19.32). *Then*

$$r_{\mathbb{C}}(E, A, D; B, C) = \left(\sup_{\mathrm{Re} s \geq 0} \|G(s)\| \right)^{-1}.$$

Furthermore, if $\|\Delta\| < r_{\mathbb{C}}(E, A, D; B, C)$ *then* (19.28) *is strangeness-free with the same blocksizes* d, a, *and* h *as for (19.1).*

Remark 5 By the maximum principle [41], the supremum of $G(s)$ over the right-half plane is attained at a finite point on the imaginary axis or at infinity. For strangeness-free DDAEs, it can be shown that it suffices to take the supremum of $\|G(s)\|$ over the imaginary axis instead of the whole right-half plane, i.e., we have

$$r_{\mathbb{C}}(E, A, D; B, C) = \left(\sup_{\mathrm{Re} s = 0} \|G(s)\| \right)^{-1},$$

see [18, Lemma A.1].

Remark 6 Perturbed systems of the form (19.28) represent a subclass of the class of systems with more general structured perturbations

$$(E + B_1\Delta_1 C_1)\dot{x}(t) = (A + B_2\Delta_2 C_2)x(t) + (D + B_3\Delta_3 C_3)x(t-\tau), \qquad (19.33)$$

where $\Delta_i \in \mathbb{C}^{p_i,q_i}$, $i = 1, 2, 3$, are perturbations, $B_i \in \mathbb{C}^{n,p_i}$ and $C_i \in \mathbb{C}^{q_i,n}$, $i = 1, 2, 3$, are allowed to be different matrices. One may formulate a structured stability radius subject to (19.33), as well, but an exact formula for it could not be expected as in the case of (19.28). For another special case that $B_1 = B_2 = B_3 = B$ and C_i are different, an analogous formulation and similar results for the structured stability radius can be obtained, cf. Theorem 1. However, due to the special row-structure of the strangeness-free form and of allowable perturbations, the consideration of perturbed systems of the form (19.28) is more reasonable. If $E = I$ and no perturbation is allowed in the leading term, then the formula of the complex stability radius by Theorem 11 reduces to that of Theorem 1 by using an appropriate matrix norm.

As a corollary we obtain the corresponding result for a special case of strangeness-free systems where already the pair (E, A) is regular with $\operatorname{ind}(E, A) \le 1$.

Corollary 2 *Consider system* (19.1) *with a regular pair* (E, A) *satisfying* $\operatorname{ind}(E, A) \le 1$ *and suppose that the system is exponentially stable and has Weierstraß-Kronecker canonical form* (19.17)*. If the system is subject to structured perturbations as in* (19.28)*, where the structure matrix* B_1 *satisfies*

$$W^{-1}B_1 = \begin{bmatrix} B_{11} \\ 0 \end{bmatrix},$$

with $B_{11} \in \mathbb{C}^{d\times p_1}$*, then the structured stability radius is given by*

$$r_{\mathbb{C}}(E, A, D; B, C) = \left(\sup_{\operatorname{Re} s = 0} \|G(s)\| \right)^{-1}.$$

For non-delayed DAEs it has been shown [17] that if the perturbation is such that the nilpotent structure in the Weierstraß-Kronecker canonical form is preserved, then one can also characterize the structured stability radius in the case that the pair (E, A) is regular and $\operatorname{ind}(E, A) > 1$.

We have seen that exponential stability is characterized by the spectrum of H if we assume that the allowable delay conditions $ND_{21} = 0$ and $ND_{22} = 0$ hold. In the following we assume that this property is preserved and that in the perturbed equation (19.28), the structure matrices B_1, B_2, B_3 satisfy

$$W^{-1}B_1 = \begin{bmatrix} B_{11} \\ 0 \end{bmatrix}, \; W^{-1}B_2 = \begin{bmatrix} B_{21} \\ 0 \end{bmatrix}, \; W^{-1}B_3 = \begin{bmatrix} B_{31} \\ B_{32} \end{bmatrix}, \; NB_{32} = 0, \qquad (19.34)$$

where $B_{j,1} \in \mathbb{C}^{d,p_j}$, $j = 1, 2, 3$, $B_{32} \in \mathbb{C}^{n-d,p_3}$, and $W \in \mathbb{C}^{n,n}$, $N \in \mathbb{C}^{n-d,n-d}$ are as in (19.17). In the following we consider structured perturbations that do not alter the nilpotent structure of the Kronecker form (19.17) of (E, A), i.e., the nilpotent matrix N and the corresponding left invariant subspace associated with eigenvalue ∞ is preserved, see [8] for the case that $\operatorname{ind}(E, A) = 1$ and $D = 0$.

Similar to the approach in [8], we now introduce the distance to the nearest pair with a different nilpotent structure

$$d^n_{\mathbb{C}}(E, A, D; B, C) = \inf\{\|\Delta\| : \text{ (19.28) does not preserve the nilpotent structure}\}.$$

Under assumption (19.34), we obtain the following result, see [17] for the case of non-delay DAEs.

Proposition 4 ([18, Proposition 4.11]) *Consider Eq.* (19.1) *with regular* (E, A) *and* $\operatorname{ind}(E, A) > 1$, *subject to transformed perturbations satisfying* (19.34). *Let us decompose* $CT = \begin{bmatrix} C_{11} & C_{12} \end{bmatrix}$ *with* $C_{11} \in \mathbb{C}^{q,r}$, $C_{12} \in \mathbb{C}^{q,n-r}$. *Then the distance to the nearest system with a different nilpotent structure is given by*

$$d^n_{\mathbb{C}}(E, A, D; B, C) = \|C_{11} B_{11}\|^{-1}.$$

Remark 7 By their definition, the blocks B_{11} and C_{11} depend on the transformation matrices W^{-1} and T, respectively. It is known that the Weierstraß-Kronecker canonical form (19.17) is not unique. However, [37, Lemma 2.10] implies that neither the product $C_{11} B_{11}$ nor the condition (19.34) depends on the choice of pair (W, T). Thus, the distance formula for $d^n_{\mathbb{C}}(E, A, D; B, C)$ obtained in Proposition 4 is indeed independent of the choice of the transformations.

We obtain an analogous formula for the complex stability radius of Eq. (19.1) with regular higher-index pair (E, A).

Theorem 12 ([18, Theorem 4.13]) *Consider an exponentially stable equation* (19.1) *with regular pair* (E, A) *and* $\operatorname{ind}(E, A) > 1$ *and assume that Eq.* (19.1) *is subject to transformed perturbations satisfying* (19.34). *Then the stability radius is given by the formula*

$$r_{\mathbb{C}}(E, A, D; B, C) = \left(\sup_{\operatorname{Re} s = 0} \|G(s)\| \right)^{-1}.$$

Moreover, if $\|\Delta\| < r_{\mathbb{C}}(E, A, D; B, C)$, *then the perturbed equation* (19.28) *has a regular pair* $(E + B_1 \Delta_1 C, A + B_2 \Delta_2 C)$ *with the same nilpotent structure in the Kronecker canonical form and it is exponentially stable.*

For an illustration of the results, see [18, Example 4.14]. Finally, we note that a similar robust stability problem can be formulated for linear time-invariant DAEs with multiple delays subject to multiple/affine perturbations and a formula of the complex stability radius can be obtained analogously. However, the problem of extending the formula of the real stability radius from delay ODEs and nondelay DAEs to delay DAEs is not solved yet.

19.3.4 Related and Other Results

In this part we discuss a related and some other robust stability results for DDAEs. The first result on the stability radius for DDAEs was given in [42], where the robust stability of a special class of problems is investigated. The so-called singularly perturbed system (SPS) of functional differential equations (FDEs)

$$\begin{aligned} \dot{x}(t) &= L_{11}x_t + L_{12}y_t \\ \varepsilon\dot{y}(t) &= L_{21}x_t + L_{22}y_t \end{aligned} \tag{19.35}$$

is considered, where $x \in \mathbb{C}^{n_1},\ y \in \mathbb{C}^{n_2},\ \varepsilon > 0$ is a small parameter;

$$\begin{aligned} L_{j1}x_t &= \sum_{i=0}^{l} A^i_{j1}x(t-\tau_i) + \int_{-\tau_l}^{0} D_{j1}(\theta)x(t+\theta)d\theta \\ L_{j2}y_t &= \sum_{k=0}^{m} A^k_{j2}y(t-\varepsilon\mu_k) + \int_{-\mu_m}^{0} D_{j2}(\theta)y(t+\varepsilon\theta)d\theta \end{aligned} \tag{19.36}$$

$j = 1, 2,\ A^i_{jk}$ are constant matrices of appropriate dimensions, $D_{jk}(.)$ are integrable matrix-valued functions, and $0 \le \tau_0 \le \tau_1 \le \ldots \le \tau_p,\ 0 \le \mu_0 \le \mu_1 \le \ldots \le \mu_m$.

A number of problems arising in science and engineering can be modeled by SPS-s of differential equations with delay, e.g., see [22] and the references cited therein. Setting $\varepsilon = 0$, the system (19.35)–(19.36) reduced to be a DAE with multiple and distributional delays.

In [42] the system (19.35)–(19.36) with coefficients subject to structured perturbations is investigated and the complex stability radius is defined in the standard way. For a nonzero ε, one may multiply both sides of the second equation in (19.35) by ε^{-1} and then obtains a DODE. Then, by using a stability criterion obtained in [14] and applying [57, Theorem 3.3], a formula of the complex stability radius is obtained without difficulty. However, in practice, this formulation is less useful because the appearance of small ε may make the computation of the stability radius ill-posed. Therefore, the asymptotic behavior of the complex stability radius of (19.35)–(19.36) as ε tends to zero is of interest. By using a similar approach as those in [15, 16], the main result of [42] is that the stability radius of the singularly perturbed system converges to the minimum of the stability radii of the reduced DAE system and of a fast boundary-layer subsystem, which is constructed by an appropriate time-scaling.

The stability and the robust stability of DDAEs can be investigated by the Lyapunov-based approach as well. By this approach, sufficient stability and robust stability criteria are usually formulated in term of linear matrix inequalities (LMIs), which can be solved numerically by computer software. In this direction, a number of results have been obtained, for example, see [20, 46, 49, 61, 65] and references therein. An advantage of the Lyapunov-Krasovskii functional method is that it can be extended to the robust stability analysis of linear systems with time-varying delays, see [1, 39, 52]. However, explicit formulas of the stability radii cannot be obtained by this approach. That is why a detailed discussion of the Lyapunov-based approach to the robust stability of DDAEs is out of the scope of this survey paper.

Stability results for linear time-varying and nonlinear DDAEs are rather limited. Asymptotic stability of Hessenberg DDAEs with index up to 3 was investigated via the direct linearization in [68]. In [44], Floquet theory was used to check the local stability of periodic solutions for semi-explicit DDAEs of index at most 2. Just very recently, a careful and complete analysis of general linear time-varying DDAEs was presented in [26]. Based on a regularization procedure, existence and uniqueness of solutions and other structural properties of DDAEs such as consistency of initial functions and smoothness requirements are analyzed. Combining with the recent spectrum-based stability results for linear time-varying DODEs [6] and for linear time-varying DAEs [43], the theory initiated in [26] would give a promisingly efficient approach to the stability and the robust stability of linear time-varying and nonlinear DDAEs.

19.4 Discussion

In this paper we have surveyed recent results on the robustness of exponential stability for linear DODEs and DDAEs. We have seen that, while for linear DODEs, most of the robustness and distance problems are well understood, many problems for DDAEs such as a characterization of the real stability radius for linear time-invariant DDAEs and for positive systems as well as the robust stability of linear time-varying DDAEs are still open. These problems are challenging and interesting research works in the future.

Acknowledgements This work is supported by Vietnam National Foundation for Science and Technology Development (NAFOSTED). The authors thank Daniel Kressner for his useful comments that led to the improved presentation of the paper.

References

1. Anh, T.T., Hien, L.V., Phat, V.N.: Stability analysis for linear non-autonomous systems with continuously distributed multiple time-varying delays and applications. Acta Math. Vietnam. **36**, 129–143 (2011)
2. Ascher, U.M., Petzold, L.R.: The numerical solution of delay-differential algebraic equations of retarded and neutral type. SIAM J. Numer. Anal. **32**, 1635–1657 (1995)
3. Baker, C.T.H., Paul, C.A.H., Tian, H.: Differential algebraic equations with after-effect. J. Comput. Appl. Math. **140**, 63–80 (2002)

4. Bellen, A., Zennaro, M.: Numerical Methods for Delay Differential Equations. Oxford University Press, Oxford (2003)
5. Brenan, K.E., Campbell, S.L., Petzold, L.R.: Numerical Solution of Initial-Value Problems in Differential Algebraic Equations, 2nd edn. SIAM, Philadelphia (1996)
6. Breda, D., Van Vleck, E.: Approximating Lyapunov exponents and Sacker-Sell spectrum for retarded functional differential equations. Numer. Math. **126**, 225–257 (2014)
7. Byers, R., He, C., Mehrmann, V.: Where is the nearest non-regular pencil. Linear Algebra Appl. **285**, 81–105 (1998)
8. Byers, R., Nichols, N.K.: On the stability radius of a generalized state-space system. Linear Algebra Appl. **188–189**, 113–134 (1993)
9. Campbell, S.L.: Singular Systems of Differential Equations I. Pitman, San Francisco (1980)
10. Campbell, S.L.: Singular linear systems of differential equations with delays. Appl. Anal. **11**, 129–136 (1980)
11. Campbell, S.L.: 2-D (differential-delay) implicit systems. In: Proceedings of the IMACS World Congress on Scientific Computation, Dublin, pp. 1828–1829 (1991)
12. Campbell, S.L.: Nonregular 2D descriptor delay systems. IMA J. Math. Control Appl. **12**, 57–67 (1995)
13. Campbell, S.L., Linh, V.H.: Stability criteria for differential-algebraic equations with multiple delays and their numerical solutions. Appl. Math. Comput. **208**, 397–415 (2009)
14. Dragan, V., Ionita, A.: Exponential stability for singularly perturbed systems with state delays. In: Makay, G., Hatvani, L. (eds.) EJQTDE, Proceedings of 6th Coll. QTDE, no. 6, pp. 1–8. Szeged, Hungary (Aug 10–14, 1999)
15. Du, N.H., Linh, V.H.: Implicit-system approach to the robust stability for a class of singularly perturbed linear systems. Syst. Control Lett. **54**, 33–41 (2005)
16. Du, N.H., Linh, V.H.: On the robust stability of implicit linear systems containing a small parameter in the leading term. IMA J. Math. Control Inf. **23**, 67–74 (2006)
17. Du, N.H., Linh, V.H., Mehrmann, V.: Robust stability of differential-algebraic equations. In: Ilchmann, A., Reis, T. (eds.) Surveys in Differential-Algebraic Equations I, DAE-F, pp. 63–95. Springer, Berlin, Heidelberg (2013)
18. Du, N.H., Linh, V.H., Mehrmann, V., Thuan, D.D.: Stability and robust stability of linear time-invariant delay differential-algebraic equations. SIAM J. Matrix Anal. Appl. **34**, 1631–1654 (2013)
19. Du, N.H., Thuan, D.D., Liem, N.C.: Stability radius of implicit dynamic equations with constant coefficients on time scales. Syst. Control Lett. **60**, 596–603 (2011)
20. Fridman, E.: Stability of linear descriptor systems with delay: a Lyapunov-based approach. J. Math. Anal. Appl. **273**, 24–44 (2002)
21. Gantmacher, F.R.: The Theory of Matrices II. Chelsea Publishing Company, New York (1959)
22. Glizer, V.Y., Fridman, E.: H_∞ control of linear singularly perturbed systems with small state delay. J. Math. Anal. Appl. **250**(1), 49–85 (2000)
23. Griepentrog, E., März, R.: Differential-Algebraic Equations and Their Numerical Treatment. Teubner Verlag, Leipzig (1986)
24. Guglielmi, N., Hairer, E.: Computing breaking points in implicit delay differential equations. Adv. Comput. Math. **29**, 229–247 (2008)
25. Ha, P., Mehrmann, V.: Analysis and reformulation of linear delay differential-algebraic equations. Electron. J. Linear Algebra **23**, 703–730 (2012)
26. Ha, P., Mehrmann, V., A. Steinbrecher: analysis of linear variable coefficient delay differential-algebraic equations. J. Dyn. Differ. Equ. (2014, to appear). See also: PREPRINT 17/2013, Institut für Mathematik, TU Berlin. http://www.math.tu-berlin.de/preprints/
27. Hairer, E., Wanner, G.: Solving Ordinary Differential Equations II: Stiff and Differential-Algebraic Problems, 2nd edn. Springer, Berlin (1996)
28. Hale, J.K.: Theory of Functional-Differential Equations. Springer, New York/Heidelberg/Berlin (1977)
29. Hale, J.K., Verduyn Lunel, S.M.: Introduction to Functional Differential Equations. Springer, Berlin/Heidelberg/New York (1993)

30. Hinrichsen, D., Pritchard, A.J.: Stability radii of linear systems. Syst. Control Lett. **7**, 1–10 (1986)
31. Hinrichsen, D., Pritchard, A.J.: Stability radius for structured perturbations and the algebraic Riccati equation. Syst. Control Lett. **8**, 105–113 (1986)
32. Hinrichsen, D., Pritchard, A.J.: Real and complex stability radii: a survey, In: Hinrichsen, D., Martensson, B. (eds.) Control of Uncertain Systems, Progress in System and Control Theory, pp. 119–162. Birkhäuser, Basel (1990)
33. Hinrichsen, D., Pritchard, A.J.: Mathematical Systems Theory I. Modelling, State Space Analysis, Stability and Robustness. Springer, New York (2005)
34. Hinrichsen, D., Son, N.K.: Stability radii of positive discrete-time systems under affine parameter perturbations. Int. J. Robust Nonlinear Control **8**, 1969–1988 (1998)
35. Hinrichsen, D., Son, N.K.: μ-analysis and robust stability of positive linear systems. Appl. Math. Comput. Sci. **8**, 253–268 (1998)
36. Hu, G., Davison, E.J.: Real stability radii of linear time-invariant time-delay systems. Syst. Control Lett. **50**, 209–219 (2003)
37. Kunkel, P., Mehrmann, V.: Differential-Algebraic Equations. Analysis and Numerical Solution. EMS Publishing House, Zürich (2006)
38. Kunkel, P., Mehrmann, V.: Stability properties of differential-algebraic equations and spin-stabilized discretization. Electron. Trans. Numer. Anal. **26**, 383–420 (2007)
39. Jiang, X., Han, Q.L.: Delay-dependent robust stability for uncertain linear systems with interval time-varying delay. Automatica **42**, 1059–1065 (2006)
40. Lamour, R., März, R., Tischendorf, C.: Differential-Algebraic Equations: A Projector Based Analysis. Springer, Berlin/New York (2013)
41. Lang, S.: Complex Analysis. Springer, New York (1999)
42. Linh, V.H.: On the robustness of asymptotic stability for a class of singularly perturbed systems with multiple delays. Acta Math. Vietnam. **30**, 137–151 (2005)
43. Linh, V.H., Mehrmann, V.: Lyapunov, Bohl and Sacker-Sell spectral intervals for differential-algebraic equations. J. Dyn. Differ. Equ. **21**, 153–194 (2009)
44. Luzyanina, T., Rose, D.: Periodic solutions of differential-algebraic equations with time-delays: computation and stability analysis. J. Bifurc. Chaos **16**, 67–84 (2006)
45. Michiels, W.: Spectrum-based stability analysis and stabilisation of systems described by delay differential algebraic equations. IET Control Theory Appl. **5**, 1829–1842 (2011)
46. Müller, P.C.: On the stability of linear descriptor systems by applying modified Lyapunov equations. Proc. Appl. Math. Mech. PAMM **3**, 136–137 (2003)
47. Ngoc, P.H.A.: Novel criteria for exponential stability on functional differential equations. Proc. AMS **141**, 3083–3091 (2013)
48. Ngoc, P.H.A., Tinh, C.T.: New criteria for exponential stability of linear time-varying differential systems with delay. Taiwan. J. Math. 18, 1759–1774 (2014)
49. Phat, V.N., Sau, N.H.: On exponential stability of linear singular positive delayed systems. Appl. Math. Lett. **38**, 67–72 (2014)
50. Qiu, L., Bernhardsson, B., Rantzer, A., Davison, E.J., Young, P.M., Doyle, J.C.: A formula for computation of the real stability radius. Automatica **31**, 879–890 (1995)
51. Riaza, R.: Differential-Algebraic Systems. Analytical Aspects and Circuit Applications. World Scientific, Hackensack (2008)
52. Hanai Abd El-Moneim Sadaka: Robust stability, constrained stabilization, observer-based controller designs for time delay systems. Ph.D. thesis, Northeastern University, Boston (2012)
53. Shampine, L.F., Gahinet, P.: Delay-differential-algebraic equations in control theory. Appl. Numer. Math. **56**, 574–588 (2006)
54. Son, N.K., Ngoc, P.H.A.: The complex stability radius of linear time-delay systems. Vietnam J. Math. **26**, 379–384 (1998)
55. Son, N.K., Ngoc, P.H.A.: Stability radius of linear delay systems. In: Proceedings of the American Control Conference, San Diego, pp. 815–816 (1999)
56. Son, N.K., Ngoc, P.H.A.: Robust stability of positive linear time delay systems under affine parameter perturbations. Acta Math. Vietnam. **24**, 353–372 (1999)

57. Son, N.K., Ngoc, P.H.A.: Robust stability of linear functional differential equations. Adv. Stud. Contemp. Math. **3**, 43–59 (2001)
58. Ngoc, P.H.A., Son, N.K.: Stability radii of positive linear functional differential equations under multi-perturbations. SIAM J. Control Optim. **43**, 2278–2295 (2005)
59. Son, N.K., Thuan, D.D.: On the radius of surjectivity for rectangular matrices and its application to measuring stabilizability of linear systems under structured perturbations. J. Nonlinear Convex Anal. **12**, 441–453 (2011)
60. Son, N.K., Thuan, D.D.: The structured distance to non-surjectivity and its applications to calculating the controllability radius of descriptor systems. J. Math. Anal. Appl. **388**, 272–281 (2012)
61. Stykel, T.: Analysis and numerical solution of generalized Lyapunov equations. Dissertation, Institut für Mathematik, TU Berlin, Berlin (2002)
62. Stykel, T.: On criteria for asymptotic stability of differential-algebraic equations. Z. Angew. Math. Mech. **92**, 147–158 (2002)
63. Tian, H., Yu, Q., Kuang, J.: Asymptotic stability of linear neutral delay differential-algebraic equations and linear multistep methods. SIAM J. Numer. Anal **49**, 608–618 (2011)
64. Wei, J.: Eigenvalue and stability of singular differential delay systems. J. Math. Anal. Appl. **297**, 305–316 (2004)
65. Xu, S., Van Dooren, P., Radu, S., Lam, J.: Robust stability and stabilization for singular systems with state delay and parameter uncertainty. IEEE Trans. Autom. Control **47**, 1122–1128 (2002)
66. Zhong, Q.C.: Robust Control of Time-Delay Systems. Springer, London (2006)
67. Zhu, W., Petzold, L.R.: Asymptotic stability of linear delay differential-algebraic equations and numerical methods. Appl. Numer. Math. **24**, 247–264 (1997)
68. Zhu, W., Petzold, L.R.: Asymptotic stability of Hessenberg delay differential-algebraic equations of retarded or neutral type. Appl. Numer. Math. **27**, 309–325 (1998)

Chapter 20
Distance Problems for Linear Dynamical Systems

Daniel Kressner and Matthias Voigt

Abstract This chapter is concerned with distance problems for linear time-invariant differential and differential-algebraic equations. Such problems can be formulated as distance problems for matrices and pencils. In the first part, we discuss characterizations of the distance of a regular matrix pencil to the set of singular matrix pencils. The second part focuses on the distance of a stable matrix or pencil to the set of unstable matrices or pencils. We present a survey of numerical procedures to compute or estimate these distances by taking into account some of the historical developments as well as the state of the art.

20.1 Introduction

Consider a linear time-invariant differential-algebraic equation (DAE)

$$E\dot{x}(t) = Ax(t) + f(t), \quad x(0) = x_0, \tag{20.1}$$

with coefficient matrices $E, A \in \mathbb{R}^{n\times n}$, a sufficiently smooth inhomogeneity $f : [0,\infty) \to \mathbb{R}^n$, and an initial state vector $x_0 \in \mathbb{R}^n$. When E is nonsingular, (20.1) is turned into a system of ordinary differential equations by simply multiplying with E^{-1} on both sides. The more interesting case of singular E arises, for example, when imposing algebraic constraints on the state vector $x(t) \in \mathbb{R}^n$. Equations of this type play an important role in a variety of applications, including electrical circuits [54, 55] and multi-body systems [48]. The analysis and numerical solution of more general DAEs (which also take into account time-variant coefficients, nonlinearities and time delays) is a central part of Volker Mehrmann's work, as witnessed by the monograph [43] and by the several other chapters of this book

D. Kressner
ANCHP, EPF Lausanne, Station 8, CH-1015 Lausanne, Switzerland
e-mail: daniel.kressner@epfl.ch

M. Voigt (✉)
Institut für Mathematik, Technische Universität Berlin, Sekretariat MA 4-5, Straße des 17. Juni 136, D-10623 Berlin, Germany
e-mail: mvoigt@math.tu-berlin.de

P. Benner et al. (eds.), *Numerical Algebra, Matrix Theory, Differential-Algebraic Equations and Control Theory*, DOI 10.1007/978-3-319-15260-8_20

concerned with DAEs/descriptor systems. A recurring theme in Volker's work is the notion of *robustness*, as prominently expressed in the papers [2, 33].

This survey is concerned with distance measures that provide robust ways of assessing properties for (20.1). Before moving on with DAEs, let us illustrate the basic idea in the simpler setting of numerically deciding whether a given matrix $A \in \mathbb{R}^{n\times n}$ is singular. In the presence of roundoff errors and other uncertainties, this problem is ill-posed: Square matrices are *generically* nonsingular and therefore the slightest perturbation very likely turns a possibly singular matrix into a nonsingular matrix. It is more sensible to ask whether A is close to a singular matrix. In the absence of additional information on the uncertainty, this naturally leads to the problem of finding the smallest perturbation $\Delta A \in \mathbb{R}^{n\times n}$ such that $A + \Delta A$ is singular:

$$\delta(A) := \min\{\|\Delta A\| : A + \Delta A \text{ is singular}\}.$$

It is well known that this *distance to singularity* coincides with $\sigma_{\min}(A)$, the smallest singular value of A, when choosing the matrix 2-norm $\|\cdot\|_2$ or the Frobenius norm $\|\cdot\|_F$ for $\|\Delta A\|$. Note that $\sigma_{\min}(A)$ is the reciprocal of $\|A^{-1}\|_2$, the (relative) 2-norm condition number of matrix inversion. This witnesses a more general relation between the condition number of a problem and its distance to ill-posedness [16]. The quantity $\delta(A)$ is robust in the sense that it is only mildly affected by perturbations of A. To see this, consider a (slight) perturbation $A \mapsto \tilde{A}$ that arises, e.g., due to roundoff errors. By the triangular inequality it holds

$$\delta(A) - \|\tilde{A} - A\| \le \delta(\tilde{A}) \le \delta(A) + \|\tilde{A} - A\|.$$

Such a robustness property holds more generally for all the distance measures discussed here.

In this survey, we will focus on distance measures for two of the most important properties of a DAE (20.1). The matrix pencil $sE - A$ associated with (20.1) is called

- *regular* if its characteristic polynomial $s \mapsto \det(sE - A)$ is not identically zero;
- *stable* if all finite eigenvalues of $sE-A$ are contained in $\mathbb{C}^-$, the open left complex half-plane .

Regularity guarantees the existence of a unique (classical) solution of the DAE for all consistent initial values [59]. On the other hand, the stability of $sE - A$ is equivalent to asymptotic stability of the homogeneous DAE (20.1) with $f(t) \equiv 0$ [18].

Numerically verifying the regularity and stability of a given pencil is a challenge; none of the straightforward approaches is guaranteed to remain robust under uncertainties in E and A. This motivates considering the distance of a given regular or stable pencil to the set of singular or unstable pencils, respectively. Then one needs to devise algorithms for computing these distances or, at least, reliable and effective bounds. As we will see below, this is by no means trivial.

20.2 The Distance to Singularity of a Matrix Pencil

In this section we will discuss the distance of a given regular matrix pencil to the nearest singular matrix pencil, that is, a matrix pencil with vanishing characteristic polynomial. To our knowledge, such a notion of distance was first introduced and discussed by Byers, He, and Mehrmann [15]. Letting $\mathbb{R}[s]$ denote the ring of polynomials with coefficients in $\mathbb{R}$, the *distance to singularity* of a matrix pencil $sE - A$ is given by

$$\delta(E, A) := \min \left\{ \left\| [\Delta E \ \Delta A] \right\|_F : s(E + \Delta E) - (A + \Delta A) \in \mathbb{R}[s]^{n \times n} \text{ is singular} \right\}.$$

As explained in [15], explicit formulas for $\delta(E, A)$ can be obtained for special cases, for example when $n \leq 2$ or E, A are scalar multiples of each other. Such explicit formulas are not known for the general case and, even worse, devising a numerical method for computing $\delta(E, A)$ or bounds thereof turned out to be an extremely difficult problem. Since the publication of [15], almost no progress has been made in this direction.

In view of the definition, one might attempt to check (nearby) singularity by inspecting the magnitudes of the coefficients of $\det(sE - A)$. This attempt is futile for most practical applications, because the polynomial coefficients exhibit a wildly different scaling as n increases except for very particular situations, e.g., when all eigenvalues are (almost) on the unit circle. To obtain more meaningful measures, one therefore needs to consider other characterizations for the singularity of a pencil. In the following, we will discuss several such characterizations and their relation to $\delta(E, A)$.

20.2.1 *Distance to Singularity and Structured Low-Rank Approximation*

The Kronecker structure of $sE - A$ is intimately related to the ranks of certain block Toeplitz matrices constructed from E, A; see [9] for an overview. Specifically, let us consider

$$W_k = W_k(E, A) := \begin{bmatrix} A & & & \\ E & A & & \\ & \ddots & \ddots & \\ & & E & A \\ & & & E \end{bmatrix} \in \mathbb{R}^{(k+1)n \times kn}. \tag{20.2}$$

It can be shown that $sE - A$ is a regular pencil if and only if $W_k(E, A)$ has full column rank for all $k = 1, \ldots, n$. Setting

$$\gamma_k := \min \left\{ \left\| [\Delta E \ \Delta A] \right\|_F : \operatorname{rank} W_k(E + \Delta E, A + \Delta A) < nk \right\},$$

it therefore holds that

$$\delta(E, A) = \min_{1 \le k \le n} \gamma_k = \gamma_n, \tag{20.3}$$

where the latter equality follows from the observation that the rank deficiency of W_k implies the rank deficiency of W_ℓ for all $\ell \ge k$.

By (20.3), computing $\delta(E, A)$ is equivalent to finding a structured perturbation that makes W_n rank deficient. Having wider applicability in signal processing, systems and control, such structured low-rank approximation problems have attracted quite some attention recently; see [45] for an overview.

To proceed from (20.3), one needs to replace the rank constraint by a simpler characterization. Clearly, the matrix $W_k(E + \Delta E, A + \Delta A)$ is rank deficient if and only if there exist vectors $x_1, \ldots, x_k \in \mathbb{R}^n$, not all equal to zero, such that

$$\begin{bmatrix} A + \Delta A & & & \\ E + \Delta E & A + \Delta A & & \\ & \ddots & \ddots & \\ & & E + \Delta E & A + \Delta A \\ & & & E + \Delta E \end{bmatrix} \begin{pmatrix} x_1 \\ x_2 \\ x_3 \\ \vdots \\ x_k \end{pmatrix} = 0. \tag{20.4}$$

Hence, we obtain the outer-inner optimization problem

$$\gamma_k = \min_{\substack{x \in \mathbb{R}^{kn} \\ x \neq 0}} f_k(x), \quad f_k(x) = \min \left\{ \left\| [\Delta E \ \Delta A] \right\|_F : W_k(E + \Delta E, A + \Delta A)x = 0 \right\}. \tag{20.5}$$

Its particular structure implies that (20.4) is equivalent to

$$[\Delta E \ \Delta A] \, Z_k = -[E \ A] \, Z_k \quad \text{with} \quad Z_k = \begin{bmatrix} 0 & x_1 & x_2 & \ldots & x_{k-1} & x_k \\ x_1 & x_2 & x_3 & \ldots & x_k & 0 \end{bmatrix}. \tag{20.6}$$

This shows that the inner optimization problem in (20.5) is a standard linear least-squares problem, admitting the explicit solution $f_k(x) = \left\| [E \ A] \, Z_k Z_k^\dagger \right\|_F$, where $Z_k^\dagger$ denotes the Moore-Penrose pseudoinverse of Z_k. Thus,

$$\gamma_k = \min_{\substack{x \in \mathbb{R}^{kn} \\ x \neq 0}} \left\| [E \ A] \, Z_k Z_k^\dagger \right\|_F$$

and, consequently,

$$\delta(E,A) = \min_{1\leq k\leq n}\ \min_{\substack{x\in\mathbb{R}^{kn}\\ x\neq 0}} \big\| \begin{bmatrix} E & A \end{bmatrix} Z_k Z_k^{\dagger} \big\|_{\mathrm{F}} = \min_{\substack{x\in\mathbb{R}^{n^2}\\ x\neq 0}} \big\| \begin{bmatrix} E & A \end{bmatrix} Z_n Z_n^{\dagger} \big\|_{\mathrm{F}}. \tag{20.7}$$

This so called *variable projection least-squares problem* can, in principle, be addressed by standard nonlinear optimization methods. Such an approach bears two major obstacles: (a) There may be many local minima. (b) The sheer number of variables (n^2 for $k = n$) restricts the scope of existing optimization methods to fairly small values of n.

Although the obstacles mentioned above have not been overcome yet, significant progress has been made in the availability of software for structured low-rank approximation problems. Specifically, the SLRA software package [46] covers the block Toeplitz structure of the matrix W_k and can hence be used to compute γ_k. We have applied this software to all examples from [15]. For the matrix pencil

$$sE - A = s\begin{bmatrix} 0 & 1/\varepsilon \\ 0 & 1 \end{bmatrix} - \begin{bmatrix} 1 & 1/\varepsilon \\ 0 & 1 \end{bmatrix}$$

and $\varepsilon = 10^{-1}$, the value $\gamma_2 = 0.0992607\ldots$ returned by SLRA approximates the true value of $\delta(E,A)$ up to machine precision. This nice behavior is also observed for smaller values of ε until around 10^{-7}, below which SLRA signals an error.

For other examples from [15], SLRA seems to deliver local minima only or exhibits very slow convergence. For the 8×8 matrix pencil arising from the model of a two-dimensional, three-link mobile manipulator [15, Example 14], SLRA returns:

γ_1	γ_2	γ_3	γ_4	γ_5	γ_6	γ_7	γ_8
0.0113	0.0169	0.0171	0.0277	0.0293	0.0998	0.0293	0.6171

These values do not reflect the fact that the exact value of γ_k *decreases* as k increases. This clearly indicates a need to further explore the potential of SLRA and related software for computing $\delta(E,A)$.

20.2.2 Distance to Singularity and Block Schur Form

For orthogonal matrices $Q, Z \in \mathbb{R}^{n\times n}$ we consider the equivalence transformation

$$Q^{\mathsf{T}}(sE - A)Z = s\begin{bmatrix} E_{11} & E_{12} \\ E_{21} & E_{22} \end{bmatrix} - \begin{bmatrix} A_{11} & A_{12} \\ A_{21} & A_{22} \end{bmatrix},$$

where we partitioned the transformed pencil such that $sE_{11} - A_{11} \in \mathbb{R}[s]^{k-1\times k}$ and $sE_{22} - A_{22} \in \mathbb{R}[s]^{n-k+1\times n-k}$ for some $1 \le k \le n$. The perturbation $s\Delta E - \Delta A$ defined by

$$Q^{\mathsf{T}}(s\Delta E - \Delta A)Z := s\begin{bmatrix} 0 & 0 \\ -E_{21} & 0 \end{bmatrix} - \begin{bmatrix} 0 & 0 \\ -A_{21} & 0 \end{bmatrix}$$

yields

$$Q^{\mathsf{T}}(s(E + \Delta E) - (A + \Delta A))Z = s\begin{bmatrix} E_{11} & E_{12} \\ 0 & E_{22} \end{bmatrix} - \begin{bmatrix} A_{11} & A_{12} \\ 0 & A_{22} \end{bmatrix}.$$

Since $sE_{11}-A_{11}$ and $sE_{22}-A_{22}$ are rectangular, and thus singular, the perturbed pencil $s(E+\Delta E)-(A+\Delta A)$ is singular as well. Partitioning $Q = \begin{bmatrix} Q_{k-1} & Q_{n-k+1} \end{bmatrix} \in \mathbb{R}^{n\times n}$ and $Z = \begin{bmatrix} Z_k & Z_{n-k} \end{bmatrix} \in \mathbb{R}^{n\times n}$ conformally, the de-regularizing perturbation satisfies

$$\left\| \begin{bmatrix} \Delta E & \Delta A \end{bmatrix} \right\|_{\mathrm{F}} = \left\| \begin{bmatrix} Q_{n-k+1}^{\mathsf{T}} E Z_k & Q_{n-k+1}^{\mathsf{T}} A Z_k \end{bmatrix} \right\|_{\mathrm{F}}.$$

Minimizing over all perturbations constructed as described above leads to the following minimax-like characterization [15, Sec. 4.2]:

$$\delta(E, A) = \min_{1\le k\le n} \Big\{ \left\| \begin{bmatrix} Q_{n-k+1}^{\mathsf{T}} E Z_k & Q_{n-k+1}^{\mathsf{T}} A Z_k \end{bmatrix} \right\|_{\mathrm{F}} :$$
$$Q_{n-k+1} \in \mathbb{R}^{n\times n-k+1},\ Q_{n-k+1}^{\mathsf{T}} Q_{n-k+1} = I_{n-k+1}, \text{ and } Z_k \in \mathbb{R}^{n\times k},\ Z_k^{\mathsf{T}} Z_k = I_k \Big\}. \tag{20.8}$$

Computing $\delta(E, A)$ via this characterization amounts to solving n optimization problems over Stiefel manifolds. Again, the possible presence of many local minima and the $O(n^2)$ degrees of freedom limit the usefulness of this characterization. A related idea was considered in [62] for computing the distance to the nearest uncontrollable system.

20.2.3 *Lower and Upper Bounds for the Distance to Singularity*

As discussed above, the characterizations (20.3) and (20.8) are of limited applicability for computing $\delta(E, A)$. It is therefore of interest to develop inexpensive lower and upper bounds, partially based on these characterizations.

Lower bound from singular values of W_k. A first lower bound proposed in [15] uses the matrix W_k defined in (20.2). Let $s\Delta E - \Delta A$ be a minimum norm de-regularizing perturbation of $sE - A$. Then $W_k(E + \Delta E, A + \Delta A)$ is rank-deficient for some $k \leq n$, and the inequalities

$$\sigma_{\min}(W_k(E,A)) \leq \|W_k(\Delta E, \Delta A)\|_{\mathrm{F}} = \sqrt{k}\delta(E,A)$$

hold for this particular value k. Consequently,

$$\frac{\sigma_{\min}(W_n(E,A))}{\sqrt{n}} = \min_{1\leq k\leq n} \frac{\sigma_{\min}(W_k(E,A))}{\sqrt{k}} \leq \delta(E,A). \tag{20.9}$$

To obtain a reliable lower bound, one needs to evaluate $\sigma_{\min}(W_k(E,A))$ for $k = n$, as its value can decrease by many orders of magnitude when increasing from k to $k+1$, see [15, Subsec 5.1]. However, the computation of the smallest singular value of the $(k+1)n \times kn$-matrix $W_k(E,A)$ gets increasingly expensive as k increases. For example, inverse iteration applied to $W_k^{\mathsf{T}} W_k$ requires $O(k \cdot n^3)$ operations and $O(k \cdot n^2)$ storage for the factorization of $W_k^{\mathsf{T}} W_k$, assuming its block tridiagonal structure is exploited.

Lower bound from one-parameter optimization. Another lower bound [15, Subsec 5.2] is obtained from the observation that $\alpha_0(E + \Delta E) - \beta_0(A + \Delta A)$ is singular for all scalars $\alpha_0, \beta_0 \in \mathbb{C}$ if the pencil $s(E + \Delta E) - (A + \Delta A)$ is singular. It follows that

$$\sigma_{\min}(\alpha_0 E - \beta_0 A) \leq \|\alpha_0 \Delta E - \beta_0 \Delta A\|_{\mathrm{F}} \leq \sqrt{|\alpha_0|^2 + |\beta_0|^2}\, \left\| \begin{bmatrix} \Delta E & \Delta A \end{bmatrix} \right\|_{\mathrm{F}}.$$

Defining

$$\mathbb{S}_{\mathbb{F}} := \left\{(\alpha,\beta) \in \mathbb{F} \times \mathbb{F} : |\alpha|^2 + |\beta|^2 = 1\right\},$$

we therefore obtain

$$\max_{(\alpha,\beta)\in\mathbb{S}_{\mathbb{R}}} \sigma_{\min}(\alpha E - \beta A) \leq \max_{(\alpha,\beta)\in\mathbb{S}_{\mathbb{C}}} \sigma_{\min}(\alpha E - \beta A) \leq \delta(E,A). \tag{20.10}$$

The first inequality is particularly suitable when E, A are real and amounts to minimizing

$$g(t) := -\sigma_{\min}(\sin(t)E - \cos(t)A), \quad t \in [0,\pi].$$

Well-known properties of singular values imply that g is piecewise smooth and Lipschitz-continuous. An efficient algorithm tailored to such a situation is described in [49]. The lower bounds (20.10) have been observed to be rather tight [15].

Upper bound from common null space. A simple upper bound is derived from the following observation. If $\operatorname{rank}\begin{bmatrix} E + \Delta E \\ A + \Delta A \end{bmatrix} < n$ or if $\operatorname{rank}\begin{bmatrix} E + \Delta E & A + \Delta A \end{bmatrix} < n$, then the pencil $s(E + \Delta E) - (A + \Delta A)$ is singular.

Thus,

$$\delta(E,A) \leq \min\left\{\sigma_{\min}\left(\begin{bmatrix} E \\ A \end{bmatrix}\right), \sigma_{\min}\left(\begin{bmatrix} E & A \end{bmatrix}\right)\right\}, \tag{20.11}$$

which becomes an equality for $n = 1$ or $n = 2$.

Upper bound from generalized Schur form. The (real) generalized Schur form [23, Thm. 7.7.2] states that $sE - A \in \mathbb{R}[s]^{n\times n}$ can be reduced to quasi-triangular form by an orthogonal equivalence transformation:

$$Q^{\mathsf{T}}(sE - A)Z = \begin{bmatrix} sE_{11} - A_{11} & \dots & sE_{1m} - A_{1m} \\ & \ddots & \vdots \\ & & sE_{mm} - A_{mm} \end{bmatrix}. \tag{20.12}$$

The diagonal blocks $sE_{ii} - A_{ii}$ are either 1×1 (corresponding to a real eigenvalue, an infinite eigenvalue or a singular block) or 2×2 (corresponding to a pair of complex conjugate eigenvalues).

Obviously, $s\tilde{E} - \tilde{A} := s(E + \Delta E) - (A + \Delta A)$ becomes singular when any of the diagonal blocks $s\tilde{E}_{ii} - \tilde{A}_{ii}$ in its generalized Schur form becomes singular. This directly gives the upper bound

$$\delta(E,A) \leq \min_{1\leq i\leq m} \delta(E_{ii}, A_{ii}) = \min_{1\leq i\leq m} \min\left\{\sigma_{\min}\left(\begin{bmatrix} E_{ii} \\ A_{ii} \end{bmatrix}\right), \sigma_{\min}\left(\begin{bmatrix} E_{ii} & A_{ii} \end{bmatrix}\right)\right\},$$

where we used (20.11).

Upper bound from singular values and vectors of W_k. Let us come back to the least-squares problem (20.7), which implies

$$\delta(E,A) \leq \left\| \begin{bmatrix} E & A \end{bmatrix} Z_k Z_k^{\dagger} \right\|_{\mathrm{F}} \leq \left\| \begin{bmatrix} E & A \end{bmatrix} Z_k \right\|_{\mathrm{F}} \left\| Z_k^{\dagger} \right\|_{\mathrm{F}}$$

for any $2n \times (k+1)$ matrix Z_k of the form (20.6), defined via vectors $x_1, \dots, x_k \in \mathbb{R}^n$. We make a particular choice of Z_k by considering the partitioning $x = \left(x_1^{\mathsf{T}} \dots x_k^{\mathsf{T}}\right)^{\mathsf{T}}$ of a (right) singular vector $x \in \mathbb{R}^{kn}$ belonging to $\sigma_{\min}(W_k)$. The structure of W_k implies that $W_k x$ is the vectorization of $\begin{bmatrix} E & A \end{bmatrix} Z_k$, and therefore

$$\left\| \begin{bmatrix} E & A \end{bmatrix} Z_k \right\|_{\mathrm{F}} = \|W_k x\|_2 = \sigma_{\min}(W_k).$$

This gives the upper bounds

$$\delta(E,A) \leq \left\| \begin{bmatrix} E & A \end{bmatrix} Z_k \right\|_{\mathrm{F}} \left\| Z_k^{\dagger} \right\|_{\mathrm{F}} \leq \sigma_{\min}(W_k)/\sigma_{\min}(Z_k),$$

which are valid for every $k = 1, \ldots, n$. In [15], it is noted that the choice of k is critical for this bound, since it might only be relatively tight for one k. As discussed above, the computation of the smallest singular value and the corresponding singular vector gets expensive, if k and/or n are not small.

In [15], further upper bounds are presented which we do not summarize here. Moreover, several examples show that none of the bounds presented above is tight for all examples. A computationally attractive way to determine or estimate $\delta(E, A)$ thus remains an open problem.

The presentation above focused on real matrices with real perturbations. Up to minor modifications, all developments directly extend to complex matrices with complex perturbations.

20.2.4 Semi-Explicit DAEs

The consideration of general unstructured perturbations $\Delta E, \Delta A$ may become inappropriate when more information on the uncertainty in the coefficients E, A if a DAE (20.1) is available. For example, for the special case of linear semi-explicit DAEs, the pencil $sE - A$ takes the form

$$sE - A = s\begin{bmatrix} I_r & 0 \\ 0 & 0 \end{bmatrix} - \begin{bmatrix} A_{11} & A_{12} \\ A_{21} & A_{22} \end{bmatrix}.$$

Since there is little reason to admit perturbations in the fixed matrix E, it makes sense to consider the modified distance to singularity given by

$$\delta_0(E, A) := \min\left\{\|\Delta A\|_F : sE - (A + \Delta A) \in \mathbb{C}[s]^{n\times n} \text{ is singular}\right\}. \qquad (20.13)$$

At least in principle, it is straightforward to incorporate such linear constraints on the perturbation structure into the structured low-rank approximation framework of Sect. 20.2.1. However, it turns out that more can be said about (20.13).

By [61, Lem. 2.2.26], the matrix pencil $sE-A$ is regular if and only if the $(n-r)\times(n-r)$ rational matrix $G(s) := A_{21}(sI_r - A_{11})^{-1}A_{12} + A_{22}$ has full normal rank. Trivially, $sI_r - A_{11}$ has full normal rank for all $A_{11} \in \mathbb{R}^{r\times r}$ and, consequently, a de-regularizing perturbation is characterized by one of the properties

$$\operatorname{rank}\begin{bmatrix} A_{12} + \Delta A_{12} \\ A_{22} + \Delta A_{22} \end{bmatrix} < n - r \text{ or } \operatorname{rank}\begin{bmatrix} A_{21} + \Delta A_{21} & A_{22} + \Delta A_{22} \end{bmatrix} < n - r.$$

In other words we obtain

$$\delta_0(E, A) = \min\left\{\sigma_{\min}\left(\begin{bmatrix} A_{12} \\ A_{22} \end{bmatrix}\right), \sigma_{\min}\left(\begin{bmatrix} A_{21} & A_{22} \end{bmatrix}\right)\right\}.$$

This improves a result from [15], where such a statement has been shown only for the case that $r = 1$ or $r = n - 1$.

20.2.5 Low-Rank Perturbations

A rather different way of imposing structure is to constrain the perturbations to be of low rank. This leads to the concept of *rank-*(κ_E, κ_A) *distance to singularity* proposed in [47]:

$$\delta_{\kappa_E,\kappa_A}(E, A) := \min\big\{ \big\|[\Delta E\ \Delta A]\big\|_F : \Delta E,\ \Delta A \in \mathbb{C}^{n\times n},$$
$$\operatorname{rank}\Delta E \le \kappa_E,\ \operatorname{rank}\Delta A \le \kappa_A,\ s(E + \Delta E) - (A + \Delta A) \text{ is singular}\big\}.$$

Here, κ_E, κ_A are supposed to be (much) smaller than n and we consider the complex case because it is the more natural setting for the developments below.

Low-rank perturbations naturally arise in situations when only a few entries of E and A are subject to uncertainties. Another motivation to consider $\delta_{\kappa_E,\kappa_A}(E, A)$ arises from the suspicion that the low dimensionality of the manifold of rank-constrained matrices could help reduce the cost of nonlinear optimization methods for computing this distance.

Following [47], we now focus on $\delta_{0,1}(E, A)$, that is, E is not perturbed at all and A is subject to rank-1 perturbations:

$$sE - \big(A + \tau u v^{\mathsf{H}}\big) \tag{20.14}$$

with $u,\ v \in \mathbb{C}^n \setminus \{0\}$ and a nonzero scalar $\tau \in \mathbb{C}$ determining the perturbation level. It is assumed that $sE - A$ itself is not singular, and therefore $\Lambda(E, A)$, the set of (finite) eigenvalues of $sE - A$, does not coincide with $\mathbb{C}$. We define the rational function $Q : \mathbb{C} \setminus \Lambda(E, A) \to \mathbb{C}$ by

$$Q(s) := v^{\mathsf{H}} R(s) u, \qquad R(s) = (sE - A)^{-1}.$$

Because of

$$\det\big(sE - A - \tau u v^{\mathsf{H}}\big) = \big(1 - \tau Q(s)\big)\det\big(sE - A\big) = \tau\big(\tau^{-1} - Q(s)\big)\det\big(sE - A\big),$$

a scalar $s \in \mathbb{C}\setminus\Lambda(E, A)$ is an eigenvalue of (20.14) if and only if $\tau^{-1} = Q(s)$. Thus, the perturbed pencil (20.14) becomes singular if and only if $\tau^{-1} - Q(s)$ vanishes on $\mathbb{C} \setminus \Lambda(E, A)$. Since $Q(s)$ is a meromorphic function on $\mathbb{C}$ with at most n poles, the latter condition is equivalent to

$$\tau^{-1} = Q(s_0), \quad 0 = \frac{\partial^j}{\partial s^j} Q(s)\bigg|_{s=s_0}, \quad j = 1, 2, \ldots, n, \tag{20.15}$$

for an arbitrary fixed $s_0 \in \mathbb{C} \setminus \Lambda(E, A)$. Note that $\frac{\partial^j}{\partial s^j} Q(s) = v^{\mathsf{H}} C_j(s) u$ with

$$C_j(s) := R(s)(ER(s))^j.$$

The smallest τ for which (20.15) holds gives the distance to singularity for *fixed* choices of u, v. Optimizing with respect to these vectors finally yields

$$\delta_{0,1}(E, A) = \min \Big\{ \big|v^{\mathsf{H}} R(s_0) u\big|^{-1} : u, v \in \mathbb{C}^n, \|u\|_2 = \|v\|_2 = 1, \\ v^{\mathsf{H}} C_j(s_0) u = 0, \ j = 1, \ldots, n \Big\} \qquad (20.16)$$

for arbitrary fixed $s_0 \in \mathbb{C} \setminus \Lambda(E, A)$; see also [47, Thm. 7]. The constraints in (20.16) can be expressed as

$$0 = \operatorname{tr}\big(v^{\mathsf{H}} C_j(s_0) u\big) = \operatorname{tr}\big(u v^{\mathsf{H}} C_j(s_0)\big) = \big\langle C_j(s_0), v u^{\mathsf{H}} \big\rangle.$$

In other words, the matrix vu^{H} is orthogonal to all matrices $C_j(s_0)$ with respect to the matrix inner product $\langle \cdot, \cdot \rangle$. Equivalently, it holds that

$$v u^{\mathsf{H}} \in \mathscr{D} := (\operatorname{span}\{C_1(s_0), \ldots, C_n(s_0)\})^{\perp} \subset \mathbb{C}^{n \times n}.$$

It turns out that the space $\mathscr{D}$ does not depend on the particular choice of s_0. Summarizing these developments, we arrive at

$$\delta_{0,1}(E, A) = \min \Big\{ |\operatorname{tr}(G R(s_0))|^{-1} : G^{\mathsf{H}} \in \mathscr{D}, \operatorname{rank} G = 1, \|G\|_F = 1 \Big\}; \qquad (20.17)$$

see [47, Thm. 13] for more details.

A difficulty in the nonconvex optimization problem (20.17) is that the two constraints $G^{\mathsf{H}} \in \mathscr{D}$ and $\operatorname{rank} G \leq 1$ need to be satisfied simultaneously. On the other hand, each of the two constraints individually constitutes a matrix manifold for which an orthogonal projection can be easily computed. This suggests the use of an alternating projection method on manifolds [44]. Starting from $G_0 = R(s_0)$ (or another suitable choice), the method proposed in [47] constructs the iteration

$$G_{2k+1} := \big(\Pi_{\mathscr{D}} G_{2k}^{\mathsf{H}}\big)^{\mathsf{H}}, \quad G_{2k+2} := \Psi(G_{2k+1}), \quad k = 0, 1, \ldots, \qquad (20.18)$$

where $\Pi_{\mathscr{D}}$ is the orthogonal projection on the linear space $\mathscr{D}$ and Ψ is the orthogonal projection on the manifold of matrices having rank at most 1 (which is simply a rank-1 truncation). It is shown in [47, Prop. 38] that *if* the sequence (20.18) converges to a matrix $G \neq 0$ then G has rank one and the pencil $sE - \big(A + \operatorname{tr}(GR(s_0))^{-1} G\big)$ is singular. General results on alternating projection methods can be applied to study the local convergence behavior of (20.18).

The paper [47] also extensively covers the distance to singularity of a Hermitian matrix pencil under Hermitian rank-1 perturbations.

20.3 The Distance to Instability

In this section we turn to the problem of computing the distance to instability, sometimes also referred to as the stability radius, and closely related problems.

20.3.1 *The Distance to Instability and Hamiltonian Matrices*

The problem of computing the distance to instability in its simplest form is formulated for an ODE system

$$\dot{x}(t) = Ax(t) \tag{20.19}$$

with $A \in \mathbb{R}^{n \times n}$. The distance to instability is the smallest norm of a perturbation Δ such that at least one eigenvalue of $A + \Delta$ does not have negative real part. If A itself is not stable, this distance is clearly 0. Assuming that A is stable, this distance satisfies

$$\begin{aligned} r_{\mathbb{F}}^{\|\cdot\|}(A) :&= \min\left\{\|\Delta\| : \Lambda(A+\Delta) \cap \overline{\mathbb{C}^+} \neq \emptyset \text{ for } \Delta \in \mathbb{F}^{n \times n}\right\} \\ &= \min\left\{\|\Delta\| : \Lambda(A+\Delta) \cap \mathrm{i}\mathbb{R} \neq \emptyset \text{ for } \Delta \in \mathbb{F}^{n \times n}\right\}. \end{aligned}$$

The appropriate choice of the field $\mathbb{F} \in \{\mathbb{R}, \mathbb{C}\}$ for the entries of Δ depends on the application. As pointed out in [58, Chapter 50], one should work with complex perturbations, even for a real matrix A, when attempting to draw conclusions about the transient behavior of (20.19). On the other hand, if one is interested in stability robust to uncertainties in the entries of a real matrix A then clearly $\mathbb{F} = \mathbb{R}$ is the preferred choice. Since $\mathbb{R} \subset \mathbb{C}$ it holds that

$$r_{\mathbb{C}}^{\|\cdot\|}(A) \leq r_{\mathbb{R}}^{\|\cdot\|}(A).$$

The norm for measuring the perturbations in the definition of $r_{\mathbb{F}}^{\|\cdot\|}(A)$ is usually chosen as $\|\cdot\| = \|\cdot\|_2$ or $\|\cdot\| = \|\cdot\|_F$. The choice between these two norms has, at most, a limited impact on the distance to instability. As we will see below, there are always minimal perturbations that have rank 1 (for $\mathbb{F} = \mathbb{C}$) or rank 2 (for $\mathbb{F} = \mathbb{R}$). Thus,

$$r_{\mathbb{C}}^{\|\cdot\|_2}(A) = r_{\mathbb{C}}^{\|\cdot\|_F}(A), \qquad r_{\mathbb{R}}^{\|\cdot\|_2}(A) \leq r_{\mathbb{R}}^{\|\cdot\|_F}(A) \leq \sqrt{2}\, r_{\mathbb{R}}^{\|\cdot\|_2}(A).$$

In the following, we will therefore simply write $r_{\mathbb{C}}(A)$.

The papers [35, 60] were among the first to consider the distance to instability and established

$$r_{\mathbb{C}}(A) = \min_{\omega \in \mathbb{R}} \sigma_{\min}(\mathrm{i}\omega I_n - A). \tag{20.20}$$

This result is constructive and gives rise to a minimal perturbation of rank 1, which is obtained from the singular vectors belonging to $\sigma_{\min}(\mathrm{i}\omega I_n - A)$ for the optimal value of ω. The characterization (20.20) is intimately related to the eigenvalues of the *Hamiltonian matrix*

$$\mathscr{H}(\alpha) := \begin{bmatrix} A & -\alpha I_n \\ \alpha I_n & -A^{\mathsf{T}} \end{bmatrix} \in \mathbb{R}^{2n\times 2n}. \tag{20.21}$$

In [14, 34], it is shown that $\alpha \geq r_{\mathbb{C}}(A)$ if and only if $H(\alpha)$ has at least one purely imaginary eigenvalue. Based on this result, Byers [14] proposed a bisection method that adapts the value of α by checking whether any of the eigenvalues of $\mathscr{H}(\alpha)$ are purely imaginary. This algorithm converges *globally* and is robust to roundoff error, provided that a structure-preserving algorithm for the Hamiltonian matrix $\mathscr{H}(\alpha)$ is used. As summarized in the Chap. 1 by Bunse-Gerstner and Faßbender, the development of such algorithms is another central theme of Volker Mehrmann's work. We also refer to [4] for an overview and [1, 3] for the corresponding software.

20.3.2 The $\mathscr{H}_\infty$ Norm and Even Matrix Pencils

The results discussed in Sect. 20.3.1 have been extended into several different directions. One particularly important extension is concerned with the computation of the $\mathscr{H}_\infty$ norm for linear time-invariant control systems of the form

$$\begin{aligned} E\dot{x}(t) &= Ax(t) + Bu(t), \\ y(t) &= Cx(t) + Du(t), \end{aligned} \tag{20.22}$$

where $sE - A \in \mathbb{R}[s]^{n\times n}$ is assumed to be regular, $B \in \mathbb{R}^{n\times m}$, and $C \in \mathbb{R}^{p\times n}$. Moreover, $u : [0,\infty) \to \mathbb{R}^m$ is an input control signal and $y : [0,\infty) \to \mathbb{R}^p$ is a measured output signal.

The transfer function of (20.22) is given by $G(s) = C(sE - A)^{-1}B + D$ and maps inputs to outputs in the frequency domain. We assume that $G(s)$ is analytic and bounded in $\mathbb{C}^+$, for which a sufficient condition is that $\Lambda(E, A) \subset \mathbb{C}^-$ and all infinite eigenvalues are geometrically simple. For such transfer functions, the $\mathcal{H}_\infty$ norm is defined as

$$\|G\|_{\mathcal{H}_\infty} := \sup_{s \in \mathbb{C}^+} \sigma_{\max}(G(s)) = \sup_{\omega \in \mathbb{R}} \sigma_{\max}(G(\mathrm{i}\omega)), \tag{20.23}$$

where $\sigma_{\max}$ denotes the largest singular value of a matrix.

Comparing (20.23) with (20.20), we obtain $\|G\|_{\mathcal{H}_\infty} = 1/r_{\mathbb{C}}(A)$ for the special case that $E = B = C = I_n$ and $D = 0$. To relate to the practically more relevant case of general B, C, we have to extend the notion of distance to instability and consider the structured complex stability radius [36, 37] given by

$$\begin{aligned} r_{\mathbb{C}}(A, B, C) &:= \min\left\{\|\Delta\|_2 : \Lambda(A + B\Delta C) \cap \overline{\mathbb{C}^+} \neq \emptyset \text{ for } \Delta \in \mathbb{C}^{m \times p}\right\} \\ &= \min\left\{\|\Delta\|_2 : \Lambda(A + B\Delta C) \cap \mathrm{i}\mathbb{R} \neq \emptyset \text{ for } \Delta \in \mathbb{C}^{m \times p}\right\}. \end{aligned} \tag{20.24}$$

This generalization accounts for perturbations of the system that have a feedback structure (that is, $y(t) = \Delta u(t)$) and thus assesses the robustness of stability with respect to external disturbances.

Provided that A is stable, it holds [36] that

$$\|G\|_{\mathcal{H}_\infty} = \begin{cases} 1/r_{\mathbb{C}}(A, B, C) & \text{if } G(s) \not\equiv 0, \\ \infty & \text{if } G(s) \equiv 0, \end{cases} \tag{20.25}$$

with $G(s) = C(sI_n - A)^{-1}B$. In other words, a small $\mathcal{H}_\infty$ norm corresponds to a large robustness of stability of the system. The definition of the structured complex stability radius has been further generalized to cover $D \neq 0$ [38, Sec. 5.2] or $E \neq I_n$ [8, 17, 18]. However, in both cases the definition and interpretation of the structured complex stability radius becomes more cumbersome. For $D \neq 0$, the radius does not depend on the perturbations in an affine but in a linear fractional way. The case $E \neq I_n$ requires to carefully treat infinite eigenvalues and to account for perturbations that make the pencil $sE - A$ singular.

The most reliable methods for computing the $\mathcal{H}_\infty$ norm are based on the following extension of the connection between $r_{\mathbb{C}}(A)$ and the Hamiltonian matrix (20.21) discussed above; see [10–12] for $E = I_n$ and [5] for general E. Provided that $\alpha > \inf_{\omega \in \mathbb{R}} \sigma_{\max}(G(\mathrm{i}\omega))$, the inequality $\|G\|_{\mathcal{H}_\infty} \geq \alpha$ holds if and only if the *even matrix pencil*

$$s\mathscr{E} - \mathscr{A}(\alpha) := \left[\begin{array}{cc|cc} 0 & -sE^{\mathsf{T}} - A^{\mathsf{T}} & -C^{\mathsf{T}} & 0 \\ sE - A & 0 & 0 & -B \\ \hline -C & 0 & \alpha I_p & -D \\ 0 & -B^{\mathsf{T}} & -D^{\mathsf{T}} & \alpha I_m \end{array}\right] \in \mathbb{R}[s]^{2n+m+p \times 2n+m+p} \tag{20.26}$$

has purely imaginary eigenvalues. Such pencils are closely related to skew-Hamiltonian/Hamiltonian pencils, for which structure-preserving algorithms are discussed in the Chap. 1; see [6, 7] for recently released software. By additionally exploiting eigenvalue and, optionally, eigenvector information, methods based on (20.26) can be implemented such that they converge globally quadratically or even faster [22].

The structured *real* stability radius $r_{\mathbb{R}}^{\|\cdot\|_2}(A,B,C)$ with respect to the matrix 2-norm is defined as in (20.24), but with the perturbation restricted to stay real: $\Delta \in \mathbb{R}^{m\times p}$. It turns out that the computation of $r_{\mathbb{R}}^{\|\cdot\|_2}(A,B,C)$ is more difficult compared to $r_{\mathbb{C}}(A,B,C)$. Provided that $G(s) = C(sI_n - A)^{-1}B$ is analytic and bounded in $\mathbb{C}^+$, the celebrated expression

$$r_{\mathbb{R}}^{\|\cdot\|_2}(A,B,C) = \left(\sup_{\omega\in\mathbb{R}} \inf_{\gamma\in(0,1]} \sigma_2\left(\begin{bmatrix} \mathrm{Re}(G(\mathrm{i}\omega)) & -\gamma\,\mathrm{Im}(G(\mathrm{i}\omega)) \\ \frac{1}{\gamma}\mathrm{Im}(G(\mathrm{i}\omega)) & \mathrm{Re}(G(\mathrm{i}\omega)) \end{bmatrix}\right)\right)^{-1} \tag{20.27}$$

holds [53], where σ_2 denotes the second largest singular value of a matrix. Again, this characterization is constructive in the sense that a minimal rank-2 perturbation can be constructed from singular vectors. The inner optimization problem in (20.27) is unimodal [53], implying that every local minimum is a global minimum and thus allowing for reliable and efficient numerical optimization. For the outer optimization problem, a numerical method similar to the stability radius and $\mathscr{H}_\infty$ norm computation is devised in [56].

The definition and computation of robust stability measures for linear delay DAEs is surveyed in the Chap. 19 by Linh and Thuan.

20.3.3 *The Distance to Instability and Pseudospectra*

The approaches discussed above have been designed for relatively small problems and it is by no means clear that they can be well adapted to large-scale problems. This is mainly because of the lack of an efficient large-scale algorithm for deciding whether a Hamiltonian matrix or an even matrix pencil has purely imaginary eigenvalues; see [41] for a more detailed discussion.

Recent work by Guglielmi and Overton [31] has initiated the development of novel algorithms that are based on pseudospectra and appear to be more suitable for large-scale problems. Moreover, as we will see in Sect. 20.3.6 below, the framework offers much more flexibility for incorporating structure.

Given a matrix $A \in \mathbb{R}^{n\times n}$ and $\varepsilon > 0$, we consider the ε-pseudospectrum

$$\Lambda_\varepsilon^{\mathbb{F},\|\cdot\|}(A) := \{\lambda \in \mathbb{C} : \lambda \in \Lambda(A+\Delta) \text{ for some } \Delta \in \mathbb{F}^{n\times n},\ \|\Delta\| < \varepsilon\}$$

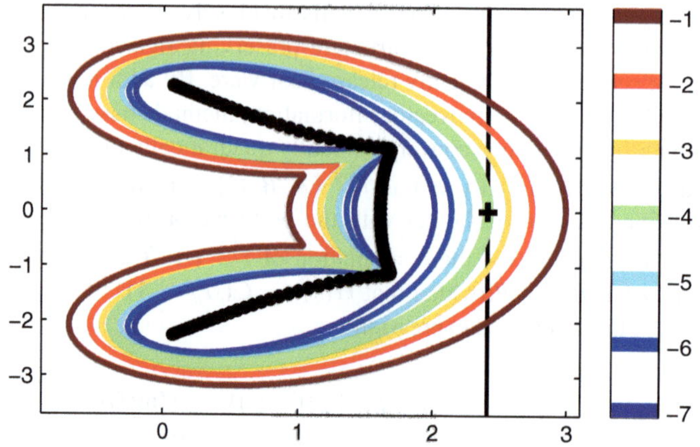

Fig. 20.1 Contour lines of the pseudospectrum $\Lambda_\varepsilon^{\mathbb{C},\|\cdot\|_2}(A)$ for the so called Grcar matrix and $\varepsilon = 10^{-1}, 10^{-2}, \ldots, 10^{-7}$. *Black crosses* denote eigenvalues and the *black vertical line* denotes the pseudospectral abscissa for $\varepsilon = 10^{-4}$

for $\mathbb{F} \in \{\mathbb{R}, \mathbb{C}\}$ and a matrix norm $\|\cdot\|$ (usually $\|\cdot\| = \|\cdot\|_2$ or $\|\cdot\| = \|\cdot\|_F$). The pseudospectrum grows as ε increases; see Fig. 20.1 for an illustration.

Basic idea. By definition, the distance to instability for a stable matrix is the smallest value of ε for which the ε-pseudospectrum touches the imaginary axis. In terms of the ε-pseudospectral abscissa

$$\alpha^{\mathbb{F},\|\cdot\|}(\varepsilon, A) := \sup\left\{\operatorname{Re}(\lambda) : \lambda \in \Lambda_\varepsilon^{\mathbb{F},\|\cdot\|}(A)\right\},$$

this means that $\varepsilon = r_{\mathbb{F}}^{\|\cdot\|}(A)$ satisfies

$$\alpha^{\mathbb{F},\|\cdot\|}(\varepsilon, A) = 0. \tag{20.28}$$

Hence, once we have an efficient method for evaluating the quantity $\alpha^{\mathbb{F},\|\cdot\|}(\varepsilon, A)$, root finding algorithms can be used to determine $r_{\mathbb{F}}^{\|\cdot\|}(A)$.

Computation of the ε-pseudospectral abscissa. In a large-scale setting, one benefits from the fact that every element of $\Lambda_\varepsilon^{\mathbb{F},\|\cdot\|}(A)$ can be realized by a low-rank perturbation. In particular, for complex pseudospectra we have

$$\begin{aligned}\Lambda_\varepsilon^{\mathbb{C}}(A) :&= \Lambda_\varepsilon^{\mathbb{C},\|\cdot\|_2}(A) = \Lambda_\varepsilon^{\mathbb{C},\|\cdot\|_F}(A)\\ &= \{\lambda \in \mathbb{C} : \lambda \in \Lambda(A + \Delta) \text{ for some } \Delta \in \mathbb{C}^{n\times n},\ \operatorname{rank}\Delta = 1,\ \|\Delta\| < \varepsilon\};\end{aligned}$$

see [58] for this and many other properties of pseudospectra. The procedure proposed in [31] attempts to construct a sequence of perturbation matrices $\Delta_k := \varepsilon u_k v_k^{\mathsf{H}}$ with $\|u_k\|_2 = \|v_k\|_2 = 1$ for $k = 1, 2, \ldots$, such that one of the eigenvalues of $A + \Delta_k$ converges to the rightmost point of the ε-pseudospectrum.

The following standard eigenvalue perturbation result is instrumental in determining these perturbations.

Lemma 1 ([57]) *Let $M_0, M_1 \in \mathbb{R}^{n\times n}$, $t \in \mathbb{R}$, and consider the matrix family $M(t) = M_0 + tM_1$. Let $\lambda(t)$ be an eigenvalue of $M(t)$ converging to a simple eigenvalue λ_0 of M_0 with corresponding right and left eigenvectors $x_0, y_0 \in \mathbb{C}^n$ for $t \to 0$. Then $y_0^{\mathsf{H}} x_0 \neq 0$ and $\lambda(t)$ is analytic near $t = 0$ with*

$$\left.\frac{\mathrm{d}\lambda(t)}{\mathrm{d}t}\right|_{t=0} = \frac{y_0^{\mathsf{H}} M_1 x_0}{y_0^{\mathsf{H}} x_0}.$$

For the initial perturbation $\Delta_1 = \varepsilon u_1 v_1^{\mathsf{H}}$ one computes the rightmost eigenvalue λ_0 of A (assuming that it is simple) with corresponding right and left eigenvectors $x_0, y_0 \in \mathbb{C}^n$. The eigenvectors are normalized such that $y_0^{\mathsf{H}} x_0 > 0$, a property that is called RP-compatibility in [31]. Then for $A_1(t) := A + t\varepsilon u_1 v_1^{\mathsf{H}}$ with an eigenvalue $\lambda_1(t)$ converging to λ_0 for $t \to 0$ we obtain

$$\operatorname{Re}\left(\left.\frac{\mathrm{d}\lambda_1(t)}{\mathrm{d}t}\right|_{t=0}\right) = \varepsilon \frac{\operatorname{Re}\left(y_0^{\mathsf{H}} u_1 v_1^{\mathsf{H}} x_0\right)}{y_0^{\mathsf{H}} x_0} \leq \varepsilon \frac{\|y_0\|_2 \|x_0\|_2}{y_0^{\mathsf{H}} x_0}. \tag{20.29}$$

Equality in (20.29) holds for $u_1 = y_0/\|y_0\|_2$ and $v_1 = x_0/\|x_0\|_2$, i.e., this choice yields the maximal local growth of the real part of λ_0. For all subsequent perturbations, consider the matrix family

$$A_k(t) := A + \varepsilon u_{k-1} v_{k-1}^{\mathsf{H}} + t\varepsilon\left(u_k v_k^{\mathsf{H}} - u_{k-1} v_{k-1}^{\mathsf{H}}\right), \quad k = 2, 3, \ldots,$$

which constitutes rank-1 perturbations of norm ε for $t = 0$ and $t = 1$. Assume that $A_k(t)$ has an eigenvalue $\lambda_k(t)$ converging to a (simple) rightmost eigenvalue λ_{k-1} of $A + \Delta_{k-1} = A + \varepsilon u_{k-1} v_{k-1}^{\mathsf{H}}$ for $t \to 0$. Moreover, let $x_{k-1}, y_{k-1} \in \mathbb{C}^n$ with $y_{k-1}^{\mathsf{H}} x_{k-1} > 0$ be the corresponding right and left eigenvectors. Analogously to (20.29) we obtain

$$\begin{aligned}\operatorname{Re}\left(\left.\frac{\mathrm{d}\lambda_k(t)}{\mathrm{d}t}\right|_{t=0}\right) &= \varepsilon \frac{\operatorname{Re}\left(y_{k-1}^{\mathsf{H}}\left(u_k v_k^{\mathsf{H}} - u_{k-1} v_{k-1}^{\mathsf{H}}\right) x_{k-1}\right)}{y_{k-1}^{\mathsf{H}} x_{k-1}} \\ &\leq \varepsilon \frac{\|y_{k-1}\|_2 \|x_{k-1}\|_2 - \operatorname{Re}\left(y_{k-1}^{\mathsf{H}} u_{k-1} v_{k-1}^{\mathsf{H}} x_{k-1}\right)}{y_{k-1}^{\mathsf{H}} x_{k-1}}.\end{aligned} \tag{20.30}$$

Again, equality in (20.30) is achieved for the choice $u_k = y_{k-1}/\|y_{k-1}\|_2$ and $v_k = x_{k-1}/\|x_{k-1}\|_2$. As a consequence, the whole process of computing the pseudospectral abscissa consists of computing the rightmost eigenvalue of a matrix with corresponding right and left eigenvectors, constructing an optimal rank-1 perturbation by using these eigenvectors and repeating this procedure for the perturbed matrix until convergence. In [31] is shown that this procedure is a fixed point iteration converging to a locally rightmost point of the ε-pseudospectrum (under a weak regularity assumption). A major weakness is that we cannot ensure to find a globally rightmost point and thus, at least in principle, only find a lower bound for the ε-pseudospectral abscissa. However, numerical examples reported in [31] indicate that this bound often attains the exact value.

Connection to low-rank dynamics. The above procedure is an iterative scheme on the manifold of rank-1 perturbations having norm ε. This naturally leads to the question whether there exists a continuous path of such matrices converging to the desired optimal perturbation. Indeed, this question has a positive answer [27, 28]. Consider the differential equations

$$\begin{aligned} \dot{u}(t) &= \frac{\mathrm{i}}{2}\,\mathrm{Im}\left(u(t)^{\mathsf{H}}x(t)y(t)^{\mathsf{H}}v(t)\right) + \left(I_n - u(t)u(t)^{\mathsf{H}}\right)x(t)y(t)^{\mathsf{H}}v(t), \\ \dot{v}(t) &= \frac{\mathrm{i}}{2}\,\mathrm{Im}\left(v(t)^{\mathsf{H}}x(t)y(t)^{\mathsf{H}}u(t)\right) + \left(I_n - v(t)v(t)^{\mathsf{H}}\right)x(t)y(t)^{\mathsf{H}}u(t), \end{aligned} \tag{20.31}$$

where $x(t)$ and $y(t)$ are the right and left eigenvectors of unit norm with $y(t)^{\mathsf{H}}x(t) > 0$ corresponding to the rightmost eigenvalue $\lambda(t)$ of the matrix $A + \Delta(t) := A + \varepsilon u(t)v(t)^{\mathsf{H}}$. It has been shown in [27] that if $\lambda(t)$ is simple and smooth, it will tend to a locally rightmost point of the ε-pseudospectrum for $t \to \infty$. In fact, the iteration discussed above corresponds to the explicit Euler method applied to (20.31). Of course, other, potentially faster and adaptive methods can be used to discretize (20.31), a major advantage of the continuous formulation apart from its elegance.

Computation of the distance to instability. Now that we know how to efficiently determine $\alpha^{\mathbb{C}}(\varepsilon, A)$ for a fixed value of ε, we apply a root-finding algorithm to the nonlinear equation (20.28) in ε, in order to determine $\varepsilon_* = r_{\mathbb{C}}(A)$. In practice, Newton and Newton-bisection schemes turned out to be effective for this purpose [8, 24]. Let $\lambda(\varepsilon)$ be the rightmost point of $\Lambda_\varepsilon^{\mathbb{C}}(A)$ with the corresponding optimal perturbation matrix $\Delta(\varepsilon) := \varepsilon u(\varepsilon)v(\varepsilon)^{\mathsf{H}}$. Furthermore, let $x(\varepsilon)$ and $y(\varepsilon)$ be the corresponding right and left eigenvectors of $A + \Delta(\varepsilon)$ with $\|x(\varepsilon)\|_2 = \|x(\varepsilon)\|_2 = 1$ and $y(\varepsilon)^{\mathsf{H}}x(\varepsilon) > 0$. Suppose that the rightmost point of $\Lambda_{\hat{\varepsilon}}^{\mathbb{C}}(A)$ is unique for a given $\hat{\varepsilon}$. Then $\lambda(\cdot)$ is continuously differentiable at $\hat{\varepsilon}$ and

$$\left.\frac{\mathrm{d}\lambda(\varepsilon)}{\mathrm{d}\varepsilon}\right|_{\varepsilon=\hat{\varepsilon}} = \frac{1}{y(\hat{\varepsilon})^{\mathsf{H}}x(\hat{\varepsilon})}.$$

We thus have the ingredients for a Newton method to determine the root ε_* of (20.28):

$$\varepsilon_{k+1} = \varepsilon_k - y(\varepsilon_k)^{\mathsf{H}}x(\varepsilon_k)\alpha^{\mathbb{C}}(\varepsilon_k, A).$$

Acceleration. The iterative procedures discussed above for computing $\alpha^{\mathbb{C}}(\varepsilon, A)$ and $r_{\mathbb{C}}(A)$ can be accelerated. In [42], subspace acceleration techniques for both quantities were proposed for which, under certain mild conditions, locally superlinear convergence was proven. In practice, these subspace methods exhibit a quite robust convergence behavior, their local convergence is observed to be even locally quadratic, and they can be much faster than the methods from [24, 31]. However, it is currently not clear how these methods can be extended to the more general situations discussed below. The vector extrapolation techniques from [51, 52], which sometimes achieve similar speedups, do not have this drawback.

Real pseudospectra. One of the beauties of the presented framework is that it seamlessly extends to real pseudospectra. In the real case, the corresponding perturbations have rank 2 and, consequently, the dynamical system (20.31) needs to be replaced by appropriate rank-2 dynamics. Such dynamics have been proposed and analyzed in [29], both for the matrix 2-norm and the Frobenius norm, yielding efficient algorithms for $\alpha^{\mathbb{R},\|\cdot\|_2}(\varepsilon, A)$ and $\alpha^{\mathbb{R},\|\cdot\|_F}(\varepsilon, A)$. These results are used in [30] to design Newton-type algorithms for the computation of $r_{\mathbb{R}}^{\|\cdot\|_2}(A)$ and $r_{\mathbb{R}}^{\|\cdot\|_F}(A)$.

20.3.4 The $\mathcal{H}_\infty$ Norm and Spectral Value Sets

The basic ideas from Sect. 20.3.3 can be generalized in a direct manner to structured complex stability radii $r_{\mathbb{C}}(A, B, C)$ with $B \in \mathbb{R}^{n\times m}$ and $C \in \mathbb{R}^{p\times n}$. Instead of $\Lambda_\varepsilon^{\mathbb{C}}(A)$, one has to consider structured complex pseudospectra of the form

$$\Lambda_\varepsilon^{\mathbb{C}}(A, B, C) = \{\lambda \in \mathbb{C} : \lambda \in \Lambda(A + B\Delta C) \text{ for some } \Delta \in \mathbb{C}^{m\times p},\ \|\Delta\| < \varepsilon\}.$$

By (20.25), this yields an approach for computing the $\mathcal{H}_\infty$ norm of a transfer function $G(s) = C(sI_n - A)^{-1}B$ with a possibly large n. However, in order to extend this to transfer functions $G(s) = C(sI_n - A)^{-1}B + D$ with nonzero D one has to consider much more complicated pseudospectral structures (called spectral value sets in [38]), given by

$$\Lambda_\varepsilon^{\mathbb{C}}(A, B, C, D) = \big\{\lambda \in \mathbb{C} : \lambda \in \Lambda\big(A + B\Delta(I_p - D\Delta)^{-1}C\big) \text{ for some } \Delta \in \mathbb{C}^{m\times p},\ \|\Delta\| < \varepsilon\big\}.$$

As shown in [24], such spectral value sets can again be realized by rank-1 perturbations, allowing for an extension of the algorithms discussed above. A further extension has been made in [8] for transfer functions of the form $G(s) = C(sE-A)^{-1}B + D$. There, an embedding of the original control system into system of larger dimension is used to eliminate D which drastically simplifies the analysis of the algorithm. However, one has to consider structured pseudospectra of a matrix pencil instead of a matrix. Special care must be taken of possible perturbations of the infinite eigenvalues or perturbations that make the pencil singular. In [8] further improvements have been made with regard to the choice of the eigenvalues to follow during the computation of the pseudospectral abscissa. In particular, it is important to not only consider the location of the eigenvalues but also their sensitivity with respect to the perturbation structure. This is related to the concepts of controllability and observability of the underlying control system (20.22).

20.3.5 The Implicit Determinant Method

Another class of algorithms goes back to Freitag and Spence [19]. For the determination of the complex stability radius $r_{\mathbb{C}}(A)$ of a stable matrix $A \in \mathbb{R}^{n\times n}$ this approach again makes use of a spectral characterization involving the Hamiltonian matrix $\mathscr{H}(\alpha)$ defined in (20.21), however in a completely different way. Denote by α_* the smallest value of α such that $\mathscr{H}(\alpha)$ has a purely imaginary eigenvalue $\mathrm{i}\omega_*$, that is,

$$(\mathscr{H}(\alpha_*) - \mathrm{i}\omega_* I_{2n})\, x_* = 0$$

for some eigenvector $x_* \in \mathbb{C}^{2n}$. This value of α_* coincides with $r_{\mathbb{C}}(A)$ and ω_* is an optimal frequency according to (20.20). Due to the spectral symmetry of $\mathscr{H}(\alpha_*)$, the eigenvalue $\mathrm{i}\omega_*$ generically forms a Jordan block of dimension two, which will be assumed throughout this section.

Motivated by related methods for determining bifurcation points in parameter-dependent nonlinear systems, the basic idea of [19] consists of setting up a (well-conditioned) system of two nonlinear equations for which (ω_*, α_*) is a regular solution. For this purpose, a normalization vector $c \in \mathbb{C}^{2n}$ is chosen such that $c^{\mathsf{H}} x_* \neq 0$. Then the genericity assumption implies that the bordered matrix

$$\mathscr{M}(\omega,\alpha) := \begin{bmatrix} \mathscr{H}(\alpha) - \mathrm{i}\omega I_{2n} & \mathscr{J} c \\ c^{\mathsf{H}} & 0 \end{bmatrix} \quad \text{with} \quad \mathscr{J} = \begin{bmatrix} 0 & I_n \\ -I_n & 0 \end{bmatrix}$$

is nonsingular at (ω_*, α_*) and hence it is also nonsingular for all pairs (ω, α) in a neighborhood of (ω_*, α_*). Thus, for all such pairs the linear system of equations

$$\begin{bmatrix} \mathscr{H}(\alpha) - \mathrm{i}\omega I_{2n} & \mathscr{J} c \\ c^{\mathsf{H}} & 0 \end{bmatrix} \begin{pmatrix} x(\omega,\alpha) \\ f(\omega,\alpha) \end{pmatrix} = \begin{pmatrix} 0 \\ 1 \end{pmatrix} \tag{20.32}$$

has a unique solution. By Cramer's rule it holds

$$f(\omega,\alpha) = \frac{\det(\mathscr{H}(\alpha) - \mathrm{i}\omega I_{2n})}{\det(\mathscr{M}(\omega,\alpha))}.$$

In particular, $f(\omega,\alpha) = 0$ is equivalent to $\det(\mathscr{H}(\alpha) - \mathrm{i}\omega I_{2n}) = 0$. From the algebraic properties of the eigenvalue $\mathrm{i}\omega_*$ it follows [19, Lem. 4(a)] that $\partial f(\omega_*, \alpha_*)/\partial\omega = 0$. In summary, (ω_*, α_*) is a solution of the nonlinear system

$$0 = g(\omega,\alpha) := \begin{pmatrix} f(\omega,\alpha) \\ \partial f(\omega,\alpha)/\partial\omega \end{pmatrix}. \tag{20.33}$$

Moreover, it can be shown [19] that this solution is regular. Hence, Newton's method applied to $0 = g(\omega,\alpha)$ converges locally quadratically to (ω_*, α_*).

To implement Newton's method, we need to evaluate $g(\omega, \alpha)$ and its Jacobian, which amounts to evaluating first- and second-order derivatives of $f(\omega, \alpha)$. By differentiating the relation (20.32), it turns out that all these derivatives can be computed by solving four linear systems with $\mathscr{M}(\omega, \alpha)$. Having these derivatives computed at the kth iterate (ω_k, α_k), the next iterate of Newton's method is determined by first solving the 2×2 linear system

$$\begin{bmatrix} \partial f(\omega_k, \alpha_k)/\partial\omega & \partial f(\omega_k, \alpha_k)/\partial\alpha \\ \partial^2 f(\omega_k, \alpha_k)/\partial\omega^2 & \partial^2 f(\omega_k, \alpha_k)/\partial\omega\partial\alpha \end{bmatrix} \begin{pmatrix} \Delta\omega_k \\ \Delta\alpha_k \end{pmatrix} = -g(\omega_k, \alpha_k),$$

and subsequently setting

$$\begin{pmatrix} \omega_{k+1} \\ \alpha_{k+1} \end{pmatrix} := \begin{pmatrix} \omega_k \\ \alpha_k \end{pmatrix} + \begin{pmatrix} \Delta\omega_k \\ \Delta\alpha_k \end{pmatrix}.$$

Numerical examples in [19] show the effectivity of this method. Since only a few linear systems have to be solved, it also has potential for large-scale systems. Similarly as for the pseudospectral approach, the method is only guaranteed to converge locally and it may converge to a solution of (20.33) that is different from (ω_*, α_*). However, the situation is less bleak in practice; numerical results reveal that the method often converges to the correct solution for the initial values proposed in [19]. As discussed in [32], there is always the possibility to check whether $\mathscr{H}(\alpha_*)$ has purely imaginary eigenvalues, but this global optimality certificate may become too expensive for large-scale systems.

As discussed in [21], the described algorithm can be extended in a rather straightforward manner to $\mathscr{H}_\infty$ norm computations, even for the general case of descriptor systems. For this purpose, one only needs to replace the $(1,1)$-block in $\mathscr{M}(\omega, \alpha)$ by a Hamiltonian matrix $\tilde{\mathscr{A}}(\alpha) - i\omega\tilde{\mathscr{E}}$ that can be derived from the even pencil $s\mathscr{E} - \mathscr{A}(\alpha)$ in (20.26); see also [61].

The extension to the real 2-norm stability radius proposed in [20] is more involved. As a basis, the 4×4 Hamiltonian matrix constructed in [56] instead of the $2n \times 2n$ Hamiltonian matrix $\mathscr{H}(\alpha)$ needs to be used. Moreover, the bordered matrix $\mathscr{M}(\omega, \alpha)$ is needs to be replaced by a matrix in three variables $\mathscr{M}(\omega, \alpha, \gamma)$ due to the fact that one has to optimize in (20.27) over two parameters ω and γ instead of ω only.

20.3.6 Structured Distances and Variations

In this section, we briefly discuss existing work on structured distances to instability for structures beyond the real perturbations and fractional perturbations (related to linear control systems) considered above. Dealing with such structures is by no means simple, even the (usually simpler) problem of the structured distance to singularity, also called structured singular value, often poses great difficulties; see [40] for summary. A notable exception are complex Hamiltonian perturbations, for which an expression based on a unimodal optimization problem, not unlike the one in (20.27), can be derived [39]. This is discussed in more detail in the Chap. 8 by Bora and Karow, which also covers Volker's work on Hamiltonian perturbations.

The lack of simple characterizations for structured singular values complicates the development of efficient algorithms that guarantee global optimality, like Byers' bisection method, for structured distances to instability. In contrast, it is fairly straightforward to incorporate structure in the pseudospectra-based algorithms from Sect. 20.3.3. This is because finding a structured perturbation that is optimal in first order is a much simpler than finding a globally optimal perturbation. Considering Lemma 1, the former problem amounts to determining a structured perturbation Δ with $\|\Delta\| \leq \varepsilon$ such that the real part of $y_0^{\mathsf{H}} M_1 x_0$ becomes maximal. For the Frobenius norm, this optimization problem has an explicit solution in terms of the orthogonal projection of $x_0 y_0^{\mathsf{H}}$ onto the set of structured matrices, which becomes particularly simple in the usual situation when this set is a subspace or a manifold. To be able to address large-scale problems, one needs to restrict the obtained iteration to low-rank matrices, that is, one needs to develop a structured extension of the dynamical system (20.19). This part is significantly more challenging and has been addressed for (complex and real) Hamiltonian matrices [25], Toeplitz matrices [13], and symplectic matrices [26] so far.

Apart from structure preservation, the flexibility of pseudospectra-based algorithms is also also witnessed by several recent extensions, for example, to nonlinear eigenvalue problems [50] and to the regularization of solvent equations discussed in the Chap. 3 by Lin and Schröder.

20.4 Summary and Conclusions

Distance measures are not only useful for quantifying the impact of uncertainty on properties of dynamical systems, they also lead to mathematically very interesting and challenging problems. They touch upon a diversity of current topics, including structured low-rank approximation, eigenvalue optimization, and low-rank dynamics. Despite the fact that distance measures are a classical topic in numerical linear algebra and control, significant progress has been made in incorporating structure and dealing with large-scale problems for robust stability calculations. On other topics, less progress has been made; in particular, the reliable estimation of the distance to the nearest singular matrix pencil remains a widely open problem.

References

1. Benner, P., Byers, R., Barth, E.: Algorithm 800: Fortran 77 subroutines for computing the eigenvalues of Hamiltonian matrices I: the square reduced method. ACM Trans. Math. Softw. **26**(1), 49–77 (2000)
2. Benner, P., Byers, R., Mehrmann, V., Xu, H.: Robust numerical methods for robust control. Tech. Rep. 06-2004, Institut für Mathematik, TU Berlin (2004)
3. Benner, P., Kressner, D.: Algorithm 854: Fortran 77 subroutines for computing the eigenvalues of Hamiltonian matrices II. ACM Trans. Math. Softw. **32**(2), 352–373 (2006)
4. Benner, P., Kressner, D., Mehrmann, V.: Skew-Hamiltonian and Hamiltonian eigenvalue problems: theory, algorithms and applications. In: Drmač Z., Marušić M., Zutek Z. (eds.) Proceedings of the Conference on Applied Mathematics and Scientific Computing, Brijuni, pp. 3–39. Springer (2005)
5. Benner, P., Sima, V., Voigt, M.: $\mathscr{L}_\infty$-norm computation for continuous-time descriptor systems using structured matrix pencils. IEEE Trans. Autom. Control **57**(1), 233–238 (2012)
6. Benner, P., Sima, V., Voigt, M.: FORTRAN 77 subroutines for the solution of skew-Hamiltonian/Hamiltonian eigenproblems – Part I: algorithms and applications. Preprint MPIMD/13-11, Max Planck Institute Magdeburg (2013). Available from http://www.mpi-magdeburg.mpg.de/preprints/2013/11/
7. Benner, P., Sima, V., Voigt, M.: FORTRAN 77 subroutines for the solution of skew-Hamiltonian/Hamiltonian eigenproblems – Part II: implementation and numerical results. Preprint MPIMD/13-12, Max Planck Institute Magdeburg (2013). Available from http://www.mpi-magdeburg.mpg.de/preprints/2013/12/
8. Benner, P., Voigt, M.: A structured pseudospectral method for $\mathscr{H}_\infty$-norm computation of large-scale descriptor systems. Math. Control Signals Syst. **26**(2), 303–338 (2014)
9. Boley, D.L.: The algebraic structure of pencils and block Toeplitz matrices. Linear Algebra Appl. **279**(1–3), 255–279 (1998)
10. Boyd, S., Balakrishnan, V.: A regularity result for the singular values of a transfer matrix and a quadratically convergent algorithm for computing its L_∞-norm. Syst. Control Lett. **15**(1), 1–7 (1990)
11. Boyd, S., Balakrishnan, V., Kabamba, P.: A bisection method for computing the H_∞ norm of a transfer matrix and related problems. Math. Control Signals Syst. **2**(3), 207–219 (1989)
12. Bruinsma, N.A., Steinbuch, M.: A fast algorithm to compute the H_∞-norm of a transfer function matrix. Syst. Control Lett. **14**(4), 287–293 (1990)
13. Buttà, P., Guglielmi, N., Noschese, S.: Computing the structured pseudospectrum of a Toeplitz matrix and its extreme points. SIAM J. Matrix Anal. Appl. **33**(4), 1300–1319 (2012)
14. Byers, R.: A bisection method for measuring the distance of a stable matrix to the unstable matrices. SIAM J. Sci. Stat. Comput. **9**, 875–881 (1988)
15. Byers, R., He, C., Mehrmann, V.: Where is the nearest non-regular pencil? Linear Algebra Appl. **285**, 81–105 (1998)
16. Demmel, J.W.: On condition numbers and the distance to the nearest ill-posed problem. Numer. Math. **51**(3), 251–289 (1987)
17. Du, N.H., Thuan, D.D., Liem, N.C.: Stability radius of implicit dynamic equations with constant coefficients on time scales. Syst. Control Lett. **60**(8), 596–603 (2011)
18. Du, N.H., Linh, V.H., Mehrmann, V.: Robust stability of differential-algebraic equations. In: Ilchmann, A., Reis, T. (eds.) Surveys in Differential-Algebraic Equations I. Differential-Algebraic Equations Forum, chap. 2, pp. 63–95. Springer, Berlin/Heidelberg (2013)
19. Freitag, M.A., Spence, A.: A Newton-based method for the calculation of the distance to instability. Linear Algebra Appl. **435**(12), 3189–3205 (2011)
20. Freitag, M.A., Spence, A.: A new approach for calculating the real stability radius. BIT **54**(2), 381–400 (2014)
21. Freitag, M.A., Spence, A., Van Dooren, P.: Calculating the H_∞-norm using the implicit determinant method. SIAM J. Matrix Anal. Appl. **35**(2), 619–634 (2014)

22. Genin, Y., Dooren, P.V., Vermaut, V.: Convergence of the calculation of $\mathscr{H}_\infty$ norms and related questions. In: Proceedings of 13th Symposium on Mathematical Theory of Networks and Systems, pp. 429–432. Padova (1998)
23. Golub, G.H., Van Loan, C.F.: Matrix Computations, 4th edn. The John Hopkins University Press, Baltimore (2013)
24. Guglielmi, N., Gürbüzbalaban, M., Overton, M.L.: Fast approximation of the H_∞ norm via optimization of spectral value sets. SIAM J. Matrix Anal. Appl. **34**(2), 709–737 (2013)
25. Guglielmi, N., Kressner, D., Lubich, C.: Low rank differential equations for Hamiltonian matrix nearness problems. Numer. Math. (2014). Also available from http://link.springer.com/article/10.1007%2Fs00211-014-0637-x
26. Guglielmi, N., Kressner, D., Lubich, C.: Computing extremal points of symplectic pseudospectra and solving symplectic matrix nearness problems. SIAM J. Matrix Anal. Appl. **35**(2), 1407–1428 (2014)
27. Guglielmi, N., Lubich, C.: Differential equations for roaming pseudospectra: paths to extremal points and boundary tracking. SIAM J. Numer. Anal. **49**(3), 1194–1209 (2011)
28. Guglielmi, N., Lubich, C.: Erratum/addendum: Differential equations for roaming pseudospectra: paths to extremal points and boundary tracking. SIAM J. Numer. Anal. **50**(2), 977–981 (2012)
29. Guglielmi, N., Lubich, C.: Low-rank dynamics for computing extremal points of real pseudospectra. SIAM J. Matrix Anal. Appl. **34**(1), 40–66 (2013)
30. Guglielmi, N., Manetta, M.: Approximating real stability radii (2014). Available from http://matematica.univaq.it/~guglielm/PAPERS/GuMaSR.pdf
31. Guglielmi, N., Overton, M.L.: Fast algorithms for the approximation of the pseudospectral abscissa and pseudospectral radius of a matrix. SIAM J. Matrix Anal. Appl. **32**(4), 1166–1192 (2011)
32. He, C., Watson, G.A.: An algorithm for computing the distance to instability. SIAM J. Matrix Anal. Appl. **20**(1), 101–116 (1999)
33. Higham, N.J., Konstantinov, M., Mehrmann, V., Petkov, P.: The sensitivity of computational control problems. IEEE Control Syst. Mag. **24**(1), 28–43 (2004)
34. Hinrichsen, D., Motscha, M.: Optimization problems in the robustness analysis of linear state space systems. In: Approximation and Optimization (Havana, 1987). Lecture Notes in Mathematics, vol. 1354, pp. 54–78. Springer, Berlin (1988)
35. Hinrichsen, D., Pritchard, A.J.: Stability radii of linear systems. Syst. Control Lett. **7**, 1–10 (1986)
36. Hinrichsen, D., Pritchard, A.J.: Stability radius for structured perturbations and the algebraic Riccati equation. Syst. Control Lett. **8**, 105–113 (1986)
37. Hinrichsen, D., Pritchard, A.J.: Real and Complex Stability Radii: A Survey. Progress in Systems and Control Theory, vol. 6, pp. 119–162. Birkhäuser, Boston (1990)
38. Hinrichsen, D., Pritchard, A.J.: Mathematical Systems Theory I: Modeling, State Space Analysis, Stability and Robustness. Texts in Applied Mathematics, vol. 48. Springer, New York (2005)
39. Karow, M.: μ-values and spectral value sets for linear perturbation classes defined by a scalar product. SIAM J. Matrix Anal. Appl. **32**(3), 845–865 (2011)
40. Karow, M., Kokiopoulou, E., Kressner, D.: On the computation of structured singular values and pseudospectra. Syst. Control Lett. **59**(2), 122–129 (2010)
41. Kressner, D.: Finding the distance to instability of a large sparse matrix. In: IEEE International Symposium on Computer-Aided Control Systems Design, Munich (2006)
42. Kressner, D., Vandereycken, B.: Subspace methods for computing the pseudospectral abscissa and the stability radius. SIAM J. Matrix Anal. Appl. **35**(1), 292–313 (2014)
43. Kunkel, P., Mehrmann, V.: Differential-Algebraic Equations. Analysis and Numerical Solution. EMS Publishing House, Zürich (2006)
44. Lewis, A.S., Malick, J.: Alternating projections on manifolds. Math. Oper. Res. **33**(1), 216–234 (2008)

45. Markovsky, I.: Low Rank Approximation: Algorithms, Implementation, Applications. Communications and Control Engineering. Springer, London (2012)
46. Markovsky, I., Usevich, K.: Software for weighted structured low-rank approximation. J. Comput. Appl. Math. **256**, 278–292 (2014)
47. Mehl, C., Mehrmann, V., Wojtylak, M.: On the distance to singularity via low rank perturbations. MATHEON-Preprint 1058, DFG-Forschungszentrum MATHEON (2014)
48. Mehrmann, V., Stykel, T.: Balanced truncation model reduction for large-scale systems in descriptor form. In: Benner, P., Mehrmann, V., Sorensen, D. (eds.) Dimension Reduction of Large-Scale Systems. Lecture Notes in Computational Science and Engineering, vol. 45, chap. 3, pp. 89–116. Springer, Berlin/Heidelberg, New York (2005)
49. Mengi, E., Yildirim, E.A., Kiliç, M.: Numerical optimization of eigenvalues of Hermitian matrix functions. SIAM J. Matrix Anal. Appl. **35**(2), 699–724 (2014)
50. Michiels, W., Guglielmi, N.: An iterative method for computing the pseudospectral abscissa for a class of nonlinear eigenvalue problems. SIAM J. Sci. Comput. **34**(4), A2366–A2393 (2012)
51. Mitchell, T.: Robust and efficient methods for approximation and optimization of stability measures. Ph.D. thesis, NYU (2014)
52. Mitchell, T., Overton, M.L.: Fast approximation of the H_∞ norm via hybrid expansion-contraction using spectral value sets. Tech. Rep. (2014)
53. Qiu, L., Bernhardsson, B., Rantzer, A., Davison, E.J., Young, P.M., Doyle, J.C.: A formula for computation of the real stability radius. Automatica **31**(6), 879–890 (1995)
54. Reis, T.: Circuit synthesis of passive descriptor systems – a modified nodal approach. Int. J. Circ. Theor. Appl. **38**(1), 44–68 (2010)
55. Riaza, R.: Differential-Algebraic Systems. Analytical Aspects and Circuit Applications. World Scientific, Singapore (2008)
56. Sreedhar, J., Dooren, P.V., Tits, A.L.: A fast algorithm to compute the real structured stability radius. In: Jeltsch, R., Mansour, M. (eds.) Stability Theory, International Series of Numerical Mathematics, vol. 121, pp. 219–230. Birkhäuser, Basel (1996)
57. Stewart, G.W., Sun, J.G.: Matrix Perturbation Theory. Academic, New York (1990)
58. Trefethen, L.N., Embree, M.: Spectra and Pseudospectra: The Behavior of Nonnormal Matrices and Operators. Princeton University Press, Princeton (2005)
59. Trenn, S.: Solution concepts for linear DAEs: A survey. In: Ilchmann, A., Reis, T. (eds.) Surveys in Differential-Algebraic Equations I. Differential-Algebraic Equations Forum, pp. 137–172. Springer, Berlin/Heidelberg (2013)
60. Van Loan, C.F.: How near is a stable matrix to an unstable matrix? Contemp. Math. **47**, 465–478 (1984)
61. Voigt, M.: On linear-quadratic optimal control and robustness of differential-algebraic systems. Dissertation, Fakultät für Mathematik, Otto-von-Guericke-Universität Magdeburg (2015). Submitted
62. Wicks, M., DeCarlo, R.A.: Computing the distance to an uncontrollable system. IEEE Trans. Automat. Control **36**(1), 39–49 (1991)

Chapter 21
Discrete Input/Output Maps and their Relation to Proper Orthogonal Decomposition

Manuel Baumann, Jan Heiland, and Michael Schmidt

Abstract Current control design techniques require system models of moderate size to be applicable. The generation of such models is challenging for complex systems which are typically described by partial differential equations (PDEs), and model-order reduction or low-order-modeling techniques have been developed for this purpose. Many of them heavily rely on the state space models and their discretizations. However, in control applications, a sufficient accuracy of the models with respect to their input/output (I/O) behavior is typically more relevant than the accurate representation of the system states. Therefore, a discretization framework has been developed and is discussed here, which heavily focuses on the I/O map of the original PDE system and its direct discretization in the form of an I/O matrix and with error bounds measuring the relevant I/O error. We also discuss an SVD-based dimension reduction for the matrix representation of an I/O map and how it can be interpreted in terms of the Proper Orthogonal Decomposition (POD) method which gives rise to a more general POD approach in time capturing. We present numerical examples for both, reduced I/O map s and generalized POD.

21.1 Introduction

To come up with a real time controller for a system of partial differential equations, the synthesis of a surrogate model of moderate size that still inherits the important system dynamics is a necessary step.

M. Baumann
Faculty EWI, Delft Institute of Applied Mathematics, Mekelweg 4, 2628 CD Delft, The Netherlands
e-mail: m.m.baumann@tudelft.nl

J. Heiland (✉)
Max Planck Institute for Dynamics of Complex Technical Systems, Sandtorstr. 1, 39106 Magdeburg, Germany
e-mail: heiland@mpi-magdeburg.mpg.de

M. Schmidt
University of Applied Sciences Offenburg, Badstraße 24, 77652 Offenburg, Germany
e-mail: schmidt@hs-offenburg.de

P. Benner et al. (eds.), *Numerical Algebra, Matrix Theory, Differential-Algebraic Equations and Control Theory*, DOI 10.1007/978-3-319-15260-8_21

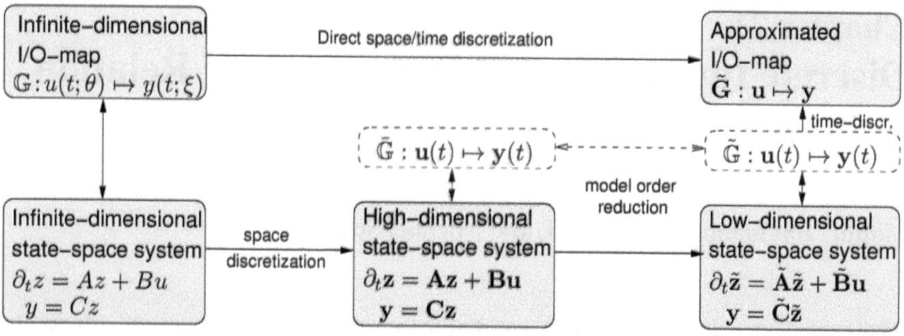

Fig. 21.1 Scheme of the classical model reduction approach and the direct discretization of the I/O map

A classical approach to obtain such surrogate models is to start with a finite but very high-dimensional state space model stemming from a spatial discretization of the system's original PDE state space model. In a next step, this model is reduced by so-called model-order reduction techniques (e.g. moment matching, balanced truncation) to a state space model of moderate size, for which current control design methods become feasible [2, 5, 15]. Often, a linearization step has to be carried out at some stage in addition. Another approach is low-order modeling POD, where the original state space model is analyzed to identify few and particularly relevant state space trajectories and which allows an approximate description of the state trajectory of the full system as a linear combination of the few ones [6, 14, 19, 20].

The classical methods mentioned above have in common that they heavily rely and focus on a state space representation. However, for control applications, only the I/O behavior of the system is of interest. Furthermore, state space representations of control systems can have simple I/O behaviors. Control engineers frequently use this insight when using black-box models from system identification methods applied to measured input/output sequences.

The high relevance of the I/O behavior has motivated the development of an alternative approach to generate surrogate models for control applications based on a direct discretization of the I/O behavior of the original infinite-dimensional system. A theoretical framework for the discretization of I/O map s of many important classes of linear infinite-dimensional systems has been established over the last years. Nonlinear systems can be treated in many cases after a linearization step [22, Ch. 3]. The result is an approximate representation of the I/O behavior via a matrix, and the approximation error is always considered with respect to the introduced error in the I/O behavior, cf. Fig. 21.1.

Our article is organized as follows: In Sect. 21.2, we will recall the essentials of the theoretical framework for the direct discretization of I/O map s established in [22] including error estimates. In Sect. 21.3, we will discuss the interesting feature of performing a special form of singular value decompositions on the I/O matrix representation, allowing to further reduce the size of the representation, but also to identify relevant locations for actuators and sensors. We will show that the well-known POD method can be interpreted as a special case of the direct I/O map discretization and a subsequent reduction. This observation gives rise to a generalized POD approach by extending the time discretization from snapshots to a wider class of time discretizations. Finally, in Sect. 21.4, we present numerical examples of applications of the direct discretization method, its application in optimal control, and of the generalized POD approach.

21.2 Direct Discretizations of I/O Maps

21.2.1 I/O Maps of Linear Systems

We follow [22] in style and notation and consider linear time-invariant (LTI) systems of first order:

$$\partial_t z(t) = Az(t) + Bu(t), \quad t \in (0, T], \tag{21.1a}$$

$$z(0) = z^0 \in D(A) \subset Z, \tag{21.1b}$$

$$y(t) = Cz(t), \quad t \in [0, T]. \tag{21.1c}$$

Here, for time $t \in [0, T]$, the state $z(t)$ takes values in a Hilbert space Z. We assume that A is a densely defined unbounded operator $A: Z \supset D(A) \to Z$ generating a C^0-semigroup $(S(t))_{t\geq 0}$ on Z.

For Hilbert spaces U and Y, we assume the controls to be functions $u \in \mathcal{U} := L^2(0, T; U)$ and the observations $y \in \mathcal{Y} \subset L^2(0, T; Y)$.

If $B: U \to Z$ and $C: Z \to Y$ are bounded, then a bounded linear I/O map $\mathbb{G} \in \mathcal{L}(\mathcal{U}, \mathcal{Y})$ can be associated with (21.1) by applying C to the unique mild solution defined as

$$z(t) = S(t)z^0 + \int_0^t S(t-s)Bu(s)\, ds, \quad t \in [0, T],$$

see, e.g., [21, Ch. 4].

For later reference, we note that we can represent $\mathbb{G}$ as a convolution with a kernel function $K \in L^2(-T, T; \mathcal{L}(U, Y))$ via

$$(\mathbb{G}u)(t) = \int_0^T K(t-s)u(s)\, ds, \quad t \in [0, T], \tag{21.2}$$

with

$$K(t) = \begin{cases} CS(t)B, & t \geq 0 \\ 0, & t < 0 \end{cases}.$$

This framework is suited for various differential equations (21.1), for instance heat equations, wave equations, transport equations, as well as linearizations of Navier-Stokes equations, cf. [22].

21.2.2 Discretization of I/O Maps in Two Steps

For the discretization of the I/O map

$$\mathbb{G}: \mathcal{U} \to \mathcal{Y}, \qquad u \mapsto y,$$

of the abstract system (21.1), we consider two steps:

1. *Approximation of signals.* For finite-dimensional subspaces $\bar{\mathcal{U}} \subset \mathcal{U}$ and $\bar{\mathcal{Y}} \subset \mathcal{Y}$ with orthogonal bases $\{u_1, \ldots, u_{\bar{p}}\} \subset \bar{\mathcal{U}}$ and $\{y_1, \ldots, y_{\bar{q}}\} \subset \bar{\mathcal{Y}}$ and corresponding orthogonal projections $\mathbb{P}_{\bar{\mathcal{U}}}$ and $\mathbb{P}_{\bar{\mathcal{Y}}}$, we find that the approximation

$$\mathbb{G}_S = \mathbb{P}_{\bar{\mathcal{Y}}} \mathbb{G} \mathbb{P}_{\bar{\mathcal{U}}}$$

is a finite-dimensional linear map which can be expressed as a matrix $\mathbf{G} \in \mathbb{R}^{\bar{q} \times \bar{p}}$ with elements $\mathbf{G}_{ij} = (y_i, \mathbb{G} u_j)_{\mathcal{Y}}$.
2. *Approximation of system dynamics.* The components $\mathbf{G}_{ij} = (y_i, \mathbb{G} u_j)_{\mathcal{Y}}$ can be obtained by computing the response of the model successively for inputs $u_1, \ldots, u_{\bar{p}}$ and by testing it against all $y_1, \ldots, y_{\bar{q}}$. For time-invariant state space systems and for bases with a space-time-like tensor structure

$$u_{(j,l)}(t) = \phi_j(t)\mu_l, \quad y_{(i,k)}(t) = \psi_i(t)\nu_k,$$

this task reduces to determining the observations $(\nu_k, C z_l(t))_Y$ from the states $z_l(t) = S(t) B \mu_l$, where μ_l, $l = 1, \ldots, p$ and ν_k, $k = 1, \cdots, q$ form bases of finite-dimensional subspaces of U and Y, where ϕ_j, $j = 1, \cdots, r$ and ψ_i, $i = 1, \cdots, s$ are bases of the time dimensions, and where $S(t)$ is the system's evolution semigroup.

Because the system's response is typically evaluated numerically, one has to consider an approximation $\mathbb{G}_{DS}$ of $\mathbb{G}_S$. The resulting total error ϵ_{DS} can be decomposed into the *signal* approximation error ϵ_S and the *dynamical* approximation error ϵ_D, i.e.

$$\underbrace{\|\mathbb{G} - \mathbb{G}_{DS}\|}_{=:\epsilon_{DS}} \leq \underbrace{\|\mathbb{G} - \mathbb{G}_S\|}_{=:\epsilon_S} + \underbrace{\|\mathbb{G}_S - \mathbb{G}_{DS}\|}_{=:\epsilon_D}, \tag{21.3}$$

in appropriate norms. In Theorem 1, it is shown that one can adjust $\bar{\mathcal{U}}$ and $\bar{\mathcal{Y}}$ and the accuracy of the numerical computations such that the errors are balanced.

Table 21.1 The considered I/O maps, their discretization, and their numerical approximation. The first row contains the operators for a general system and for a space-time like tensor structure of the discrete signal spaces $\bar{\mathcal{U}} = \mathcal{U}_{h_1,\tau_1} = \mathcal{R}_{\tau_1} \cdot U_{h_1}$ and $\bar{\mathcal{Y}} = \mathcal{Y}_{h_2,\tau_2} = \mathcal{S}_{\tau_2} \cdot Y_{h_2}$. The second line lists their matrix representations and the third line contains the low-dimensional approximation of $\tilde{\mathbf{G}}$

∞-dimensional system	Discrete I/O spaces	Numerical approximation
$\mathbb{G}: \mathcal{U} \to \mathcal{Y}$	$\mathbb{G}_S: \bar{\mathcal{U}} \to \bar{\mathcal{Y}}$	$\mathbb{G}_{DS}: \bar{\mathcal{U}} \to \bar{\mathcal{Y}}$
	$\mathbb{G}_S: \mathcal{U}_{h_1,\tau_1} \to \mathcal{Y}_{h_2,\tau_2}$	$\mathbb{G}_{DS}: \mathcal{U}_{h_1,\tau_1} \to \mathcal{Y}_{h_2,\tau_2}$
	$\mathbf{G}: \mathbb{R}^{pr} \to \mathbb{R}^{qs}$	$\tilde{\mathbf{G}}: \mathbb{R}^{pr} \to \mathbb{R}^{qs}$
	$\mathbf{H} = \mathbf{M}_{\bar{\mathcal{Y}}} \mathbf{G}: \mathbb{R}^{pr} \to \mathbb{R}^{qs}$	$\tilde{\mathbf{H}}: \mathbb{R}^{pr} \to \mathbb{R}^{qs}$
		$\hat{\mathbf{G}}: \mathbb{R}^{\hat{p}\hat{r}} \to \mathbb{R}^{\hat{q}\hat{s}}$

In what follows, we consider discrete I/O maps and their numerical approximation. We have summarized the symbols used for the definition of the I/O maps on the different levels of approximation in Table 21.1.

21.2.3 Discretization of Signals

21.2.3.1 Space-Time Discretization and Matrix Representation

Following the notation used in [17, 22], we recall the definitions and notions for direct discretization of I/O map s. In order to discretize the input signals $u \in \mathcal{U}$ and $y \in \mathcal{Y}$ in space and time, we choose four families $\{U_{h_1}\}_{h_1>0}$, $\{Y_{h_2}\}_{h_2>0}$, $\{\mathcal{R}_{\tau_1}\}_{\tau_1>0}$, and $\{\mathcal{S}_{\tau_2}\}_{\tau_2>0}$ of subspaces $U_{h_1} \subset U$, $Y_{h_2} \subset Y$, $\mathcal{R}_{\tau_1} \subset L^2(0,T)$ and $\mathcal{S}_{\tau_2} \subset L^2(0,T)$ of dimensions $p(h_1) = \dim(U_{h_1})$, $q(h_2) = \dim(Y_{h_2})$, $r(\tau_1) = \dim(\mathcal{R}_{\tau_1})$, and $s(\tau_2) = \dim(\mathcal{S}_{\tau_2})$ and define

$$\mathcal{U}_{h_1,\tau_1} := \mathcal{R}_{\tau_1} \cdot U_{h_1} \quad \text{and} \quad \mathcal{Y}_{h_2,\tau_2} = \mathcal{S}_{\tau_2} \cdot Y_{h_2}.$$

We denote the orthogonal projections onto these subspaces by $\mathbb{P}_{\mathcal{U},h_1,\tau_1} \in \mathcal{L}(\mathcal{U})$ and $\mathbb{P}_{\mathcal{Y},h_2,\tau_2} \in \mathcal{L}(\mathcal{Y})$. To approximate $\mathbb{G}$, we define

$$\mathbb{G}_S = \mathbb{G}_S(h_1,\tau_1,h_2,\tau_2) = \mathbb{P}_{\mathcal{Y},h_2,\tau_2} \mathbb{G} \mathbb{P}_{\mathcal{U},h_1,\tau_1} \in \mathcal{L}(\mathcal{U},\mathcal{Y}).$$

In order to obtain a matrix representation of $\mathbb{G}_S$, we introduce families of bases $\{\mu_1,\dots,\mu_p\}$ of U_{h_1}, $\{\nu_1,\dots,\nu_q\}$ of Y_{h_2}, $\{\phi_1,\dots,\phi_r\}$ of $\mathcal{R}_{\tau_1}$ and $\{\psi_1,\dots,\psi_s\}$ of $\mathcal{S}_{\tau_2}$ and corresponding mass matrices $\mathbf{M}_{U,h_1} \in \mathbb{R}^{p\times p}$, $\mathbf{M}_{Y,h_2} \in \mathbb{R}^{q\times q}$, $\mathbf{M}_{\mathcal{R},\tau_1} \in \mathbb{R}^{r\times r}$ and $\mathbf{M}_{\mathcal{S},\tau_2} \in \mathbb{R}^{s\times s}$, for instance via

$$[\mathbf{M}_{U,h_1}]_{l_1 l_2} = (\mu_{l_1}, \mu_{l_2})_U, \qquad l_1, l_2 = 1,\dots,p.$$

These mass matrices induce weighted scalar products and corresponding norms in the respective spaces, which we indicate by a subscript w, like $\mathbb{R}^p_w$ with $(\cdot,\cdot)_{p;w}$ and $\|\cdot\|_{p;w}$. We write signals $u \in \mathcal{U}_{h_1,\tau_1}$ and $y \in \mathcal{Y}_{h_2,\tau_2}$ as

$$u(t) = \sum_{l=1}^{p}\sum_{j=1}^{r} \mathbf{u}^l_j \phi_j(t)\mu_l, \qquad y(t) = \sum_{k=1}^{q}\sum_{i=1}^{s} \mathbf{y}^k_i \psi_i(t)\nu_k, \tag{21.4}$$

where $\mathbf{u}^l_j$ are the elements of a block-structured vector $\mathbf{u} \in \mathbb{R}^{pr}$ with p blocks $\mathbf{u}^k \in \mathbb{R}^r$, and the vector $\mathbf{y} \in \mathbb{R}^{qs}$ is defined similarly.

We obtain a matrix representation $\mathbf{G}$ of $\mathbb{G}_S$ by setting

$$\mathbf{G} = \mathbf{G}(h_1,\tau_1,h_2,\tau_2) = \kappa_{\mathcal{Y}}\mathbb{P}_{\mathcal{Y}}\mathbb{G}\mathbb{P}_{\mathcal{U}}\kappa_{\mathcal{U}}^{-1} \in \mathbb{R}^{qs\times pr},$$

where some dependencies on h_1,τ_1,h_2,τ_2 have been omitted. With the norms $\|\cdot\|_{pr;w}$ and $\|\cdot\|_{qs;w}$ in the product spaces with the mass matrices

$$\mathbf{M}_{\mathcal{U},h_1,\tau_1} = \mathbf{M}_{U,h_1} \otimes \mathbf{M}_{\mathcal{R},\tau_1} \in \mathbb{R}^{pr\times pr}, \quad \mathbf{M}_{\mathcal{Y},h_2,\tau_2} = \mathbf{M}_{Y,h_2} \otimes \mathbf{M}_{\mathcal{S},\tau_2} \in \mathbb{R}^{qs\times qs},$$

the isomorphisms $\kappa_{\mathcal{U},h_1,\tau_1} \in \mathscr{L}(\mathcal{U}_{h_1,\tau_1},\mathbb{R}^{pr}_w)$ and $\kappa_{\mathcal{Y},h_2,\tau_2} \in \mathscr{L}(\mathcal{Y}_{h_2,\tau_2},\mathbb{R}^{qs}_w)$ associating functions with coefficient vectors are unitary mappings and we can define the discrete $\mathscr{L}(\mathcal{U},\mathcal{Y})$-norm as

$$\|\mathbf{G}(h_1,\tau_1,h_2,\tau_2)\|_{qs\times pr;w} := \sup_{\mathbf{u}\in\mathbb{R}^{pr}} \frac{\|\mathbf{G}\mathbf{u}\|_{qs;w}}{\|\mathbf{u}\|_{pr;w}} = \|\mathbf{M}^{1/2}_{\mathcal{Y},h_2,\tau_2}\mathbf{G}\mathbf{M}^{-1/2}_{\mathcal{U},h_1,\tau_1}\|_{qs\times pr}.$$

For later reference, we define $\mathbf{H} = \mathbf{H}(h_1,\tau_1,h_2,\tau_2) := \mathbf{M}_{\mathcal{Y},h_2,\tau_2}\mathbf{G} \in \mathbb{R}^{qs\times pr}$, which is a matrix of $q\times p$ blocks $\mathbf{H}^{kl} \in \mathbb{R}^{s\times r}$ with block elements

$$\mathbf{H}^{kl}_{ij} = [\mathbf{M}_{\mathcal{Y}}\kappa_{\mathcal{Y}}\mathbb{P}_{\mathcal{Y}}\mathbb{G}(\mu_l\phi_j)]^l_j = (\nu_k\psi_i,\mathbb{G}(\mu_l\phi_j))_{\mathcal{Y}}. \tag{21.5}$$

We have the following convergence result:

Lemma 1 (Lem. 3.2, [22]) *For all $(h_1,\tau_1,h_2,\tau_2) \in \mathbb{R}^4_+$, we have*

$$\|\mathbf{G}(h_1,\tau_1,h_2,\tau_2)\|_{qs\times pr;w} = \|\mathbb{G}_S(h_1,\tau_1,h_2,\tau_2)\|_{\mathscr{L}(\mathcal{U},\mathcal{Y})} \le \|\mathbb{G}\|_{\mathscr{L}(\mathcal{U},\mathcal{Y})}.$$

If the subspaces $\{\mathcal{U}_{h_1,\tau_1}\}_{h_1,\tau_1>0}$ and $\{\mathcal{Y}_{h_2,\tau_2}\}_{h_2,\tau_2>0}$ are nested, i.e.

$$\mathcal{U}_{h_1,\tau_1} \subset \mathcal{U}_{h_1',\tau_1'}, \quad \mathcal{Y}_{h_2,\tau_2} \subset \mathcal{Y}_{h_2',\tau_2'}, \quad \text{for } (h_1',\tau_1',h_2',\tau_2') \le (h_1,\tau_1,h_2,\tau_2),$$

then $\|\mathbf{G}(h_1,\tau_1,h_2,\tau_2)\|_{qs\times pr;w}$ is monotonically increasing and bounded, i.e. convergent, for $(h_1,\tau_1,h_2,\tau_2) \searrow 0$.

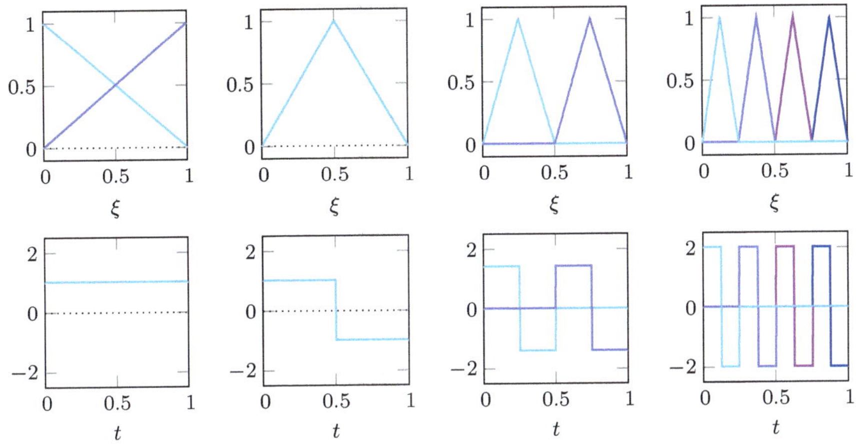

Fig. 21.2 Four levels of bases for nested $L^2(0,1)$-subspaces: Hierarchical basis for piecewise linear functions (*above*) and Haar wavelets for piecewise constant functions (*below*)

21.2.3.2 An Example for Signal Discretizations

Let $U = Y = L^2(0,1)$ and let U_{h_1} and Y_{h_2} be spanned by continuous piecewise linear functions and let $\mathcal{R}_{\tau_1}$ and $\mathcal{S}_{\tau_2}$ be spanned by piecewise constant functions. For equidistant grids one can easily construct nested bases as illustrated in Fig. 21.2.

If we denote the orthogonal projections onto U_{h_1} and $\mathcal{R}_{\tau_1}$ by P_{U,h_1} and $P_{\mathcal{R},\tau_1}$, respectively, one can show that there exist $c_U = 1/2$ and $c_{\mathcal{R}} = 1/\sqrt{2}$, independent of h_1, τ_1 and T, such that

$$\|u - P_{U_{h_1}} u\|_{L^2(0,1)} \leq c_U h_1^2 \|\partial_\xi^2 u\|_{L^2(0,1)} \quad \text{for } u \in H^2(0,1),$$

$$\|v - P_{\mathcal{R}_{\tau_1}} v\|_{L^2(0,T)} \leq c_{\mathcal{R}} \tau_1 \|\partial_t v\|_{L^2(0,T)} \quad \text{for } v \in H^1(0,T),$$

see e.g. [7]. From *Fubini*'s theorem, one can infer that a similar estimate holds for time space tensorized bases, i.e., the projection $\mathbb{P}_{\mathcal{U},h_1,\tau_1}$ satisfies

$$\|u - \mathbb{P}_{\mathcal{U},h_1,\tau_1} u\|_{\mathcal{U}} \leq (c_U h_1^2 + c_{\mathcal{R}} \tau_1)\|u\|_{\mathcal{U}_s} \quad \text{for all } u \in \mathcal{U}_s = H^{1,2}((0,T) \times (0,1)).$$

The same is true for the similarly defined projection $\mathbb{P}_{\mathcal{Y},h_2,\tau_2}$.

Similar estimates also hold for higher spatial dimensions as they are classical results from the interpolation theory in Sobolev spaces, see e.g. [7]. For spaces of higher regularity, using basis functions of higher polynomial degree, higher approximation orders are also possible.

21.2.3.3 Signal Approximation Error

As shown in [22, Lem. 3.3], the error in the signal approximation can be decomposed via $\epsilon_s := \|\mathbb{G} - \mathbb{G}_S\|_{\mathscr{L}(\mathcal{U},\mathcal{Y})} = \epsilon_{s,inp} + \epsilon_{s,outp}$ with

$$\epsilon_{s,inp} := \sup_{u \in \ker \mathbb{P}_{\mathcal{U},h_1,\tau_1}} \frac{\|\mathbb{G}u\|_{\mathcal{Y}}}{\|u\|_{\mathcal{U}}}, \quad \epsilon_{s,outp} := \max_{u \in \mathcal{U}_{h_1,\tau_1}} \frac{\|(I - \mathbb{P}_{\mathcal{Y},h_2,\tau_2})\mathbb{G}u\|_{\mathcal{Y}}}{\|u\|_{\mathcal{U}}}.$$

As laid out in the following remarks, a good approximation in $\|\cdot\|_{\mathscr{L}(\mathcal{U},\mathcal{Y})}$ can only be achieved, if the subspaces $\mathcal{U}_{h_1,\tau_1}$ and $\mathcal{Y}_{h_2,\tau_2}$ are chosen *specifically* for $\mathbb{G}$. In short, the response of the chosen inputs needs to be well captured by the chosen output discretization. On the other hand, the discretization of the input space has to catch the major output modes of the system.

Remark 1 The usual requirement for families of approximating subspaces $\mathcal{U}_{h_1,\tau_1}$ and $\mathcal{Y}_{h_2,\tau_2}$ that they become dense in their superspaces is sufficient for pointwise convergence of $\|(\mathbb{G} - \mathbb{G}_S)u\|_{\mathcal{Y}} \to 0$ for every $u \in \mathcal{U}$, i.e. for convergence in the *strong operator topology*, but not for *uniform* convergence $\|\mathbb{G} - \mathbb{G}_S\|_{\mathscr{L}(\mathcal{U},\mathcal{Y})} \to 0$, cf. [22, Rem. 3.11].

Remark 2 For compact $\mathbb{G} \in \mathscr{L}(\mathcal{U},\mathcal{Y})$, there exist orthonormal systems $\{\hat{u}_1, \hat{u}_2 \dots\}$ of $\mathcal{U}$ and $\{\hat{y}_1, \hat{y}_2, \dots\}$ of $\mathcal{Y}$ and nonnegative numbers $\sigma_1 \geq \sigma_2 \geq \dots$ with $\sigma_k \to 0$ such that $\mathbb{G}u = \sum_{k=1}^{\infty} \sigma_k (u, \hat{u}_k)_{\mathcal{U}} \hat{y}_k$ for all $u \in \mathcal{U}$, see e.g. [26]. Thus, if we choose $\mathcal{U}_{h_1,\tau_1}$ and $\mathcal{Y}_{h_2,\tau_2}$ as the span of $\hat{u}_1, \dots, \hat{u}_r$ and $\hat{y}_1, \dots, \hat{y}_s$, respectively, with $s = r$ and $r \in \mathbb{N}$, we obtain an efficient approximation $\mathbb{G}_S$ of $\mathbb{G}$ with $\|\mathbb{G} - \mathbb{G}_S\|_{\mathscr{L}(\mathcal{U},\mathcal{Y})} \leq \sigma_{r+1}$. However, we must point out that the I/O map of a linear system is not compact unless it is the zero map. This can be deducted from the common fact that the I/O map of a causal linear system is a *Toeplitz* operator [23] which is compact only if it is the zero operator [10, Rem. 7.15].

21.2.4 System Dynamics Approximation

We discuss the efficient computation of the discrete I/O map $\mathbf{G} = \mathbf{M}_{\mathcal{Y}}^{-1}\mathbf{H}$ of a state space system, via the approximation of the associated convolution kernel $K \in L^2(0, T; \mathscr{L}(U, Y))$.

21.2.4.1 Kernel Function Approximation

If we insert (21.2) in (21.5), after a change of variables, we obtain

$$\mathbf{H}_{ij}^{kl} = \int_0^T \int_0^T \psi_i(t)\phi_j(s)(\nu_l, K(t-s)\mu_k)_Y \, ds \, dt = \int_0^T \mathbf{W}_{ij}(t)\mathbf{K}_{kl}(t) \, dt$$

where $\mathbf{W}\colon [0,T] \to \mathbb{R}^{s\times r}$ and $\mathbf{K}\colon [0,T] \to \mathbb{R}^{q\times p}$,

$$\mathbf{W}_{ij}(t) = \int_0^{T-t} \psi_i(t+s)\phi_j(s)\,ds, \qquad \mathbf{K}_{kl}(t) = (\nu_k, K(t)\mu_l)_Y,$$

so that

$$\mathbf{H} = \mathbf{M}_{\mathcal{Y}}\mathbf{G} = \int_0^T \mathbf{K}(t) \otimes \mathbf{W}(t)\,dt.$$

Remark 3 For piecewise polynomial ansatz functions $\psi_i(t)$ and $\phi_j(t)$, $\mathbf{W}(t)$ can be calculated exactly. For particular choices, $\mathbf{W}(t) \in \mathbb{R}^{r\times r}$ is a lower triangular Toeplitz matrix for all $t \in [0,T]$. Hence, the matrices $\mathbf{H}_{ij} = \int_0^T \mathbf{W}_{ij}(t)\mathbf{K}(t)\,dt \in \mathbb{R}^{q\times p}$ satisfy $\mathbf{H}_{ij} = \mathbf{H}_{i-j}$ for $1 \le i,j \le r$ and $\mathbf{H}_{ij} = 0$ for $1 \le i < j \le r$. This implies that $\mathbf{H}_{ij}$ are *Markov parameters* of a discrete-time linear time-invariant causal *MIMO* system [22, Ch. 3.4.3].

For state space systems (21.1), the matrix-valued function $\mathbf{K}$ reads

$$\mathbf{K}_{kl}(t) = (\nu_k, CS(t)B\mu_l)_Y = (c_k^*, S(t)b_l)_Z,$$

where $c_k^* = C^*\nu_k \in Z$ and $b_l = B\mu_l$ for $k = 1,\ldots,q$ and $l = 1,\ldots,p$. Accordingly, one can obtain $\mathbf{K}$ by solving p homogeneous systems

$$\dot{z}_l(t) = Az_l(t), \qquad t \in (0,T], \tag{21.8a}$$

$$z_l(0) = b_l, \tag{21.8b}$$

recalling that (21.8) has the mild solution $z_l(t) = S(t)b_l \in C([0,T];L^2(\Omega))$. Typically, $z_l(t)$ is numerically approximated by $z_{l,\mathtt{tol}}(t)$, which gives only an approximation $\tilde{\mathbf{K}}$ of $\mathbf{K}$ and, thus, an approximation $\tilde{\mathbf{H}}$ of $\mathbf{H}$. Here, the subscript $\mathtt{tol}$ refers to a parameter that controls the error $z_l - z_{l,\mathtt{tol}}$ as we will specify later. Then, the approximation $\mathbb{G}_{DS}$ of $\mathbb{G}_S$, depending on h_1, h_2, τ_1, τ_2 and $\mathtt{tol}$, is given by

$$\mathbb{G}_{DS} = \kappa_{\mathcal{Y}}^{-1}\tilde{\mathbf{G}}\kappa_{\mathcal{U}}\mathbb{P}_{\mathcal{U}}, \quad \text{with } \tilde{\mathbf{G}} = \mathbf{M}_{\mathcal{Y}}^{-1}\tilde{\mathbf{H}}.$$

Note that it might be preferable to consider an adjoint system and approximate the kernel functions via an adjoint system, cf. [22, Rem. 3.14].

21.2.4.2 System Dynamics Error

The approximation error in the system dynamics is related to the approximation error of the states via (21.8) as follows:

Proposition 1 (Thm. 3.6, [22]) *The system dynamics error ϵ_D satisfies*

$$\begin{aligned}\|\mathbb{G}_S - \mathbb{G}_{DS}\|_{\mathscr{L}(\mathcal{U},\mathcal{Y})} &\leq \sqrt{T}\|\mathbf{K} - \tilde{\mathbf{K}}\|_{L^2(0,T;\mathbb{R}_w^{q\times p})} \\ &\leq p\sqrt{T}\sqrt{\frac{\lambda_{\max}(\mathbf{M}_{Y,h_2})}{\lambda_{\min}(\mathbf{M}_{U,h_1})}} \max_{1\leq l\leq p} \|\mathbf{K}_{:,l} - \tilde{\mathbf{K}}_{:,l}\|_{L^2(0,T;\mathbb{R}^q)}.\end{aligned}$$

Here $\mathbf{K}_{:,l}$ and $\tilde{\mathbf{K}}_{:,l}$ denote the l'th column of $\mathbf{K}(t)$ and $\tilde{\mathbf{K}}(t)$ that are defined via the exact and the approximate solutions of (21.8)*, respectively, $\lambda_{\max}(\mathbf{M}_{Y,h_2})$ is the largest eigenvalue of $\mathbf{M}_{Y,h_2}$ and $\lambda_{\min}(\mathbf{M}_{U,h_1})$ the smallest eigenvalue of $\mathbf{M}_{U,h_1}$. $\mathbb{R}_w^{q\times p}$ denotes the space of real $q \times p$-matrices equipped with the weighted matrix norm $\|\mathbf{M}\|_{q\times p;w} = \sup_{\mathbf{u}\neq 0} \|\mathbf{M}\mathbf{u}\|_{q;w}/\|\mathbf{u}\|_{p;w}$.*

21.2.4.3 Error in the State Approximation

As laid out in Sect. 21.2.4.1, the approximation of the system dynamics is realized via numerically solving homogeneous PDEs (21.8) for p different initial values.

In view of the error estimates, we will focus on the approximation properties of the chosen discretization, i.e. whether

$$\|\mathbf{K}_{:,l} - \tilde{\mathbf{K}}_{:,l}\|_{L^2(0,T;\mathbb{R}^q)} < \texttt{tol} \tag{21.9}$$

is *guaranteed* for a given $\texttt{tol} > 0$ rather than on efficiency.

For the concrete system (21.8), the handling of the initial values b_l is an issue, as they in general only belong to Z but not necessarily to $D(A)$. For the heat equation, this typically leads to large but quickly decaying gradients in the exact solution $z_l \in C^1((0,T], H^2(\Omega) \cap H_0^1(\Omega))$.

This is reflected in the analytical bound for the general case

$$\|\partial_t z(t)\|_{L^2(\Omega)} = \|\Delta z(t)\|_{L^2(\Omega)} \leq \frac{c}{t}\|z^0\|_{L^2(\Omega)} \quad \text{for all } t \in (0,T],$$

with some constant $c > 0$ independent of z^0 and T, cf. [18, p. 148].

To capture large gradients in the simulation, adaptive space and time discretizations are the methods of choice [11]. As for the guaranteed accuracy of the state approximation, the combination of discontinuous Galerkin time discretizations with standard Galerkin space discretizations comes with (a priori and a posteriori) error estimators, that also work for adapted meshes [12, 18, 24]. We distinguish two types of error estimates.

1. *Global state error estimates* that measure the global error $(z_l - z_{l,\mathtt{tol}})$, see [12] for a priori and a posteriori estimates for parabolic problems that guarantee (21.9) by ensuring that

$$\|\mathbf{K}_{:,l} - \tilde{\mathbf{K}}_{:,l}\|^2_{L^2(0,T;\mathbb{R}^q)} \leq \|C\|^2_{\mathscr{L}(Z,Y)} \sum_{i=1}^{q} \|v_i\|^2_Y \|z - z^{(l)}_{\mathtt{tol}}\|^2_{L^2(0,T;Z)}. \tag{21.10}$$

2. *Goal-oriented error estimates* that measure the error $\|\mathbf{K}_{:,l} - \tilde{\mathbf{K}}_{:,l}\|_{L^2(0,T;\mathbb{R}^q)}$ directly. This is advantageous when the error in the *observations* $\mathbf{K}_{:,l}$ is small although the error in the states is large, see, e.g., [1].

Thus, for typical applications, via a suitable choice of the approximation schemes, we can fulfill the following assumption:

Assumption 1 *For a given tolerance* $\mathtt{tol}$*, the approximations* $z_{l,\mathtt{tol}}$ *to the solutions* z_l *of* (21.8) *can be computed such that*

$$\|\mathbf{K}_{:,l} - \tilde{\mathbf{K}}_{:,l}\|_{L^2(0,T;\mathbb{R}^q)} < \mathtt{tol}, \qquad l = 1, \ldots, p.$$

21.2.5 Total Error Estimates

We show how the previously introduced error estimates sum up to an estimate of the total error in the approximation of $\mathbb{G}$.

Theorem 1 (Thm. 3.7, [22]) *Consider the I/O map* $\mathbb{G} \in \mathscr{L}(\mathcal{U}, \mathcal{Y})$ *of the infinite-dimensional linear time-invariant system* (21.2) *and assume that* $\mathbb{G}|_{\mathcal{U}_s} \in \mathscr{L}(\mathcal{U}_s, \mathcal{Y}_s)$ *with spaces* $\mathcal{U}_s \subset U$ *and* $\mathcal{Y}_s \subset Y$ *such that for* $\alpha_1, \beta_1, \alpha_2, \beta_2 \in \mathbb{N}$*, the families of subspaces* $\{\mathcal{U}_{h_1,\tau_1}\}_{h_1,\tau_1}$ *and* $\{\mathcal{Y}_{h_2,\tau_2}\}_{h_2,\tau_2}$ *satisfy*

$$\|u - \mathbb{P}_{\mathcal{U},h_1,\tau_1} u\|_{\mathcal{U}} \leq (c_{\mathcal{R}} \tau_1^{\alpha_1} + c_U h_1^{\beta_1}) \|u\|_{\mathcal{U}_s}, \qquad u \in \mathcal{U}_s, \tag{21.11a}$$

$$\|y - \mathbb{P}_{\mathcal{Y},h_2,\tau_2} y\|_{\mathcal{Y}} \leq (c_{\mathcal{S}} \tau_2^{\alpha_2} + c_Y h_2^{\beta_2}) \|y\|_{\mathcal{Y}_s}, \qquad y \in \mathcal{Y}_s, \tag{21.11b}$$

with positive constants $c_{\mathcal{R}}$, $c_{\mathcal{S}}$, c_U *and* c_Y*. And assume that the error in solving for the state dynamics can be made arbitrarily small, i.e. Assumption 1 holds.*

Let $\delta > 0$ *be given. Assume, the chosen subspaces* $\mathcal{U}_{h_1^*,\tau_1^*}$ *and* $\mathcal{Y}_{h_2^*,\tau_2^*}$ *fulfill*

$$\tau_1^* < \left(\frac{\delta}{8 c_{\mathcal{R}} \|\mathbb{G}\|_{\mathscr{L}(\mathcal{U},\mathcal{Y})}}\right)^{1/\alpha_1}, \quad h_1^* < \left(\frac{\delta}{8 c_U \|\mathbb{G}\|_{\mathscr{L}(\mathcal{U},\mathcal{Y})}}\right)^{1/\beta_1}, \tag{21.12a}$$

$$\tau_2^* < \left(\frac{\delta}{8 c_{\mathcal{S}} \|\mathbb{G}\|_{\mathscr{L}(\mathcal{U}_s,\mathcal{Y}_s)}}\right)^{1/\alpha_2}, \quad h_2^* < \left(\frac{\delta}{8 c_Y \|\mathbb{G}\|_{\mathscr{L}(\mathcal{U}_s,\mathcal{Y}_s)}}\right)^{1/\beta_2}, \tag{21.12b}$$

and that one can solve numerically the homogeneous systems (21.8) *for* $k = 1, \dots, p(h_1)$ *such that*

$$\|\mathbf{K}_{:,l} - \tilde{\mathbf{K}}_{:,l}\|_{L^2(0,T;\mathbb{R}^q)} < c_{K,l} := \frac{\delta}{2\sqrt{T}\,p(h_1^*)}\sqrt{\frac{\lambda_{\min}(\mathbf{M}_{U,h_1^*})}{\lambda_{\max}(\mathbf{M}_{Y,h_2^*})}}.$$

Then,

$$\|\mathbb{G} - \mathbb{G}_{DS}\|_{\mathscr{L}(\mathcal{U}_s,\mathcal{Y})} < \delta.$$

Moreover, the signal error $\epsilon_S' := \|\mathbb{G} - \mathbb{G}_S\|_{\mathscr{L}(\mathcal{U}_s,\mathcal{Y})}$ *and the system dynamics error* $\epsilon_D := \|\mathbb{G}_S - \mathbb{G}_{DS}\|_{\mathscr{L}(\mathcal{U},\mathcal{Y})}$ *are balanced in the sense that* $\epsilon_S', \epsilon_D < \delta/2$.

Proof The proof of [22, Thm. 3.7] is readily extended to the considered situtation noting that [22, Thm. 3.5] is valid for any $\mathcal{U}_s$ and $\mathcal{Y}_s$ for which $\mathbb{G}_{|\mathcal{U}_s} \in \mathscr{L}(\mathcal{U}_s, \mathcal{Y}_s)$ and (21.11) holds.

Remark 4 Considering subspaces $\mathcal{U}_s$ and $\mathcal{Y}_s$ allows for possibly better error estimates and does not exclude the case that $\mathcal{U}_s = \mathcal{U}$ or $\mathcal{Y}_s = \mathcal{Y}$. In particular, for modelling distributed control and observation, one can choose $Z = L^2(\Omega)$ for a spatial domain Ω and defines $U = L^2(\Theta)$ and $Y = L^2(\Sigma)$ as function spaces over domains of control $\Theta \subset \Omega$ and observation $\Sigma \subset \Omega$. Then $\mathcal{U}_s$ and $\mathcal{Y}_s$ can be chosen as subspaces of higher regularity like

$$\mathcal{U}_s = H^{\alpha_1,\beta_1}((0,T) \times \Theta), \quad \mathcal{Y}_s = H^{\alpha_2,\beta_2}((0,T) \times \Xi), \quad \alpha_1, \beta_1, \alpha_2, \beta_2 \in \mathbb{N},$$

for which (21.11) is obtained using standard approximation schemes [7].

21.3 Higher Order SVD for I/O Maps and POD

21.3.1 Dimension Reduction of the I/O Map via SVDs

An accurate resolution of the signal spaces by general basis functions may lead to large dimensions of the discrete I/O map $\tilde{\mathbf{G}}$. We show how to employ a *Tucker decomposition* or *higher order singular value decomposition* (HOSVD) [8] to reduce the degrees of freedom in input and output space while preserving accuracy and the tensor structure.

We consider the discrete spaces as introduced in Sect. 21.2.3.1 with their dimensions p, q, r, s and their indexing via j, i, l, k as in (21.4).

For $\tilde{\mathbf{G}} \in \mathbb{R}^{qs \times pr}$ considered as a fourth-order tensor $\tilde{\mathbf{G}} \in \mathbb{R}^{s \times r \times q \times p}$ with $\tilde{\mathbf{G}}_{ijkl} = \tilde{\mathbf{G}}_{ij}^{kl}$ there exists a HOSVD

$$\tilde{\mathbf{G}} = \mathbf{S} \times_1 \mathbf{U}^{(\psi)} \times_2 \mathbf{U}^{(\phi)} \times_3 \mathbf{U}^{(\nu)} \times_4 \mathbf{U}^{(\mu)}, \tag{21.13}$$

with the *core tensor* $\mathbf{S} \in \mathbb{R}^{s\times r\times q\times p}$ satisfying some orthogonality properties and unitary matrices $\mathbf{U}^{(\psi)} \in \mathbb{R}^{s\times s}$, $\mathbf{U}^{(\phi)} \in \mathbb{R}^{r\times r}$ and $\mathbf{U}^{(\nu)} \in \mathbb{R}^{q\times q}$, $\mathbf{U}^{(\mu)} \in \mathbb{R}^{p\times p}$. Here, $\times_1, \ldots, \times_4$ denote tensor-matrix multiplications. We define a *matrix unfolding* $\tilde{\mathbf{G}}^{(\psi)} \in \mathbb{R}^{s\times rqp}$ of the tensor $\tilde{\mathbf{G}}$ via

$$\tilde{\mathbf{G}}^{(\psi)}_{im} = \tilde{\mathbf{G}}_{ijkl}, \quad m = (k-1)ps + (l-1)s + i,$$

which is putting all elements belonging to $\psi_1, \psi_2, \ldots, \psi_s$ into one respective row. Similarly, we define the unfoldings $\tilde{\mathbf{G}}^{(\phi)} \in \mathbb{R}^{r\times qps}$, $\tilde{\mathbf{G}}^{(\nu)} \in \mathbb{R}^{q\times psr}$ and $\tilde{\mathbf{G}}^{(\mu)} \in \mathbb{R}^{p\times srq}$. Then we can calculate $\mathbf{U}^{(\psi)}$, $\mathbf{U}^{(\phi)}$, $\mathbf{U}^{(\nu)}$ and $\mathbf{U}^{(\mu)}$ in (21.13) by means of four SVDs like

$$\tilde{\mathbf{G}}^{(\psi)} = \mathbf{U}^{(\psi)}\Sigma^{(\psi)}(\mathbf{V}^{(\psi)})^{\mathsf{T}},$$

with $\Sigma^{(\psi)}$ diagonal with entries $\sigma_1^{(\psi)} \geq \sigma_2^{(\psi)} \geq \ldots \sigma_s^{(\psi)} \geq 0$ and $\mathbf{V}^{(\psi)}$ column-wise orthonormal. The $\sigma_i^{(\psi)}$ are the *n-mode singular values* of the tensor $\tilde{\mathbf{G}}$.

From (21.13), we derive an approximation $\hat{\mathbf{G}} \in \mathbb{R}^{s\times r\times q\times p}$ of $\tilde{\mathbf{G}}$ by discarding the smallest n-mode singular values $\{\sigma_{\hat{s}+1}^{(\psi)}, \ldots, \sigma_s^{(\psi)}\}$, $\{\sigma_{\hat{r}+1}^{(\phi)}, \ldots, \sigma_r^{(\psi)}\}$, $\{\sigma_{\hat{q}+1}^{(\nu)}, \ldots, \sigma_q^{(\nu)}\}$ and $\{\sigma_{\hat{p}+1}^{(\mu)}, \ldots, \sigma_p^{(\mu)}\}$, i.e. by setting the corresponding parts of $\mathbf{S}$ to zero. Then we have

$$\|\tilde{\mathbf{G}} - \hat{\mathbf{G}}\|_F^2 \leq \sum_{i=\hat{s}+1}^{s} \sigma_i^{(\psi)} + \sum_{j=\hat{r}+1}^{r} \sigma_j^{(\phi)} + \sum_{k=\hat{q}+1}^{q} \sigma_k^{(\nu)} + \sum_{l=\hat{p}+1}^{p} \sigma_l^{(\mu)},$$

see [8]. Unlike the matrix case, this approximation needs not be optimal in a least square sense, see e.g. [9] for best norm approximations.

Finally, expressing the signals in terms of the corresponding leading singular vectors contained in $\mathbf{U}^{(\psi)}$, $\mathbf{U}^{(\phi)}$, $\mathbf{U}^{(\nu)}$, and $\mathbf{U}^{(\mu)}$, we obtain a low-dimensional representation of $\hat{\mathbf{G}}$ in the smaller space $\mathbb{R}^{\hat{q}\hat{r}\times\hat{p}\hat{s}}$.

Apart from being used for defining a reduced model, the major HOSVD modes can be interpreted as the most relevant input and output signals - an insight that can be exploited for sensor and actuator design and placement.

21.3.2 I/O Maps and the Classical POD Method

To illustrate the relation of POD and HOSVD-reduced I/O map s, we consider the finite-dimensional LTI system

$$\dot{v} = Av + f$$

with the output $y = v$ and the sole (one-dimensional) input $Bu = f$.

Consider the grid $\mathcal{T} = \{t_i\}_{i=1}^s$ of time instances

$$0 = t_1 < t_2 < \cdots < t_s = T$$

and let $\bar{\mathcal{Y}}$ be the span of the nodal vectors of $v(t)$ taken at $t \in \mathcal{T}$. Then, the corresponding $\tilde{\mathbf{G}}$ for a single input dimension is a $s \times 1 \times q \times 1$ tensor that can be unfolded into the matrix

$$\tilde{\mathbf{G}}^{(v)} = \mathbf{X} = \begin{bmatrix} v_1(t_1) & \dots & v_1(t_s) \\ \vdots & \ddots & \vdots \\ v_q(t_1) & \dots & v_q(t_s) \end{bmatrix} \in \mathbb{R}^{q \times s}. \tag{21.14}$$

As laid out in the discussion of the truncated HOSVD in Sect. 21.3.1, a reduced basis for the space dimension of the state space can be obtained via a truncated SVD of $\tilde{\mathbf{G}}^{(v)}$. We observe that this reduced basis would be the well-known reduced basis used in POD, see [25] for an introduction.

21.3.3 A Generalized Approach in Time Capturing for POD

In POD, one uses samplings from the time evolution to compress the spatial state dimension. Instead of considering the matrix of snapshots (21.14) from discrete time instances, we propose considering the measurement matrix

$$\mathbf{X}_{gen} = \begin{bmatrix} (v_1, \psi_1)_{\mathcal{S}} & \dots & (v_1, \psi_s)_{\mathcal{S}} \\ \vdots & \ddots & \vdots \\ (v_q, \psi_1)_{\mathcal{S}} & \dots & (v_q, \psi_s)_{\mathcal{S}} \end{bmatrix} \in \mathbb{R}^{q \times s}, \tag{21.15}$$

of the same dimension that is obtained by testing the spatial components of the state v against the basis functions of a discrete $\mathcal{S}_{\tau_2} = \operatorname{span}\{\psi_1, \cdots, \psi_s\} \subset L^2(0,T)$. Note that for smooth trajectories, the standard snapshot matrix is obtained from (21.15) by testing against *delta distributions* located at $t \in \mathcal{T}$.

The L^2 orthogonal projection of the state vector $\mathbf{v}$ onto the space spanned by the measurements is given as

$$\hat{\mathbf{v}}(t) = \mathbf{X}_{gen}\mathbf{M}_{\mathcal{S}}^{-1}\boldsymbol{\psi}(t),$$

where $\boldsymbol{\psi} := [\psi_1, \dots, \psi_s]^{\mathsf{T}}$ and where $\mathbf{M}_{\mathcal{S}}$ is the mass matrix of $\mathcal{S}_{\tau_2}$, i.e. $[\mathbf{M}_{\mathcal{S}}]_{i_1 i_2} = (\psi_{i_1}, \psi_{i_2})_{\mathcal{S}}$. As for the classical POD, cf. [25, Thm. 1.12], where $\mathbf{M}_{\mathcal{S}} = I$, one can show that the generalized POD modes are the eigenvalues corresponding to the largest eigenvalues of the operator

$$\mathbf{R} = \int_0^T \hat{\mathbf{v}}\hat{\mathbf{v}}^\mathsf{T}\, dt = \int_0^T \mathbf{X}_{gen}\mathbf{M}_\mathcal{S}^{-1}\boldsymbol{\psi}(t)\boldsymbol{\psi}(t)^\mathsf{T}\mathbf{M}_\mathcal{S}^{-1}\mathbf{X}_{gen}^\mathsf{T}\, dt$$
$$= \mathbf{X}_{gen}\mathbf{M}_\mathcal{S}^{-1}\underbrace{\int_0^T \boldsymbol{\psi}(t)\boldsymbol{\psi}(t)^\mathsf{T}\, dt}_{=\mathbf{M}_\mathcal{S}}\mathbf{M}_\mathcal{S}^{-1}\mathbf{X}_{gen}^\mathsf{T} = \mathbf{X}_{gen}\mathbf{M}_\mathcal{S}{}^{-1}\mathbf{X}_{gen}^\mathsf{T}.$$

Therefore, in an implementation, the generalized POD modes can be obtained via an SVD of

$$\mathbf{X}_{gen}\mathbf{M}_\mathcal{S}^{-1/2}. \tag{21.16}$$

21.4 Numerical Examples and Applications

By means of four numerical examples we present numerical convergence results, an application of the discretized I/O map for the solution of optimization problems including the use of SVD, as well as numerical tests comparing the I/O map s motivated POD variant to the standard approach.

21.4.1 Convergence of the I/O Map for a Heat Equation

We consider a heat equation with homogeneous Dirichlet boundary conditions, which for a domain Ω with a C^2-boundary and for $Z = L^2(\Omega)$ becomes a system of type (21.1) with the Laplace operator

$$A = \Delta : D(A) = H^2(\Omega) \cap H_0^1(\Omega) \subset Z \to Z. \tag{21.17}$$

Since A is the infinitesimal generator of an analytic C^0-semigroup of contractions $(S(t))_{t\geq 0}$, the mild solution z of (21.1) exhibits the following stability and regularity properties, see e.g. [21, Ch. 7] and [13].

(i) If $z_0 = 0$ and $u \in \mathcal{U}$, then $z \in H^{1,2}((0,T)\times\Omega)$ with

$$\|z\|_{H^{1,2}((0,T)\times\Omega)} \leq c\|u\|_\mathcal{U}. \tag{21.18}$$

(ii) Assume that $u \equiv 0$. For $z_0 \in D(A)$ we have $z \in C^1([0,T]; D(A))$, but for $z_0 \in Z$ we only have $z \in C^1((0,T]; D(A))$.

For the numerical tests, we will consider domains $\Omega \subset \mathbb{R}^2$ and define control and observation operators as follows: For given points $a_c, b_c, a_m, b_m \in \bar{\Omega}$, let $\Omega_c = (a_{c,1}, a_{c,2}) \times (b_{c,1}, b_{c,2})$ and $\Omega_m = (a_{m,1}, a_{m,2}) \times (b_{m,1}, b_{m,2})$ be rectangular subsets of Ω where the control is active and the observation takes place, respectively. Let $U = Y = L^2(0,1)$ and define $C \in \mathscr{L}(L^2(\Omega), Y)$ and $B \in \mathscr{L}(U, L^2(\Omega))$ via

$$(Cz)(\xi) = \int_{a_{m,1}}^{b_{m,1}} \frac{z(x_1, x_2(\xi))}{b_{m,1} - a_{m,1}} \, dx_1, \tag{21.19}$$

$$(Bu)(x_1, x_2) = \begin{cases} u(\theta(x_1))\omega_c(x_2), & (x_1, x_2) \in \Omega_c \\ 0, & (x_1, x_2) \notin \Omega_c \end{cases}, \tag{21.20}$$

where $\omega_c \in L^2(a_{c,2}, b_{c,2})$ is a weight function and $\theta : [a_{c,1}, b_{c,1}] \to [0,1]$ and $x_1 : [0,1] \to [a_{m,1}, b_{m,1}]$ are affine-linear transformations.

Since $C_{|H^2(\Omega)} \in \mathscr{L}(H^2(\Omega), H^2(0,1))$, we have $\mathbb{G} \in \mathscr{L}(\mathcal{U}, \mathcal{Y}_s)$ as well as

$$\mathbb{G}_{|\mathcal{U}_s} \in \mathscr{L}(\mathcal{U}_s, \mathcal{Y}_s), \quad \text{with } \mathcal{U}_s = H^{1,2}((0,T) \times \Theta), \quad \mathcal{Y}_s = H^{1,2}((0,T) \times \Xi).$$

Also, for $u \in \mathcal{U}_s$, we have $\|u\|_{\mathcal{U}} \leq \|u\|_{\mathcal{U}_s}$, and for $u \in \mathcal{U}$, we have $\|\mathbb{G}u\|_{\mathcal{Y}_s} \leq c'\|z\|_{H^{1,2}((0,T)\times\Omega)} \leq c\,c'\|u\|_{\mathcal{U}}$, where c is the constant used in the stability estimate (21.18) and $c' = \max\{\|C\|_{\mathscr{L}(L^2(\Omega), L^2(\Xi))}, \|C\|_{\mathscr{L}(H^2(\Omega), H^2(\Xi))}\}$.

For the concrete test case, we set $T = 1$ and $\Omega = (0,1)^2$ and choose $\Omega_c = \Omega$, $\Omega_m = (0.1, 0.2) \times (0.1, 0.9)$, and $\omega_c(x_2) = \sin(\pi x_2)$, see Fig. 21.3. For this choice, for inputs of the form $u(t;\theta) = \sin(\omega_T \pi t)\sin(m\pi\theta)$ with $\omega_T, m \in \mathbb{N}$, there is an analytic expression for the outputs in terms of the eigenfunctions of the Laplace operator (21.17).

For the finite-dimensional approximation of the I/O map $\mathbb{G}$, we choose hierarchical linear finite elements in U_{h_1} and Y_{h_2} and Haar wavelets in $\mathcal{R}_{\tau_1}$ and $\mathcal{S}_{\tau_2}$ and compute $\mathbb{G}_{DS}(h_1, \tau_1, h_2, \tau_2, \texttt{tol})$. The tolerance `tol` refers to the accuracy of the numerical approximation of the system dynamics, cf. Assumption 1.

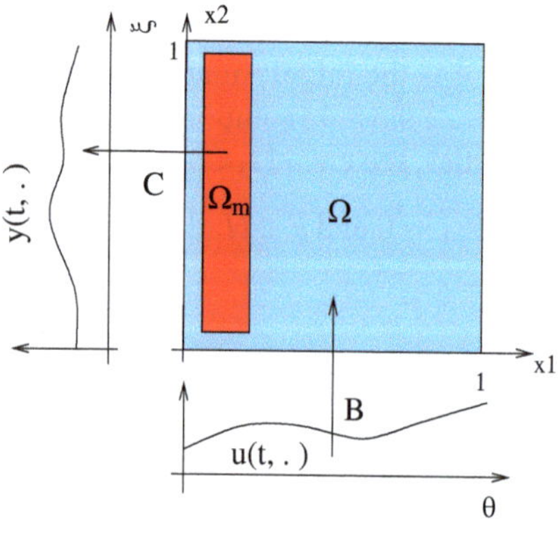

Fig. 21.3 Setup for the heat equation with Dirichlet boundary conditions and known solutions

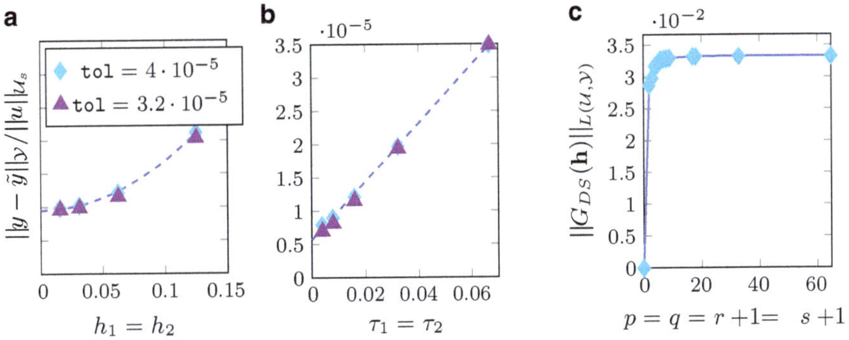

Fig. 21.4 The relative output errors for input $u(t;\theta) = \sin(10\pi t)\sin(5\pi\theta)$ for varying $h_1 = h_2$ and fixed $\tau_1 = \tau_2 = 1/64$ (**a**) and for varying $\tau_1 = \tau_2$ and fixed $h_1 = h_2 = 1/17$ (**b**). The dashed lines are the linear and the quadratic fit. (**c**): the norm of the discrete I/O map $\|\mathbb{G}_{DS}(\mathbf{h})\|_{\mathscr{L}(\mathcal{U},\mathcal{Y})}$ versus approximation space dimensions $p = q = r + 1 = s + 1$ for a fixed tolerance $\texttt{tol} = 4.0e-5$

Convergence of single outputs. For the present setup, with inputs $u(t;\theta) = \sin(\omega_T \pi t)\sin(m\pi\theta)$, and known outputs $y = \mathbb{G}u$, we investigate the relative error $\|y - \tilde{y}\|_{\mathcal{Y}}/\|u\|_{\mathcal{U}_s}$, with $\tilde{y} = \mathbb{G}_{DS}(h_1, \tau_1, h_2, \tau_2, \texttt{tol})u$, for varying discretization parameters h_1, τ_1, h_2, τ_2 and tol. For $m = 5$ and $\omega_T = 10$, we observe a quadratic convergence with respect to decreasing $h_1 = h_2$ (cf. Fig. 21.4a) and a linear convergence for $\tau_1 = \tau_2$ (cf. Fig. 21.4b). It is notable, however, that due to the system dynamics error, the error converges to a positive plateau value depending on the tolerance tol.

Convergence of the norm $\|\mathbb{G}_S(h_1, \tau_1, h_2, \tau_2)\|_{\mathscr{L}(\mathcal{U},\mathcal{Y})}$ for nested subspaces. Successively improving the signal approximation by adding additional basis functions, the norm $\|\mathbb{G}_S(h_1, \tau_1, h_2, \tau_2)\|_{\mathscr{L}(\mathcal{U},\mathcal{Y})}$ converges, cf. Lemma 1. We approximate $\|\mathbb{G}_S\|_{\mathscr{L}(\mathcal{U},\mathcal{Y})}$ by $\|\mathbb{G}_{DS}\|_{\mathscr{L}(\mathcal{U},\mathcal{Y})}$, where $\mathbb{G}_{DS}$ has been calculated with $\texttt{tol} = 4.0e-5$. In Fig. 21.4c, the approximations $\|\mathbb{G}_S(h_1, \tau_1, h_2, \tau_2)\|_{\mathscr{L}(\mathcal{U},\mathcal{Y})} = \|\mathbb{G}_S(\frac{1}{p-1}, \frac{1}{r}, \frac{1}{q-1}, \frac{1}{s})\|_{\mathscr{L}(\mathcal{U},\mathcal{Y})}$ are plotted for increasing subspace dimensions $p = q = r + 1 = s + 1 = 2, 3, \ldots, 65$.

21.4.2 I/O Maps and Higher Order SVD for Optimal Control

We demonstrate the use of discrete I/O map s in optimization problems like

$$J(u, y) = \frac{1}{2}\|y - y_D\|_{\mathcal{Y}}^2 + \frac{\alpha}{2}\|u\|_{\mathcal{U}}^2 \to \min, \quad \text{s. t. } y = \mathbb{G}u, \ u \in \mathcal{U}_{ad}, \tag{21.21}$$

where $\mathcal{U}_{ad} \subset \mathcal{U}$ is the subset of admissible controls, $y_D \in \mathcal{Y}$ is a target output signal, and $\alpha > 0$ is a regularization parameter. With $\mathbf{y}_D = \kappa_{\mathcal{Y},h_2,\tau_2}\mathbb{P}_{\mathcal{Y},h_2,\tau_2}y_D$ and

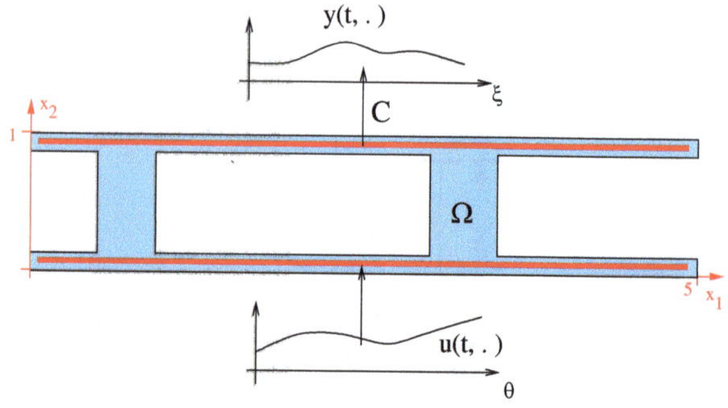

Fig. 21.5 Setup of the test case for the heat equation with homogeneous Neumann boundary conditions

$\bar{U}_{ad} = \{\mathbf{u} \in \mathbb{R}^{pr} : \mathbf{u} = \kappa_{\mathcal{U},h_1,\tau_1}\mathbb{P}_{\mathcal{U},h_1,\tau_1}u,\ u \in \mathcal{U}_{ad}\}$, we define the discrete approximation to (21.21) as,

$$\bar{J}_{\mathbf{h}}(\mathbf{u},\mathbf{y}) = \frac{1}{2}\|\mathbf{y}-\mathbf{y}_D\|^2_{qs;w} + \alpha\|\mathbf{u}\|^2_{pr;w} \to \min \quad \text{s. t. } \mathbf{y} = \tilde{\mathbf{G}}\mathbf{u},\ \mathbf{u} \in \bar{U}_{ad} \tag{21.22}$$

For an optimization problem without control constraints, i.e. $\mathcal{U}_{ad} = \mathcal{U}$ and $\bar{U}_{ad} = \mathbb{R}^{pr}$, the solution $\bar{\mathbf{u}}$ of (21.22) is readily defined by

$$(\tilde{\mathbf{G}}^{\mathsf{T}}\mathbf{M}_{\mathcal{Y}}\tilde{\mathbf{G}} + \alpha\mathbf{M}_{\mathcal{U}})\bar{\mathbf{u}} = \tilde{\mathbf{G}}^{\mathsf{T}}\mathbf{M}_{\mathcal{Y}}\mathbf{y}_D. \tag{21.23}$$

As the test case for optimal control, we consider a model for the heat conduction in two infinitely long plates of width 5 and height 0.2 which are connected by two rectangular bars and which are surrounded by an insulating material. That is, for $t \in (0,1]$ we solve a heat equation with homogeneous Neumann boundary conditions on a domain Ω as in Fig. 21.5. We assume, that we can heat the bottom plate and measure the temperature distribution in the upper plate, i.e. to model the control and observation we set $\Omega_c = (0.05, 4.95) \times (0.05, 0.15)$, $\Omega_m = (0.05, 4.95) \times (0.85, 0.95)$ and $\omega_c(x_2) = \sin(\pi(x_2 - 0.05)/0.1)$ and define B and C similarly to (21.19).

As the target output, we choose $y_D = \mathbb{G}u_0$, i.e. the output for the input $u_0 \equiv 1$. Then, the solution of the optimal control problem (21.21) will give an optimized input u_* that leads to a similar output as u_0 but at lower costs.

As the reference for a subsequent HOSVD based reduction, we solve (21.23) with an approximated I/O map $\tilde{\mathbf{G}} \in \mathbb{R}^{17\cdot 64 \times 65\cdot 64}$ and $\alpha = 10^{-4}$ for an approximation $\bar{u}$ to u_*. This took 0.33 s on a desktop PC. The norm of the input was reduced by 27.9 % with a relative deviation of $\mathbb{G}\bar{u}$ from y_D of 9.4 %. Having employed a HOSVD to reduce the I/O map to $\hat{\mathbf{G}} \in \mathbb{R}^{3\cdot 5 \times 3\cdot 5}$, i.e. having truncated all but the 3 most relevant spatial and the 5 most relevant temporal input and output modes, the calculation of $\bar{u}$ took less than 0.0004 s. The optimal control on the base of $\hat{\mathbf{G}}$ came with a norm reduction of 27.4 % if compared to u_0 while admitting a deviation from the target of 9.5 %. See Fig. 21.6 for an illustration of the optimization results.

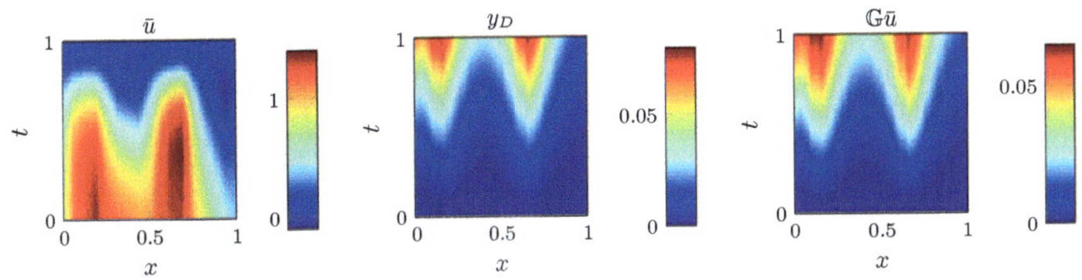

Fig. 21.6 Application of the SVD-reduced approximated I/O map $\hat{\mathbf{G}} \in \mathbb{R}^{3\cdot5\times3\cdot5}$ in an optimization problem. From left to right: the optimal control $\bar{u}$, the target output $y_D = \mathbb{G}u_0$, and the output $\mathbb{G}\bar{u}$

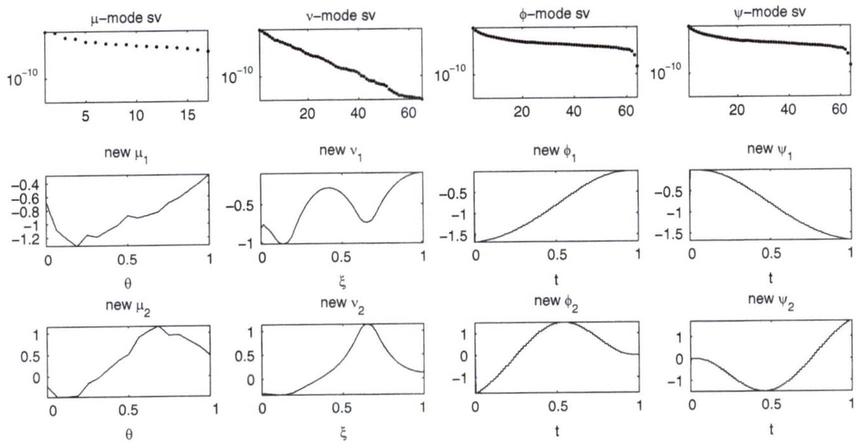

Fig. 21.7 Results of an HOSVD applied to the I/O map of the test case illustrated in Fig. 21.5. In the first row the n-mode singular values in a semilogarithmic scale are plotted. In the 2nd and 3rd row the respective most relevant modes are shown

The insights provided by an HOSVD are illustrated in Fig. 21.7, namely the distribution of the n-mode singular values indicating how many degrees of freedom are needed to capture the dynamics and the most relevant input and output modes, cf. Sect. 21.3.1. From the modes one can draw conclusions on effective actuations. In the given example, the locations of the connecting bars are clearly visible in the spatial input modes μ_1 and μ_2.

21.4.3 *Driven Cavity and Generalized POD*

We consider a driven cavity flow at *Reynolds number* $Re = 2{,}000$ in the unit square. For the spatial discretization we use *Taylor-Hood* elements on a uniform triangulation of the domain by 50^2 triangles, see Fig. 21.8 for an illustration of the flow and the discretization. We linearize the equations about the velocity solution α

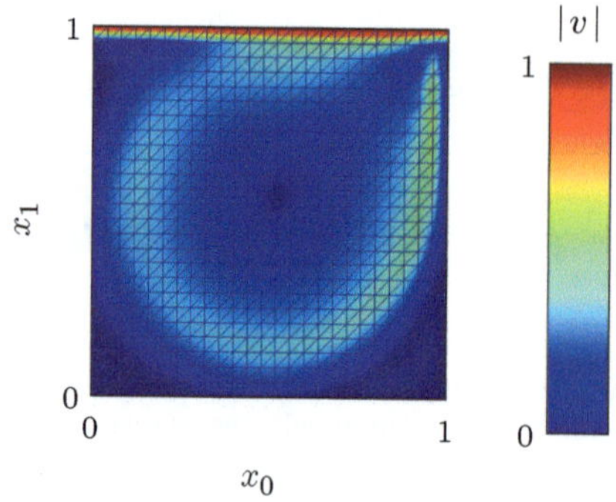

Fig. 21.8 Illustration of the driven cavity at the steady state for $Re = 2{,}000$ and the triangulation of the domain

of the associated steady state Stokes problem, and consider the time evolution of the velocity $v(t) \in \mathbb{R}^{N_v}$ and the pressure $p(t) \in \mathbb{R}^{N_p}$, N_v, $N_p \in \mathbb{N}$, modeled by

$$M\dot{v}(t) = -A(\alpha, Re)v(t) + J^{\mathsf{T}}p(t) + f(t), \tag{21.24a}$$

$$0 = Jv(t), \tag{21.24b}$$

$$v(0) = \alpha, \tag{21.24c}$$

in the time interval $(0, 5]$. Here, $M \in \mathbb{R}^{N_v,N_v}$ is the velocity mass matrix, $J \in \mathbb{R}^{N_p,N_v}$ and its transpose J^{T} are discretizations of the divergence and the gradient operator, $A \in \mathbb{R}^{N_v,N_v}$ models the diffusion and the linearized convection depending on Re and v_0, and $f \in \mathbb{R}^{N_v}$ is a source terms that arises from the linearization and the incorporation of the boundary conditions. See [16] for details of the modeling.

We compare the I/O map motivated variant of Sect. 21.3.3 of POD to the standard approach that we refer to as POD. To the generalization we refer as gmPOD. To the columns of the measurement matrices (21.14) or (21.15) we will refer to as *measurements* which is more general than the standard term *snapshot*. By s, we denote the number of measurements, i.e. the number of snapshots or the dimension of the test space, respectively. By $\hat{k}$, we denote the number of modes, i.e. leading singular vectors of the measurement matrix, that are taken for the reduced order model (often called *POD-dimension*).

Say $U_{\hat{k}}$ is the matrix of the chosen $\hat{k}$ modes obtained via a truncated SVD of (21.14) for POD or (21.16) for gmPOD. Then, a reduced model of (21.24) is given as

$$\hat{M}\dot{\hat{v}}(t) = \hat{A}\hat{v}(t) + \hat{f}(t, U_{\hat{k}}\hat{v}(t)), \tag{21.25a}$$

$$\hat{v}(0) = \hat{\alpha}, \tag{21.25b}$$

where $\hat{M} := U_{\hat{k}}^{\mathsf{T}}MU_{\hat{k}}$, $\hat{A} := U_{\hat{k}}^{\mathsf{T}}AU_{\hat{k}}$, and $\hat{f} = U_{\hat{k}}^{\mathsf{T}}f$, with M, A, and f being the mass matrix, coefficient matrix, and a nonlinearity or source term. The solution $\hat{v}$ is related to the actual solution v via the ansatz $v = U_{\hat{k}}\hat{v}$. Accordingly, the initial value is typically chosen as $\hat{\alpha} = U_{\hat{k}}^{\mathsf{T}} v(0)$. We consider only the velocity component for the measurements for (21.24a). Since the measurements fulfill $JX = 0$, we have that $U_{\hat{k}}^{\mathsf{T}} J^{\mathsf{T}} = 0$.

As the basis for the (generalized) time capturing, we use the hierarchical basis of piecewise linear functions illustrated in Fig. 21.2.

The time integration for the full order system was implemented using the implicit trapezoidal rule on a uniform time grid. The corresponding reduced order systems were numerically integrated using *Scipy*'s (for the current test case) and *Matlab*'s (for the following test case) built-in ODE solvers `integrate.odeint` and `ODE45`. The source code of the POD and gmPOD tests is available from the author's *Github* account [4].

We consider the error $e_{s,\hat{k}} \approx \left(\int_0^T \|v(t) - U_{\hat{k}}\hat{v}(t)\|^2_{L^2(\Omega)} dt\right)^{1/2}$ where v is the solution of the full order system and $\hat{v}$ solves the reduced order system constructed by means of s measurements and $\hat{k}$ POD modes. Here, T is the endpoint of the considered time interval and Ω is the spatial domain of the considered test case. The error is evaluated numerically using the piecewise trapezoidal rule in time and the finite element space norm on the chosen discretization of the full order model.

The conducted tests were designed to evaluate the evolution of the error $e_{s,\hat{k}}$ with respect to the number of measurements s and the dimension of the reduced model $\hat{k}$. Both parameters are limiting factors: the amount of needed memory increases linearly with s and the computational complexity to compute the POD modes via an SVD increases like $\hat{k}$ times the square of the spatial dimension. Also, the effort for the numerical solution of the reduced systems scales with the number of POD modes necessary for the required accuracy.

For the test case with the linearized flow equations (21.24), we find that gmPOD significantly outperforms POD. Throughout the investigated range of measurements and number of chosen POD modes, the approximation error after the gmPOD reduction is much smaller than after a POD reduction of the same dimension, see Fig. 21.10b.

21.4.4 Nonlinear Burgers' Equation and Generalized POD

As the last test case, we consider a system stemming from a spatial discretization of the nonlinear Burgers' equation,

$$\partial_t z(t,x) + \partial_x\Big(\frac{1}{2}z(t,x)^2 - \nu\partial_x z(t,x)\Big) = 0, \tag{21.26}$$

with the spatial coordinate $x \in (0,1)$, the time variable $t \in (0,1]$, and the viscosity parameter $\nu = 0.01$, completed by zero Dirichlet boundary conditions and a step function as initial conditions as illustrated in Fig. 21.9a. See [3] and [4] for details on the implementation.

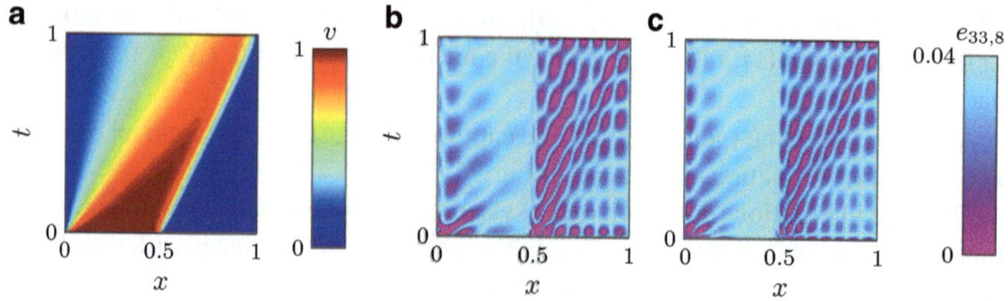

Fig. 21.9 Full order approximation of Burgers' equation (**a**) and the error made by POD (**b**) and gmPOD (**c**) using $s = 33$ snapshots and a reduced model of dimension $\hat{k} = 8$

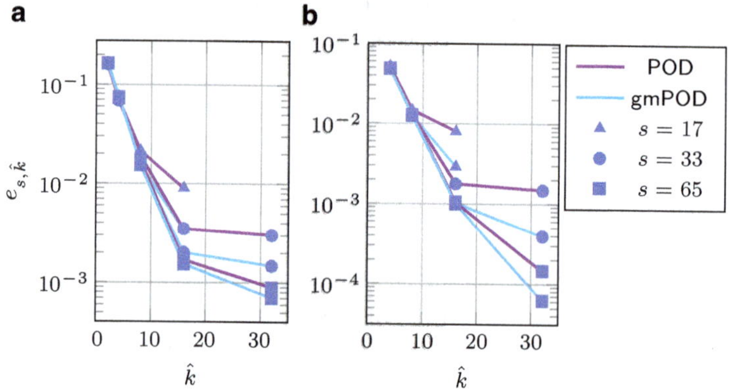

Fig. 21.10 The time and space approximation error $e_{s,\hat{k}}$: (**a**) for the semi-discrete Burgers' equation (Sect. 21.4.4) and (**b**) for the linearized flow equations (Sect. 21.4.3) for varying numbers s of measurements and varying numbers $\hat{k}$ of POD modes used for the reduced model

We compute the gmPOD-reduced model and compare it to the standard POD reduction as explained in Sect. 21.4.3. Again, the generalized POD approach gmPOD outperforms the classical variant in terms of accuracy versus number of snapshots and POD modes, see Fig. 21.10a for the error plot and Fig. 21.9b–c for an illustration of the approximation error.

21.5 Final Remarks and Outlook

The presented framework is suitable to provide direct discretizations of I/O map s of linear infinite-dimensional control systems with *spatially distributed* inputs and outputs. This allows for tailoring numerical approximations to the I/O behavior of systems which is particularly important in control setups. The provided methodology, thus, comes with error estimates in the relevant $\mathcal{L}(\mathcal{U}, \mathcal{Y})$ norm. We have illustrated the approximation properties in a numerical example.

The discrete I/O map can be expressed as a tensor and reduced by higher order SVDs to the most relevant input and output modes. Apart from the reduction, a HOSVD also gives insights for controller and sensor design.

We have shown that the idea of sampling trajectories in time to compress spatial information is related to POD reduced order modeling. The freedom to choose the test functions for the sampling can be used to define POD reduced models on the basis of measurements that are more general than snapshots. The newly introduced generalized POD variant has shown to significantly outperform classical POD reduction for a linear and a nonlinear test case.

The proposed variant of POD seems promising in further aspects that are subject to future work. Unlike in the standard approach, the sampling is done in a subspace of the state space which can be exploited for improved error estimates. Secondly, the nonlocalized measurements have smoothing properties that might be of advantage for noisy measurements. Finally, the freedom in bases choice allows for problem specific measurement functions which may reduce the number of necessary snapshots.

References

1. Ainsworth, M., Oden, J.T.: A Posteriori Error Estimation in Finite Element Analysis. Wiley-Interscience, New York (2000)
2. Antoulas, A.C.: Approximation of Large-Scale Dynamical Systems. Society for Industrial and Applied Mathematics (SIAM), Philadelphia (2005)
3. Baumann, M.: Nonlinear model order reduction using POD/DEIM for optimal control of Burgers' equation. Master's thesis, Delft University of Technology (2013)
4. Baumann, M., Heiland, J.: genpod – matlab and python implementation with test cases. https://github.com/ManuelMBaumann/genpod.git, September (2014)
5. Benner, P., Mehrmann, V., Sorensen, D., (eds.): Dimension Reduction of Large-Scale Systems. LNSCE, vol. 45. Springer, Heidelberg (2005)
6. Berkooz, G., Holmes, P., Lumley, J.L.: The proper orthogonal decomposition in the analysis of turbulent flows. In: Annual Review of Fluid Mechanics, vol. 25, pp 539–575. Annual Reviews, Palo Alto (1993)
7. Ciarlet, P.G.: The Finite Element Method for Elliptic Problems. Classics in Applied Mathematics, vol. 40. Society for Industrial and Applied Mathematics (SIAM), Philadelphia (2002)
8. De Lathauwer, L., De Moor, B., Vandewalle, J.: A multilinear singular value decomposition. SIAM J. Matrix Anal. Appl. **21**(4), 1253–1278 (2000)
9. De Lathauwer, L., De Moor, B., Vandewalle, J.: On the best rank-1 and rank-$(R_1, R_2, \cdots, R_N)$ approximation of higher-order tensors. SIAM J. Matrix Anal. Appl. **21**(4), 1324–1342 (2000)
10. Douglas, R.G.: Banach Algebra Techniques in Operator Theory. Academic, New York (1972)
11. Eriksson, K., Estep, D., Hansbo, P., Johnson, C.: Introduction to adaptive methods for differential equations. Acta Numer. vol. 4, pp 105–158. Cambridge University Press, Cambridge (1995). http://journals.cambridge.org/action/displayFulltext?type=8&fid=2604116&jid=ANU&volumeId=4&issueId=-1&aid=1771172
12. Eriksson, K., Johnson, C.: Adaptive finite element methods for parabolic problems. II. Optimal error estimates in $L_\infty L_2$ and $L_\infty L_\infty$. SIAM J. Numer. Anal. **32**(3), 706–740 (1995)
13. Evans, L.C.: Partial Differential Equations. Graduate Studies in Mathematics, vol. 19. American Mathematical Society, Providence (1998)

14. Gerhard, J., Pastoor, M., King, R., Noack, B.R., Dillmann, A., Morzynski, M., Tadmor, G.: Model-based control of vortex shedding using low-dimensional galerkin models. AIAA-Paper 2003-4262 (2003)
15. Gugercin, S., Antoulas, A.C.: A survey of model reduction by balanced truncation and some new results. Int. J. Control **77**(8), 748–766 (2004)
16. Heiland, J., Mehrmann, V.: Distributed control of linearized Navier-Stokes equations via discretized input/output maps. Z. Angew. Math. Mech. **92**(4), 257–274 (2012)
17. Heiland, J., Mehrmann, V., Schmidt, M.: A new discretization framework for input/output maps and its application to flow control. In: King, R. (ed.) Active Flow Control. Papers contributed to the Conference "Active Flow Control II 2010", Berlin, May 26–28, 2010, pp 375–372. Springer, Berlin (2010)
18. Johnson, C.: Numerical Solution of Partial Differential Equations by the Finite Element Method. Cambridge University Press, Cambridge (1987)
19. Lehmann, O., Luchtenburg, D.M., Noack, B.R., King, R., Morzynski, M., Tadmor, G.: Wake stabilization using POD Galerkin models with interpolated modes. In: Proceedings of the 44th IEEE Conference on Decision and Control and European Conference ECC, Invited Paper 1618 (2005)
20. Pastoor, M., King, R., Noack, B.R., Dillmann, A., Tadmor, G.: Model-based coherent-structure control of turbulent shear flows using low-dimensional vortex models. AIAA-Paper 2003-4261 (2003)
21. Pazy, A.: Semigroups of Linear Operators and Applications to Partial Differential Equations. Applied Mathematical Sciences, vol. 44. Springer, New York (1983)
22. Schmidt, M.: Systematic discretization of input/output Maps and other contributions to the control of distributed parameter systems. PhD thesis, TU Berlin, Fakultät Mathematik, Berlin (2007)
23. Staffans, O.J.: Well-posed Linear Systems. Cambridge University Press, Cambridge/ New York (2005)
24. Thomée, V.: Galerkin Finite Element Methods for Parabolic Problems. Springer, Berlin (1997)
25. Volkwein, S.: Model reduction using proper orthogonal decomposition. Lecture Notes, Institute of Mathematics and Scientific Computing, University of Graz, Austria (2011)
26. Werner, D.: Funktionalanalysis. Springer, Berlin (2000)

MIX
Papier aus verantwortungsvollen Quellen
Paper from responsible sources
FSC® C105338

If you have any concerns about our products, you can contact us on
ProductSafety@springernature.com

In case Publisher is established outside the EU, the EU authorized representative is:
Springer Nature Customer Service Center GmbH
Europaplatz 3, 69115 Heidelberg, Germany

Printed by Libri Plureos GmbH
in Hamburg, Germany